The International Exhibition of 1862

The Illustrated Catalogue of the Industrial Department

VOLUME 1: BRITISH DIVISION 1

ANONYMOUS

CAMBRIDGE
UNIVERSITY PRESS

CAMBRIDGE
UNIVERSITY PRESS

University Printing House, Cambridge, CB2 8BS, United Kingdom

Published in the United States of America by Cambridge University Press, New York

Cambridge University Press is part of the University of Cambridge.
It furthers the University's mission by disseminating knowledge in the pursuit of
education, learning and research at the highest international levels of excellence.

www.cambridge.org
Information on this title: www.cambridge.org/9781108067287

© in this compilation Cambridge University Press 2014

This edition first published 1862
This digitally printed version 2014

ISBN 978-1-108-06728-7 Paperback

This book reproduces the text of the original edition. The content and language reflect
the beliefs, practices and terminology of their time, and have not been updated.

Cambridge University Press wishes to make clear that the book, unless originally published
by Cambridge, is not being republished by, in association or collaboration with, or
with the endorsement or approval of, the original publisher or its successors in title.

CAMBRIDGE LIBRARY COLLECTION

Books of enduring scholarly value

Technology

The focus of this series is engineering, broadly construed. It covers technological innovation from a range of periods and cultures, but centres on the technological achievements of the industrial era in the West, particularly in the nineteenth century, as understood by their contemporaries. Infrastructure is one major focus, covering the building of railways and canals, bridges and tunnels, land drainage, the laying of submarine cables, and the construction of docks and lighthouses. Other key topics include developments in industrial and manufacturing fields such as mining technology, the production of iron and steel, the use of steam power, and chemical processes such as photography and textile dyes.

The International Exhibition of 1862

Replete with detailed engravings, this four-volume catalogue was published to accompany the International Exhibition of 1862. Held in South Kensington from May to November, the exhibition showcased the progress made in a diverse range of crafts, trades and industries since the Great Exhibition of 1851. Over 6 million visitors came to view the wares of more than 28,000 exhibitors from Britain, her empire and beyond. Featuring explanatory notes and covering such fields as mining, engineering, textiles, printing and photography, this remains an instructive resource for social and economic historians. The exhibition's *Illustrated Record*, its *Popular Guide* and the industrial department's one-volume *Official Catalogue* have all been reissued in this series. Including a floor plan of the main buildings, Volume 1 begins with a concise history of the exhibition written by John Hollingshead (1827–1904). It then catalogues and illustrates impressive examples of British manufacturing.

Cambridge University Press has long been a pioneer in the reissuing of out-of-print titles from its own backlist, producing digital reprints of books that are still sought after by scholars and students but could not be reprinted economically using traditional technology. The Cambridge Library Collection extends this activity to a wider range of books which are still of importance to researchers and professionals, either for the source material they contain, or as landmarks in the history of their academiwc discipline.

Drawing from the world-renowned collections in the Cambridge University Library and other partner libraries, and guided by the advice of experts in each subject area, Cambridge University Press is using state-of-the-art scanning machines in its own Printing House to capture the content of each book selected for inclusion. The files are processed to give a consistently clear, crisp image, and the books finished to the high quality standard for which the Press is recognised around the world. The latest print-on-demand technology ensures that the books will remain available indefinitely, and that orders for single or multiple copies can quickly be supplied.

The Cambridge Library Collection brings back to life books of enduring scholarly value (including out-of-copyright works originally issued by other publishers) across a wide range of disciplines in the humanities and social sciences and in science and technology.

THE

ILLUSTRATED CATALOGUE

OF THE

INTERNATIONAL EXHIBITION.

G. Sykes, Del.

J. H. Le Keux, Sc. 30 Argyle St. W.C.

INTERNATIONAL EXHIBITION OF 1862.

South Front.— Entrance to Picture Galleries

THE INTERNATIONAL EXHIBITION *of* 1862.

THE ILLUSTRATED CATALOGUE

OF THE

INDUSTRIAL DEPARTMENT.

BRITISH DIVISION—VOL. I.

PRINTED FOR HER MAJESTY'S COMMISSIONERS.

Printed for Her Majesty's Commissioners by

CLAY, SON, & TAYLOR, Bread Street Hill. | EDMUND EVANS, Raquet Court.
CLOWES & SON, Stamford Street. | PETTER & GALPIN, La Belle Sauvage Yard.
SPOTTISWOODE & Co., New Street Square.

CONTENTS.

VOLUME I.

A CONCISE HISTORY OF THE INTERNATIONAL EXHIBITION OF 1862.

THE ILLUSTRATED CATALOGUE OF THE INDUSTRIAL DEPARTMENT.

CLASS I. MINING, QUARRYING, METALLURGY, AND MINERAL PRODUCTS.

    II. CHEMICAL SUBSTANCES AND PRODUCTS, AND PHARMACEUTICAL PROCESSES.
        Section a. Chemical Products.
            b. Medical and Pharmaceutical Products and Processes.

CONTENTS.

VOLUME II.

CONTENTS

A CONCISE HISTORY

OF THE

INTERNATIONAL EXHIBITION

of 1862,

ITS RISE AND PROGRESS, ITS BUILDING AND FEATURES,

AND A SUMMARY OF ALL FORMER EXHIBITIONS,

BY JOHN HOLLINGSHEAD.

LIST OF ILLUSTRATIONS

TO

HOLLINGSHEAD'S CONCISE HISTORY.

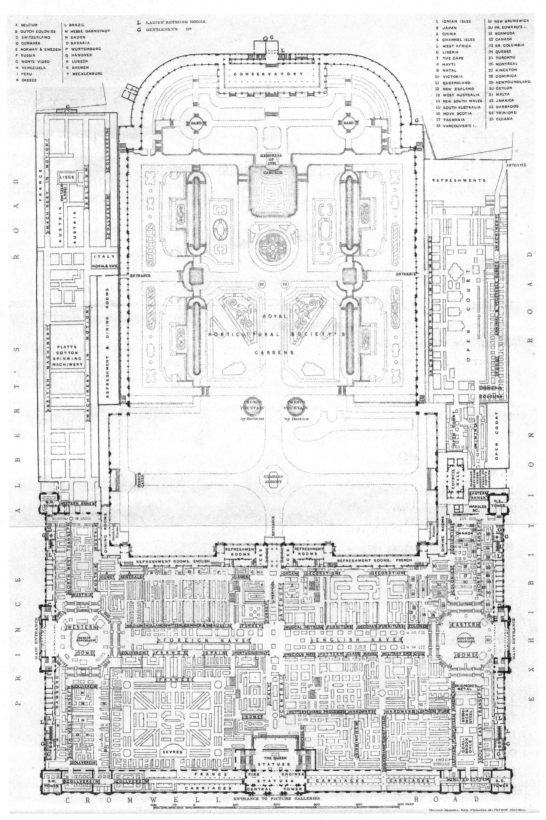

The material originally positioned here is too large for reproduction in this reissue. A PDF can be downloaded from the web address given on page iv of this book, by clicking on 'Resources Available'.

INTERNATIONAL EXHIBITIONS.

CHAPTER I.

INTRODUCTION.

WHEN the Great Exhibition of 1851 was first put in motion, its promoters knew little of the probable success of such a display—of the extent to which it would be supported by exhibitors or visited by the public. They could only be encouraged by the records of certain exhibitions which had been merely national in character and design. They were fed upon statistics, more or less reliable, which sometimes led them to hope, sometimes to despair. They had to overcome the apathy of many supporters, and to check the wild enthusiasm of others. Their administrative mechanism—with the exception of the Society of Arts—was all new, and it creaked and occasionally stuck fast, until all the parts settled down in their appointed places, and were smoothed by action and hard work. They had set themselves a difficult and novel task. They wished to attract exhibitors from the remotest corners of the earth, and to provide a palace for them—a temple dedicated to the worship of trade—without the aid of a government grant. They met with assistance where they least expected it, and opposition where they expected assistance. They had to feel their way, step by step; to send out travelling commissioners to solicit aid in the great centres of industry; to appoint committees and then teach them their duties; and to do thousands of things, unfortified by precedents and in doubt as to results. The world was all before them where to choose, and they confined themselves to no nation and to no class. Never was such a broad appeal made to the trading instincts of mankind. It seemed as if the country, conscious of its own strength, was anxious to enter into an industrial contest with the whole world. Wherever any handicraft was practised, any package shipped, any bill of exchange drawn, the challenge was sent. Some thought the appeal was too broad, and even dangerous. The doctrine of free markets, notwithstanding the recent partial abolition of the corn-tax, was not as popular then as it is now, and our tariff, instead of being pared down to thirty-five articles, including varieties, was one of more than five hundred articles, excluding varieties. Those were the days when the Custom-house looked sharply after "*bonâ-fide* nutmegs," and sweet-meats played an important part in the national finance. Public opinion, in its surly moods, accused the Exhibition promoters of giving up their country into the hands of invited savages. The memorable May-day of 1851 was looked forward to with dread by many honest people, who regarded it as the turning-point in England's fate. They expected that London would be ravaged at will, and planted with many varieties of new disease. The tomahawk was looked for in Hyde Park, the stiletto in Cheapside, and dirt, strange costumes, and stranger manners everywhere. Unmanageable crowds were pictured assembling in the chief thoroughfares to make the Exhibition a stalking-horse for riot and plunder. Wild fears produced over-caution in the laying out of plans, and the police and army were concentrated as if for an internal war. When the statistics of 1851, however, came to be gathered together, it was found that there had been less crime, less disorder, and fewer accidents than the annual average.

The building of course was not free from panic-stricken criticism. Its light-ness was regarded as an evident sign of weakness, and its size was held to increase the danger. Although its strength was tested in every mechanical way, a broad margin was left for doubts, and not one half of the numbers were at first carefully admitted who swarmed into it unchecked later in its brief life.

Like boys who have hesitated long on the bank of a clear stream, but who leaping in full of dread, are surprised to find how harmless and pleasant the water is, we can now afford to smile at our fears of twelve years ago. The Exhibition came and went. Strange nations were brought together, and learned to know each other better. Though all the good that was once expected from this gathering, in maintaining the peace of the world, cannot, with our painful after-knowledge, be claimed now, still we need not be ashamed of the Exhibition and its results. Unless friendly intercourse, hard work, and industrial rivalry are hollow mockeries, it is impossible that 1851 can have left no good mark upon the world.

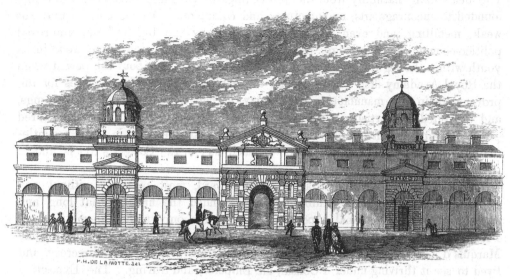

ROYAL MEWS, CHARING CROSS, 1828.

CHAPTER II.

EARLY EXHIBITIONS.

NDUSTRIAL exhibitions, after a stormy existence of more than a century, have established a claim to the honours of history. In their early youth, in the dim old times, they may have been content with the pedler's pack, the travelling show-van, or a booth at a fair; but now they have grown rich and important, and have settled down into something like a permanent institution. At first, when they gave up their gipsy life, and started in regular business, they began as national, almost parochial, displays; and it was long before the growing free-trade spirit of the age allowed them to become international. Museums occasionally dabbled in the products of foreign industry, but with evident wonder and distrust. A catalogue of rarities exhibited at the Public Theatre of Leyden, in 1699, gives us a curious account of these early exhibitions. There was a Norway house, built of beams, without mortar or stone, side by side with a mermaid's hand, a crocodile, and several thunderbolts. There were a pair of Laplan's breeches and a pair of Polonian boots, mixed up with the chair of a midwife, and a model of a murdering knife found in England, "whereon was written, Kil the males, rost the females, and burn the whelps." There were a Roman lamp, "which burnes alwayes under ground," and a Persian tobacco-pipe, in companionship with the stomach of a man, and a mushroom said to be a hundred years old. Arabian jewels, East Indian coral trees, Egyptian linen, Chinese songs on Chinese paper, and a pot of China beer, had to be taken in connection with such delicacies as the snout of a sawfish, the skin of a woman, "prepared like leather," or the ears and tongue of a thief who had been hanged.

The Society of Arts may claim the credit of originating national exhibitions. The idea sprang naturally from the proceedings of a Society which was ostensibly founded to encourage arts, manufactures, and commerce. Amongst much that was weak, meddling, and even ridiculous, regarded by the light of our improved politico-economical knowledge, the Society succeeded in doing some good in its youth within the legitimate scope of its labours. In 1756—about the period when the Royal Academy first began its fine art exhibitions—it offered prizes for improvements in the manufacture of tapestry, carpets, porcelain, and other things, and exhibited the articles which were offered for competition. It also offered prizes for improvements in agricultural and other machines, and in 1761, a gentleman was paid to attend an exhibition of machinery in the Society's rooms, and to explain the models exhibited.

A few years after this, France came forward, most probably without any knowledge of the English exhibitions, and founded the first of that long and successful series of national expositions, which were only made international in France in 1855. The first French Exposition was opened in Paris, in 1797, by the Marquis d'Avèze, who originated the idea in the stormy days of the Directory, and lived to see it thriving under a Consul, an Emperor, and a King. The Exposition of 1798 remained open only three days, and the articles exhibited were of an aristocratic and costly, rather than of a popular character. The exhibitors numbered only one hundred and ten, and a jury of nine men was appointed to decide upon their merits. The second Exposition took place in 1801, when the exhibitors reached two hundred and twenty-nine. This display was considered so successful, and the preparation for it had been found so effective in keeping distressed workpeople employed, that the third Exposition was fixed to take place in 1802, after the short period of one year. The exhibitors, notwithstanding this short breathing-time, had increased to five hundred and forty, and their productions showed an extraordinary improvement in every way. Mechanical science had made manufacture easier, and had reduced the price of all articles in popular demand.

The fourth French Exposition opened, after a longer interval, in 1806, supported by the largely increased number of one thousand four hundred and twenty-two exhibitors. The fifth Exposition took place in 1819, after an interval of thirteen years, and showed the moderately increased number of one thousand six hundred and sixty-two exhibitors. It displayed, however, a marked improvement in many branches of popular manufacture. The sixth Exposition, in 1823, showed a slight decrease in the exhibitors, who, from the former number, had fallen to one thousand six hundred and forty-eight. On the other hand, the jury rewards were increased from eight hundred and nine to one thousand and ninety-one.

The seventh Exposition, in 1827, had one thousand seven hundred and ninety-five exhibitors. Steam-power in manufacture now began to be felt; goods had improved; prices had diminished; and the foundation was laid of a large export trade. The eighth Exposition was held in 1834, when a steady progress was shown in every branch of industry, and the exhibitors had increased to two thousand four hundred and forty-seven. The ninth Exposition, in 1839, had three thousand two hundred and eighty-one exhibitors, and was remarkable for its display of raw produce and a purer taste in design. The tenth Exposition, in 1844, was supported by three thousand nine hundred and sixty exhibitors, of whom no less than three thousand two hundred and fifty-three were honourably recognized by the jury. The eleventh Exposition, in 1849—the last of the purely national displays in France before our Great International Exhibition of 1851—was supported by four thousand four hundred and ninety-four exhibitors; and its great and predominating attraction was machinery.

The progress of like exhibitions in England during the early part of this period was not by any means so marked and steady. Such industrial displays had to fight their way against a vast amount of apathy and prejudice. The first project set on foot for commencing an annual public exhibition of this kind was coldly received, and even denounced by the mouthpieces of public opinion.

This exhibition, however, was formed in 1828, under the patronage of King George IV., on the plan which had been found successful in France, the Netherlands, and the United States, and the place fixed upon for the display was a royal stable. The King's Mews at Charing Cross, which was pulled down in 1833, and which stood on the site of Trafalgar Square, was fitted up to receive the few productions sent in for exhibition, and the Committee of management, consisting of the Hon. G. Agar Ellis as chairman, and a number of distinguished men, issued the following manifesto :—" It appears to the Committee that it has long been a desideratum among our most intelligent merchants and manufacturers, that an Annual Exhibition of specimens of new and improved productions of our artisans and manufacturers, conducted on a scale that should command the attention of the British public resident in and annually visiting the metropolis, would be highly conducive to the interests of the foreign commerce as well as the internal trade of the United Kingdom. In the opinion of the Committee such an exhibition will not only prove a powerful stimulus in promoting the farther improvement of our already successful manufacturers, but will also bring into notice the latent talents of many skilful artisans and small manufacturers now labouring in obscurity, and sacrificing inventions valuable alike to the country and to themselves, from wanting such an opportunity of introducing them to the British public."

Public opinion professed to wish well to any plan having the promotion of English arts and manufactures, and the encouragement of English inventive talent for its object, but it doubted whether the proposed exhibition was not extremely at variance with the established tastes and habits of British artisans and manufacturers. In a spirit of narrow national pride, it thought that exhibitions of the kind were only suited to countries where the arts were still in a state of infancy, and where they stood in need of every sort of adventitious aid; but that people like the British, who had eclipsed all other nations in the variety and excellence of their manufactures, could get on very well without such stimulating projects. The Boltons, the Wedgwoods, the Strutts, the Arkwrights, and the Bramahs of the time were almost exhorted not to support such a scheme, and no labour was spared to nip this, our first really national exhibition, in the bud.

The promoters of the exhibition, notwithstanding this opposition, worked very energetically in carrying out their plan. This Committee, or Board, as it was called, consisted of the Hon. G. Agar Ellis, M.P., before mentioned; Dr. Birkbeck; Mr. John Hales Calcraft; John Earl of Clare; Mr. Henry Drummond; Hugh Viscount Ebrington, M.P.; the Hon. G. M. Fortescue, M.P.; George Granville

Earl Gower; Lord Francis L. Gower, M.P.; Mr. John Labouchere; George Viscount Morpeth, M.P.; the Hon. Granville Ryder; Dudley Viscount Sandon, M.P.; Mr. C. Baring Wall, M.P.; Mr. Alexander R. Warrand; and the Hon. J. Stuart Wortley, M.P. The Treasurer was Mr. J. Labouchere; Mr. T. S. Tull acted as Secretary, and Dr. Birkbeck was the chairman of the Committee of Inspection.

The exhibition was opened on Monday, June the twenty-third, 1828, and it was described by the following long title:—"The National Repository for the Exhibition of Specimens of New and Improved Productions of the Artisans and Manufacturers of the United Kingdom, Royal Mews, Charing Cross." The public never accept a long name for a book or a building, and much in the same fashion as the "Great International Exhibition of 1851" found itself re-named "The World's Fair," or the "Crystal Palace," this forerunner of industrial exhibitions found itself simply called the "National Repository."

The outline of plan put forward by the managing Committee consisted of the following divisions:—Under the "first class" were brought in any "entirely new and ingenious constructions where a new principle is discovered, or one before known, but never practically adopted, is brought into operation." Under the "second class" were arranged any "new adaptation of some known principle, but in a manner essentially different from all that has been done before in that line of manufacture or mechanical workmanship." Under the "third class" were brought in "all improvements upon a discovery already made, by which the preparation of any article is facilitated, or its utility increased." Permission was also given to exhibit in this class all objects which were highly finished, or which "distinguished themselves by exquisite taste; likewise every description of elaborate workmanship, such as would not find a place in an exhibition of arts."

The under-committees of inspection were five in number; one governed by a chairman who was by profession a civil engineer; another governed by a chairman well acquainted with chemistry and the chemical arts; a third governed by a chairman well acquainted with silk, cotton, and woollen manufactures; a fourth governed by a chairman who was a mathematical instrument maker; and the fifth governed by a chairman who was well acquainted with workmanship in all kinds of metals.

It was resolved that the decision of the under-committees, with regard to the selection of articles submitted to their inspection, as well as to the class to which they might belong, should be final when signed by the chairman of the General Committee of Inspection. It was also resolved that the Presidents, Vice-Presidents, and Secretaries of all the Mechanics' Institutions in the United Kingdom should be invited to take on them the office of Committee of Inspection, with power to add to their number in their respective districts, and with the same power of deciding upon the admission of articles as the London Under-Committees of Inspection.

No charge for space in the building was made on any articles approved of by the committees of inspection, but all specimens had to be left under the control of the board of management until the close of the exhibition; then they were returned to their owners, unless they had been sold by request, in which case the exhibitor received his money instead of his goods.

When the building was thrown open on the appointed day, a great number of persons of all ranks hastened to inspect the articles exhibited. An extensive gallery, which ran from end to end of the King's Mews, had been neatly fitted up for the display, and various specimens of curious and highly-wrought manufactures, models of looms, bridges, &c., &c., and samples of useful and improved articles for domestic comfort or foreign commerce were arrayed with labels descriptive of the peculiar qualities which obtained them admission. A model of a chapel, and of a number of weavers in the act of weaving a piece of *Gros de Naples*, were also exhibited, to the great horror of a rather unfriendly critic. "If," he exclaimed, " the exhibition at Romani's cheap hosiery shop, Cheapside, of a stocking-weaver working at his loom is quackery, what is this ?"

There were no " special correspondents " of newspapers—no graphic reporters, in 1828, and therefore the accounts we get of this exhibition are dry and business-like. We are told of "beautifully executed works in chasing and cutlery," of "weaving in silks of remarkable patterns," of "models of engines and machinery for many purposes," of " little-known manufactures," and of " a multitude of curiosities;" but no word-pictures of the display are attempted, and when the reporter enters into more minute details, it is generally in no very friendly spirit that he does so.

Public opinion, in one of its surly moods, rose up and called the " National Repository " a " toy-shop." As nicknames go a long way in an argument, this was considered to be a severe hit, and public opinion was encouraged to renew the attack. It walked round the unfortunate exhibition, selected all the weakest points, probed them without mercy and without judgment, knocked the exhibitors down, leaped upon their models, admitted the respectability of the managing committee while it accused them of being fools; and, in fact, behaved in that overbearing way too common with public opinion when asked to tolerate a novel experiment. It accused the exhibition of being paltry, and said if it was indeed national, the nation was never before shown in so contemptible a light. It affected to remember that such exhibitions were had recourse to in France and Holland, in order to excite a manufacturing spirit in the people of those countries, and to enable them to dispense with the wares of England. Bonaparte—the great bugbear—was held to have been at the bottom of it all,—striving to close the Continent against us, and to bring ruin on our trade and manufactures. Public opinion shook the British manufacturer by the hand, and congratulated him on the envied ascendency he had maintained in spite of all his supposed rivals and enemies had done or could possibly do. It was not surprised that the English people should brood over these facts as they were called, and should frown on any attempt to introduce what it erroneously styled a foreign institution. It thought that no person could be so blind as not to see, that to foster such exhibitions would only be to adopt the stale device of an enemy. It boldly announced that France had turned her back at last upon these displays, and had resolved to hold no more, being convinced that they were more injurious than useful. The assertion was incorrect, and we may therefore regard the inference as unreliable. Public opinion is not always right, and in this case it was eminently wrong.

The " National Repository," thus hunted down by those who ought to have been its friends, could not boast of a very profitable existence. It struggled through with some four exhibitions of decreasing merit, and, like an actor who persists in keeping the stage too long, its last appearance gave its enemies fair material for banter. It was called the "exotic thing," and although its critics professed to have no desire to exult over the failure of good intentions, still they did exult considerably. The "exotic thing" was shown up in a good deal of the three-notes-of-admiration style of writing, and was broken on the arithmetical wheel. As it had only collected sixty specimens of industrial art for this fourth exhibition, and there were five hundred and fifty manufacturing towns in England, Scotland, and Ireland, it was easily shown that this only gave one invention to every nine towns. The keen-eyed critic was much shocked, on looking a little deeper into the exhibition, to find that at least eight of the sixty specimens exhibited were the production of foreigners; but he abstained from asking what the country was coming to. He was ungallantly severe upon certain "young misses of the Minories and the Gravel Pits," who had contributed "scissors-and-pencil work" to the exhibition—"pretty imitations of Nature," as it was called in the catalogue. The shilling a head charged for admission was considered dear, and the "exotic thing" was consigned to oblivion as a fourth-rate bazaar.

When it left the "King's Mews" in the following year (1833), and carried on its withered business for a short time at a room in Leicester Square, it was still followed by a few barking enemies. It was contrasted disadvantageously with the "National Gallery of Practical Science"—the "Adelaide Gallery" in the Strand —which started with the attraction of many electrical machines, a noisy steam gun, and an electrical eel, gradually sunk into a casino, and is now an echoing desert.

SOCIETY OF ARTS' HOUSE.

CHAPTER III.

THE SOCIETY OF ARTS.

URING this time the Society of Arts had kept the lamp burning. In 1829, the Secretary of the Society read papers on several of the leading industries of the country, and from this date specimens of raw materials, manufactures, and new inventions were frequently collected in the old rooms in the Adelphi, for the instruction of the members and the public. Then followed local Trade Exhibitions, held at Manchester, Birmingham, Leeds, Dublin, and other places; and the Exhibition of Manufactures at the Free-Trade Bazaar, held in Covent Garden Theatre, in 1845. In that year the Society of Arts tried to revive the idea of forming periodical exhibitions of industrial products in England on the plan of the French Expositions. A Committee of the Council of the Society was appointed to make the necessary inquiries as to the willingness of manufacturers to contribute their productions to such an exhibition, and a fund was subscribed for the purpose of meeting the preliminary expenses, but owing to the want of sympathy on the part of the manufacturers the project was not then proceeded with. The English people were then very imperfectly acquainted with the value of such exhibitions. They required to be educated on this point, and education had to be provided.

The Society of Arts had been losing strength for many years, when it was aroused to a new course of life in 1846 by its Royal President. The Prince Consort advised its Council, that the action of the Society most likely to prove immediately beneficial to the public and itself, would be that which would encourage most efficiently the application of the Fine Arts to our Manufactures.

To carry out this idea, the Council at once established a Special Prize Fund, and offered premiums and medals for the production of manufactured articles of simple form—for colours to be used in porcelain, and capable of resisting the action of acids, but not then used in England, and for excellence in combined form and colour. The object of these prizes was to promote a love for the beautiful, by supplying articles of elegant form, made of cheap materials, and suited to the uses of every-day life. The first competitive designs were to be sent in to the Society on or before the fifteenth of May, 1846, and amongst the articles received at that date, was a tea-service in one colour, manufactured by the Messrs. Minton, to which the Society of Arts awarded its special prize. It might be said that the Great Exhibition of 1851 was founded on a teacup, for upon this tea-service, and the jugs, mugs, and other articles rewarded with prizes in 1846, the whole subsequent action of the Society relative to exhibitions was based.

The articles rewarded with prizes in 1846, together with those sent in for competition in 1847, formed the basis of the first exhibition of "Select Specimens of British Manufactures and Decorative Art," which was opened at the house of the Society of Arts in March, 1847. The introduction to the catalogue sets forth the object of the exhibition in the following words:—"The Exhibition of Select Specimens of British Manufactures and Decorative Art is the commencement of a series of annual exhibitions, by means of which the Society hopes to contribute essentially to the progress of those objects for the encouragement of which it was originally instituted." . . . . "The first step in the improvement of an art or manufacture is the knowledge of what has already been done in that art or manufacture." . . . . "To make improvements with advantage we should begin at the very summit of that perfection which has already been attained. It is for this reason that the Society of Arts have now thought it to be their duty to exhibit each year in some department of arts or manufactures, the degree of perfection that has already been attained." . . . .

"We have no doubt that after the eyes of the public are familiarized with specimens of the best decorative art, they will prefer them to subjects which are vulgar and gaudy; and that after a series of such annual exhibitions, no manufacturer will have to complain that his best productions are left on his hands, and his worst preferred."

Manufacturers were slow in agreeing with the Society of Arts about the value of these exhibitions. Very few competitors came forward in 1846, and it was with difficulty the judges could find subjects worthy of reward. The exhibition of 1847 would have been a total failure but for two individuals, who made it a point of personal favour with a few great manufacturers, to be permitted to select from their stores a sufficient number of articles to make a show. The result was highly satisfactory. Twenty thousand people visited the exhibition, and the Council arranged a third display, which was opened in March, 1848. This time the contributions from manufacturers were sent in unsolicited, and even forced upon the Society, and upwards of seventy thousand persons visited the Society's rooms. So great and genuine was the success of this exhibition considered, that in a paper read to the Society on the eighth of March, 1848, it was proposed to obtain the loan of a selection of the articles exhibited, and to circulate them through the country wherever Schools of Design existed. It was also proposed, "with the co-operation of the Board of Trade, that the Society of Arts should, every fourth year, make a collected exhibition of the principal subjects exhibited in the previous three years, and of others expressly prepared for the special purpose, and that such national exhibition should take place in some large building purposely provided, if not at the cost of the Government, at least with the government sanction."

The eyes of the Society were still directed affectionately to Trafalgar Square, as a spot flavoured with exhibition facts and associations, and an open space which could afford to bear a large temporary building. Some thought that the square of Somerset House would be the best site, and another party opened negotiations with the proprietor of the Baker Street Bazaar. It was proposed that the Society of Arts should collect the articles to be exhibited, and manage the money details of the exhibition. The admission was to be partly free and partly by payment, and the receipts were to be applied to the expenses incurred in paying for honorary medals and rewards, and to form a fund for future exhibitions.

A deputation from the Society of Arts waited upon the President of the Board of Trade to submit this scheme. Mr. Lefevre, the Secretary of the Board of Trade, formed one of the deputation, and, as a member of the Society of Arts, he pointed out the advantages which the Schools of Design would derive from the liberal offer of the Society. Mr. Labouchere, the President, immediately accepted this offer, and promised the deputation the co-operation of the Board in carrying out the proposed plan for giving the provincial Schools of Design the benefit of the periodical exhibitions of the Society of Arts. He also expressed a deep interest in the proposed national exhibition, and referred the deputation to the Chairman of the Woods and Forests (Lord Morpeth) to arrange the site for the necessary building.

The Society's exhibition of manufactures in 1848 was followed by an exhibition of pure Art—known as the "Mulready Exhibition"—in June of the same year, and at the opening of the Society's Session in November, 1848, its first exhibition of models of machinery was announced to take place in January, 1849. This formed the beginning of a series of exhibitions of inventions, which have been held annually from that time in the Society's rooms, each year proving, by the increased applications for space, that the ordinary resources of the Society of Arts were becoming less and less sufficient to meet the growing interest of the public in such collections.

In the spring of 1848, the third general "Exhibition of Recent Specimens of British Manufactures and Decorative Art" was held at the old house in the Adelphi. It was far more successful than the former exhibition; the visitors were more numerous, the articles sent in were of a superior quality, and the public taste was better educated to appreciate their excellence. This exhibition was closely followed by a second art display—known as the "Etty Exhibition"—which took place in the same rooms in June, 1849.

Eighteen hundred and forty-nine was the eventful year in which the industrial scheme began to show signs of unmistakable life. In that year an exhibition of manufactures and art in connection with the meeting of the British Association for the Advancement of Science was held at Birmingham with very encouraging results. In France, the National Exposition of that year also met with unusual success, and coming as it did just as this country had embarked on its career of partial free-trade, it gave a fresh impulse to the idea of holding a Great National Exhibition of British Industry.

On the occasion of the opening of the third exhibition of British manufactures at the rooms of the Society of Arts, on the seventh of March, 1849, we find the following remarks in the address of the Council read to the members:—

" The Society is aware that the exhibitions, necessarily limited each year to certain classes of manufactures, are only parts of a series of displays which it is proposed shall culminate every fifth year in a Great National Exhibition embracing *all* manufactures.

" The revolution of the first fifth year will arrive in 1851, and the Council feel that it will be necessary forthwith to mature those arrangements for giving due effect to this event, which have already been successfully instituted and carried to a certain point with the President of the Board of Trade and the Chief Commissioner of Woods and Forests. The Board of Trade has already promised co-operation, and the Chief Commissioner of Woods, a suitable site for the building in which the exhibition may be made. It only remains for the Government to take the risk of providing a temporary building of dimensions sufficiently ample for the purpose. The Society of Arts, having practically demonstrated the means of establishing such exhibitions, and educated most successfully a numerous public of all classes of society to appreciate them and crowd to see them ; having induced the noble designers, most eminent manufacturers, ingenious mechanics, skilled workmen, and men of science, to assist in these exhibitions ; having been aided by the active co-operation and good-will of the most distinguished nobles,— the Council feel that they shall be warranted in preferring a request to Her Majesty's Government to do its part in this great object, and to provide once in every fifth year a suitable building in which national exhibitions, duly representing the best productions in all branches of manufactures, may be found."

At the distribution of prizes awarded by the Society, and presented by the Prince Consort on the fourteenth of June, 1849, just after the exhibition of British manufactures had closed, so great had been the success of that exhibition over all the preceding exhibitions, that it was stated that the great object of a national exhibition of industry was more likely than it had ever before appeared to be carried out to a successful issue.

The promoters of this scheme only contemplated a national exhibition, and they asked for pecuniary aid from Government to enable them to carry it out.

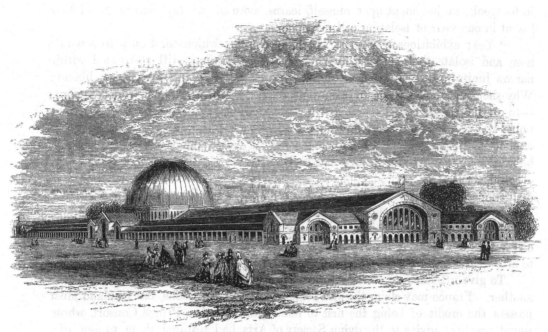

DESIGN OF BUILDING COMMITTEE, 1851.

CHAPTER IV

EARLY STRUGGLES OF THE FIRST INTERNATIONAL EXHIBITION.

MANY have advanced claims since 1851 to be considered the originators of the proposition for holding universal or international exhibitions. There have not been quite so many claimants for the honour as came forward to assert a parental right over the new rifle movement, but they have been sufficiently numerous and self-confident. Amongst the mass of claims—some of them of the wildest kind—there is one which must be generally admitted. M. Boucher de Perthes, President of the *Société Royale d'Emulation* of *Abbeville*, boldly recommended the holding of an " Exposition Universelle " in the year 1834, in an address which he then delivered to that Society. It may be found in the published records of that Institution (at page 517), and is reprinted in *Le Petit Glossaire* of the same author in 1835. The following is a close translation of this remarkable speech, so full of enlarged views, too liberal for the time :—

" Let us work, but let us be tolerant, and look upon every workman as a brother, whatever be his name, his colour, or his country. Let us not quarrel with competition. That alone will enlighten us, and point out the true path of our industry. It is competition which makes the good workman, because it is competition which indicates to each one his proper task. Exhibitions which have so beneficial an influence upon industry are but an embodiment of this rivalry. It is here that the producer brings the fruit of his labour side by side with that of his neighbour,—takes the measure of his efforts, estimates the merits of his productions, and sitting, so to speak, in judgment upon himself, learns more in one day than he could have learnt in one year of isolation and monopoly.

" Yes; exhibitions are better than prohibitions, which tend only to separate men and isolate them. Why then are these exhibitions still restrained within narrow limits? Why are they not instituted upon a scale truly large and liberal? Why should we be afraid to open our halls to manufacturers whom we call foreign; —to the Belgians;—to the English;—to the Swiss;—to the Germans? How noble would be a European Exhibition, and what a mine of instruction it would offer for all! Do you imagine that the country in which it should take place would be a loser by it? Do you believe that if the *Place de la Concorde,* opened this first of May, 1834, to the productions of French industry, should be opened to pro- ductions of the whole world,—do you believe, I say, that Paris, that France, would suffer, or that we should, in consequence, produce less, or become inferior? No, Gentlemen: France would not suffer any more than the foreigner; nor our city more than the capital. Exhibitions are always beneficial, and beneficial to all."

To give utterance to an idea is one thing, but to create the thing thought of is another. France may claim the idea of international exhibitions, but England must possess the credit of being the first to realize them. The Prince Consort, whose sound practical advice to the dying Society of Arts had restored them to new life, was the first to take the Society's plan for an enlarged national display in hand, and to mould it into a universal exhibition. His views were broader than the European circle in which M. Boucher de Perthes wished to confine his scheme, and all the nations of the earth were invited to free competition.

At a meeting held at Buckingham Palace on the twenty-ninth of June, 1849— the minutes of which are given in Mr. Henry Cole's Introduction to the Illustrated Catalogue of 1851—the Prince Consort suggested the four great divisions—of *Raw Material—Machinery and Mechanical Inventions—Manufactures—and Sculp- ture and Plastic Art*—of which he proposed the Great Exhibition should consist. He likewise disposed of all questions about the site for a building by suggesting the occupation of that vacant ground in Hyde Park, which afterwards contained the renowned "Crystal Palace." The questions of premiums, or prizes, to manufac- turers,—of a Royal Commission to give weight and authority to the plan and to aid in carrying it out,—and of a broad popular subscription to be organized by the Society of Arts, were also discussed and settled. Above all, the international character of the proposed exhibition was fixed in the following words:—" It is con- sidered that, whilst it appears an error to fix any limitation to the productions of machinery, science, and taste, which are of no country, but belong as a whole to the civilized world, particular advantage to British industry might be derived from placing it in fair competition with that of other nations."

A second meeting was held at Osborne on the fourteenth of July in the same year, at which a general outline of a plan of operations was submitted, and the details were more fully discussed with the President of the Board of Trade. On the thirty-first of July, 1849, the Prince Consort, in his capacity of President of the Society of Arts, addressed a letter to the Home Secretary, Sir George Grey, in order to bring the subject officially under the notice of Her Majesty's Govern- ment, and the answer received was prompt and satisfactory.

The next step was taken by the Society of Arts, acting under the direction of the Prince Consort. As funds were required to set the machinery in motion, and the ordinary resources of the Society were not available for such a purpose, it became necessary to provide for an estimated outlay in building and preliminary expenses of seventy thousand pounds. In this position they were compelled at the commencement of their proceedings to make an arrangement with a firm— Messrs. James and George Munday—willing to advance the sums likely to be required, in consideration of a share in the contingent profits. The proportion of this share was at first fixed, but afterwards, at the request of the Society of Arts, it was left to be decided at the close of the exhibition by arbitrators chosen on both sides. A clause was introduced in the agreement, giving the Society of Arts power to cancel it if requested to do so by the Lords of Her Majesty's Treasury within a specified period, provision being made for the repayment to the Messrs. Munday of any sum that might have been advanced by them, together with a fair compensation for the outlay and risk which they might have incurred.

At this period so little was known of the general feeling of manufacturers and agriculturists towards such displays in this country, that a commission, consisting of Mr. Henry Cole and Mr. Francis Fuller, was appointed to visit the principal towns in England, Ireland, and Scotland, and collect opinions from the leading men. The result was most satisfactory. On the point of the *general expediency of such periodical exhibitions* they met with the most perfect unanimity in all parts of the country, and expressions of surprise, if not of regret, that England should have been so slow in instituting such an exhibition. It was considered that the benefits of the exhibition would be great, individually and nationally; that great good had been done on the Continent by such displays; and that the broader the competition the better would it be for all. A willingness to exhibit was found to be general; and on the point of *whether prizes should be awarded*, although it was considered that the best prize was commercial success, it was admitted that the wholesome rivalry caused by worthy prizes would be beneficial if a thoroughly impartial distribution could be secured.

On the important point of *whether such exhibitions should be supported by a national grant or by voluntary subscriptions*, the preponderance of opinions was in favour of the voluntary principle; and it was generally considered that if the financial aid of Government were sought, the public would feel themselves relieved in a great measure from the necessity of assisting.

Fifty towns were visited after this, and meetings were held, at all of which favourable resolutions were passed; and by January, 1850, the names of sixty thousand influential persons had been obtained as supporters of the great plan.

While the Society of Arts, through their commissioners, were thus canvassing the country, they judged it to be wise to put themselves in possession of what the French people had done in promoting such undertakings. Mr. Digby Wyatt therefore, acting under their instructions, prepared an elaborate analytical report upon the French Exposition of 1849, with an historical sketch of all previous French expositions.

Before the plan however was adopted by the country, a dinner was given at the Mansion House in May, 1850, by Mr. Farncomb, then Lord Mayor of London, to promote the exhibition. At this banquet the Prince Consort placed the subject with great force and clearness before the municipal authorities assembled, and, through the press, before the whole country the next morning. In his own words, the collection and exhibition in one building of the Works of Industry of All Nations was " to give a true test and a living picture of the point of development at which the whole of mankind had arrived in this great task, and a new starting-point from which all nations will be able to direct their further exertions."

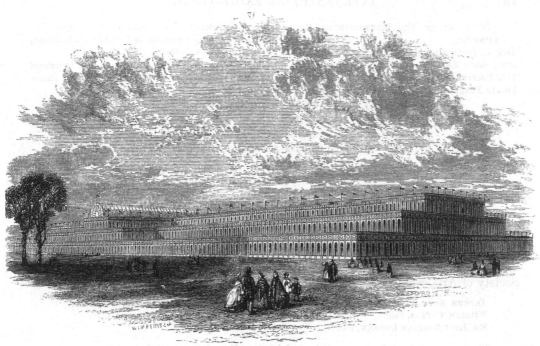

EXHIBITION, 1851.—END VIEW.

CHAPTER V.

THE ROYAL COMMISSION.

UPON the presentation of the reports prepared under the direction of the Society of Arts, the Queen was pleased to issue a Royal Commission, which was published in the London Gazette of the third of January, 1850. It provided—that a full and diligent inquiry should be made into the best mode by which the productions of English colonies and foreign countries might be introduced into the kingdom, as to the site for and the general conduct of the proposed exhibition, and as to the best mode of determining the nature of the prizes, and of securing the most impartial distribution of them. It also provided for the appointment of local commissioners where they were required in England or abroad, and for their removal if necessary, and a power to summons witnesses for examination. Under this Commission the following appointments were made:—

HER MAJESTY'S COMMISSIONERS.

PRESIDENT, HIS ROYAL HIGHNESS PRINCE ALBERT, K.G., F.R.S.

HIS GRACE THE DUKE OF BUCCLEUCH, K.G., F.R.S.	RT. HON. W. E. GLADSTONE, M.P.
RT. HON. THE EARL OF ROSSE, K.P., Pr. of R.S.	SIR RICHARD WESTMACOTT, R.A.
RT. HON. THE EARL OF ELLESMERE, F.S.A.	SIR CHARLES LYELL, F.R.S.
RT. HON. THE EARL GRANVILLE.	SIR CHARLES LOCK EASTLAKE, P.R.A., F.R.S.
RT. HON. LORD STANLEY.	THOMAS BARING, ESQ., M.P.
RT. HON. LORD OVERSTONE.	CHARLES BARRY, ESQ., R.A., F.R.S.
RT. HON. LORD JOHN RUSSELL, M.P., F.R.S.	THOMAS BAZLEY, ESQ.
RT. HON. HENRY LABOUCHERE, M.P.	RICHARD COBDEN, ESQ., M.P.

C

THE PRESIDENT OF THE INSTITUTION OF CIVIL ENGINEERS.
THOMAS FIELD GIBSON, ESQ.
JOHN GOTT, ESQ.
THE PRESIDENT OF THE GEOLOGICAL SOCIETY.
PHILIP PUSEY, ESQ., M.P., F.R.S.

JOHN. SHEPHERD, ESQ., Chairman of the Hon. East India Company (Successor to Sir Archibald Galloway, K.C.B., &c., who died in 1850).
ROBERT STEPHENSON, ESQ., M.P., F.R.S. (appointed on his resignation of the Chairmanship of the Executive Committee).
ALDERMAN THOMPSON, M.P.

J. SCOTT RUSSELL, Esq., F.R.S.,
SIR STAFFORD HENRY NORTHCOTE, BART., C.B., } Secretaries.
EDGAR A. BOWRING, *Acting Secretary.*

The TRUSTEES appointed by the Society of Arts for the Prize Fund of twenty thousand pounds, were:—

THE RIGHT HON. THE EARL OF CLARENDON, K.G.

SIR JOHN PETER BOILEAU, BART.
JAMES COURTHOPE PEACHE, ESQ.

The TREASURERS for all receipts arising from donations, subscriptions, or any other source on behalf of or towards the proposed exhibition, appointed by the Society of Arts, were:—

ARTHUR KETT BARCLAY, ESQ.
WILLIAM COTTON, ESQ.
SIR JOHN WILLIAM LUBBOCK, BART.

SAMUEL MORTON PETO, ESQ.
BARON LIONEL DE ROTHSCHILD.

These appointments were confirmed by the Royal Commissioners.

The TREASURERS for payment of all executive expenses, appointed by the Society of Arts, were:—

PETER LE NEVE FOSTER, ESQ.
JOSEPH PAYNE, ESQ.

THOMAS WINKWORTH, ESQ.

The Executive Committee, four members of which were originally appointed by the Society of Arts before the issuing of the Commission, and one member of which, Mr. George Drew, was appointed to watch the interests of the money contractors, Messrs. Munday, were:—

HENRY COLE, ESQ.
CHARLES WENTWORTH DILKE, ESQ. (the younger).
FRANCIS FULLER, ESQ.

ROBERT STEPHENSON, ESQ. (subsequently replaced by LIEUT.-COL. REID).
GEORGE DREW, ESQ. (who subsequently withdrew).

MATTHEW DIGBY WYATT, ESQ., *Secretary.*

The duties of this Committee were " to carry the exhibition into effect, under the directions of the Prince Consort and the Commissioners."

A Special General Meeting of the Members of the Society of Arts, in conformity with a requisition presented to the Council of that body, was held on the eighth of February, 1850, " to consider the position of the Society with respect to the Industrial Exhibition proposed to be held in 1851." At this meeting Mr. Scott Russell read a statement of the preliminary steps which had been taken, and the Society resolved that the scheme of the Prince Consort was worthy of him, and of the hearty co-operation of the members, and it pledged itself to promote the success of the undertaking in every way. It resolved that a subscription list for this purpose should be opened at the house of the Society of Arts, and that the members should be requested to use their exertions to get it filled.

At the first meeting of the Royal Commission the subject of the contract with the Messrs. Munday was taken into consideration. It was thought that the exhibition would stand on a much firmer, because more popular, basis, if the wishes of the leading traders of England, as set forth in the report of Messrs. Cole and Fuller, were acted upon, and the whole burden of the scheme thrown upon voluntary contributions. In arriving at this conclusion the Commissioners announced that they did not intend to cast any slur upon the contract. They were fully convinced that it had been entered into with the best intentions by the Society of Arts, and with a most liberal spirit by the Messrs. Munday, and that its conditions were strictly reasonable and, indeed, favourable to the public. They judged, however, that the maintenance of any contract giving to a great national undertaking the appearance of a private speculation would not be agreeable to the public, and would endanger the success of the exhibition both at home and abroad. Upon these grounds they addressed a request to the Lords of the Treasury, that their Lordships would exercise the power—before alluded to—reserved to them in the contract between the Society of Arts and the Messrs. Munday, of requesting the former body to give immediate notice of their intention to determine the contract. In order to make this request valid, it was necessary that the Lords of the Treasury should intimate their willingness to take upon themselves the responsibility of repaying to the Messrs. Munday the sums already advanced by them, together with such amount of compensation as arbitrators might award to them. As it had been understood, however, from the first issue of the Commission that no part of the funds for carrying on the exhibition was to come out of Her Majesty's Exchequer, the Lords of the Treasury required that, before making such an intimation to the Society of Arts, they should receive from the Royal Commission an assurance that the necessary sum should be forthcoming when wanted, which assurance the Commissioners gave.

The amounts advanced by the Messrs. Munday were twenty thousand pounds sterling invested for the Prize Fund, and two thousand five hundred pounds advanced for general purposes. This sum was repaid, with interest, on the twenty-second of November, 1850; and the question of compensation for loss of time, personal services, and risk having been referred, by mutual consent, to Mr. Robert Stephenson, that gentleman, after a full inquiry into the circumstances, and after hearing counsel on the case, ultimately awarded five thousand one hundred and twenty pounds and the costs to the Messrs. Munday, which sum was paid.

The Commission defined the functions of the Commissioners to be only those of inquiry and general direction, while the pecuniary responsibilities and the performance of the executive duties were to fall on the Society of Arts; but the cancelling of Messrs. Munday's contract altered this, and placed the whole responsibility on the shoulders of the Commissioners. The Executive Committee therefore felt it becoming to tender their resignation, which was not accepted, and arrangements were made to meet the altered circumstances of the case. Her Majesty was advised to issue supplementary commissions, appointing Mr. Robert Stephenson, M.P., a Commissioner, upon his resignation as Chairman of the Executive Committee (before alluded to), and Lieutenant-Colonel W. Reid, R.E., in his place. Messrs. Munday, and their representative, Mr. George Drew, from this time ceased to attend the meetings of the Executive Committee.

The Commissioners now appealed directly to the public to contribute to this great national undertaking. A subscription list was opened, and they announced to the public that they were exclusively responsible for the application of the funds, and would insure an effectual control over the expenditure, and a proper audit of the accounts. The Council of the Society of Arts felt, with the Government, that the broad scheme suggested by the Prince Consort merited being made a national undertaking, and they therefore resigned the work of realization to the Royal Commission without a murmur. In determining the contract with the Messrs. Munday, they made no conditions with the Royal Commissioners as to the appropriation of any probable surplus arising from the exhibition. They said, in an address to the general body of Members on the twenty-second of July, 1850, " the exhibition is now being carried out under the Prince Consort even on a larger scale than the Society had originally meditated. The satisfaction with which the extension of plan, so peculiarly their own, has been viewed by the Society, is testified to by the fact that the members have contributed in their several localities no less a sum than six thousand one hundred and eighty-seven pounds, in addition to a further sum of eleven hundred and one pounds paid into the hands of the Treasurers at the Society's house in answer to an appeal made to them in February." This amount may appear small, compared with the sum readily subscribed by Members of the Society under the Guarantee Deed for the Exhibition of 1862; but public confidence in such undertakings has grown even more rapidly than the undertakings themselves.

The subscriptions promised to the Commissioners were made public from time to time. The total amount reported was seventy-nine thousand two hundred and twenty-four pounds thirteen shillings and fourpence, of which sum sixty-seven thousand eight hundred and ninety-six pounds twelve shillings and ninepence had been actually paid to the credit of the Commission on the twenty-ninth of February, 1850. A portion of the subscriptions received in some of the provincial districts was retained to defray the expenses of collection and local management. This amount was only gathered after many laborious meetings in town and country, and the collection occupied so much time, that it was clear some other plan must be adopted, or the International Exhibition be given up.

"At the commencement of the Commissioners' proceedings" (says the First Report of the Commissioners), "while they were incurring no expenses beyond those of the remuneration of their officers and the necessary outlay on printing, advertising, and other comparatively small items, the subscriptions received from time to time were amply sufficient for their wants; and they did not experience any inconvenience from the want of a more definite legal position than that of a mere Commission of Inquiry. When, however, in the month of June, 1850, the plan of a building to cost seventy-nine thousand eight hundred pounds had been approved, and it became necessary that a contract should be made for its erection, questions naturally arose as to the power of the Commission to enter into and to enforce such a contract,—as to the person or persons by whom such a contract should be signed, and the individual responsibility which, by so signing it, they would incur,—and as to the mode in which the money that should be required beyond the amount of the subscriptions received was to be provided."

It was suggested at the last moment to raise money on a guarantee for a much larger sum than was likely to be required, and this sum was fixed at a quarter of a million sterling. A few individuals (amongst whom Sir Samuel Morton Peto stood prominent) boldly stepped forward to incur the responsibility of signing a guarantee deed to this amount, and the Prince Consort, amongst the few, put his name down for ten thousand pounds. The Commissioners, acting on their doubts about their position, obtained a Royal Charter of Incorporation from Her Majesty, dated August the fifteenth, 1850, and, with the Guarantee Deed in their hands, they were able to obtain what advances they required from the Bank of England. The sums so borrowed, amounting in the whole to thirty-two thousand five hundred pounds, were repaid, with interest, on the twenty-second of May, 1851, out of the receipts at the doors, after the Exhibition had been open for three weeks. The direct control over the whole expenditure of the Commission (subject to the approval of the Commissioners) was exercised by a FINANCE COMMITTEE, consisting of Lord Granville (Chairman), Lord Overstone, Mr. Labouchere, Mr. Gladstone, Sir Alexander Spearman, Mr. T. F. Gibson, Mr. Thos. Baring, Mr. Cobden, and Sir S. Morton Peto, with Mr. Edgar A. Bowring as Secretary.

The business of the Commission was divided amongst several Committees, but the largest share of the work fell upon the Executive Committee. As Mr. Drew withdrew, although invited to remain, and Mr. Fuller stated that he was unable to devote the whole of his time to the exhibition, this Committee was practically reduced to three members,—Sir William Reid, Mr. C. Wentworth Dilke, and Mr. Henry Cole. Their duty required them to exercise continued watchfulness over every part of the vast undertaking. Sir W. Reid more particularly undertook the task of communicating with the public departments, Mr. Henry Cole settled the troublesome questions of space and arrangement, and Mr. C. Wentworth Dilke took charge of the correspondence, and was the general superintendent. To take the letter-work of this Committee alone ;—from October, 1849, to December, 1851, inclusive, fifty-one thousand nine hundred and thirteen letters were received, one hundred and sixty-one thousand six hundred and thirty-one letters were sent out, and the amount paid for postage and parcels was nearly fifteen hundred pounds. The greatest number of letters received on one day was five hundred and twenty-two, on the first of March, 1851, during the correspondence relative to space and arrangement. The greatest number despatched on one day was seven thousand eight hundred and thirty-five, on the ninth of October, 1851, when sending out cards of admission to exhibitors and others for the closing of the Exhibition. The papers printed by the chief Committees in carrying on the business of the Exhibition were eleven hundred and thirty-three in number, and they were circulated to the extent of a million and a half of copies.

The local organization begun by the Society of Arts, by which sixty-five district committees were formed, was much extended by the Royal Commissioners. The local committees were increased to two hundred and ninety-seven, and about four hundred and fifty local commissioners were nominated. To insure uniformity of action, and do away with a vast amount of letter-writing, two special travelling commissioners—Dr. Lyon Playfair, and Lieut.-Colonel Lloyd—were appointed to communicate with these local committees. Commissioners were also appointed, or committees formed, in eleven British colonies and thirty foreign countries.

EXHIBITION, 1851—TRANSEPT.

CHAPTER VI.

THE EXHIBITION OF 1851.

WITH all this elaborate organization it was impossible to make any extensive inquiries as to the amount of articles likely to be tendered for exhibition. There was no time to be lost in commencing the building, and it was therefore necessary to lay down some arbitrary rule with regard to space. The Commissioners accordingly fixed the area of the building at eight hundred thousand square feet, or a little more than eighteen acres,—a space between three and four times as large as that occupied by any previous exhibition abroad. The space thus fixed was increased during the carrying out of the undertaking by additional galleries to rather more than one million of square feet.

At the third meeting of the Commissioners, held on the twenty-fourth of January, 1850, they appointed a Building Committee, consisting of:—

THE DUKE OF BUCCLEUCH.
THE EARL OF ELLESMERE.
MR. (afterwards SIR CHARLES) BARRY, R.A.
MR. (afterwards SIR WILLIAM) CUBITT, PR. I.C.E.   } All Members of the Royal
MR. STEPHENSON, M.P.   Commission.
MR. BRUNEL.
MR. COCKERELL, R.A.
MR. DONALDSON.

This Committee held thirty-eight meetings between February the fifth and July the twenty-third, 1850.

The first debate was about a site, and the Commissioners selected the spot in Hyde Park (with the permission of the Crown) which had been originally suggested by the Prince Consort. Out of the twenty acres of land chosen, the Building Committee proposed to cover sixteen acres with buildings, the plan and arrangement of which were to be determined by public competition. An invitation, dated March the thirteenth, 1850, was therefore published, calling upon architects and others for designs.

The eighth of April was the last day fixed for the receipt of these plans, giving barely a month for their execution; but, short as the time was, no less than two hundred and thirty-three designs and specifications were sent in by the appointed date. Twenty-seven of these came from France, two from Belgium, three from Holland, one from Hanover, one from Naples, two from Switzerland, one from Prussia, and one from Hamburgh, making altogether thirty-eight as contributed by foreigners. One hundred and twenty-eight were sent in by Londoners, fifty-one by designers living in the country; six came from Scotland; three from Ireland; and seven were anonymous. One was sent in by "a lady with great diffidence," but the majority were sent in with great confidence. Twelve additional plans were received after the eighth of April, and the whole were publicly exhibited for one month from the tenth of June, at the Institution of Civil Engineers. Some of them sprang from that imaginative school which loves to turn all our black river-side wharves into marble palaces on paper, and dreams of reviving the glories of ancient Babylon at Fleet Ditch. One presented a striking picture of a great central dome depressed at the top, which looked very much like a smashed egg.

The Building Committee reported on the merits of all, dividing the competitors into two lists. One they considered "entitled to favourable and honourable mention;" the other to "further higher honorary distinction." They selected none of the designs, however, and announced that they considered "no single plan so accordant with the peculiar objects in view, either in the principle or detail of its arrangement, as to warrant them recommending it for adoption."

Some of the designs taught the Building Committee what to avoid, while others gave them valuable information. From the careful examination of the various plans many practical conclusions as to the mode of arranging the building were derived. Taking a hint from one and a feature from another, the Committee prepared a design, for the realization of which a complete set of working drawings, specifications, and quantities were prepared under the immediate superintendence of Mr. M. D. Wyatt, Mr. Owen Jones, and Mr. C. H. Wild. Invitations were issued for tenders upon the basis of the plans so prepared, and for any suggestions that might show a way of reducing the estimates of cost. Nineteen tenders were received at the appointed time, the tenth of July, only eight of which were for the entire work.

Objections were raised at this period, chiefly on the part of the public, to the erection of any building in Hyde Park that was not of a light and temporary character. The subject was debated in the House of Commons on the first of July, 1850, and full explanations were given by individual members of the Commission in both Houses. A feeling, however, still prevailed against the employment of durable materials, and particularly of brickwork, in the erection of the building; and the existence of this feeling induced Mr., now Sir Joseph, Paxton to turn his attention to the subject, and led him to submit a plan for a structure chiefly of glass and iron, similar to those which had been successfully tried by him at Chatsworth. Messrs. Fox, Henderson, and Co. submitted estimates for the construction of the Building suggested by Sir Joseph Paxton, and the Commissioners, after much deliberation, adopted this plan on the twenty-sixth of July, and accepted the tender of the contractors to carry it out for the sum of seventy-nine thousand eight hundred pounds.

The duty of superintending the construction of the building fell upon Sir William Cubitt, who acted on behalf of the Commissioners. The chief gentlemen employed under him were, Mr. Wild, who was responsible for the engineering details; Mr. Owen Jones, who was responsible for the decoration; and Mr. Digby Wyatt, who attended to the general building arrangements, the contracts, and monthly accounts. Mr. Wyatt from that time ceased to act as Secretary to the Executive Committee, although his name was retained in all official documents, so as to avoid inconvenience. Possession of the site was obtained on the thirtieth of July, and a hoarding was immediately erected, enclosing it. The Contract Deed was not actually signed till the thirty-first of October, but the commencement of the works was not delayed for this preliminary, and the first column of the building was fixed as early as the twenty-sixth of September. On the fourteenth of November the Commissioners, to meet the requirements of the Lords of the Treasury and the Commissioners of Woods and Forests, entered into a Deed of Covenant with Her Majesty, binding them to remove the building and to restore the site to the Crown before the first of June, 1852.

Great ingenuity was bestowed upon the adaptation of mechanical contrivances to shorten work during the progress of the building, an elaborate description of which, by Mr. Digby Wyatt, may be found in the first volume of the Illustrated Catalogue for 1851. Numerous tests were applied to the work in its different stages to prove its stability, and so calm the fears of the public on this head. Every cast-iron girder when brought on the ground was weighed, and tried in an hydraulic press. The wrought-iron trusses were carefully examined, and some of the most questionable points in the foundations were tested by being loaded with extraordinary weights. The gallery floors were proved both by fixed and moving loads, and a careful observance of the effect of storms upon every part of the building convinced those intrusted with its charge that there was no chance of its being taken up by a strong wind, as predicted by popular ignorance, and blown like a balloon over Kensington and Chelsea.

The Crystal Palace at Sydenham still shows us an outline of what the Exhibition building was, being mainly built of the materials removed from Hyde Park when the " World's Fair " was over, and the stern provisions of the covenant had to be carried out. Its general plan was a parallelogram, eighteen hundred and forty-eight feet long, and four hundred and eight feet wide; the greatest length running from east to west. There was also a projection on the north side, forty-eight feet wide, and nine hundred and thirty-six feet long. This area, the measure of which in acres has been before stated, was subdivided into twelve avenues of various widths, extending from east to west. The chief passage was seventy-two feet wide and sixty-three feet high, and this occupied the centre. It was flanked on both sides by passages alternately twenty-four feet and forty-eight feet wide. The first of these side avenues on either side of the centre passage was sixty-three feet high, the next two on either side were forty-three feet high, and the remainder were all twenty-three feet high. About the middle of the building, taking it lengthways, these avenues were cut through at right angles by a transept seventy-two feet wide, the half-circular roof of which rose to a height of one hundred and eight feet, enclosing a row of huge elm-trees.

Two other groups of trees on the ground gave rise to open courts which were enclosed within the building, and thus Nature stood on terms of good-fellowship with Art.

The total area roofed over was equal to about eighteen acres and a quarter, and nearly two hundred and ninety-four thousand panes of glass were used, the bulk of them being forty-nine inches long by ten inches broad. The avenues just mentioned were formed by rows of hollow cast-iron columns, eight inches in diameter, which acted as supports for the building, and rain-water drains. They were placed in lines twenty-four feet from each other, and rose in one, two, and three tiers to uphold the roof at the different levels before named. In the lower tier these columns were nineteen feet long, and in the two upper tiers they were seventeen feet. Between each of them were inserted short pieces, each three feet long, of such a shape that they served to support girders in horizontal tiers at three different levels; the bases of the columns were also separate pieces, and they varied in length to suit the different levels of the site. Three thousand three hundred columns were fitted up altogether, and reckoning the different articles made of cast and wrought iron which helped to form the building, there were five hundred and thirty-seven thousand and eighty-two separate pieces, representing over four thousand four hundred and eighty-six tons. Nearly two thousand yards of four and five-inch gas-pipes were laid down; the wrought timber used amounted to two hundred and sixty-five thousand pieces, or more than a million and a third of lineal feet, and the rough timber to nearly four hundred and thirteen thousand cubic feet.

The girders, part of which were of cast and part of wrought iron, were all three feet deep (with the exception of four in the roof at the intersection of the nave and transept, which were six feet deep), thus producing unbroken horizontal lines through every part of the building. These girders were all alike, and they formed a lattice-work combining great strength with lightness of appearance. All the twenty-four feet girders were of cast iron, and of these there were two thousand one hundred and fifty. The roof trusses, of greater length, three hundred and seventy-two in number, were made chiefly of wrought iron, the general lines being the same as in the cast-iron girders.

The lower tier of girders in parts of the building more than one story in height, formed the support for the floor of the galleries, which were twenty-four feet wide, and extended the whole length of the palace in four parallel lines, two on each side of the centre avenue, interrupted only by the transept, round the ends of which they were continued. Many cross galleries connected the long lines, and the additional space thus obtained was over two hundred and seventeen thousand square feet. The floor of the galleries consisted of cross beams, under-trussed, so as to distribute the whole weight brought upon the floor pretty equally upon the eight points at which the ends of the beams rested on the cast-iron girders. Upon this foundation were fixed the ordinary floor-joists and floor. The galleries were reached by ten double staircases, with flights eight feet wide, so arranged as to communicate equally with either of the two lines of gallery between which they were placed. In those parts of the building which were more than two stories in height there was a second horizontal tier of girders twenty feet above the gallery, which served to give stiffness to the columns between which they were fixed. The upper tier of girders and trusses in all cases supported the roof, which was the most novel and interesting part of the building. In its general form it was flat, but in detail it consisted of a series of ridges and furrows, the rise and fall of which was not very great. As the roof, girders, and trusses were twenty-four feet apart lengthways, they were made to carry the main gutters on their upper edge in the transverse direction of the building. The space between these was spanned by light wood beams or rafters, contrived so as to support the glass roof, and to carry into the main gutters the rain-water and the condensed vapour formed under it at the same time. The total length of the gutters used was nearly twenty-four miles. Between these rafters the glass roof was supported by light wooden sash-bars sloping upwards at an inclination of two and a half to one. The advantage of this form of roofing for large areas was its great lightness and economy. The glass of the roof was fixed into the sash-bars, which were grooved to receive it. About two hundred miles of sash-bars and eight hundred and ninety-six thousand square feet of glass were required for the roof; and the whole weight of glass used was about four hundred tons.

The outer enclosure of the building was formed by dividing the twenty-four feet spaces between the iron columns into three panels; those on the lower story were filled in with boarding, those in the upper story with glazed sashes. Metal louvres, fixed in frames three feet high, were introduced at the top of each story round the entire circuit of the building, and in the lower story similar ventilating frames formed a plinth four feet high immediately above the floor. The total ventilating surface thus obtained amounted to nearly forty-one thousand square feet, or very nearly one acre. Each story was crowned outside with a cornice and cresting ornament, and over the columns posts were carried up, to which flagstaffs were fixed.

Three entrances were provided, one in the centre of the south side and one at each end of the building; and in order to make the departure of large crowds easy, seventeen other doors were provided for exit only. The floor was entirely boarded, and on the ground floor spaces of about half an inch were left between the boards to allow the dirt to pass through. Much money and jewelry passed between these boards along with the dirt, and the right of searching for these valuables was bought from the contractors when the building was taken down. The gallery floor, unlike the ground floor, was closely boarded and tongued, to prevent the passage of dust.

The roof of the transept has been mentioned as having been half-circular instead of flat, like that of the remainder of the building. This roof was supported by arched timber ribs placed twenty-four feet apart, or one over every column, the tops forming sockets into the end of which the feet of the ribs were fixed. Horizontal timbers between these supported minor ribs at distances of eight feet, and upon these a ridge and furrow roof was constructed in a manner similar to that already described, but following the curve of the arched ribs instead of being worked on a horizontal plane. A narrow gallery was constructed along the ridge of the arched roof, that workmen might go up to do the necessary repairs. The ends of the transept were filled in with fan-like tracery and glazed sashes. The only portion of untransparent roofing in the whole building was on both sides of the arched roof just described, where there was a lead flat twenty-four feet wide, which afforded the opportunity of giving some additional strength to resist any tendency in the arched ribs to spread outwards at the springing. During most of the time while the work was in full activity, more than two thousand men were employed on the ground, with four powerful steam-engines.

Many important additions were made to the building as originally undertaken by Messrs. Fox and Henderson for the sum of seventy-nine thousand eight hundred pounds. The ventilating contrivances were increased, the galleries were doubled in extent; the outer railing, the gas-lighting, both external and internal, extra offices, staircases, and refreshment accommodation were provided, and a considerable extent of additional area was enclosed. The planing of the floor, the ornamental painting, both inside and outside, the boiler-house and its connections with the main building, provision of water for fountains, and increased provision for safety from fire, the entire enclosure and separation of the department of machinery in motion, and many other important additions served greatly to increase the difficulty of completing the work within the given time, as well as to considerably raise the cost. The total amount at which the whole of the bills for the building on use and waste terms were settled, after careful examination on the part of Sir William Cubitt and the officers responsible to him, was one hundred and seven thousand seven hundred and eighty pounds seven shillings and sixpence. The contractors, however, on making up their prime cost accounts, discovered that they were heavy losers by their contract. They explained this by pointing to the speed at which the work had been done, leaving no time to make the most economical arrangements with under-contractors and others. Their statement was carefully inquired into, and every voucher examined, and in consideration of the important services they had rendered to the Exhibition by their punctuality in preparing the building, a further sum of thirty-five thousand pounds was paid to them by the Commissioners on the seventh of November, 1851, on their signing an agreement to abide by such terms and conditions as the Commissioners might afterwards prescribe with regard to the verification and settlement of the accounts, and the occupation and sale of the Exhibition building. The ultimate sale of the building to the "Crystal Palace Company" of Sydenham for seventy thousand pounds, about Midsummer, 1852, placed the contractors in a very different financial position. Under their contract they were to receive all money arising from the sale of the building as "old materials," and this was estimated at a little over thirty-three thousand pounds. The receipt of the additional thirty-seven thousand pounds relieved them from all loss, and their accounts with the Commissioners were finally closed in a mutually satisfactory manner.

The division of the space provided by the Exhibition building was a labour which taxed the ingenuity and energy of the Committees and officers. The million of square feet covered by the "Crystal Palace," when the deductions for passages were made, only gave half a million of square feet for the display of goods, besides the vertical space. The whole available space, vertical and horizontal, was divided in two; one-half was given to England and her colonies, and the other half to foreign countries.   No rent was charged for space, and no prices were allowed to be fixed to the goods displayed.   The number of exhibitors in the United Kingdom alone demanding space were eight thousand two hundred and thirteen ; the horizontal space demanded by them was four hundred and sixteen thousand three hundred and fifty-four square feet, and the vertical space one hundred and ninety-four thousand eight hundred and eighty-six square feet.   These demands, as far as the horizontal space was concerned, would have required the whole Palace if they had been complied with ; but all the claims were sifted and reduced, and a fair proportion of room was given to each claimant.

The division of the goods was made upon the plan originally laid down by the Prince Consort, and the Exhibition therefore had its four great departments : Raw Materials, subdivided into four classes ; Machinery, subdivided into six classes ; Manufactures, subdivided into nineteen classes ; and Fine Arts, which formed a class by themselves.   The British articles occupied the western half of the building, according to a geographical arrangement ; and the foreign articles occupied the eastern half.   The foreign and colonial divisions were arranged according to their latitudes, the countries lying nearest to the Equator being placed nearest to the transept.

The reception of the goods commenced on the twelfth of February, 1851, and nearly the whole of the British goods were received and completely arranged before the day of the opening.   On the foreign side great progress was also made, but some of the packages from abroad did not arrive till a later period.   The Custom-house regulations and charges were not enforced against any goods intended for exhibition, and the building was placed upon the footing of a bonded warehouse. The British packages and articles received and unpacked in the building amounted to nearly twenty-one thousand ; and the foreign and colonial packages to twelve thousand five hundred and fifty.

Contrary to all popular expectation, the Exhibition was opened by Her Majesty punctually on Thursday the first of May, 1851, with all the advantage of fine weather, state patronage, and the good wishes of everybody, in the presence of twenty-five thousand spectators.   The day was kept as a general holiday in London, and the persons who collected outside the building and in the adjoining parks could not have been much less than a million.

The Exhibition remained open one hundred and forty-one days.   Its foreign exhibitors numbered six thousand five hundred and fifty-six, and the exhibitors of the United Kingdom and dependencies seven thousand three hundred and eighty-two (exclusive of India), forming a grand total of thirteen thousand nine hundred and thirty-eight.   The estimated value of the articles exhibited (excluding the famous Koh-i-noor diamond) was :—

	£	s.	d.
United Kingdom	1,031,607	4	9
Dependencies of ditto	79,901	15	0
Foreign Countries	670,420	11	7
Total	£1,781,929	11	4

The number of prize medals was two thousand nine hundred and eighteen, and of Council Medals, one hundred and seventy, awarded by Juries selected with great care on a representative principle.

The whole daily admissions by payment reached five million two hundred and sixty-five thousand four hundred and twenty-nine, and by season tickets, seven hundred and seventy-three thousand seven hundred and sixty-six;—together six million thirty-nine thousand one hundred and ninety-five. The average number of visitors present on each day appears from these figures to have been forty-two thousand eight hundred and thirty-one. It is not possible to say what proportion of this number consisted of visits paid by distinct individuals, and what of repeated visits by the same persons. The greatest number present on any one day was on Tuesday, the seventh of October, when one hundred and nine thousand nine hundred and fifteen persons were counted by the police. The numbers on the Monday and Wednesday of the same week (the last week of the Exhibition) were very little less, having been one hundred and seven thousand eight hundred and fifteen, and one hundred and nine thousand seven hundred and sixty, respectively. The greatest number of persons present in the building at any one time was ninety-three thousand two hundred and twenty-four, on the seventh of October. The ventilating arrangements had been slightly altered from the original design, but the presence of great crowds seems to have had very little effect upon the temperature of the building.

The gross receipts from daily admissions were three hundred and fifty-six thousand two hundred and seventy-eight pounds three shillings and sevenpence; from season tickets, sixty-seven thousand five hundred and fourteen pounds one shilling; —together, four hundred and twenty-three thousand seven hundred and ninety-two pounds four shillings and sevenpence. The gentlemen's season tickets were three guineas, the ladies' two guineas, not transferable. The admission was one pound each day on the second and third of May, and five shillings from the fifth to the twenty-fourth of May inclusive. On and after the twenty-sixth of May, Mondays, Tuesdays, Wednesdays, and Thursdays were shilling days, Fridays were half-crown days, and Saturdays were five-shilling days. The only alterations made in these rates were a reduction in the price of season tickets to thirty shillings and twenty shillings for gentlemen's and ladies' tickets respectively, and a reduction in the rate of admission on Saturdays to half a crown, both of which changes came into operation at the beginning of August. The Commissioners' receipts from all sources, to the twenty-ninth of February, 1852, including subscriptions, but excluding the loans from the Messrs. Munday and the Bank of England (together fifty-five thousand pounds), were five hundred and six thousand one hundred pounds six shillings and elevenpence. The expenditure between the same dates, excluding the repayment of the above two loans, was two hundred and ninety-two thousand seven hundred and ninety-four pounds eleven shillings and threepence, leaving a balance in the hands of the Commissioners of two hundred and thirteen thousand three hundred and five pounds fifteen shillings and eightpence. The money receipts at the doors exhibit several curious facts. The loss on light gold was two hundred and eighteen pounds four shillings and eightpence. The loss on defaced and foreign coin was two hundred and thirty-one pounds sixteen shillings and tenpence. The spurious coin amounted to twelve bad crowns, two hundred and sixty bad half-crowns, one thousand and thirty-four bad shillings, ninety bad sixpences, and three bad fourpenny pieces;—making a loss under this head of ninety pounds and five shillings, and a total loss of five hundred and forty pounds six shillings and sixpence.

The amount expended by the visitors in refreshments, according to a return furnished by Messrs. Schweppe and Co., the contractors, was seventy-five thousand five hundred and fifty-seven pounds fifteen shillings, giving a general average of about threepence per head.   Buns and effervescing drinks—such as soda-water, lemonade, and ginger-beer—seem to have been the chief favourites, if we take numbers instead of quantities or values;—the first having reached a sale of one million eight hundred and four thousand seven hundred and eighteen, the second of one million ninety-two thousand three hundred and thirty-seven, bottles.   The consumption of water in and for the building was estimated at from one hundred thousand to two hundred and seventy thousand gallons a day.

The Exhibition was closed to the public on Saturday, the eleventh of October, 1851.   The following Monday and Tuesday were set aside for the free admission of exhibitors and their friends; and on Wednesday, the fifteenth of October, the final closing ceremony took place in the presence of the exhibitors, jurors, foreign and local commissioners, and others.   After the presentation of the Jury Reports to the Commissioners, the Prince Consort, on behalf of the Commission, took leave of all those who had given their assistance towards conducting the Exhibition to its prosperous issue.   The removal of the goods immediately commenced.   By the fourteenth of November the British side was cleared, and by the fifteenth of January, 1852, the last straggling foreign exhibitor had disappeared with all that belonged to him.

The special literature and art which sprang out of this display was vast and curious.   "Aunt Mavor," and many other imitators, came forward with "alphabets" of the Exhibition.   Religious tracts, and sermons improving the occasion, were published in waggon-loads.   Catnach issued street-ballads, some sentimental, some comic, and some stated to be "by authority."   French dramatists constructed plays showing the adventures of helpless foreigners in England, and their British imitators adapted these productions to our native stage.   Projectors, more or less mad, amongst other suggestions proposed that houses in bleak situations should be covered with crystal palaces, to convert them into little Madeiras.   Stories and songs about the display were written in various dialects,—in French, German, Italian, Russian, Dutch, Spanish, Portuguese, Swedish, Arabic, Turkish, and a dozen other tongues; and the great Mansion House speech of the Prince Consort (before alluded to) was translated into at least five languages.   Engravings of all degrees of merit on every material, from paper to gelatine, appeared in shoals, and musical composers loaded us with various compositions, from airs to symphonies.

No one who looked upon this Exhibition when it was full of life could help feeling that it was a great creation.   It was like nothing that had been seen before, but like much that had been dreamed of.   The long passages lined with the most precious products of industry,—the rich tapestries hanging from lofty galleries, the hum of voices, the strange faces, the mixed costumes, the perfumes of scented water, the trees springing up in the midst of loaded bazaars, the running fountains, and the light glittering roof bending over all, contributed to form a world such as had only been imagined in Eastern fables.   When it had melted away, and the green turf began to grow again in its place, no man living thought to look upon its like again.

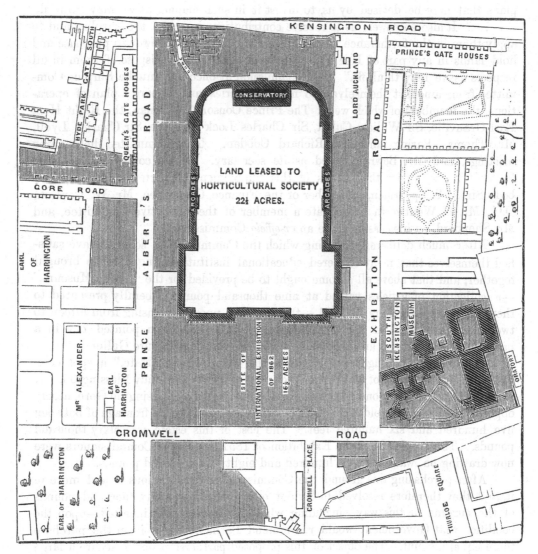

PLAN OF KENSINGTON GORE ESTATE.

CHAPTER VII.

THE SURPLUS, AND ITS APPLICATION.

THE first steps which the Royal Commissioners took on ascertaining the existence of a surplus, was to apply to Her Majesty for a Supplemental Charter, empowering them to dispose of the money in accordance with the expectations held out to the subscribers at the time their aid was solicited. This Charter was dated the second of December, 1851, and the Commissioners thus describe it: " It (the Charter) gives us the power to dispose of the surplus in the furtherance of any plans that may be devised by us, to invest it in such manner as we may think fit. . . . . It also empowers us to receive contributions in aid of the surplus, and to apply them in the furtherance of our plans; and it gives us power to purchase and hold lands in any part of Her Majesty's dominions, and to dispose of them in all respects as we may think fit." The Commission at once appointed a Surplus Committee from amongst themselves, to make inquiries and prepare a plan of operations, the members of which were : The Prince Consort, Chairman, the Right Hon. Earl Granville, Sir William Cubitt, Sir Charles Lock Eastlake, Sir Charles Lyell, Mr. Thomas Baring, and Mr. Richard Cobden. This Committee held three meetings, and Mr. Bowring acted as its secretary. In the course of 1852 the Commissioners, under the powers of this Supplemental Charter, added to their body Sir A. Y. Spearman, a member of the Finance Committee, Mr. W. Coulson, Q.C., Mr. C. Wentworth Dilke, late a member of the Executive Committee, and Mr. Shepherd, on his ceasing to be an *ex-officio* Commissioner.

After much deliberation, during which the Commissioners seem to have satisfied themselves that many scattered educational institutions ought to be brought together, and that above all a home ought to be provided for the " Trade Museum," —a collection of articles valued at nine thousand pounds, liberally presented to them by many exhibitors in 1851,—they decided upon purchasing from fifteen to twenty acres of land at South Kensington, which had been pointed out in a parliamentary report as an eligible site for a new National Picture Gallery.

The first estate bought was known as the "Gore House Estate," a spot celebrated as the residence of Mr. Wilberforce, and afterwards of Lady Blessington. It contained about twenty-one acres and a half, was situated at Kensington Gore, nearly opposite the site of the old Exhibition building, and possessed a frontage of between five hundred and six hundred feet. The cost of this estate was sixty thousand pounds. The surplus from the Exhibition of 1851, which the Commissioners were now drawing on, was about one hundred and eighty-six thousand pounds.

After purchasing this ground the Commissioners were anxious to add more to it, and they therefore resolved to lay out one hundred and fifty thousand pounds of the surplus in this way—including what they had already invested—upon the condition that Government would recommend Parliament to join in the purchase to an equal amount. The object of this proposed partnership was to secure a large block of cheap land in London before the spread of building placed it out of their reach, to which some of the overcrowded national exhibitions might be removed if necessary, and on which an educational institution might be erected for the improvement of designing art in connection with manufactures. The Government having pledged themselves to this scheme, the Commissioners bought another estate of forty-eight acres from the trustees of the Baron de Villars for one hundred and fifty-three thousand five hundred pounds.

The promised parliamentary vote of one hundred and fifty thousand pounds towards the South Kensington land purchases of the Royal Commissioners was obtained in the session of 1852-53, and the formal partnership between the Government and the Commissioners then commenced. This union compelled the Commissioners to increase their members, and the following state officers, the Lord President of the Council, the First Lord of the Treasury, the Chancellor of the Exchequer, the President of the Board of Trade, and the First Commissioner of

Works, were added to the Commission. The Right Hon. B. Disraeli and Sir Roderick Murchison were also elected members of the same body.

With a capital of three hundred and forty-two thousand five hundred pounds (one hundred and sixty-five thousand pounds of which were contributed by the Commissioners out of the surplus, while the remainder was supplied by Parliament in two distinct sums) the South Kensington estate was gradually secured. Seventeen acres more land were purchased of the Earl of Harrington, making, with the former purchases, about eighty-six acres—a plot larger by several acres than St. James's Park—and various small changes were effected with adjoining proprietors to render the estate more compact. The cost of the eighty-six acres was two hundred and eighty thousand pounds, or an average of three thousand two hundred and fifty pounds an acre.

As an instance of the cost of making improvements in the metropolis, and to show the comparative cheapness of the land purchased by the Commissioners, it may be well to mention the outlay on some of the more important improvements undertaken of late years. The line of street from Oxford Street to Holborn contained two hundred and twenty thousand one hundred and fifty-one square feet, and its total cost was two hundred and ninety thousand pounds, or an average of more than fifty-seven thousand pounds an acre. The new thoroughfare from Bow Street to Charlotte Street, Bloomsbury, contained sixty-one thousand six hundred and fifty-three square feet, and its total cost was ninety-six thousand pounds, or an average of nearly sixty-eight thousand pounds an acre. Again, the new line from Coventry Street to Long Acre, which contains sixty-five thousand four hundred and ten square feet, cost one hundred and eighty thousand pounds, or an average of nearly one hundred and twenty thousand pounds an acre.

Much of the South Kensington land when the Commissioners bought it was laid out as market-gardens, and the neighbourhood had been famous for nurseries for more than two centuries. The bold speculation was at first looked upon with great distrust, and the obvious joke about sinking money in a cabbage-garden was freely indulged in at the expense of the Commissioners. The sum they were dealing with had many claimants with a supposed title, even after setting aside the government contribution and the pure Exhibition, or "shilling surplus," which no one could own. There was the subscription list of nearly sixty-eight thousand pounds, made up of various sums collected from all parts of the country. Although each contribution was sent in on the clear understanding that it was to be "absolute and definite," many local bodies considered that this part of the surplus ought to have been returned for the direct benefit of local institutions. The original scheme of the Commissioners for a great Central Industrial University would have benefited these local institutions, and particularly the local schools of design. The suggestions sent in to the Commissioners from all sides for the disposal of the whole surplus were very numerous. One correspondent proposed that the Great Exhibition building should be bought and turned into a winter residence for invalids; another proposed that the fund should go to alleviate Irish and Highland destitution; and another that the Exhibition building should be purchased for a great public reading-room. The Commissioners duly registered all the propositions, but followed none of them, and by these means turned disappointed friends into active enemies.

The joke about the "cabbage-garden" would have been very severe, if half

London had not once been a cabbage-garden or a brick-field.   Those who made it were ignorant of the laws which govern metropolitan progress, and the Commissioners were soon able to show that their investment was commercially wise.   They obtained an Act of Parliament for stopping up certain lanes and by-ways which cut through their property, and they formed nearly two miles of new roadway from eighty to one hundred feet broad, the chief lines of which went round the best part of their estate.   These roads were the Cromwell Road, the Exhibition Road, and the Prince Albert Road, forming with the main Kensington Road four sides of a square.

When the Government in 1856 lost its Bill for removing the National Gallery of Pictures from Charing Cross to South Kensington, the Commissioners proposed to dissolve partnership with the State, and this dissolution was effected in the early part of 1858.   The sums advanced by the Government were repaid by the Commissioners, subject to a deduction for the ground and buildings of the Department of Science and Art, popularly known as the South Kensington Museum.   The connection of the Commissioners with this Department ceased at this point, and they became nothing more than Trustees for the surplus, buying and selling land.   They disposed of some outlying parts of their estate—about twelve acres—in building leases on very advantageous terms, for the ground, with the exception of a small corner in the Gore House estate, where the London clay crops up, is red gravel to a depth of more than twenty feet.   The new roads, the letting of the upper part of the great centre square—about two-and-twenty acres, —to the Horticultural Society for an Italian exhibition garden, the general improved tone given to the neighbourhood, and the march of time and population have so improved the property, that the Commissioners have now nearly doubled their original capital.   Under the arrangement with the Horticultural Society the Commissioners have expended about fifty thousand pounds sterling in the erection of architectural arcades in the new gardens, and the Society have expended an equal amount in terraces, fountains, the great conservatories, and the laying out of the grounds.

CORK EXHIBITION, 1852.

CHAPTER VIII.

INTERMEDIATE EXHIBITIONS.

HE great financial and general success of the Exhibition of 1851 naturally encouraged the repetition of such displays all over the world. Many worthy imitators rose up at short intervals; some being merely local or national displays, others being international. There was the Cork Exhibition in 1852, the daily admissions of which reached seventy-four thousand and ninety-five, the season-ticket admissions fifty-four thousand nine hundred and thirty-six, and the receipts respectively two thousand eight hundred and seventy-four pounds, and one thousand five hundred and forty-five pounds. Two were started simultaneously in 1853; one in New York, and the other in Dublin, both of which were universal exhibitions. The New York Exhibition was a private speculation, and was not a commercial success, owing chiefly to a long delay in the opening caused by the building not being finished. It was visited by a Government appointed Commission of six members—Lord Ellesmere, Sir Charles Lyell, Mr. Wentworth Dilke, Mr. Wallis, Mr. Whitworth, and Professor J. Wilson—whose elaborate reports were laid before Parliament in

DUBLIN EXHIBITION, 1853.

1854. The Dublin Exhibition in the same year, also universal in its design, owed its existence to the patriotic energy of Mr. Dargan, who made himself responsible for the pecuniary part of the undertaking, and paid the losses, which were something considerable, out of his own pocket. This exhibition was very popular with the visitors and the native population, and although it drew so heavily upon its spirited promoter's resources, it showed more visits, compared with the amount of local population, than the Great Exhibition. The population of the metropolis in 1851 was two million three hundred thousand, and the admissions recorded at the Hyde Park building give little more than two and a half visits to each of these individuals. The lowest rate of admission in 1851 was one shilling, but in Dublin it was sixpence. The daily visitors at the latter place reached six hundred and thirty-four thousand five hundred and twenty-three; the season-ticket visitors three hundred and sixty-six thousand seven hundred and forty-five, and the receipts respectively twenty-eight thousand nine hundred and eighty one pounds, and eighteen thousand three hundred and eighty-two pounds. With the local population of two hundred and fifty-four thousand, this gave nearly four visits to each individual.

The Munich Exhibition of 1854 came next, and it was made the subject of a special report by Consul-General Ward. This display was not International in the broadest sense, though it was not confined to the products of Bavaria. The whole of Germany was allowed to take a part in the competitive struggle.

NEW YORK EXHIBITION, 1853.

The Twelfth Exhibition in Paris followed this, in 1855, being the first great French International Exhibition. It imitated very closely the plan of 1851. The exhibitors, although showing a decrease upon those of 1851 in London, showed a marked increase upon those of the eleventh French Exposition in 1849. The French exhibitors numbered nine thousand seven hundred and ninety, without Algeria, as against four thousand four hundred and ninety-four in 1849, an increase greatly attributable to the admission of international exhibitors. The total number of exhibitors in 1855—national and international—was twenty thousand eight hundred and thirty-nine, more than fifteen hundred of whom were from the United Kingdom, and nearly eleven hundred from the British colonies. The value of the British goods exhibited was estimated at six hundred and ninety-three thousand six hundred and twenty-seven pounds, in addition to the fine arts collection, valued at one hundred and thirty-seven thousand five hundred and sixty pounds. Out of the British exhibitors in the industrial classes, nine hundred and thirty-one received honorary awards from the juries, fifteen obtained the grand medal of honour, thirty-two the medal of honour, three hundred and one the first-class medal, three hundred and fifty-three the second-class medal, and two hundred and thirty honourable mention. The whole British nation, in fact, through its exhibitors, obtained honourable mention—in some cases more than it deserved. One enthusiastic writer in the *"Visite à l'Exposition Universelle,"* produced under the direction of M. Tresca, describing the English mechanical models exhibited,

MUNICH EXHIBITION, 1854.

gives us credit for our magnificent Britannia tubular bridge, which was the first to join England to Scotland!

The lowest rate of admission to this French Exhibition was four *sous* (there were many days on which the building was open free), and the total number of visitors between the fifteenth of May, the date of its opening, and the first of December, when it was finally closed to the public, was four million five hundred and thirty-three thousand four hundred and sixty four; more than nine hundred thousand being visitors to the palace of the Fine Arts. Of the whole number, forty thousand were British subjects, including two thousand seven hundred and sixty-eight furnished with workmen's passports free of charge. The total receipts are stated as a little under one hundred and eighteen thousand pounds sterling.

After this great international display came the Manchester Fine Art Exhibition in 1857; a collection of ancient and modern pictures and works of art never before equalled. To furnish it hundreds of cautious collectors gave up their priceless treasures for a time, as they have again given them up in 1862. The daily admissions at this exhibition reached one million fifty-three thousand five hundred and thirty-eight; the season-ticket admissions two hundred and eighty-three thousand one hundred and seventy-seven, and the receipts respectively, sixty thousand five hundred and six pounds, and twenty-three thousand and fourteen pounds.

To make this list complete, though departing for a moment from strict chronological order, we must not forget that partial revival of some of the old City Trade Guilds which has lately given us the exhibitions of the Ironmongers' and Painter-Stainers' Companies. The Society of Arts' displays have of course taken place year after year, and in 1861 we have had the Dublin Art Exhibition, the Edinburgh Art Treasures' Exhibition, and the very hopeful Italian National Exhibition at Florence.

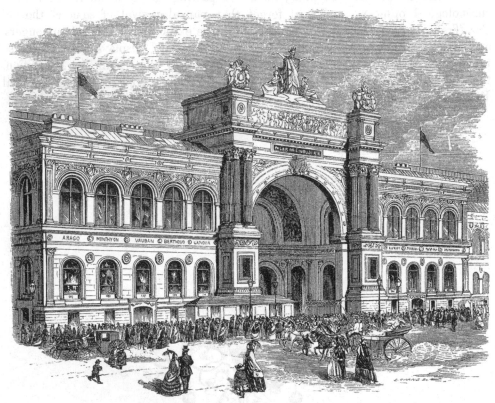

PARIS EXHIBITION, 1855.

CHAPTER IX.

BEGINNING OF THE SECOND GREAT EXHIBITION.

THE doctrine of finality was not regarded by the friends of industrial exhibitions with any more favour when it was applied to their cherished schemes than it was by certain politicians when applied to politics. The Society of Arts always held a belief that the world never stands absolutely still, and that where there is no decay there must be material progress. With an increase of capital comes an increase of population in an equal degree, and this is called the advance of civilization. We may not be more moral, more imaginative, nor better educated than our ancestors, but we have steam, gas, railways, and power-looms, while there are more of us, and we have more money to spend.

This belief sustained the Society of Arts when they first proposed, in the early part of 1858, to repeat the Great Exhibition of 1851. They considered that if such reviews of industry were to be anything beyond a mere show in a gigantic bazaar, they ought to be periodical. In none of the Society's communications on the subject of the first Great International Exhibition, whether to the Royal Commissioners, the different manufacturers throughout the country, or the supporters of the scheme in London, was the word "periodical" ever omitted. They never contemplated a single gathering of half the products of the earth under one roof—the creation of an Ark of Industry—which after a six months' life was to be shattered at a blow, and the pieces never collected again. Literary sentimentality and English composition may have invested the display of 1851 with this character, but certainly not the practical minds of its originators, who had before them not only the results of the homely annual displays (before described) at the old rooms in the Adelphi, but the records of what had been done in France since 1797 at average periods of about five years.

The year chosen by the Society of Arts for the second great exhibition was 1861, and during 1858 resolutions and facts favourable to such a repetition were industriously circulated amongst the commercial public. At the close of the same year the Society brought the matter officially under the notice of the Commissioners of 1851, inquiring whether they were willing to undertake the management of the proposed exhibition.

The reply sent to this communication by the Commissioners, asked for further information as to the prospects of such a scheme, and the support it was likely to receive from manufacturers and the public. The Commissioners added that they had no funds to meet the expenses of such an exhibition, but that if the Society's report should be favourable, they were willing to consider how they could most effectively help the undertaking.

The Society of Arts replied to this letter of the Commissioners on the eleventh of March, 1859, stating that they would do all in their power to furnish the required information, and would try to obtain subscriptions to a proposed guarantee fund of two hundred and fifty thousand pounds. They considered that if they succeeded in this last endeavour they should give the best proof of a probability of success. The letter also set forth various reasons why, in the opinion of the Society, the proposed exhibition was likely to produce still more definite and valuable results than the Exhibition of 1851.

The war in Italy, which broke out in the early part of 1859, and threatened to be of long continuance, together with the disturbed state of the Continent, gave little hope that the exhibition, so far as its international character was concerned, could be held with success. The Council of the Society of Arts therefore passed a resolution at the beginning of June, 1859, to put off the proposed exhibition until a more favourable period. This resolution met with the entire concurrence of the Commissioners, and the original proposal was accordingly abandoned.

The Italian war having been unexpectedly brought to an early conclusion, the exhibition project was revived. As it was too late to make the necessary arrangements for holding the exhibition in 1861, the following year was chosen for the display. The Society of Arts forwarded to the Commissioners on the eighth of March, 1860, a copy of a proposed guarantee agreement for securing the means of holding the exhibition, and asked to be informed whether the Commissioners were willing to grant a site on their Kensington Gore estate for the purposes of that and future exhibitions, and if so, on what terms. The following were the principal conditions of the draft guarantee agreement in question :—No subscriber was to be liable unless the deed was signed to the extent of at least two hundred and fifty

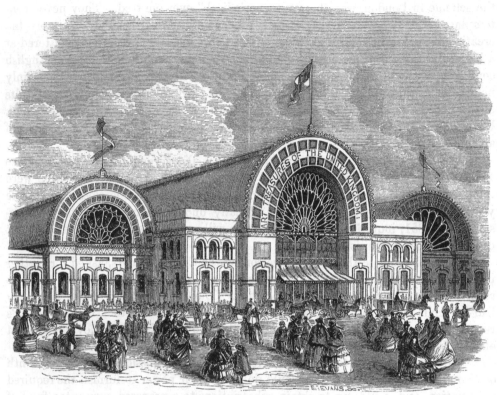

MANCHESTER EXHIBITION, 1857.

thousand pounds, and in the event of loss attending the exhibition, each subscriber
was to contribute in rateable proportion to his subscription to liquidate such loss.
Earl Granville, the Marquis of Chandos, Mr. Thomas Baring, Mr. Dilke, and Mr.
Thomas Fairbairn, were to be invited to be the five Trustees of the exhibition.
Application was to be made to the Commissioners for a site on the South Kensington estate, and the Trustees were to erect whatever buildings, whether
permanent or temporary, they might think necessary for the exhibition, but on the
express condition that at least one-third of the sum so expended by them should be
employed in erections of a permanent character, suitable for decennial or other
exhibitions, and when not so used, suitable for other purposes tending to the
encouragement of arts, manufactures, and commerce, such permanent buildings
to be vested in the Society of Arts.   In the event of there being a surplus, it was
to be applied to the encouragement of arts, manufactures, and commerce in such
a way as the subscribers might determine upon at a meeting to be called for the
purpose.   In the event of there being a loss which the Society of Arts declined to
liquidate, the permanent buildings above mentioned were to be sold, and if, after
such sale, there still remained a deficit, the ultimate loss was to be borne by the
subscribers rateably, as already stated.

The facts collected by the Society of Arts in support of this second exhibition
came chiefly from Mr. William Hawes, the Registrar-General, Mr. C. M. Willich.

Colonel Owen, and Sir Cusack P. Roney.   They showed that the population of Great Britain, which was twenty-five millions one hundred and eighty thousand five hundred and fifty-five in 1851, would be about twenty-nine millions in 1862, and that London then would contain half a million more people than it did at the period of the first Great Exhibition.   They showed that one half of this population would consist of persons between the ages of fifteen and fifty, and that one-fourth would consist of persons who were too young to benefit by the Exhibition of 1851. They showed that the length of railways in England alone would be nearly eleven thousand miles in 1862, compared with six thousand seven hundred and fifty-five in 1851, and that the general system of railway management would be much improved.   The continental managers have now learned to appreciate through-booking, return-tickets, and excursion traffic at reduced rates, which they would not look at a few years back.   Many continental lines have been opened since the year of the Great Exhibition, all more or less converging towards this country, and several others of great importance in shortening existing routes, and putting us in communication with new districts, have been completed during 1859 and 1860. The steam passages between America and Europe have been more than quad-rupled, and the fares lowered at least thirty per cent.   The chain of railways now joining New York, Boston, Portland, and Quebec, has been tripled since 1851; the distance between London and India has been decreased twenty-five per cent., and between England and Australia fifty per cent.; the time taken for passages to and from our West Indian colonies has been diminished one-third, and we have a well-organized steam communication with South America and Africa, which did not exist in 1851.

Adopting the statements of Mr. Hawes, which were enlarged and repeated in 1861, we were told that " the people are better employed, crime is decreasing, and their social and intellectual condition is improved.   Most important discoveries have been made in the preparation of colours for printing and dyeing, producing what are called the 'Aniline' series; great economy has been effected in the manufacture of glass, and a process has just been made perfect for transferring photographs to that material.   The manufacture of agricultural implements, and especially the application of steam power to them, has been so improved and extended that it is now a highly important branch of trade.

" Photography, hardly known in 1851, has become an important branch of art and industry, used alike by the artist, the engineer, the architect, and the manufacturer.

" Marine telegraphy, only just accomplished in 1851—the public communica-tion with Dublin having been opened in June, and that with Paris in November, 1852—has now become almost universal, linking together distant countries.

" The electric telegraph has become universal, and in every direction facilities for communication have been increased.   We have repealed the duties on soap and paper, the only manufactures the prosperity of which was then thwarted by Excise restrictions.

" We have abolished all taxes on the dissemination of knowledge, and have given increased facilities for the circulation of knowledge by post.

" We have repealed the Import duties, or very nearly so, on raw materials, the produce of foreign countries.

" We have admitted free of duty, confident in our strength, the manufactures of foreign countries to compete with our own.

" Old industries have been stimulated and improved. New industries have arisen.

" In fine art, painting, and sculpture it is hardly possible, except in very extraordinary periods, that a marked change can be observed in a single ten years, but this country certainly holds its own as compared with the productions of other countries.

" In the manufacture of iron, improvements have also been made; new bands of ore have been discovered; and day by day we are economizing its production; and a metal between iron and steel is now produced at one process, which heretofore required two or more processes alike expensive and difficult.

" In steam power, especially that applied to railroads and to ocean steam navigation, economical appliances have advanced rapidly.

" The use of coal for locomotives in place of coke, and super-heating steam and surface-condensing in ocean steamers, tend to increase the power and economize the cost of these powerful engines of civilization.

" In shipbuilding, the past ten years have produced great changes.

" Our navy and mercantile marine have alike advanced in scientific construction and in mechanical arrangements. The ocean steamers which were then employed in the postal service included but one of two thousand tons; now there are many of nearly double that tonnage, with corresponding power and speed—increasing the facilities and decreasing the risk of communication with our colonies and foreign countries.

" In printing great advances have been made. By the perfection of chromatic printing, views of distant countries, copies of celebrated pictures, most beautifully coloured, have been brought within the reach of almost every class, displacing works which neither improved the taste nor gave useful information; and by the application of most expensive and most beautiful machinery to the printing of our daily journals, we have been enabled profitably to meet the increased demand caused by the cheapness of our newspapers. Invention and mechanical contrivance have thus kept pace with the requirements of intellect and the daily increasing love of knowledge."

So energetic were the Society of Arts in putting many of these facts before the public, and in obtaining promises to sign the necessary guarantee deed, that on the eighth of June, 1860, they were able to address another letter to the Commissioners, stating that four hundred and fifty-five persons had already intimated their intention to subscribe sums to the guarantee fund, amounting altogether to three hundred and eight thousand three hundred and fifty pounds. They also informed the Commissioners that the amount which it was intended to invest in the permanent buildings already spoken of was to be at least fifty thousand pounds, and called attention to the liberality with which the Society had surrendered its claim to a share of the surplus profits of the Exhibition of 1851.

" In reply to this application " (the Commissioners say in their fourth report) " we informed the Society of Arts that we should be happy to grant rent free until the thirty-first of December, 1862, for the purposes of the Exhibition of 1862, the use of the whole of the land on the main square of our estate lying on the south

FLORENCE EXHIBITION, 1861.

side of the arcades and entrances to the gardens of the Horticultural Society, estimated at sixteen acres, on the understanding that all the buildings to be erected for the exhibition, whether permanent or temporary in their character, should be subject to our approval, and that all the temporary buildings should be removed within six months after the close of the exhibition if required by us; the Trustees of the exhibition being at liberty, on the other hand, to remove the buildings termed permanent if the exhibition should be attended with pecuniary loss. We further expressed our readiness to grant to the Society, in recognition of their long-continued services in advancing the interests of the arts and manufactures, and especially in preparing the way for the Great Exhibition of 1851, a lease for ninety-nine years at a moderate ground rent of those permanent buildings if retained on our ground, on condition of not less than the sum of fifty thousand pounds, named in the Society's letter of the eighth of June, being expended on them by the Trustees, and of their not covering more than one acre of ground; and also on condition of their being used solely for holding exhibitions and for purposes connected with the promotion of arts and manufactures. With respect to the Society's application, that we should appropriate a portion of our estate for the purpose of future exhibitions analogous to the proposed Exhibition of 1862, we informed them that with the view of meeting their wishes as far as was consistent with our public duty, and at the same time bearing in mind our obligations to our mortgagees, we would undertake, in the event of the payment to us of the sum of ten thousand pounds out of the profits (if any) of the Exhibition of 1862, to reserve for the purposes of another International Exhibition in 1872, to be conducted by such body as might be approved by us, the remainder of the land now proposed to be lent by us for the Exhibition of 1862 that was not covered by the permanent buildings already referred to, such reservation not interfering in any way with the free use by us of that land in the intervening period."

The Society of Arts accepted these terms, and at once entered into correspondence with the five gentlemen named in the draft guarantee deed as the proposed Trustees of the exhibition, to obtain their definite acceptance of the trust thus offered to them. These gentlemen thought that the offer of management should be made to the Commissioners of 1851, and the Society, therefore, made another application to the Commissioners, asking whether they would undertake the management of the exhibition within the conditions expressed in the guarantee agreement. The Commissioners were unwilling to take the management of the undertaking with the restrictions imposed, even had there not been many legal difficulties in the way of their doing so, but they expressed their general approval of the object which the Society of Arts had in view in organizing the proposed exhibition, and their readiness when the trust was accepted by the five gentlemen named to afford assistance in advising the Trustees on certain important principles and financial points, and to elect as Commisioners those two of the five Trustees —Lord Chandos and Mr. Fairbairn—who were not already members of their body.

Upon this reply being communicated by the Society to the proposed Trustees, those gentlemen announced their willingness to accept the trust in the following letter :—

                                                  " London, November 22nd, 1860.
" SIR,
    " We have to acknowledge the receipt of your letter of yesterday, enclosing the copy of a communication from Her Majesty's Commissioners for the Exhibition of 1851 to the Council of the Society of Arts, in which the Commissioners express their general approval of the object which the Society has in view in organizing the Exhibition of 1862, and their willingness to render such support and assistance to the undertaking as may be consistent with their position as a chartered body, and with the powers conferred upon them by their Charter of Incorporation.

    " Under these circumstances we have to request that you will intimate to the Council of the Society of Arts our willingness to accept the trust which the Council and the Guarantors have in so flattering a manner expressed a wish to repose in us, on the understanding that the Council will forthwith take measures for giving legal effect to the Guarantee, and for obtaining a Charter of Incorporation satisfactory to us.

                    " We have the honour to be, Sir,
                        " Your obedient servants,
                            (Signed)   GRANVILLE,
                                        *CHANDOS,
                                        THOMAS BARING,
                                        †C. WENTWORTH DILKE,
                                        THOMAS FAIRBAIRN."

" P. LE NEVE FOSTER, Esq.,
        Secretary to the Society of Arts."

---

    * Now the Duke of Buckingham and Chandos.   † Now Sir Charles Wentworth Dilke, Bart.

RAISING THE FIRST RIB.

CHAPTER X.

THE EXHIBITION OF 1862.—ITS ORGANIZATION.

THE earliest act of the Trustees, on accepting office, during the interval required for the preparation of the necessary legal powers, was to obtain a modification of certain conditions made between the Society of Arts and the Commissioners of 1851. In the month of January, 1861, they informed the Council of the Society of Arts, that, with a due regard to the interests of the exhibition, they would not be justified in expending more than twenty thousand pounds on that part of the site intended to be leased to the Society, especially as the average cost per acre of the whole exhibition building was estimated not to reach twelve thousand pounds.

On receiving this information, the Society of Arts requested the Commissioners of 1851 to reconsider the agreement; and the Commissioners consented to accept the new proposition of the Trustees, upon condition that if the necessary surplus should exist at the close of the exhibition, as much money should be laid out in improving the architectural character of the permanent portion of the building, as would bring the original sum expended up to fifty thousand pounds. "Our reason," the Commissioners say in their fourth report, "for making this stipulation was, that we felt that if we allowed any permanent buildings to be retained on the site, it was important that they should be sufficiently handsome in an architectural point of view, which could not be the case if no larger amount were expended upon them than the limited sum proposed by the Trustees. By the arrangement agreed to by us the retention of the buildings would not disfigure the estate or offer any impediment to the ornamental execution of any future building scheme on the property, which might otherwise be seriously interfered with if so small a sum as that desired by the Trustees were expended on so large an area as that intended to be covered by these buildings.

"It is obvious that the above alteration in the terms originally agreed to between the Society of Arts and ourselves, combined with the alteration which it involved in the stipulations contained in the draft guarantee agreement, was one greatly to the benefit of the subscribers to the guarantee, and greatly tending to lessen the chance of their being eventually called upon to contribute any portion of the sums respectively guaranteed by them. By the original stipulations the Trustees were required to expend *at least* one-third of the total sum expended by them on buildings for the exhibition in the erection of permanent buildings to be leased to the Society of Arts, at the same time that there was no limit whatever upon the amount which the Trustees might choose so to spend. By the altered arrangements, however, the total sum that the Trustees are empowered to expend on the buildings in question, under any circumstances, is a *maximum* of fifty thousand pounds; and even a considerable portion of that sum (that to be expended on the architectural completion of the permanent buildings) is not to be spent in the first instance, but only in the event of the exhibition being successful. The liability of the guarantors was thus, in fact, reduced from an unlimited *minimum* to a limited *maximum*."

While this matter was under discussion, an application was made by the Trustees to the Commissioners, representing that additional space was imperatively required, and applying for the loan of an unoccupied portion of land lying between the western arcades of the Horticultural Gardens and Prince Albert's Road—the same land, in fact, which the Commissioners are prepared to devote, if desired, to the British Museum. The Trustees also applied for the loan of the south arcades of the Horticultural Gardens, to be fitted up as refreshment rooms. Both these requests were complied with; the latter with the liberal consent of the Horticultural Society, but upon the condition, made by the Commissioners as some balance for a sacrifice of rent, that the Trustees should provide a permanent wall and roof to the south arcades, and a permanent wall to the west arcades. The Commissioners felt it to be necessary to make this stipulation, in order to prevent the stability of the arcades being endangered by the manner in which it was proposed to employ them in connection with the exhibition. A subsequent request for more space on the opposite side of the ground—to form another annexe—was also complied with.

Various meetings with the Trustees and their solicitor, and also with the counsel and solicitor of the Bank of England, were attended by the chairman and secretary of the Society of Arts, to adjust the conditions of the Charter of Incorporation, as well as the Guarantee Deed; and the forms of both instruments having been settled in a satisfactory manner, a petition to the Crown for a Charter was presented by the Society. The Charter, having received the approval of the law officers of the Crown, was sealed on the fourteenth of February, 1861, and under it the five Trustees received their legal title of "The Commissioners for the Exhibition of 1862." The Charter, the Guarantee Deed, and a full list of the Guarantors —the main props of the exhibition—are given in an appendix.

The plans of the Society of Arts having thus received the gracious approval of Her Majesty, the Council transmitted the Charter to the Commissioners for the Exhibition of 1862, and received the following letter from Mr. Sandford, their secretary:—

SIR,   Council Office, Feb. 20, 1861.

I AM directed to acknowledge the receipt of Mr. Foster's letter of the sixteenth of February, enclosing the Charter which has been granted to Earl Granville, K.G., the Marquis of Chandos, Mr. Thomas Baring, M.P., Mr. C. Wentworth Dilke, and Mr. Thomas Fairbairn, incorporating them as The Commissioners for the Exhibition of 1862.

The Commissioners, on the twenty-second of November last, agreed to act, after a guarantee had been promised to such an extent as to show a strong opinion in the public mind that the time for holding a second international exhibition had arrived; after the guarantors had expressed an opinion that the absolute control of the undertaking ought to be intrusted to five gentlemen, named by them; and after the Commissioners of the Exhibition of 1851 had intimated their approval of the project, and their confidence in the proposed mode of management, and had promised their support and assistance. The Commissioners, therefore, gladly accept a Charter which conveys to them Her Majesty's gracious assurance that she is earnestly desirous to promote the holding of an international exhibition of industry and art in the year 1862, and that she is pleased to sanction the proposed arrangements.

The powers conveyed by the Charter will, however, be practically inoperative until the deed of guarantee has been executed. When this has been done, the Bank of England has agreed to advance the necessary loan of money on liberal terms. The Commissioners therefore desire me to request that you will represent to the Council the necessity of having the deed signed as soon as possible.

The Commissioners, unwilling to lose valuable time, have, during the interval required for the preparation of the requisite legal powers, taken such provisional steps as their position permitted.

The most pressing point was the building required for the exhibition. In 1850, notwithstanding the possession of considerable funds, and the assistance of the most eminent architects and engineers, seven months elapsed before a design was adopted. The Commissioners therefore felt that if they postponed the consideration of this subject until they were a legally constituted body, the cost of the building would be greatly increased, and a serious risk incurred of its non-completion by the appointed time.

The arrangements made by the Society of Arts, when negotiating for a site on the estate of the Commissioners of 1851, and their arrangement that the exhibition was to include pictures, a branch of art not exhibited on the former occasion, rendered it necessary to contemplate the erection of a building in some parts of a more substantial character than that of 1851.

A plan was submitted to the Commissioners by Captain Fowke, R.E., who had been employed by Her Majesty's Government, in the British Department of the Paris Exhibition of 1855. This design was adapted to the proposed site, and was intended to meet the practical defects which experience had shown to exist both in the buildings in Hyde Park and in the Champs Elysées. It appeared well adapted for the required purposes, and its principal features were of a striking character, and likely to form an attractive part of the exhibition. The Commissioners submitted the design to the competition of ten eminent contractors, four of whom took out the quantities. Three tenders (one a joint one from two of the contractors invited) were sent in on the day named in the invitation, but all were greatly in excess of the amount which the Commissioners could prudently spend, with a due regard to the interests of the guarantors.

The Commissioners have, therefore, had under their consideration modifications of the plan, which, without destroying its merits, would materially reduce its cost.

The Commissioners having learnt that the French government had applied, on the third of November last, to the Foreign Office, to know whether it was intended to hold an international exhibition in England in 1862, entered into private communication with that government, from whom they have received satisfactory assurances of support, accompanied by a statement that it had been the intention of the Emperor to hold an international exhibition in Paris in 1862, had the project not been entertained in England.

The Commissioners also requested the Duke of Newcastle, the Secretary of State for the Colonies, to announce the design entertained of holding an exhibition, and the intention of the promoters to apply to the Crown for a Charter; and the Commissioners have been informed that his Grace has addressed a communication to that effect to all the governors of Her Majesty's colonies.

The Commissioners have had under their consideration the revision of the rules laid down in 1851, respecting the award of prizes, the constitution of juries, the affixing of prices, the distribution of space, the mode of classification, and also the organization of the additional department of the Fine Arts.

E

When, therefore, the guarantee deed has been executed, the Commissioners hope to be able to proceed at once with the construction of the buildings, and to announce the rules and regulations for the arrangement of the exhibition.

I have the honour to be, Sir,
Your obedient servant,
F. R. SANDFORD.

To obtain signatures to the Guarantee Deed from persons residing in almost every portion of the kingdom, was no light labour; but the work was undertaken by the officers of the Society of Arts with a degree of energy which insured the early completion of the task, and the Commissioners were enabled, on the fifteenth of March, to publish in the 'London Gazette' a notice that the Guarantee Deed had been signed for two hundred and fifty thousand pounds, which sum it had been arranged should be subscribed before the instrument would become binding on the Guarantors.

Her Majesty's Commissioners, with as little delay as possible, now began their work. In the early part of March, 1861, they took possession of offices in the Strand, and at some of their earliest meetings they arrived at the following

DECISIONS ON POINTS RELATING TO THE EXHIBITION.

1. Her Majesty's Commissioners have fixed upon Thursday the first day of May, 1862, for opening the Exhibition.

2. The Exhibition building will be erected on a site adjoining the gardens of the Royal Horticultural Society, and in the immediate neighbourhood of the ground occupied in 1851, on the occasion of the first international exhibition.

3. The portion of the building to be devoted to the exhibition of pictures will be erected in brick, and will occupy the entire front towards Cromwell Road; the portion in which machinery will be exhibited will extend along Prince Albert's Road, on the west side of the gardens.

4. All works of industry to be exhibited should have been produced since 1850. The decision whether goods proposed to be exhibited are admissible or not, must in each case eventually rest with Her Majesty's Commissioners.

5. Subject to the necessary limitation of space, all persons, whether designers, inventors, manufacturers, or producers of articles, will be allowed to exhibit; but they must state the character in which they do so.

6. Her Majesty's Commissioners will communicate with foreign and colonial exhibitors only through the commission which the government of each foreign country or colony may appoint for that purpose; and no article will be admitted from any foreign country or colony without the sanction of such commission.

7. No rent will be charged to exhibitors.

8. Every article produced or obtained by human industry, whether of
Raw materials,
Machinery,
Manufactures, or
Fine arts,
will be admitted to the exhibition, with the exception of
Living animals and plants,
Fresh vegetable and animal substances liable to spoil by keeping,
Detonating or dangerous substances.
Copper caps, or other articles of a similar nature, may be exhibited, provided the detonating powder be not inserted; also lucifer matches, with imitation tops.

9. Spirits or alcohols, oils, acids, corrosive salts, and substances of a highly inflammable nature, will only be admitted by special written permission, and in well-secured glass vessels.

10. The articles exhibited will be divided into the following classes:—

SECTION I.

Class 1. Mining, quarrying, metallurgy, and mineral products.
„ 2. Chemical substances and products, and pharmaceutical processes.
„ 3. Substances used for food.
„ 4. Animal and vegetable substances used in manufactures.

SECTION II.

Class 5. Railway plant, including locomotive engines and carriages.
    „   6. Carriages not connected with rail or tram roads.
    „   7. Manufacturing machines and tools.
    „   8. Machinery in general.
    „   9. Agricultural and horticultural machines and implements.
    „ 10. Civil engineering, architectural, and building contrivances.
    „ 11. Military engineering, armour and accoutrements, ordnance and small arms.
    „ 12. Naval architecture, ships' tackle.
    „ 13. Philosophical instruments, and processes depending upon their use.
    „ 14. Photographic apparatus and photography.
    „ 15. Horological instruments.
    „ 16. Musical instruments.
    „ 17. Surgical instruments and appliances.

SECTION III.

Class 18. Cotton.
  „ 19. Flax and hemp.
  „ 20. Silk and velvet.
  „ 21. Woollen and worsted, including mixed fabrics generally.
  „ 22. Carpets.
  „ 23. Woven, spun, felted, and laid fabrics, when shown as specimens of printing or dyeing.
  „ 24. Tapestry, lace, and embroidery.
  „ 25. Skins, fur, feathers, and hair.
  „ 26. Leather, including saddlery and harness.
  „ 27. Articles of clothing.
  „ 28. Paper, stationery, printing, and bookbinding.
  „ 29. Educational works and appliances.
  „ 30. Furniture and upholstery, including paper-hangings and papier mâché.
  „ 31. Iron and general hardware.
  „ 32. Steel and cutlery, and edge tools.
  „ 33. Works in precious metals and their imitations, and jewelry.
  „ 34. Glass.
  „ 35. Pottery.
  „ 36. Manufactures not included in previous classes.

SECTION IV.—MODERN FINE ARTS.

(See Decisions 111—123.)

Class 37. Architecture.
  „ 38. Paintings in oil and water colours, and drawings.
  „ 39. Sculpture, models, die-sinking, and intaglios.
  „ 40. Etchings and engravings.

11. Prizes, or rewards for merit, in the form of medals will be given in Sections I., II., III.

    (*a.*) These medals will be of one class, for merit, without any distinction of degree.

    (*b.*) No exhibitor will receive more than one medal in any class or sub-class.

    (*c.*) An international jury will be formed for each class and sub-class of the exhibition, by whom the medals will be adjudged, subject to general rules which will regulate the action of the juries.

    (*d.*) Each foreign commission will be at liberty to nominate one member of the jury for each class and sub-class, in which staple industries of their country and its dependencies are represented.

Her Majesty's Commissioners have resolved that an industry shall be ranked as a staple one which has twenty exhibitors in a class, or fifteen exhibitors in a sub-class. But Her Majesty's Commissioners will give to each foreign commission the alternative of sending a specified number of jurors, determined by the experience of past exhibitions, and by the relative spaces allotted to the several countries.

Note.—In the nomination, each sub-class is to be considered a separate jury. Should it happen that a foreign commission is not represented by fifteen exhibitors in any of the sub-classes of a general class, the fact that they have an aggregate of twenty exhibitors in the whole class would not entitle them to a juror. The sub-classes will act as separate juries, only to be united for confirmation of awards, and for general purposes of administration.

    (*e.*) The names of the foreign jurors must be sent to Her Majesty's Commissioners before the tenth of March, 1862; at the same time, the class or sub-class on which each juror is to serve must be specified.

(*f.*) The British jurors will be chosen in the following manner:

Her Majesty's Commissioners will take steps to secure a certain number of jurors on behalf of India and the colonies; and every exhibitor in the United Kingdom will propose the names of three persons to act on the jury for the class in which he exhibits.

From the persons so named, Her Majesty's Commissioners will select the requisite number of jurors.

(*g.*) The names of all the jurors will be published in April, 1862.

(*h.*) The juries will be required to submit their awards, with a brief statement of the grounds of each, to Her Majesty's Commissioners before the fifteenth day of June, 1862.

(*i.*) The awards will be published in the Exhibition building, at a public ceremony.

(*j.*) They will immediately afterwards be conspicuously attached to the counters of the successful exhibitors, and the grounds of each award will be very briefly stated.

(*k.*) If an exhibitor accepts the office of juror, no medal can be awarded in the class or sub-class to which he is appointed, either to himself individually, or to the firm in which he may be a partner.

(*l.*) The medals will be delivered to the exhibitors on the last day of the exhibition.

12. Prices may be affixed to the articles exhibited in sections I., II., III.

13. Her Majesty's Commissioners will be prepared to receive all articles which may be sent to them on or after Wednesday the twelfth of February, and will continue to receive goods until Monday the thirty-first of March, 1862, inclusive.

14. Articles of great size or weight, the placing of which will require considerable labour, must be sent before Saturday the first of March, 1862; and manufacturers wishing to exhibit machinery or other objects that will require foundations or special constructions, must make a declaration to that effect on their demands for space.

15. Any exhibitor whose goods can properly be placed together will be at liberty to arrange such goods in his own way, provided his arrangement is compatible with the general scheme of the exhibition and the convenience of other exhibitors.

16. Where it is desired to exhibit processes of manufacture, a sufficient number of articles, however dissimilar, will be admitted for the purpose of illustrating the process; but they must not exceed the number actually required. (17–25). [Several numbers were left blank with a view of incorporating future decisions.]

26. Exhibitors will be required to deliver their goods at such part of the building as shall be indicated to them (see Article 153), with the freight, carriage, porterage, and all charges and dues upon them paid.

27. The vans will be unloaded, and the articles and packages taken to the places appointed in the building, by the officers of Her Majesty's Commissioners.

28. Upon receipt of notice from Her Majesty's Commissioners, that the articles are deposited in the building, exhibitors or their representatives, or agents, must themselves unpack, put together, and arrange their goods.

29. Packing-cases must be removed at the cost of the exhibitors or their agents as soon as the goods are examined and deposited in charge of the Commissioners. If not removed within three days of notice being given, they will be disposed of, and the proceeds, if any, applied to the funds of the exhibition.

30. The following regulations have been adopted by the principal railway companies, with the view of affording facilities for the conveyance of goods to and from the Exhibition, and for warehousing packing-cases :—

(*a.*) The Railway Clearing House classification will be adopted.

(*b.*) Subject to the owners bearing all risks, the charge for conveyance will be one half the ordinary rates, with an addition of 5s. per ton each way, for the extra delivery to and from the Exhibition—the site of the building being beyond the usual delivery boundary.

(*c.*) All charges for goods going to the Exhibition must be prepaid, but for the return journey prepayment will be optional.

(*d.*) No less additional charge for delivery or collection than 6d. for a single consignment will be made in addition to the freight.

(*e.*) On application by exhibitors, empty packages will be warehoused by the railway companies up to the end of 1862, at the following scale of charges, including cartage from and to the Exhibition :—

5s. per package not exceeding 3 feet in its greatest dimensions.
7s. 6d.   „   „   4 feet   „   „
10s.   „   „   5 feet   „   „
20s.   „   • „   8 feet   „   „

Above that size, Special agreement. (31–34.)

35. No counters or fittings will be provided by Her Majesty's Commissioners. Exhibitors will be permitted, subject only to the necessary general regulations, to erect, according to their own taste, all the counters, stands, glass frames, brackets, awnings, hangings, or similar contrivances which they may consider best calculated for the display of their goods.

36. Exhibitors, or their representatives, should provide whatever light temporary covering may be requisite (such as sheets of oiled calico) to protect their goods from dust; and, in the case of machinery and polished goods, should make the requisite arrangements for keeping the articles free from rust during the time of the exhibition. (37–42.)

43. Exhibitors must be at the charge of insuring their own goods, should they desire this security. Every precaution will be taken to prevent fire, theft, or other losses, and Her Majesty's Commissioners will give all the aid in their power for the legal prosecution of any persons guilty of robbery or wilful injury in the exhibition, but they will not be responsible for losses or damage of any kind which may be occasioned by fire or theft, or in any other manner.

44. Exhibitors may employ assistants (male or female) to keep in order the articles they exhibit, or to explain them to visitors, after obtaining written permission from Her Majesty's Commissioners; but such assistants will be forbidden to invite visitors to purchase the goods of their employers. (45–49.)

50. Articles once deposited in the building will not be permitted to be removed without written permission from Her Majesty's Commissioners. (51–54.)

55. Her Majesty's Commissioners will provide shafting, steam (not exceeding thirty pounds per inch), and water, at high pressure, for machines in motion.

56. Persons who may wish to exhibit machines, or trains of machinery, in motion, will be allowed to have them worked, as far as practicable, under their own superintendence, and by their own men. (57–70.)

70. Intending exhibitors in the United Kingdom are requested to apply without delay to the secretary to Her Majesty's Commissioners, for a *Form of Demand for Space*, stating at the same time in which of the four sections they wish to exhibit.

71. The following is the form which has to be filled up :—

1. Name and Christian name of applicant (or name of firm) ....
2. Nature of business carried on ....
3. Address ⎰ No. of street or square, &c. ....
   ⎱ and
   ⎱ Name of Town ....
4. Nature of articles to be exhibited ....
5. Number of Class in which they are to be exhibited ....

Floor Space.

6. Probable space that will be required for articles or case in which they will be shown .. ..

⎰ Length .... feet.
⎰ Breadth .... feet.
⎱ Height .... feet.

Hanging or Wall Space.

⎰ Height .... feet.
⎱ Width .... feet.

100. Foreign and colonial exhibitors should apply to the commission, or other central authority, appointed by the foreign or colonial government, as soon as notice has been given of its appointment.

101. Her Majesty's Commissioners will consider that to be the central authority in each case which is stated to be so by the government of its country, and will only communicate with exhibitors through such central body.

102. No articles of foreign manufacture, to whomsoever they may belong, or wheresoever they may be, can be admitted for exhibition, *except with the sanction of the central authority of the country of which they are the produce.* Her Majesty's Commissioners will communicate to such central authority the amount of space which can be allowed to the productions of the country for which it acts, and will also state the further conditions and limitations which may from time to time be decided on with respect to the admission of articles. All articles forwarded by such central authority will be admitted, provided they do not require a greater aggregate amount of space than that assigned to the country from which they come; and provided also, that they do not violate the general conditions and limitations. It will rest with the central authority in each country to decide upon the merits of the several articles presented for exhibition, and to take care that those which are sent are such as fairly represent the industry of their fellow-countrymen.

103. Separate space will be allotted to each foreign country, within which the commissioners for that country will be at liberty to arrange the productions intrusted to them in such manner as they think best,

subject to the condition that all machinery shall be exhibited in the portion of the building specially devoted to that purpose, and all pictures in the fine art galleries, and to the observance of any general rules that may be laid down by Her Majesty's Commissioners for public convenience.

104. By arrangements made with Her Majesty's Government, all foreign or colonial goods intended for exhibition, sent and addressed in accordance with the regulations laid down by Her Majesty's Commissioners of Customs, as set forth in article 108, will be admitted into the country, and allowed to be transmitted to the Exhibition building without being previously opened, and without payment of any duty. But all goods which shall not be re-exported at the termination of the exhibition will be charged with the proper duties, under the ordinary Customs' regulations.

106. Every article sent separately, and every package, must be legibly marked with the name of the foreign country or colony of which it is the produce or manufacture, and, as far as practicable, with the name of the exhibitor or exhibitors.

107. The following is the form of address which should be adopted:—

To the Commissioners for the Exhibition of 1862.

BUILDING, SOUTH KENSINGTON, LONDON.

From [state Country, and Exhibitor's name.]

To prevent loss, miscarriage, or mislaying, articles, or packing-cases containing them, which occupy less bulk than two cubic feet, should not be sent separately; but packages under such size containing, as far as possible, the same classes of articles, should be transmitted in combination.

108. Her Majesty's Commissioners of Customs have laid down the following regulations upon the importation of goods intended for the Exhibition:—

a. All packages containing goods intended for the International Exhibition of 1862 shall be specially reported as such, and shall be addressed to the Commissioners of the International Exhibition, or to one of their officers, and be consigned to a duly-accredited agent, and shall be accompanied with a specification of their contents and value. They shall be separately entered as intended for the International Exhibition, and the agents in passing their entries shall specify the full contents of the packages, together with the value.

b. Such packages as may be landed in London shall be forwarded unopened to the Exhibition in charge of an approved licensed carman, accompanied by a cart note from the landing officer, giving a description of the packages and the marks and numbers thereon; and in cases where there may be reason to suppose they contain other goods than those for the Exhibition, they shall also be accompanied by a revenue officer.

c. Packages landed at the outports shall be forwarded with a similar note by railway or other public conveyance, under seals of office, direct to the Exhibition, the officers at the respective ports taking care that the packages bear no private address, and that the documents relating thereto be immediately forwarded to the proper officers of Her Majesty's Customs stationed at the Exhibition.

d. On the arrival of the goods at the Exhibition, no package shall be opened without the knowledge and consent of the officer of Customs, and if the goods be found to agree with the entry or specification, they will, if free, be at once considered as out of charge of the Customs, the entry or declaration being deemed sufficient for all statistical purposes.

e. In the case of all dutiable goods, an account will be taken by the officers of the Crown at the time of the first opening of the packages, but such deficiencies as may occur within the building from any legitimate or unavoidable cause, the officers being fully satisfied thereof, shall not be charged with duty.

f. That the building be considered, for all practical purposes, a "bonded warehouse;" and that in all cases where dutiable goods shall not be exported, but retained for use in this country, the duty shall be assessed by the officer in charge at the building (and received in the Exhibition by a clerk duly appointed for the purpose), in accordance with the practice now existing in regard to articles found in "passengers' baggage."

g. In the case of dutiable goods for exportation, an entry shall be passed in the long room, and bond given for their due exportation; and on the receipt of this entry by the officer in charge of the building the goods shall be packed in his presence, and if for shipment at an outport, placed under seal, and forwarded in charge to a railway or other public company; but if for shipment at London they shall then be sent in charge of Customs officers, at the expense of the exporter, to be delivered into the charge of the searcher of the station from which they are to be shipped, without further examination, under the regulations applicable to goods shipped direct from the warehouse.

109. It is not the intention of Her Majesty's Commissioners to take any steps in reference to the protection of inventions or designs, by patent or registration, the law on these points having been materially simplified since 1851.

DECISIONS SPECIALLY APPLICABLE TO

SECTION IV.—MODERN FINE ARTS.

Class 37. Architecture.
  ,,   38. Paintings in oil and water colours, and drawings.
  ,,   39. Sculpture, models, die-sinking, and intaglios.
  ,,   40. Engravings and etchings.

110. The object of the exhibition being to illustrate the progress and present condition of *modern art*, each country will decide the period of art which in its own case will best attain that end.

111. The exhibition of British art in this section will include the works of artists alive on or subsequent to the 1st of May, 1762.

112. It is not proposed to award prizes in this section.

113. Prices will be not allowed to be affixed to any work of art exhibited in this section.

114. One half of the space to be allotted to Section IV. will be given to foreign countries, and one half will be reserved for the works of British and colonial artists.

115. The subdivision of the space allotted to foreign countries will be made after consideration of the demands received from the commission, or other central authority, of each foreign country. It is therefore important that these demands should be transmitted to Her Majesty's Commissioners at the earliest possible date.

116. The arrangement of the works of art within the space allotted to each foreign country will be entirely under the control of the accredited representatives of that country, subject only to the necessary general regulations.

117. For the purposes of the catalogue it will be necessary that the central authority of each foreign country should furnish Her Majesty's Commissioners, on or before the 1st of January, 1862, with a description of the several works of art which will be sent for exhibition, specifying in each case the name of the artist, the title of the work, and (when possible) the date of its production.

118. The space at the disposal of Her Majesty's Commissioners for the display of British art being limited, and it being at the same time desirable to bring together as careful and perfect an illustration as possible, a selection of the works to be exhibited will be indispensable.

119. The selection of exhibitors, the space and number of works to be allowed to each, and the arrangement of them, will be intrusted to committees to be nominated by Her Majesty s Commissioners.

120. In the case of living artists, Her Majesty's Commissioners would desire to consult the wishes of the artists themselves as to the particular works by which they would prefer to be represented. The selection of works so made by the artists will not necessarily be binding upon Her Majesty's Commissioners, but in no case will any work by a living artist be exhibited against his wish, if expressed in writing, and delivered to the Commissioners on or before the 31st of March, 1862.

121. Her Majesty's Commissioners will avail themselves of the following eight art institutions of this country in communicating with artists who are members of those institutions :—

    The Royal Academy,
    The Royal Scottish Academy,
    The Royal Hibernian Academy,
    The Society of Painters in Water Colours,
    The Society of British Artists,
    The New Society of Painters in Water Colours,
    The Institute of British Artists,
    The Institute of British Architects.

122. Intending exhibitors in the British division of Section IV., who are not members of any of the preceding institutions, may at once receive forms of demand for space, by applying to the secretary to Her Majesty's Commissioners. These forms must be filled up and returned before the 1st of June, 1861.

ADMISSION OF VISITORS.

123. Her Majesty's Commissioners have adopted the following regulations with respect to the admission of visitors to the Exhibition :—

    (*a.*) The Exhibition will be open daily (Sundays excepted) during such hours as the Commissioners shall from time to time appoint.

(*b.*) The Royal Horticultural Society having arranged a new entrance to their Gardens from Kensington Road, the Commissioners have agreed with the Council of the Society to establish an entrance to the Exhibition from the Gardens, and to issue a joint ticket, giving the owner the privilege of admission both to the Gardens and to the Exhibition on all occasions when they are open to visitors, including the flower shows and fêtes held in the Gardens up to the 18th of October, 1862.

(*c.*) There will, therefore, be four principal entrances for visitors :—
    (1.) From the Horticultural Gardens, for the owners of the joint tickets, Fellows of the Society, and other visitors to the Gardens.
    (2.) In Cromwell Road.
    (3.) In Prince Albert's Road.
    (4.) In Exhibition Road.

(*d.*) The regulations necessary for preventing obstructions and danger at the several entrances will be issued prior to the opening.

(*e.*) Admittance to the Exhibition will be given only to the owners of season tickets, and to visitors paying at the doors.

SEASON TICKETS.

(*f.*) There will be two classes of season tickets :—
The first, price 3*l.* 3*s.*, will entitle the owner to admission to the opening and all other ceremonials, as well as at all times when the building is open to the public.
The second, price 5*l.* 5*s.*, will confer the same privileges of admission to the Exhibition, and will further entitle the owner to admission to the Gardens of the Royal Horticultural Society at South Kensington and Chiswick (including the flower shows and fêtes at these gardens) from the 1st of February to the 18th of October.

(*g.*) Season tickets must be signed before presentation. The owners must produce them, and must write their names in a book at the door each time they enter the building or Gardens.

(*h.*) Season tickets are not transferable, and if presented by any other persons than the registered owners, will be forfeited, and the names of the offenders will be published. If lost, they will not be replaced.

PRICES OF ADMISSION.

(*i.*) On the 1st of May, on the occasion of the opening ceremonial, the admissions will be restricted to the owners of season tickets.

(*j.*) On the 2nd and 3rd of May the price of admission will be 1*l.* for each person ; and the Commissioners reserve to themselves the power of appointing three other days, when the same charge will be made.

(*k.*) From the 5th to the 17th of May, 5*s.*

(*l.*) From the 19th to the 31st of May, 2*s.* 6*d.*, except on one day in each week, when the charge will be 5*s.*

(*m.*) After the 31st of May the price of admission on four days in each week will be 1*s.*

ADMISSION OF PERSONS AND RECEPTION OF ARTICLES DURING THE ARRANGEMENT OF THE EXHIBITION.

ADMISSION OF PERSONS.

125. No person whatever will be admitted into the building unless he is the bearer of a pass or day ticket.

126. Passes, for the period specified thereon, available only at a particular entrance, and for a particular part of the building, will be issued as follows :—
To commissioners of colonies and foreign countries, on application to the Secretary to Her Majesty's Commissioners ;
To Custom-house officers, and their attendants, on application to Her Majesty's Commissioners through the superintendent, British side ;
To exhibitors, and others not before specified, obtaining a pass available for more than one day for British side ;
To exhibitors and others, not before specified, obtaining a pass available for more than one day for Foreign side.

127. An exhibitor, British side, desiring a pass must apply for a form to be obtained at the office for passes, at least two days before it is required. The application must be addressed to the superintendent

of the class to which the exhibitor belongs, who will forward it to the superintendent of the British side, with his recommendation attached. If granted, the name will be entered on the registry, and the pass forwarded to the applicant.

128. Passes for workmen in the employ of exhibitors, British side, will be issued, at the discretion of their class superintendent, to exhibitors themselves, on application to him, and superintendents of classes will state in their daily reports any irregularity or misconduct they may observe.

129. No application for a pass will be received from a foreign exhibitor except through the commissioner of his own country.

130. Lists of foreign exhibitors requiring passes, signed by the commissioner of the country to which they belong, will be considered as vouchers, on which the number of passes required will be delivered to such commissioners.

131. Passes for workmen, Foreign side, will be issued on the application of the commissioner of the foreign country requiring their admission.

132. Day tickets will be issued, on application, by the superintendents of the several classes, and by the superintendent of the Foreign side.

133. An exhibitor, British side, requiring a day ticket, will be furnished with it by the superintendent of the class to which he belongs, on application at the building.

134. Applications for passes or day tickets on account of exhibitors from India or the colonies must be addressed to the superintendents of the Indian and colonial departments at the building.

135. An exhibitor, Foreign side, requiring a day ticket, will be furnished with it on application to the superintendent of the Foreign side.

136. Passes and day tickets must be shown on entering and leaving, and whenever demanded, within the building.

137. Passes and day tickets are not transferable; the transfer of a pass or ticket will, on discovery, subject the holder to expulsion from the building, and prevent the person to whom it was originally issued from obtaining another admission.

138. If the possessor of a pass is found in a part of the building beyond the specified limits of his pass, it will be taken from him.

139. All persons holding passes will be considered as agreeing to conform to all regulations from time to time issued.

140. Infringement of rules will be followed by expulsion and forfeiture of pass.

141. If a pass is lost by the owner, he must pay ten shillings before obtaining its renewal.

142. The door-keepers will be held responsible for any person found in the building without a pass.

143. The officer in charge of a class space will be held responsible for the presence of any unauthorized person within it.

144. Every person, not properly authorized, found in the building, or handling or conveying or removing any package or article from one part of the building to another, will be liable to be given into custody.

145. No person will be allowed to carry any bundle or parcel of any size or kind whatever out of the building before it has been opened and examined.

146. All persons using tow or cotton waste must provide slate or metal boxes for containing the refuse.

147. Smoking is strictly forbidden.

148. The introduction of lucifer matches into the building is strictly forbidden.

149. Officers and servants of the Commissioners must attend punctually at the appointed hours, and enter their names in the books provided.

150. Intoxication or disobedience of orders will subject the offender to immediate dismissal.

151. The receipt of fees or payments of any kind by any officers or servants of Her Majesty's Commissioners will subject the receivers to dismissal.

MODE OF PASSING ARTICLES OF THE UNITED KINGDOM INTO THE BUILDING.

152. All articles must be delivered at the building with all charges and dues whatever upon them paid.

153. The entrance for the reception of goods into the building will be as follows :—

No. 1. The entrances to the Eastern Annexe in Exhibition Road for Classes 1, 2, 3, 4, and 9, and part of 5 and 7, marked Door A.

No. 2. The central gates in the Cromwell Road for fine arts and general manufactures, English and foreign, marked Door B (Fine Arts), Door C (General Manufactures).

No. 3. The entrance by Gore Lane for English and foreign machinery, marked Door D.

No. 4. The entrance by the Horticultural Society's Road in Prince Albert's Road for English machinery only, marked Door E.

154. Goods and machinery will be received during such hours only as may from time to time be fixed.

155. No persons but the carmen will be allowed to enter with a waggon; they will not be permitted to leave the waggon while within the building.

156. Every article sent separately, and every package, must be legibly marked with the class number, and the name of the exhibitor or exhibitors, and special labels will be accordingly sent to them.

157. An officer will be appointed to superintend the unloading and transporting of the articles to their respective places, and the following regulations will be observed in the reception and distribution of goods :—

    1st. Every package on its delivery into the building to be entered in a register, with the name and address of the sender, the class to which it belongs, and whether received in good condition or damaged.

    2nd. A rotation number to be marked distinctly on each package; the same number to be entered against the sender's name in the register.

    3rd. All packages to be removed from the landing stage, and deposited in charge of the class superintendents, or their deputies, at places appointed for such purposes, as soon as possible after their reception into the building.

    4th. Packages accidentally delivered or received at the wrong door, or from which the name of the exhibitor may have become obliterated, to be deposited in a place set apart for such packages.

    5th. In order to facilitate the answering of inquiries, the rotation numbers of all packages received during the day will be entered in a book containing the names of the exhibitors, alphabetically arranged, to be made up every evening.

158. Each class will have a superintendent, a deputy superintendent, and the requisite number of attendants attached to it.

159. The space for each exhibitor will be distinctly marked in the building.

160. Trucks will be numbered for each class, and wheeled off to their respective classes by the officers of the Exhibition.

161. All packing-cases, &c., must be removed by the exhibitors as soon-as they receive orders from the Commissioners to do so. Packing-cases not removed within three days after notice will be sold, and the proceeds applied to the funds of the Exhibition.

MODE OF PASSING FOREIGN AND COLONIAL GOODS INTO THE BUILDING.

162. The receipt of all foreign and colonial articles will be subject to the control of the officers of the Customs, and in case any difficulty should arise, application is to be made to the senior Customs officer at the building.

163. The officers of the Customs and their servants will be provided with passes, and will be privileged to enter all portions of the building in which they may have business.

164. All carts and waggons bringing foreign goods will enter the building at Cromwell Road, Door C, excepting machinery, which will enter the machinery department by Gore Lane, Door D.

165. Goods and machinery will be received during such hours only as may from time to time be fixed.

166. All articles and packages must be delivered at the building with all charges and dues whatever paid on them.

167. Every article sent separately, and every package, must be legibly marked with the name of the foreign country or colony of which they are the produce or manufacture, and as far as practicable with the name of the exhibitor or exhibitors.

168. No person but the carmen and the officers of Customs in charge will be allowed to enter the building with a cart or waggon. Neither the carmen nor the officer of Customs in charge will be permitted to quit the waggon whilst in the building.

169. The carts or waggons will be unloaded in rotation by the officers of the Exhibition, when rotation numbers will be affixed to each package by the officers of the Customs.

170. The packages must be produced on arrival to the officer of Customs at the Exhibition, who will see that the Customs number as well as the name of the foreign country, is affixed to each package, which will then be placed on trucks, and conveyed to its destination in the building.

171. The officer in charge of each division will see that the packages belong to that division, that the Customs rotation number is marked thereon, and that the goods are then properly stored within it. When the articles of each country are deposited within the space assigned to them, the commissioner and agents appointed by foreign commissions and colonial committees, or the exhibitors, must themselves unpack, put together, and arrange them. In the case of foreign and colonial productions, as they must be necessarily unpacked for a considerable time before they are finally arranged for exhibition, it is suggested

that the consignees or agents should be authorized to provide proper temporary coverings, such as glazed calico, to protect the articles from dust, &c., and in the case of machinery and polished goods make the requisite arrangements for keeping the articles free from rust.

172. The officers of the Exhibition in charge of each division will cause all packages properly certified as empty to be arranged in places hereafter to be determined upon.

173. All packing-cases, &c., must be removed by exhibitors, or their agents, as soon as they receive notice from Her Majesty's Commissioners to do so.

174. All packing-cases not removed within three days after. notice, will be sold, and the proceeds applied to the funds of the Exhibition.

175. Exhibitors intending to introduce foreign articles upon which duty has been paid, with the view of exhibiting them amongst the goods of the country of which they are the produce or manufacture, must have a ticket attached to each with the words "duty paid" thereon; and to prevent difficulty in their delivery at the close of the Exhibition they should be brought under the notice of the principal officer of Customs at the time they are brought in.

The organization of a working staff followed next, and the formation of the following divisional

COMMITTEES OF ADVICE.

BUILDING COMMITTEE.

EARL OF SHELBURNE.
WILLIAM BAKER, ESQ.

WILLIAM FAIRBAIRN, ESQ., LL.D., F.R.S.

HON. EDWIN B. PORTMAN, *Secretary.*

FINANCE COMMITTEE.

RIGHT HON. R. LOWE, M.P.
SIR A. SPEARMAN, BART.
EDGAR A. BOWRING, ESQ.

T. F. GIBSON, ESQ.
HENRY THRING, ESQ.

LORD FREDERICK CAVENDISH, *Hon. Secretary.*

INDUSTRIAL DEPARTMENT.

GENERAL COMMITTEE ON ORGANIZATION.

MARQUIS OF HARTINGTON, M.P.
LORD STANLEY, M.P.
LORD NAAS, M.P.
LORD STANLEY OF ALDERLEY.
RIGHT HON. LORD MAYOR OF LONDON.
RIGHT HON. T. MILNER GIBSON, M.P.
RIGHT HON. W. HUTT, M.P.
THOMAS BAZLEY, ESQ., M.P.
T. F. GIBSON, ESQ.
DR. LYON PLAYFAIR, C.B., F.R.S.
HENRY COLE, ESQ., C.B.

WILLIAM DARGAN, ESQ., Dublin.
PRESIDENT OF THE ROYAL AGRICULTURAL SOCIETY
CHAIRMAN OF THE SOCIETY OF ARTS.
CHAIRMAN OF THE ROYAL DUBLIN SOCIETY.
PRESIDENT OF THE INSTITUTION OF CIVIL ENGINEERS.
PRESIDENT OF THE INSTITUTION OF MECHANICAL ENGINEERS.
PRESIDENTS OF THE CHAMBERS OF COMMERCE IN THE UNITED KINGDOM.

EDGAR A. BOWRING, ESQ., C.B., *Hon. Secretary.*

CLASS COMMITTEES.

MINING COMMITTEE.—CLASS I.—MINING, QUARRYING, METALLURGY, AND MINERAL PRODUCTS.

SIR R. I. MURCHISON, D.C.L., F.R.S.
I. RAWSON BARKER, ESQ.
SAMUEL BLACKWELL, ESQ.
R. B. GRANTHAM, ESQ., C.E., F.G.S.
PROFESSOR MASKELYNE.
PROFESSOR PERCY, M.D.; F.R.S.

PROFESSOR RAMSEY, F.R.S.
W. S. RODEN, ESQ.
PROFESSOR WARINGTON SMYTH, F.R.S.
THOMAS SOPWITH, ESQ., F.R.S.
NICHOLAS WOOD, ESQ.
HUSSEY VIVIAN, ESQ., M.P.

ROBERT HUNT, ESQ., F.R.S., *Superintendent.*

CHEMICAL COMMITTEE.—CLASS II.—CHEMICAL SUBSTANCES AND PRODUCTS, AND PHARMA-CEUTICAL PROCESSES.

W. T. Brande, Esq., F.R.S.
Warren De la Rue, Esq., Ph.D., F.R.S.
Professor Faraday, D.C.L., F.R.S.
Thomas Graham, Esq., F.R.S.
A. W. Hofmann, Esq., LL.D., F.R.S.

W. A. Miller, Esq., M.D., F.R.S.
Dr. Lyon Playfair, C.B., F.R.S.
Theophilus Redwood, Esq., Ph.D.
J. Stenhouse, Esq., LL.D., F.R.S.

C. W. Quin, Esq., *Superintendent.*

MACHINERY COMMITTEE.—
CLASS V.—RAILWAY PLANT, INCLUDING LOCOMOTIVE ENGINES AND CARRIAGES.
VII.—MANUFACTURING MACHINES AND TOOLS.
VIII.—MACHINERY IN GENERAL.
X.—CIVIL ENGINEERING, ARCHITECTURAL AND BUILDING CONTRIVANCES.

The Duke of Sutherland, *Chairman.*
The Earl of Caithness.
Lord Richard Grosvenor, M.P
John Anderson, Esq.
Frederick J. Bramwell, Esq.
Joseph Cubitt, Esq.
James Fenton, Esq.
John Fernie, Esq.

Capt. Douglas Galton, R.E., F.R.S.
Thomas Elliot Harrison, Esq.
George Willoughby Hemans, Esq.
Sampson Lloyd, Esq.
John Robinson M'Clean, Esq.
Henry Maudslay, Esq.
John Penn, Esq.
John Scott Russell, Esq., F.R.S.

D. K. Clark, Esq., *Superintendent.*

CLASS Xв.—SUB-COMMITTEE.—SANITARY APPLIANCES.

Earl of Shaftesbury.
Earl of Ducie, F.R.S.
Earl Fortescue.
Right Rev. Lord Bishop of Bath and Wells.
Right Hon. Lord Mayor.
Sir Joseph Olliffe, M.D.
J. F. Campbell, Esq.
E. Chadwick, Esq., C.B.
William Fairbairn, Esq. LL.D., F.R.S.

Captain Douglas Galton, R.E., F.R.S.
G. Godwin, Esq., F.R.S.
P. H. Holland, Esq.
Owen Jones, Esq.
Dr. Letheby.
R. Rawlinson, Esq., C.E.
J. Simon, Esq., F.R.S.
J. Sutherland, Esq., M.D.
T. Twining, Esq.

Hon. Edwin B. Portman, *Secretary.*

AGRICULTURAL AND HORTICULTURAL COMMITTEE.—CLASS IX.—AGRICULTURAL AND HORTI-CULTURAL MACHINES AND IMPLEMENTS.

Earl of Clancarty.
Earl of Erne.
Lord Portman.
Lord Talbot de Malahide, F.R.S.
J. Easton, Esq.
B. T. Brandreth Gibbs, Esq., *Superintendent.*
John Gibson, Esq.

Chandos W. Hoskyns, Esq.
Charles Lee, Esq.
J. Hall Maxwell, Esq., C.B.
James Stirling, Esq., C.E.
H. S. Thompson, Esq., M.P.
Professor Wilson.

Hon. Edwin B. Portman, *Secretary.*

MILITARY COMMITTEE.—CLASS XI.—MILITARY ENGINEERING, ARMOUR AND ACCOUTREMENTS, ORDNANCE AND SMALL ARMS.

Major Gen. The Hon. James Lindsay, M.P., *Chairman.*
Colonel Shafto Adair, F.R.S.
J. Anderson, Esq.
T. Graham Balfour, Esq., M.D., F.R.S.
Lieut.-Col. R. S. Baynes.
W. Fairbairn, Esq., LL.D., F.R.S.
Capt. Fowke, R.E.

Lieut.-Col. A. Lane Fox.
Col. J. H. Lefroy, F.R.S.
J. G. Logan, Esq., M.D.
Major Porter, R.E.
Sir Sibbard Scott, Bart.
Capt. A. C. Tupper.
Capt. Tyler, R.E.
Capt. Sir W. Wiseman, Bart., R.N.

Major Moffatt, *Superintendent.*

NAVAL COMMITTEE.—CLASS XII.—NAVAL ARCHITECTURE, SHIPS' TACKLE.

CAPT. FREDERICK ARROW.
REAR-ADMIRAL C. R. D. BETHUNE, C.B.
CAPT. MARK C. CLOSE.
CAPT. RICHARD COLLINSON, C.B.
CAPT. E. G. FISHBOURNE, C.B., *Chairman.*

JOHN LAIRD, ESQ.
CAPT. SIR FREDERICK NICOLSON, BART., C.B.
CAPT. WASHINGTON, F.R.S.
CLIFFORD WIGRAM, ESQ.

MAJOR MOFFATT, *Superintendent.*

PHILOSOPHICAL COMMITTEE.—CLASS XIII.—PHILOSOPHICAL INSTRUMENTS AND PROCESSES DEPENDING UPON THEIR USE.

SIR DAVID BREWSTER, F.R.S.
PROFESSOR B. C. BRODIE, F.R.S.
CHARLES BROOKE, ESQ., F.R.S.
DR. CARPENTER, F.R.S.
DR. FRANKLAND, F.R.S.
FRANCIS GALTON, ESQ., F.R.S.

J. P. GASSIOT, ESQ., F.R.S., *Chairman.*
PROFESSOR TYNDALL, F.R.S.
PROFESSOR WHEATSTONE, F.R.S.
COLONEL SIR H. JAMES, R.E., F.R.S.
WARREN DE LA RUE, ESQ., F.R.S.
JAMES GLAISHER, ESQ., F.R.S.

C. R. WELD, ESQ., *Superintendent.*

PHOTOGRAPHIC COMMITTEE.—CLASS XIV.—

EARL CAITHNESS.
EDWARD KATER, ESQ., F.R.S.

HUGH DIAMOND, ESQ., M.D.

P. LE NEVE FOSTER, ESQ., M.A., *Superintendent.*

SURGICAL COMMITTEE.—CLASS XVII.—SURGICAL INSTRUMENTS AND APPLIANCES.

J. MONCRIEFF ARNOTT, ESQ., F.R.C.S., F.R.S.
T. BELL, ESQ., V.P.R.S.
ALEXANDER BRYSON, ESQ., M.D., F.R.S., Inspector of General Hospitals and Fleets.
ARTHUR FARRE, ESQ., F.R.C.S.
F. SEYMOUR HADEN, ESQ., F.R.C.S.
CÆSAR H. HAWKINS, ESQ., F.R.C.S., F.R.S.

WILLIAM LAWRENCE, ESQ., F.R.C.S., F.R.S.
THOMAS LONGMORE, ESQ., F.R.C.S., Deputy Inspector General.
JAMES LUKE, ESQ., F.R.C.S., F.R.S.
JAMES PAGET, ESQ., F.R.C.S., F.R.S.
JOHN F. SOUTH, ESQ., F.R.C.S.

J. R. TRAER, ESQ., F.R.C.S., *Superintendent.*

EDUCATION COMMITTEE.—CLASS XXIX.—EDUCATIONAL WORKS AND APPLIANCES.

VISCOUNT ENFIELD, M.P.
RIGHT HON. C. B. ADDERLEY, M.P., *Chairman.*
RIGHT HON. W. COWPER, M.P.
HON. AND REV. S. BEST.
SIR JOHN ACTON, BART., M.P.
R. R. W. LINGEN, ESQ.
THE VERY REV. THE DEAN OF HEREFORD.
THE VERY REV. THE DEAN OF CHRIST CHURCH, OXFORD.
THE MASTER OF TRINITY COLLEGE, CAMBRIDGE.
THE REV. DR. TEMPLE.
M. ARNOLD, ESQ., H.M., Inspector of Schools.
REV. SAMUEL CLARK.

REV. F. C. COOK, H.M. Inspector of Schools.
REV. B. M. COWIE, H.M. Inspector of Schools.
EDWARD HAMILTON, ESQ.
JOSEPH D. HOOKER, ESQ., M.D., F.R.S.
REV. M. MITCHELL, H.M. Inspector of Schools, *Vice-Chairman.*
REV. M. PATTISON.
W. PORTAL, ESQ.
J. S. REYNOLDS, ESQ.
REV. W. ROGERS.
PHILIP S. SCLATER, ESQ., M.A., F.R.S.
E. C. TUFNELL, ESQ., H.M. Inspector of School.
REV. A. WILSON.

J. G. FITCH, ESQ., M.A., *Superintendent.*

FINE ART DEPARTMENT.

GENERAL COMMITTEE.

THE DUKE OF BUCCLEUCH, K.G.
THE MARQUIS OF LANSDOWNE, K.G.
THE MARQUIS OF HERTFORD, K.G.
EARL SPENCER.

EARL STANHOPE, F.R.S.
EARL OF MALMESBURY.
EARL SOMERS.
EARL OF DUDLEY.

LORD ASHBURTON, F.R.S.
LORD OVERSTONE.
LORD TALBOT DE MALAHIDE, F.R.S.
LORD LLANOVER.
LORD TAUNTON.
LORD ELCHO, M.P.
THE LORD CHIEF BARON.
SIR STAFFORD NORTHCOTE, BART., M.P.
SIR FRANCIS SCOTT, BART.
THOMAS ASHTON, ESQ.
R. H. CHENEY, ESQ.
REV. E. COLERIDGE.
E. C. FIELD, ESQ.
R. S. HOLFORD, ESQ., M.P.
H. T. HOPE, ESQ.
JOHN RUSKIN, ESQ.
WM. STIRLING, ESQ., M.P.

S. J. STERN, ESQ.
TOM TAYLOR, ESQ.
JOHN WALTER, ESQ., M.P.
W. WELLS, ESQ.
THE PRESIDENT OF THE ROYAL ACADEMY.
THE PRESIDENT OF THE ROYAL SCOTTISH ACADEMY.
THE PRESIDENT OF THE ROYAL HIBERNIAN ACADEMY.
THE PRESIDENT OF THE OLD SOCIETY OF PAINTERS IN WATER COLOURS.
THE PRESIDENT OF THE NEW SOCIETY OF PAINTERS IN WATER COLOURS.
THE PRESIDENT OF THE SOCIETY OF BRITISH ARTISTS.
THE PRESIDENT OF THE INSTITUTE OF BRITISH ARTISTS.
THE PRESIDENT OF THE ROYAL INSTITUTION OF BRITISH ARCHITECTS.

P. LE NEVE FOSTER, ESQ., M.A., *Secretary:*

COMMITTEE FOR CLASS XXXVII.—ARCHITECTURE.

ARTHUR ASHPITEL, ESQ.
JAMES BELL, ESQ.
PROFESSOR DONALDSON.
JAMES FERGUSSON, ESQ.
T. HAYTER LEWIS, ESQ.
A. J. BERESFORD HOPE, ESQ., *Chairman.*

G. G. SCOTT, ESQ., R.A.
SYDNEY SMIRKE, ESQ., R.A.
WILLIAM TITE, ESQ., M.P., F.R.S., President of the Institute of British Architects.
M. DIGBY WYATT, ESQ.

J. B. WARING, ESQ., F.R.I.B.A., *Superintendent.*

COMMITTEE FOR CLASS XXXVIII.—PAINTINGS IN OIL AND WATER COLOURS, AND DRAWINGS.

SIR CHARLES EASTLAKE, F.R.S., President Royal Academy, *Chairman.*
SIR J. W. GORDON, President Royal Scottish Academy.
S. CATTERSON SMITH, ESQ., President Royal Hibernian Academy.

F.Y. HURLSTONE, ESQ., President Society of British Artists.
F. TAYLOR, ESQ., President Society of Painters in Water Colours.
H. WARREN, ESQ., President New Society of Painters in Water Colours.

R. REDGRAVE, ESQ., R.A., *Professional Adviser.* R. J. SKETCHLEY, ESQ., *Assistant.*

SUB-COMMITTEE.—CLASS XXXVIIIA.—ART DESIGNS FOR MANUFACTURES.

MARQUIS OF SALISBURY, K.G.
EARL OF DUDLEY.
SIR J. P. BOILEAU, BART., F.R.S.
H. A. BOWLER, ESQ.
HENRY COLE, ESQ., C.B.
C. D. FORTNUM, ESQ.

D. MACLISE, ESQ., R.A.
R. MONCKTON MILNES, ESQ., M.P.
GODFREY SYKES, ESQ.
T. WINKWORTH, ESQ.
M. DIGBY WYATT, ESQ.

J. LEIGHTON, ESQ., *Superintendent.*

COMMITTEE FOR CLASS XXXIX.—SCULPTURE, MODELS, DIE SINKING, AND INTAGLIOS.

THE MARQUIS OF LANSDOWNE, K.G.
THE EARL OF GIFFORD, M.P.
J. H. FOLEY, ESQ., R.A.

A. H. LAYARD, ESQ., M.P.
R. WESTMACOTT, ESQ., R.A., F.R.S.

EDMUND OLDFIELD, ESQ., M.A., *Superintendent.*

COMMITTEE FOR CLASS XL.—ETCHINGS AND ENGRAVINGS.

W. H. CARPENTER, ESQ.
D. COLNAGHI, ESQ.
G. T. DOO, ESQ., R.A., *Chairman.*

R. J. LANE, ESQ., A.R.A.
W. SMITH, ESQ., *Superintendent.*

C. W. FRANKS, ESQ., *Secretary.*

These committees suggested the appointment of local committees, and the following circular was accordingly drawn up and sent, with a copy of the Charter, to every mayor, provost, and chamber of commerce in the United Kingdom :—

SIR,

I AM directed to inform you that at a meeting of a committee appointed for the purpose of advising Her Majesty's Commissioners as to the arrangements to be made for obtaining an instructive display of the industrial products of the United Kingdom at the approaching exhibition, the following resolutions were adopted, and have been submitted to the consideration of the Commissioners :—

1st. That it is the opinion of this Committee that, in order to obtain an adequate representation of the various interests concerned, and to excite local sympathy in the objects of the exhibition, it is expedient that a system of local organization should be adopted in the first instance.

2nd. That in the opinion of this Committee it is expedient that all persons desirous of being exhibitors should send in demands for space to Her Majesty's Commissioners, who should refer such demands either to trade committees or local committees, as may seem in each case most expedient.

Her Majesty's Commissioners have accordingly, in the "Decisions," of which I enclose a copy, invited all persons in this country who wish to take part in the exhibition to make early demands for space. These demands have already reached them in very considerable numbers, but no final decision with respect to the organization of the committee to whom they will be referred for consideration can be arrived at till further information is obtained as to the amount of space that will probably be required in each class of the exhibition, the districts from which the demands for space will mainly proceed, and the local organization that will be adopted in those districts.

The assurances which have already been received by Her Majesty's Commissioners from foreign governments, of the interest which is felt abroad in the progress of the undertaking, leave no doubt but that every effort is being made by foreign exhibitors in preparing to take part in the exhibition.

Her Majesty's Commissioners trust that equal zeal will be shown in this country by all classes of the community ; and with a view to the adoption of the local organization declared to be necessary by the first of the resolutions above recited, they desire me to request that you will have the goodness to take such measures as you may deem expedient for the purpose of stimulating local action in your town and neighbourhood, and for encouraging the production of suitable articles for exhibition. If it should appear to you to be advisable that a local committee should be formed, Her Majesty's Commissioners would be glad to learn that this had been done ; the committee being selected with express reference to the various industries of the town and surrounding districts. If such a committee should be formed, it would in the first instance be made the channel of communication with the exhibitors in your district, and the secretary would be furnished from time to time with such information as might be requisite for that purpose.

I have to request that you will inform me of the name and address of the members and secretary of the local committee, if such a committee should be appointed.

I have the honour to be, Sir,
Your obedient servant,
F. R. SANDFORD,
Secretary.

The duties of the local committees were mainly as follows :—

1. To act as the channel of communication between exhibitors and Her Majesty's Commissioners, and to give publicity in their districts to such information as may be useful to intending exhibitors, and others interested in the exhibition.

2. To encourage by every means in their power the production of articles suited for exhibition.

3. To examine the lists of proposed exhibitors, in order to see that they fairly represent the industries of the districts ; and that the principal producers appear in them.

4. To enter into communication with the leading manufacturers who have not sent any demand for space, with the view of urging them to do so at an early date, and to furnish proper forms of application for this purpose, which will be supplied by Her Majesty's Commissioners.

5. To examine the lists of applicants with the view of limiting the demands of those exhibitors who may have formed extravagant ideas of the worth of their goods, and of the space they should occupy.

6. In cases where applicants for space have made demands under more than one class, to ascertain the exact amount of space they will require in each class.

7. At a somewhat later date to superintend, in accordance with such general regulations as may be laid down, the selection or rejection of articles proposed for exhibition, and the allotment of space to such as may be declared admissible.

8. To take such steps as may appear expedient for the purpose of encouraging a desire to visit the exhibition; and for the systematic organization of such visits by all classes of the community.

Circulars were also sent to all the exhibitors who had exhibited in London in 1851, and in Paris in 1855, and the following communications, similar in spirit, were sent to the various governors of our colonies:—

SIR,

WITH reference to the printed "Decisions" of Her Majesty's Commissioners, which have already reached your Excellency through the Secretary of State for the Colonies, I am now instructed to enter into some further explanations for the information of the gentlemen in the colony under your government, who may undertake the duty of forming a collection for the International Exhibition in 1862.

In the first place, it would facilitate the arrangements here if the appointment of the commission or central authority referred to in the "Decisions" were to take place at as early a period as practicable.

Your Excellency will have remarked, that no article will be admitted from the colonies without the sanction of such commission or central authority, and it is important that Her Majesty s Commissioners should know with whom they can officially correspond. Upon this point I am instructed further to observe, that it is highly desirable, in the interests of the colonies, that whoever may be nominated as agent in this country, should be a man of business, well acquainted with the resources of the colony he represents.

The lists of articles admissible are so ample as to include every kind of produce, raw or manufactured, with only the three specified exceptions. With regard to one of these, viz., "fresh vegetables and animal substances liable to spoil by keeping," it seems desirable to explain that the term "fresh" is to be literally interpreted; therefore, articles of export, in whatever manner prepared, so as to keep without undergoing change, will be admissible.

Produce, such as wine, or other articles the result of fermentation, now admissible, although excluded from the Exhibition of 1851, will be submitted to the judgment of a special jury, who will decide upon their respective merits.

In the article of timber, the specimens should in all cases be converted into plank or scantling, of such a size as to show its mercantile value. If possible they should be four inches thick, and cut so as to show the "sap" on both edges. Moreover, since there is great uncertainty as to the origin of much colonial timber, it will be very desirable that each kind of timber should be accompanied by a few twigs, showing its leaves and flowers, when procurable. If the latter are pressed between sheets of paper enclosed in boards, they will furnish the evidence required.

Each colony will have a separate space assigned to it in which to exhibit its products, distinct from that of other colonies. It is, however, the wish of Her Majesty's Commissioners also to classify colonial raw produce, bringing all textile materials, all minerals, and so on, into one general comparative view; and they therefore invite exhibitors to furnish, when practicable, duplicate specimens for that purpose.

Her Majesty's Commissioners not having as yet information as to the number, size, and kind of articles which it is intended to exhibit, are unable at present to assign any fixed amount of space to each colony; but they will be prepared to act with the greatest possible liberality in this respect.

In estimating the probable area which the objects of exhibition may be expected to occupy, it is wished that each colony should mention the superficial area in square feet that its contributions will actually cover; and if wall surface should also be required, then the height and width of the articles for which such space is needed must also be specified. The large number of colonies to be provided for renders the earliest possible information upon this subject indispensable.

Her Majesty's Commissioners desire me to add, that they trust your Excellency will cause the particular attention of intending exhibitors to be drawn to that paragraph in the "Decisions" in which it is announced that the latest period at which goods can be received is March the thirty-first, 1862.

Her Majesty's Commissioners have appointed Dr. Lindley to assist the various colonial committees, if they require any advice from England.

I have the honour to be, Sir,
Your obedient servant,
F. R. SANDFORD,
Secretary.

The foreign exhibitors were stimulated and advised through our Foreign Office, and the following letter from the Commissioners set this machinery in motion:—

Offices, 454 West Strand, London, W.C.
May 18, 1861.

Sir,

I AM directed by Her Majesty's Commissioners for the Exhibition of 1862 to acquaint you, for the information of Lord John Russell, that, having taken into their consideration the question of the amount of space which may be allotted to the productions of each country in the building to be erected for the exhibition, they have decided upon the apportionment detailed in the accompanying schedule, so far as the industrial sections (Sections I., II., III.) of the exhibition are concerned. The amount of space to be allotted to the fine art section (Section IV.) will be the subject of a further communication.

The Commissioners are anxious to ascertain at as early a period as possible the extent to which each country is likely to avail itself of the space placed at its disposal; and I am accordingly directed to request that you will move Lord John Russell to bring under the notice of the governments of the several countries a particular statement of the amount of space allotted to the country to which each respective communication may be addressed, in order that it may be laid by them before the commission, or such other central authority as may be, or has been, appointed for the purpose.

Lord John Russell will perceive that, of the total space, it is intended to reserve one half for the productions of Great Britain and her dependencies, the other half being devoted to the productions of other countries.

The Commissioners are, however, of opinion that it will be advisable, in the various communications which they are now requesting his Lordship to transmit to those countries, that they should be informed that, of the gross space so allotted, about one-half will be required to be given up to passages; and each state should be particularly cautioned that, if it desires to appropriate more space to passage-room than half the gross amount allotted to it, it must find such excess out of its own allotment.

As it will be absolutely necessary to receive timely notice of the extent to which each country will take advantage of the offer thus made to it, Her Majesty's Commissioners conceive that the first of November of this year might be specified as the date by which the various commissions or other central authorities should be requested to give full information on this point.

Whilst it is the intention of the Commissioners to preserve as much as possible the national features of the groups of objects contributed by each country, they propose to adopt the principle of local classi-fication in the building, which was successfully carried out in the Paris Exhibition, and, as a general rule, to keep distinct each section of the exhibition named in their decisions. That is to say, Classes 1 to 4 will be in one division of the building; Classes 5 to 17, relating chiefly to machinery, in a second division; Classes 18 to 36, being manufactures, in a third; whilst Classes 37 to 40, being fine arts, will be in a distinct and special building.

Further, the Commissioners propose with the concurrence of each country, to bring together into international groups, the articles exhibited by each, under the following classes:—No. 15, Photographic apparatus and photography, and No. 29, Educational works and appliances; and to devote to each of these two classes a separate division or apartment in the building.

The Commissioners therefore request that each state will also return, by the first of November, an approximate division of the space which it considers will be occupied by each of the several classes.

They are also anxious that arrangements should be made by which the various descriptions of articles may be packed like only with like, and may be delivered at the building with the Class to which they belong, distinctly marked on such packages.

I have the honour to be, Sir,
Your obedient Servant,
F. R. SANDFORD,
Secretary.

To the Under Secretary of State for Foreign Affairs.

SCHEDULE.

THE total amount of space in the building devoted to the exhibition of objects in the first three, or industrial, sections, is eight hundred thousand square feet gross, of which it is proposed to reserve one-half, or four hundred thousand square feet, to Great Britain and her dependencies. The remaining half will be appropriated to foreign countries.

It is understood that such space is available to the extent of one-half for exhibiting purposes, the other half, at least, being reserved for passage-room.

F

The space mentioned in this schedule, covered and uncovered, has been subsequently increased by the annexes and the galleries to one million two hundred and thirty-one thousand square feet; and the same proportion—one-half—in the industrial section has been reserved for the passages.

The local committees in some cases, instead of observing the rules laid down for their guidance, acted upon the principle of giving each proposing exhibitor a little space, without much reference to his trading position, or the value of the articles which he proposed to exhibit. This was an error which, when discovered, considerably increased the labour of the Commissioners, who had to go through each local list, and arrange the claims for space upon a sounder system.

Without attempting to compete with the first Exhibition, the Commissioners thus decided that works should be selected for their excellence; that they should be arranged in classes, and not according to countries; that music, painting, and photography should be included in the display; and that foreigners should be admitted to contribute on the same conditions as British exhibitors. The Commissioners devoted considerable care to the fine arts department, that being the leading untried feature in connection with English international exhibitions. They decided that the display of the British school of art should be limited to the works of artists living within the century prior to 1862, but that foreigners should have liberty to select their art specimens without any such chronological restriction. In order to insure a full and honourable exhibition under this head, they appealed to art collectors as well as to artists, and took every precaution for the preservation of the pictures intrusted to their care. The selection of works has been made under competent professional advice, and the exceptional course has been adopted of paying all charges for conveyance to and from the building. Sculpture, models, die-sinking, and intaglios, have been included under the head of modern fine arts, as well as architecture, and their selection has been confided to special committees acting on certain rules laid down in the decisions of the Commissioners.

VIEW OF BUILDING FROM THE HORTICULTURAL GARDENS, 1862.

CHAPTER XI.

THE EXHIBITION OF 1862.—ITS WORK AND FEATURES.

NEARLY nine thousand eight hundred and sixty-two applicants sent in claims for space, of whom all but two thousand were in the industrial department.

The work of the Commissioners has been heavy, responsible, and troublesome; although they have not nearly so many official records to show as the managers of the first exhibition. Their letters received in thirteen months have been about forty-four thousand; their letters posted have numbered about eighty-four thousand; but their printed documents have only reached about two hundred and fifty.

Anything which stirs the public mind to the same extent as the present exhibition is sure to produce a host of eccentric proposals, some of them verging on the confines of madness, others displaying, through all their extravagance, a certain calculating selfishness. The first exhibition had a great forcing power in bringing forth these curious proposals and suggestions, but the second exhibition has had a greater.

The first odd communication in writing which the Commissioners received came direct from a lady who concealed her name. She was anxious that the building, even in its earliest stage of erection, should be secured from fire, and she hoped that the Commissioners had not neglected to insure it in some respectable office. She was led to write in this strain, because her sister had recently had a fire at her house, and was not covered by any policy of insurance.

The office of the exhibition had not been open many months before an American gentleman called to make a proposition. He was the fortunate possessor of the embalmed body of Julia Pastrana—a poor creature—half baboon, half woman, who created a sensation in England a few years ago; and he thought that arrangements might be made with the Commissioners to show this dead wonder at sixpence a head.

Another gentleman, not half so worldly-minded, but enthusiastic about the art of flying, wished to exhibit an aërial machine in action under one of the great domes, where he thought he could spring up and down like an acrobat in a gigantic baby-jumper. When his offer was politely declined, he as politely thanked the Commissioners, feeling that their object in refusing him permission to exhibit was only to save him from making a very great personal sacrifice in preparing his machine.

Another scientific exhibitor wished to send "Evidences of one general metallic root," and the following is his communication on this subject :—

"Hard labour and multiple experiments has proved to me the evidence of one general root metallique. Out of the fundamental principle, and by the developpement of the primitive formations often natural influences interferring, various mixtions are produced, but when the actives and passive agents are settled to a more or less neutral state and a homogene equilibrum of their parts of atomes is constituted, a homogene characteristic individualité is, or can be produced, and a so-called Simple Element is etablished, this Element, inseparable from his special Character and Indidual Unité, cannot be divided further by the ordinary Official Chemical Rule and Methode—from the Bar Metal to the Oxides from the Oxides to the fluide State and again (vice-versa)—but, when higher and most exalted Affinitys produced in a Philosophical Way might be known and applied, then, the homogene Equilibrum of the Individual Unité, affected by a higher affinité then its constituent Atoms posess itself, consequently chemical combinaison will follow on one side, reduction to a more primitive State on the other, and the parts of the so called Unité of the pretended Simple Element returned to the primitive Root."

One exhibitor had a scheme for showing coffins; another one for showing widows' caps; another one for the display of peculiar wigs; and another one for the exhibition of a patent moustache guard, to protect the moustache from soup while the wearer is dining.

One gentleman—a native of France—of a poetical turn of mind, wished to put the whole official catalogue into flowing verse, and to work up all the minutes, documents, and decisions of the Commissioners into an epic poem.

Another thoughtful friend of the Commissioners sent a number of small physic powders for each member of the staff, all the way from Baden Baden. They were as carefully directed as medicine packets usually are, and were intended to repair the exhausted frames of the over-worked officials.

The smallest contribution which was declined was a penny loaf of the year 1801. The applicant for space to exhibit this loaf, stated that he believed it to be the oldest piece of bread in the world. He had offered it to the Commissioners for the Exhibition of 1851, and he now offered it to the Commissioners for 1862. It was purchased by the applicant's father sixty years ago, when wheat was selling at a guinea a bushel, and for the purpose of preserving it as a specimen of very dear bread, a string net was made, in which it has been imprisoned ever since.

A project was submitted to the Commissioners for securing the money receipts of the exhibition by a system of astronomical checks based on the signs of the zodiac. The sun's radiations were to do a great deal in keeping the money, and ticket-takers honest; crowning honesty with a glory, and scorching dishonesty with the mark of the beast. The whole scheme was elaborate and confused; and though put forward as a serious business proposition, it read like one of those head-strong allegories written in imitation of John Bunyan.

The demands of the eight thousand industrial would-be exhibitors, if they had been complied with, would have swallowed up seven times the contents of the building, and, therefore, even the most sensitive exhibitor will admit that a little cutting down was necessary. Some of the classes had their demands pared more than others, the Commissioners and the committees being guided in their decisions by the relative importance of different branches of industry. The greatest number of applications was in connection with the class for iron and general hardware; and here one hundred and four claims were rejected. In the class for steel and cutlery there were only one hundred and twenty applications. Glass and pottery produced very few applications, and the number of exhibitors registered under this head is therefore small. There were two hundred and eighty-nine applicants for space to exhibit agricultural implements, and of these about one hundred and fifty were accepted. In mining and metallurgy there were not more than twenty rejections, and the number of exhibitors accepted was about three hundred and sixty. The result of all this winnowing has been that nearly five thousand five hundred British industrial exhibitors have been chosen under the divisions in the following table, to occupy about three hundred and eighty-six thousand and seven hundred square superficial feet of horizontal space, including passages:—

SECTION I.—RAW MATERIALS.

	Approximate No. of Exhibitors.	Approximate No. of sq. ft.
Class 1. Mining, quarrying, metallurgy, and mineral products ...	360	8,400
„ 2. Chemical substances and products, and pharmaceutical processes	202	5,100
„ 3. Substances used for food, including wines	163	4,500
„ 4. Animal and vegetable substances used in manufactures ...	247	7,500
	972	25,500

SECTION II.—MACHINERY.

	Approximate No. of Exhibitors.	Approximate No. of sq. ft.
Class 5. Railway plant, including locomotive engines and carriages ...	· 83	
„ 6. Carriages not connected with rail or tram roads	116	
„ 7. Manufacturing machines and tools	241	113,532
„ 8. Machinery in general	242	
„ 9. Agricultural and horticultural machines and implements ...	150	33,800
„ 10. Civil engineering, architectural, and building contrivances ...	164	13,962
„ 11. Military engineering, armour and accoutrements, ordnance, and small arms	130	12,610
„ 12. Naval architecture, ship's tackle	150	
„ 13. Philosophical instruments and processes depending upon their use	149	7,625
„ 14. Photography and photographic apparatus	165	2,966
„ 15. Horological instruments	130	2,700
„ 16. Musical instruments	91	5,870
„ 17. Surgical instruments and appliances	134	2,475
	1,945	195,540
Carried forward	2,917	221,040

SECTION III.—MANUFACTURES.

	Approximate No. of Exhibitors.	Approximate No. of sq. ft.
Brought forward	2,917	221,040
Class 18. Cotton ...	63	4,684
„ 19. Flax and hemp ...	81	6,483
„ 20. Silk and velvet ...	64	4,722
„ 21. Woollen and worsted, including mixed fabrics generally	235	21,093
„ 22. Carpets	44	(Vertical space)
„ 23. Wove, spun, felted, and laid fabrics, when shown as specimens of printing or dyeing ...	51	3,546
„ 24. Tapestry, lace, and embroidery ...	85	5,307
„ 25. Skins, fur, feathers, and hair ...	68	1,316
„ 26. Leather, including saddlery and harness ...	135	4,583
„ 27. Articles of clothing ...	201	7,402
„ 28. Paper, stationery, printing, and bookbinding ...	223	6,250
„ 29. Educational works and appliances ...	234	4,344
„ 30. Furniture and upholstery, including paper-hangings and papier-mâché ...	258	25,272
„ 31. Iron and general hardware ...	409	25,522
„ 32. Steel and cutlery ...	127	13,316
„ 33. Works in precious metals, and their imitations, and jewelry ...	84	7,968
„ 34. Glass ...	81	15,580
„ 35. Pottery ...	62	5,475
„ 36. Manufactures not included in previous classes ...	31	2,800
	2,536	165,663
Total ...	5,453	386,703

It must not be supposed that the rejected contributors bore their rejection very patiently. One gentleman said, that "If Diogenes were alive, he would find abundant use for his lantern in guiding the Commissioners in their search for truth." Another, more indignant, wrote to say:—"I am determined to exhibit, and shall petition all the Commissioners, even to the Prince of Wales himself, should this application be unsuccessful. If all means fail, I shall inquire through the press—the leading daily and literary journals—for an explanation of the system of preference which dictates refusal to one and the acceptance of another exhibitor."

In Class one, the department for mining, quarrying, metallurgy, and mineral products—which is under the management of Mr. Robert Hunt, F.R.S.—the commercial motive to exhibit is weaker than in any other class. The transport of the specimens has been costly, and few of the exhibitors can expect any reward except thanks for having contributed to an instructive and interesting collection.

The National Committee for this class drew up a scheme with the view of suggesting, to some extent, the nature of the objects which it was thought desirable should be exhibited in this department of the Exhibition.

The committee suggested the desirability of keeping the size of mineral specimens within moderate limits, as being more convenient for display, and better adapted for exhibiting special peculiarities than unwieldy masses. This did not of course apply to any remarkable examples, such as sections of Lodes, or peculiar and illustrative phenomena. They also recommended that where building stones were exhibited in the form of cubes, the uniform size of eight inches should be preserved, and it was deemed essential that two surfaces of the cube should be left in the natural state—undressed. These and all mineral specimens were to have labels attached, carefully giving the locality from which they were obtained, and, if possible, the geological formation to which they belong.

The committee desired to see models or drawings of the most approved methods adopted in working our mines and collieries, of the machinery employed for draining, for ventilation, and for winding, and also of the improvements which have been introduced for preparing minerals for the market. They also hoped to see good illustrations of the metallurgical processes employed, and of the commercial results obtained.

The following is the outline plan sent to each exhibitor in this department, and it may be taken to represent the general features of the collection :—

MINING AND QUARRYING OPERATIONS.

    DRAWINGS and SECTIONS showing the relations of the Minerals to the rocks in which they occur.
    PLANS and SECTIONS and MODELS of the workings of MINES and COLLIERIES.
    MODELS, or DRAWINGS, of MACHINERY employed for—
        1. VENTILATION.
        2. DRAINING.
        3. RAISING Minerals.
        4. LOWERING and RAISING Miners.
        5. STAMPING and CRUSHING Ores.
        6. WASHING and DRESSING Ores.
    Tools and other Appliances.
GEOLOGICAL and MINERALOGICAL MAPS.   PLANS, SECTIONS, or MODELS.
MINERAL PRODUCTS.

Non-Metalliferous Minerals.

    COALS and MINERALS used as Fuel.
        1. Bituminous Coal.
        2. Cannel Coal, and Torbanite.
        3. Anthracite.
        4. Lignite.
        5. Peat.
        6. Bituminous Shales.
        7. Native Naphtha, Pitch, Bitumen, &c.

The more important commercial varieties are exhibited with chemical analyses, statements of heating power, and physical peculiarities.

    CLAYS and FELSPATHIC Minerals :—
        1. Porcelain Clay or Kaolin.
        2. China Stone.
        3. Potters' Clay.
        4. Pipe Clay.
        5. Brick Clay, and Brick Earth, &c.
    BUILDING STONES of all varieties.
    SLATES and SLABS.
    PAVING STONES, &c.
    SANDS for GLASS-MAKING, &c.
    CEMENT STONES and CEMENTS, LIMESTONES, &c.
    ROTTEN STONE.
    FULLERS' EARTH.
    FLUOR SPAR.
    BARYTES, STRONTIAN, and other Minerals employed in the Arts.
    COPROLITES and other Mineral Manures.
    SALT.
    GEMS.
    STONES used for ornament.
    MILLSTONES, GRINDSTONES, HONESTONES, &c.

Metalliferous Minerals.   Some Localities.

IRON :—

    MAGNETIC OXIDE .   .   .   .

Devonshire.
Cornwall.
Ireland, &c.
Yorkshire.

HEMATITE (Anhydrous Red Oxide)	Cumberland. Lancashire. Isle of Man. Flintshire. South Wales. Somersetshire. Scotland. Ireland.
SPECULAR IRON ORE	Brendon Hills.
BROWN HEMATITE (Hydrated Oxide)	Forest of Dean. Peak District of Derbyshire. Brendon Hills, Devonshire. Cornwall. Llantrissant, S. Wales. Carnarvonshire. Flintshire. Isle of Man. Ireland, &c.
SPATHOSE ORE	Exmoor. Cornwall. Devonshire. Durham.
HYDRATED OXIDES	Yorkshire. Lincolnshire. Northampton. Oxford. Warwick. &c. &c.

CARBONATES :—

Argillaceous Carbonate . . . . Black Band . . . . Hydrated Oxides of the Carboniferous formations .	Coal Fields generally.
COAL BRASSES of Coal Measures of	South Wales.
MIXED CARBONATES and Hydrated Oxides . . .	Cleveland. Whitby. Westbury. Seend. &c. &c.
PISOLITIC Iron Ores	Carnarvonshire.

COPPER :—

NATIVE	Cornwall. Devonshire.
OXIDE	Cornwall. Devonshire. Ireland.
CARBONATE	Cornwall. Cheshire. Shropshire.

SULPHIDES :—
    Gray Ore . . . .
    Yellow Ore, &c. . . . .
	Cornwall.
	Devonshire.
	Lancashire.
	Cumberland.
	Shropshire.
	Derbyshire.
	Isle of Man.
	Wales.
	Scotland.
	Ireland.

Other VARIETIES which enter into Commerce . .	Do.

Tin :—

Oxide	{ Cornwall.
Tin Pyrites	{ Devonshire.

Lead :—

	Cornwall.
	Devonshire
	Somerset.
	Derbyshire.
	Shropshire.
Carbonates.	Yorkshire.
Sulphides	Cheshire.
Other Varieties used in the Arts	Westmoreland.
	Durham.
	Northumberland.
	Cumberland.
	Wales.
	Scotland.
	Ireland.

Silver :—

Native	{ Cornwall.
Sulphides	{ Devonshire.
Chlorides	
Argentiferous Gossans, &c.	Cornwall.

Gold :—

	Merionethshire.
Gold Quartz	Cornwall.
	Devonshire.

Zinc :—

	Cumberland.
Carbonate	Isle of Man.
Silicate	Cardiganshire.
Sulphide	North Wales.
	Cornwall.
	Devonshire.

Sulphur Ores :—

	Cornwall.
Pyrites of Metalliferous veins	Devonshire.
Do. of Coal measures—" Coal Brasses "	Cumberland.
	Carnarvonshire.
	Wicklow, Ireland, &c.

Cobalt

{ Cornwall.
{ Cumberland.

Nickel

{ Cornwall.
{ Cumberland.

Uranium . . . . Cornwall.

Tungsten (Wolfram)

{ Cornwall.
{ Devonshire, &c.

Arsenic.

Manganese

{ Cornwall.
{ Devonshire.
{ Warwickshire.

This large collection is chiefly placed in the eastern annexe—a companion to the show of agricultural implements. The court is largely fitted up with ornamental marbles.

The Lead Hills of Lanarkshire, once the scene of a wild rush after gold, and always a district producing rare materials, have sent a series of their treasures. Lead ores and their products have come from Durham, Northumberland, and Cumberland. The mountain limestone regions of Yorkshire have furnished lead and copper, and Cardiganshire and Flintshire exhibit varieties of their galenas and blendes. The local committees of Exeter, Tavistock, and Redruth have carefully looked up the more remarkable examples of the minerals of Devonshire and Cornwall. That tin ore which has been worked in Cornwall and on Dartmore for more than two thousand years, and which is now being obtained at the rate of ten thousand tons per annum, is shown as obtained from the stream and from the mine. Several varieties of copper ore, many of rich argentiferous lead ores of the West, the zinc ores of these counties, and sulphur ores (iron pyrites) and many others, not quite so important, but remarkable as being rare, are also amongst the illustrations of Nature's bounty to us. Ireland, too, sends over her lead, copper, and pyritic ores.

Although the Exhibition of 1862 does not exhibit such a complete collection of iron ores as the industry of Mr. S. Blackwell brought together in 1851, yet most of the new discoveries, and they are many and important, find a place. The iron ores of Weardale, those of Cleveland, and the recent discoveries of the West Riding of Yorkshire, the remarkable deposits of Lincolnshire, those of the Midland Counties, and the best examples from Somerset, Devon, and Cornwall, form an interesting series.

Coal specimens are abundant. There is scarcely a useful seam from which an example has not been sent. In 1851 there were some gigantic blocks, many of them of little interest, but cubical pieces of about eighteen inches are now the rule.

The building and ornamental stones of the kingdom are fully represented. The sandstones of the coal measures of the North; the Dolomites of Yorkshire and the neighbouring counties; the Liassic limestones, the Oolites, especially those from the neighbourhood of Bath; Portland stone, Purbeck marble, Stonesfield slate, and numerous other varieties illustrate our richness in constructive materials.

In metallurgy, the examples of iron are numerous, and, at the same time, as nearly all varieties of pig iron are to be found in the Exhibition, there are some of the most remarkable examples of rolled iron that have ever yet been shown. Rails from sixty to upwards of one hundred feet in length, without a weld; bars of remarkable size and length; sheets of iron of most unusual dimensions; and armour plates which have resisted the battering power of Armstrong's guns, are there; with cranks, one weighing above twenty-five tons, and beams of singular size and strength, which prove the capabilities of British forges.

The specimens of gold from Devonshire and Wales will probably be general objects of interest. Two monolithic obelisks, belonging to this department, are erected as trophies in the nave—one made of Cheesewring granite twenty-seven feet high; the other made of gray granite from the Ross of Mull, thirty-five feet high. In the mineral court there are also a beautiful serpentine obelisk, fifteen feet high; a fine collection of Derbyshire marbles, principally black; and a few fluor-spar ornaments, the celebrated fluor-spar being nearly exhausted. One exhibitor has sent a richly inlaid table that may be regarded as a perfect work of art, while other exhibitors have contributed fine examples of Florentine work in British stones.

The only metal exhibited which has become an article of commerce since 1851 is aluminium; and of this there are numerous specimens, both manufactured and unmanufactured. It has a fine white colour, slightly inclining to blue especially after being well hammered; when cold, like silver, it is susceptible of a very fine matting, is easily burnished or polished or drawn into wire, and can easily be beaten out, either hot or cold, to the same extent, and as perfectly as gold and silver. It is much lighter than ordinary metals: its elasticity and tenacity are about the same as those qualities in silver, and it is easily run into sand and métallic moulds.

Perhaps the most generally interesting objects in the mineral department will be the model of the celebrated Barrow works near Ulverstone, a working model from the Durham mines, set in motion by water power, and other models from the Midland Counties, and from Lancashire, showing every feature of this underground branch of industry.

The chemical and pharmaceutical section of the Exhibition, Class two, can boast of a collection which has never been equalled in variety and excellence. Amongst the exhibitors are over two hundred of the first manufacturers in the country.

Hardly a name of any eminence is missing in any branch of chemical manufacture, from magenta and borax, down to matches and blacking. A complete series of drugs, systematically arranged, has been prepared by a committee of the most eminent members of the Pharmaceutical Society, assisted by Professor Redwood, who is on the National Committee of Class two. This collection is peculiarly instructive and interesting to pharmacists; and its value is greatly enhanced by the Society allowing the specimens to be handled and examined under the supervision of a curator. A very interesting collection of chemicals, illustrating the improvements made in calico-printing and dyeing since 1851, has been formed, at the suggestion of Dr. Lyon Playfair, by Mr. Robert Rumney of Manchester, whose combined knowledge, as a chemist and a manufacturer, rendered him peculiarly fitted for the work. In no department of applied chemistry have such strides been made within the last ten years as in dyeing and calico-printing; and in the history of these useful arts, perhaps no similar period has been so fruitful in good results. A collection of products, illustrating the discovery of the coal-tar dyes formed by the first workman in this fertile field, Mr. W. H. Perkin, is also exhibited: in fact the various dyes are particularly well represented. The coal-tar series is most fully represented, and numerous specimens of the lichen and madder dyes are also exhibited. Altogether the specimens exhibited will tend to show that England has now become the dye-producing nation of Europe, instead of having to depend on Holland, France, and other countries for the supply of lichen and madder dyes wherewith to ornament the produce of her millions of silk, woollen, and cotton looms.

The larger and coarser kind of chemicals, such as alum, soda, copperas, the prussiates, &c., in the manufacture of which this country has always been preeminent, are here as a matter of course. Some splendid specimens of salts, in a high state of purity, are exhibited by many well-known firms, and the more delicate materials of absolute purity for laboratory use, show that English manufacturers can compete most satisfactorily with those of the Continent in this respect.

Fine pharmaceutical chemicals, such as the cinchona alkaloids, the opium alkaloids, the valerates, strychnia, &c., receive adequate representation; several collections of the first-named substances being particularly complete. The rarer products of the laboratory, interesting only to the scientific chemist, are well represented by the numerous specimens exhibited by many eminent scientific men and traders. A collection of raw pigments for the manufacture of artists' colours, including specimens of real ultramarine, valued at more than one thousand pounds, deserves great attention; and the coarser colours for painters' use, including specimens of varnishes and the gums from which they are made, although not so interesting to the casual observer, are nevertheless most complete and valuable. Besides the splendid collection of drugs and pharmaceutical products exhibited by the Pharmaceutical Society, there are many very valuable specimens to be seen in the different cases. A series of products, illustrating the manufacture of starch for laundry use, and a good sprinkling of blacking, matches, manures, vegetable and animal blacks, blue and black-leads, complete this interesting collection.

Class three, consisting of food products, including beverages and tobacco, is well represented by several well-known firms. Many, however, have hung back, fearing that although the general public were well contented to consume these substances in the ordinary way, yet they might not care to see them ranged on counters as specimens for exhibition. The collections of dried fruits, tea, coffee, and cocoas, show that by the exercise of a little taste and judgment these substances can be made as gratifying to the eye and mind as they are to the palate. Although somewhat uninteresting to look at, the specimens of preserved meat are very numerous, and indicate how much the trade in this article has increased of late years. A pig cured whole forms a prominent object in this department. Series of specimens contributed by various exhibitors illustrate the process of sugar-refining in a very complete and interesting manner. Great brewers strive for mastery not only in the quality of their beers, but in the ingenious way they have each adopted for showing this article. Handsome cases of pickles and preserves, and cases of confectionery, form conspicuous objects, round which the younger portion of the visitors to the building will form longing groups. The collections of corn and seeds will afford valuable information to the foreign and colonial agriculturists.

Class four consists of animal and vegetable substances used in manufactures. To commence with animal substances, the large tallow and soap series of products are well represented. These are very important branches of British manufacture, and we have every reason to be proud of the display made. The perfumery stalls, with their tasteful decorations, will form points of attraction for lady visitors, more especially as there are several beautifully designed scent fountains in full play. Further down we come to some exquisite specimens of ivory turning; and the wax-flower show, which exceeds in extent any that has hitherto taken place, will be one of the lions of the Exhibition. Some splendid specimens of woods are shown; gutta percha, india-rubber, and ebonite are exhibited by most of the well-known houses; the fishing-rod and tackle-makers have an excellent display, and some very beautiful specimens of canes and sticks are to be seen in the cases of various exhibitors. The manufacture of combs, although not a large one, is well represented, and the same may be said of glue and gelatine. Fibre manufactures are led by specimens of cocoa-nut matting covered with brilliant patterns that one could hardly have expected from so coarse a material. The walls of this class are covered with some magnificent specimens of veneers of large size, which divide the space with leather-cloth, cocoa-nut matting, and kamptulicon. Some graceful specimens of basket-work will surprise even our French friends, who have carried this manufacture to such perfection. In miscellaneous manufactures belonging to this class, we find some fine specimens of sponge, cork, glue, and gelatine, and some interesting applications of gold-beaters' skin. The magnificent collection of fleeces shown by the Royal Agricultural Society cannot fail to attract great attention, and the collection of cottons by the Cotton Supply Association is most valuable and interesting.

These three Classes are under the management of Mr. C. W. Quin, F.C.S.

Classes five, seven, eight, and ten—comprising railway plant, locomotive engines and carriages, manufacturing machines and tools, machinery in general, and civil engineering, architectural and building contrivances—are under the management of Mr. D. K. Clark, C.E. The first three classes, which include machinery in motion and at rest, are exhibited in the western annexe. This department is substantially fitted up to provide steam power, and the means of transport for heavy materials. A single line of railway runs from end to end on each side; six double-flue boilers, thirty feet long by six feet and a half in diameter, are built in at the north end, communicating with a chimney which is seventy-five feet high, and which has a diameter of ten feet at the base. Two hundred elegant iron columns, of the Doric order, have been raised at intervals ten feet apart, and ten feet high above the floor, supporting two thousand feet of shafting, two inches and a half in diameter. Two thousand feet of steam pipe, having a graduated diameter from fifteen to eight inches, are laid down in a bricked subway, or pipe culvert, side by side with two thousand feet of exhaust pipe, eighteen inches in diameter.

In this department many of the very heavy goods were delivered, and were unloaded by two travelling steam cranes capable of lifting five tons each. The steam power supplied here is from four to five hundred horses; and two pumps are placed in this annexe to work the two great French fountains that are exhibited in the Horticultural Gardens. There is also a travelling crane capable of lifting twenty-four tons, in two twelves. The largest steam engines are in the French section, where two are exhibited of sixty horse power each.

Mr. Ashton's ingenious steam winch—like a veteran worker who has earned an honourable place—is also exhibited in this annexe.

Perhaps the greatest novelties in Class five are Ramsbottom's self-feeding tender, to supply water to express engines without stopping, and several smoke-consuming locomotives.

The pile carpet and worsted looms will excite general interest in Class seven (a); and in Class seven (b) there are many new varieties of steam-hammers, variations upon Mr. J. Nasmyth's original plan. Amongst others is an interesting "steam smith," or radial hammer, exhibited by Messrs. Neilson and Co. of Glasgow.

In Class eight, amongst the marine engines, parts of Messrs. Penn's large engines made for the "Achilles" (a sister ship to the "Warrior"), with a crank shaft weighing seventeen tons, are worthy of special notice. Several important varieties of traction engines are exhibited, made to run on common roads; and the washing and mangling machines will surely not be despised, when we know that the annual washing bill of the United Kingdom is now estimated at six millions sterling.

The following is a classified list of the articles exhibited in this important department, with approximate details of the allotted space :—

MACHINERY DEPARTMENT.

CLASS V.

ALLOTMENTS.

	Superficial Square Feet.
1. Locomotive Engines and Tenders	2,993
2. Ditto Models—Fittings and Pieces	11
3. Railway Carriages and Waggons	999
4. Ditto Carriages—Models	12
5. Ditto Wheels and Axles	318
6. Ditto Carriage and Waggon Fittings and Pieces ...	62
7. Ditto Brakes	48
8. Permanent Way	600
9. Switches and Crossings	300
10. Weighing Machinery	1,000
11. Road Signals	150
12. Train Signals	40
13. Sundries	493
Total allotments	7,026

CLASS VIIa.

MACHINERY FOR PREPARING, SPINNING, AND WEAVING, COTTON, WOOL, FLAX, SILK, &c.

	Superficial Square Feet.
1. Cotton Spinning	15,249
2. Cotton Looms	1,372
3. Calico and Fancy Looms	136
4. Pieces of Cotton Machinery	46
5. Woollen and Worsted Spinning	253
6. Ditto Ditto Looms	292
7. Pile Carpet and Worsted Looms	350
8. Knitting Machinery	82
9. Flax Spinning	1,198
10. Flax and Jute Weaving	1,760
11. Silk Throwing and Weaving	99
12. Sundry	226
	21,063

CLASS VIIb.

MACHINES AND TOOLS (FOR WOOD AND METAL.)

	Superficial Square Feet.
1. Forging	25
2. Forging-hammers	1,004
3. Machine Tools	7,217
4. Small Tools	240
5. Wood-working Tools	2,447
6. Sundry Cutting-tools	214
7. Drawing and Rolling Mills	120
8. Hand-tools	151
9. Paper-making	2,729
10. Paper-working	360
11. Printing	902
12. Type	348
13. Lithographic Presses	250
14. Stamping, Embossing Presses	84
15. Brick, Tile, and Pipe Machinery	756
16. Bread and Biscuit Machinery	650
17. Confectionery and Grocery Machinery	97
18. Sewing-machines	218
19. Needle-making, Pen-making	136
20. Boot-making, Leather-making	113
21. Sundry	320
	18,381

CLASS VIII.

ALLOTMENTS.

	Superficial Square Feet.
1. Steam-Engines ...	2,216
2. Ditto ditto Marine ...	2,003
3. Ditto Boilers ...	50
4. Engine and Boiler Fittings ...	516
5. Electro-Magnetic and Air Engines ...	413
6. Hydraulic Machinery ...	2,841
7. Fire Engines ...	127
8. Water Meters ...	26
9. Air Machinery ...	110
10. Soda Water and Ice Machines ...	382
11. Washing and Mangling Machines ...	310
12. Domestic Machines ...	22
13. Brewing and Distilling Machinery ...	411
14. Beer Engines ...	80
15. Sugar Mills and Machinery ...	3,190
16. Flour Mills ...	287
17. Oil, Coffee, and other Mills ...	580
18. Stone Mills ...	169
19. Weighing Machines ...	160
20. Measuring Instruments ...	45
21. Cranes and Windlasses ...	690
22. Leather Bands ...	120
23. Traction Engines for Common Roads ...	771
	15,519

Besides making arrangements for showing machinery in motion, and illustrating it by processes, Her Majesty's Commissioners reserved space for the exhibition of manufacturing processes in certain handicrafts which could be carried on without danger to the building. Invitations were issued to steel-pen makers, pin and needle makers, button makers, medal strikers, gold chain manufacturers, engine turners for watches, type casters, type printers, lithographic and copper-plate printers, earthenware and porcelain printers, potters, brick and drain-tile makers, glass-blowers on a small scale, turners in metal, ivory, and wood; glove makers, and pillow-lace makers of various kinds. Applications to exhibit were not received from all these manufacturers by the date fixed, but the result is that visitors will see illustrations of the following processes:—

Needle making.	Lithographic printing.
Medal striking.	Copper-plate printing.
Gold chain making.	A potter's wheel.
Type casting.	Brick and drain-tile making.
Type printing by hand.	Wood carving.

Class ten in this department includes furnaces, iron and paper pipes, drainage contrivances, baths and other sanitary contrivances, bridges, aqueducts, viaducts, models of docks and harbours, diving apparatus, models of iron buildings, and improved cottage roofs, window-sashes, ornamental building contrivances in marble, granite, slate, and other materials; stones, bricks, cement, tiles, and pipes, including a fine collection of Shropshire ornamental tile-work, contrivances for preserving wood, warehouse machinery, fire-escapes, contrivances for ventilation, smoke-curing, and improved filters and gas-measuring apparatus. Many useful novelties are exhibited in this department, and amongst the most interesting models are those of the Boyne Viaduct—illustrating the principle of lattice-construction—the great Saltash Viaduct, and Rennie's docks at Cadiz.

The display of agricultural implements in Class nine—which is placed in the long covered and uncovered eastern annexe—is under the management of Mr. Brandreth Gibbs, and is substantially the annual exhibition of the Royal Agri-

WESTERN ANNEXE.

cultural Society.   This society has held twenty-three annual exhibitions in different towns from 1839 to 1861 inclusive, and always with a hopeful increase of exhibitors and visitors.   Its great display in Hyde Park in 1851 was eminently successful; satisfactory to exhibitors and useful to agriculture.   These shows of the society have led to such improvements in the design and manufacture of agricultural implements that a large export trade has been created in this branch of industry, which figures prominently in the government returns.

In horticultural exhibitions a progress even more rapid has been shown. Not only have they led to improvements in the culture of native, and the naturalization of foreign plants, but they have taken botany out of its abstract existence in books and pictures, and have placed it practically before the eyes of the public, to aid in a useful, beautiful, and what is gradually becoming a necessary part of education. The annual displays at 'Chiswick have been the creators of the Botanic Gardens in the Regent's Park, and of the Royal Horticultural Society; and the display of fruits and flowers in the gardens of this society, attached to the international building will, doubtless form one of the most attractive features of the Exhibition.

In the agricultural section gardening and farming are jointly represented; and the collection of implements and models includes horticultural buildings, wire netting, fencing, and gardening apparatus.

During the last twenty years great improvements have been made in the construction of agricultural machinery and implements. Before that period a large proportion of the implements of the farm were the production of the village black-smith; regular implement manufacturers were but few, and the want of railway communication prevented the better class of the machines then made being ex-tensively used beyond the immediate neighbourhood in which they were constructed. Of late years, however, vast 'strides have been made, and the wants of agriculture are now provided with machinery, which, for its ingenuity and excellence of work-manship, will bear favourable comparison with any other branches of our national industry.

It is unnecessary here to enter into any elaborate description of the various kinds of machinery which are now used as aids to the agriculturist, partly because it would occupy too much valuable space, and secondly, because not only the general character, but the individual merits of the different implements hitherto invented from time to time, during the last twenty years, have been annually recorded in the Reports of the Royal Agricultural Society of England.

The report on Class nine in the Great Exhibition of 1851, written by the late Mr. Pusey, M.P, has brought the history of agricultural machinery down to that period, and it therefore only remains for us now to allude in general terms to some of the principal inventions and improvements that have since appeared.

It was impossible for Her Majesty's Commissioners on the present occasion to set apart more than a limited area for Class nine; the demands of exhibitors had therefore to be cut down to a small fraction of the space they originally applied for, and it was stipulated in granting the allotments, that certain im-plements specified by the Committee of Selection should be exhibited in each. By this means a tolerably complete collection of agricultural machinery has been secured, but from the frequency of new inventions it was impossible to anticipate what novelties might be forthcoming at the International Exhibition of 1862. It was therefore considered desirable to give the exhibitors the oppor-tunity of placing within their allotments any additional machines they might wish to exhibit.

No doubt the most important introduction since 1851 has been the application of steam to the cultivation of the soil.

It is true that an engine and apparatus were exhibited in the Great Exhibi-tion in Hyde Park, but the practical adaptation of steam power to the cultivation of the soil has taken place since that period, and therefore it claims special notice on this occasion.

There will be found in the building examples of most of the different systems that have lately been before the public, and these will be best understood by an examination of the respective exhibitors' stands which contain in some instances not only the implements themselves, but also models and diagrams illustrating the manner of working them.

The next great improvement has been that of reaping machines for grain crops and mowing machines for natural and artificial grasses. And here again the invention of reaping machines dates back earlier than the year 1851. Long before that period one was known in Scotland, but the attention of manufacturers and agriculturists was first seriously turned to this class of machine when M'Cormick and Hussey contributed one to the Great Exhibition of 1851. Since then various improvements have been made, and probably one of the most important has been that of the "self-delivery." The original machines exhibited by both M'Cormick and Hussey required manual labour to deposit the corn, whereas a large proportion of those now made deliver by mechanical means.

The mowing machines for cutting meadow and other grasses and clover for hay, &c., may be regarded as a natural sequence to the use of reaping machines, and they form an important addition since 1851. Several examples will be found in this class.

We must now allude to the great improvements which have been made in the threshing machines. These do not now simply supersede the old hand flail, but in many cases thresh, winnow, and prepare the grain ready for market, at one operation. They are known as "combined finishing machines." There are others which prepare the grain ready for the last operation, viz., final dressing.

We may also refer to the elevators, for raising corn in the straw, or straw only, to the rick, and they effect a considerable saving of labour.

The straw thatch-weaving machine is also new. It is a modification of the loom, working the straw into a portable covering.

The root-pulpers now in use may be regarded as carrying the operations of the turnip-cutters one step further.

As the above enumerated are the chief new introductions, it will at once be evident that the advancement made in agricultural mechanism is not to be estimated by the number of new kinds of machinery, *i. e.*, machines constructed to carry out some new process, but rather by the comparative perfection to which those already known have been brought.

To subdivide the different classes of machinery, or to enumerate separately all the minor inventions and implements, would far extend the limits assigned to this brief introductory notice; and we therefore merely classify the different specimens comprised in Class nine, under the following general heads :—

I. TILLAGE AND DRAINAGE OF THE SOIL.
    Applications of steam-power to the cultivation of the soil.
    Ploughs, various, subsoil ditto, pulverizers, &c.
    Cultivators, scarifiers, and grubbers.
    Clod crushers, land rollers, and Norwegian harrows.
    Harrows, ordinary and chain.
    Potato diggers.

II. THE CULTURE OF THE SOIL AND HARVESTING OF CROPS.

  Drills for sowing corn, seeds, and manures, and dibbling machines.
  Manure distributors for dry and liquid manures.
  Horse-hoes, turnip-thinning machine for thinning the growing plants, weed extirpators, and thistle
    destroyers.
  Turnip-fly machines for destroying that insect.
  Haymakers or tedding machines.
  Reaping machines.
  Mowing machines, simple and combination with reapers, and combined hedge-clipper.
  Horse-rakes.
  Carts, waggons, farm railways (tramways), stone-collecting cart not requiring manual labour.

III. PREPARATION OF GRAIN FOR MARKET AND OF FOOD FOR CATTLE.

  Steam-engines, fixed and portable, and self-propelling.
  Threshing-machines, fixed and portable, including combined finishing.
  Straw elevators, for raising the straw to the stack.
  Winnowing or dressing machines, corn screens, riddles, and separators.
  Smut machines, for removing smut from grain.
  Barley humellers or avellers, for removing the awns off barley.
  Clover drawer, for drawing or threshing the seed from the plant.
  Chaff-cutters and gorse-bruisers.
  Mills of sorts for crushing, grinding, &c.
  Oil-cake breakers.
  Horse gear for giving motion to horse threshing and other machines.
  Turnip and root cutters and pulpers, and turnip tailer, for removing roots, &c., off turnips.
  Root-washers
  Steaming apparatus for cooking food for cattle.

IV. MISCELLANEOUS.

  Straw thatch-weaving machines.
  Sack trucks and sack elevators.
  Churns and dairy utensils, cheese press, curd mill.
  Dynamometer for testing the power required to work various machines.
  Cowhouse fittings and feeding cribs.
  Hand tools, drainage ditto, levels.
  Apple mills and cider presses.
  Liquid manure pumps and hydraulic apparatus.
  Gates, fencing, &c.

HORTICULTURAL.

  Conservatories, garden-engines, tools, &c.

Classes eleven and twelve—military engineering, armour and accoutrements, ordnance and small arms, and naval architecture and ships' tackle—are under the superintendence of Major Moffatt. In the first of these classes—under Sub-Class *a*—there is a praiseworthy contrivance in the application of the study of the centre of gravity to the suspension of soldiers' belts and ammunition. There are also many specimens of improved cartouch boxes, &c.

Under Sub-Class *b* are exhibited many improvements in tents for field service.

In Sub-Class *c* may be studied an extensive range of every engine of destruction which the human mind has yet devised, from swords to great guns, from the old Brown Bess—more praiseworthy for carrying the queen of British weapons at its muzzle than for its accuracy of aim with the projectile of that date—down to our death-dealing, repeating weapons, our breech-loading arms, and our long-range rifles.

Amongst the more formidable engines of destruction are the Armstrong and Whitworth one-hundred pounders, Blakeley's five-hundred pounders, and the Mersey Steel Company's six-hundred pounder. As to their respective merits, the visitors must judge for themselves, for the opinions of scientific men on this head are divided.

The trophy of great guns by Mr. Anderson of the Royal Carriage Department will at once attract the close scrutiny of the scientific, being a development of the process of construction of the great Armstrong gun, with a microscopic lens arrangement.

War in all its branches may be studied in an inspection of Colonel Shafto Adair's model of London, with its projected lines of defence for the great city, and also in the model of Fort Torr by Colonel Harness, with Captain Ducane's details of the same fortification. There are other fine models, all deserving minute inspection; not the least interesting is one of the Crimean Monument by Captain Brine, Royal Engineers.

In Class twelve there is a beautiful series of models, lent by the Lords of the Admiralty, illustrative of the progress in the construction of naval architecture from the time of the "Great Harry," 1514, to the iron-clad frigate the "Warrior." There is another large model of the "Warrior," placed in the nave. Next to the "Queen" line-of-battle ship is to be seen a fine model of the "Northumberland," illustrating prospective progress—a splendid iron-cased frigate of that name being now under construction on a larger scale than even the present wonder of the day, the "Warrior." In this class are to be found many new inventions contrived to meet the apparent difficulty of keeping these leviathan ships' heads in the desired direction in rough weather.

The National Life-Boat Institution exhibits some interesting models of boats for improving this humane branch of the naval service.

The Commissioners of the Trinity House, of the Northern Lights, Edinburgh, and of the Ballast Board, Dublin, are exhibiting some very interesting, handsome models of the danger beacons of our isles, many of them originating in the scientific zeal of His late Royal Highness the Prince Consort.

There is also exhibited a splendid model of the American system of boat building by machinery. This is a process by which we can obtain a boat ready for use, of any dimensions, turned out complete from the log of timber at a few hours' notice.

There are models of ships for steam under as well as over the water; gun-boats and floating batteries, penetrable and impenetrable; ships for commerce, and ships for war—in fact, naval architecture with all its appliances may be here studied by the most critical and scientific.

Classes thirteen, twenty-five, and twenty-six are under the superintendence of Mr. C. R. Weld. The first class (philosophical instruments, and the processes depending on their use) comprises all descriptions of philosophical and optical instruments, electric telegraph apparatus, including M. Caselli's pantagraph, which transmits autographic messages, thereby avoiding all risk of error, models of engines, &c. Amongst the exhibitors are all the celebrated makers of instruments, and many private gentlemen. The great improvements and numerous inventions in electric telegraphy during the past ten years are strikingly shown by the various new instruments exhibited by the Universal Private Telegraph, the British and Irish, the Submarine and other Telegraph Companies. Mr. Wheatstone's ingenious and beautiful domestic telegraphs are shown in working order, and many inventions and contrivances to utilize this valuable discovery are to be seen in this class.

Microscopes, as might be expected, form a prominent feature : the great makers of these important instruments exhibit microscopes of all kinds, from that intended to assist the researches of the most scientific physiologist to the more humble instrument made for the student. Among them are many binocular microscopes, which are rapidly advancing in the estimation of microscopists. The exquisitely adjusted balances and delicate weights are well worthy of minute inspection, and among the optical instruments the excellence of modern glass is particularly deserving of notice. In this class is shown a very interesting series of photographic views of the late total eclipse of the sun as seen in Spain. This class also contributes several trophies to the Exhibition, among the most important of which are several large telescopes and lighthouse lanterns with all the recent improvements.

The Kew Observatory, under the management of a Committee of the British Association for the Advancement of Science, contributes a beautiful series of instruments for the automatic registration of the variations of magnetometers. These instruments present some convenient, but not very important modifications of Mr. Charles Brooke's very ingenious instruments as employed in the Greenwich and Paris observatories.

Class twenty-five (skins, fur, feathers, and hair) is very full and complete, every branch of those special trades being well represented. The central object of attraction in this section will doubtless be the great ostrich-feather trophy, and another smaller trophy of hair work.

Class twenty-six comprises all kinds of undressed and dressed leather, saddlery, harness, and whips ; and in this section one firm exhibits a handsome trophy.

The photographic section—Class fourteen—which is under the direction of Mr. P. Le Neve Foster, M.A., divides the space of the central tower with the educational department. It may almost be said to form a new class, for although the Talbotype paper process was shown in 1851, the collodion process was not invented until the middle of that year.

The old difficulty of defining the position of photography has been felt by the Commissioners and the Committee, and the result is that it is allowed to hover over the confines of art, more as an attendant than as a companion.

This department of the Exhibition comprises a large and perfect collection of photographic apparatus and appliances—a great variety of tents and out-door contrivances ; with lenses exhibited by Mr. Ross, and some novel lenses sent by Mr. Dallmeyer. Amongst the leading attractions are the art designs photographed in ceramic materials, and burnt in by M. F. Joubert. This gentleman also exhibits several specimens of the same work in which he has succeeded in combining many colours.

Mr. Fox Talbot has sent some specimens of photographic engraving ; Sir Henry James some samples of photozincography, or photographs on zinc—a principle now

successfully applied to the printing of the Ordnance maps.   Mr. Field exhibits
some specimens of photolithography, and Mr. Pouncey of Dorchester has sent
some examples of printing in carbon, or photographs which are as permanent as engravings.   The difficulty in printing in this style is to get a gradation
of tone, and in this M. Joubert with his phototypes has met with considerable
success.

Mr. Thurston Thompson and Mr. Caldesi exhibit gigantic photographs of
Raffaele's cartoons, which are wonderful examples of the manipulation of large plates.
The former gentleman also exhibits several photographs of Turner's pictures.

The most striking objects for general observers will doubtless be the full-
length portraits enlarged to the size of life; and of these several exhibitors have
sent specimens.

The collection of coloured photographs and miniatures is very fine, and also the
portraits of "men of the time."   Rejlander's composed pictures—or subjects where
a figure is taken from one canvas, a tree from another, and a building from a third,
and combined in one high art whole—are very interesting, and show what may be
done with photography, either in a serious or a comic vein.

Classes fifteen, twenty-eight, and thirty-eight (a), comprising horological instruments, paper, stationery, printing and bookbinding, and art designs for manufactures
—the latter illustrative of British art from 1762 to 1862—are under the superintendence of Mr. J. Leighton, F.S.A.

In horology, the greatest novelties are improved designs rather than new
movements, though several clocks are exhibited constructed to go more than
twelve months.   One exhibitor sends an astronomical clock, impelled by gravita-
tion, which requires no oil to the escapement; and another exhibitor has sent a
clock which shows the time and longitude at important places.   A large number
of electro-magnetic clocks are exhibited, with some mercurial time-pieces, some
steam or speed clocks, a geographical clock, showing the time throughout the
world, and a new balance constructed to resist all extremes of temperature.   Under
what may be called the fine arts of watchmaking, many manufacturers exhibit
some very rich examples of heraldic enamelling and engraving.

In paper are shown many efforts to obtain that important material without
the aid of cotton or linen rags—from such substances as straw and other vegetable
fibres—while the capabilities of the material are shown in water-pipes made of
paper, fancy boxes, stationery, &c.

The Bank of England exhibit their valuable "Notes for the prevention of
forgery," and the great seal of England is also exhibited by Mr. Wyon.

Printing and type-founders are represented in great strength.   Messrs. Bradbury exhibit their "nature-printing;" a thing quite new in England since 1851;
and specimens are also exhibited of the new art of autotypography, the invention
of Mr. George Wallis, by which drawings are so executed that they can be engraved direct in a few seconds on a metal plate for printing from at the ordinary
copper-plate printing-press.   Designs direct from the artist's hand are thus produced, suited to a variety of purposes in fine and industrial art.

In printing processes may also be seen that ingenious toy of the Electro-
Block Company by which engravings may be reduced or enlarged to any
extent.

In chromo-lithography great progress has been made during the same period in the imitation—almost in the reproduction—of water-colour drawings; and the improvement in steam printing and in wood blocks is almost as great.

The publishers of London show together in a collective case, where may be seen some of the most luxurious volumes produced by the leading houses. The reprint of the first edition of Shakespeare; the Bibles of Bagster and Spottiswoode, are also in this collection.

The specimens of bookbinding, both in hand-tooling and blocking (for which England is unrivalled), are very fine; particularly publishers' bookbinding, which has had the advantage of the best art.

The designs of Mr. John Leighton, the superintendent, are very numerous in this department.

The art designs for manufactures comprise a fine collection of drawings and models in all departments of art industry capable of reproduction. Designs for glass and ceramic wares, precious and other metals, furniture and carving, plastic decorations, and other objects in relief; also designs for textile fabrics, paper-hangings, mural decorations, tiles, mosaics, inlays, stained painted and decorated glass, &c., are freely exhibited, and also a number of original illuminations. The following is a list of the deceased artists, specimens of whose works have in most cases been obtained:—

CHIPPENDALE,	BACON (Sculptor),
CHAMBERS,	GANDY,
ADAMS,	FLAXMAN,
SOANE,	T. HOPE,
JEFFERY WYATT,	HOLLAND,
STOTHARD,	PITTS (Sculptor),
BRIDGENS,	B. WYATT,
TATHAM,	WYON (Sculptor),
PUGIN,	BARRY,
JAMES WYATT,	COTTERILL (Sculptor).

Class sixteen, comprising musical instruments, is under the management of Mr. C. Boosé, and it promises to be as popular in 1862 as it was in 1851.

Grand pianos are exhibited, some remarkable for the beauty of their cases, others for the richness of their tone, and they are well supported by every variety of stringed and wind instruments. An oak piano of the time of Charles the First is shown, and an historical series of pianos, from the old harpsichord—the favourite of our great-grandmothers—down to the improved instrument of the present day. In one part of this collection is a self-blowing harmonium, the wind for which is supplied by clockwork; in another part is a group of Æolian harps, and in another corner is a double bass with a remarkably ingenious apparatus for producing enharmonic scales of harmonics. Many valuable improvements in the mechanism of pianos are exhibited, and the inner machinery of these domestic instruments is shown from the first stage to the last. An oblique piano, with a new action, is a novelty in its way, and also some metal bag-pipes suitable for tropical climates. A new music time-keeper is worthy of notice; also an omnitonic flute, adjustable at will to any key; and the central object of attraction to listeners rather than to sight-seers is the orchestrion, or self-acting organ, which fairly imitates the melody of a full orchestra.

Class seventeen is devoted to the exhibition of surgical instruments and appliances. In the list of exhibitors in this class will be found the names of nearly all the important instrument makers of England; and the objects exhibited will show how much ingenuity and skill can be exercised in the invention of instruments to aid the surgeon in removing disease and in diminishing suffering and distortion. When the mind has fairly grasped the immense benefit which the surgeon is capable of conferring on his fellow-creatures, it can realize the importance to mankind of such instruments as are exhibited in this class; and the International Exhibition of 1862 will fairly demonstrate how much originality in invention, and how much perfection in manufacture have been reached during the past eleven years. Surgical instruments generally may be subdivided into those used, first, for general surgical purposes, secondly, in ophthalmic surgery, thirdly, in orthopædic surgery, fourthly, in aural surgery, fifthly, dental surgery, and sixthly, in obstetric surgery; so that in all and each of these subdivisions the professional man will find examples of the more important instruments which have been devised for the relief or cure of those diseases and injuries to which the human body is liable.

The superintendence of this class has been given to Mr. Traer, F.R.C.S.; and it may be advisable to state that whereas in 1851 surgical instruments were comprehended under " philosophical instruments," they now worthily form a special class of their own.

The textile division of the Exhibition—comprising Classes eighteen, nineteen, twenty, twenty-one, twenty-two, twenty-three, twenty-four, and twenty-seven—is under the management of Mr. George Wallis.

These classes include cotton, flax and hemp, silk and velvet, woollen, worsted, and mixed fabrics generally; carpets, woven, spun, fitted and laid fabrics shown as specimens of printing and dyeing; tapestry, lace and embroidery, and articles of clothing.

The textile fabrics of the United Kingdom are located in the south-eastern gallery, and thus occupy one-half of the space on the south side of the great nave, the other half being devoted to foreign productions. The arrangements commence with the class for printing and dyeing (Class twenty-three), and in this the productions of Manchester and Glasgow form the leading features, and in combination with the printed table-covers and bandannas of the London houses, constitute the staple of the exhibits, the illustrations of dyeing not being very numerous. Cotton manufactures succeed, and in this class, too, the manufactures of Manchester and Glasgow with the sewing threads of Leicester, Paisley, and Huddersfield, constitute the leading features.

In the cross gallery leading from the gallery next the nave to that along the north wall of the picture gallery, lace, tapestry, and embroidery (Class twenty-four) are placed. Nottingham, London, and Dublin are here effectively represented by a remarkable display of lace goods of all kinds and qualities. The productions of hand lace-workers of Buckinghamshire, Bedfordshire, and Northamptonshire are also located here.

The important class of woollens, worsted, and mixed fabrics (Class twenty) is arranged along the remainder of the gallery next the nave, and continued to the eastern dome, and thence along the cross gallery next the south-eastern transept. The manufactures of the West Riding of Yorkshire, the West of England, and the Metropolis occupy space to the staircase at the side of the dome : Norwich fabrics being placed across the angle of the gallery, and at the back the poplins of Dublin and the shawls of Paisley. Glasgow manufactures in this class are placed next the transept, and the coarser woollens and mixtures of Scotland, together with the blankets and flannels of Rochdale and Witney, and a collection of yarns from various quarters, complete the arrangements of this department.

Class nineteen—linen, flax and hemp manufactures—follows woollens ; the coarser fabrics, with cordage, mats, &c., being placed in the remaining portion of the cross gallery, and a portion of the gallery against the north wall of the picture saloons. Here, and on the western side just named, the fine fabrics of linen, damasks, &c., are to be found, finishing with a very extensive display from the north of Ireland.

The silk manufactures of Great Britain have been located against the north wall of the picture gallery ; and a complete collective display of nearly every class of silk goods manufactured in London, Macclesfield, and Manchester, together with the ribbons of Coventry and the spun silks and thrown silks of Leek and Derby, forms one of the great features of the textile division.

The class for clothing (Class twenty-seven)—divided into four sub-classes, viz., hats, millinery and artificial flowers, general clothing, and boots and shoes—has been arranged in the extreme angle of the south-eastern transept gallery.

Passing from the flax and hemp manufactures, boots and shoes are placed along the south end of the transept ; next to these, gloves and hosiery, and the variety of articles coming under the head of general clothing. Millinery and artificial flowers, and finally, hats and caps, complete the arrangements to the end of the gallery next to the eastern dome.

It may be, perhaps, well to remark here that the most defective parts of the textile division are those of calico printing and cotton manufactures. Neither of the displays convey a fair idea of the extent and importance of these industries ; and although every exertion was made consistent with the functions imposed upon Her Majesty's Commissioners, to insure a proper representation in these two branches of our national industry, printers and manufacturers did not move sufficiently in the matter until it was too late, as, according to the early demands in each class, the spaces had to be allotted in the mass at a comparatively early date. This practically excluded many producers who at a later period made proposals to exhibit.

The educational section—Class twenty-nine—has been placed under the superintendence of Mr. J. G. Fitch, M.A., who was from the first a member of the National Committee of Advice. The original scheme which was drawn up by that committee was a very comprehensive one, and although it has not been altogether realized, we think it may be interesting to place it on record here :—

BUILDINGS, FITTINGS, AND FURNITURE.

BUILDINGS. Plans, Sections, Elevations, Drawings, Photographs, and Models of—

Schools :
    Infant.
    Primary.
    Secondary.
    Industrial.
    Sunday.
    Adult.
    Trade.
    Art.
    Technical.
    Swimming.
    Riding.

Schools—*cont.*
    Fencing, &c.
    Higher.
    Lecture Rooms.
    Institutes.
    Public Libraries.
    Museums.
    Private Studies.
    Dormitories.
    Training Colleges.
    Universities.

FITTINGS AND FURNITURE. Specimens, Models, Drawings, &c. of—

Desks.
Galleries.
Forms and seats.
Black boards and easels.
Inkstands and wells.
Tables, Work-tables, &c.

Teachers' and pupil-teachers' desks and boxes.
Beds and cribs for infants.
Timepieces.
Curtains for Schools.
Cases and Stands for Maps and Diagrams.
Receptacles for hats, cloaks, &c.

SANITARY ARRANGEMENTS, SPECIALLY SUITED FOR COLLEGES, SCHOOLS, AND INSTITUTES.
    Apparatus for heating, lighting, and ventilation.
    Play and exercise grounds.
    Lavatories, &c.

FURNISHED or FITTED MODELS, and collections of furniture, &c. requisite for schools and other educational
    institutions.

BOOKS AND INSTRUMENTS OF TEACHING GENERALLY.

READING AND SPELLING.
    Books. Primers, reading books, works on elocution, &c.
    Tabular Lessons. Alphabets, spelling exercises, &c.
    Materials. Boxes of letters, &c.

WRITING.
    Books. Manuals for teachers. Copy-books, &c.
    Copies and models for imitation. Diagrams of forms and proportions of letters.
    Materials. Slates, pencils, pencil-holders. Pens, pen-holders, pen-menders. Ink. Rulers.
    Mechanical Expedients for directing the hand or otherwise assisting the pupil in learning to write.

ARITHMETIC.
    Books. Theory or practice of arithmetic, mensuration, or book-keeping.
    Tabular Lessons. Elementary illustrations of number. Sheet exercises and sums.
    Pictorial Illustrations and diagrams of weights and measures; illustrations of the various systems
        of weights and measures in the United Kingdom.
    Mechanical Appliances. Ball-frames, cubes, &c.

RELIGIOUS INSTRUCTION.
    Books. Bible manuals. Compendiums of Scripture or ecclesiastical history. Catechisms. Books
        for Sunday-school use, &c.
    Bible Pictures. Illustrations of Eastern life and manners, &c.
    Maps, Charts, and Models illustrating the chronology, history, or geography of the Bible.

HISTORY (Secular).
    Books. Manuals of ancient or modern history. Biographies. Reading books.
    Chronological Charts and diagrams. Systems of mnemonics applied to chronology, &c.
    Pictures, in series or singly, exhibiting historical events.

GEOGRAPHY.
    Books and Atlases.
    Maps, Charts, Models, and Diagrams. Outline maps. Simple projections.
    Globes, plain or in relief.
    Miscellaneous Appliances. Slate globes, maps in relief, models and pictures of physical phe-
        nomena, &c.
    National Surveys.

LANGUAGE.

>*Books.* Works on composition, the analysis of sentences. The philosophy and structure of language. Dictionaries and grammars of ancient or modern languages. Editions of classic authors. Courses of reading and instruction.
>
>*Tabular Lessons* in parsing, etymology, or logical analysis.

MATHEMATICS.

>*Books.* Treatises and exercises on pure or applied mathematics,
>
>*Illustrations.* Geometrical diagrams; models and drawings for elementary lessons on form and quantity, &c.
>
>*Mathematical Instruments.* Simple and cheap instruments for school use, singly or in cases. Mariners' compasses. Sextants, theodolites, levelling instruments, &c.

PHYSICAL SCIENCE.

>*Books.* Text books and manuals on astronomy, mechanics, electricity, chemistry, mineralogy, &c.
>
>*Drawings and Diagrams* illustrating scientific truths.
>
>*Models and Apparatus* employed in teaching.
>
>*Cheap Collections of Objects* adapted for chemical, electrical, or other scientific experiments.

NATURAL HISTORY.

>*Books.* Manuals or reading books on botany, zoology, and geology.
>
>*Drawings and Pictures.* Illustrations of structure, appearance, relative sizes, or local distribution of plants and animals.
>
>*Charts and Diagrams* to simplify or exhibit systems of classification.
>
>*Elementary Collections* of natural history.

MUSIC.

>*Books.* Theory or practice of vocal or instrumental music. Exercises:
>
>*Compositions.* Chants, part songs, school songs, &c.
>
>*Diagrams and Tabular Lessons* showing scales, systems of musical notation, &c.
>
>*Instruments of Instruction.* Black boards for music lessons. Tuning-forks, pitch-pipes, metrenomes. Cheap musical instruments for schools, juvenile bands, &c.

DRAWING, PAINTING, AND DESIGN.

>*Books.* Handbooks of instruction for teachers, exercises for pupils, &c.:
>
>*Copies.* Drawings and pictures, models, casts, &c.
>
>*Materials.* Paper, pencils, rubbers, chalks, brushes, easels, colours, canvas, palettes, &c.
>
>*Diagrams and Models.* Illustrations of theory of perspective, laws of vision, &c.

DOMESTIC ECONOMY.

>*Books.* Text books and reading books, adapted for school use, on needlework, cooking, choice of food, materials for dress, management of a house, &c.
>
>*Illustrations.* Pictures, diagrams, models, and specimens of household implements, furniture, &c. suitable for educational use.

INDUSTRIAL EDUCATION GENERALLY.

>*Books.* Manuals of gardening, agriculture, or other industrial work done in schools, or other institutions of technical instruction, whether for children or for adults.
>
>*Instruments and Illustrations* employed therein.

SOCIAL AND ECONOMIC SCIENCE.

>*Books.* Manuals, and reading books, on wages, capital, labour, the conditions of industrial success, &c.
>
>*Tabular Lessons* or other visible illustrations of such subjects.

PHYSIOLOGY AND THE LAWS OF HEALTH.

>*Books.* Text books, and reading books, on animal physiology, functions of the skin, cleanliness, food, ventilation, respiration, general conditions of health.
>
>*Diagrams and Drawings.*
>
>*Anatomical Models* for teaching.

GENERAL KNOWLEDGE.

>*Books.* Text books on common things, the philosophy of every-day life, &c. Lessons on objects. Courses of miscellaneous instruction.
>
>*Drawings and Diagrams* exhibiting the structure and use of familiar things, as a watch, a door-lock, tools and simple machines, weights, lengths, &c.
>
>*Models and Specimens* used in teaching.

SCHOOL REGISTERS.

>Roll-books, registers of attendance, payments, progress, &c. Expedients for facilitating the collection of educational statistics.

TABLETS AND PICTURES FOR WALL USE, including contrivances for rendering schoolrooms cheerful and ornamental.

TEACHING FOR THE BLIND, THE DEAF AND DUMB, IDIOTS, OR OTHERS MENTALLY OR PHYSICALLY DEFICIENT.
  Books. Embossed for the blind. Alphabets for the deaf and dumb, &c.
  Treatment of defective utterance.
  Instruments and Apparatus adapted for these purposes.

SPECIAL AND PROFESSIONAL EDUCATION.
  Books. Manuals of military, naval, legal, medical, engineering, or other professional instruction.
  Instruments and Apparatus used in such instruction.

THEORY AND PRACTICE OF TEACHING.
  Books. Methods and systems of teaching. Model lessons. Teachers' manuals. Courses of
    pædagogy. Schemes of examinations. Histories of education. Reports, &c., of Committee of
    Council on Education, Boards, and societies of education. Statistics of education, histories,
    reports and regulations of public libraries, book-hawking societies for promoting the sale of
    pure literature, literary and scientific societies, institutes, &c.

LIBRARIES.
  Collections, lists, or specimens of books adapted for school libraries, either by their cheapness, or
    by arrangement or classification.

APPLIANCES FOR PHYSICAL EDUCATION.—TOYS AND GAMES.

MANUALS OF DRILL, military, naval, or general, for boys, girls, or adults. Exercise books, &c., for the
    use of teachers.

GYMNASTIC APPARATUS.
  For Playgrounds. Specimens, models, and diagrams of swings, poles, parallel bars, inclined
    planes, &c.
  For Indoor Use. Dumb-bells, chest expanders, &c.

APPARATUS EMPLOYED IN INFANT SCHOOLS.
  Articles used in *Kinder-Garten* occupations.
  Models, puzzles, and expedients for educating the eye or hand.
  Specimens of the commoner tools used by workmen, as smiths, carpenters, gardeners, &c., and
    models of articles of household furniture generally.
  Picture books and cards.
  Instructive games and toys.

MATERIALS USED IN, AND PHOTOGRAPHS AND PICTURES, illustrative of national and other games and
    exercises of strength or agility.

Miscellaneous toys and games.

SPECIMENS OF SCHOOL-WORK.

WRITING, plain, ornamental, or illuminated.

DRAWING AND DESIGN, plain and coloured drawings from maps, copies, models, nature, memory.
    Modelling in clay, wax, &c. Cutting out paper, from copies or invention.

NEEDLEWORK.
  Ordinary. Sewing, knitting, darning, &c.
  Artistic. Embroidery, lace-work, worsted-work, &c. Specimens of dressed dolls, &c.

INDUSTRIAL WORK GENERALLY.
  Basket Work, Artificial Flowers, Matting, &c.
  Floral and other *decorative* work for school fêtes, &c.

MUSEUMS.

MUSEUMS.
  National.
  Local.
  Trade.
  Itinerating.
  Classified collections, of small cost, for educational use, to illustrate common objects, specific
    sciences or studies, or particular book or courses of instruction.
  Special floras or faunas, &c.

TAXIDERMY.
  Methods of mounting, labelling, and preserving objects from dust, insects, &c.
  Specimens.

The plan here sketched represents the conception formed by the National Committee of a possible educational collection on an extensive scale. It was prepared on the assumption that the court would be an international one, and that means could be taken for bringing into comparison illustrations of the methods and processes of teaching from all parts of the world. The committee also hoped that a very large space would be at their disposal, and therefore sought to make their list as nearly as possible an exhaustive one; but the event has not justified these anticipations. A strong wish was expressed on the part of the foreign commissioners to keep all educational contributions in the courts devoted to the several countries; and the idea of an international exhibition in this department was therefore abandoned. Moreover, the space allotted by the Commissioners to the Education Committee necessarily amounted only to one-fifteenth part of the total area demanded by British exhibitors alone. In these circumstances it became necessary to sacrifice much of the unity and symmetry of the original plan. For example, it was very early determined by the committee that results or specimens of school-work were altogether inadmissible, since in so small a space it was simply impossible to represent the work done in different classes of schools with even approximate fairness. In the department of school-buildings it is to be regretted that so few models are to be found illustrative of the great improvements which have taken place recently in the structure, fitting, and internal arrangement of elementary schools.

Nevertheless the collection, though somewhat heterogeneous, contains some noteworthy objects. One recess on the north side of the room is devoted to the exhibition of a collection of the most recent text-books and manuals employed in teaching. Another contains—besides many miscellaneous articles of school furniture—an interesting collection of models illustrative of the system pursued in Reformatory schools, and a beautiful model of one of the school-houses on the Philanthropic Society's farm at Redhill. In the adjacent bay, which is wholly occupied by materials and models employed in teaching drawing, the Department of Science and Art exhibits a complete series of the copies employed in the instruction of pupils in the Government Schools of Design.

The small court in the north-east corner of the educational department is entirely devoted to the illustration of methods employed in teaching the Blind. Here will be found specimens of all kinds of embossed or raised type, of writing and ciphering frames, of musical notation, and of maps in relief. A very beautiful embossed globe used in the School for the Indigent Blind adorns the centre of this recess, while specimens of the work done in various Blind asylums are displayed on the walls and on the screens. In the arrangement of the articles in this division, the superintendent has had the advantage of the assistance of Viscount Cranborne, Mr. Edmund C. Johnson, and the Rev. B. G. Johns, Chaplain of the Blind Asylum, who have taken great interest in the selection of the articles, and to whom all the credit of the arrangement is due.

On the south side of the room, the principal education societies—the National, the British and Foreign, the Home and Colonial, the Christian Knowledge societies, and others—exhibit specimens of the articles employed in fitting up schools or in elementary teaching generally. A very beautiful collection of models of school fittings, executed in miniature, and exhibited by the National Society, deserves especial notice. The Home and Colonial School Society also displays an interesting series of articles used in the "Kinder Garten" system, which has been so successfully adopted by that society in its infant schools. Mr. Myers's collection of pictures and other objects adapted for use in the nursery and in the schoolroom, is especially worthy of attention.

Of the apparatus employed in teaching elementary science the collection of globes, the orrery, and several other collections, will probably attract the eye of the visitor first; but the scientific collection, the curious case of small birds, on the staircase, and the very systematic and beautiful geological collections will be found, on close examination, to possess great interest and merit. The large cases of stuffed birds, though not specially educational in their character, are remarkable evidences of the perfection which the art of taxidermy has reached in England.

Some surprise has been expressed that toys should have a place in the educational court. Perhaps we shall be less disposed to question the wisdom of this arrangement when we consider that of modern toys many have a definite educational purpose, and are designed expressly to beguile the little ones into efforts of observation, of counting, of construction, or of invention. Moreover, it must not be forgotten that everything which helps a little child to open its eyes and look about him, which teaches him to use his fingers, which fills his dreams and fancies with happy images, or even which only gives him an hour of innocent enjoyment, plays a part in his education which none but a pedant can afford to disregard.

The walls of the apartment are covered with maps and diagrams, of which the most conspicuous is a gigantic map of the British Isles.

Classes thirty, thirty-three, thirty-four, and thirty-five, under the management of Mr. J. B. Waring, consist of furniture, paper-hanging, and other wall decorations; works in the precious metals and jewelry; pottery and glass of every description. In Class thirty, the well-known names of all the leading manufacturers guarantee a first-rate exhibition. About two thousand five hundred square feet are set apart for contributors in the Mediæval style, and are filled with a series of remarkable works, illustrating the great advance made in this particular branch of Art since the formation of the Mediæval Court, in the first Exhibition of 1851.

The arrangement of this court was carried out by Mr. Burges and Mr. Slater, and they are able to show, amongst other things, a reredos by Mr. Street, executed by Mr. Earp, and a portion of that for Waltham Abbey, with the cartoon of the rest. There is likewise a cast of the sculpture in the Bedminster reredos. Mr. Redfern contributes casts of his sculptures of the Ascension, for the Digby Mortuary Chapel at Sherborne, and for the Westropp monument in Limerick Cathedral, —the latter being arranged in connection with a portion of the actual carved work of the monument. Several exhibitors contribute fonts, and there is a cast of the renaissance font at Witley, carved by Mr. Forsyth.

A cast of Dr. Mill's monument (designed by Mr. Scott) and effigy at Ely has been sent. Mr. Nichol sends another effigy arranged on a high tomb in connection with some subjects in relief. The late Lord Cawdor's high tomb is also exhibited; and one of the circular panels, with a cut subject, for the Lichfield pavement. In woodwork, there are the stalls of Chichester Cathedral; a rich bureau, which was shown about two years ago at the Architectural Exhibition; and a decorated organ.

In metal-work, one exhibitor sends a rich iron font cover; the Ecclesiological Society exhibits the frontal which it is going to present to St. Paul's Cathedral, designed according to the "Cologne" method; and the Dean of Peterborough has kindly lent the new frontal for his cathedral, executed by the Ladies' Ecclesiastical Embroidery Society.

Two well-designed groups of furniture, one in the mediæval style, the other of the modern school, are placed at the east entrance of the main avenue. The Goldsmiths' Court is filled with a noble display of works in the precious metals; three richly grouped trophies in the main avenue attest the artistic excellence of the designs executed by the prominent exhibitors, and for the first time also the working jewellers of Birmingham come forward in a body to challenge comparison with the London, Scotch, and Irish jewellers.

In Class thirty-five the leading firms are bent on surpassing all former efforts; some with majolica and Palissy ware, encaustic tiles, and a frieze by Lucca della Robbia; others with parian and china, from designs by Durham, Marochetti, Monti, Foley, Gibson, and Marshall. One of the fountains is a triumph of ceramic art, and the china trophy in the main avenue is a worthy pendant to one by another exhibitor. One firm contributes the main feature of the court itself, which contains evidence also of the energy and advancing taste of the principal firms of the Staffordshire potteries. Every description of glass is well represented in Class thirty-four, and in the stained-glass gallery, and the great northern windows of the east and west transepts, the old and the younger rising firms bear witness to the great progress made during the last ten years in elaborate stained glass compositions, secular as well as ecclesiastical. Two handsome trophies of glass are placed in the nave. Classes thirty-seven and ten (c), the one consisting of architectural designs and models, the other of large works remarkable for architectural merit, such as church doors, altars, pulpits, &c., both under the charge of Mr. Waring, will perhaps, more than any other classes in the Exhibition, sustain the well-deserved reputation of our countrymen in the art of architecture. If this art is not so attractive to the public generally as the more easily appreciated arts of painting, sculpture, and decoration, it will be acknowledged by all connoisseurs foreign and English, to be that in which we exhibit works of pre-eminent excellence.

In Classes six, thirty-one, thirty-two, and thirty-six,—comprising carriages, iron and general hardware, steel and cutlery, and unclassed manufactures,—which are under the management of Mr. T. A. Wright—the display has much of an art character. The prominent features in Classes thirty-one and thirty-two—placed in the south end of the south-eastern transept—are the screen for Hereford Cathedral; Bessemer's specimens of his patent steel (a new product since 1851); a new peal of bells, which are so made that they may be chimed by a child ignorant of music; and a large class trophy, to which two firms have contributed elaborate specimens of ornamental metal work; and two others equally elaborate specimens of mediæval metal work. Several separate courts and collections in this section of are worthy centres of attraction.

Some beautiful iron-work gates—called the Norwich gates—are exhibited in the nave. In the carriage section (Class six) of this department there is a trophy-drag—a specimen of what may be considered as our national carriage; and Class thirty-six, comprising dressing-cases and travelling-bags—which were not honoured with a class in 1851, has gratefully erected a magnificent trophy-case designed by Mr. J. B. Waring, filled by four exhibitors with dressing-cases, despatch boxes, &c.

The Birmingham contribution to this collection contains a very complete representation of the various articles in metal for which the town and district has so long been celebrated. The court more particularly appropriated to the display of articles in Class thirty-one—the general hardware court—contains the productions of more than one hundred exhibitors. At its entrance will be found a display of metallic bedsteads, raw materials of every kind for the use of brass-founders, also gas-fittings of all kinds, and stamped and general brass-foundry. Tin-plate working and japan wares, wire-drawing, fire-irons, stoves and grates, iron safes, hollow wares in copper and iron, tinned and enamelled, tubing of all kinds, scales and weighing machines, medals and dies, hooks and eyes, pins, steel pens, locks and general hardware, knife-cleaning machines, steel toys, metal mountings for the use of bookbinders, fire-guards, coffin furniture, saddlers' ironmongery, ornamental panels of various kinds, screws, nails, hinges, &c., &c., are also exhibited; and, in fact, as far as Birmingham is concerned, all its leading branches of industry are fully represented by all its leading manufacturers. The majority of the exhibitors in 1851 are again present, with other houses which now for the first time have entered the industrial field. In other parts of the building will be found the contributions of its gun-makers and sword-cutlers; its glass; its horological instruments; its leather, saddlery, harness, and whips; its buttons; its manufactures in papier-mâché; its steel and cutlery; and the results of the workings of its jewellers, electro-metallurgists, and silver-plate workers.

The collection in the Sheffield and Rotherham court—under Class thirty-one—comprises stove grates, fenders, fire-irons, hot-air stoves, kitchen ranges, ornamental iron-work consisting of hat-stands, ballusters, &c., &c., manufactured by most of the firms who exhibited in the Exhibition of 1851.

A portion of this court is occupied by Britannia metal goods, brass goods, consisting of high pressure taps, cocks, hydrants, &c., &c.

Adjoining Class thirty-one are exhibited in cases round the walls, edge tools, joiners tools, files, steel table-knives, scissors, sheep shears; and above, in vertical cases, saws, scythes, sickles, &c., and the general class of Sheffield manufactures.

On one block are exhibited goods of a similar character to those on the counters round the walls, but including steel, springs, &c.

Two other blocks are occupied by goods of a new class of manufacture, consisting of heavy castings of steel, crank axles for locomotives, driving wheels, axles, tyres, points for crossings, bells, railway-carriage springs, buffers, &c., &c., which it is anticipated will prove of a superior quality to those shown by German manufacturers. The above articles are now being made in large quantities, and have, since the last Exhibition, become one of the most extensive branches of the Sheffield trade.

The counter under the gallery is covered with cases filled with fine cutlery —scissors, tailors' shears, table-knives, small edge-tools, bowie-knives, hatchets, &c., being samples of goods for which Sheffield holds so high a position.

The goods displayed by Walsall exhibitors represent all the important branches of trade for which that town is justly celebrated. There is nearly every description of saddlery, harness, and bridle work; also saddlers' ironmongery, comprising all kinds of bits, stirrups, silver-plated and brass harness, carriage furniture, &c. &c.

The collection sent by Wolverhampton comprises a variety of locks, and amongst them a new patent keyless lock, based upon the permutation principle, which has been manufactured for this department of the Exhibition, having two hundred and forty-four millions one hundred and forty thousand one hundred and twenty-five combinations, to open all of which would take a man—supposing he could live so long—one hundred and thirty years. Safes, hollow-ware, general hardware, and choice examples of japanned goods complete the contributions from this important manufacturing centre.

The following is a list and plan of the main trophies fixed throughout the British half of the building:—

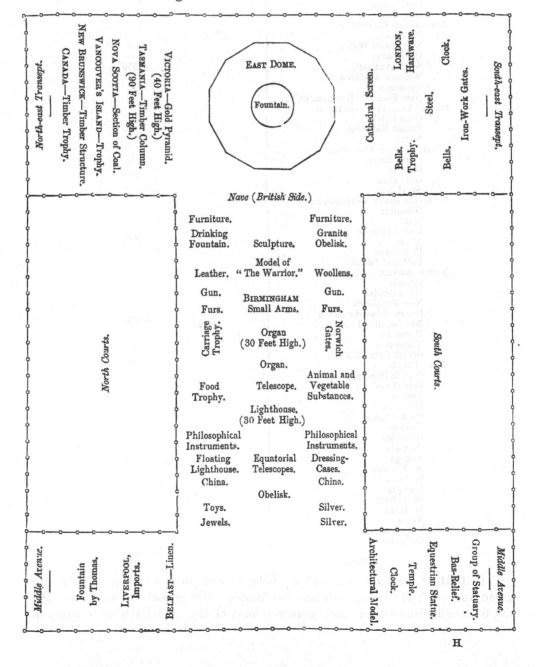

This completes the industrial portion of the British side, and we may now turn to the colonial and Indian collections.

The colonial and foreign exhibitors—being mostly at a safe distance—were not so difficult to deal with as the home exhibitors. Twelve thousand eight hundred and twenty-two superficial feet were detached from the British half of the space for the British colonies, and apportioned as follows:—

EAST INDIAN COLONIES (exclusive of India) :—

	Superficial sq. ft.
Ceylon	
Mauritius and Seychelles	
Straits Settlements	300
Hong Kong	
Labuan	

AUSTRALIAN COLONIES :—

Victoria	
New South Wales	
Queensland	
South Australia	4,550
Western Australia	
Tasmania	
New Zealand (European only)	

SOUTH AFRICAN COLONIES :—

Cape Colony	
British Kaffraria	640
Natal	

WEST AFRICAN COLONIES :—

Sierra Leone	
Gambia	
Gold Coast	200
St. Helena	

MEDITERRANEAN COLONIES, &c. :—

Gibraltar	
Malta	
Ionian Islands	
Aden	400
Heligoland	
Falkland Islands	

NORTH AMERICAN GROUP :—

Canada	
Nova Scotia	
New Brunswick	
Prince Edward's Island	
Newfoundland	5,895
Bermuda	
Vancouver	
British Columbia	

WEST INDIES :—

British Honduras	
British Guiana	
Jamaica	
Bahamas	
Turk's Island	
Trinidad	
Barbados	
Grenada	
Tobago	837
St. Vincent	
St. Lucia	
Antigua	
Montserrat	
St. Kitts	
Nevis	
Virgin Islands	
Dominica	

Total	12,822

In 1851, the colonies were, as a whole, almost unrepresented. The notice given was too short; the undertaking was hurried; the project was quite new, and not thoroughly understood; and, moreover, most of the colonies were scarcely in a position to go to much expense for contributions. The East India Company, however, made a noble display, and some few of the British colonies a respectable appearance in 1851, and also at Paris in 1855.

According to the latest official returns, the aggregate population of the colonies and possessions under British rule exceeds one hundred and ninety-five millions, of which the great bulk, one hundred and eighty-five millions, are distributed over British India. In these colonies a total revenue is raised of about forty-four million pounds, and the yearly value of the external trade, imports and exports, is upwards of one hundred and seventy-six million pounds.

It appears, from the official reports,'that out of twenty-three thousand five hundred and seventy-five superficial foot of horizontal net space allotted to the British colonies in 1851, but six thousand one hundred and eighty feet were occupied. The only colonies then specially represented were—Canada, which made a good display; a few objects indirectly sent for exhibition from Nova Scotia, New Brunswick, Newfoundland, and Bermuda. From the West Indies a small collection was sent from the Bahamas, and a few odds and ends from Antigua and Barbados. Trinidad and British Guiana were well represented. Of the African colonies, the Cape was the only one that sent a collection; a few objects illustrating the products of St. Helena and the West Coast of Africa were shown by London merchants and individual exhibitors in England.

Of the Eastern colonies, the Mauritius sent but little; but a fair collection was transmitted from Ceylon.

In 1851, the Australasian colonies were but poorly represented, although a few made some efforts to put in an appearance. The New South Wales and Tasmanian collections were creditable, and a few things were sent from South Australia and New Zealand. With the exception of a small collection from Malta, this formed the aggregate of the colonial efforts.

At the Paris Exhibition in 1855, the few colonies that did send articles made a very satisfactory display. Canada, especially, obtained honour for its varied collections, which occupied upwards of three thousand feet of space, contributed by about three hundred and fifty exhibitors. Jamaica covered an area of about five hundred feet, and British Guiana three hundred and fifty, whilst Barbados and the Bahamas were the only other West Indian colonies that sent. Ceylon occupied nearly as much space as Demerara, and the Mauritius sent a small collection. The Australasian colonies on that occasion were very well represented, although one or two did not show; two hundred and fifty-one exhibitors from New South Wales occupied eight hundred and seventy-one square feet; one hundred and eighteen exhibitors from Tasmania, four hundred and twenty-nine feet; thirty-six from Victoria, two hundred and eighty-nine feet; and ten from Newfoundland, one hundred and seventeen feet. The official returns show that the twelve British colonies which exhibited products at Paris in 1855 filled about five hundred feet more space than all the colonies which were represented in 1851.

The contributions from the colonial possessions and many of the miscellaneous and outlying countries which have no special government, or where no commission has been formed, are under the superintendence of Dr. Lindley, F.R.S., who, besides his well-known scientific attainments, brings to the work the business knowledge and experience gained in the same department at the Exhibition of 1851.

The following Table gives the population, revenue, debt, aad foreign commerce of the dependencies of the British Empire, chiefly for the year 1859, the latest date for which complete statistics can be obtained:—

COLONY.	Population according to the last return.	Revenue.	Debt.	Imports.	Exports.	Total external commerce.
		£	£	£	£	£
EASTERN—						
India	185,908,277	36,060,788	97,851,807	34,545,650	30,532,298	65,077,948
Ceylon	2,000,000	747,037	None.	3,474,487	2,524,752	5,902,239
Mauritius and Seychelles . .	313,047	609,634	None.	2,025,890	2,544,793	4,570,683
Straits Settlements . .	273,774	125,453	None.	7,811,698	7,422,855	15,834,553
Hong Kong	86,941	70,000	None.	Not stated.	Not stated.	..
Labuan	1,774	6,707	None.	30,724	6,358	37,082
AUSTRALASIAN—						
Victoria	540,322	3,257,724	8,000,000	15,622,891	13,867,860	29,490,751
New South Wales . .	336,572	1,511,964	2,500,000	6,597,053	4,768,049	11,365,102
Queensland . . .	25,000	160,000	None.	521,695	609,794	1,131,489
South Australia . .	122,735	511,927	830,200	1,507,494	1,655,876	3,163,370
Western Australia . .	14,837	57,945	None.	125,315	93,037	218,352
Tasmania	86,596	429,425	345,260	1,163,907	1,193,898	2,357,805
New Zealand (European only) .	71,508	459,649	500,000	1,551,030	551,484	2,102,514
AFRICAN SETTLEMENTS—						
Sierra Leone . . .	38,318	31,432	None.	169,727	247,261	416,988
Gambia	6,939	15,599	None.	76,150	110,364	186,514
Gold Coast . . .	151,000	8,286	None.	114,596	118,563	233,159
St. Helena . . .	5,490	20,736	None.	120,181	21,465	141,646
Cape Colony . . .	267,096	650,925	868,711	2,579,359	2,021,371	4,590,720
British Kaffraria . .	..	..	None.	..	..	..
Natal	160,170	50,905	165,000	199,917	110,415	310,332
MEDITERRANEAN POSSESSIONS, &c.—						
Gibraltar	17,750	32,500	None.	Not stated.	Not stated.	..
Malta	145,802	147,385	None.	2,428,909	1,775,794	4,204,703
Ionian Islands . . .	233,973	130,262	300,000	1,306,303	649,057	1,955,360
Aden	..	..	..	..	45,297	..
Heligoland . . .	2,800	..	..	..	..	..
Falkland Island . . .	540	7,657	None.	13,890	6,892	20,782
NORTH AMERICAN COLONIES—						
Canada (Census 1861) . .	2,501,370	1,947,829	11,661,010	4,953,396	6,711,032	11,664,428
Nova Scotia . . .	277,117	139,788	200,000	1,620,191	1,377,826	2,998,017
New Brunswick . . .	193,800	160,107	226,025	1,416,034	1,073,422	2,489,456
Prince Edward's Island .	80,872	27,402	28,966	234,698	178,680	413,378
Newfoundland . . .	122,638	133,735	182,500	1,323,288	1,357,113	2,680,401
Bermuda	10,982	16,765	None.	160,914	41,420	202,334
Vancouver	18,000	..	..	..	..	..
British Columbia . . .	6,000	50,000	None.	177,219	168,000	345,219
WEST INDIAN POSSESSIONS—						
British Honduras . .	29,000	27,982	None.	175,293	288,161	463,454
British Guiana . . .	127,695	275,618	449,802	1,179,901	1,228,844	2,408,745
Jamaica	377,433	279,935	913,607	853,015	961,007	1,814,022
Bahamas	27,619	30,727	None.	213,166	141,896	355,062
Turk's Island . . .	3,300	11,067	900	42,655	33,488	76,143
Trinidad	68,600	180,174	232,417	734,902	820,606	1,555,508
Barbados	135,939	87,595	None.	1,049,237	1,225,572	2,274,809
Grenada	35,517	16,948	9,400	124,660	131,307	255,967
Tobago	16,363	9,100	..	57,691	77,897	135,588
St. Vincent . . .	30,128	19,911	..	131,451	178,990	310,441
St. Lucia	30,000	12,832	15,000	103,973	101,879	205,852
Antigua	36,000	34,446	47,500	203,997	289,063	493,060
Montserrat . . .	7,053	3,513	..	19,718	16,746	36,464
St. Kitts	20,741	17,845	None.	110,835	136,511	247,346
Nevis	9,571	4,721	None.	34,748	48,186	82,934
Virgin Islands . . .	6,053	19,993	None.	10,075	11,789	22,864
Dominica	25,023	14,211	8,000	66,506	96,861	163,367

The colonies and outlying dependencies exhibiting, arranged alphabetically, with a rough estimate of their exhibitors, are:—

	Approximate Number of Exhibitors.		Approximate Number of Exhibitors.
Bahamas	5	Natal	Commission.
Barbados	Commission.	New Brunswick	36
Bermuda	„	Newfoundland	22
British Columbia . . .	„	New South Wales . . .	335
British Guiana	„	New Zealand	114
Channel Islands	5	Nova Scotia	65
Canada	Commission.	Prince Edward's Island .	Commission
Cape of Good Hope . .	„	Queensland	98
Ceylon	41	South Australia	77
Dominica	Commission.	St. Helena	Commission.
Grenada	Nothing.	St. Vincent	4
Hong Kong	Commission included in China.	Tasmania	650
		Trinidad	1
		Vancouver	6
Jamaica	·195	Victoria	Commission.
Malta	Commission.	Western Australia . . .	69
Mauritius	22		
		Total . .	1,745

Most of the industrial divisions in the Exhibition are well filled by nearly all the colonies exhibiting; and the collection of raw produce is particularly rich and interesting.

The Eastern colonies begin with Hong Kong, which shows a small but interesting collection of Chinese produce and manufactures, and takes under its wing a number of British merchants, officers, and some exhibitors who show choice and very attractive specimens of Chinese industry and silk, rich velvets, silk and gold embroidered carpets, porcelain, china, enamels, and articles of raw produce.

Ceylon has forwarded a very large and interesting collection of colonial products—coffee and cinnamon, woods and fibres, pearls, &c. Mauritius sends beautiful specimens of sugar, fruits, and vegetable substances.

The Australian colonies exhibit one of the most extensive and finest collections of the whole group, and on the collecting, arranging, and despatching of these a very large amount of money has been expended.

New South Wales has a beautifully arranged collection of its gold products from all the principal fields, in the several shapes of nuggets, quartz, grain gold, washing stuff, coin from the Sydney mint, &c. It sends an excellent assortment of Australian wines, the best of its wools and fleeces, and cloth made from them, and stuffed alpacas and the shorn fleeces of the flocks now in the colony; coal, minerals, native woods, and various agricultural produce and manufactures.

Queensland, which appears for the first time in Europe, has come forward most creditably with its ornamental and useful woods, wool, cotton, and tropical products.

South Australia is principally strong in its rich mineral products of copper and lead, and malachite manufactures, and its wheat and flour, for which it has always been noted.

Western Australia also sends specimens of woods, in which it is especially rich, some of the spars and planks being very fine. Its other products assimilate to those already mentioned.

Victoria has gone to great expense to forward an enormous collection; the only difficulty being to find room for one-half the goods sent. One of the most striking objects is a gilded obelisk representing the actual amount of gold found in the colony since 1851—about eight hundred tons, or one hundred and three millions sterling. Its manufactures and general industry are well represented, and a more extensive and varied collection has never before been sent from any British colony to Europe.

Tasmania sends, besides its wool, manufactures and agricultural produce, a noble trophy, rising ninety or one hundred feet, made of its native woods, with a circular staircase in the interior. Two whale-boats with all their gear are slung from it, and a fine native spar, surmounted with a flag, rises from the centre.

New Zealand sends from several of its provinces wool, wood, coal, gold, and agricultural produce.

The African settlements which exhibit are:—Natal, which though a comparatively young colony, has taken great pains to get together a fitting representation of its indigenous wealth and native industry. Tropical industry, agriculture, and the products of the chase are chiefly represented, and a large counter or carved side-board of native wood, with glazed panels, covering charts, photographs, and water-colour drawings of natives and scenery, forms a striking object. The Cape Colony is unrepresented, except by a few individual exhibitors. St. Helena and one or two of the other West African settlements have sent small contributions.

The Mediterranean possessions which exhibit are Malta and the Ionian Islands. The former shows stone, lace, silver work, and other products of industry, with some agricultural specimens. The Ionian Islands have a fine collection, not only of agricultural but of manufacturing industry, sent by about one hundred and seventy exhibitors. The embroidery and silver filigree work, the silks, and other articles, are very elegant.

The North American colonies generally have sent a large collection, and all are well represented. Canada has been rather tardy; but the Lower Provinces have taken great pains to send such collections as may give a fair idea of their chief products. Timber, minerals, the products of the fisheries, agricultural and other implements, hardware, and homespuns form their main products.

The West Indian possessions exhibit, if not much variety, still many interesting articles. Their staples of coffee, sugar, rum, arrowroot, and cocoa possess, it is true, little novelty; but their woods and fibres are valuable, and their gums, oils, drugs, and other raw materials, will be examined with interest by many manufacturers.

We next come to India proper, which has ten thousand superficial feet of space allotted to it, a deduction from the British half of the building. The management of this large and important part of the Exhibition has been given to Dr. J. Forbes Watson, the reporter on Indian products, who has devoted so much attention to a due development of the staples of India, and has lately been so closely occupied in re-arranging the valuable East India Museum at Fife House, Whitehall Yard.

The collection is more varied and extensive than the former one in 1851, when India had twenty-four thousand feet assigned to it. The articles for which space cannot be given have been removed to the India Museum, Fife House, Whitehall Yard, there to constitute a supplementary collection.

The India Museum itself contributes from its varied resources a collection of very considerable interest; and several firms in this country, in their capacity of importers, as well as a few private individuals, exhibit valuable specimens.

Some examples showing the aptitude of cloth made of Indian cotton for taking dyes are worthy of special attention.

Glancing at the productions of special importance consigned direct from India, we have first those from Bengal.

The collection from Bengal, the North Western Provinces, Oude, the Punjab, Burmah, the States of Ulwar, and the Tenasserim and Martaban Provinces, including the Straits Settlements, is of great interest It amounts to nearly six thousand specimens, equal to more than double those sent from the same parts in 1851.

First stand the collections of the products of the soil, and among these particularly that of oil seeds, oils, gums, resins, medicinal substances, fibres, and timbers. Under this head tea occupies a prominent position, the tea districts of Assam, Cachar, Sythet, Darjeeling, Dehra Dhoon, Gurhwal, Kumaon, and Kangra being all represented. This division of the Indian collection, whilst containing a number of specimens entirely new or never before exhibited, comprises a large variety of products indigenous to different localities.

Among the articles contributed are several which show the great improvement which has undoubtedly taken place in a variety of manufactures, and even of works of art.

Amongst the latter, the paintings on ivory from Delhi show a great advance in that branch of native talent.

Among the specimens forwarded on the present occasion, there are two—one a landscape, the other the interior of a temple—which will attract attention.

Several of the paintings are faithful copies of photographs taken at Lucknow.

The Calcutta "Chickun" work, or needle embroidery, is deserving of special notice. The assortment comprises specimens of extraordinary cheapness, as well as of the highest finish.

The floss silk embroideries on Cashmere cloth and net from Delhi are particularly fine. Though greatly superior to what has been sent on former occasions, they have the further advantage of greatly improved patterns, with extraordinary cheapness.

Carpets show a considerable advance in workmanship, and the blending of colours is remarkably fine.

Of Cashmere shawls the collection is very large. The greater portion consists of private contributions, and comprises varieties of every description. Some sent from Sirinuggur, the capital of Cashmere, and from Umritsur, are considered very superior.

The "kuftgori" work, or steel inlaid with gold, from Goojerat, in the Punjab, will attract the attention of those interested in this kind of manufacture.

The collection contains a variety of articles under the name of papier-mâché, those made at Cashmere being deserving of notice. The specimens sent show considerable improvement in comparison with those contributed in 1851 and 1855.

The assortment of silk cloths from various parts of India is very large. Some from the Punjab and other parts are excellent, and of brilliant dye.

The embroidered silks and brocades from Benares stand unrivalled. The workmanship is of the highest finish, whilst the interweaving of the gold and silver threads with the silk shows exquisite taste. The kinkobs or brocades are very rich.

The greater portion of the collection under the head of manufactures in cotton and hemp consists of specimens of cloth, &c., made by prisoners in the various jails; those from the Punjab and from Meerut, as well as the hemp fabrics from Barnagore, near Calcutta, are worthy of special notice.

The specimens of paper forwarded with the present collection form a large variety. There is the famous paper made of the daphne plant at Nepal, illustrated by the fibre of the fresh shrub and specimens of all the stages of its preparation.

The manufactures in straw from Monghyr are good, and their merit is enhanced by their cheapness.

The lac-work sent from Bareilly shows a great improvement in comparison with the specimens sent to the Paris Exhibition.

The turnery from Lahore, called "pack puttan" work, commends itself on this occasion by its superior workmanship.

In addition to a collection of clay figures from Kishnagur, are some excellent specimens of plastic models from Oude.

Of sculpture there is a large variety of specimens in stone, ivory, marble, and various kinds of woods, all of them much superior to any hitherto forwarded, the ivory carvings from Berhampore being especially remarkable.

From Cuttack the silver filigree work upholds the reputation of native workers in the precious metals for minuteness and accuracy of design and execution.

The muslins, plain and embroidered, of Dacca manufacture will as usual command admiration from their peculiar delicacy of texture.

The collection from Bombay is interesting and important.

Cotton, its preparation and manufacture, is well represented. His Highness the Rao of Kutch contributes samples of raw and manufactured cotton from Bhooj, Kutch, and Mandanee, and the cloth as made largely at the last-named place for exportation to Muscat and the Arabian coast. There are also specimens from Seebee, Jacobabad, and Shikarpoor, of the raw material; and from Khyrpoor His Highness Meer Ali Morad sends cotton cloth and chintz as applied to clothing purposes. The districts of Broach, Khandeish, Rewa Kanta, Belgaum, Dharwar, Guzerat, Kattiawar, South Mahratta, Berar, Pahlunpoor, Thurad, Wurryee, and Ahmedabad, in the manufacture of cotton, are well represented; the towellings, canvas, and duck from Broach being also varied and remarkable.

Rich samples of raw, manufactured, and dyed silks have been forwarded from Khyrpoor, Musher, Shikarpoor, Kutch, Zeyd, Bokhara, Cabul, Kishur, Candahar, Mandanee, Dharwar, Ahmedabad, and Belgaum, while most of these places furnish specimens of mixed silk and cotton fabrics.

In wool and woollens, Bombay sends examples of the produce and manufacture of Cashmere, Cabul, Herat, Musher, Khorassan, Kelat, Kutch, Dharwar, and also Shikapoor and Upper Scinde, whence come specimens of fibre, and goats' and camels' hair.

The plain and embroidered muslins from Bombay give a fair idea of the quality of manufacture under this head at Khyrpoor, Bhooj, Kutch, Surat, and other places, while the embroideries generally, on velvet, silk, leather, and cloth, are of a high degree of excellence, both in design and execution.

In works of art or skill, Bombay is more than adequately represented, as the elaborate carvings in black-wood, sandal-wood, and ivory, and in a material termed "ratanglee," from Malabar, will fully testify. The articles of inlaid ivory, &c., are also of a high degree of excellence, while the same description of work and enamelling in the precious metals is well represented by arms and armour contributed by their Highnesses Meer Ali Morad of Khyrpoor and the Rao of Kutch. The last named also presents costly specimens of skill in the gold and silversmiths' art. The shield exhibited was manufactured at Mandanee.

The models are mostly copies of agricultural implements, showing the local mode of preparing the land for the culture of cotton, grain, pulses, &c.

In other manufactures and arts there are samples of cutlery, jewelry, lac-ware, &c., from Bombay and Scinde, and in mineral products and works in stone, &c.; alum, saltpetre, subcarbonate of soda, from Scinde; indigo from Jacobabad; sulphur, lead, copper ores, and "galena," from Beloochistan; agates and car-nelians from the Ruttimpoor mines in Rewa Kanta; and a substitute for marble from Bhooj.

The collection from Madras, if not so large as those from the other divisions of India, yet consists of a great variety of specimens. A valuable portion of the whole consignment from this presidency is that which, under the head of raw pro-ducts, includes timbers, oils, seeds, gums, resins, dyes, &c.

There is a valuable assortment of woods from Vizagapatam, South Canara, and Bangalore, of oils from South Canara, of dyes, gums, saltpetre, sugar, and candies from Salem and North Arcot; mineral products and shells from Vizaga-patam and the Andaman and Cocos Islands. Of grain and seeds, Madras sends a good assortment, as also does South Canara, whence there are fair samples of condi-ments, spices, medicinal substances and distillations, fruits, &c.

Of fibres and rope, Burmah, Chinglepet, Vizagapatam, Bangalore, and Sálem each send examples, the last named and Cuddapah furnishing indigo and coffee.

Passing to the important article of cotton, we find Oopum and Bourbon cotton from Salem in small quantities, and raw and manufactured specimens from Arnee, Chinglepet, North Arcot, Bangalore, and other localities.

Of woollens there seems to be but a limited supply, and of silks, raw and manufactured, the best samples are those from Salem and Madras (town); a rug of this material from Tanjore will attract attention.

The muslins are chiefly from Madras, Arnee, and Salem, and include a specimen woven from a peculiar thread at Máderpak, North Arcot. This collection, though small, is particularly good.

In shawls, embroideries, and silks, those forwarded from Madras are like the specimens from other parts of India, unrivalled in variety and beauty of design. The lace work from Tinnevelly and Vizagapatam is also worthy of high com-mendation.

Glancing at the few rich feather ornaments from Kurnool and Vizagapatam, the large straw mattings from Pulghaut and North Arcot, the writing-paper from Madras and Salem, and the large assortment of carvings and constructions in ivory, sandalwood, buffalo-horn, and porcupine-quills deserve notice. Specimens of this work are sent from all parts of the presidency,—and through Madras from the Malay peninsula,—from the town of Madras, North Canara, Vizagapatam, Malacca, Burmah, &c., while Bangunpully, Hyderabad, and Burmah contribute a limited number of specimens of lac-ware.

The models of useful appliances include a portable kitchen and compendious sandwich-box from the Andaman Islands; while as works of art and skill, an exquisite temple from Burmah, a curious model in pith from Tanjore, and some copper figures from North Arcot are worthy of notice.

Of arms, cutlery, &c., the collection for exhibition is comparatively small, Salem, Malabar, and Burmah supplying the best specimens.

The fine arts in Madras are represented by a collection of engravings, photographs, and drawings. Some miniatures on ivory especially illustrate the delicacy of finish, minuteness of detail, and brilliancy of colour which have ever been remarkable in native drawings, while in these, as in some of the Bengal drawings, a marked improvement will be noted with regard to shade and perspective.

Among the varieties from Madras may be especially mentioned the pottery from the School of Industrial Arts in Madras, as well as from Bangalore and North Arcot; these, with most of the examples of native manufacturing and artistic skill, affording abundant evidence of the existing will and ability to adapt the wealthy resources and produce of India to the requirements of an advanced stage of civilization.

The samples of raw produce received for exhibition from Mysore include cotton and silk from Bangalore and Chittledroog, grain, oil, pulses, and substances used for food from Bangalore, Astagram, Chittledroog, and Nugger, honey from Coorg, and sugar from Paulhully.

Mineral substances have been sent from Chittledroog, Astagram, and Bangalore.

In manufactures, the most noticeable collections are in cotton, silk, and wool (plain and embroidered), from Bangalore, Astagram, and Chittledroog. There are also carpets from Bangalore, and models and carvings in sandal-wood, of which with some inlaid work, there are several good specimens.

From Burmah, in addition to the many specimens forwarded through the Bengal Committee at Calcutta, a large collection has been contributed by Messrs. Halliday, Fox, and Co. This extensive assortment comprises a large series of grains, woods, medicinal roots, seeds, spices, pulses, and cotton, and specimens of metal, ivory, precious stones and gold dust (from Ava).

The samples of Burmese manufactures, &c., include musical instruments, jewelry, models of boats, implements, &c., and some curious Burmese pictures.

An extensive collection of woods has been forwarded from Rangoon for the purpose of illustrating the timber growth of the province of Pegu.

Of the raw produce and manufactures of Singapore, Penang, Malacca, &c., a well-selected variety has been forwarded for exhibition, especial care having been taken to fairly illustrate the natural resources of the Malay Peninsula, and the adjacent islands.

The mineral classes are represented by coal from Borneo and Malacca, iron from Saigoor (Cochin China), earths and clays from Singapore, Malacca, and Penang. The tin ore of Kassang, Malacca, Penang, &c., is stated to yield from forty-five to sixty per cent. of metal after a rude process of washing and smelting.

Vegetable substances are fully represented by a choice collection of woods from Malacca, Penang, &c., each contributing a variety of specimens.

There are also samples of fibres from Singapore, Malacca, Manilla, Penang, Siam, the Eastern Archipelago, &c.; rope from Penang, tanning substances from Singapore and Rhio, dyes from Cambodia, Penang, and Siam, vegetable tallow from Cochin China, Siac, and Borneo, gutta percha from the Malay Peninsula and the Eastern Archipelago, and tobacco from Bally and Penang.

Of grain specimens are forwarded from Singapore, Malacca, and Province Wellesley, Penang, whence comes an excellent assortment of rice.

Siam, Java, Sambawa, and Malacca furnish seeds and pulses. Singapore, Malacca, Sarawak, and Penang send sago, arrowroot, and spices; and from Singapore and Penang fair samples of sugar and coffee.

Another important feature in this consignment is the collection of medicinal substances and processes from Singapore and Malacca; while the oils and gums from these places, and from Siam, Sumatra, Java, Cochin China, and Malacca are unusually interesting.

Of cotton, the principal specimens are from Singapore, Malacca, and Penang, while the manufactures consist principally of arms from Tringanu and Malacca; embroideries and silks from Singapore and Penang, and lace from Malacca.

The works in metals, &c. (including implements), come from Malacca, Penang, and Singapore, while the lac-ware of Tringanu and Singapore, and the work-boxes made of the pandanus palm will be found curious of their kind.

There is also a collection of photographic portraits of natives. The importance and value of this consignment rest on the very interesting series of samples forwarded, and which well illustrate the productive character of the soil from which they have sprung.

This completes the industrial portion of Great Britain and her dependencies, and we now just touch upon the fourth British section of the exhibition.

The division of modern fine arts—comprising paintings, drawings, sculpture, and engravings—is fully described in its special catalogues, but we cannot pass it over without a few words.

The main portion of the building set apart for the reception of works of art stretches to the right and left of the principal entrance, and occupies the upper part of the whole frontage in Cromwell Road. Branching off from each extremity smaller galleries, on the same level, run at right angles to the principal galleries, along the Exhibition Road and Prince Albert's Road; and, as far as the domes, the walls are entirely covered by paintings in oil and water-colours, and drawings, an important addition to the attraction on this occasion, paintings not being included in the International Exhibition of 1851. These have been grouped together in Class thirty-eight. The space has been divided into two equal parts, and while the south-western angle has been appropriated to foreign artists, the south-eastern portion is devoted to the exhibition of works by masters of the British school.

To convey any adequate notion of the British treasures of art collected here, or of the machinery by which they have been brought together, would be impossible within the limits, and foreign to the purpose of this summary. This will easily be

PICTURE GALLERY, 1862.

understood when it is known that fourteen hundred pictures are here exhibited, and that the greatest care has been taken to prevent the admission of any but works of the highest character.

The period of art represented is that between 1762 and 1862. This includes Hogarth, who died in 1764, and with very few exceptions may be considered to embrace every artist of eminence who has flourished in the United Kingdom.

In the catalogue of a collection gathered from public and private galleries—and liberally lent by many hundreds of art patrons—it is almost unnecessary to say that the names of Sir Joshua Reynolds, Richard Wilson, Bonington, Nasmyth, Morland, Etty, Hilton, Turner, and Leslie occur repeatedly, or that a due prominence has been given to such artists as Crome, Callcott, Raeburn, and Allan. Gainsborough and Sir Thomas Lawrence occupy, of course, distinguished places; and the list would be incomplete indeed that did not include many specimens of Collins, Constable, the Chalons, and Sir David Wilkie. Nor must notice be omitted of our truly national school of water-colour painting; founded by Cosens Girtin, and above all by Turner, whose works illustrate both the commencement and the full perfection of the art; those who laboured with them to extend its scope have not been overlooked, and Edridge, Robson, Davint, Fielding, Prout, and Varley, are well represented; while the works of Stothard, Barrett, and Cox lead up to the living representatives of this beautiful art. These specimens have been selected by Mr. Redgrave, R.A., Inspector-General for Art at the South Kensington Museum, and the collection must speak for itself of the care and judgment here displayed.

With the works of living artists—occupying as nearly as possible one-half of the hanging space in the British division—a different course has been followed. While the same supervision has been exercised to prevent the admission of inferior pictures, it was considered only just that the artists should themselves have a voice in deciding what—in the limited number for which room could be found—should be the particular specimens of their art by which each should be represented. A committee was accordingly formed in the early part of 1861, consisting of the presidents of the several art societies of the United Kingdom, under whose direction a list was drawn up of those artists who should be invited to contribute. To each of these an invitation was issued, and the order in which their works were entered upon the lists they sent in, has been observed as nearly as possible in the applications made for the loan of them to the several proprietors. If an equally satisfactory representation has not in all cases been the result, this has arisen from causes quite independent of the machinery adopted to produce it. The hanging has been the joint work of Mr. Redgrave (assisted by his brother, Mr. J. Redgrave), and Mr. Creswick, R.A.

The collection of sculpture, medals, and intaglios,—Class thirty-nine,—under the management of Mr. Edmund Oldfield, will be an important addition to the fine art section. For the first time perhaps in the history of such exhibitions, sculpture holds a prominent and recognized position side by side with painting. For many years at the Royal Academy it was thrust into an ill-lighted room, the most appropriate inscription for which would have been that well-known line from Horace,—"The perfection of art is to conceal art." Here it has been carefully arranged about the building, to add to the general effect of the display, with the exception of one room devoted to the productions of deceased sculptors. This room contains fine examples of Banks, Flaxman, Chantrey, Westmacott, and Wyatt. All the modern artists have come forward to exhibit, and the result is that we have a fine collection numbering about one hundred and twenty statues of different sizes, forty or fifty busts, and twenty or thirty bas-reliefs. Mr. Gibson's celebrated tinted Venus, which has been for some years past the admiration of foreign connoisseurs, is lent by its owner to the Commissioners of the International Exhibition. The judges of art will thus have an opportunity of deciding for themselves whether colour enhances the beauty of sculpture. It is erroneously supposed that Mr. Gibson has tinted his statues to represent life, whereas he has only endeavoured by colour to soften the general effect, and to give the appearance of ivory, a material much used by the ancients. Apart from the colour, this statue is undoubtedly the finest work of modern sculpture. Mr. Gibson has represented his Venus as the Goddess of Marriage, a dignified and beautiful matron, with a tortoise at her feet. This statue was executed ten years ago for its present owner, and by his permission it remained in Mr. Gibson's studio at Rome until 1859. After that time it was removed to its owner's house, and it has never before been publicly exhibited. Nearly thirty of the principal pieces of statuary are arranged in the picture galleries and towers.

A small collection of intaglios, cameos, and medals—the latter containing many fine works of the Wyons, father and son—completes the general features of this important department.

The collection of British etchings and engravings—Class forty—is divided into two sections, the first containing the works of deceased, and the second those of living engravers. The former is divided into—Etchings, Line engravings, Mezzotints, Stipple, and Wood. The arrangement of each subdivision is a chronological one, not calculated according to the dates of the births or deaths of the artists, but as nearly as possible on the times when their principal works were published. The number of these represented is eighty-four; their engravings are about four hundred and fifty, and they occupy three hundred and twenty-seven frames. They are, with few exceptions, early proof impressions of the very highest quality.

The etchings do not call for any especial remark; but among the line engravings will be found the most celebrated works of Hogarth (those engraved entirely by himself), Browne, Woollett, Sir R. Strange, W. Sharp, the Heaths, Raimbach, the Cookes, and others of nearly equal importance.

The mezzotints are peculiarly interesting, and the wonderful productions of MacArdell, Pether, Dixon, Earlom, J. R. Smith, V. Green, and others, are sufficient to prove the unrivalled excellence of the British school in this branch of art.

The sections of stipple and wood engravings include many of the best specimens of Bartolozzi, Haward, Collyer, Scriven, Bewick, Clennell, Williams, &c.

The department of living engravers is similarly subdivided, with an additional section of lithographs, but the arrangement of each subdivision is an alphabetical one. It contains about three hundred works, by sixty-eight engravers, which fill nearly one hundred and eighty frames. With very few exceptions, the proofs are contributed by the artists themselves. The whole arrangement of this class has been undertaken by Mr. William Smith, assisted in the hanging by Mr. Colnaghi.

We now come to the foreign half of the Exhibition. It is impossible at present, and within the limits of this book, to do more than give an outline of this important section.

The following Table shows the different foreign countries (with their population, imports, and exports) to which half the available space in the present building was originally allotted :—

COUNTRY.		Population.	Value of Imports from in 1860.	Value of Exports to in 1860.
			£	£
Arabia		..	..	..
Belgium	Kingdom.	4,671,187	4,070,866	3,964,670
CENTRAL AMERICA :—				
Mexico	Republic.	8,137,853	490,221	538,949
Costa Rica	"	135,000		
Guatemala	"	850,000	224,909	196,091
Honduras	"	350,000		
Salvador	"	600,000		
China	Empire.	415,000,000	9,323,764	2,915,542
Denmark	Kingdom.	2,752,500	2,642.877	1,594,050
Egypt	Viceroyalty.	5,125,000	10,352,574	2,598,912
Feejee Islands		.	..	..
France and Algeria	Empire.	39,500,000	17,895,210	15,759,258
GERMANY :—				
Austria	Empire.	35,000,000	986,349	1,488,098
Northern Germany (The Two Mecklenburgs, Hanse Towns)		1,100,000	7,524,016	13,850,705
Zollverein (Baden, Bavaria, Brunswick, Frankfort-on-the-Maine, Hanover, Hesse Cassel, Hesse Darmstadt, Nassau, Oldenburg, Prussia, Saxony, Thuringian Union, Wurtemburg)		33,543,000	7,920,511	4,846,283
Greece	Kingdom.	1,100,000	677,342	374,211
Holland (and Colonies)	"	18,200,000	8,713,952	10,247,151
Italy	"	21,729,000	2,748,525	5,277,720
Japan	Empire.	50,000,000	167,511	..
Liberia	Republic.	..	..	..
Morocco	Empire.	6,000,000	280,424	214,510
Norway	Kingdom.	1,500,000	1,160,992	630,773
Persia	"	10,000,000	..	31,970
Portugal (and Colonies)	"	6,349,000	2,281,844	2,225,495
Rome	Pontificate.	3,125,000	..	..
Russia	Empire.	75,149,000	16,201,498	5,446,879
Sandwich Islands	Kingdom.	70,000	298	35,373
Siam	"	..	75,240	13,556
SOUTH AMERICA :—				
Argentine Confederation	Republics.	1,200,000	1,101,428	1,820,935
Brazil	Empire.	7,700,000	2,269,130	4,571,308
Chile	Republic.	1,559,000	2,582,448	1,737.929
Ecuador	"	1,041,000	107,033	76,271
New Granadan Confederation	Republics.	2,224,000	24,940	854,500
Peru	"	2,500,000	2,581,138	1,428,172
Uruguay	"	301,000	867,328	944,002
Venezuela	"	1,565,000	..	327,357
Spain	Kingdom.	21,307,000	8,026,600	5,078,551
Sweden	"	3,734,240	3,193,308	940,613
Switzerland	Republic.	2,535,000	..	..
Tunis	Empire.	1,000,000	13,954	4,845
Turkey	"	35,600,000	3,253,246	5,206,566
Moldavia and Wallachia	"	4,000,000	2,252,242	201,273
United States	Republic.	31,500,000	44,724,312	22,907.681
Western Africa		..	1,776,565	1,145,434
		857,752,780	166,512,595	120,495,633

Some of these states declined to exhibit, or were so tardy in accepting the space placed at their disposal, that they were unavoidably shut out from the Exhibition. The following list shows the exhibiting countries, with the space which they each fill in the building :—

Countries.	Horizontal space. Superficial Square Feet.			Vertical space. Super. Sq. Feet.				Approximate Number of Exhibitors.
	Ground.	Gallery.	Annexe.	Large Picture Gallery.		Small Picture Gallery.		
				Feet.	Inches.	Feet.	Inches.	
France and Colonies . .	94,419	28,350	24,750	350	0	250	0	4,000
Belgium	21,930	11,267	15,750	215	0	...		863
Austria	36,000	10,408	6,000	66	0	62	0	1,410
The Zollverein	49,500	13,562	20,250	196	0	451	0	2,875
Hanse Towns . . . } The two Mecklenburgs .}	6,225	3,400	...	...		...		254
Russia.	10,800	3,250	...	30	0	40	0	659
Italy	8,906	6,875	2,000	63	6	58	9	2,070
Rome	3,469	...	...	15	0	10	0	46
Switzerland	9,000	4,836	2,000	37	0	40	0	482
Holland	7,200	...	...	100	0	40	0	385
Denmark.	4,500	1,250	300	53	6	37	0	299
Sweden . . . . . } Norway . . . . . }	7,200	1,750	900	35 0 32 0		30 0 36 0		608 213
Spain	4,000	1,875	...	37	0	27	9	1,133
Portugal and Colonies .	3,531	1,250	...	...		...		1,065
Turkey . . . . . } Egypt . . . . . . } Tunis . . . . . . }	6,250	8,050	...	...		...		15 ... ...
GENERAL SPACE.								
Greece	800	1,250	...	...		...		252
Costa Rica	600	...	...	...		...		11
Guatemala	124	...	...	...		...		...
Brazil	1,250	...	...	...		...		230
Ecuador	1,000	...	...	...		...		...
Venezuela	300	...	...	...		...		...
Peru	200	...	...	...		...		38
Uruguay	224	...	...	...		...		34
GENERAL SPACE TAKEN FROM BRITISH HALF OF BUILDING.								
United States . . .	5,250	...	...	...		...		60
China and Japan . . .	1,350	...	...	...		...		35
India	10,000	...	...	...		...		251
Liberia. . . . . . } West and Central Africa.} Ionian Isles . . . . } Madagascar . . . . } Hayti . . . . . . } Siam and Fejee Isles . }	1,750	...	...	...		...		... 7 177 10 1 3
						Total ...		17,486

The French court and its approaches form a perfect exhibition by themselves, undisturbed by any jarring tints or effects produced by unsympathetic neighbours. Everything that human ingenuity, money, and taste can produce is exhibited in this court, and it will be from no want of energy if they fail to carry off the first prizes. Our great industrial rivals are determined to give us a royal entrance to their kingdom within a kingdom. The portico to their court, which they have erected in the nave, is of cast iron, and manufactured by M. Barbezat. It forms a corridor seventeen feet in breadth, twenty-five feet in height, and extending east and west one hundred and fifty feet, the entire length of the French Court. The portico in the centre has three divisions between four leading columns. The east and west divisions, ten feet in width, are entrances to the exhibition. The centre division, twelve feet wide, is backed by a noble sheet of spotless plate glass perfectly transparent, manufactured at St. Gobain. The dimensions of this glass are ten feet and a half broad by sixteen feet and a half high. At the foot of the glass is a richly ornamented couch or divan, for public use. The sides are protected by lions of cast iron, the size of life, placed on pedestals of granite. Its centre division is surmounted by the imperial arms, and is decorated with flags. The entrances on either side have chandeliers in bronze and gilt, inlaid with crystals. The diameter of these chandeliers is six feet. The four main columns of the portico entrance are hung with carpets of various exhibitors from Aubusson and Paris.

Along the entire length of the corridors, on either side, is space for exhibitors' goods, divided off by a light fencing of cast iron. Here are exhibited carpets of various kinds, and furniture in ebony and other woods—some of particularly light and fantastic design. The corridors are in some parts enriched with paintings, and at intervals are monumental mantelpieces in carved wood, bronze, and marble, which occupy the central space of the corridors, like trophies in the nave.

The French Exhibition generally will show a great advance upon 1851, and a considerable advance upon 1855. It is probable that the Treaty of Commerce will not have been without influence on this display. Many manufacturers who make no sign are those who, right or wrong, believe the treaty to be against their interests, while those to whom the treaty is undoubtedly beneficial have made the greatest exertions to be worthily represented.

About four thousand exhibitors have come forward from the empire and its colonies (the colonies claiming nearly a thousand), of whom nearly one half show under the head of raw produce.

The machinery section is well filled by at least seven hundred exhibitors, and manufactures are in the hands of more than fourteen hundred exhibitors. The whole display is remarkable for the number of collective exhibitions, especially in the wine and food class, more than forty of these being sent from various departments of the country. Amongst the special exhibitions are the forest products of the Landes; a collection of essays on the food of birds; a collection of French mammalia and birds both useful and mischievous; an exhibition of oak wood for ships; a geological and botanical collection; the results of the labours of the Society of Acclimatization; with a number of paintings of animals and plants for forage.

The agricultural implements of France are shown in a collective exhibition, and also the wines of Champagne and Burgundy, the wines and brandies of Montauban, and cereals and farinaceous grain. A collective display of ornamental and general leather-work is very interesting; the textile fabrics are rich and numerous; art-furniture, china, and glass fill prominent places; and the Gobelin and Beauvais tapestries have not been forgotten.

The French machinery, which is distinguished for its mechanical finish, includes all known types of machines. It has much beauty of form, and shows a judicious distribution of parts in relation to the strain to be supported. As regards vessels, the great marine engines used in the Mediterranean ships are also remarkable for the above qualities, and they include a direct application of the screw. The railway department presents some remarkable models of locomotives—the inventions of engineers whose object has been to utilize the adherence of the wheels on the curves, to increase the power of the steam, and to introduce largely the employment of coal as a combustible. The travelling carriages are more noticeable for their beauty than for the increased comfort which they promise to travellers.

Several improvements in boilers are exhibited, showing how these machines may be made so as to be easily taken to pieces and cleaned. The means are also shown of purifying water from matters likely to form incrustations. The Giffard feeder, as a piece of ingenious mechanism, is, perhaps, the most striking invention of all.

The locomotives are numerous, and present many details worthy of the study of engineers.

The hydraulic machines show great progress since 1851, and also the ice-producing apparatus.

In mechanics and metallurgy there are some new improvements both in form and in construction; and some important machines are exhibited for making sugar from cane or beet-root on a large or small scale.

Belgium is well represented by a large collection spread through all the thirty-six industrial classes. It is strong in iron and textile fabrics; and while the Ministry of Public Works of Brussels has shown the mineral products of the country, about seventy-six apprentice schools have sent examples of cotton and linen manufactures. The Belgian Government also exhibits silk and velvet goods, and woollen and mixed fabrics, produced in the Flemish apprentice schools; and the show of Brussels lace is of course particularly attractive. Manufacturing machines and tools form a considerable part of this collection, and also machinery in general. The section of manufactures is particularly well filled by more than five hundred exhibitors. The food products form a large class, also the animal and vegetable substances; but the classes that are the best filled are those representing flax and hemp, and woollen and worsted manufactures.

The Austrian collection is furnished by about fourteen hundred exhibitors, who show in all the thirty-six industrial classes, and the strength of their display lies chiefly in the multiplicity of raw products exhibited.

The various kinds of coal found throughout the Austrian empire are represented by several complete collections, and also by some geological maps belonging to the Imperial Geological Society of Vienna.

Both table and rock salts (of which Austria produces four hundred thousand tons annually), as well as mercury, sulphur, and saltpetre, are exhibited from the Imperial mines.

Much interest will be excited by the collection of vegetable products; grain —of which twelve million six hundred thousand quarters are produced annually in the provinces—forming an important part.

Beet-root sugar is also another interesting feature of this section, as the hundred and ten millions of pounds that are manufactured yearly entirely supply the national demand for that commodity to the total exclusion of the colonial produced sugar.

The display of wines is another important feature, as the annual produce of Austria reaches four hundred and twelve millions of imperial gallons, and nearly all the provinces of the empire contribute to this department.

Specimens of flax, raw silk, and woods of all kinds are shown amongst the raw products, Austria producing annually of the latter two thousand four hundred and ten millions of cubic feet. Wool is another important national product, which amounts to seventy-seven millions of pounds-weight yearly.

Amongst the productions in Classes three and four Hungary holds the highest position.

The great distance has unfortunately prevented the transport of much machinery; nevertheless some of the Austrian exhibitors show to advantage in this branch. There is a locomotive exhibited which is made to be used only on mountains, having four cylinders, with a rate of speed equal to ninety English miles an hour. This is shown by the Staats Eisenbahn Geschellschaft.

The musical instruments are very fine, particularly those from Vienna, which are distinguished by their variety, cheapness, and good workmanship.

The woollen stuffs and cloths from Brünn and Reichenberg, which have already earned for themselves a good name at the Paris Exhibition of 1855, will now show what further progress has been made since that period.

In the iron industrial department Upper Austria stands prominent, knives, tools, and general cutlery being shown in a "collective exhibition."

Vienna sends a rich collection of her manufactures—works in leather, fashionable stationery, and turnery, most of which are already favourites in the London markets.

The Austrian exhibitors of paper will excite some attention, on account of the cheapness of their products; and the glass manufacturers of Bohemia, who have so long held such a high position in the commercial world, show the very decided progress which they have made in their particular industry.

The display in Class twenty-nine, made by the Austrian Government, faithfully represents the present state of public education in that country.

The result of the tour of the 'Novara' round the world is also exhibited; the beautiful anatomical preparations of Professor Hyrte; and the representation of human growth made by Doctor Litrarzick.

The "Zollverein," or German Free-Trade Union, section includes a considerable collection of mining products from Prussia and other states, arranged in a systematic order. Amongst them is a pillar of rock salt from Stassfurth, near Magdeburg, remarkable for its purity and small degree of hydroscopical quality.

There is also a series of chemical substances and products, and amongst them the wines and wools are worthy of special notice.

The machinery and iron and steel manufactures include some remarkable castings from Krupp's factory in Westphalia, a cannon of cast steel from Westphalia, and a locomotive engine from one of the best workshops in Berlin.

Pantographs are also exhibited, which are used for executing "guillochée" works on printing rollers and copper plates; needle-pointing machines from Tserlohn, and philosophical instruments and apparatus from Berlin and Cassel (Electorate of Hesse). There is a good collection of arms and cutlery from Rhenish Prussia, and amongst them the sword presented to His Majesty the King of Prussia by Mr. Lueneschloss, in Solingen. A collection of models of ploughs of every century, from a professor of the Academy at Hohenheim (Würtemberg), are very interesting.

Amongst the musical instruments is an orchestrion belonging to the Grand Duke of Baden, a number of pianofortes from Leipsic and Berlin, and a large number of harmoniums from Stuttgart.

Amongst the woven articles, the silk goods from Crefeld and Bielefield take high rank, and form a most attractive show. Woollen and worsted, linen and cotton goods from the different States of the Zollverein, especially Prussia, Saxony, and Würtemberg, and likewise hosiery from the same countries, form a leading feature of this Exhibition.

The display of varnished leather work, principally from Mayence and Worms (Grand Duchy of Hesse), and Rhenish Prussia, is very important; and likewise the show of fancy leather goods from Offenbach (Grand Duchy of Hesse), Frankfort-on-the-Maine, and Berlin.

The exhibition of works in precious metals is very interesting. There is a splendid collection of chased and embossed silver works from Berlin, and amongst them the wedding present given to His Royal Highness the Crown Prince of Prussia and the Princess Royal by the city of Berlin.

Jewelry and ornaments in gold from Hanau (Electorate of Hesse), Pforzheim (Grand Duchy of Baden), and Frankfort-on-the-Maine are exhibited; the works in amber from the Prussian coast of the Baltic are very remarkable.

In porcelain, the royal manufactories at Berlin and Meissen (Saxony) send a rich collection of specimens.

The iron foundries at Berlin and Hanover send some large ornamental works in bronze, and two colossal couching lions come from the latter town to add to the effect of the German courts.

Amongst the industrial branches connected with science and art specimens will be exhibited from the printing-offices of Berlin, Brunswick, Gotha, and Leipsic; with globes and maps from different States of the Zollverein.

A good collection of photographs is exhibited, chiefly from Munich; and the portrait, life-size, is shown of His Royal Highness the Prince of Wales, taken on his journey to the Holy Land. Bavaria also sends a collection of very excellent drawing pencils.

Colours for printing and lithography are sent from Frankfort-on-the-Maine and Hanover; toys principally from the duchies of Saxony and the city of Nürnberg; clocks from the Grand Duchy of Baden, and straw plaiting from the Black Forest.

The models of the new exchange building at Berlin, the Derschauer railway bridge, and the sloping plain of the Oberlaendische Canal, in East Prussia, will be peculiarly interesting to architects.

The Zollverein exhibitors number about two thousand nine hundred, spread through the whole thirty-six industrial classes. The kingdom of Prussia claims nearly fourteen hundred of these exhibitors, and is particularly strong in the mining, chemical products, food, woollen manufactures, and hardware classes.

The first of the Hanse Towns—Hamburgh—exhibits soap, fancy models, philosophical instruments, furniture made of carved wood, basket-work and walnut-tree furniture, fancy furniture made of harts' horn, for sportsmen, and various folding beds, for railway travelling.

Bremen exhibits a friction balance, or frictiometer, for ascertaining the laws of friction and testing lubricating substances; some chased and embossed silver work, tea and coffee services in the genuine old German style, and silver table services.

Lübeck exhibits specimens of marchpane, a peculiar bread, and preserved food used much amongst sailors; the two Mecklenburgs exhibit a collection of raw products, of which wool is the staple; and Mecklenburg-Schwerin displays some handsome inlaid work, and folding doors from the castle of the Grand Duke.

Russia is well represented in furs and cereals, and in a collection of minerals sent by one of the public administrations of the country. Amongst other curiosities it has sent a valuable seal-skin carpet. Its largest division of exhibitors show in the wine and food class, and in the class for animal and vegetable substances used in manufactures. Under machinery it has a good display of carriages, and its strong numerical point in manufactures will be found in the skins, furs, feathers, leather, and articles of clothing. It also sends a valuable collection of works in mosaic, marbles, "pietra-dura," paintings on china from the Imperial factory, and glass, plain, coloured, and jewelled. Most of the articles are luxurious and ornamental rather than useful, but they are of rare execution, value, and beauty. Two colossal china vases form part of the collection, on which have been copied, in a large size, a picture of Inigo Jones, from the original painting by Vandyke, and a picture of John Locke, from the original painting by Kneller. After the Exhibition, we believe it is the intention of the Emperor to present these vases to the Royal Society, and also to present a collection of precious marbles to the Geological Society.

Its exhibitors are spread over about thirty-four classes, and they number upwards of six hundred.

Italy is represented in a large and valuable collection, spread through all the industrial classes, with the exception of the one devoted to naval architecture. Its exhibitors number at least two thousand, and the display is particularly strong in mining and metallurgy, food and wine, vegetable products, and silk and velvet manufactures. The collection of pictures in embroidery is very rich and interesting; and the promise given at the Florence exhibition is more than redeemed.

The city of Rome shows through about forty-six exhibitors in eighteen industrial classes. The chief part of its display consists of natural and artificial stones for building decoration, inlaid stone tables, pavements, a few richly bound books, and some lace and tapestry made by the inmates of the state prisons. The photographs and specimens of silk manufactures are valuable and interesting.

Switzerland has nearly five hundred exhibitors, who make a display in about thirty-three classes. The food class is particularly well represented; and of course Class fifteen—horological instruments—is well filled, and by about seventy exhibitors. The show of silk and velvet goods by this country is rich and extensive.

Holland—or the Netherlands, as it is most frequently called—makes a good show, spread over about thirty-three classes. Its greatest division of exhibitors is in the food class; and it shows well in animal and vegetable substances used for manufactures, in paper and printing, in furniture, and in iron and general hardware. A collection of pipes and tobacco forms a noticeable feature in the display from this country.

Denmark is represented in about thirty-two classes by about three hundred exhibitors, who show well in raw materials and manufactures. It has sent a small collection of musical instruments, and another small collection of domestic and ornamental furniture. Amongst the clothing exhibited are some interesting Esquimaux dresses.

The strong point in the contribution from Sweden is the collection of woods and metals. Nearly half the exhibitors from this country—or two hundred and thirty—show under the raw material section. In cotton and woollen fabrics, and general hardware (under the head of manufactures), it makes a fair display; and also in agricultural implements and civil engineering.

Its exhibitors are spread thinly over at least thirty-four out of the thirty-six industrial classes.

Norway has sent a small but interesting collection, spread over about thirty-two classes. Its largest divisions of exhibitors are under the heads of mining and general manufactures. It shows very little in food, very little in machinery, except naval architecture and ship's tackle, but its manufactures, especially its articles of clothing, are fairly represented. Specimens of iron, copper, lead, and silver ores are sent, the latter from the government silver-mine of Kongsberg. In Class three some curious cereals from Finmarken are shown, grown in the latitude of seventy degrees north; and in Class thirty-six are shown objects illustrating the life and industry of the Norwegian and Lapland peasants, and some carvings in wood.

Spain, as might be expected, makes the greatest show under the first section —that of raw materials—having about one hundred and fifty-six exhibitors of mining, quarrying, metallurgy, and mineral products, seventy or eighty exhibitors of animal and vegetable substances used in manufactures, and nearly six hundred exhibitors—or more than one-half of its whole number—who show food and wines. Its display, therefore, consists chiefly of wheat, fruits, oil, preserves, and wine, including wine made from dried grapes, and peculiar products, such as acorn coffee. In manufactures it has twelve exhibitors, who show cotton; in machinery it only numbers about thirty exhibitors, some of the classes, such as that for the machines and tools, having only one exhibitor.

This collection is scattered through about fifteen classes.

Portugal, like Spain, displays its strength in food and wine. In that class its exhibitors are over six hundred, or about three-fifths of its whole number. It makes a good show in the mining classes, and also in the animal and vegetable substances used in manufactures, particularly in woods. It shows wine, oil, cheese, cereals, poncho cloth—a peculiar production like blanketing—cotton fabrics, and silk fabrics, including a satin towel richly embroidered with gold. In Class thirty-six it has a good show of miscellaneous manufactures, such as straw cloaks, and wax flowers and fruits, and it spreads its collection over about twenty-nine classes.

Turkey is represented by its government, and a few private contributors, principally in a large collection which has been gathered by Mr. Hyde Clarke at the request of the Ministry of Commerce, and by the Governor-General of Turkey. The collection includes food, fruits, woods, wool, silver articles and filigree work from Thessaly; textile fabrics, silkworms' eggs, imported in large quantities to France and Italy; an alarm lock, which rings a bell when it is opened; and a copy of the Turkish national jest-book.

Egypt stands alone with a small and valuable collection, Tunis having refused to exhibit under the management of the Turkish Commissioners; and Guatemala, Ecuador, and Venezuela have each sent a late but interesting contribution.

Greece has sent, through about two hundred and fifty exhibitors, a collection n which food and vegetable products stand prominent. The manufactured articles are not numerous, and the display altogether is in about twenty-two classes.

Costa Rica is represented by its government in ten classes. Its chief display is under the head of mining, chemical substances, food, raw materials used in manufactures, skins, and gold, silver, and filigree work.

The Brazilian collection includes oils, a great variety of native woods, vegetable products, the celebrated patchouli scent, a fine collection of minerals from the Imperial Museum, with photographs, oil paintings, silver work, and watchwork, and a valuable display of gold and diamonds.

The Imperial Government has spared no expense to make this collection large and perfect, and it has given instructions to the agents here to get a special report drawn up upon it by some eminent scientific writer.

Peru is represented by about thirty-eight exhibitors, whose display consists chiefly of silver, silver and plated articles, and mercury, wools, including alpaca, cotton, and wine. Amongst antiquities are shown some silver and other articles belonging to the ancient inhabitants of the country, and some blankets or coverlets made from cotton of great antiquity, which were recently found in the ruins of an Indian city.

The Oriental republic of Uruguay shows through about thirty exhibitors, who have sent specimens of wool, roots and skins, timber and wheat, lead, iron, copper and coals, with some wine and food, and a few coloured marbles.

The United States are represented by about sixty exhibitors in about twelve of the industrial classes. The most ingenious pieces of mechanism exhibited, are a machine for making tufted carpets, and one for setting up and distributing type.

The Japanese collection is large and particularly interesting, because it comes from a country about which very little is known. The works of industry and art in which the Japanese most excel are very varied. Many of them will not only bear comparison with the best workmanship of Europe, but in many points they cannot be rivalled. Manchester and Birmingham, London and Paris, will each find in a Japanese collection articles that either cannot be produced in their workshops, or only at a cost that would make them practically unsaleable. Many of these articles, however, with all their delicacy of workmanship and perfection of material, such as the finer kinds of eggshell porcelain, and china ; the inlaid, enamelled, and chiselled metal-work ; the crape silk fabrics, and the lacquered ware, are procurable in Japan, especially by the native purchasers, at very moderate prices. Others again are very costly, and can only be obtained at prices which in Europe would probably be thought far beyond their value. These are chiefly specimens of old lacquer, old bronzes—the finer kinds of ivory carving, swords, and armour, of which latter class the armed retainers of the Daimios, and the feudal chiefs themselves, are extravagant admirers and collectors. When wealthy they will give any price for an approved weapon by a maker of great repute.

As the object of this Japanese collection is to exhibit, as far as limits of space and means will allow, a fair sample of the Japanese, and their capabilities of production in rivalry with the nations of the West, all the articles selected will be found to throw some light on this question of competitive power of production as well as on the progress in civilization of a people who have been nearly wholly unaided by contact with the European race.

The various objects are thus classified :—

SPECIMENS OF LACQUER WARE.—Lacquering on wood; lacquer and inlaid woods mixed; lacquer on other materials, shells, ivory, tortoise-shell, &c.

SPECIMENS OF STRAW-BASKET WORK, and lacquer, and lacquer combined in articles of use and ornament; basket and ratan work.

SPECIMENS OF CHINA AND PORCELAIN of every variety, enamelled, lacquered, and plain; also of pottery, and quaint forms of earthenware.

SPECIMENS OF METALLURGY AND MINERAL PRODUCTS.—Bronzes, simple and inlaid with other metals; medallions and intaglios in pure and mixed metals; brooches, medals, buttons, &c.; cutlery and workmen's tools; arms and armour.

MANUFACTURES OF PAPER.—Raw materials; paper for rooms, for writing, for handkerchiefs, for waterproof coats, &c.; imitation leather.

TEXTILE FABRICS. — Silk crapes, silks, tapestry; printed cottons; fabrics from the bark of a creeper.

WORKS OF ART.—Carvings in ivory, wood, paintings, illustrated works, lithochrome prints, &c.

EDUCATIONAL WORKS AND APPLIANCES.—Books of science, scientific models and instruments (chiefly copied from the Dutch), Japanese shells, toys, &c., and a miscellaneous collection of specimens of lacquer ware, lacquering on wood, inlaid wood and lacquer mixed.

LACQUER ON THE MATERIALS, as ivory, shells, and tortoiseshell, &c.; and inlaid woods.

No exact list of exhibitors can be given here, but the articles number more than six hundred, the bulk of which have been gathered together by Mr. R. Alcock, C.B., Her Majesty's minister at the court of the Tyckoon.

The foreign display of paintings, drawings, engravings, sculpture, and intaglios is numerically as strong as the British exhibition. The space accorded to it, as we have before said, is equal to the British space, and is quite as well filled. Many visitors may be disappointed at not seeing numerous specimens of their favourite old masters; but they must remember that this is chiefly a fine art exhibition of the modern school, gathered from all the leading cities of the world.

France leads off with about two hundred oil paintings—specimens of the great French School by about one hundred and thirty artists; nearly forty water-colour pictures—the works of about twenty artists; fifty groups of sculpture, by about forty-four artists; and about one hundred and thirty engravings, lithographs, and architectural drawings. In oil paintings it sends six specimens of Paul Delaroche, five of Meissonnier, one of Doré in water-colour (the illustrations of Dante); and it may be said generally to be represented by about four hundred and twenty works of art by about two hundred and sixty artists.

Switzerland sends more than a hundred oil pictures by about fifty artists, amongst which are fifteen specimens of Jules Hébert, with a few pieces of sculpture and engravings.

Belgium contributes about one hundred and thirteen oil paintings by fifty-three artists, about twenty-six groups of sculpture by sixteen artists, and two engravings. It has sent nine pictures by L. Gallait and eight by Madou.

The great German school of painting is well represented by the Zollverein, and other German states. Prussia has sent nearly two hundred works of art in architectural designs, oil paintings, sculpture, and engravings, which are the productions of about one hundred and twenty artists. Berlin and Dusseldorf have each sent a fine collection of paintings, and the engravings from the former place are choice and numerous. Amongst the sculpture from Berlin is a case of medals by Carl Fischer. Only one specimen has been sent of Peter Von Cornelius, and two of Oswald Achenbach ; and amongst the oil paintings from Dusseldorf is a series of eleven pictures by Heinrich Mücke, representing the life of St. Meinrad. In sculpture Robert Cauer and Gustav Eichler have sent the greatest number of groups, and most of the works of art from Prussia generally are exhibited by the artists.

Bavaria has sent about forty-four works of art—chiefly oil paintings—by about twenty-seven artists, amongst which are seven pictures by Carl Wilhelm Müller. Saxony is represented by about twenty artists in thirty paintings, drawings, and groups of sculpture ; the Grand Duchy of Baden has sent a small fine art collection ; and the Duchy of Brunswick, Frankfort-on-the-Maine, the Electorate of Hesse, the Duchy of Saxe-Coburg and Gotha, and Würtemburg are represented by about forty works of art—the productions of about thirty artists. North Germany and the Hanse Towns have also sent a small collection ;—Hamburgh having contributed about twenty works by about twelve artists. Amongst these are three pictures by F. Heimerdinger—one being a picture of fairy-life from a tale by L. Tieck, and three pictures by B. Mohrhagen.

The Austrian school—as it is called—is represented by about eighty oil paintings, sixteen water-colour pictures, nineteen pieces of sculpture, five engravings, and about ten architectural sketches. These are the works of about ninety-seven artists.

Holland exhibits about one hundred and twenty oil paintings by fifty-nine artists, and two engravings. Amongst the pictures are seven by D. Bles, and eight by P. Van Schendel.

The (so denominated) Spanish school is represented by about twenty-three artists and thirty works of art in oil-painting, sculpture, and engraving. Velasquez and Murillo are only represented by the engravers.

Russia sends more than a hundred works of art by about sixty artists. Eighty of them are oil paintings, five are groups of sculpture and medals, three are architectural sketches, and seventeen are engravings. Amongst them is a collection of forty-seven medals exhibited by the Academy of Fine Arts of St. Petersburgh, and a monument representing the Empress Catherine II., by Felix Chopin. The earliest picture exhibited is one by Anthony Lodsuiko, who died in 1773 ; and there are seven specimens of Axenfeldt, and five portraits by Levitsky Demetrius.

Italy, as may naturally be expected, is well represented in the fine art classes. About forty architectural designs of various degrees of merit by about thirty artists, and eighty oil paintings and drawings by about sixty artists, form the display of pictures. About seventy groups of statuary and busts—some of them by English artists —have been sent in the sculpture class ; and the engravings reach at least fifty, by about twenty-two engravers. G. G. Battista and Luigi Marchesi are represented by the most numerous specimens amongst the paintings ; and the engravings include twelve plates by Luigi Calamatta of Rome, and some works by Giuseppe Longhi, Raffaelle Morghen, and Leonardo da Vinci.

Rome has sent about fifty-seven pieces of sculpture by about thirty-six artists, including eight specimens of G. M. Benzoni ; a great number of valuable cameos ; a few fine drawings ; a good many engravings ; a small collection of medals ; and a large collection of mosaics by seventeen artists, including a contribution from the Vatican. Its oil paintings number about forty-five by about twenty-four artists, —chiefly modern painters.

Norway is represented by about sixty works of art by about twenty-one artists. Amongst some fifty oil paintings are nine specimens of Boe—pictures of flowers, fruit, birds, and jewelry—six landscapes by Gude, and eleven specimens of Tidemand—figure pictures, in two of which the landscapes are painted by the former artist. Several portraits in ivory are in the small collection of sculpture.

Sweden is represented by about forty works of art in oil painting and sculpture by about twenty-four artists. The rustic scenes of J. F. Häckert and Miss A. Lindegrin are the most numerous amongst the pictures.

Denmark has sent about one hundred and ten works of art by about sixty-four artists, including six groups of sculpture by Thorwaldsen and five by J. A. Jerichau. Amongst the oil paintings E. Jerichau is the artist most largely represented.

For the first time in the history of all exhibitions Turkey fills a place as an exhibitor of pictures. Five paintings have been sent by M. Paul Musurus Bey, and they comprise portraits and sketches of still life. The artist, who is the son of the Russian Commissioner, is only twenty years of age, and therefore these works must be judged leniently.

Greece has sent two oil paintings, five groups of sculpture, and eight engravings—the works of about fifteen artists; and the Ionian Islands are represented by about twelve pictures—the productions of seven artists. The sculpture from Greece consists chiefly of statues by L. and G. Phytalae.

The American fine art display is small, numbering about a dozen pictures and engravings, the chief of which is Mr. Cropsey's "Autumn on the Hudson."

The articles—and groups of articles—exhibited throughout the building in the industrial classes have been estimated at one hundred thousand; the British fine art exhibitors at two thousand, and the foreign fine art exhibitors at about the same number.

The following is the list and plan of the foreign trophies which have been fixed in the building:—

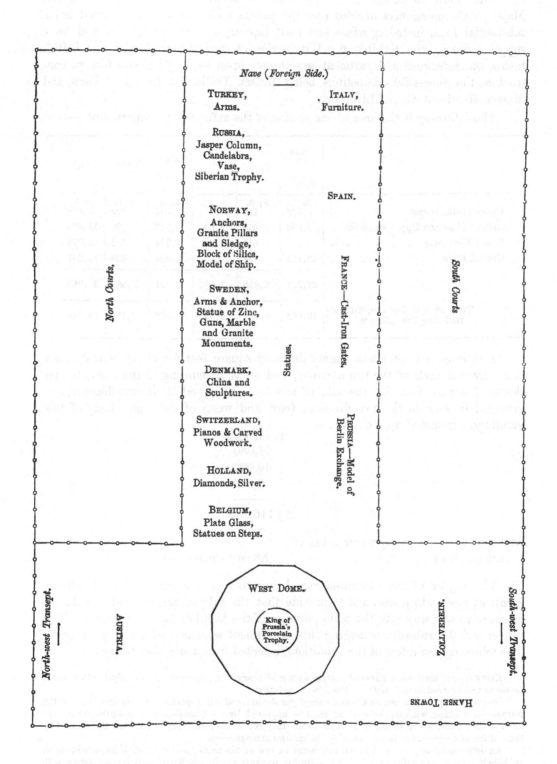

After amusement and instruction comes a demand for refreshment, and Her Majesty's Commissioners decided that the public would be best accommodated if substantial food, including wines and malt liquors, as well as lighter food, were supplied during the Exhibition. With a view of suiting both foreign and British tastes, the refreshment department was thrown open to English and foreign contractors, the successful competitors being Messrs. Veillard and Co., of Paris, and Messrs. Morrish & Co., of Liverpool.

The following is the area of one section of the refreshment department:—

FLOOR.	Dining and Refreshment Rooms.	Kitchens.	Store Rooms.	Vestibules, Shafts, Stairs, &c.	Retiring Rooms.	Total.
	Sq. ft.	Sq. ft.	Sq. ft.	Sq. ft.	Sq. ft.	Sq. ft.
Upper Dining-rooms	5,510	450	..	1,515	195	7,670
Gallery Floor and Upper Mezzanine .	5,896	660	2,695	2,318	330	11,899
Lower Mezzanine	..	858	..	1,784	132	2,774
Ground Floor	10,125	858	..	3,284	430	14,697
	21,531	2,826	2,695	8,901	1,087	37,040
Total of both Sections, without including the cellarage . .	43,062	5,652	5,390	17,802	2,174	74,080

A covered area of about twenty thousand square feet for cheap refreshments is set apart in each of the two annexes; and areas, amounting in the whole to one thousand square feet, for the sale of ices, tea, coffee, and light refreshments, are provided in each of the two divisions (east and west) of the main body of the buildings. Summed up, we have:—

Square Feet.
74,080
40,000
2,000
—————
2) 116,080
—————

And this gives, as the approximation of
each contract . . . . . 58,040 square feet.

The object of the Commissioners being to secure a supply of good refreshments at moderate prices, and to provide that the sale of refreshments should not interfere in any way with the main purposes of the Exhibition, the conditions of the tender and the rules for managing this department were necessarily very stringent. The following are a few of the conditions, selected from more than thirty:—

Every tender must have annexed thereto a scale of charges in respect of the principal refreshments proposed to be served to the public by the person tendering.

No cooking apparatus can be allowed except gas stoves; and all apparatus or arrangements for the purpose of cooking, warming, or washing must be approved by the Commissioners in writing under the hand of their Secretary or General Manager, before they can be used, and may at any time be removed by them if found dangerous or inconvenient in the general arrangements.

An office must be provided by each contractor in one of his areas; and there shall be present there, on behalf of the contractor, at all times while his servants are in the building, a representative, with whom any communications from the Commissioners may be left; and any notice or communication left at the office of the contractor shall be deemed to have been duly served on or made to the contractor himself.

The contractor and his servants shall be subject to all bye-laws and regulations that may be made by the Commissioners for the orderly conduct of the Exhibition, and of the persons employed therein.

Any servant of any contractor who may misconduct him or herself, by overcharge, incivility, or in any other way, shall be immediately discharged by the contractor, on his being required to do so in writing by the Secretary of the Commissioners.

The admissions into the building of servants and other persons on business connected with the refreshment department shall be regulated by the Commissioners; but no provisions or materials for cooking will be allowed to be introduced into the building, except between the hours of six and eight A.M., unless in special cases, and then only with the written permission of their General Manager.

No cooking will be allowed on the premises, except the cooking, by gas stoves, of dishes to be served hot.

The refreshment areas are to be kept open throughout the hours during which the public are admitted to the Exhibition; and the contractor engages to keep at the area let to him, on every day on which the Exhibition is open, a sufficient supply of all refreshments specified in the scale of prices, and to sell the same at the prices therein specified: such refreshments to be of the best quality.

The contractor shall exhibit copies in English, French, German, and Italian, each printed in specified colours, of the scale of prices in such places at or near his refreshment areas as the Commissioners may determine. Additions in writing to the list of refreshments, with the prices annexed, may be made from time to time. The cost of the attendance of waiters shall be included in the price of refreshments. A sufficient number of waiters, speaking the four above-named languages, must be provided, having distinctive marks on their collars, of colours corresponding with those of the scale of prices.

The contractor shall supply fresh filtered water, and, so long as the Commissioners deem it necessary, iced water, in glasses, gratis to all persons who ask for the same, at all times when the Exhibition is open to the public, in such manner as the Commissioners may require.

The Commissioners will not be responsible for any losses or damage which may happen to the property of a contractor from any cause whatever.

A few words about the three leading Catalogues will now close this long Chapter.

The Illustrated Catalogue of the present Exhibition is published by the direct authority and at the risk of the Commissioners, who will doubtless derive a considerable profit from it. It has been produced under the superintendence of Mr. Joseph Cundall. Each exhibitor is allowed two lines of description, and all matter beyond that, or space for woodcuts, is charged at the rate of five pounds per page for each edition of ten thousand copies. It will be issued in parts, and is meant to be something more than a pretty drawing-room book to please a few idle loungers before or after dinner. It is a work got up in a utilitarian spirit for the benefit of the exhibitors, and it will be a lasting and substantial record of the Exhibition.

The Illustrated Catalogue of the former display—brought out in a lump at the close of 1851—ran to four volumes of about six hundred pages each. The cost of the woodcuts was about six thousand six hundred pounds sterling, and of this the contractors, Messrs. Spicer and Clowes, furnished five thousand three hundred pounds, the exhibitors finding the other thirteen hundred pounds. The present catalogue will probably reach three large volumes of seven hundred pages each, and the exhibitors this time will contribute woodcuts costing at least from five to six thousand pounds. All the important trading and manufacturing firms have secured pages, and the volumes will be particularly rich in illustrations of machines. One eminent maker fills thirty pages with woodcuts and descriptions of cotton-spinning machines, showing the various processes, from opening the raw cotton to the final production of the web; two other makers have forty pages; and three leading agricultural houses have taken forty-four pages between them, which they have filled with engravings of farming implements. The illustrations of pottery and works in the precious metals are of a very superior class. They have been produced under the care of Mr. P. H. Delamotte and Mr. Dudley, and it is certain that the part in which they will appear will be by far the most attractive. The purely advertising sheets in each of the parts are let at ten pounds for an inner page, twenty-five pounds for a page nearest the type, or for the last page, and fifty pounds for the back cover. These charges are based on a guaranteed circulation of ten thousand; and when the parts reach that number the advertisement will have to be renewed.

The Shilling, or General Catalogue, has been gradually built up under the superintendence of Mr. Sydney Whiting. The "editing" of this volume has not only involved drudgery unknown even to the sub-editor of a daily newspaper, but has required the faculty for arrangement, and a large store of miscellaneous knowledge. Wonderful specimens of French and German English were constantly dropping in, and the names and addresses of exhibitors, with the briefest possible description of articles to be exhibited (all confined, by official command, to sixteen words) were not always very easy to read. Of the difficulty of deciphering the handwriting of several of these correspondents many instances might be given more or less curious, but the following will suffice as a unique example of how wide may become the divergence from a supposition in reference to handwriting and the fact. The written words supposed to be "glass, china, and crystal," were so read, and placed in Class thirty-four, but were ultimately translated into "glacè, chinè, and crystallisè," referring to silk goods in Class twenty. Occasionally the slightest printed or written error in the formation of a letter might change an article which has absorbed perhaps years of labour into something for which the exhibitor has a profound contempt. The words "electric chain" passed muster through the hands of compiler, printer, and reader, but an accident made it appear that the article was an electric chair. The names of compound firms are troublesome when one partner has a surname like a Christian name, and Howell, James, and Co., if not so well known, might have been put James Howell and Co. Indistinct writing will convert the letter u into two ll's, with the most ludicrous effect on certain names. We are all very sensitive about our names, whether we are celebrated as authors or inventors of a patent roasting-jack. It would, however, be ungracious not to admit that, generally speaking, the required forms have been correctly filled up, and all necessary information given with good-will and readiness; and we only refer to the foregoing facts to show the impossibility of excluding errors from such a compilation. The guaranteed edition of this catalogue is to reach two hundred and fifty thousand copies; its ordinary advertising pages are let at fifty pounds each; the page next to the type has been let for three hundred pounds; and the back page of the cover for five hundred pounds. The Shilling Fine Arts Catalogue—compiled by Mr. C. W. Franks—has the same guaranteed edition, and its advertising pages have been let at about the same prices.

INTERNATIONAL EXHIBITION OF 1862

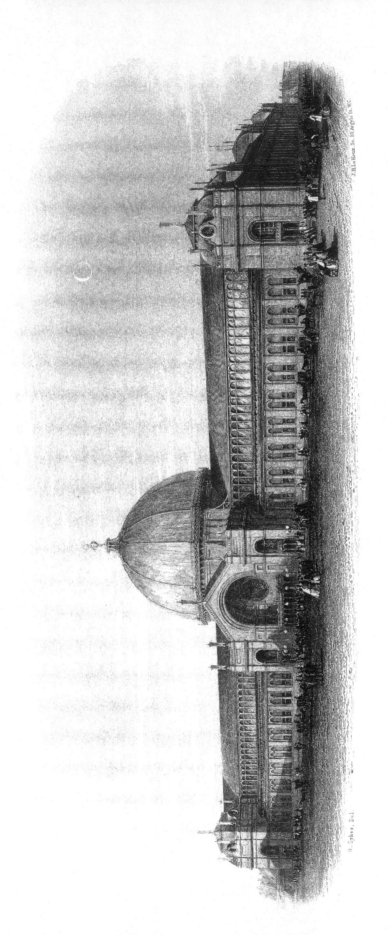

G. Sykes, Del.

J. H. Le Keux Sc. 30 Argyle St. W.C.

INTERNATIONAL EXHIBITION OF 1862.

West Front — Main Entrance.

SOUTH GALLERY IN PROGRESS.

CHAPTER XII.

THE BUILDING FOR THE EXHIBITION OF 1862.

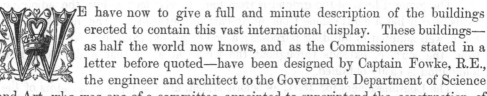

E have now to give a full and minute description of the buildings erected to contain this vast international display. These buildings— as half the world now knows, and as the Commissioners stated in a letter before quoted—have been designed by Captain Fowke, R.E., the engineer and architect to the Government Department of Science and Art, who was one of a committee appointed to superintend the construction of the whole of the works on the estate of the Commissioners of 1851, and was secretary to the British Department of the Paris Exhibition in 1855. While superintending the erection of the southern arcades, which he had planned for the Royal Horticultural Society's gardens, he carefully considered a design for a building on the Kensington Gore estate which should appropriately close up the open side of the great central square, and should provide ample space and protection for industrial displays as well as a noble gallery for pictures and engravings. This design was well matured during the progress of the Horticultural Society's arcades, and was therefore ready to be submitted to the Royal Commissioners for the Exhibition of 1862 as soon as they were appointed.

On the arrangements for the ground being completed, the Commissioners therefore had before them a design founded on a large experience and an intimate knowledge of the site. As competition in the case of the 1851 Exhibition had only resulted in disappointment and delay, they carefully considered Captain Fowke's plans, and eventually accepted them, subject, however, to some modification on account of the cost of their execution, which was estimated at five hundred and ninety thousand pounds. The leading feature of the original design was the great hall, which was to have been five hundred feet long, two hundred and fifty feet wide, and two hundred and ten feet high. Its proposed position was immediately behind the central entrance of the south front, above which it would have towered. This feature, however, was suppressed on account of its cost; but there will be no architectural or constructive difficulties to prevent its being added subsequently should the Commissioners approve of it and the necessary funds be available.

The ground on which the main building stands is about sixteen acres in extent, and occupies the southern portion of the land purchased at South Kensington by the Commissioners for the Exhibition of 1851. It is nearly rectangular in shape, and it measures about twelve hundred feet from east to west, by five hundred and sixty feet from north to south. It lies immediately south of the Royal Horticultural Society's gardens, the southern arcade of which has been lent to the Exhibition Commissioners for refreshment rooms. The Cromwell Road forms the southern boundary, while on the east it is shut in by the Exhibition Road, and on the west by Prince Albert's Road.

The whole of this ground is covered by buildings of a permanent character, and to secure as much additional space as possible, the two long strips of ground between the east and west arcades and the adjacent roads are roofed in by means of temporary sheds, to give ample space for the exhibition of machinery and other large and heavy objects, which cannot be conveniently shown in the main building. The additional area afforded by these two annexes is about seven acres, which makes the total extent of the Exhibition buildings twenty-four and a half acres.

The general level of the ground is from four to six feet below the adjoining roads, and it has a gentle slope from north-east to south-west. The nature of the soil is well adapted to building purposes; a bed of gravel, from four to twelve feet from the surface, extends throughout the whole area, so that a dry and firm foundation is easily obtained.

The Commissioners for 1851, as before stated, are the legal proprietors of the site, but to secure the greater portion of it for the intended exhibition in 1872, they have agreed to reserve about sixteen acres of it for that purpose on receiving ten thousand pounds by way of ground rent. It is already agreed that a lease shall be granted to the Society of Arts of the central portion of the picture gallery, one acre in extent, along the Cromwell Road, for ninety-nine years, on condition that ground rent to the amount of two hundred and forty pounds per annum be paid to the 1851 Commissioners, and that the building be given up unreservedly for the use of the exhibition in 1872.

In the general design of the building, a ground plan of which is given, its suitability for future international exhibitions has been keep steadily in view, and it has a much more permanent character than the famous Crystal Palace erected for the exhibition in 1851.

Glass and iron are no longer the main features of the design, but are succeeded by lofty walls of brickwork, which surround the ground on all sides, and form the walls of the fine arts galleries.   The east end and west sides, by being continued past the southern arcade of the gardens, have a frontage of seven hundred and fifty feet, and the frontage on the south is eleven hundred and fifty feet.   The north front is the lower arcade of the gardens, which has had a permanent upper story added to it.   The interior space thus enclosed is entirely covered in by roofs of various heights, and is divided into nave, transepts, aisles, and open courts; the latter, occupying comparatively a very small portion of the space, are roofed with glass as in 1851, but the other parts have opaque roofs, and are lighted by clerestory windows.

The interior supports are hollow cast-iron columns, as in 1851, of somewhat larger dimensions, being a foot wide, with an inch of metal in them.   They are so arranged as to come at intervals of twenty-five or fifty feet from centre to centre; in fact, twenty-five is the unit here as twenty-four was in 1851, and nearly all the leading dimensions, both vertical and horizontal, are multiples of that number. The exceptions to this rule are the nave and transepts, which are eighty-five feet wide; the former runs east and west, and terminates in the centre of those fronts, having its central line eighty-one feet north of the centre line of the building; the latter extend north and south from the ends of the nave throughout the whole width.   At the intersection of the nave and transepts are the great domes.   The aisles are continued all round the nave and transepts, and the space enclosed by them forms the open or glass courts.

The columns are not supported as they were in 1851.   On that occasion they were attached to connecting pieces, which, terminating in a large flat base plate, rested on concrete laid flush with the ground: these connecting pieces of course varied in height to suit the slope of the ground.   This plan has been avoided in the present building by bedding the columns themselves on York slabs laid on brick piers, which are founded on concrete; the slabs being all adjusted to the same level throughout by varying the height of the brickwork, only one length of column is used, and the facility of setting them up is thus greatly increased.

The total area roofed in is nine hundred and eighty-eight thousand square feet, or sixty millions of cubic feet; it is therefore considerably larger than the covered part of the 1851 Exhibition, which only occupied seven hundred and ninety-nine thousand square feet.   It has also, when actual covered space is alone considered, slightly the advantage of the Paris Exhibition, which had a covered area of nine hundred and fifty-three thousand square feet.   If, however, we compare the total space, covered and uncovered, occupied by each, Paris is considerably larger, for the more favourable character of its climate for out-of-door display enabled the authorities of that exhibition to increase the area of ground given up to exhibiting space by five hundred and forty-seven thousand square feet, while, with our variable climate, it has not been thought advisable to have more than thirty-five thousand feet of ground unroofed.   The total areas, covered and uncovered occupied by the two exhibitions are one million five hundred thousand square feet for Paris, and one million twenty-three thousand square feet for 1862.

The French Exhibition therefore considerably exceeded ours in size, but it was not nearly so compact in form, and its temporary annexes made up a very large portion of it, occupying six hundred thousand of the nine hundred and fifty-three thousand square feet, while our two annexes do not amount to more than one-third of the total area.

The plans of this huge building, as before stated, were submitted to ten leading contractors: three tenders for construction were sent in, and that furnished by Messrs. Kelk and Messrs. Lucas for the modified building, being the lowest, was at once accepted.

These two eminent firms joined their resources together, and became partners in the work.

The nature of the contract was peculiar. The whole responsibility for the execution of the works rested with the contractors, and the amount they are to receive is contingent on the receipts of the exhibition. The Commissioners have the option of purchasing the building out and out, or of merely paying for the use of it. For the rent of the building a sum of two hundred thousand pounds is guaranteed absolutely; if the receipts exceed four hundred thousand pounds, the contractors are to be paid one hundred thousand pounds more for rent, and if the sum is fully paid, then the centre acre of the great picture galleries is to be left as the property of the Society of Arts. The contractors are also bound, if required, to sell the whole for a further sum of one hundred and thirty thousand pounds, thus making its total cost four hundred and thirty thousand pounds.

All proceedings were submitted to Captain Fowke, R.E., who acted for Her Majesty's Commissioners. He conferred with a Building Committee, consisting of the Earl of Shelburne, Mr. W. Fairbairn, and Mr. W. Baker; and the Commissioners reserved to themselves the final approval of everything. Captain Fowke was assisted by Captain Philpotts and Lieutenant Brooke, and certain non-commissioned officers of the Royal Engineers. The date agreed upon for the building to be completed and given up by the contractors was the twelfth of February, 1862, and on that day, although much of the decoration, many of the minor details, and part of the western dome had not been completed, the Commissioners took formal and practical possession of their property. The galleries and stairways had been previously tested by a body of four hundred men marching over them, and the greatest deflection of the iron beams was only one-eighth of an inch, and of the iron and wooden trusses only three-eighths to five-eighths of an inch. The following report on this subject was addressed to Her Majesty's Commissioners by the professional members of the Building Committee:—

MY LORDS AND GENTLEMEN,

FEELING that it would be a source of satisfaction to the Commissioners, as well as to ourselves, as members of the Building Committee, and also a due precaution for the public safety, that the gallery and other floors of the International Exhibition Building at South Kensington should be thoroughly proved, we undertook a series of experiments on Monday last.

We have to report that, in carrying out these experiments, the various floors and stairs were put to a more severe test than they would be subjected to with the largest number of people that could possibly be assembled upon them at any other time during the Exhibition. The result of these experiments fully bears out our calculations on the strength of the different parts of the structure, and we feel perfectly satisfied as to the stability of the building for the purpose for which it was intended.

The two large domes, in the strength of which we have taken great interest, were eased from their temporary support last week, and no observable settlement took place.

The following are the particulars of the tests:—The first caused a large body of men, about four hundred in number, to be closely packed upon a space twenty-five feet by twenty-five feet, on one lay of flooring; we then moved them in step, and afterwards made them run over the different galleries and down each staircase; at the same time we caused the deflections of the girders carrying these floors to be carefully noted at several places, and had the satisfaction of finding that, in each case, the deflections were very nearly the same, thus exhibiting a remarkable uniformity in the construction. The cast-iron girders, with twenty-five feet bearings, deflected only one-eighth of an inch at the centre, and the timber trussed beams of the same bearing placed between these girders deflected half an inch at the centre. In every instance the girders and trusses recovered their original position immediately on the removal of the load.

We are, my Lords and Gentlemen, yours faithfully,

WILLIAM FAIRBAIRN, C.E.
WILLIAM BAKER, C.E.

London, Feb. 13.

The general outline of the south front presents an elevation eleven hundred and fifty feet long, and fifty-five feet high in the brickwork, with two projecting towers at each end, rising sixteen feet above the general outline, and a larger tower in the centre, in which is the main entrance to the picture galleries. Semicircular headed panels, separated by pilasters, are built at central intervals of twenty-five feet throughout the whole length; a high plinth extends all round, and between the arches are circular niches. In the lower portion of each panel is a window, to admit light and air to the ground floor, and to ventilate the picture gallery above.

The exterior is chiefly built in plain brickwork, and with no more ornament than such work admits of. The panels are plastered in cement, and it is proposed to ornament them with English mosaics. The exterior decoration will eventually depend on the funds, and the way in which they are applied. At present the building is very incomplete in external ornament, but any amount of architectural beauty that can be paid for may be added hereafter. The Exhibition Charter provides that fifty thousand pounds shall be spent in the architectural completion of the central portion of the building out of contingent profits.

The two great domes, being each three hundred feet from the south front, can never in any way enter into its effect. If the middle hall, with its great central dome, should ever be built, then the Cromwell Road front will not be without this ornament. Each dome keeps its place as the centre of its own front, and its effect is utterly independent of its fellow, which is one thousand feet from it. The upper terrace of the Horticultural Gardens is the only point from which the two present domes appear simultaneously; and when thus viewed, so completely does the building carry on the symmetrical lines of arcades and terraces, that the duality of the domes is at once accepted as the natural complement of the system which has governed the laying out of the entire quadrangle.

The main entrance to the southern portion of the building is through three arches in the central tower, twenty feet wide and fifty feet high, resting on piers fourteen feet thick, decorated with terra-cotta columns. Above the arches is the cornice and frieze, on the top of which, and above the middle porch, is an ornamental clock dial. Passing through the archways, the visitor enters a large vestibule and hall, one hundred and fifty feet long and one hundred and ten feet wide, leading to the industrial courts and galleries; and a flight of steps on either side, twenty feet wide, conduct him to the picture gallery.

The chief requisites in a picture gallery are a solid building capable of resisting every change of weather, well ventilated, and an equally distributed light throughout, admitted in such a way as will prevent its rays being directly reflected from the surface of the picture to the eye of the visitor. A light, therefore, satisfying this condition when the observer is standing at a convenient distance, is the only one which can be called perfect.

No one can have observed pictures lighted by ordinary windows without experiencing the unpleasant effect produced by the improper reflection of the rays, or glitter from the pictures, as it is called. It is for this reason that one is often puzzled where to find a place from which to see the whole of a large picture to advantage. This defect exists in many of the finest galleries, both in this country and on the Continent, and the result is that some pictures can scarcely be seen at all, while others can only be observed from one or two points, which are always more or less crowded, according to the merits of the subject.

This is obviated by admitting the light at a particular angle from the roof, by means of a skylight extending along its entire length, and which, in the present case, measures thirty-one feet in width, that is, fifteen feet six inches from the ridge on either side. The entire width of the opening, measured on a horizontal plane, is twenty-nine feet two inches. Each room is fifty feet wide, and at a height of thirty-two feet nine inches a cove, springing from a cornice on either side, reaches the height of the tie-bar of the principals (forty-two feet ten inches above the floor), twelve feet four inches from the wall, thus leaving a space twenty-five feet four inches between the coves. In this space a transparent calico ceiling (hereafter to be replaced by ground glass) is introduced, which, however, is raised two feet four inches above the highest point of the cove, or forty-five feet seven inches from the floor. The space between the highest point of the cove and the eave of the calico is occupied by louvres for ventilation.

These proportions give the gallery as much light as possible, and glitter from the surface of the pictures is avoided. As regards the quantity of light admitted, it may be briefly stated that the opening for admission is exactly half the floor area of the gallery, the former being twenty-five feet wide, the latter fifty. In dealing with the quantity of light, another important point must not be lost sight of, namely, the height of the opening from the floor, and its consequent distance from the picture. In this gallery this is reduced to a minimum consistently with the avoidance of glitter, being only forty-five feet seven inches from the floor.

The following illustration will explain the question of glitter, or reflection of light, from the varnish of pictures:—Supposing a mirror to be hung against the entire surface of the wall. It will be seen, by referring to the diagram on the opposite page, that a ray of light from the skylight, at its extremity furthest from either wall, striking that wall at A, at a height of twenty-three feet three inches above the floor, will be reflected so as to reach the eye at E of a beholder (say five feet three inches above the floor) standing five feet on the other side of the centre of the room, or thirty feet from the mirror, and, consequently, all the rays striking below that point will fall below his eye; or, in other words, he will not be able to

see the image of the skylight in the mirror at any point below twenty-three feet
three inches from the floor, and, as a matter of course, there will be no glitter
on the wall, or on pictures hung against it, below that point. Consequently, to
see pictures without glitter hung higher up, it will be necessary for the spectator
to retire still further from the centre of the gallery.

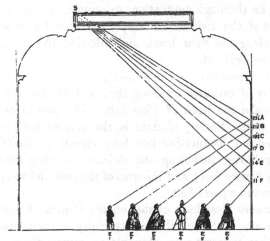

Transverse Section of Picture Gallery, showing the way of admitting the light to avoid glitter.

It will be seen from the diagram that this point, which is called the glitter
point, alters with the position of the beholder. For instance, at E, five feet from
the wall, the glitter point is at F, eleven feet from the floor, while, on coming
closer, it will descend in proportion. On the other hand, by receding to a distance
of ten feet, the wall may be seen without glitter to a height of fourteen feet. Looking again to the same diagram, it will be seen that, apart from all considerations of
reflection, a person desiring to see a picture at a height of fourteen feet would
naturally retire ten feet, if not more, from it; and the same may be said of the other
heights and positions shown on the sectional diagram, so that in any position in
which a person can conveniently examine a picture he may be sure of having its
surface free from glitter.

This system of lighting increased the difficulty of successfully treating the exterior of the building, for it prevented any windows being placed in the upper part
of the side walls; but after the very successful application of these principles of
lighting to picture galleries which have been constructed within the last few years
at South Kensington, it was wisely determined to forego all other considerations,
and apply the same principles to the rooms destined to receive the choicest works
of art of the present age.

On ascending the stairs the visitor enters a vestibule of similar proportions to
the one below, from which he obtains one unbroken vista throughout the whole
extent of the main gallery. This gallery is about as long as the Louvre at Paris.

Entering the first room on either side, he finds himself in a spacious hall,
three hundred and twenty-five feet long, fifty feet wide, and forty-three feet high.
Passing through this, he enters one of the wing towers, which forms a room
fifty-two feet by forty-five feet, and sixty-six feet high; he then enters another

room seventy-five feet long, and of the same width and height as the first, from which he may pass into the end tower, whence he will have another uninterrupted view of the whole main gallery.

The interior architectural decorations of these rooms are very simple, and may be briefly described as a plain cove extending to each side of the skylight, and resting on a moulded cornice.

Arrangements for thorough ventilation, so essential to the preservation of the pictures and comfort of the visitors, are amply provided for by admitting fresh air through apertures along the floor level, and allowing the vitiated air to escape through louvres in the skylight.

Descending to the ground floor, the same sized rooms are repeated; but as they are lighted by means of ordinary windows, they will be devoted to other objects than those coming under the head of Fine Arts. The part of the picture gallery which is to revert to the Society of Arts is the central hall and the two large rooms, three hundred and twenty-five feet long, on either side of it. The area of the south picture·galleries, including the staircase, is sixty-three thousand nine hundred and twenty square feet; and the area of the east and west picture galleries is twenty-seven thousand square feet.

So many erroneous ideas have existed on the Continent respecting the space devoted to the fine arts in this exhibition, especially in comparison with the space in Paris in 1855, that it seems very desirable to correct them. A writer in the 'Athenæum' last year said: "The building for the fine arts in England is of a far more substantial character than that in the Allée Marbœuf. The one is of brick and iron, intended to be permanent; the other was of timber and plaster, and only temporary. On this account, if on no other, there will be much less space in England than in Paris. According to official documents, the Rez de Chaussée, in Paris, devoted to oil pictures only, contained twelve thousand five hundred and three square metres of hanging space, or about one hundred and seventy-one thousand six hundred and thirty-three square feet, of which France, according to the account of 'Prince Napoleon's visits,' retained seven thousand four hundred and forty-five square mètres, or full three-fifths. The upper gallery at Paris contained about three thousand one hundred square mètres, of which France retained four-fifths. In England, the whole space devoted to the fine arts, including staircase, contains only ninety thousand nine hundred and twenty square feet of hanging space, or little more than half that in Paris; and this England equally divides with foreign nations, retaining only forty-five thousand four hundred and sixty square feet, or exactly one-half. England asked in Paris for twelve thousand square feet of wall space for paintings in oil and water-colour engravings and works of architecture; and, according to Mr. Redgrave's accurate report, occupied ten thousand four hundred and ninety square feet, being considerably less than one-sixteenth of the whole French space. France has been accorded ten thousand square feet of space for 1862, which is more than one-tenth of the whole space for the fine arts. And when it is remembered that she has no school of water-colour painters to provide for, the proportion in her favour is still greater. In 1855, England took a small side gallery, whilst the large saloons were appropriated by France and other continental nations. In 1862 all the space will be equally excellent, and equally divided, not only as to area, but as to the nature and structure of the galleries, between England and the countries she has invited. Moreover, the United Kingdom is to be represented by the works of a hundred years, whilst in France the works were those of living artists."

Before concluding the description of the picture gallery, it may be interesting to add its constructive details. The foundations throughout are carried down to the gravel, here from six to twelve feet below the surface of the ground, in concrete, on which ordinary brick footings are laid. In the front wall the piers carrying the semicircular arches are twelve feet wide by three feet two inches thick, and the intervening panel having merely its own weight to support, is only nine-inch work. The back wall is of rather a different construction. This is a plain wall from top to bottom, with numerous arches through it on the ground floor; it is built for the most part hollow, with piers so placed that the weight of the floor and roof will come on them. This system of hollow walling gives the greatest strength with the least amount of material, and secures a straight face at either side. The floor of the picture gallery has been constructed of great strength, so as to bear with perfect safety the greatest load which can be brought on it. It is carried on girders thirteen and a half by twelve feet, resting on the side walls and intermediately supported by two cast-iron columns. These girders cross the building at central intervals of twelve and a half feet, and over them are laid joists about thirty-six feet by about eight feet four inches, two feet apart, to carry the floorboards. A portion of this floor has been proved to one hundred and forty pounds to the foot, which exceeds the greatest load it can have to bear when densely crowded with visitors. The walls in the Picture Gallery are lined throughout with wood, which is kept at a short distance from the brickwork, so as to guard against damp. The design of the roof is the same as that already employed by Captain Fowke in one of the South Kensington galleries, and also in the Irish National Gallery in Dublin, and seems well adapted for its purpose; the principals which support it consist of two strongly trussed double timber rafters, connected by an iron tie-bar four feet above the level of the wall-plate. The coved ceiling is thus made four feet higher than it could have been with an ordinary tie-beam roof.

The principals are placed at central intervals of twenty-five feet; they rest on flat stones built into the walls, and strongly trussed purlins, carrying the skylight rafters and upper portion of the cove, are suspended to them. The skylight is glazed with sixteen-ounce glass, and the rest of the roof is covered with slates.

The east and west fronts, though differing from the south, are not less imposing. They are in all respects similar to each other in their general effect. Here the huge domes, rising to a height of two hundred and sixty feet, show to most advantage, and the transept roof, with its lofty clerestory windows, is in full view. To the observer below the form of each dome appears nearly that of a semicircle: this effect is obtained by making its height eleven feet more than its semi-diameter, which fully allows for the loss by perspective diminution.

From the crown of each dome rises the finial to a height of fifty-five feet. Each dome is in the middle of each façade; its centre is the point formed by the intersection of the centre lines of the nave and transepts, and the front of the building is advanced from it one hundred and eight feet. Under each noble arched recess is the main entrance to the industrial courts, the effect of which forms one of the most pleasing exterior parts of the building. The porches are each one hundred and sixty-two feet in extreme width, and they each contain a deep semicircular arched recess of sixty-eight feet span and eighty feet high, with a deep covering, capable of receiving an almost endless variety of decoration if such be desired hereafter.

In the tympan of the recess is the great rose window, which will be visible from end to end within, the window in one closing the vista as the spectator looks from a standing point beneath the other. Minor porches on either side, thirty-six feet wide, forming wings, support a pedimental gable, which rises to a level with the ridge of the nave and transept roofs, and is finished with a bold line of balustrade. The entrances beneath are enclosed by an arcaded framing filling up the recess for one quarter of its height, and having a balcony above. The flat brickwork of the wings is relieved by pilasters, one on each side of the minor porch; these carry a light cornice moulding, surmounted by an attic.

On either side of the central entrance, recessed fifteen feet from it, the exterior walls of the building extend two hundred and thirty-five feet to the north and south; these enclose the auxiliary picture galleries. There is a high plinth from end to end, and immediately above are panels formed by a series of coupled semi-circular arched recesses, with bold pilasters between. Over all is an appropriate cornice, supported by corbels. By the wall being reduced to the height of thirty-six and a half feet, the lofty clerestory windows of the transept, which rise immediately behind, come into the composition. As in the south front, the lower portions of the panels are occupied by windows to give light and ventilation to the offices and retiring rooms, with which the ground floor on these sides are occupied. The upper floor is used as an auxiliary picture gallery, and is therefore lighted on the same principles as the rooms on the south front.

A visitor can enter the auxiliary galleries, independent of the main gallery, by means of stairs on either side of the east and west entrances; or he may have access to them from the end towers of the latter, already described. They form four distinct rooms, two hundred and forty-seven feet long, twenty-five feet wide, and seventeen feet high. The same principles of lighting and ventilating being observed in these galleries as in the larger one, their construction is similar, subject, of course, to the alterations rendered necessary by their smaller size.

The main and auxiliary picture galleries of the fine arts department afford four thousand six hundred lineal feet of hanging space, from seventeen to thirty feet high; yet all this amount, large as it may seem, is required, and even more would have been desirable could it by any possibility have been obtained. An idea of their extent may be formed by the fact that, in walking once up and down the galleries, the visitor will have to traverse a mile all but sixty yards; and, presuming the moderate allowance of seventy-five per cent. of the available wall space to be actually covered by pictures, the aggregate area will equal seven thousand six hundred square yards, or about one and a half acres.

To complete the survey of the exterior, we must examine the north front, for which purpose it is necessary to enter the Royal Horticultural Society's gardens. The large space here afforded admits of a connoisseur criticising the building from several points of view. For our purpose, however, it will suffice to imagine our station to be on the central walk, one or two hundred yards from the south arcade.

From no other point can a better view of the building be obtained. The south arcade forms the basement of the north front, to which an upper story has been added. The façade is divided into two floors, except the central portion, which has a mezzanine interposed. The ground floor, consisting of the southern arcade of the gardens, with its pleasing arrangement of twisted terra-cotta columns, is doubtless familiar to many people. The whole front is divided into five faces, in different lines of advance. By subdividing the centre mass into three sections a very great variety and relief of design is obtained. The middle of the front is occupied by the entrance from the gardens, through three ten-feet ornamented brick archways, supported by coupled stone columns; these are immediately opposite the southern entrance, from which point will be obtained one unbroken vista across the whole building up to the cascade and conservatory at the north end of the gardens.

In examining the five divisions of the façade, we find that the centre (seventy feet high) presents three levels,—the arcade of the gardens, the shallow mezzanine (interrupted by the central arches before mentioned), and the upper floor. On each side of the centre are the lights of the arcade, consisting of tripled ornamental brick arches on terra-cotta columns, separated by pilasters; the upper lights are similarly arranged, and the whole is surrounded by a panelled frieze of appropriate design, with openings for ventilation. Over this is seen the roof, of good pitch, following the line of the ground plan.

The two corresponding recesses on each side are thrown back twenty-five feet, and extend in an unbroken line for two hundred feet, with a height of sixty feet: the level of the upper floor here corresponds with that of the centre mezzanine, and the lighting, both above and below, is effected by eleven sets of tripled arches, similar to those in the centre division.

At each end are the returns into the garden; and in the fifty feet which completes the length is an entrance archway, ten feet wide, on the far side of which the tripled arched light is repeated.

The upper and lower floors on this front occupy an area of twenty-six thousand eight hundred square feet: the whole of this space has been given up to the refreshment rooms and offices connected with the Exhibition.

No matter what may be the financial results of the great display, these rooms at least will remain after its close, and form part of the Horticultural Gardens.

In making a survey of the interior, it is hardly necessary to go minutely into the construction of the whole, and we shall therefore only dwell on those parts which present any novelties. It would be unprofitable to do otherwise, for there are certain portions which differ in no essential from many ordinary iron structures. The nave and transepts are similar in all respects.

Entering by the east or west front, the visitor ascends two steps until he comes to the level of the daïs under each dome. From this point, six feet above the rest of the floor, he may in one view command the interior of the whole building. A very serious defect in the ground has been remedied by this arrangement. The roads surrounding the site are about five feet above the level of the ground on which the building stands. Had this contour been rigidly followed, the visitor on entering would have gone down into a pit.

This immediate descent, apart from its inconvenience, would have totally marred the interior aspect of the building.   Had the whole area been raised to suit the road level, it will be obvious that the cost would have been considerably increased.   From the daïs three flights of steps, eighty feet wide, conduct the visitor into the nave and transept on either side.   The nave is eight hundred feet long, eighty-five feet wide, and one hundred feet high to the ridge of the roof.

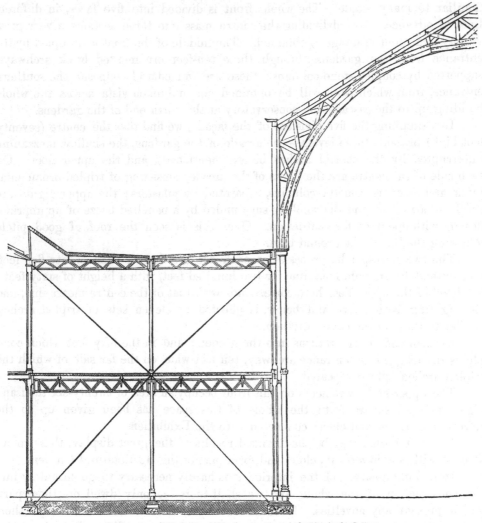

Section of one-half of a Nave Rib, showing the cross-bracing in the Gallery.

The supports on either side consist of square and round cast-iron columns, coupled together ; the former carry the gallery floor, and the latter, advancing into the nave, receive the principals of the roof.   These columns are fifty feet high, in two lengths of twenty-five feet each, and from their capitals spring the roof frames, which consist of three thicknesses of plank, from eighteen inches to two feet six inches deep, firmly nailed and bolted together, and so arranged that their ends break joint.   The centre plank is four inches thick, and each of the outer ones is three inches; the lower edges are tangents to an imaginary semicircle, round which they form half of a nearly regular polygon. From the springing rise the posts of the clerestory windows, twenty-five feet high. The principal rafters of the roof frames rise from the top of these posts, and are carried up, after passing a tangent, to the back of the arch, to meet at the ridge in a point twenty-five feet above the top of the clerestory. The angles over the haunches and crown of the arch are firmly braced together, so as to reduce the thrust as much as possible.

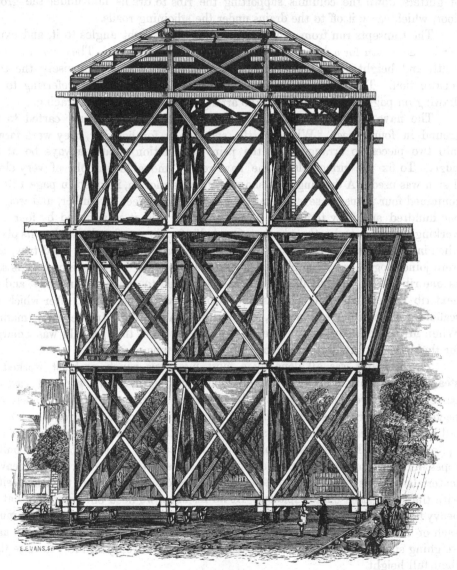

TRAVELLING SCAFFOLDING.

The rib is repeated thirty times in the length of the nave, and from its graceful curve and lightness it produces a fine effect. Between every roof principal is a clerestory light twenty-five feet high, consisting of three arches springing from intermediate mullions. The roof is covered with felt and zinc, on one and a half inch plank, which is laid diagonally, so as to brace the whole together. The nave is therefore entirely dependent for its light on the clerestory windows; but this arrangement is found to be satisfactory, and a substantial water-tight covering is thus insured, having the advantage of obviating all chance of that unpleasant glare which the experience of 1851 proved to be unavoidable with a glass covering. The building at the same time is thus made cooler in summer and warmer in winter than a simple glass structure. The rain-water from the roof is conducted, by means of gutters, down the columns supporting the ribs to drains laid under the ground floor, which carry it off to the drains under the adjoining roads.

The transepts run from each end of the nave, at right angles to it, and extend north and south for a length of six hundred and fifty feet. They are the same width and height as the nave, and the ribs of its roof are of precisely the same construction. This construction will be more fully understood by referring to the drawing on page 138, which shows the arrangement of planks and bracing.

The nave ribs were made at Mr. Kelk's yard in Pimlico, and carted to the ground in four pieces. When on the spot ready for hoisting, they were formed into two pieces, so arranged that the point of junction should always be at the ridge. To fix the principals in the nave roof, a movable scaffold of very clever design was used. A drawing of it, showing its construction, is given in page 139. It contained four thousand seven hundred and forty cubic feet of timber, and weighed one hundred and forty tons; but, notwithstanding this, it was moved by four men working crowbars under the wheels. One half of a rib was first hoisted to its place; when in position the other half was raised, and as soon as both were fixed true, they were joined together by completing the arch and bracing over its crown. As soon as one rib was up the travelling scaffold was moved to the adjoining bay, and the next rib completed. The purlins and boarding were then fixed, after which the scaffold was again moved forward, and another bay covered in the same manner. When Mr. Crace's workmen came in to paint the roof, this scaffold was enlarged for their use until its weight reached two hundred tons.

The hoisting was all done by a most ingenious winch, or hoist, worked by steam, the invention of Mr. Ashton. This machine has two grooved cast-iron barrels, which are made to revolve by means of a system of toothed wheels connected with a portable steam-engine. A rope is passed round the grooves. On the fall being manned and the barrels set in motion, the coils of the rope are gathered up, and a great hoisting power obtained. By means of snatch-blocks and pulleys ropes were led from this simple machine to all parts of the building, and the heaviest materials, such as girders, columns, scaffold beams, &c., were hoisted to their position with the greatest ease and rapidity. As an instance, we need only mention that the heavy floor girders, weighing about one ton and a quarter, were raised in two minutes, each of the columns in about the same time, and the ponderous ribs of the nave, weighing six tons and a half, required only from ten to twenty minutes to raise them their full height.

The only portion of the Crystal Palace of 1851 which can be compared with the nave is the great central transept, the height of which was one hundred and eight feet, or four feet more than that of the nave; but it was narrower by thirteen feet, being only seventy-two feet wide, while the total length of the nave of the present building is very nearly three times as long.

The ribs of the transept were fixed over a standing scaffold all through, which alone consumed thirty thousand three hundred and thirty-six cubic feet of timber. The reason for this was, that as the domes divide the transepts into four separate lengths, four travellers would have been necessary, and though these would not have taken nearly so much material as the standing scaffolds, yet the contractors thought that the difference would not have compensated for the additional labour required in building the former.

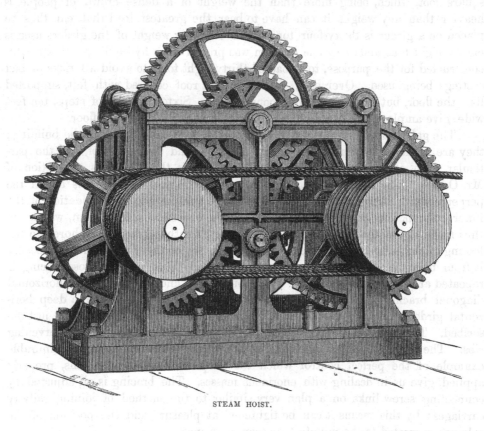

STEAM HOIST.

The general plan shows three large buildings parallel to each other,—the refreshment rooms, the nave, and the picture gallery. These are connected at their ends by transepts, and thus two vast oblong spaces are enclosed, one to the north and the other to the south of the nave.

On both sides of the nave and the inner sides of the transepts are aisles fifty feet wide. Another aisle, twenty-five feet wide, is carried along the outer sides of the transepts, and along the back wall of the south front.

After deducting the space occupied by all these aisles from the oblongs above referred to, we have remaining two smaller ones, that north of the nave, seven hundred and fifty by eighty-seven feet, and that at the south of it, seven hundred and fifty by two hundred feet. Each of these is subdivided into three courts by two fifty-feet aisles. The centre courts are one hundred and fifty feet long, and those at the ends two hundred and fifty feet. The dividing aisles on the north lead to the refreshment rooms, on the south to the entrance vestibule.

Twenty-five feet above the ground floor are the galleries, following the same line as the aisles; they give an additional exhibiting space of two hundred and eight thousand square feet. Particular care, as the testing has shown, has been taken to make these galleries amply strong for the heavy moving loads they may have to bear. The floors are supported on cast-iron girders fixed to the columns; over them are laid two strong suspended trusses, which carry the joists and boarding.

Supposing a floor to be loaded with one hundred and forty pounds to the square foot, which, being more than the weight of a dense crowd of people, is heavier than any weight it can have to bear, the greatest load that can thus be placed on a girder is thirty-four tons. The breaking weight of the girders used is eighty-eight tons, and every one of them was proved in an hydraulic press, specially constructed for the purpose, to a load of thirty-eight tons, to avoid all risks of bad castings being used. Over each gallery is a flat roof covered with felt, supported like the floor, but of much lighter construction. Sixteen flights of steps, ten feet wide, give ample means of ascending from the ground to the upper floor.

The galleries play a very important part in the construction of the building; they are made to form an abutment to the nave and transept roof, and the particular form of bracing by which this is effected is the ingenious suggestion of Mr. Ordish. The roof thrusting outwards tends to throw the columns out of the perpendicular; strong iron braces are therefore anchored to the foundation of the inner column, and carried up to the top of the opposite outer column, which are thus made to counteract the thrust of the roof. Another bracing, anchored to the footing of the outer column, is carried up to the top of the inner column, to secure it from being acted on by the force of the wind. This vertical cross-bracing is repeated at every hundred feet, or every fourth bay, and by introducing horizontal diagonal bracing under the roof flats, they are turned as it were into a deep horizontal girder, supported at two ends by the columns vertically braced as just described. This horizontal girder therefore takes the thrust of the three intervening ribs. The way in which the bracing is introduced is very clever, and is an admirable example of the perfect control which the simplest mechanical means, properly applied, give us in dealing with enormous masses. The bracing is all adjusted by connecting screw links on a plan very similar to the method of joining railway carriages; by this means it can be tightened at pleasure, and the position of the columns corrected to the minutest fraction of an inch.

The drawing of the rib in the illustration on page 138 shows the vertical cross-bracing.

The aisles and galleries enclose six courts, three north of the nave, two of which are two hundred and fifty by eighty-seven feet, and the other one hundred and fifty by eighty-seven feet,—three south of the nave, two of which are two hundred and fifty by two hundred feet, and the other one hundred and fifty by two hundred feet. These form the open or glass covered courts, and are the only portions of the building which in this particular resemble the Crystal Palace of 1851. They have only a ground floor, and the roof, which is on the ridge and valley plan, but in spans of fifty feet, is entirely covered with glass. The roof is carried on square iron columns fifty feet apart each way, at the top of which, fifty feet above the ground, wrought-iron trellis girders are fixed on lines running east and west. The columns and trellis girders carry the principals of the roof, which are all of iron, on the trussed-rafter plan, eight feet apart. The roofs are drained by channels in the valleys conducting the water down the hollow iron columns. The effect of these courts, with their light glass roofs admitting floods of light into the building, gives a pleasing variety to the interior, and affords most valuable exhibiting space.

The great domes, from their huge size, form one of the most prominent features in the building.

It has been before stated that they are situated at the intersection of the nave and transepts. Their form and position have been thus determined. The intersection of the lines of columns in the nave and transept aisles forms two octagons, which, though not mathematically regular, are regular in this one respect—their opposite sides are parallel and equal, the length of the sides being alternately eighty-five feet and thirty-five feet five inches. The points at the angles of these octagons are the chief supports of the domes. For this purpose there is a column at each angle, two feet in diameter, and for architectural effect, as well as for carrying the groined ribs, the object of which will be presently explained, the lower portion of these two-feet columns is clustered with two round columns and one square column of smaller dimensions.

Though the chief points of support, however, are at the eight angles of the octagon, the dome is a dodecagon, the other four points being thus obtained:—The last bay of the nave and transept, instead of having a roof resting on wooden principals going straight across, has two iron diagonal ribs crossing it, forming as it were a groined arch, whose apex is a point in the centre of the bay and in a line with the roof ridge. By joining the apices of these groins and the points in the octagon already determined, we get a nearly regular dodecagon, having its opposite sides parallel and equal, and with eight sides in pairs, each equal to forty-three feet nine inches, and the four remaining sides coming between these pairs, each equal to thirty-five feet five inches. This dodecagon forms the base of the dome, which will thus have eight sides over the nave and transepts, and four sides over the corners of the aisles, equal respectively to the dimensions just given, and a diameter of one hundred and sixty feet.

Each groined rib transmits the weight on it to two columns outside the octagon, so that the dome may be said to rest on sixteen points, its pressure on the angles of the octagon being nearly five times $(4\frac{9}{25})$ as much as it is on the adjacent columns of the nave and transepts.

By the ingenious and novel plan of the groined roof-ribs a dodecagon dome is made to seem to stand on an octagon; no additional columns of support but those actually coming in the sides of the nave and transepts are used, and thus an uninterrupted vista is obtained through both these channels, and a beautiful architectural effect is produced.

Each rib is two feet deep, with a web of three-eighth-inch plate iron, to the edge of which is riveted a top and bottom flange formed of angle iron in such a way as to give the top flange an area of nearly ten inches and three quarters, and the bottom flange an area of nearly nineteen inches and three quarters. The principal rafter and its upright are also made of wrought iron, having a web twelve inches deep, with an equal top and bottom flange of angle iron riveted to it so as to give it a sectional area of about twenty inches and a quarter. Radial pieces of iron eight inches by five-sixteenths of an inch, connecting the upright and principal rafter with the circular portion of the rib, are introduced every five feet. At the intersections the ribs are strengthened by additional plates of iron, and here, for a short distance, they assume the form of a box girder.

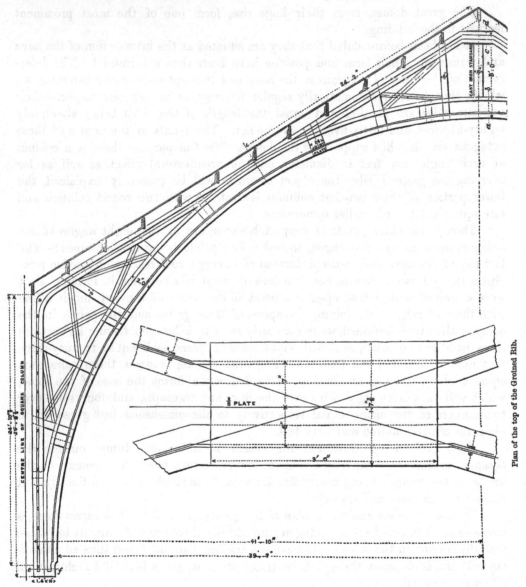

Elevation of Diagonal Ribs, supporting the Rib of Dome over Nave and Transepts.

The intersections of the principal rafters and semi-ellipses are connected by a cast-iron standard, which is continued up above the ridge of the roof to a point one hundred and seven feet from the nave floor line, this being the level of the bed on which the dome ribs rest.

The large columns at the angles of the octagon are two feet in external diameter with seven-eighths of an inch of metal, and they were raised in three lengths to a height of ninety-five feet, their ends being joined together by flanges and screwed nuts on the inside. To fasten the bolts, a man was lowered down inside the columns, the diameter of which is sufficient to give him room to screw up the nuts. The columns are thus kept perfectly smooth on the outside, and appear like one casting ninety-five feet long. To the top of each two-feet column is bolted a cast-iron stanchion twelve feet high, whose summit is therefore just one hundred and seven feet above the nave floor. On the tops of these stanchions, and resting on ornamental brackets, a gallery three feet wide is carried round the outside and inside of the drum. It is not, however, accessible to the public, but only to men employed in opening the louvres which are here placed for ventilation. To the upper side of the gallery, and through it to the stanchions, the double wrought iron tie-plate acting as the

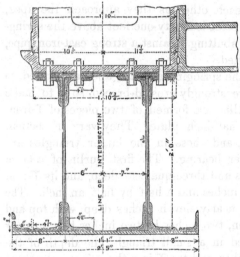

Section on Line *a b* in elevation, showing junction of Diagonal Ribs at crown of Arch.

Top Plate to receive Column.

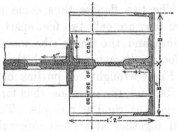

Section on Line *e f* in elevation, showing Section of Seat of Diagonal Rib on corner column of Dome.

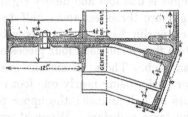

Section showing Seat of Square and Diagonal Ribs on cap of Column of Nave.

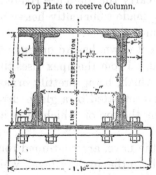

Section on Line *c d* in elevation, showing junction of Top of Diagonal Ribs.

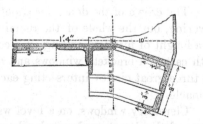

Section of Seat of Diagonal Rib, separated from Square Ribs.

dome's hoop, is securely bolted. It consists of an inner plate of iron six inches by three-eighths of an inch, which is connected with an outer plate ten inches by three-eighths of an inch, so that both these plates take the thrust of the dome. The dome ribs are bedded on the top of them, with their feet bolted through to the heads of the stanchions.

L

Each dome rib is an iron girder made of boiler plate and angle iron. The top and bottom flanges are nearly equal in section, the former being nineteen square inches and five-eighths, the latter twenty square inches and three-quarters. There is no continuous web between the two flanges, but they are joined at eight-feet intervals by two pieces of boiler plate, having a three-inch wood spacing piece between. The first seven feet of each rib is vertical, and the girder is here three and a half feet deep. At the summit of the vertical portion, which is one hundred and fourteen feet above the nave floor level, is the springing line The top flange follows the curve traced by a radius of ninety-one feet nine inches and seven-eighths, the centre being a point twelve feet and nearly three inches and a quarter beyond the centre of the dome ; the bottom flange is on a curve whose radius is ninety feet one and a quarter inches, and centre fourteen feet and half an inch beyond the dome's centre.

The two flanges thus come nearer each other as they approach the apex, where they are only two feet apart. This point is ninety-one feet above the springing line, and the twelve ribs meet there, abutting against a strong cast-iron pipe, one foot in diameter, to which they are bolted.

Eight wrought-iron purlins between the springing and the apex are bolted to the ribs, and the divisions thus formed are strongly crossed-braced, so as to make the whole as rigid as possible. These purlins are formed of two pieces of T-iron, joined together at six-feet intervals by a half-inch plate. They vary in section, decreasing as they approach the summit, and those in the larger triangles are slightly heavier, on account of their longer bearing. The first purlin of a large triangle is one foot and nearly eight inches and three-quarters deep, and its T-iron flanges are three feet and a half by three inches and a half by half an inch. The upper purlin has a continuous web-plate nearly eleven inches deep, with top and bottom flanges made of four pieces of L-iron, two inches by two inches by a quarter of an inch. These dimensions are altered in a small triangle, the depth of the lower purlin being nearly one foot nine inches, and the T-iron three inches by three inches by half an inch ; the upper purlin is here eleven inches deep, but the L-iron is the same as before. Wrought-iron sash-bars to carry the glass are riveted to the purlins every eighteen inches, every fifth bar being made sufficiently heavy to assist in the cross-bracing, and prevent the purlins twisting.

The crown of the dome, for about thirty-two feet down, has an ornamental zinc covering, but the whole of the remainder is glazed. From the apex rises the finial to a height of fifty feet, resting on a concave base, which, being prettily ornamented with cast-iron brackets, windows, and mouldings, is terminated by a globe surrounded by three great circles intersecting each other, from the top of which rises a gilded pinnacle.

Clerestory windows, on a level with those of the nave, are continued round the drum of the dome ; and panels above these reach to the gallery level, which has an ornamental railing all round.

These domes are the largest that have ever yet been executed, being one hundred and sixty feet in external diameter, and they have been the most difficult part of the works at Kensington. The dome of St. Peter's is one hundred and fifty-seven feet and a half in diameter, and that of St. Paul's one hundred and twelve feet ; but although the Exhibition domes are in themselves larger in every way than any yet constructed, they do not rise to so great a height above the ground as either of those with which they have been compared. The Exhibition domes spring from a height of one hundred and fourteen feet, and the top of the finial is two hundred and sixty feet above the ground, while the cross on St. Peter's is four hundred and thirty-four feet, and that of St. Paul's three hundred and forty feet, above the pavement. The height of the dome inside, from the top of the raised floor, is two hundred feet.

The scaffolds for the construction of these domes were on a greater scale than anything of the kind ever executed. They were literally forests of timber, looking at a distance like fine lace-work. They occupied nearly the whole interior space of the domes, and were cross-braced and bolted together in every possible way, so as to give them sufficient strength, for they had to bear the weight of the whole of the iron in the domes, one hundred and twenty tons in each.

The scaffold was carried up in eight different stages, between which were horizontal beams. The central portion was a square of twenty-four feet, rising to a height of two hundred feet. As this ascended each stage was cross-braced vertically. From the centre a scaffold radiated into each triangle of the dome, being triangular in shape, though not quite so large. These radiating scaffolds had independent vertical bracing, while at each stage they were cross-braced horizontally, and connected with the central scaffold as well as with each other. The main timbers in the scaffold were from fourteen inches to twelve inches square, while the cross-bracing was on an average twelve inches by six inches. This work was put up by Mr. Clemence, the contractor's clerk of works, and was looked upon as a triumph in scaffolding. It was of immense strength, and so skilfully constructed, that very little of the timber in it was spoiled by cutting, so that when taken down, every particle of wood used, amounting to forty thousand six hundred and seventy-two cubic feet in each scaffold, was as available for any other work as if it had just come from the builder's yard.

These scaffolds were completed in eight weeks, and every beam in them was hoisted by the steam-winch before described, without the aid of which they would have required at least double the time, and have been far more costly to execute.

The annexes, or temporary buildings adjoining the Exhibition, next require description.

The plan of having detached buildings for machinery will be a great improvement on the 1851 Exhibition, where everything was under the same roof; for, admirably arranged and ventilated as that building was, yet the smell of oil and grease inseparable from machinery, occasionally intruded itself on the general visitors.

The western annexe is nine hundred and seventy-five feet long; for a length of seven hundred and twenty feet it is two hundred feet wide, the remaining two hundred and fifty-five feet being one hundred and fifty feet wide. The east side is enclosed by the back wall of the west arcade of the gardens, and the west side, which adjoins the road, has a plain lath and plaster front. It is covered by a ridge and valley roof, supported on most ingeniously constructed light wooden ribs of fifty feet span, placed at fifteen-feet intervals. These ribs are similar in construction to those of the nave, that is, they are formed of planks nailed together, but they are very much lighter. The circular portion springs at a height of ten feet above the ground line. Its elevation is nearly half of a regular polygon, described

GENERAL VIEW OF THE BUILDING IN PROGRESS, BEFORE ADDITIONAL STORY AND REFRESHMENT COLONNADE.

about a semicircle whose diameter is fifty feet; it consists of three planks nine inches wide, the centre plank is one and a quarter inches thick, and has nailed to it on either side a three-quarter inch plank, the ends breaking joint all through.   The principal rafters, which are composed of two three-quarter inch planks, rise from a point twenty-eight feet above the ground, and meet above the curved ribs, so as to make the ridge five feet above the crown of the arch.   The upright, which has its foot morticed into a sleeper resting on piles, is formed of one and a quarter inch centre-plank, with a three-quarter inch plank on each side, having a strengthening piece four inches by three inches spiked to it on either side to prevent its bending. The principal rafter and upright are connected with the curved rib by radial pieces of one and a quarter inch plank, which are brought rather below the inner line of the curve, and finished off, for the sake of ornament, by a spear-head.   The roof-frames are therefore merely planks nailed together, and so disposed that the weight comes on their edge.   One half of the roof is covered with boards and felt, and the other half has a glazed skylight, with louvres for ventilation throughout the whole length. The span of each rib is fifty feet, so that in the two hundred feet width there are four spans, and in the hundred and fifty feet, three.

The western annexe, as before mentioned, is devoted to the exhibition of machinery in motion, for which purpose steam-pipes, water-pipes, and shafting are led through it.   There is a boarded floor all through, but the heavy machinery is of course bedded on the ground, independent of the floor, which is only used for passages.   The entrance to it is through the north end of the west transept, from which point the successive ribs of the roof afford a fine perspective view from end to end.

The superficial extent of the western annexe is one hundred and eighty-four thousand square feet, or about four and a half acres.   It is of itself a perfect exhibition of its kind, and contains the most ingenious mechanical contrivances of the age.

The building itself is in many respects worthy of its contents, for its ingenuity, economy, and simplicity.   It required no bolting or framing, and any person of ordinary intelligence, able to drive a nail, could have constructed the ribs, which have nothing in them but nails and sawn planks.   Each rib was made in a horizontal position, over a full-sized drawing, marked on a platform, and, when complete, it was hoisted vertically by means of a derrick.   To prevent it from wabbling, which, from its extreme thinness, it was very liable to do, it was stiffened while being raised by having scaffold-poles tied across the angles, which themselves formed the scaffolding for finishing the roof.

The frames are braced together at the top of the uprights, and the ribs are strutted from the wall-plate to prevent buckling.

The rain-water is let off by pipes attached to every third rib, to drains under the floor.

The eastern annexe is exactly similar to the western in its construction, but by having a large open court three hundred and fifty feet by one hundred feet left in it, its covered area is only ninety-six thousand square feet.   Its total length is seven hundred and seventy-five feet, and it is entered from the east transept by means of a covered communication or tunnel under the porch of the Horticultural Gardens.

This annexe is built for large agricultural implements, and any other heavy machines which do not require to be put in motion to show them off, and many large metallurgic, mineralogical, and geological specimens are also placed here.

The laying out of the works was commenced on the ninth of March, 1861, by three independent agencies—Mr. Marshall acted on the part of the contractors, while Mr. Wakeford, and Serjeant Harkin of the Royal Engineers, acted for the Commissioners.

Great care had to be taken with the measurements, for the slightest error would have thrown the work out considerably, and have occasioned great difficulty in fitting the girders. In the three separate measurements made, the mean variation was only three-eighths of an inch, a difference quite imperceptible in a piece of ground one thousand two hundred feet long by six hundred feet wide. A glance through any of the aisles will show how accurately the work has been conducted'; and whether they be examined on the square or diagonally, the columns will be found to range in line as perfectly as they would show in a plan.

About two weeks were occupied in making the measurements, so that the building may be said to have been actually commenced in the beginning of April, 1861, and from that date its progress was uninterrupted and rapid.

It is always interesting, and at the same time it gives a good idea of the size of any building, to state the quantities of the chief materials used in its construction.

There are seven millions of bricks used in the Exhibition building, and these have all been supplied by Messrs. Smeed, of Sittingbourne. Nearly all the cast-iron work was supplied from the Staveley iron-works in Derbyshire, and Mr. Barrow, the owner, superintended its manufacture. There are upwards of four thousand tons of this metal in the building, and to show what care was taken with the castings, we may mention that only four girders proved defective, by breaking in the hydraulic press.

There are upwards of eighty-two thousand and twenty-five feet of columns, equal in length to four miles, and if the one thousand two hundred and sixty-six girders used were placed end to end, they would reach a distance of six miles. The wrought iron was chiefly supplied by the Thames Iron Company, the builders of the "Warrior." This firm undertook the supply of all the iron for the domes, the groined ribs, the fifty-feet roofs, and the iron trellis girders which support them; but at the close of 1861, finding themselves behindhand, they gave up the western dome, and the iron-work for that was supplied and fixed by Messrs. Kelk and Lucas. The total quantity of wrought iron in connection with these parts amounts to one thousand two hundred tons. Mr. Ashton, who fixed the iron-work for Sir Joseph Paxton's two glass buildings, was charged with the same duty here.

The timber-work was executed partly at the works of Messrs. Lucas, at Lowestoft, and partly at Mr. Kelk's works at Pimlico. The former prepared all the window-sashes, &c., &c., by machinery; and the latter constructed the heavy ribs of the nave and transepts. Upwards of one million three hundred thousand superficial feet of floor have been laid down.

To cover the roofs four hundred and eighty-six thousand three hundred and eighty-six square feet of felt is used, equal to eleven acres; and to complete the whole of the glazing, five hundred and fifty-three thousand superficial feet of glass was required, which weighs two hundred and forty-seven tons, and would cover twelve acres and three-quarters.

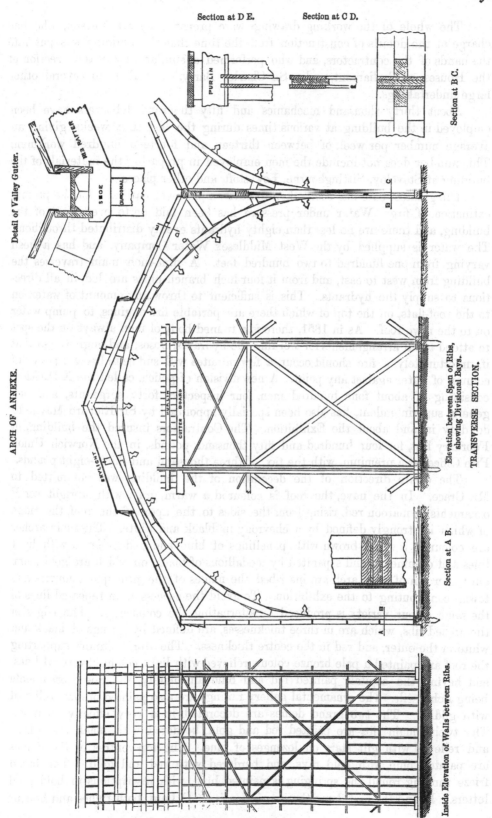

The whole of the working drawings were prepared by Mr. Meeson, who had charge of the details of construction from the time that the building was put into the hands of the contractors, and who performed a similar duty in the erection of the Houses of Parliament, under Sir Charles Barry, as well as in several other large undertakings.

About thirty thousand mechanics and fifty thousand labourers have been employed in the building at various times during the last sixty weeks, giving an average number per week of between thirteen and fourteen hundred workmen. This number does not include the men employed in preparing the materials of the building at Staveley, Sittingbourne, Lowestoft, and other places.

The greatest care has been taken to have ample means at hand for the prompt extinction of fire. Water under pressure has been laid on to every part of the building, and there are no less than eighty hydrants evenly distributed throughout. The water is supplied by the West Middlesex Water Company, and has a head varying from one hundred to two hundred feet. A nine-inch main traverses the building from west to east, and from it four-inch branch-pipes are led in all directions to supply the hydrants. This is sufficient to throw any amount of water on to the roof-flats, on the top of which there are portable fire-engines, to pump water on to the nave roof. As in 1851, there is a trained body of men always on the spot to attend to all arrangements for working the hydrants, hoses, and engines; so that if, unfortunately, a fire should occur, a few minutes will suffice to bring a powerful column of water against any point. A new division of police, called the X Division, consisting of about four hundred men, four inspectors, forty sergeants, and one general superintendent, has also been specially appointed by Sir Richard Mayne to do duty in and about the Exhibition. The Contractors insured the building, in February last, for four hundred and fifty thousand pounds, in the Norwich Union Fire Office, at a premium, with the tax, of three thousand and thirty-eight pounds.

The entire direction of the decoration of the building was committed to Mr. Crace. In the nave, the roof is coloured a warm gray, with upright scroll ornaments in maroon red, rising from the sides to the apex of the roof, the ridge of which is strongly defined by a chevrony in black and white. The main arches are coloured a warm brown with panellings of blue and red, relieved with light lines and ornaments, and separated by medallions of black, on which are gold stars. On the crown of each arch are inscribed the names of the principal countries and towns contributing to the exhibition. To avoid the succession of repeated lines of the same colour, variety is produced by alternating the colourings. The edges of the arched ribs, which are in three thicknesses, are defined by springs of black and white in the outer, and red in the centre thickness. The iron columns supporting the roof are painted a pale bronze colour, relieved with light coloured vertical lines, and having the capitals painted red and blue alternately, the raised ornaments being richly gilt. The ornamental gallery railings are also painted bronze, relieved with gilding. The two grand domes are decorated in a very effective manner. The twelve main ribs are painted red and gold, bordered with black and white, and relieved with gilt stars on lozenges of blue. The top centres of the domes are painted blue, with gold rays, and bordered with red and gold. The broad frieze running round the springing is painted blue, with inscriptions in bold gold letters. Round the east dome the inscription is: "O Lord, both riches and honour

PERSPECTIVE VIEW OF NAVE.

To face page 152.

come of thee, and thou reignest over all; and in thine hand is power and might; and in thine hand it is to make great." Round the western dome is the following inscription, taken from the first book of Chronicles: "Tua est Domine magnificentia, et potentia, et gloria, atque Victoria: et tibi laus: cuncta enim quæ in cœlo sunt, et in terra tua sunt, tuum Domine regnum." At the west end of the nave is: "Gloria in excelsis Deo et in terra pax;" and at the east end of the nave is: "The wise and their works are in the hand of God." The cornice above is painted principally in a red and gold colour. The walls above the arches under the freize are richly ornamented in red panelling. In the four smaller compartments are Europe, Asia, Africa, and America, and in the spandrils of the large arches are painted medallions containing figures representing arts, sciences, and manufactures. The walls at the ends of the nave and transepts are also richly ornamented and inscribed with appropriate legends. At the south-west end of the western transept is: "Deus in terram respexit et implevit illam bonis suis;" and at the north-west end is: "Domini est terra et plenitudo ejus." At the north-east end of the eastern transept is the following line from Cowper:—

"Alternately the nations learn and teach;"

and at the south-east end is another line from the same poet:—

"Each climate needs what other climes produce."

The picture galleries have their walls painted a sage green as a background for the pictures, and the cove is tinted to correspond, the cornices and soffits being vellum colour, relieved with maroon lines and ornaments. The wall round the arches is also ornamented.

To show with what energy it was necessary to carry out the work, we may mention that the whole of the picture galleries on the east side were painted and decorated in five days. Mr. Crace was asked on a Saturday if his designs were ready—he decided on the colours on Monday morning, the work was commenced at midday, and completed at midday on the following Saturday.

It was indispensable in all the designs for the decorations, that they should be so arranged as to be easy of execution; that the important principle of "the greatest effect at the least cost" should be strictly attended to. All ornament therefore had to be done by stencilling, and all the colouring on woodwork in distemper. Distemper is a very ready means of colouring surfaces, because one coat of it bears out and gives a result more solid and more luminous than four coats of oil paint, but it has the disadvantage of not being preservative like the last, and it cannot be washed. Perhaps no one will ever discover the very rough state of the principals of the roof of the nave, which are simply saw-cut, besides being blemished by the process of carting and lifting them. The coat of distemper conceals all that. It is composed of whiting and size, made of any tint required by adding the usual colours.

Stencilling is performed by cutting out the pattern required in stout, strong paper, which is then varnished over to strengthen and preserve it; taking care also to leave proper ties to keep the pattern together. In the progress of the work more than one hundred men were employed at this stencilling, out of whom, we have reason to believe, scarcely half a dozen ever did it before, and yet the task has been very well executed, and reflects much credit on all those engaged upon it.

It is proposed to raise sufficient funds to execute two large mosaic pictures, twenty-three feet high by thirteen feet wide, as experiments for decorating the panels of the outside walls, following out the plan before alluded to. The mosaics will be made of pottery, in geometric forms, by the pressure of dry powder. Various experiments in laying the mosaics have been made by Messrs. Minton (Stoke-upon-Trent) with mosaics of their own manufacture, and by Messrs. W. B. Simpson and Sons, of West Strand, with mosaics manufactured by Messrs. Maw. The experiments are very promising; and they prove that mosaic pictures may be as easily worked and used in England as in ancient Greece and Rome, or mediæval Italy. They will be as imperishable as the hardest and most perfect terra-cottas. They will create a new branch of industry, which may be worked in any locality, and probably by women as well as men.

The designs will illustrate Industry, Science, and Art. Some cartoons have been already prepared by Mr. Cope, R.A., Mr. J. C. Hook, R.A., Mr. Godfrey Sykes, and Mr. Townroe; two of these will be executed in mosaics as soon as the funds are provided.

The ornamental borders will be designed and the mosaics worked out under the superintendence of Mr. Godfrey Sykes and his assistants.

When two panels have been done, and all the necessary arrangements have been made, after the close of the Exhibition of 1862, for filling the others, designs from other subjects will be sought from the artists named below.

The following are the principal subjects which, at present, it is proposed should be executed, and the artists named are those who have already kindly consented to undertake to make designs for them, when the proper period arrives.

I. SUBJECTS ILLUSTRATING THE PRODUCTION OF RAW MATERIALS.

1. Agriculture, Holman Hunt; 2. Chemistry, W. Cave Thomas; 3. Fishing, J. C. Hook, R.A.; 4. Hunting, Frederick Leighton; 5. Metallurgy, Eyre Crowe; 6. Mining, F. Barwell; 7. Planting, &c., Michael Mulready; 8. Quarrying, G. F. Watts; 9. Sheep Shearing, C. W. Cope, R.A.; 10. Vintage, F. R. Pickersgill, R.A.

II. SUBJECTS ILLUSTRATING MACHINERY.

1. Astronomy, S. Hart, R.A.; 2. Engineering (reserved); 3. Horology (reserved); 4. Mechanics (reserved); 5. Navigation, J. E. Millais, A.R.A.; 6. Railways, R. Townroe.

III. SUBJECTS ILLUSTRATING MANUFACTURES AND HAND LABOUR.

1. Bricklaying, D. Maclise, R.A.; 2. Carpentry, R. Burchett; 3. China Painting, H. A. Bowler; 4. Glass Blowing (reserved); 5. Iron Forging, Godfrey Sykes; 2. Jewelry, D. G. Rossetti; 7. Lace Making, R. Redgrave, R.A.; 8. Metal Casting, A. Elmore, R.A.; 9. Printing, R. Redgrave, R.A.; 10. Straw Plaiting, C. W. Cope, R.A.; 11. Weaving, Octavius Hudson; 12. Pottery, Godfrey Sykes.

IV. SUBJECTS ILLUSTRATING FINE ARTS.

1. Architecture, W. Mulready, R.A.; 2. Painting, W. Mulready, R.A.; 3. Sculpture, W. Mulready, R.A.; 4. Music, J. C Horsley, A.R.A.

The designs, before they are executed, will be approved by a Committee of the Artists.

The Marquis of Salisbury, K.G., Mr. Layard, M.P., and Mr. Cole, C.B., act as a Committee of Management for carrying out the experiments, and all communications should be addressed to G. F. Duncombe, Esq., Secretary, South Kensington Museum, London, W.

The building can be viewed only as a utilitarian structure for the present. Although it thoroughly provides for the wants of the Exhibition, much remains to be done to render it complete and perfect. Perhaps no building in the world, twenty-four and a half acres in extent, has ever been erected at so low a liability as two hundred thousand pounds, capable of being sold for four hundred and thirty thousand pounds. This charge is only at the rate of twopence per cubic foot, whilst the rate for first-class dwelling-houses is one shilling and fourpence. The Houses of Parliament cost three shillings per cubic foot, and ordinary public buildings generally may be taken at from ninepence to a shilling per cubic foot. Economy has reigned paramount, and we can hardly expect one pound to do as much as six or eight pounds have done in other cases. When the Exhibition has succeeded, and the guarantors are free to deal with a surplus—when the Society of Arts becomes as much the proprietor of the whole, as it is now virtually the proprietor of a large portion of the Cromwell Road front—then we may venture to hope that that Society will be proud of its property, and have the means, as well as the desire, to render it, both inside and outside, a complete building, worthy of being the home of international exhibitions.

POSTSCRIPT.

THE opening of this Temple of Industry will teach many of us, if we need teaching, that the great scheme of creation never stands still. Many youths will have risen into manhood since 1851, many ripe men will have shrunk into old age, and many honoured and once familiar faces will have vanished from the crowd. We shall look in vain for one kingly presence—the keystone of the ceremony—and shall be reminded with much sorrow but with more hope that we are raising a monument to his memory. When the master dies the ship is not left a spectral hulk upon the stocks; she is launched to carry his name into the remotest corners of the earth.

We cannot easily overrate the courage and intellectual originality of a Prince who has taught thousands that there is no wisdom in despising the followers of trade. He did nothing to soil his Royal dignity when he shook hands with Commerce, and pointed out to Art-industry the road it should take. He guided his private conduct by the highest moral principles, and his public acts by that suggestive saying of his own great countryman, Goethe:—

"We should do our utmost to encourage the beautiful, for the useful encourages itself."

We mourn his loss, but only in a selfish spirit, and envy him his end, dying in the lap of his peaceful victories, in the full glory of a spotless and consistent life.

APPENDIX.

CHARTER OF INCORPORATION OF THE COMMISSIONERS FOR THE EXHIBITION OF 1862.

VICTORIA, by the grace of God of the United Kingdom of Great Britain and Ireland Queen, Defender of the Faith, to all to whom these presents shall come greeting: Whereas the Society for the Encouragement of Arts, Manufactures, and Commerce, incorporated by Charter under our Great Seal, bearing date at Westminster the 10th day of June, in the 10th year of our reign, and whereof our most dearly beloved Consort is President (and which Society is hereinafter referred to as the Society of Arts), did previously to the year 1851 establish and cause to be held from time to time, exhibitions of the products of industry and art, which exhibitions resulted in, or conduced to, the holding of the Exhibition of the Works of Industry of all Nations in the year 1851, and which last-named exhibition was attended with great success and public advantage: And whereas the said Society, in order to promote the objects for which it was in-corporated, is desirous that facilities should be afforded for holding from time to time international exhibitions of the products of industry and art, and it hath been represented to us by the said Society that many of its members and others of our loving subjects are desirous that such an international exhibition should be holden in the metropolis in the year 1862, or so soon after as conveniently may be, and the said Society is desirous that the entire control and management of such exhibition shall be confided to the Right Honourable Granville George Earl Granville, Lord President of our Council, and Knight of our most noble Order of the Garter; the Right Honourable Richard Plantagenet Campbell Temple Nugent Brydges Chandos Grenville, com-monly called Marquis of Chandos; Thomas Baring, Esq., M.P.; Charles Wentworth Dilke the younger, Esq.; and Thomas Fairbairn, Esq.; who are willing to undertake the duty of conducting such exhibition, provided that the holding thereof be approved by us, and that we should be willing to grant to them our Charter of Incorporation, to enable them to conduct and manage the same: And whereas it hath also been represented to us that it is essential to the success of such undertaking that we give our sanction thereto, in order that it may have the confidence not only of all classes of our subjects, but of the subjects of foreign countries, and for such objects, as well as for other the purposes herein appearing, the said Society hath besought us to authorize the said Earl Granville, the Marquis of Chandos, Thomas Baring, Charles Wentworth Dilke, and Thomas Fairbairn to carry into effect such undertaking, and to grant to them our Charter ot Incorporation: And whereas it hath been further represented to us that, with a view to the arrangements for the said exhibition, it will be necessary for the corporation to be hereby created forthwith to borrow sums not exceeding in the whole two hundred and fifty thousand pounds, and that the Governor and Company of the Bank of England, or other persons, will be willing to advance that sum, on having the repayment thereof secured by the covenant of the corporation, to be hereby created, and by the covenant of a sufficient number of other persons: And whereas it hath been further represented to us that, with a view of forwarding the undertaking, many of our loving subjects are willing to enter into proper covenants to effect such purpose, the covenants to be so framed as in the event of any payment being made thereunder, as far as practicable to subject the covenantors to bear such payment rateably, according to the amounts by them subscribed, but not exceeding in each case the amount of the subscription: And it hath also been represented to us that it is essential to the well-conducting of the affairs connected with the undertaking, and with the view of preventing disputes and litigation hereafter in reference thereto, that the general nature of the undertaking, as sanctioned and approved by us, and of the duties, rights, and powers of the persons conducting the same, shall, so far as conveniently may be, be defined, and shall be notified to all whom it may concern, by means of such charter: And whereas it is further represented to us, that under arrangements made between the said Society and the Commissioners for the Exhibition of 1851, incorporated by our Royal Charter bearing date the 15th day of August, in the fourteenth year of our reign, and continued and endowed with further powers by our Royal Charter bearing date the 2nd day of December, in the fifteenth year of our reign, those Commissioners have agreed to grant, rent free, the use of a certain site for the said Exhibition of 1862, subject to certain regulations relating to the approval by them of the buildings to be erected thereon, and with a provision that, in case the persons having the conduct of that exhibition should, before the 31st day of December 1862, give notice of the desire of the said Society to retain certain permanent buildings intended to be erected for the Exhibition of 1862, the Commissioners for the Exhibition of 1851 would grant to the said Society a lease of the site to an extent not exceeding one acre, whereon those permanent buildings should be erected, with a view, amongst other things, to assist the holding of future exhibitions; and in case the same persons shall, out of the profits of the undertaking, pay to the same Commissioners a sum of ten thousand pounds, those Commissioners have agreed to reserve (subject to certain conditions) a certain site for an exhibition to be held in the year 1872.   Now know ye that we, being earnestly desirous to promote the holding of an international exhibition of industry and art in the year 1862, do, by these presents, for us, our heirs and successors, give, grant, and ordain that the said Earl Granville, the Marquis of Chandos, Thomas Baring, Charles Wentworth Dilke, and Thomas Fairbairn, and the survivors and survivor of them, and such other persons, if any, as shall be appointed, in manner hereinafter provided, to be Commissioners, in lieu of them or any of them, shall be one body politic and corporate, by the name of " The Commissioners for the Exhibition of 1862," and by that name shall and may sue and be sued, implead and be impleaded, and shall have perpetual succession and a common seal, with full power to alter, vary, break, or renew the same at their discretion: And we will and ordain that the corporation hereby incorporated, hereinafter referred to as " Our Commissioners," is incorporated for the purpose of conducting and managing an international exhibition of the products of industry and art of all nations, such exhibition to be held in or near the metropolis in the year 1862, or within such further time as is hereinafter provided in that behalf; and we will and ordain that our Commissioners shall have the entire conduct, control, and management of the said exhibition, and of the funds that may arise from that undertaking, and that such exhibition may be carried on either in accordance with the precedent afforded by the Exhibition of 1851, or in such other mode or manner as our Commissioners shall in their discretion think fit, but subject to such special directions as are hereinafter contained.   And we will and ordain that our Commissioners shall have power to borrow and take up at interest for the purposes of the said undertaking, such sum or sums of money as they may think fit, and may from time to time for such purpose mortgage or pledge the funds or other property of the said corporation, and may under their common seal execute any deed or deeds of covenant or other deed or deeds for securing repayment of any sum or sums so to be borrowed, with interest, and may also procure any persons willing to guarantee the repayment of any such sum or sums or any part thereof, to execute a deed of covenant for payment of such sums as the covenantors may be willing to become liable for, so as to guarantee the due repayment of any sum or sums which may be so borrowed with interest, and all costs, charges, and expenses caused by the non-payment thereof, and that the said deed or deeds of covenant shall contain all necessary and proper provisions, and in particular provisions to insure, as far as practicable, that none of the covenantors shall ultimately bear more than his fair and proper proportion of the sums which they may respectively covenant to pay, and the several persons who shall make and enter into such covenants are hereinafter referred to as the guarantors, and the sum or sums of money which shall be so borrowed and secured to be paid are hereinafter referred to as " The Guaranteed Debt of the Corporation."   And we will and ordain that each of the several persons hereby incorporated, and any person who may, as hereinafter provided, be appointed in the place of any of them, may execute the said deed of guarantee in his own individual capacity for such sum as he may think fit.   And we do hereby direct and authorize our Commissioners to make and enter into such arrangements as they and the Commissioners for the Exhibition of 1851 may mutually agree upon, for holding the exhibition on a portion of the estate of those Commissioners at Kensington Gore, in accordance with the arrangements already made with them by the Society of Arts, or which may hereafter be made by our Commissioners with the Commissioners for the Exhibition of 1851, so as ;such other or further arrangements shall not, without the approval of the Society, be inconsistent with the arrangements already made between the Society and those Commissioners; or they may choose and contract for the occupation of any other site for holding the intended exhibition, provided such site be situate within ten miles from St. Paul's Cathedral, in the city of London, measured in a direct line. And we will and direct that in case the exhibition shall be held on any part of the lands of the Commissioners for the Exhibition of 1851, then that our Commissioners shall cause a sum not exceeding fifty thousand pounds to be expended on buildings of a permanent character, and such as may be adapted for the purposes for which the Society of Arts may require to have a lease of the site of such buildings, under the arrangements now made or contemplated between them and the Commissioners for the Exhibition of 1851, and which buildings are hereinafter referred to as " The Permanent Buildings." And we will and ordain that our Commissioners may contract for, erect, and, subject to such special directions as are herein contained, may remove, or may leave standing at the close of such exhibition, any building or buildings erected for the same in accordance with such arrangements as have been or shall be lawfully made in that behalf; and may, if they think fit, distribute prizes to exhibitors, and may do all matters and things connected with such distribution; and shall have full power to receive and take such sums of money as they may direct for entrance to the exhibition or for the rent of any part of the buildings to be erected or otherwise relating to the premises, and to dispose of all moneys which shall come to their hands as they shall think fit, for and towards the purposes of the said exhibition or otherwise, in the execution of the powers hereby given to them, including the payment of all expenses, charges, and liabilities which they may incur or become subject to; and that they shall have full power to give effectual discharges to any persons paying any moneys to them, and to settle and adjust any accounts relating thereto; and generally, to do all matters and things that may be necessary, or may appear to them to be expedient, for promoting the ends and designs of the said exhibition. And we do hereby ordain that it shall be lawful for our Commissioners, and they shall have full power and authority, from time to time, to depute or choose any persons, and to give to them all or any of the powers and authorities hereby given to our Commissioners as they shall think fit, for managing and conducting all or any of the matters and things hereby authorized to be done by our Commissioners, and which may be necessary for conducting, or in any manner relate to or concern the said exhibition. And we do hereby ordain that it shall be lawful for our Commissioners from time to time to appoint one or more secretaries and such other officers as they may think fit, and to remove all persons appointed by them, and to appoint others or not, as they see fit. And we do hereby ordain that our Commissioners may elect one member of their corporation to be the chairman thereof, and from time to time may vary such chairman as they think fit; and also that our Commissioners may elect such other person· or persons as they may think fit to be Commissioners in lieu of any one or more of them who may die or desire to be discharged from or become incapable to act in the execution of the office of Commissioner before the duties of such office shall be fully performed. And we will and ordain that such appointment of a Commissioner or Commissioners shall be made by a resolution to be passed at a meeting specially to be called for that purpose, but no appointment shall be effectual and valid unless and until the person or persons appointed shall be approved by us, such approval to be testified by a minute in writing, to be signed by one of our principal secretaries of state, and published in the ' London Gazette.' And we order and direct that our Commissioners shall meet when and at such place or places as from time to time they shall direct or determine, and that all and every the powers hereby given to our Commissioners may be exercised at any meeting of any two or more of the Commissioners, and that the decision of the majority attending at any meeting shall be binding, and determine any question proposed; and that when the votes shall be equal the chairman of the corporation for the time being, if present, shall, in addition to his vote as a member, have the casting vote, and that our Commissioners shall and may from time to time make and repeal or alter such rules, orders, regulations, and bylaws for the management of the business of the undertaking as they may think fit, so as the same be not contrary to the laws of this our realm, and such rules, orders, regulations, and bylaws shall when made, and till the same shall be repealed or altered, be as effectual as if they were contained in this our Royal Charter: Provided always, and we will and ordain that in case it shall appear to our Commissioners, from any cause not now foreseen, expedient to postpone the holding of such exhibition until some time in the year 1863, it shall be lawful for them, with the consent in writing of any one of our principal secretaries of state, to do so, by inserting in the London Gazette,' on or before the 1st day of March, 1862, notice that the said exhibition is to be so postponed, and in that case they shall and

M

may hold such exhibition accordingly in the year 1863 ; and in case after making any contracts or engagements for the holding of such exhibition, they shall from like cause see fit to abandon it altogether, they may, with the like consent, so do, giving like notice thereof, upon and subject to their making compensation to persons with whom they may have entered into any contracts in relation to the holding thereof, or incident thereto, which in such case we require and authorize them to make. And we do will and ordain that so soon as conveniently may be after the closing or abandonment of the exhibition, our Commissioners shall sell, dispose of, and convert into money, all property and effects belonging to them which can be so sold and converted, particularly all the buildings erected by them for the purposes of the undertaking, save and except "The Permanent Buildings." And we will and ordain that immediately after such sale and conversion into money, our Commissioners shall, out of the moneys to arise by such sale and conversion, or of which they shall be otherwise possessed, proceed, after payment of all costs, charges, and expenses incident to the undertaking, to pay and discharge, so far as such moneys will extend, in such order and priority as the law may require or our Commissioners see fit, all their debts and liabilities, save and except the guaranteed debt of the corporation ; and after payment of all such debts and liabilities, except as aforesaid, and providing and setting apart a reasonable sum for the payment of future expenses incident to the completion of their duties, our Commissioners shall apply the surplus of such moneys, if any, in or towards the payment and satisfaction of the guaranteed debt of the corporation, or in case the guarantors, or any of them, shall have been called upon to pay, and have paid, any moneys in respect of the guaranteed debt of the corporation, then in repaying to them, so far as the moneys applicable for such purposes will extend, the amount which the guarantors shall have so paid, in such manner, as far as practicable, as to secure that none of the guarantors shall pay more than his just and fair proportion of the sum which he shall have bound himself to contribute. And we will and ordain that as soon as may be after such sale and conversion as aforesaid, our Commissioners shall cause a statement of the accounts relating to the undertaking to be made up, and shall submit for examination the vouchers for the receipt and expenditure to the Governor of the Bank of England, the Deputy-governor of the Bank of England, and the Comptroller-general of the National Debt, or such person or persons as such Governor, Deputy-governor, and Comptroller-general, or any two of them, shall appoint to make such examination, and shall submit a duplicate of such statement to the Society of Arts for their information ; and our Commissioners shall then proceed to ascertain whether or not (having reference, if necessary, to the value of the permanent buildings, and calculating such value according to the amount such buildings are likely to realize if taken down and the materials sold) there has been a gain or loss attendant upon the undertaking, and shall forthwith certify, under their common seal, whether there shall have been a gain or loss, and, as near as may be, the estimated amount of such gain or loss, having reference to the value of the permanent buildings, and shall cause their certificate to be forthwith published in the 'London Gazette.' And in case, irrespective of the value of the permanent buildings, there shall have been a loss attending the said undertaking, then if the Society of Arts shall, with a view to obtain a lease of the permanent buildings in accordance with such arrangement as hereinbefore in that behalf mentioned, be willing out of their corporate funds to bear and sustain that loss, it shall be incumbent upon our Commissioners, if so required by the Society of Arts, by notice in writing under the hand of their secretary, to be delivered within one calendar month from the publication of such certificate, to make and enter into such arrangement with the Society as may secure to them the benefits of such lease, subject to the Society bearing such loss and undertaking to provide sufficient funds to enable our Commissioners to pay and satisfy all the remaining debts and liabilities of the said corporation, including the guaranteed debt of the corporation, or so much thereof as shall remain unpaid, and the Society undertaking to indemnify the guarantors from all loss and liability in respect thereof ; but in default of the said Society serving such notice in due time, or of their duly and effectually performing all acts to carry out such arrangement as provided for by the clause last hereinbefore contained, then our Commissioners shall forthwith, or so soon as conveniently may be, sell the permanent buildings, and out of the proceeds thereof, after payment of all cost incident to such sale, or otherwise incident to the undertaking and remaining unpaid, shall discharge all debts and liabilities, if any, attending the undertaking, remaining unpaid, except the guaranteed debt of the corporation, and shall apply the surplus, if any, in or towards satisfaction of the guaranteed debt of the corporation, or in case the guarantors, or any of them, shall have been called upon to pay, and have paid, any moneys in respect of the guaranteed debt of the corporation, then in repaying to them, so far as the moneys applicable for such purposes will extend, the amount which the guarantors shall have so paid, in such manner, as far as practicable, as to secure that none of the guarantors shall pay more than his just and fair proportion of the sum which he shall have bound himself to contribute; and if any surplus shall remain after all such payments, then such surplus shall be disposed of in manner hereinafter directed as to and concerning the ultimate disposable profit of the undertaking in case of there being a gain attending the undertaking. And we will and ordain that in case, after payment of all the debts and liabilities attending the undertaking, it shall be found that irrespective of the permanent buildings, there shall have been a gain attending the undertaking, then the permanent buildings shall be left standing for the Society of Arts, in accordance with the aforesaid arrangements, and out of such gain our Commissioners shall firstly pay to the Commissioners for the Exhibition of 1851, if desired by the Society of Arts, as hereinbefore recited, a sum not exceeding ten thousand pounds as a consideration for their reserving a site containing sixteen acres or thereabouts for an exhibition of the products of industry and art to be held in the year 1872, on the lands belonging to such Commissioners, and shall, secondly, apply in completing the permanent buildings in an architectural manner, and in a manner suitable for the objects for which they are to be employed by the Society of Arts, so much of the unexpended portion of the sum hereinbefore mentioned to be intended to be expended on the permanent buildings not exceeding fifty thousand pounds, as in the judgment of our Commissioners jointly with that of the Commissioners for the Exhibition of 1851 may be requisite for that purpose. And we will and ordain that if there shall remain any surplus of such gain arising from the same undertaking after all the payments hereinbefore provided for, such gain shall be considered as the ultimate disposable profit of the undertaking, and shall be disposed of as hereinafter in that behalf provided, viz., We will and ordain that our· Commissioners shall apply the ultimate disposable profit of the undertaking for such purposes connected with the encouragement of arts, manufactures, and commerce as shall be determined by the guarantors at a meeting to be called for the purpose, at such time and place, and in such manner, by advertisement or otherwise, as our Commissioners shall think fit, and whereof twenty-eight days' notice at the least shall be given, at which meeting the question to be determined shall be decided and settled by the votes of guarantors representing the majority in value of the subscriptions of the persons actually present and voting: Provided further, that before proceeding to ascertain the amount of each subscription, for the purpose of such decision, it shall be lawful for the chairman of the meeting to take a show of hands on any question to be submitted to the meeting, and his decision, if not objected to as to such show of hands, shall be considered conclusive and binding without the actual necessity of ascertaining the exact amount for which each guarantor shall have signed the agreement. And we will and ordain that the services of our said Commissioners shall be rendered gratuitously: but we direct that our Commissioners may, out of the corporate funds, allow and pay to their secretaries and officers, and other persons who may aid them in the conduct of such exhibition, such salaries and gratuities or other remuneration as they may think fit; and they may thereout also pay the costs, charges, and expenses incurred, or to be incurred, by the Society of Arts in promoting the said undertaking, and in getting the requisite instruments made and executed by the guarantors: Provided always, that when and as soon as any sum or sums of money which may have been borrowed by our said Commissioners under the powers aforesaid, and all interest thereon shall be fully paid, and all other the matters and things intrusted to be done by this our charter by the said Commissioners hereby incorporated shall be fully performed, or become incapable of being executed, and when the same shall have been certified under the corporate seal to one of our principal secretaries of state, then these presents, and every matter and thing herein contained, shall be absolutely void.

In witness whereof we have caused these our letters to be made patent. Witness ourself at our palace at Westminster, this fourteenth day of February, in the twenty-fourth year of our reign.

DEED OF GUARANTEE GIVEN TO SECURE AN ADVANCE TO BE MADE BY THE BANK OF ENGLAND TO THE COMMISSIONERS FOR THE EXHIBITION OF 1862.

THIS INDENTURE made the fifteenth day of February in the year of our Lord 1861, between the several persons who have subscribed their names in the first column of the second schedule hereto, and who have affixed their seals hereto, of the first part; and Sir Alexander Spearman, of the National Debt Office, London, Baronet; Matthew Marshall, of the Bank of England, Esquire; and George Earle Gray, of the Bank of England, Esquire; hereinafter called the said Trustees of the second part. Whereas the Commissioners for the Exhibition of 1862, incorporated by Her Majesty's Letters Patent, dated the fourteenth day of February, 1861, have applied to the Governor and Company of the Bank of England, and have requested them to advance the sum of two hundred and fifty thousand pounds, to be repaid with interest on the first of January, 1863, or if the exhibition should be postponed until the year 1863, then on the first day of January, 1864, which the said Governor and Company have consented and agreed to do, on the security of a covenant by the corporation (to be embodied in a separate deed) and of the covenants hereinafter contained. And whereas for the convenience of execution of this indenture by the parties hereto of the first part, the same has been prepared in ten parts. Now this indenture witnesseth that each party hereto, of the first part, for himself, his heirs, executors, and administrators, hereby covenants with the said Trustees, their executors and administrators, that the covenantor, his or her heirs, executors, or administrators, shall at any time, or from time to time when required so to do by the Trustees or Trustee for the time being of these presents, or any two of the Trustees for the time being (if more than two), at any time after the period when the repayment of the said loan shall become due, pay to the said Trustees, or the survivors or survivor of them, or the executors or administrators of the survivor of them, any sum or sums of money, not exceeding in the whole the sum set after his or her seal in the third column of the second schedule hereto: Provided nevertheless, and it is hereby agreed as follows:—

1. That when and so soon as the aggregate of the sums expressing the limits of the liability of the persons who shall become covenantors (by affixing their seals to any part of these presents) shall amount to two hundred and fifty thousand pounds at least, the Commissioners, or any two of them, shall sign and publish in the 'London Gazette' an advertisement in the form contained in the first schedule hereto, and unless and until an advertisement in such form or to the like effect shall be published, none of the parties hereto of the first part, shall be liable under the covenant hereinbefore contained.

2. That whenever any sum of money shall be required to be paid under the preceding covenant, a notice, signed by one at least of the Trustees for the time being, of the sum required to be paid, and of the day (not being less than twenty-eight clear days after the date of the notice) at which the same is to be payable, shall be delivered or sent by post to the usual or last known place of abode of the covenantor, or the executors or administrators of any covenantor, from whom such payment is required, or to the address mentioned in the fifth column of the second schedule for that purpose; and such notice shall also name a bank or banks into which, or one of which, the required payment may be made.

3. That the calls to be made by the Trustees or Trustee for the time being shall, subject to the provisions of the fourth stipulation, be made rateably on each person who shall become a covenantor hereunder by executing any part of these presents, according to the amounts set opposite to his or her seal in the second schedule to the part hereof which he or she shall so execute, the intention of the parties being, that as far as is practicable, and subject to the fourth stipulation, each covenantor and his estate shall ultimately bear his rateable proportion, and no more than his rateable proportion, of the entire sum to be raised.

4. That the Trustees or Trustee for the time being may abstain or desist from enforcing any requisition for payment against any covenantor or the estate of any covenantor, if they or he shall be satisfied that an attempt to enforce it will be useless or unproductive, and may compromise any such requisition against any covenantor or his estate, on such terms as to reduction of amount,

postponement of payment (with or without security), or otherwise as they or he shall think proper, and in determining the amount to be called for, may take into consideration the probability that some of the covenantors may be unable or cannot be compelled to pay their proportion of the sums to be paid.

5. That the moneys recovered under these presents shall be applied first in paying all costs and expenses incurred in the execution of the trusts, next in payment to the Governor and Company of the Bank of England of whatever sum may be due to them from the Commissioners, and any costs and expenses incurred by them by reason of the non-payment thereof, and the balance shall be returned by the Trustees or Trustee for the time being, to the persons who have contributed it, in such manner as they or he may think just, and calculated to give effect, as far as practicable, to the principle of rateable contribution between the covenantors, and that the Trustees or Trustee for the time being, may, after full satisfaction of the Commissioners' debt to the said Governor and Company, enforce the preceding covenant against any covenantor or his estate, for the purpose of setting right any discrepancy between what has been, and what ought according to that principle to have been, contributed by any contributor or contributors, and may apply for that purpose the money raised by so enforcing the covenant.

6. That no persons sued under the preceding covenant shall be entitled to resist the claim on the ground that the requisition for payment is improperly made, or unnecessary, or excessive in amount, or that no such notice as is required by the second stipulation has been given.

7. That if the said Commissioners shall keep a banking account with the Bank of England, and while the said sum of two hundred and fifty thousand pounds, or any part thereof, shall be due and owing by the said Commissioners to the Governor and Company of the Bank of England, any sum shall be standing or paid to the credit of such banking account, the said Governor and Company shall not be bound or required to retain or apply any such sum as last aforesaid in or towards payment of what may be due in respect of the said sum of two hundred and fifty thousand pounds. In witness whereof the said parties hereto have hereunto set their hands and seals, the day and year first above written.

The FIRST SCHEDULE above referred to.

We hereby certify that the aggregate of the sums, expressing the limits of the liability of the persons who have executed the deed of guarantee for enabling the Commissioners for the Exhibition of 1862 to obtain advances from the Bank of England, amounts to the sum of two hundred and fifty thousand pounds.

Signatures of the Commissioners, or any two of them.

The SECOND SCHEDULE above referred to.

1st Column.	2nd Column.	3rd Column.	4th Column.	5th Column.	6th Column.
The Signature of the Guarantor.	The Seal of the Guarantor.	Sum for which the Guarantor renders himself liable.	Christian and Surname (or the usual Name) of the Guarantor to be written in full length by the Witness who attests the Signature at, or immediately after, the time at which the Signature of the Guarantor is affixed.	Address of the Guarantor, to which address all Notices may be sent, Notices so sent being hereby declared sufficient for all purposes connected with these presents or the intended Exhibition.	Name and Address of the Witness attesting the Signature and sealing of the Agreement by the Guarantor whose Name is set opposite to that of the Witness in the fourth Column of this Schedule.

DECISIONS REGARDING JURIES.

HER MAJESTY'S COMMISSIONERS:

THE EARL GRANVILLE, K.G., Chairman.

THE DUKE OF BUCKINGHAM AND CHANDOS.

SIR C. WENTWORTH DILKE, Bart.

THOMAS BARING, Esq., M.P.

THOMAS FAIRBAIRN, Esq.

F. R. SANDFORD, *Secretary and General Manager.*

Special Commissioner for Jury Department . . DR. LYON PLAYFAIR, C.B., F.R.S.

Secretary of Juries . . . . J. F. ISELIN, M.A.

GENERAL CONDITIONS.

1. Objects exhibited in the Industrial Department of the Exhibition have been divided into thirty-six head classes, which, for the convenience of the Juries, have in several cases been subdivided into sections. The classes, with their sections, amount to sixty-five in number.

2. Each class, when not subdivided, and each section of a divided class, will have a separate Jury, but the sectional Juries of a class will be associated as one united Jury in the election of a deputy-chairman and in the confirmation of awards. (Vide §§ 7 and 8.)

3. The following is the list of classes and sections for which Juries are appointed:—

Class 1. Mining, quarrying, metallurgy, and mineral products.
  „ 2. Chemical substances and products, and pharmaceutical processes.
      Section *a.* Chemical products.
      „   *b.* Medical and pharmaceutical products and processes.
  „ 3. Substances used for food.
      Section *a.* Agricultural produce.
      „   *b.* Drysaltery, grocery, and preparations of food as sold for consumption.
      „   *c.* Wines, spirits, beer, and other drinks, and tobacco.
  „ 4. Animal and vegetable substances used in manufactures.
      Section *a.* Oils, fats, and wax, and their products.
      „   *b.* Other animal substances used in manufactures.
      „   *c.* Vegetable substances used in manufactures, &c.
      „   *d.* Perfumery.
  „ 5. Railway plant, including locomotive engines and carriages.
  „ 6. Carriages not connected with rail or tram roads.
  „ 7. Manufacturing machines and tools.
      Section *a.* Machinery employed in spinning and weaving.
      „   *b.* Machines and tools employed in the manufacture of wood, metal, &c.
  „ 8. Machinery in general.
  „ 9. Agricultural and horticultural machines and implements.
  „ 10. Civil engineering, architectural, and building contrivances.
      Section *a.* Civil engineering and building contrivances.
      „   *b.* Sanitary improvements and constructions.
      „   *c.* Objects shown for architectural beauty.
  „ 11. Military engineering, armour and accoutrements, ordnance and small arms.
      Section *a.* Clothing and accoutrements.
      „   *b.* Tents, camp equipages, and military engineering.
      „   *c.* Arms and ordnance.
  „ 12. Naval architecture—ships' tackle.
      Section *a.* Ships for purposes of war and commerce.
      „   *b.* Boats, barges, and vessels for amusement, &c.
      „   *c.* Ships' tackle and rigging.

Class 13. Philosophical instruments, and processes depending upon their use.
„ 14. Photographic apparatus and photography.
„ 15. Horological instruments.
„ 16. Musical instruments.
„ 17. Surgical instruments and appliances.
„ 18. Cotton.
„ 19. Flax and hemp.
„ 20. Silk and velvet.
„ 21. Woollen and worsted, including mixed fabrics generally.
„ 22. Carpets.
„ 23. Woven, spun, felted, and laid fabrics, when shown as specimens of printing or dyeing.
„ 24. Tapestry, lace, and embroidery.
„ 25. Skins, furs, feathers, and hair.
    Section a. Skins and furs.
      „ b. Feathers, and manufactures from hair.
„ 26. Leather, including saddlery and harness.
    Section a. Leather and manufactures generally made of leather.
      „ b. Saddlery, harness.
„ 27. Articles of clothing.
    Section a. Hats and caps.
      „ b. Bonnets and general millinery.
      „ c. Hosiery, gloves, and clothing in general.
      „ d. Boots and shoes.
„ 28. Paper, stationery, printing, and bookbinding.
    Section a. Paper, card, and millboard.
      „ b. Stationery.
      „ c. Plate, letterpress, and other modes of printing.
      „ d. Bookbinding.
„ 29. Educational works and appliances.
    Section a. Books and maps.
      „ b. School fittings, furniture, and apparatus.
      „ c. Appliances for physical training, including toys and games.
      „ d. Specimens and illustrations of natural history and physical science.
„ 30. Furniture and upholstery, including paper-hangings and papier-mâché.
    Section a. Furniture and upholstery.
      „ b. Paper-hangings and general decoration.
„ 31. Iron and general hardware.
    Section a. Iron manufactures.
      „ b. Manufactures in brass and copper.
      „ c. Manufactures in tin, lead, zinc, pewter, and general braziery.
„ 32. Steel cutlery and edge tools.
    Section a. Steel manufactures.
      „ b. Cutlery and edge tools.
„ 33. Works in precious metals, and their imitations, and jewelry.
„ 34. Glass.
    Section a. Stained glass, and glass used in buildings and decorations.
      „ b. Glass for household use and fancy purposes.
„ 35. Pottery.
„ 36. Dressing-cases, despatch-boxes, and travelling-cases.

4. The articles in the building are arranged as much as possible in the sixty-five classes and sections, so as to be coincident with the field of action of each Jury, and to facilitate its labours.

5. If exhibitors accept the office of Jurors, or Associate Jurors or Experts, they cease to be competitors for prizes in the class or section to which they are appointed, and these cannot be awarded either to them individually, or to the firms in which they may be partners.

6. Juries may take evidence when a majority of the Jury deem it advisable, and name the persons to be consulted as Experts: Jurors of another class may also be called in aid by a Jury, when a knowledge involved in that class is required. The Juries, should they desire to secure such aid permanently, may appoint the persons so called in to be "Associate Jurors," but they will not possess a vote. Various nations having returned the names of persons fitted to act as Supplementary Jurors or Deputies to the several classes, lists of these will be published with the view of enabling Juries to obtain their aid, should the Juries so desire.

7. The awards made by the Sectional Juries must be submitted to, and be confirmed by, the Head Jury of the class, which consists of all the Sectional Juries united.

8. Before a Jury can finally make their awards, they must be submitted to a council, consisting of the chairmen of the Juries, in order to secure uniformity of action, and a compliance with the regulations originally laid down by that body.

9. The awards of a Jury, when reported by the Council of Chairmen as being made in conformity to the rules, are final.

10. The Juries will commence their duties on Wednesday, the 7th of May, at ten o'clock, and will be aided in the general transaction of their business by Dr. Lyon Playfair, C.B., who has been appointed Special Commissioner in charge of the Department of Juries. The Special Commissioner, either personally or by deputy, may be present at their deliberations, but he will not have a vote, or interfere in the adjudication of awards.

CONSTITUTION OF JURIES.

11. The Juries consist of the nominees of the Foreign Commissions, with the addition of a certain number of British and Colonial Jurors.

12. Each of the thirty-six Head Juries will be presided over by a chairman to be nominated by Her Majesty's Commissioners, and he will be aided by a deputy-chairman, to be elected at the first meeting of the Jury.

13. The Sectional Juries will elect presidents to take the chair at their meetings; but these Sectional Presidents will not have a seat in the Council of Chairmen.

14. Juries will appoint one of their own body as a reporter or secretary to record the results of their deliberations.

15. Any foreign nation which may not have a representative on a Jury, may nominate one of the actual Jurors as a person through whom they can convey official information in regard to the exhibits and interests of that nation.

16. In the event of any nominee of foreign commissions being obliged to leave the country before the completion of the work of the Juries, a Deputy Juror may be substituted in his place. Such deputies, when announced as likely to be required, may be elected as associates, if desired by the Juries.

COUNCIL OF CHAIRMEN.

17. The chairmen of the thirty-six Head Juries will be associated as a body, to be called the " Council of Chairmen," and will be presided over by the Right Honourable Lord Taunton, appointed by Her Majesty's Commissioners the President of the Council.

18. In the absence of a chairman, the deputy chairman of a Jury will take his seat at the Council.

19. The Council of Chairmen will be constituted of British subjects and of at least an equal number of foreigners.

20. The Council of Chairmen will have to determine the general mode of action of the Juries, and to define the general principles to which it will be advisable to conform in making the awards in the several departments of the Exhibition. It is the wish of Her Majesty's Commissioners that medals should be awarded to articles possessing decided superiority, of whatever nature that superiority may be, and not with reference to a merely individual competition.

21. The Council of Chairmen must see that the awards of the several Juries are in accordance with the rules laid down, before they are considered final.

22. It will be desirable that the Council of Chairmen should hold a meeting before the . assembling of the Juries; and this meeting will be held on Saturday the 3rd of May, at twelve o'clock at noon.

23. In order to represent the wishes of Her Majesty's Commissioners, and to explain its rules, the Special Commissioner will attend the meetings of the Council, and aid it in the transaction of business; but he will not possess a vote, or act as a member of the Council.

MEETING OF JURIES.

24. The Jurors, on being appointed, will receive immediate notice of appointment, and their names will be published.

25. The Juries will meet for the transaction of business on Wednesday the 7th of May. The place, day, and hour of each meeting of the Juries will be fixed by the chairman, or in his absence

by the vice-chairman. The meetings of the Sectional Juries will in like manner be determined by the president of these Sectional Juries, to be elected at their first meeting. Notice of the days and hours of meeting will be given at the office of the Secretary of Juries, from whence summonses will be issued. The days of meetings will also be posted in the place where the Jury usually meet, and near the secretary's office.

26. In the event of an equality of votes at any meeting of a Jury or Sectional Jury, the chairman for the time being shall have a casting vote.

27. Although it will be impossible to set apart special days in which the Juries alone can examine the articles exhibited, to the exclusion of the public, arrangements will be made to carry on these examinations with as little inconvenience as possible.

28. *Jurors, immediately on their arrival in London, will report themselves to J. F. Iselin, Esq., M.A., the Secretary of Juries, at the Jury Office, on the north side of the main entrance to the Exhibition Building from Prince Albert's Road, where they will obtain their Juror's Ticket, which entitles them to admission, and receive all necessary information.*

29. *All correspondence relating to the work of the Juries is to be addressed to J. F. Iselin, Esq., M.A., Secretary of Juries, Jury Office, Exhibition Building, South Kensington, W.*

POSITION OF CLASSES IN THE BUILDING.

No. of Class.	Nature of Articles Exhibited.	Position in Building.
1.	Mining, quarrying, &c.	South Court, Eastern Annexe.
2.	Chemical substances	Eastern Annexe, South-east Passage.
3.	Substances used as food	} Eastern Annexe, East Side.
4.	Animal and vegetable substances	
5.	Railway plant	Western Annexe.
6.	Carriages	South-east Court.
7.	Manufacturing machines and tools	Western Annexe.
8.	Machinery in general	Western Annexe.
9.	Agricultural implements	West side of Eastern Annexe.
10.	Civil engineering	South Court.
11.	Military engineering	South Court.
12.	Naval architecture	South Court.
13.	Philosophical instruments	Gallery, North Court.
14.	Photography	Central Tower, and Gallery, North Court.
15.	Horological instruments	Gallery, North Court.
16.	Musical instruments	North Court.
17.	Surgical instruments	Gallery, North Court.
18.	Cotton	Gallery, South.
19.	Flax and hemp	Gallery, South-east Transept.
20.	Silk and velvet	South-east Gallery.
21.	Woollen and worsted	South-east Gallery.
22.	Carpets	Under North-east Gallery and on Gallery walls.
23.	Woven, spun, felted, and laid fabrics	South-east Gallery.
24.	Tapestry, lace, and embroidery	South-east Gallery.
25.	Skins, fur, feathers, and hair	Transept South Court.
26.	Leather	Transept South Court.
27.	Articles of clothing, &c.	South-east Angle.
28.	Paper, stationery, &c.	Gallery, North Court.
29.	Educational works	Central Tower.
30.	Furniture and upholstery	North Court.
31.	Hardware	South Court.
32.	Steel, cutlery, and edge tools	Transept, South Court.
33.	Precious metals	South Court, Central Division.
34.	Glass	South Court, Central Division.
35.	Pottery	North Court, Central Division.
36.	Manufactures not included in previous Classes	Gallery, North Court.

POSITION OF FOREIGN COUNTRIES IN THE BUILDING.

Name.	Ground.	Gallery.
France	S.W. Court.	S.W. Gallery.
Zollverein	S.W. Transept.	S.W. Transept. Gallery.
Austria	N.W. Transept.	N.W. „
Belgium	N.W. Court, No. 1.	N.W. Gallery, No. 1."
Holland	„ „ 2.	
Switzerland	„ „ 3.	„ „ 2.
Denmark	„ „ 4.	„ „ 3.
Norway and Sweden . .	„ „ 5.	„ „ 4.
Russia	„ „ 6.	„ „ 5.
Costa Rica	N.E. Court, No. 1.	
Peru	„ „ 2.	
Uruguay	„ „ 3.	
Venezuela	„ „ 4.	
Ecuador	„ „ 5.	
Argentine Republic . .	„ „ 6.	
Brazil	„ „ 7.	
Turkey	„ „ 8.	
Spain	S.W. Court, No. 2.	Spain, S.W., No. 2.
Portugal	S.C. Court, No. 1.	Portugal, S.C., No. 1.
Italy	„ „ 2.	Italy, S.C., No. 2.
Rome	„ „	
America	S.E. Court.	
Hanse Towns . . .	S.W. Transept, No. 2.	
Canada	⎫	
New Brunswick . . .	⎪	
Prince Edward's Island .	⎪	
British Columbia . . .	⎪	
Vancouver's Island . .	⎪	
Nova Scotia	⎪	
Tasmania	⎪	
Victoria	⎪	
Bermuda	⎬ N. E. Transept, East Side.	
Newfoundland . . .	⎪	
Ceylon	⎪	
Malta	⎪	
Jamaica	⎪	
Dominica	⎪	
St. Vincent	⎪	
Trinidad	⎪	
Barbados	⎭	
British Guiana	⎫	
New South Wales . . .	⎪	
Queensland	⎪	
Victoria	⎪	
South Australia . . .	⎪	
Hayti	⎪	
Bahamas	⎪	
Natal	⎬ Nave, North Side.	
Western Australia . . .	⎪	
New Zealand	⎪	
Ionian Islands	⎪	
Japan, China	⎪	
St. Helena, West Africa	⎪	
Siam, Liberia	⎭	

References :—S.W., South-west ; N.W., North-west ; N.C., North Centre ; S.C., South Centre.

INTERNATIONAL EXHIBITION OF 1862.

LIST OF GUARANTORS.

Those marked with an Asterisk (*) are Members of the Society of Arts.

*HIS ROYAL HIGHNESS THE PRINCE CONSORT, K.G. £10,000

*Abel, Frederick Augustus, F.R.S., Royal Crescent, Woolwich, s.e. £100	
*Acland, Sir Thomas Dyke, Bart., F.R.S., Killerton, Exeter 200	
*Acland, Henry W., M.D., F.R.S., Oxford ... 100	
*Adam, James Skipper, 8 Philpot Lane, e.c. ... 200	
*Adams, George G., 126 Sloane Street, s.w. ... 100	
Adams, Wm. Salkeld (W. S. Adams and Sons), 57 Haymarket, s.w. 100	
Agnew, Thomas, Manchester 100	
Agnew, Thomas, jun., Manchester 100	
Agnew, William, Manchester 100	
Ainsworth, Thomas, Cleator, Whitehaven ... 500	
*Akroyd, Edward (Jas. Akroyd and Son), Halifax 500	
*Aldam, William, Frickley Hall, near Doncaster 100	
*Alexander, H. B., The Laurels, Barnes, s.w. ... 100	
Alexander, James, 10 Porchester Terrace, Bayswater, w. 500	
*Aley, Frederick William, 8 Thurloe Place, s.w. 100	
*Alger, John, 16 Oakley Square, n.w. ... 100	
Allen, C. Bruce, Architectural Museum, South Kensington, s.w. 100	
*Allen, Edward Ellis, 2 Brunswick Place, Brompton, s.w. 100	
Allin, Thomas C., 23 Onslow Square, s.w. ... 100	
*Allhusen, Christian, Gray Street, Newcastle-on-Tyne 1000	
Allsopp, Henry (Samuel Allsopp and Sons, Burton-on-Trent), Hinslip Hall, Worcester ... 1000	
Amies, Nathaniel Jones, Manchester 100	
*Anderson, Arthur, 122 Leadenhall Street, e.c. ... 500	
*Anderson, Sir James, Glasgow 100	
Anderson, Thomas, M.D., Glasgow ... 100	
*Anderton, James, 20 New Bridge Street, e.c. ... 200	
*Andrew, W. P., 26 Montagu Square, w. ... 500	
*Angell, Joseph, 10 Strand, w.c. 500	
*Ansted, David Thomas, Bon Air, Guernsey ... 100	
*Antrobus, Sir Edmund, Bart., 146 Piccadilly, w. 2000	
Anworth, W. S. (Middleton and Anworth), Norwich 100	
*Appold, John George, F.R.S., 23 Wilson Street, Finsbury, e.c. 1000	
Archer, Professor J. C., Edinburgh 100	
Armitage, Benjamin (Elkanah Armitage and Sons), Manchester 1000	
Armitage, William (Armitage and Rigbys), Manchester 500	

*Armstrong, Sir Wm. George, C.B., Newcastle-on-Tyne £500	
*Artingstall, George, Warrington 500	
*Ashburton, Lord, Bath House, Piccadilly, w. ... 3000	
Ashford, Wm. (W. and G. Ashford and Winder), Birmingham 100	
Ashton, Thomas, Manchester 500	
*Asprey, Charles, 166 New Bond Street, w. ... 1000	
*Atkinson, Wm., 47 Gordon Square, w.c. ... 100	
*Austin, George, 7 London Street, e.c. ... 100	
*Bagnall, Charles, Tootherley Hall, Lichfield ... 500	
Bagley, Thos. (Henderson and Co.), Durham ... 300	
*Bake, Henry, 8 Philpot Lane, e.c. 200	
Baker, Anthony Kington, Cheltenham ... 100	
*Baker, William, 30 Cranbourne Street, w.c. ... 100	
Balderson, Henry, Corner Hall, Hemel Hempstead 100	
Balfour, George Edmund, Manchester ... 250	
*Ball, John, 3 Moorgate Street, e.c. 500	
Balleras, Guillermo Esteban, Seville Villa, Carlton Hill, St. John's Wood, n.w. 500	
Baring, Thos., M.P., 41 Upper Grosvenor St., w. 3000	
Barker, T. Herbert, M.D., F.R.S., Bedford ... 100	
Barlow, Fredk. Pratt (John Dickinson and Co.), 61 Old Bailey, e.c. 1000	
Barlow, Edward, Manchester 100	
Barnard, John Ansley Louis, 8 Alfred Villas, Albert Road East, Dalston, n.e. 100	
Barnett, Sampson, 23 Forston Street, n. ... 100	
Barrer, Victor, 1 Ironmonger Lane, e.c. ... 100	
*Barrett, Henry (R. Barrett and Sons), Beech Street, Barbican, e.c. 200	
Barruchson, Arnold, The Downs, Waterloo, Liverpool 200	
*Bartholemew, Charles, Broxholme, Doncaster ... 100	
*Bartlett, Wm. Edward, 8 King William St., e.c. 100	
Bass, Michael Thomas, M.P., Burton-on-Trent ... 1000	
*Bateman, J. F., F.R.S., 16 Great George Street, Westminster, s.w. 200	
*Bateman, Joseph, LL.D., 24 Bedford Place, Kensington, w. 100	
*Bates, Joshua, 21 Arlington St., St. James's, s.w. 3000	
*Batty, George, Pavement, Finsbury, e.c. ... 100	
*Bax, Edward, 1 Charing Cross, s.w. 100	
*Bazley, Thomas, M.P., Manchester 1000	
Beale, Sam., M.P., 19 Park St., Westminster, s.w. 1000	
Beaumont, John, Dalton, Huddersfield ... 500	

Beaumont, Joseph, jun., Huddersfield ...	£250
*Beckwith, Edward Lonsdale (Boord, Son, and Beckwith), Bartholomew Close, E.C. ...	500
*Begbie, Thos. S., 4 Mansion House Place, E.C. ...	500
Behrens, Solomon Levy, Manchester	300
*Belcher, Sir Edward, Union Club, S.W. ...	100
*Bell, John, 15 Douro Place, Victoria Road, Kensington, W.	100
Bellville, W. G., 64 Red Lion Street, Holborn, W.C.	100
*Bendon, George, 50 High Holborn, W.C. ...	100
*Benedict, Jules, 2 Manchester Square, W. ...	300
Benham, Frederick, 19 Wigmore Street, W. ...	500
Bennett, Henry, 166 Gresham House, City, E.C.	200
*Bennett, John, 65 Cheapside, E.C.	500
*Bennett, John, 50 Westbourne Park Villas, W. ...	500
Bennett, Wm. Cox, 62 Cornhill, E.C. ...	100
*Bentall, Edward Hammond, Heybridge, Maldon, Essex	100
Bentley, Joseph, 13 Paternoster Row, E.C. ...	100
*Benyon, Richard, M.P., 34 Grosvenor Square, W.	1000
Besley, Frederick, Nicholson's Wharf, E.C. ...	200
Besley, Robert, Fann Street, E.C.	200
Bessborough, Earl of, 40 Charles Street, Berkeley Square, W.	300
*Best, William, Gresham Club, E.C.	100
*Beveridge, Erskine, Dunfermline	500
Bevington, James B., Neckinger Mills, Bermondsey, S.E.	100
Bevington, Samuel B. (Bevington and Sons), Bermondsey, S.E.	100
*Beyer, Charles Frederick (Beyer, Peacock, and Co.), Manchester	1000
Bicknell, Henry Sandford, 3 High Street, Newington Butts, S.	500
*Biddle, Daniel, 81 Oxford Street, W.	500
Billinge, James, Ashton, near Wigan ...	100
Bingley, Alfred William, Bath Hotel, Arlington Street, Piccadilly, S.W.	100
*Bird, George, 58 Edgeware Road, N.W. ...	500
Bird, Stephen, Hornton Villa, Kensington, W. ...	250
*Bird, William, 2 Laurence Pountney Hill, E.C....	100
*Birkbeck, George Henry (Birkbeck and *Tongue), 34 Southampton Buildings, W.C.	300
*Birley, Richard, Sudley, Pendleton, Manchester	500
*Black, Henry, 1A Berners Street, Oxford Street, W.	100
Blacklock, William Thomas, Manchester ...	500
Blackwell, Samuel, 259 Oxford Street, W.C. ...	200
*Blackwell, Thomas (Crosse and Blackwell), Soho Square, W.	300
*Blair, Harrison, Manchester	200
*Blake, H. Wollaston, F.R.S., 8 Devonshire Pl., W.	100
Blandy, Jno. Jackson, Highgrove, Reading, Berks	100
*Blashfield, John Marriott, Stamford Pottery, Stamford	100
Blews, W. H. M. (Wm. Blews and Son), Birmingham	100
*Bodkin, William Henry, West Hill, Highgate, N.	500
*Boileau, Sir John P., Bart., F.R.S., 28 Upper Brook Street, W.	500
Bowker, Charles Hardy, Manchester	100
*Bowley, Robert Kanzor, 53 Charing Cross, W.C.	500
*Bowring, Edgar Alfred, Board of Trade, S.W. ...	200
*Braby, Fredk., Fitzroy Works, Euston Road, N.W.	100
*Bradley, J. W., 47 Pall Mall, S.W.	100

*Bragg, John (T. and J. Bragg), Birmingham ...	£150
*Branston, Robt. E., Denmark Hill, Camberwell, S.	200
*Brassey, Thomas, 4 Great George Street, Westminster, S.W.	2000
Breach, J. G., Burlington Hotel, Cork Street, W.	250
*Breffit, Edgar, 61 King William Street, E.C. ...	1000
Brett, John W., 2 Hanover Square, W. ...	500
Brewster, Sir David, K.H., University, Edinburgh	100
*Bridson, Henry, Harwood, near Bolton ...	250
Brigg, John Fligg (J. F. Brigg and Co.), Huddersfield	150
Briggs, George, 45 Wigmore Street, W. ...	100
Brock, William, Exeter	250
*Brocklehurst, Thos, Unett, The Fence, Macclesfield	100
Brodie, John Lamont, Manchester	100
*Brook, Charles, Huddersfield	200
*Brook, Charles, jun. (Jonas Brook and Bros.), Meltham Mills, Huddersfield	1000
*Brooke, Charles, F.R.S., 16 Fitzroy Square, W.	100
Brooke, Edward, Norton Lodge, Timperley, Cheshire	250
Brooke, George (Starkey Bros.), Huddersfield ...	1000
*Brooke, Thomas (John Brooke and Sons), Huddersfield	1000
*Brooks, Vincent, 1 Chandos Street, W.C. ...	200
*Brooman, Richard Archibald, 166 Fleet St., E.C.	250
*Brough, Geo. (Stratton and Brough), 3 Coventry Street, W.C.	200
Brown, Mrs. H., at Miss Burdett Coutts's, Stratton Street, W.	500
Brown, John (John Brown and Co.), Shirl Hill, Sheffield	250
*Brown, Michael Lewis, 47 St. Martin's Lane, W.C.	100
Browning, Jno. (Spence and Browning), Minories, E.C.	250
*Brunlees, James, 5 Victoria St., Westminster, S.W.	1000
Brunswick, George, 72 Newman St., W. ...	100
*Brunswick, Myrthil, 26 Newman Street, Oxford Street	100
*Buccleuch, Duke of, 37 Belgrave Square, S.W. ...	5000
Buck, Joseph, 124 Newgate Street, E.C. ...	100
*Buckingham and Chandos, Duke of, Wootton, Aylesbury	1000
Buckley, John Arthur, Girdlers Hall, Basinghall Street, E.C.	200
Buckley, Robt. Orford, 19 Cleveland Square, W.	1000
Buckley, Nathaniel, Ashton-under-Lyne ...	100
Buckton, Joshua, Leeds	100
Bunning, James Bunstone, Guildhall, E.C. ...	500
Burgess, James Reeve, 47 Brewer Street, Golden Square, W.	100
Burrell, Charles, Thetford, Norfolk	500
*Burt, Henry Potter, Littlecot, Streatham Common, S.	500
*Burzorjee, Dr., Northwick Lodge, St. John's Wood, N.W.	100
*Butter, Henry, 4 Minerva Pl., Barnsbury Pk., N.	100
Cadbury, James, South Place, Grimsbury, Banbury	250
Caley, Nathaniel Henry, London Lane, Norwich	100
Caley, F. G. (Caley Brothers), Windsor ...	100
Callaghan, Wm., 23A New Bond Street, W. ...	200
*Callow, James Wm. (Callow and Son), 8 Park Lane, W.	100
Calvert, Henry, Manchester	100

*Cama, Muncherzee Hormusjee, 21 Gresham House, E.C. £200
*Campbell, C. Minton, Potteries, Stoke-upon-Trent 2000
*Campbell, James, 158 Regent Street, W. ... 100
*Candy, Charles, Watling Street, E.C. 500
Cane, Thomas, Hereford 200
Cannan, Herbert Harris, 36 Basinghall Street, E.C. 100
*Carstairs, Peter, The Green, Richmond, S.W. ... 200
Carver, John (Carver Bros.), Manchester ... 100
*Cassell, John (Cassell, Petter, and Galpin), La Belle Sauvage Yard), E.C. 250
Cave, Stephen, M.P., 35 Wilton Place, S.W. ... 500
*Chadwick, David, 56 Pall Mall, Manchester ... 100
*Chadwick, Edwin, C.B., 5 Montague Villas, Richmond, S.W. 100
*Chadwick, John, 12 Mosley Street, Manchester ... 200
*Challoner, Col. Thomas Challoner Bisse, Portnal Park, Surrey 500
Chambers, Geo. Wilton (Geo. Wright and Co.), Burton Weir, Sheffield 500
*Chambers, Thos. King, M.D., 22B Brook Street, Grosvenor Square, W. 300
Chambers, Thomas, Common Serjeant, 3 Pump Court, Temple, E.C. 100
Chance, J. Timmins (Chance, Bros.), Birmingham 1000
*Chantrill, George Frederick, Liverpool ... 100
*Chapman, Edward (Chapman and Hall), 193 Piccadilly, W. 300
*Chappell, Thomas (Chappell and Co.), 50 New Bond Street, W. 500
Chapple, Frederick, Hayton Hall, near Prescot ... 500
*Charley, William, Seymour Hill, Belfast ... 100
*Charlton, Henry, Edgbaston, Birmingham ... 100
*Chater, Joseph, St. Dunstan's Hill, E.C. ... 100
Chatfield, Chas., Broad Green House, Croydon, S. 100
*Chawner, Richard Croft, The Abnalls, Lichfield 100
*Chester, Harry, 63 Rutland Gate, S.W. ... 300
*Christie, William, 1 Sussex Terrace, King's Road, Chelsea, S.W. 100
*Christy, Henry, 35 Gracechurch Street, E.C. 1000
Christy, Richard (W. M. Christy and Sons), Manchester 500
*Churchward, J. G., Admiralty House, Dover ... 500
*Clabburn, William H., Thorpe, near Norwich ... 100
*Clare, Charles Leigh, Manchester ... 250
Clark, Charles, Mayor of Wolverhampton ... 100
*Clark, Henry, M.D., F.S.A., Southampton ... 100
*Clarke, Cyrus, Street, Somerset 100
Clarke, Sir James, Bart., Bagshot Park, Surrey ... 200
Clarke, Joseph, North Hill Cottage, Highgate, N. 100
*Clarke, J. P., King Street Mill, Leicester ... 100
*Clarke, Robert, London Coffee House, Ludgate Hill, E.C. 100
Clarke, Somers, 57 Regency Square, Brighton ... 100
Clanricarde, Marquis of, 2 Carlton Terrace, S.W. 500
*Claudet, A., F.R.S., 107 Regent Street, W. ... 100
*Clay, Richard, Bread Street Hill, E.C. ... 300
*Clegg, Thomas, Corporation Street, Manchester ... 500
*Clennell, John E., London Fields, Hackney, N.E. 100
*Clifford, Charles, Temple, E.C. 200
Clouston, Peter, Lord Provost of Glasgow ... 100
*Clowes, George (Clowes and Sons), Duke Street, Stamford Street, S. 500
*Clutton, John, 3 Sussex Square, Hyde Park, N.W. 500

*Cobden, George Long, 13 Leonard Place, Kensington, W. £300
Cobbett, Arthur, 18 Pall Mall, S.W. 100
*Cobbett, Richard, 25 Northumberland Street, Strand, W.C. 100
*Cock, John, jun., South Molton, North Devon ... 100
Cockerill, Wm. James, Chelsea Villas, Brompton, S.W. 100
*Cohen, H. L., 2 Cleveland Terrace, Hyde Park, W. 100
*Cole, Henry, C.B., 17 Onslow Square, S.W. ... 300
*Cole, Thomas, 6 Castle Street, Holborn, E.C. ... 100
Coleman, Richard, Chelmsford 100
Colquhoun, John C., 8 Chesham Street, S.W. ... 100
Coles, Richard, Mayor of Southampton ... 100
*Collard, Charles Lukey (Collard and Collard), 16 Grosvenor Street, W. 1000
Collins, Thomas Samuel, M.P., Knaresborough ... 200
*Collyer, Robert, M.D., 8 Alpha Road, St. John's Wood, N.W. 500
*Colman, Edward (J. and J. Colman), 26 Cannon Street, E.C. 1000
*Colnaghi, D. (Colnaghi, Scott, and Co.), 13 Pall Mall East, S.W. 1000
Conolly, Thomas, M.P., 19 Hanover Square, W ... 1000
*Conybeare, H., 20 Duke Street, Westminster, S.W. 1500
*Cook, Thomas W., 8 Clifford Street, New Bond Street, W. 250
Cooke, Christopher, 58 Pall Mall, S.W. ... 250
*Cooke, Hindley, and Law, 11 Friday Street, E.C. 1000
*Cooke, William, 26 Spring Gardens, S.W. ... 1000
Coope, Octavius Edward, Rochetts, Brentwood, Essex 1000
Copestake, Sampson (Copestake, Moore, Crampton, and Co.), 5 Bow Churchyard, E.C. ... 1000
*Corbett, John, Stoke Works, Bromsgrove ... 100
*Corderoy, Edward, Clapham Park, S. 100
*Cornforth, John, Birmingham 100
Cottam, Edward (Robinson and Cottam), 7 Parliament Street, S. W. 300
Cottam, Louis (Cottam and Co.), 2 Winsley Street, W. 300
*Coulthurst, Wm. M., 59 Strand, W.C. ... 1000
*Cousens, Frederick Wm., 16 Water Lane, E.C. ... 300
*Coutts, Miss Angela Georgina Burdett, Stratton Street, W. 3000
*Cowie, Thomas S., 24 George Street, Hanover Square, W. 500
Cowlishaw, Wm. George (James Houldsworth and Co.), Manchester 200
Cowper, Henry, Banbury 100
*Cowper, the Right Hon. Wm., M.P., 17 Curzon Street, W. 100
*Crace, John Gregory, 14 Wigmore Street, W. ... 200
*Crampton, T. R., 15 Buckingham Street, Strand, W.C. 1000
*Creed, Henry, 33 Conduit Street, W. 100
*Cremer, Wm. Henry, jun., 210 Regent Street, W. 200
*Cremer, W. H., 27 New Bond Street, W. ... 200
Crisp, Thos. Dawson (Clabburn, Son, and Crisp), Norwich 100
*Croll, Alexander Angus, 10 Coleman Street, E.C. 1000
Crossfield, Henry, Liverpool 500
Cubitt, Lewis, 52 Bedford Square, W.C. ... 500
*Cubitt, Wm. (Lord Mayor), Mansion House, E.C. ; Ponton Lodge, Andover 1000

*Cundall, Joseph, 168 New Bond Street, w.   ...   £100
*Cunningham, H. D. P., R.N., Bury, Gosport   ...   100
Cuthell, Andrew, 63 Warwick Square, s.w.   ...   1000
Curt, Joseph, 33 Great Portland Street, w.   ...   100

*Daniell, Richard Percival (Daniell and Co.), 129
   New Bond Street, w.   1000
*Darbishire, Sam. Duckinfield, Pendyffryn, Conway   1000
Davies, John, Grove Hill, Woodford, Essex   ...   100
*Davis, Frederick, 100 New Bond Street, w.   ...   500
*Davison, Frederic (Gray and Davison), 370 Euston
   Road, N.W.   200
Davison, Robert, 8 London Street, E.C.   ...   100
*Davy, Charles, 100 Upper Thames Street, E.C. ...   100
Day, C. A. (C. A. Day and Co.), Southampton ...   1000
*Day, Wm. (Day and Son), 6 Gate Street, Lincoln's
   Inn, w.c.   1000
*Deacon, Henry, Appleton, near Warrington   ...   100
*Debenham W. J. (Deacon, Son, and Freebody),
   44 Wigmore Street, w.   500
*De la Rue, Warren (Thos. De la Rue and Co.),
   110 Bunhill Row, E.C.   1000
Dent, William, Bickley Park, Bromley, s.E.   ...   200
*Denton, John Bailey, Stevenage   100
*Derham, James, Wrington Villa, Cothan Road,
   Bristol   100
*Derham, Samuel, Nelson Street, Bristol   ...   100
*Devonshire, Duke of, Devonshire House, Picca-
   dilly, w.   2000
Dewhirst, George Charnley, Manchester   ...   200
*Dickins, Thomas, Edgemoor House, Higher
   Broughton, Manchester   200
Dickinson, James (Wm. Dickinson and Sons),
   Blackburn   500
*Dickson, Peter, 28 Upper Brook Street, w.   ...   1000
*Dilke, C. Wentworth, 76 Sloane Street, s.w. ...   1000
*Dillon, John (Morrison, Dillon, and Co.), Fore
   Street, E.C.   1000
*Dixon, George, Broad Street, Birmingham   ...   100
*Dixon, Thomas, 7 St. James's Place, Hampstead
   Road, N.W.   100
Dixon, Wm. Hepworth, Essex Villa, Queen's Road,
   St. John's Wood, N.W.   100
Dobree, Bonamy, 1 Broad Sanctuary, Westminster,
   s.w.   500
*Dobson, Benjamin, Mere Hall, Bolton, Lancashire   100
*Docker, Frederick William, 24 Denbigh Street,
   Pimlico, s.w.   100
Dolby, John Edward Adolphus, Western House,
   Earl's Court, Old Brompton, s.w.   ...   100
*Donald, William, 69 Regent Street, w.   ...   100
Donkin, Bryan (Bryan Donkin and Co.), Bermond-
   sey, s.E.   500
Douglas, Francis Brown, Lord Provost of Edin-
   burgh   100
*Doulton, Henry (Doulton and Co.), Lambeth, s.   200
Doveston, George (Doveston, Bird, and Hull),
   Manchester   300
*Driver, Henry, Mayor of Windsor   400
*Drax, J. S. W. Sawbridge Erle, M.P., Chaboro'
   Park, Blandford   100
*Ducie, Earl, 30 Prince's Gate, s.w.   500
Dudley, John Crews, Broad Street, Oxford   ...   100
Dugdale, James, Manchester   1000
Dunlop, John Macmillan, Manchester   100

Dunlop, Walter, Bradford, Yorkshire £200
*Dunn, Thomas, Richmond Hill, Sheffield   ...   250
*Dunn, Thomas, Windsor Bridge Iron Works, Man-
   chester   150
Dyte, Hen., 6 King's Bench Walk, Temple, E.C.   100

Eardly, Sir Culling Eardley, Bart., Bidnell Park,
   Hadfield   1000
Eastlake, Sir Charles L., President of the Royal
   Academy, 7 Fitzroy Square, w.   200
Easton, James (Easton, Amos, and Sons), 25
   Russell Square, w.c.   500
*Easton, James, jun. (Easton, Amos, and Sons), 42
   Tavistock Square, w.c.   1000
*Eavestaff, Wm. G., 60 Great Russell Street, w.c.   100
*Ebury, Lord, 107 Park Place, Grosvenor Square, w.   500
Ecroyd, Wm. F. (W. Ecroyd and Sons), Burnley   500
*Edgar, W. S. (Swan and Edgar), 10 Piccadilly, w.   1000
*Edgington, Benjamin, 2 Duke Street, Southwark, s.   500
Edmeston, James, 5 Crown Court, Old Broad
   Street, E.C.   100
*Edwards, J. Passmore, 166 Fleet Street, E.C. ...   100
*Edwards, Morton, 5 George Street, Hanover
   Square, w.   100
Edwards, Richard (J. Edwards and Son), Burslem   500
*Edwardes, Thomas Dyer, 5 Hyde Park Gate, Ken-
   sington Gore, w.   250
*Elkington, Frederick (Elkington and Co.), Bir-
   mingham   2000
Elliott, Chas. (Elliott Brothers), 30 Strand, w.c.   300
*Elliott, George Augustus, 13A Belgrave Square,
   s.w.   100
*Ellis, James, M.D., Sudbrook Park, Petersham,
   s.w.   200
Elwell, Edward, Forge, Wednesbury   200
*Emanuel, Harry, 70 and 71 Brook Street, w.   ...   1000
*Emmens, William, 5 Lothbury, City, E.C.   ...   100
England, George, Crystal Palace, Sydenham, s.E.   1000
*Ernest, Henry (Coleman, Ernest, and Row), 2
   Old Swan Lane, E.C.   250
Evans, Edward, Boveney Court, Windsor   ...   100
*Evans, E. Bickerton (Hill and Evans), Green Hill,
   Worcester   100
Evans, Frederick Mullett (Bradbury and Evans),
   Whitefriars, E.C.   1000
Evans, Jeremiah (J. Evans, Sons, and Co.), 33
   King William Street, E.C.   500
*Evans, S. Lavington, 12 High Street, Oxford   ...   100
*Evill, Wm., jun., Lyncome House, Battersea, s.w.   300
*Ewart, Wm., M.P., 6 Cambridge Square, w.   ...   500

*Fairbairn, Andrew, Woodsley House, Leeds   ...   500
*Fairbairn, Thomas, 17 Park Lane, w.   ...   1500
*Fairbairn, Wm., F.R.S., Manchester   1000
Fairbairn, Wm. Andrew, Manchester   500
Falmouth, Lord, 2 St. James's Square, s.w.   ...   500
Farlow, Charles, 191 Strand, w.c.   100
Farquhar, Thomas Newman, Crystal Palace, Sy-
   denham, s.E.   1000
Farrow, Charles (Farrow and Jackson), 18 Great
   Tower Street, E.C.   100
*Faulkner, David, 3 Brydges Street, Strand, w.c.   100
*Fauntleroy, Robert Thomas (Robert Fauntleroy
   and Co.), 100 Bunhill Row, E.C.   1000
*Fawell, Thomas, Stourbridge   100

Fearon, John Peter, Cuckfield, Sussex ...	£100
Fenton, Francis Henry, 63 St. James's Street, s.w.	100
Ferguson, John F., Belfast	200
Ferrabee, James, Stroud, Gloucestershire ...	100
*Feversham, Lord, 1 Great Cumberland Street, w.	100
*Field, John, Dornden, Tunbridge Wells ...	1000
*Field, William, 224 Oxford Street, w. ...	200
*Filmer, Thomas H., 28 Berners Street, w. ...	100
*Finnis, T. Q., Alderman, 79 Great Tower Street, E.C.	1000
Findlay, C. B., Glasgow	100
*Firmin, George J., 80 Borough Road, s.E. ...	100
*Fisher, Anthony Lax, M.D., 14 York Place, Portman Square, w.	100
Fisher, Richard, Midhurst	500
*Fisher, Robert, 13 Highbury Park, N. ...	250
Fisher, Rev. Samuel, Hope Parsonage, Hanley ...	200
*Fisher, Samuel, 33 Southampton Street, Strand, W.C.	100
*Fladgate, W. M., 40 Craven Street, Strand, w.c.	500
*Fletcher, John Bowman, 17 New Burlington Street, w.	100
Forrest, James Alexander, Liverpool ...	100
Forster, John, 46 Montagu Square, w. ...	250
*Forster, Sampson Lloyd, The Five Ways, Walsall	100
Foster, Thomas (Elsmore and Foster), Burslem...	250
*Fortescue, The Hon. Dudley F., M.P., 17 Grosvenor Square, w.	100
Fortnum, Charles Drury Edward, Stanmore Hill, Middlesex	500
*Foster, Charles Finch, Garden House, Cambridge	100
*Foster, John Porter (Foster, Porter, and Co.), 47 Wood Street, E.C.	1000
Foster, Wm. Orme, M.P., Stourton Castle, Stourbridge	1000
*Fothergill, Benjamin, 65 Cannon Street, E.C. ...	100
*Fowke, Capt. Francis, R.E., Park House, South Kensington, s.w.	300
*Fowler, John, 2 Queen's Square Place, s.w. ...	1000
Fowler, Charles, Totridge House, High Wycombe, Bucks	100
Fowler, Wm. Cave, 16 Aldersgate Street, E.C. ...	100
Fownes, Edward (Fownes, Brothers, and Co.), 41 Cheapside, E.C.	100
*Fox, Edwin (Halliday, Fox, and Co.), 4 Cullum Street, E.C.	200
Foxwell, Thomas S., Shepton Mallett	100
Francis, Charles Larkin (Francis, Brothers, and Pott), Nine Elms, s.	500
Franklin, Abraham Gabáy, 14 South Street, Finsbury, E.C.	100
Franklin, Jacob Abraham, 14 South Street, Finsbury, E.C.	100
*Franklyn, George Woodroffe, M.P., Lovell Hill, Windsor, and Carlton Club, s.w.	500
*Frith, J. G., 13 Wimpole Street, Cavendish Square, w.	1000
Froggort, William, Manchester	100
*Fussell, Rev. J. G. C., Privy Council Office, s.w., and 16 Cadogan Place, s.w.	200
*Galpin, Thos. Dixon (Petter and Galpin), La Belle Sauvage Yard, E.C.	250
Gardener, Robert, Manchester	100
Gardner, Samuel (John Kenyon and Co.), Sheffield	200
Garfit, Thomas, Boston	£100
Garraway, Frederick, 94 Inverness Terrace, Kensington, w.	100
*Garrett, Richard, Leiston Works, Suffolk ...	500
Gask, Charles (Grant and Gask), 59 Oxford Street, w.	500
*Gaskell, John, St. Nicholas-at-Wade, Margate ...	100
*Gassiot, J. P., F.R.S., 77 Mark Lane, E.C. ...	1000
*Gibbs, Henry Hucks, St. Dunstan's, Regent's Park, N.W.	500
*Gibson, The Right Hon. Thos. Milner, M.P., 3 Hyde Park Place, Cumberland Gate, s.w.	1000
*Gifford, Wm. James, Ford, Wellington, Somerset	200
*Gilhee, Wm. Armand (L. M. Fontainemoreau and Co.), 4 South Street, E.C.	100
Gilbey, Walter (W. and A. Gilbey), 357 Oxford Street, w.	1000
*Girdwood, John, 49 Pall Mall, s.w.	100
Girdwood, William, Old Park, Belfast ...	100
*Glaisher, James, F.R.S., 13 Dartmouth Terrace, Lewisham, s.E.	100
*Glass, Thomas, 24 Somerset Street, Kingsdown, Bristol	100
*Glover, Thomas, 8 Upper Chadwell Street, Pentonville, E.C.	1000
Glyn, George Carr, M.P., 67 Lombard Street, E.C.	500
Glyn, Sir R. P., Bart., Lombard Street, E.C. ...	500
*Glynn, Joseph, F.R.S., 28 Westbourne Park Villas, Bayswater, w.	100
*Godwin, George, F.R.S., 24 Alexander Square, Brompton, s.w.	200
*Goff, Joseph, jun., 25 Grosvenor Place s.w. ...	100
Gooddy, Edward (Barlow, Gooddy, and Jones), Manchester	250
*Goode, Thomas (Goode and Co.), 19 South Audley Street, w.	250
*Gooden, James Chisholm, 33 Tavistock Square, W.C.	200
Goodman, John Dent, Minories, Birmingham ...	100
*Goore, Wm. Henry P., Camden Villas, Moscow Road, Palace Gardens, w.	100
Gordon, Lewis Dunbar Brodie, 2½ Abingdon Street, s.w.	1000
*Gotto, Henry (Parkins and Gotto), Oxford St., w.	1000
*Gower, The Hon. E. F. Leveson, M.P., 16 Bruton Street, w.	200
*Graham, Foster (Jackson and Graham), 37 Oxford Street, w.	250
*Graham, Peter (Jackson and Graham), 37 Oxford Street, w.	1000
*Graham, William, 31 Threadneedle Street, E.C., and 17 Cleveland Square, Hyde Park, w. ...	1000
*Graham, William, Manchester	100
*Grant, Alexander, 2 Clement's Court, Wood St., E.C.	250
*Granville, The Earl, K.G., 16 Bruton Street, w.	1500
*Gray, Capt. Wm., M.P., 10 St. James's Place, s.w.	1000
*Greaves, Richard, The Cliff, Warwick ...	100
*Green, Daniel, jun., 11 Finsbury Circus, E.C. ...	100
Green, James, 35 Upper Thames Street, E.C. ...	100
Greenall, Gilbert, Walton Hall, Warrington ...	500
Gregory, Charles, 212 Regent Street, w. ...	500
*Gregson, Samuel, M.P., 37 Upper Harley St., w.	100
*Grew, Nathaniel, 8 New Broad Street, E.C. ...	100
*Griffiths, Robert, 69 Mornington Road, N.W. ...	250

Grissell, Henry, Regent's Canal Iron Works, N.... £500
Grove, George, Crystal Palace, Sydenham, S.E. ... 100
*Gruneisen, Chas. Lewis, 16 Surrey Street, Strand, W.C. 100
*Guedella, Henry, Gresham Club, E.C. ... 500
Gundry, Wm. (Gundry and Sons) 1 Soho Sq., W. 100
Gunter, Richard, Lowndes St., Lowndes Sq., S.W. 300
*Gurney, Samuel, M.P., 65 Lombard Street, E.C. 1000
*Gwynne, J. E. Anderson (Gwynne and Co.), Essex Street Wharf, Strand, W.C. ... 500

Hacking, Richard, Bury, Lancashire 1000
*Haden, F. Seymour, 62 Sloane Street, S.W. ... 100
*Hadfield, William, Manchester 100
*Hale, Warren Stormes, Alderman, 71 Queen Street, E.C. 1000
*Hall, S. C., Boltons, West Brompton, S.W. ... 200
*Hall, Walter, 10 Pier Road, Erith 200
*Hammond, William Parker, 74 Camden Road Villas, N.W. 100
*Hancock, Charles Frederick, 39 Bruton Street, Bond Street, W. 1000
*Hanhart, N. (M. and N. Hanhart), 64 Charlotte Street, Rathbone Place, W. 100
*Hankey, Thomson, M.P., 45 Portland Place, W. 500
*Hannington, C. S., North Street, Brighton ... 100
Hargreaves, W., 34 Craven Hill Gardens, W. ... 200
*Harris, James, Hanwell, W. 200
*Harrison, Robert (Harrison, Lloyd, and Co.), 19 Friday Street, E.C. 500
*Harrison, Thos. E., 27 Great George Street, S.W. 100
*Harrison, W., F.G.S., Blackburn 500
Hart, Ernest, 69 Wimpole Street, W. 100
*Hart, Charles (Hart and Son), Cockspur St., W.C. 100
Harter, James Collier, Manchester 1000
*Hartley, James (J. Hartley and Co.), Sunderland 1000
Hatch, Henry, Park Town, Oxford 100
Haward, W. G., Haverstock Hill, N.W. ... 100
*Hawkins, George, 88 Bishopsgate Street Without, E.C. 200
*Hawkshaw, John, F.R.S., 43 Eaton Place, S.W. 1000
Haworth, J., Blackburn 100
Hayes, Henry Wm. (Hayes and Co.), 4 Great Marlborough Street, W. 1000
*Headley, Richard, Stapleford, Cambridge ... 100
Heald, Nicholas, Manchester 100
*Healey, Elkanah, Oakfield, Gateacre, near Liverpool 100
*Heath, T. Vernon, 43 Piccadilly, W. 100
*Heather, James, The Crescent, Camden Road Villas, N.W. 100
*Hemming, Frederick H., 104 Gloucester Place, Portman Square, W. 100
*Henderson, Geo. Wm. Mercer, 103 Eaton Place, S.W. 500
*Henderson, John (Banks, Bros., Henderson and Co.), Wigton... 200
Henty, Robert, 40 Brunswick Square, Brighton 500
Hepworth, Wm., 108 Market Street, Manchester 100
Heron, Joseph, Manchester... 250
Heugh, John, Manchester 500
*Hewitt, Henry John, 19 Alexander Square, Brompton, S.W. 200
Hewitt, Thomas, Summer Hill House, Cork ... 100
*Heymann, Lewis (Heymann and Alexander), Nottingham 1000

*Heywood, J. Sharp (Wilkinson, Heywood, and Clark), Battle Bridge, N. £500
Hibbert, George, 21 Queen Street, Mayfair, W. ... 500
Hibbert, John Tomlinson, Manchester ... 100
*Hill, Charles, 29 Threadneedle Street, E.C. ... 300
*Hill, Thomas Rowley (Hill and Evans), Catherine Hill House, Worcester 100
*Hilton, Thomas, Great Suffolk St., Southwark, S. 100
*Hindley, Charles Hugh (Charles Hindley and Sons), 134 Oxford Street, W. 1000
*Hinstin, Ernest (Hinstin, Bros.), 22 Milk Street, Cheapside, E.C. 150
*Hinxman, J. H., M.D., Lee Terrace, Blackheath S.E. 500
Hirst, Joseph, Huddersfield 500
Hirst, Wm. Edwards, Huddersfield 250
Hitchcock, Geo. Chas., 45 Lime Street, E.C. ... 250
Hoare, Capt. Dean John, 45 Great Marlborough Street, W. 100
*Hobbs, Ashley, and Co. 76 Cheapside, E.C. ... 1000
Hodge, Wm. (Leighton and Hodge), 13 Shoe Lane, E.C. 500
Hodges, John Francis, Mayor of Dorchester ... 100
Hodgkinson, Grosvenor, Windthorpe Hall, Newark 100
Hodgson, Kirkman Daniel, M.P., 36 Brook St., W. 500
Hodgson, William Nicholson, Newby Grange, Carlisle 250
Holdsworth, Wm. Bailey (Wm. B. Holdsworth and Co.), Hunslet, Leeds 100
Holland, Sir Henry, Bart., 25 Brook Street, W. 200
Holland, James (Holland and Sons), 23 Mount Street, Grosvenor Square, W. 1000
*Hollins, M. Daintry, Potteries, Stoke-upon-Trent 2000
Holland, Robert, Stanmore, Middlesex ... 500
*Holmes, Herbert M., Derby 200
*Holmes, James, 4 New Ormond Street, W.C. ... 300
Hoole, Henry E. (Mayor of Sheffield), Green Lane Works, Sheffield 500
*Hooper, George N. (Hooper and Co.), Haymarket, S.W. 250
*Hope, Henry Thomas, 116 Piccadilly, W. ... 2000
Hopkinson, James (J. and J. Hopkinson), 225 Regent Street, W. 200
*Hopkinson, Jonathan, 40 Grosvenor Place, S.W. 500
*Horn, James, 14 High Street, Whitechapel, E. ... 100
Hornblower, Jethro (Hornblower, Fenwick, and Co.), 50 Mark Lane, E.C. 500
Horsfall, Jas. (Webster and Horsfall), Birmingham 500
Horsley, J. Callcott, A.R.A., 1 High Row, Kensington, W. 100
*Horton, Isaac, 16 Clapham Rise, S. 500
*Hoskyn, Chandos Wren, Harewood, Ross, Herefordshire 500
*Houghton, F. Burnett, 6 Clarendon Terrace, Kensington, W. 100
*Houghton, George, 4 St. John's Park Villas, Haverstock Hill, N.W. 100
Hovendon, Robert, Crown Street, Finsbury, E.C. 100
*Hubert, Samuel Morton (John Woollams and Co.), 69 Marylebone Lane, W. 500
Hughes, James, 9 Crescent, Oxford 100
*Humby, George, 2 Aberdeen Place, N.W. ... 100
Hunt, Henry A., 54 Eccleston Square, S.W. ... 500
Hunt, T. N., 2 Upper Portland Place, W. ... 500
*Hunt and Roskell, 156 New Bond Street, W. ... 2000
Hunter, Michael, jun., Master Cutler of Sheffield 100

*Hutchinson, John (Hutchinson and Earle), Appleton House, Widnes, Lancashire £1000
*Hutt, The Right Hon. William, Gibside, near Gateshead 1000
*Hutton, Thomas, J. P., Elm Park, Dublin ... 100
Hyam, David (Davis, Albert, and Co.), 60 Houndsditch, E.C. 1000
Hynam, John, 7 Wilson Street, E.C. 200

*Ibbotson, Thos. Hamer (Ibbotson and Langford), Manchester 250
Ionides, Alex. Constantine, Tulse Hill, Surrey, s. 100
*Isaacs, Saul (S. Isaacs, Campbell, and Co.), 71 Jermyn Street, w. 1000

*Jackson, John, jun. (Geo. Jackson and Sons), 49 Rathbone Place, w.c. 200
*Jackson, John, jun., 49 Rathbone Place, w.c. ... 100
*Jackson, R. M., 45 Piccadilly, w. 250
*Jackson, Samuel, 66 Red Lion Street, E.C. ... 200
Jacques, Richard Machel, Easby Abbey, Richmond, Yorkshire 100
James, D. D. (Wm. Cory and Son), Commercial Road, Lambeth, s. 500
*James, Jabus Stanley (Powis, James, and Co.), 26 Watling Street, E.C, ... 100
*Jarrett, Griffith, 37 Poultry, E.C. 500
Jay, George (G. Jay and Son), King Street, Norwich 100
*Jeanes, John (Johnstone and Jeanes), 67 New Bond Street, w. 500
*Jeffery, Wm. S. (Howell, James, and Co.), 5 Regent Street, s.w. 1000
Jenkins, Leonard (Jenkins, Hill, and Jenkins), Birmingham 500
*Joel, Joseph, Brompton Hall, Brompton, s.w. ... 1500
Johnson, Edmund, 10 Castle Street, Holborn, w.c. 100
*Johnson, Frederick, 12 North Street, Westminster, s.w. 500
*Johnson, Henry, 39 Crutched Friars, E.C. ... 1000
Johnson, Henry, Mayor of Stamford 100
*Johnson, J. M., 3 Castle Street, Holborn, w.c. ... 100
*Johnson, Richard, Oak Bank, Fallowfield, Manchester 100
Johnston, James, Newmill, Elgin, N.B. ... 100
*Jones, David Morgan (John Morgan and Co.), Amen Corner, E.C. 100
*Jones, Edward, The Larches, Handsworth, Birmingham 100
*Jones, Frederick John, 10 Aldermanbury, E.C. ... 100
*Jones, John, Clock House, Wandsworth, s.w. ... 500
*Jones, Owen, 9 Argyle Place, w. 100
*Jones, Rev. Wm. Taylor, Sydenham College, s.E. 100
*Joubert, Ferdinand, 36 Porchester Terrace, w. ... 200

*Keeling, Henry Ling, Monument Yard, E.C. ... 250
Keighley, William, Huddersfield 250
*Keith, Daniel, 124 Wood Street, Cheapside, E.C. 500
*Kelk, John, 13 South St., Grosvenor Square, w. 3000
*Kelly, Sir Fitzroy, M.P., 32 Dover Street, w. ... 1000
Kelsall, Henry (Kelsall and Kemp), Rochdale ... 200
*Kent, George, 199 High Holborn, w.c. ... 200
*Kimber, Thos., Holland House, Blackheath, s.E. 200

*Kimpton, Thomas, 5 Bath Street, Newgate Street, E.C. £250
*Kinder, Arthur, 18 Great George Street, Westminster, s.w. 100
*Kisch, Simon Abraham, 8 Lancaster Place, w.c. 100
Kitson, James, Mayor of Leeds 500
*Knight, George, 2 Foster Lane, E.C. 100
*Knill, Stuart, The Crossletts in the Grove, Blackheath, s.E. 250
Knowles, John, Manchester 100

*Ladd, William, 11 Beak Street, Regent Street, w. 100
*Lambert, Charles (Lambert and Butler), 141 Drury Lane, w.c. 500
*Lambert, Thomas (Thos. Lambert and Son), Short Street, Lambeth, s. 500
*Landon, James, 91 Inverness Terrace, Bayswater, w. 100
Lang, Robert, Redland, Bristol 100
*Langton, Wm. H. Gore, M.P., 2 Prince's Gate, Hyde Park, w. 500
*Lankester, Edwin, M.D., 8 Saville Row, w. ... 100
*Langworthy, Edward Reilly, Victoria Park, Manchester 250
*Lansdowne, Marquis of, K.G., Lansdowne House, w. 1000
*Lavanchy, John R., 6 New Burlington Street, w. 100
*Lawrence, Frederick, 94 Westbourne Terrace, w. 200
*Lawson, A. M. (Peter Lawson and Son), Edinburgh 1000
*Lea, John W. (Lea and Perrins), Upper Wick, Worcester 100
*Leaf, Sons, and Co., 39 Old Change, E.C. ... 1000
*Leather, J. Towlerton, Leeds 500
*Le Breton, Francis, 21 Sussex Place, Regent's Park, N.W. 200
*Le Couteur, Col. J., F.R.S., Bellevue, Jersey ... 100
*Ledger, Robert Goulding, St. John's, Southwark, s.E. 100
*Leeks, Edward Frederick, 73 Warwick Square, s.w. 100
*Leeman, George, Lord Mayor of York, York ... 200
*Leigh, Evan, Miles Platting, Manchester ... 200
*Leighton, George Cargill, Milford House, Strand, w.c. 250
*Leighton, John, 12 Ormond Terrace, Regent's Park, N.W. 250
Leppoc, Henry Julius, Manchester 100
*Lethbridge, J. C., 25 Abingdon Street, Westminster, s.w. 500
*Letts, Thomas, 8 Royal Exchange, E.C. ... 100
*Letts, Thos., jun., Sydenham, s.E. 100
*Leuchars, William, 18 Piccadilly, w. ... 500
*Levinsohn, Louis, 7 Finsbury Square, E.C. ... 100
Levy, John, 9 Orchard Place, Southampton ... 250
Levy, Levin (Lee, Brothers), 27 Wood Street, E.C. 100
*Lewis, Arthur (Lewis and Allenby), 195 Regent Street, w. 1000
*Lewis, Harvey, M.P., 24 Grosvenor Street, w. ... 2000
*Lewis, Jas., 6 Bartlett's Buildings, Holborn, E.C. 100
*Lewis, Waller, M.D., General Post Office, E.C. 100
Lewis, Wm., Alderman, The Mount, Rainbow Hill, Worcester 100
*Lézard, Joseph (Baume and Lézard), 21 Hatton Garden, E.C. 200

N

Liebert, Bernhard, Manchester £500
Lindley, Dr. John, F.R.S., Acton Green, Turnham
   Green, w. 100
Line, William, Daventry 100
Lings, Thomas, Manchester 100
*Little, Thomas, 43 Oxford Street, w. 200
Lloyd, Edward Rigge, Albion Tube Works, Bir-
   mingham 100
Lloyd, George, 70 Great Guildford Street, S.E. ... 100
*Lloyd, Sampson (Lloyd and Lloyd), Wednesbury 200
Lloyd, Samuel, jun., Old Park Iron Works, Wed-
   nesbury 100
*Loader, R. A. C., 23 and 24 Finsbury Pavement,
   E.C. 500
*Lock, Samuel Robert (Lock and *Whitfield), 178
   Regent Street, w. 500
*Lockwood, Ben., Huddersfield 250
Lorimer, George (Master of the Merchant Co.),
   Edinburgh 200
Losada, J. R., 105 Regent Street, w. 200
*Lovegrove, Samuel, Blackheath, S.E. 100
*Lowe, George, F.R.S., Finsbury Circus, E.C. ... 200
Lowe, J. Stanley, 31 Corn Market Street, Oxford 100
*Loysel, Edward, C.E., 92 Cannon Street, E.C. ... 1000
*Lucas, Charles (Lucas Bros.), Belvidere Road, s. 1500
*Lucas, James H., 13 Upper Woburn Place, N.W. 1000
Lucas, Philip, Manchester 300
*Lucas, Thomas (Lucas Bros.), Belvidere Road, s. 1500
Ludlam, Jeffery, 174 Piccadilly, w. 300
Lumley, Wm. Golden, 10 Sussex Place, Regent's
   Park, N.W. 100
*Lutwidge, R. W. S., 19 Whitehall Place, s.w. ... 500
*Lycett, Francis, 2 Highbury Grove, N. ... 100
*Lyell, Sir Charles, 53 Harley Street, w. ... 300
*Lyle, James Grieve, 20 Little Moorfields, E.C. ... 200
*Lyon, Arthur, 32 Windmill Street, Finsbury, E.C. 100
*Lyons, Morris, Birmingham 100
*Lyte, F. Maxwell, Florian, Torquay 100

*Macadam, Chas. Thos., 109 Fenchurch Street, E.C. 100
*Macarthur, Major-Gen. Edwd., 134 Piccadilly, w. 1000
*Mackenzie, Rev. Charles, Westbourne College, w. 100
*Mackintosh, R. J., M.A., 2 Hyde Park Terrace,
   Kensington Gore, w. 100
*Maclea, Chas. G., 17 Blenheim Terrace, Leeds ... 200
Maclean, Miss M., 3 Edwardes Place, Kensington,
   w. 100
Maclean, Lt.-Col. H. D., Lazenby Hall, Penrith 250
*Mackintosh, R. F., 2 Hyde Park Terrace, W. 100
Mac Leod of Mac Leod, 9 Cambridge Square, w. 500
*Malcolm, Major-General George Alexander, 67
   Sloane Street, s.w. 200
*Malcolm, John W., M.D., 7 Great Stanhope
   Street, w. 500
*Mackrell, W. T., Abingdon Street, s.w. ... 500
Mallinson, Thomas, Huddersfield 500
*Manby, Charles, F.R.S., 50 Harley Street, Caven-
   dish Square, w. 100
Mansfield, George (Wright and Mansfield), 3 Gt.
   Portland Street, w. 300
Mappin, Frederick Thorpe (Thos. Turton and Sons),
   Sheaf Works, Sheffield 250
Mappin, Joseph Charles (Mappin Bros.), Baker's
   Hall, Sheffield 200

Mappin, John Newton (Mappin and Co.), 77 Ox-
   ford Street, w. £500
Marjoribanks, Dudley Coutts, M.P., 29 Upper
   Brook Street, w. 1000
*Marjoribanks, E., 34 Wimpole Street, w. ... 2000
Marrian, James Pratt, Birmingham 100
*Marsh, Matthew Henry, M.P., 43 Rutland Gate,
   s.w. 500
Marshall, Thos. R. (W. Marshall and Co.), Edin-
   burgh 100
Marshall, William, Penworthan Hall, near Preston 200
Marshall, W. S., 20 Strand, w.c. 1000
Martin, George Wm., 68 Gloucester Crescent,
   N.W. 100
*Martin, George Wm., 14 and 15 Exeter Hall, w.c. 100
Martin, Richard (Martin, Hall, and Co.), Sheffield 200
Martin, W. H., Burlington Arcade, w. ... 100
Martineau, Joseph, Basing Park, Alton ... 1000
*Martyn, Silas Edward, 46 Thurloe Square, s.w.... 200
*Maw, George, Benthall Park, Broseley ... 250
Maxwell, Wm. James, Richmond, s.w. ... 100
May, Walter (Walter May and Co.), Birmingham 100
*Maynard, Joseph, 57 Coleman Street, E.C. ... 1000
McClure, Wm., Manchester 100
McConnel, James, Best Hill, Prestwich, Man-
   chester 100
McConnel, W., Best Hill, Prestwich, Manchester 100
McConnel, Henry, Cressbrook, Bakewell ... 1000
McCormick, William, M.P., 16 Cambridge Ter-
   race, Regent's Park, w. 2000
McCracken, James John (J. and R. McCracken
   and Co.), 7 Old Jewry, E.C. 200
*McFarlane, Walter, Saracen Foundry, Glasgow ... 100
McGarel, Charles, 2 Belgrave Square, s.w. ... 1000
*McLean, John Robinson, 2 Park Street, West-
   minster, s.w. 2000
McQueen, Wm. Benjamin (McQueen Bros.), 184
   Tottenham Court Road, w. 500
*Mechi, J. J., Alderman, 4 Leadenhall Street, E.C. 1000
*Meekins, Thomas Mossom, LL.D., 44 Chancery
   Lane, w.c. 100
Mellor, Wright, Huddersfield 250
*Messenger, Sam. (Messenger and Co.), Birming-
   ham 500
*Metchim, Wm. Paul, 20 Parliament Street, s.w. 100
*Metzler, George (Metzler and Co.), 137 Great
   Marlborough Street, w. 300
Meyers, Barnett, 2 Mill Lane, Tooley Street, s.E. 100
Michell, Richard, 93 Oxford Street, w. ... 300
Micholls, Henry, Manchester 100
*Middleton, Sir George N. Broke, Bart., Shrubland
   Park, Ipswich 500
*Miles, Alfred Webb, 73 Brook Street, Hanover
   Square, w. 300
Miles, Henry, The Dounfield, Kington, Hereford-
   shire 500
*Miles, Pliny, 169 King's Road, Chelsea, s.w. ... 200
*Millar, John, M.D., Bethnal House, Bethnal
   Green, N.E. 100
Mills, Charles, 67 Lombard Street, E.C. ... 500
Mills, Edward W., 67 Lombard Street, E.C. ... 500
*Mitchell, Rev. M., 15 St. James's Square, s.w.... 100
*Moate, Charles, R., 65 Old Broad Street, E.C. ... 1000
*Montgomerie, H. E., 17 Gracechurch Street, E.C. 100
Moon, Richard, Bevere, Worcester 250

Moreland, John Brogden, 76 Old Street, E.C. ...	£100
*Moreland, Joseph, 76 Old Street, St. Luke's, E.C.	100
Moreland, Richard, Eagle House, Holloway, N. ...	100
*Moreton, John, Wolverhampton	100
Morgan, John, Amen Corner, E.C.	100
*Morgan, William Vaughan (Patent Plumbago Crucible Company), Battersea, s.w.	500
*Morley, Samuel, 18 Wood Street, E.C. ...	1000
*Morrish, Francis Edward, Lancaster Buildings, Liverpool	100
Mosley, Thomas (Thomas Mosley, Huish, and Co.), Manchester	100
*Mouat, Frederick, M.D., Athenæum Club (Messrs. R. C. Lepage and Co.), 1 Whitefriars Street, Fleet Street, E.C.	100
*Muir, William, Britannia Works, Manchester ...	100
*Munn, Major W. Augustus, Faversham, Kent ...	100
*Murchison, John Henry, Surbiton Hill, Kingston-on-Thames	100
*Murchison, Sir Roderick Impey, F.R.S., 16 Belgrave Square, s.w.	500
·Murray, Eugene, Glebe House, St. Mary Street, Woolwich, s.E.	200
*Murray, John, 7 Whitehall Place, s.w. ...	1000
*Myers, George, Lambeth, s.	1000
*Napier, Robt., West Shandon, Glasgow ...	2000
*Napier, Hon. Wm., 2 Old Palace Yard, s.w. ...	500
*Navroji, Dádábhái, 32 Great St. Helen's, E.C. ...	100
*Needham, William (Needham and Kite), Phœnix Iron Works, Vauxhall, s.	200
*Neighbour, Geo. L. (Neighbour and Sons), 127 High Holborn, w.c.	500
Neild, William (Thos. Hoyle and Sons), Manchester	1000
*Neilson, Walter M., Glasgow	300
Newbold, Robert (Joseph Rodgers and Sons), Norfolk Street, Sheffield	250
Newen, George, 1 Hyde Park Terrace, w. ...	1000
Newmarch, William, 7 Cornhill, E.C. ...	100
Newton, Fredk. (Newton and Co.), 3 Fleet Street, E.C.	100
*Nicholay, J. A., 82 Oxford Street, w. ...	1000
*Nicholls, G. P. (J. and G. Nicholls), Aldine Chambers, Paternoster Row, E.C.	200
Nichols, Robert Cradock, 5 Westbourne Park Place, w.	100
Nicholson, William Newzam, Newark-on-Trent ...	100
Nickols, Richard, Joppa, Leeds	100
*Nicoll, Donald (H. J. and D. Nicoll), 114 Regent Street, w.	1000
*Nightingale, Charles (W. and C. Nightingale), 64 Wardour Street, w.	250
*Nind, Philip, 30 Leicester Square, w.c. ...	300
*Noble, Matthew, 13 Bruton Street, w. ...	200
*Nolan, Edward Henry, Ph.D., LL.D., 29 Abingdon Villas, Kensington, w.	200
North, David (Wright and North), Wolverhampton	100
*North, Frederick, M.P., The Lodge, Hastings ...	250
*Northcote, Stafford H. (S. Northcote and Co.), 29 St. Paul's Churchyard, E.C.	250
Novelli, Augustus Henry, 69 Grosvenor Street, w.	1500
*Obbard, Robert, Paragon, Blackheath, s.E. ...	100

*Odams, James, 109 Fenchurch Street, E.C ...	£200
Olivier, Charles Henry (Olivier and Carr), 37 Finsbury Square, E.C.	200
Oppenheim, John Moritz, 85 Cannon Street, West, E.C.	1000
Ordish, R. M. (Ordish and Co.), 18 Great George Street, s.w.	100
Osborne, Charles, Whitehall Street, Birmingham	100
Osler, Clarkson (F. and C. Osler, Birmingham ...	500
Other, Christopher (Other and Robinson), Wensleydale, Bedale, Yorkshire	500
*Owen, Lt.-Col. H. Cunliffe, R.E., Devonport ...	200
Owen, Rev. Joseph Butterworth, 40 Cadogan Place, Chelsea, s.w.	100
*Pakington, the Right Hon. Sir John S., Bart., M.P., 41 Eaton Square, s.w.	200
*Palk, Sir Laurence, Bart., M.P., 47 Rutland Gate, s.w.	200
*Palmer, George (Huntley and Palmers), Reading, Berks	200
*Palmer, Philip, 118 St. Martin's Lane, w.c. ...	100
Panizzi, Antonio, British Museum, w.c. ...	200
Panmure, Lady, 19 Chesham Street, Belgrave Square, s.w.	500
Parker, Charles, Binfield, Berks	300
*Parker, James, Great Baddow House, Chelmsford	100
Parsons, Thomas, 92 Regent Street, w. ...	100
*Part, John Cumberland, 186 Drury Lane, w.c.	100
*Paterson, John, 104 Wood Street, E.C. ...	200
*Paxton, Sir Joseph, M.P., Rockhills, Sydenham, s.E.	1000
*Payne, James, Plough Bridge Works, Rotherhithe, s.E.	250
*Pearce, John (Halling, Pearce, and Stone), 2 Cockspur Street, s.w.	1000
*Pease, Henry (Henry Pease and Co.), Darlington	250
*Pease, Joseph (J. and J. W. Pease), Darlington...	1000
*Peake, Thomas, The Tileries, Tunstall ...	250
Pedler, George Stanbury, 199 Fleet Street, E.C.	100
*Peel, Geo., Soho Iron Works, Manchester ...	200
*Pender, John (Pender and Co.), Mount Street, Manchester	500
*Penn, John, The Cedars, Lee, Kent, s.E. ...	1000
Pepper, John Henry, Morton House, Kilburn Priory, Edgeware Road, N.W.	100
Perry, Stephen (Jas. Perry and Co.), 37 Red Lion Square, w.c.	100
Philips, Robert Nathaniel, Manchester... ...	100
Phillips, Frederick D. (Phillips and Samson), 40 High Holborn, w.c.	100
*Phillips, George (W. P. and G. Phillips), 359 Oxford Street, w.	500
Phillips, Mark, Snitterfield, Stratford-on-Avon ...	100
*Phillips, Robert, 23 Cockspur Street, s.w. ...	1000
*Phillips, Sir Thomas, 11 King's Bench Walk, Temple, E.C.	300
*Phillips, William Phillips (W. P. and G. Phillips), 155 New Bond Street, w.	500
*Phythian, Thomas, 430 West Strand, w.c. ...	100
*Pickstone, Wm., Manchester	100
Pike, Robert, St. Aldate Street, Oxford ...	100
*Pillischer, M., 88 New Bond Street, w. ...	100
*Pinches, T. R., 27 Oxendon Street, Haymarket, s.w.	150
*Pitts, Samuel, 14 Catherine Street, w.c. ...	100

Playfair, Lyon, Dr., C.B., Edinburgh	£200
*Platt, John (Platt, Bros., and Co.), Oldham ...	500
Plowman, Joseph, St. Aldate Street, Oxford ...	100
*Pollard, George, 10 Walbrook, E.C.	100
*Poole, Henry, Saville Row, w.	1000
Poole, Thomas, 25 Princes Street, Cavendish	
Square, w.	100
Pope, W. A., 53 Charles Street, Berkeley Square,	
w.	100
*Portal, Wyndham S., Malshanger, Basingstoke ...	100
Potter, Alan, 28 Falkner Square, Liverpool ...	1500
*Potter, Edmund, F.R.S., Manchester	500
Potter, J. G. (E. E. and J. G. Potter), Darwen,	
Lancashire	500
*Potter, Wm. Simpson, 1 Adam Street, Adelphi,	
w.C.	500
Poulter, James, Dover	100
Powell, William (John Hardman and Co.), Great	
Charles Street, Birmingham	500
Power, Bonamy Mansell, 19 Chesham Street, s.w.	1000
*Prescott, W. G., 62 Threadneedle Street, E.C. ...	1000
Price, David, 10 York Terrace, Regent's Park,	
N.W.	1000
Price, Dr. David S., Crystal Palace, Sydenham,	
S.E.	100
Price, George, Cleveland Safe Works, Wolverhamp-	
ton	100
*Pritchard, John, M.P., 89 Eaton Square, s.w. ...	200
Privett, Harry, 47 Brewer Street, Golden Square,	
s.w.	100
*Purssell, Alfred, 80 Cornhill, E.C.	100
Purvis, Prior, M.D., Blackheath, s.E.	100
Quilter, William, 3 Moorgate Street, E.C. ...	100
Quin, Frederick F., M.D., 111 Mount Street, w.	200
Ramsay, Rear-Admiral W., 23 Ainslie Place,	
Edinburgh	100
*Ransford, Henry, Huron Lodge, Boltons, West	
Brompton, s.w.	100
*Ratcliff, Charles, Wyddrington, Birmingham ...	500
Ravenscroft, Francis, Birkbeck Lodge, Boundary	
Road, St. John's Wood, N.W.	100
*Rawlinson, Robert, 17 Ovington Square, s.w. ...	100
*Read, Reginald, M.D., 1 Guildford Place, w.C. ...	200
*Redgrave, Alexander, Eagle Lodge, Old Brompton,	
s.w.	100
Redgrave, Richard, 18 Hyde Park Gate South, w.	200
*Redgrave, Samuel, 17 Hyde Park Gate South,	
Kensington, w.	200
*Reed, Charles, F.S.A., Paternoster Row, E.C. ...	100
Reed, Thomas Allen, 41 Chancery Lane, w.C. ...	100
*Reiss, James, 110 Cross Street, Manchester ...	500
*Reynolds, Charles William, 2 Eaton Place, s.w.	500
Rich, Sir Charles, Bart., 12 Nottingham Place,	
Regent's Park, N.W.	200
Richards, Westley, High Street, Birmingham ...	200
*Richardson, Francis, Park Lodge, Blackheath, s.E.	100
Richardson, G. B., 23 Cornhill, E.C.	100
Richardson, James (Richardson Brothers), Edin-	
burgh	500
Richardson, John, 40 King William Street, E.C.	100
*Richardson, Thomas, Newcastle-on-Tyne ...	200
Rickards, Francis Philip, Manchester	100

*Rideout, Wm. Jackson, Farnworth, Manchester	£250
*Rimmel, Eugène, 96 Strand, w.c.	100
*Robb, Alexander, 79 St. Martin's Lane, w.c. ...	500
*Roberts, Daniel, 16 Northampton Place, Old	
Kent Road, s.E.	100
Roberts, Edward, F.S.A., 25 Parliament St., s.w.	100
Robertson, David (Robertson, Brothers, and Co.),	
Glasgow	100
Robinson, F. (Robinson and Cottam), 7 Parlia-	
ment Street, s.w.	300
*Robinson, Henry Oliver, 16 Park Street, West-	
minster, s.w.	500
Robinson, J. C., 33 Alfred Place, West Brompton,	
s.w.	100
*Robinson, James (Rigby and Robinson), 7 Park	
Lane, Piccadilly, w.	500
Robinson, John Henry, New Grove, Petworth,	
Sussex	100
Robinson, John (Sharp, Stewart, and Co.), Man-	
chester	500
*Robinson, Joseph, Berkhampstead	1000
Rock, James, jun., 6 Stratford Place, Hastings ...	100
*Roe, George, Nutley, Dublin	250
Roebuck, Samuel, Salebank, Manchester ...	100
Rogers, Francis, 2 Arundel Place, Barnsbury	
Park, N.	100
*Rolls, Jesse Gouldsmith, C.E., 4 Church Court,	
Clement's Lane, E.C.	200
*Rolt, Peter, St. Michael's House, Cornhill, E.C.	1000
*Roney, Sir Cusack P., 15 Langham Place, w. ...	1000
Rose, Hugh, Chairman of Chamber of Commerce,	
Edinburgh	200
*Rose, J. Anderson, Salisbury Street, Strand, w.C.	100
*Rose, Wm. Anderson, Alderman, Queenhithe, E.C.	500
*Rosse, The Earl of, Rosse Castle, Parsonstown ...	1000
Rothery, H. Cadogan, 94 Gloucester Terrace,	
Hyde Park, w.	100
Round, Joseph, 33 Beaumont Street, Oxford ...	100
*Routledge, Thomas, jun., Eynsham, Oxford ...	100
Rowbotham, Samuel, Bradford Street, Birmingham	100
*Rumbold, Wm. Henry, The Grange, Tunbridge	
Wells	100
*Rumney, Robert, Manchester	100
Runtz, John, Burlington House, Milton Road,	
Stoke Newington, N.	100
*Ryland, Arthur, Mayor of Birmingham ...	100
*Rylands, John (Rylands and Sons), Manchester ...	500
Sacred Harmonic Society (by Treasurer, R. K.	
Bowley), Exeter Hall	1000
*Sadler, Charles James, Broad Street, Oxford ...	100
*Salisbury, Marquis of, K.G., Hatfield	1000
*Salamons, Aaron, Old Change, E.C.	500
*Salomons, David, M.P., Gt. Cumberland Place, w.	250
*Salt, Titus, Saltaire, Bradford	3000
*Samuel, James, C.E., 26 Great George Street,	
Westminster, s.w.	1000
Samson, Henry, Manchester	100
Sandbach, Henry R., Hafodunos, Llanrwst ...	100
Sandbach, Wm. R., Willesbourne Hall, Warwick	100
Sandeman, George G., 15 Hyde Park Gardens, w.	1000
*Sandford, Francis Richard, 5 Gloucester Terrace,	
Hyde Park, w.	100
Sangster, John (W. and J. Sangster), 75 Cheap-	
side, E.C.	100

*Sassoon, S. David, 17 Cumberland Terrace, Regent's Park, N.W. £1000
*Saul, George Thomas, Bow Lodge, Bow, E. ... 100
*Saunders, William Wilson, F.R.S., Lloyd's, E.C. 500
Savory, John, 143 New Bond Street, W. ... 500
Sawyer, Frederick, "The London,' Fleet Street, E.C. 2500
*Schlesinger, Julius, Walmer Villas, Bradford, Yorkshire 100
*Schofield, Wm. F., Aldborough, Borough Bridge 1000
*Schuster, Leo, 18 Cannon Street, E.C. ... 3000
Schwabe, Adolphe (Salis, Schwabe, and Co.), Manchester 500
*Scott, Sir Francis E., Bart., 97 Euston Square, S.W. 300
Scott, Walter, Manchester 100
Scott, William (*Rogerson and Co.), Newcastle-upon-Tyne 1000
*Seaman, William Mantle, 199 Sloane Street, S.W. 100
*Sedgwick, John Bell, 1 St. Andrew's Place, Regent's Park, N.W. 100
*Shanks, Andrew, Robert Street, Adelphi, W.C. ... 250
*Sharples, Joseph, Hitchin 200
Shaw, Charles Thomas, 66 Great Hampton Street, Birmingham 100
*Shearer, B. P., Levanmore House, Bishop's Waltham 200
Shelley, Sir John Villiers, Bart., M.P., Maresfield Park, Sussex 100
*Shepperson, Allen Thos., Dulwich Hill, Surrey, S. 100
*Sheriff, Alex. Clunes, Shrubs Hill, Worcester ... 100
Shilson, William, Neithrop, Banbury 200
*Shove, W. S., Lee Terrace, Lee, S.E. 500
*Shuttleworth, Joseph (Clayton, Shuttleworth, and Co.), Lincoln 1000
*Siemens, Charles W., 3 Great George Street, S.W. 100
*Silk, Robert (Silk and Sons), 8 Long Acre, W.C. 100
Siltzer, John (Siltzer and Co.), Manchester ... 500
*Simon, George (Lightly and Simon), 123 Fenchurch Street, E.C. 500
*Simpson, W. B., 456 West Strand, W.C. ... 100
*Slaney, Robert A., M.P., 5 Bolton Row, Mayfair, S.W. 100
Slade, Felix, Walcot Place, Lambeth, S. ... 500
*Smirke, Sydney, R.A., 79 Grosvenor Street, W. ... 1000
Smith, George (Wm. Smith, Son, and Co.), Leeds 500
*Smith, George Henry (Wrigley and Smith), Manchester 100
Smith, George Robert, 73 Eaton Square, S.W. ... 500
*Smith, James, Seaforth, Liverpool 500
Smith, John, 1 Great George Street, S.W. ... 1000
Smith, John (Beckett and Co.), Bankers, Leeds 1000
Smith, Mark (Wm. Smith and Bros.), Heywood, Manchester 300
*Smith, R. M., Edinburgh 250
*Smith, Wm., C.E., 19 Salisbury Street, Adelphi, W.C. 150
Smith, Wm., 20 Upper Southwick Street, Cambridge Square, W. 100
Smith, W. H., 186 Strand, W.C. 500
*Snelgrove, John (Marshall and Snelgrove), 11 Vere Street, W. 1000
*Solly, S. R., F.R.S., 10 Manchester Square, W. ... 300
Solomon, Henry, 134 and 31 Houndsditch, E.C. 250
*Solomon, Joseph, 22 Red Lion Square, W.C. ... 100
Solomon, Leon, 69 Grosvenor Street, W. ... 1000

*Somes, Joseph, M.P., National Club, S.W. ... £1000
*Sopwith, Thomas, F.R.S., 43 Cleveland Square, S.W. 200
Sotheby, S. Leigh, Buckfastleigh House, South Devon 500
Sowler, John, Manchester 100
Sowler, Thomas, Manchester 100
*Spark, Henry King, Greenbank, Darlington ... 500
*Sparks, William, Crewkerne 200
*Sparrow, Charles, 11 New North Street, Red Lion Square, W.C. 100
Spence, James, Liverpool 250
*Spicer, William R. (Spicer Bros.), 19 Bridge Street, E.C. 500
*Spiers, Richard James, Alderman, Oxford ... 250
*Squire, William, 5 Coleman Street, E.C. ... 500
Stainton, Jas. Joseph, Meadows, Lewisham, S.E. 100
Standen, Richard Spiers (Standen and Co.), 5 Park Street, Oxford 100
Standish, John, Bagshot, Surrey 200
Standring, James, Mayor of Margate 100
*Stanley, the Right Hon. Lord., M.P., 23 St. James's Square, S.W. 500
Stanley of Alderley, Lord, 40 Dover Street, W. 500
*Stanton, George, Coton Hill, Shrewsbury ... 100
*Staples, Joseph, 10 South Street, Brompton, S.W. 200
Starey, Thomas, Rawstorn, Nottingham ... 200
*Starr, Henry (Wheatley, Starr, and Co.), 156 Cheapside, E.C. 1000
Steane, James S. (Oxon Wine Co.), 42 Corn Market Street, Oxford 100
Stebbing, Joseph Rankin, F.R.A.S., Southampton 100
Steers, Spencer James, Halewood, Prescot, Lancashire 100
Steinthal, Henry Michael, Manchester ... 100
Stephenson, Henry (Stephenson, Blake, and Co.), Allen Street, Sheffield 100
Stern, Sigismond James, Manchester 500
*Stevenson, John, Canal Foundry, Preston ... 100
Stillwell, Edward Swift (Stillwell, Son, and Ledger), 25 Barbican, E.C. 250
Stewart, Charles E., 30 Upper Harley Street, W. 250
Stirling, Wm. (Wm. Stirling and Sons), Glasgow 500
Stock, T. O., 18 Austin Friars, E.C. 300
*Story, George Marvin, 2 Coleman Street, E.C. ... 100
Straker, Samuel (Straker and Son), 81 Bishopsgate Street Within, E.C. 100
Stubbs, Henry, Manchester 200
*Sulivan, Laurence, Right Hon., Broom House, Fulham, S.W. 100
*Sutherland, Duke of, Stafford House, S.W. ... 500
Sutton, M. H. (Sutton and Sons), Reading, Berks 250
Sylvester, J. J., Professor, F.R.S., Royal Military Academy, Woolwich, S.E. 100
*Symonds, John, 3 Ingram Court, Fenchurch Street, E.C. 100
*Taber, John, 39 Crutched Friars, E.C. ... 100
*Tagg, William, 49 Chichester Place, W.C. ... 100
Tamplin, F. A., Liverpool 250
Tannett, T. (Smith, Beacock, and Tannett), Leeds 500
*Tapling, Thos (T. Tapling and Co.), Gresham Street, E.C. 1000
Taunton, Lord, 27 Belgrave Square, S.W. ... 1000
Tayler, J. Fred., 11 Upper Phillimore Gardens, S. 200

*Taylor, Thomas George, Nelson Lodge, Stoke Newington Road, N. £200
*Taylor, William, Newport Pagnell 200
*Telford, Charles, Widmore, Bromley, Kent ... 500
Temple, Henry, 3 Elm Court, Temple, E.C. ... 100
*Terrell, William, 7 Apsley Place, Redland, Bristol 100
*Teulon, Seymour, Teachley's Park, Limpsfield, Surrey 100
Thackeray, W. M., 36 Onslow Square, s.w. ... 100
*Thomas, J. Evan, 7 Lower Belgrave Place, Pimlico, s.w. 100
*Thompson, Harry S., M.P., Kirby Hall, York ... 250
Thompson, Richard A., South Kensington Museum, s.w. 100
Thompson, Samuel W., Thingwall, Liverpool ... 500
Thompson, William, Thurnbury Lodge, Park Town, Oxford 250
*Thring, Henry, 5 Queen's Gate Gardens, South Kensington, s.w. 100
*Thurston and Co., 14 Catherine Street, Strand, w.c. 500
Tighe, Right Hon. Col. W. F. Fownes, Woodstock Park, Inistioge, Ireland 200
Tillett, Samuel, 6 Wellington Terrace, Bayswater, w. 100
*Todé, Edward Henri, 8 Took's Court, Lincoln's Inn Fields, E.C. 300
Tod-Heatly, Grant H., 5 Berkeley Square, w. ... 300
*Tootal, Edward, The Weaste, Manchester ... 500
*Topham, John, 32 King William Street, E.C. ... 150
Tottie, Charles, 2 Alderman's Walk, E.C. ... 500
*Towle, John, Hincksey Mills, Oxford ... 100
Tregelles, Nathaniel (Tregelles and Taylor), 54 Old Broad Street, E.C. 100
*Treggon, Wm. Thomas (Treggon and Co), 22 Jewin Street, E.C. 100
Trehonnais, F. R. de la, Oak Villa, Norwood, S. 100
*Trower, George S., 33 Hyde Park Square, w. ... 200
Truscott, Francis Wyatt (Truscott, Son, and Simmons), Suffolk Lane, E.C. 500
Truss, Thomas Seville, 53 Gracechurch Street, E.C. 100
Tubbs, Robert, 62 Harley Street, w. 100
Tuely, Nathaniel Clissold, 8 Spencer Villas, Southfields, Wandsworth, s.w. 100
*Tulloch, James, 16 Montague Place, w.c. ... 100
*Turner, B. B. (Brecknell, Turner, and Sons), 31 Haymarket, s.w. 500
*Twining, Thomas, jun., Perryn House, Twickenham, s.w. 1000
*Tylor, Alfred (Tylor and Sons), Warwick Lane, E.C. 1000
*Tysoe, John, Manchester 100

Underhay, Fredk. George, 23 Arundel Square, Barnsbury, N. 100
*Underwood, Joseph, 5 Hyde Park Gardens, w. ... 1000
Uzielli, Mrs., Hanover Lodge, Regent's Park, N.W. 5000
*Uzielli, Theodosius, 21 Threadneedle Street, E.C. 3000

*Vallentin, James, Shearn Lodge, Walthamstow, N.E. 200
*Veitch, James, jun., King's Road, Chelsea, s.w. 100
*Venning, James M., 7 Petersham Terrace, Queen's Gate, w. 200

Vernon, G. H., Grove, Retford £100
Viccars, Richard, Padbury 250
Vickers, Henry, Mayor of Sheffield 100
*Vieweg, Augustus Julius, 82 Wood Street, E.C. 200
*Vigers, Edward, 12 Chepstow Villas West, w. ... 100
*Vignoles, Charles, 21 Duke Street, Westminster, s.w. 1000
*Virtue, James Sprent, 294 City Road, E.C. ... 500

*Walker, James, F.R.S., 28 Great George Street, Westminster, s.w. 1000
*Walker, Joseph William, 27 Francis Street, Tottenham Court Road, w.c. 250
Wallis, John, Wood Green, Tottenham, N. ... 100
*Walter, Capt. Edward, Army and Navy Club, s.w. 500
Walters, Edward, Manchester 100
*Ward, John, 5 and 6 Leicester Square, w.c. ... 500
*Watkins, William, 52 Lime Street, E.C. ... 100
Watson, James (James Nisbet and Co.), 21 Berners Street, w. 100
*Webb, John, 11 Grafton Street, Bond Street, w. 1000
Weeks, John, 54 Baker Street, w. 500
*Welch, John Kemp, 51 Berners Street, w. ... 500
*Welch, John Kemp (J. Schweppe and Co.), 51 Berners Street, w. 500
Wenlock, Lord, Escrick Park, York 200
*Wertheimer, Samson, 154 New Bond Street, w. 300
*Westhead, Joshua P. Brown, M.P. (J. P. and E. Westhead and Co.), Manchester 1000
*Westley, W. (Carpenter and Westley), 24 Regent Street, s.w. 200
*Westmacott, Richard, R.A., 1 Kensington Gate, w. 100
*Wetter, Conrad, 67 Myddelton Square, E.C. ... 100
*Whatman, James, F.R.S., Vintners, Maidstone ... 1000
Whelon, William, Mayor of Lancaster ... 100
*Whichcord, John, F.S.A., 16 Walbrook, E.C. ... 500
*Whishaw, James, 16 York Ter., Regent's Park, N.W. 100
White, Arthur Bernard, 83 Inverness Ter., Kensington Gardens, w. 100
*White, George Frederick (J. B. White and Bros.), Millbank Street, s.w. 500
*White, Henry, 5 Queen Street, E.C. 100
*White, Henry Clarence, 38 Great Tower St., E.C. 250
*Whitehead, James Heywood, Royal George Mill, Manchester 250
*Whittingham, Charles, 14 Richmond Villas, Barnsbury, N. 200
*Whittington, Rev. R. 18 Guildford Street, Russell Square, w.c. 100
Wildes, George, Manchester 300
*Wilkinson, David (Molineaux, Webb, and Co.), Manchester 200
*Wilkinson, John, jun. (J. Wilkinson, Son, and Co.), Hunslet, Leeds 500
Willans, Wm., President of the Chamber of Commerce, Huddersfield 500
*Williams, H. R., Board of Trade, s.w. ... 100
*Williams, J. W. Hume, 3 Dr. Johnson's Buildings, Temple, E.C. 100
*Willet, John, 35 Albyn Place, Aberdeen ... 100
*Williams, Wm., Crosby Hall, E.C. 100
Williamson, Robert, Scarborough 100

Willis, George (Willis and Sotheron), 136 Strand, w.c. £500
Willoughby d'Eresby, Lord, 142 Piccadilly, w. ... 1000
*Wilson, Erasmus, F.R.C.S., F.R.S., 17 Henrietta Street, Cavendish Square, w. 200
*Wilson, Geo. F., F.R.S., Belmont, Vauxhall, s. ... 500
*Wilson, Professor John, University, Edinburgh ... 300
*Winkworth, Thos., 7 Sussex Pl., Canonbury, N. ... 100
*Winsor, Wm. (Winsor and Newton), 38 Rathbone Place, w. 500
*Withers, George, 8A Baker St. Portman Sq., w. 100
*Wodderspoon, James, 7 Serle Street, w.c. ... 1000
Wodehouse, Lord, 48 Bryanstone Square, w. ... 100
Wolfenden, James Rawsthorne, Mayor of Bolton 500
*Wood, John, Theddon Grange, Alton, Hants ... 100
Wood, Joseph, Mayor of Worcester 100
*Wood, Nicholas, Durham 100
*Wood, Vice-Chancellor, Sir W. Page, 31 Great George Street, s.w. 100
Woodcock, William, Manchester 100
*Woodd, Basil T., M.P., Conyngham Hall, Knaresborough 250

*Woodd, Robert Ballard, 108 New Bond Street, w. £500
*Woodhouse, John Thos., Ashby-de-la-Zouch ... 500
*Woollams, Henry, 110 High Street, Manchester Square, w. 250
*Woollcombe, Thomas, Kerr Street, Devonport ... 100
*Woolloton, Charles, 246 High St., Borough, s.e. 500
Wright, T. B., Birmingham 100
*Wright, Joseph, Saltley Works, Birmingham ... 500
Wrigley, Joseph, jun. (J. and P. C. Wrigley and Co.), Huddersfield 250
Wrigley, Thomas, 32 Princess Street, Manchester 500
Wyatt, M. Digby, 37 Tavistock Place, w.c. ... 100
Wyld, William, 45 Rue Blanche, Paris ... 100

*Yolland, Colonel Wm., R.E., 17 Westbourne Park, w. 100
Younghusband, Joseph T., 53 Clifton Road, St. John's Wood, N.w. 500

*Zanzi, Alexander, 30 Brompton Crescent, s.w. 100

TOTAL AMOUNT GUARANTEED, £446,850.

LONDON:
PRINTED BY WILLIAM CLOWES AND SONS, STAMFORD STREET,
AND CHARING CROSS.

THE

ILLUSTRATED CATALOGUE

OF THE

INTERNATIONAL EXHIBITION.

SECTION I.

CLASS I.

MINING, QUARRYING, METALLURGY, AND MINERAL PRODUCTS.

[1]

AARON, E. & W., *Liverpool*.—1. Halkyn hydraulic limestone. 2. Halkyn Chirt stone. 3. Holywell Roman cement stone.

[2]

ABERDARE COAL COMPANY, *Cardiff, Glamorganshire*.—Specimens of Aberdare Company's Merthyr steam coal, from the four-feet and nine-feet seams.

The coals, raised by this company, and shipped at Cardiff, are used extensively for marine purposes by the English and French Governments, and are entered in the naval contracts of both countries. They are also very largely used, by all the great steam navigation companies in England, and on the Continent.

Evaporative power	...	...	...	15·852
Specific gravity	...	...	...	1·323
Coke ...	...	...	...	86·8
Moisture	...	...	...	·93
Frangibility { Large	...	·...	86·4
{ Small	...	...	13·6
Ash ...	...	...	...	...	1·90
Carbon	...	...	...	...	89·33
Hydrogen	...	...	...	...	4·23
Nitrogen	...	...	...	...	1·57
Sulphur	...	...	...	...	·67
Oxygen	...	...	...	...	1·60

[3]

ABERDARE IRON COMPANY, *Aberdare, Glamorganshire*.—Coal; iron ore, pigs, refined metal, and railway iron.

[4]

ABERDARE STEAM FUEL COMPANY (Limited), *White Lion Court, London; Cardiff, and Aberdare*.—Patent steam fuel.

The ABERDARE PATENT STEAM FUEL, is made by compressing the small of the best South Wales steam coal, until its density somewhat exceeds that of the coal itself. From the regular form, and uniform size of the blocks, a larger quantity by weight can be stowed in a given space, to the extent of one-third, than of the coal; and at the same time the evaporating power is maintained; facility in firing is in no degree diminished; all risk of spontaneous ignition is removed; and much advantage is gained in convenience and cleanliness of trans-shipment, especially in foreign ports. Moreover, while coal, when stored abroad, is subject to a depreciation of from 20 to 30 per cent., this fuel is absolutely incapable of injury, even by a lengthened exposure to the full effects of tropical heat and rain. Weight, per cubic foot, 80 lbs. Space occupied by one ton, 28 cubic feet. Supplied in blocks 9 in. by 6 in. square, of 14 lbs. each.

[5]

ADAIR, JOHN G., *Bellgrove, Ballybrittas*.—Coal from exhibitor's colliery, Ballylehane, Queen's County, Ireland. Marbles, minerals, and building materials, natural productions of County Donegal, Ireland.

[6]

AYTOUN, ROBERT, 3 *Fettis Row, Edinburgh*.—Safety-cage for miners, and hoist.

[7]

BANKART, F., & SONS, *Red Jacket Works, Briton Ferry, Glamorganshire; and 9 Clement's Lane, Lombard Street, London.*—Copper and ores.

[8]

BARBER, WALKER, & CO., *Eastwood, Nottinghamshire.*—Coals.

[9]

BARKER, RICHARD, *Wood Bank, Egremont, Cumberland.*—Hematite iron ores, and spars associated with them.

[10]

BARKER, RAWSON, & CO., *Sheffield.*—Leads: white, red, and refined.

[11]

BARNES, THOMAS ADDISON, *Grosmont Iron Works, Whitby, Yorkshire.*—Grosmont iron-stone and pig iron, from the Whitby Cleveland district.

[12]

BARRINGER & CARTER, *Mansfield, Nottinghamshire.*—A remarkably fine red moulding sand, found only at Mansfield.

[13]

BARROW, BENJAMIN, *Clifton House, Ryde, Isle of Wight.*—Mineral products of the Isle of Wight.

[14]

BARROW, RICHARD, *Stavely Works, near Chesterfield.*—Coal, ironstone, and iron.

[15]

BATSON, ALFRED, *Ramsbury, Wilts.*—Devonshire madrepores collected by the exhibitor; inlaid by John Thomas, Babbicombe, Devon.

[16]

BAYLY, J., *Plymouth.*—Ores of copper, tin, and lead.

[17]

BEADON, W., *Otterhead, Honiton, Devon.*—Siliceous sands for stuccoing, plastering, &c. Mineral black, natural pigment. Fine clays, iron ores, &c.

[18]

BELL BROTHERS, *Newcastle-on-Tyne.*—Aluminium and its compounds. Pig iron and iron ores from Cleveland.

[19]

BENNETT, THOMAS, 11 *Woodbridge Street, Clerkenwell, London.*—Specimens of leaf gold.

[20]

BENNETTS, WILLIAM, *Camborne.*—Safety-skip, adapted for raising men and minerals, from coal and copper mines.

[21]

BENTHOLL, H., 14 *Chatham Place, Blackfriars.*—Porphyry.

[22]

BENTLEY, JOHN F., *Stamford.*—Specimens of the building stones, &c., of this district, worked in design.

[23]

BICKFORD, SMITH, & CO., *Tuckingmill, Cornwall.*—Safety-fuse: a small column of gun-powder enclosed in fibrous material or metal, for conveying fire to the charge in blasting.

[24]

BIDDULPH, J., & CO., *Swansea.*—Minerals, iron ore, and coal.

[25]

BIDEFORD ANTHRACITE MINING COMPANY, *Bideford, Devon.*—Mineral black paint and culm.'

This superior paint has been exclusively used in her Majesty's dockyards and arsenals, for the last forty years. Its properties are thus spoken of in a certificate bearing date Aug. 12th, 1847:—"Its superiority is observable in the preservation of wood, iron, and canvas. It covers the work well; dries quick and hard; is more durable, and does not blister like other blacks; and has a body inferior only to white lead.—William Smith, master-painter. Attested. R. Dundas." The paint may be procured from the company, in any quantity, at a moderate price.

[26]

BIRD, EDWARD, *Matlock Bath, Derbyshire.*—Copy of Egyptian obelisk in black marble; paperweights, &c., engraved and etched.

[27]

BIRD, WM., & CO., 2 *Laurence Pountney Hill, London.*—Specimens of British iron, steel, and tin-plates.

[28]

BIRLEY, SAMUEL, *Ashford, Derbyshire.*—Black marble table inlaid with arabesques, &c.

[29]

BLAENAVON IRON AND COAL COMPANY, *Monmouthshire, and Cannon Street, London.*—Iron angle bars, tee rails, weldless tires, girders, and pigs.

[30]

BLAENCLYDACH COAL COMPANY, *Neath.*—Samples of coal.

[31]

BOLCKOW & VAUGHAN, *Middlesborough, Tees.*—Coal, coke, ironstone, pig iron, rail, plate, bar, and other manufactured irons.

[32]

BOUNDY, T., *Swansea.*—Arsenic.

[33]

BOWLING IRON COMPANY, *near Bradford, Yorkshire; London, 5 Bankside.*—Boiler plates, tyres, bars, angles, &c.

[*Obtained a First Class Medal of 1851 Exhibition, and the Silver Medal of the Paris Exhibition, 1855.*]

In order to give a general idea of the nature and scope of their operations, the Bowling Iron Company subjoin a list of the various branches of the iron trade in which they are engaged.

Iron-masters, engineers, millwrights, boiler-makers, &c., manufacturers of plates, tyres, bars, sheets, hoops, angle and tee iron, steam hammers, and forgings.

The sole agents for London, France, Germany, &c., are Messrs. Macnaught, Robertson, and Craig, whose offices in London are at 14 Cannon Street, E.C., and 5 Bankside, S.E.; and in Paris at 55 Rue de Douai.

[34]

BOXALL, JOHN JAMES, *Pulborough, Sussex.*—Green sandstone. Pulborough church, which is 600 years old, is built of this stone.

[35]

BRADLEY, CHRISTOPHER L., *Prior House, Richmond, Yorkshire.*—Copper and lead ore from the mountain limestone, Yorkshire.

[36]

BREWER, ROBERT, *Rudloe Firs, Corsham, Chippenham.*—Stone vase, and two cubes of Bath stone.

[37]

BRIGHT, S., & CO., *Buxton.*—Fine black marble vases, and inlaid mosaic work.

[38]

BRISTOL AND FOREST OF DEAN COMPANY, *Princess Royal Colliery, near Lydney.*—Coal, from the Yorkley and Whittington seams.

[39]

BROWN, J., & CO., *Sheffield.*—Samples of steel manufacture.

[40]

BROWN & JEFFCOCK, Civil and Mining Engineers, *Barnsley.*—Coals and ironstones from South Yorkshire coal fields; geological and mining maps and sections.

Specimens of the following coals are exhibited, viz., Melton Field, or Wathwood, or Wood Moor; Cannel coal from same bed; Woolley Silkstone, or Abdy, or Winter; High Hazel, or Kent's thick coal; Barnsley, or Elsecar, or Darnall; Flockton; Parkgate; Thorncliffe thin; Silkstone four feet; Silkstone; Halifax or Ganister bed.

Ironstones from the South Yorkshire coal-field, as used at the Milton and Elsecar, Parkgate, and Thorncliffe Iron Works. At the Parkgate Iron Works, near Rotherham, armour plates for the new ships of war are made in large quantities.

MAP of the SOUTH YORKSHIRE COAL-FIELD, showing the outcrops of the coals, directions of the faults, and the situation of the various collieries and iron works.

Sections showing the relative position of the coals and ironstones worked in this district.

One class of coals is very valuable for iron and steel making, and for locomotive and marine steam-engines; other kinds are suitable for gas-making and domestic purposes, and are well known in the London and other markets, as Flockton, Silkstone, Barnsley House coal, &c.

[41]

BROWN & RENNIE, *Kilsyth, by Glasgow.*—Coal and coke.

[42]

BROWNE, WILLIAM, *St. Austell, Cornwall.*—China clay of every description, china stone, and red hematite iron ore.

The exhibitor has on sale at his various works, china clay of the purest descriptions, suitable for every purpose; and can supply china stone in any quantity for pottery and other manufactures. He offers, also, a large supply of exceedingly rich red and black hematite iron ores from his mines in Devonshire.

[43]

BRUNTON, J. D., *Barge Yard, Bucklersbury.*—Condensed peat, and peat charcoal.

[44]

BRUNTON, W., & Co., *Penhellick Safety-Fuse Works, near Camborne.*—Safety-fuse for blasting in mining, quarrying, and submarine operations.

W. Brunton & Co. are manufacturers of every description of safety-fuse; and the inventors of the gutta-percha fuse, which has been supplied to the Royal Arsenal, Woolwich, to the Arctic expedition, and which is in use in every part of the globe. The branch works of the firm are at Brymbo, near Wrexham.

[45]

BUDD, J. P., *Swansea.*—Iron, and tin-plates.

[46]

BULL, GEORGE, D.D., Dean of Connor, *Redhall, Co. Antrim.*—Large quartz crystal, or Irish diamond; weight 83½ lbs.

[47]

BUTLIN, THOMAS, & Co., *East End Iron Works, Wellingborough.*—Iron and its ores.

[48]

BUTTERLEY COMPANY, *Butterley Iron Works, Alfreton.*—Section of coal-pit. Armour plates, deck beams, rolled girders, joists, and other iron.

[49]

BWLCH Y GROES SLATE COMPANY (Limited), *Llanberis, Carnarvon.*—Roofing slate; the green a fine specimen.

[50]

BYERS, JOSHUA, & SON, Producers and Manufacturers, *Stockton-on-Tees, Durham.*—Lead ore from Grasshill mine, Teesdale. Silver and litharge. Refined, common, and slag leads, sheet lead; lead pipe; and thin sheet lead.

[*Obtained Prize Medal at the Exhibition of 1851.*]

[51]

CAITHNESS, EARL OF, 17 *Hill Street.*—Caithness flags.

[52]

CALOW, JOHN THOMAS, *Staveley, Derbyshire.*—Patent safety apparatus for shafts of mines, &c.

[53]

CAMPBELL BROTHERS, *William Street, Blackfriars.*—Pig and bar iron, manufactured at Calder and Govan Iron Works.

[54]

CANNAMANNING CHINA CLAY COMPANY, *Newton Abbott.*—Pipe, potters', and china clays.

[55]

CASE & MORRIS, Proprietors, *Rose Bridge, Three Hall Collieries, Ince, Wigan.*—Section of actual strata of Rose Bridge and Ince Hall collieries, coal, &c.

[56]

CHAFFER, THOMAS, *Burnley, Lancashire, and* 14 *Great Howard Street, Liverpool.*—Worsthorne, Hambleton, and Portsmouth stone.

[57]

CHAMBERS, J., *Alfreton.*—Coal.

[58]

CHEESEWRING GRANITE COMPANY, 6 *Cannon Street, Cornwall.*—Design by John Bell for memorial of the Exhibition of 1851, one-fifth full size. (*Nave.*)

[59]

CHILD, W. J. & T., *Hull, Leeds, and Grindleford Bridge, Derbyshire.*—French and Derbyshire Peak millstones.

[60]

CLAY CROSS COMPANY, *Clay Cross, near Chesterfield.*—Samples of coal, lime, limestone, ironstone, and pig iron.

[61]

COAL OWNERS OF NORTHUMBERLAND AND DURHAM, *Newcastle-on-Tyne.*—Map and section of coal-fields.

[62]

COCHRANE & CO., *Woodside, Dudley, and Ormesley Iron Works, Middlesborough-on-Tees.*—Iron pipes and pig iron.

[63]

COLLES, ALEXANDER, *Marble Mills, Kilkenny.*—Black Kilkenny marble chimney-piece, made by machinery.

[64]

COLLEY, GEORGE, 8 *Upper Dorset Street, Belgrave Road, Pimlico.*—Vase in freestone.

[65]

CONNORREE MINING COMPANY, *Connorree Mines, Ovoca, Ireland.*—Sulphur pyrites, precipitate of copper, and sulphur and copper ores.

[66]

COPELAND, GEORGE ALEXANDER, *Carwythenack House, Constantine.*—A series of patent waterproof blasting cartridges.

[67]

CORBETT, W. F., *Great Charles Street, Birmingham.*—Apparatus to prevent over-winding at pits.

[68]

CORBETT, JOHN, *Stoke Prior Salt Works, Bromsgrove, Worcestershire.*—Refined table salt, butter salt, and provision salt.

[69]

COURAGE, ALFRED, & CO., *Bagillt, Flintshire.*—Lead smelting, and manufacturing patent sanitary pipes. Zinc spelter making.

[70]

COWPEN COAL COMPANY, *Cowpen Colliery, Blyth.*—Black of Cowpen Hartley steam coal.

[71]

COX, BROTHERS, & Co., *Derby.*—Red, white, and orange lead, shot, lead pipes, plates of Derbyshire silver, &c.

[72]

CRAIG, GEORGE, & SON, *Caithness Pavement Works, Thurso.*—Specimen of Caithness flags for tables, shelving, and pavement.

[73]

CRAWLEY, C. E., 17 *Gracechurch Street.*—Improved miners' safety-lamp, combining greater safety with increased light.   (*See page* 7.)

[74]

CRAWLEY, G. B., *Neath.*—Samples of coal.

[75]

CRAWSHAY, H., & Co., *Lightmoor Collieries, Cinderford.*—Rocky vein coal.

[76]

CRAWSHAY, H., & Co., *Abbot's Wood Mines, Cinderford.*—Black Brush iron ore.

[77]

CROWN PRESERVED COAL COMPANY (Limited), 62 *Moorgate Street, London.*—Preserved coal.

[78]

CWMORTHIN SLATE COMPANY (Limited), *Merionethshire.*—Slates and slabs.

[79]

DABBS, JOHN, Agent for LORD NORTHWICK, *Stamford.*—Freestone from Ketton Quarries, Rutland.

[80]

DAGLISH, JOHN, *Hetton Collieries, Durham.*—Model of ventilating furnace for coal-mines; self-registering water-gauge.

[81]

DAVIS, DAVID, *Bute Crescent, Cardiff.*—Sample of Davis's upper four feet and Blaengwawr Merthyr steam coals.

These coals are on the English, French, and Spanish Government lists, and are largely consumed in steam ships, locomotive engines, and manufactories throughout the world. The following companies (as well as the London contractors, and consumers in every country) will testify to their superior quality :—

The Peninsular and Oriental Steam Packet Company
The West India Royal Mail Steam Packet Company.
The Montreal Mail Packet Company.
The Philadelphia and New York Transatlantic Company.
The Cunard Royal Mail Company.

REPORT of William Allen Miller, Esq., M.D., F.R.S., King's College; of W. Hoffman, Esq., LL.D., F.R.S., Royal College of Chemistry; and E. Frankland, Esq., Ph. D., F.R.S., Saint Bartholomew's Hospital.

DAVID DAVIS'S MERTHYR AND STEAM COALS.	Nine-feet Vein.	Upper Four-feet Vein.	CHEMICAL ANALYSIS OF 100 PARTS OF DRIED COAL.	Nine-feet Vein.	Upper Four-ft. Vein.
Theoretic and exportative power of 1 lb. of this coal . . . . . }	15·882	15·895	Ash	1·83	2·50
			Carbon	89·52	89·27
Specific gravity	1·328	1·356	Hydrogen	4·31	4·42
Coke	88·05	85·60	Nitrogen	1·20	1·32
Moisture	0·68	0·83	Sulphur	0·89	0·78
Frangibility { Large . . . . Small	75·2 24·8	78·8 21·2	Oxygen	2·25	1·71

CRAWLEY, C. E., 17 *Gracechurch Street*.—Improved miners' safety-lamp, combining greater safety with increased light.

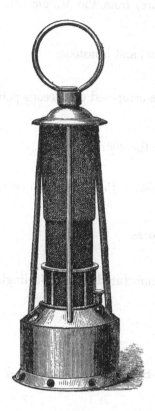 

This lamp combines several important advantages, viz.,

1st. Without the use of glass it gives from three to four times the light of the common "Davy lamp."

2nd. It never requires snuffing, thus not only keeping the inside of the lamp from getting foul, but giving less trouble to the miner, and at the same time producing a more even light.

3rd. It will, on account of its peculiar construction, consume, while burning with a good flame, from one to two cubic feet of gas per minute (the light being thereby improved), thus tending, though in a small degree, to prevent the accumulation of gas, and so to some extent to lessen the risk of explosion.

4th. Owing to the fact of the gas passing into the lamp principally from below, the wire gauze that covers the flame does not readily become red hot.

5th. It can be instantly extinguished, if required, without trouble; a matter of great importance in case of a sudden irruption of gas.

6th. The lock is very simple, though entirely differing in principle from all others now in use, and is rendered perfectly secure, by means of a seal placed over the lock and completely concealing it, in such a manner as to render it absolutely impossible to open the lamp without breaking the seal, thus forming a perfect detector. This seal consists of a very small thin metal disc, having any kind of device stamped upon it, which could be varied from day to day; none but the one authorized person knowing beforehand what seal would be used on any particular day.

7th. The great increase, however, in the light would of itself remove the chief temptation to open the lamp, added to which, it gives, if anything, less light when opened.

8th. It is also, under ordinary circumstances, impossible to light a pipe by drawing the flame through the gauze, as is the case with the "Davy."

9th. Nor, for the same reason, can the flame be driven through the gauze by a current of air; which being the case in the "Davy," has been supposed to be the cause of so many explosions.

10th. There is also another patent improvement in this lamp which will be hailed by the miner as a great boon, viz., an insulated handle, which enables it to be carried at all times without inconvenience, however hot the rest of the lamp may become.

These lamps are made entirely by machinery, and the manufacture is carefully superintended by the patentee, so as to insure perfect accuracy in the fitting of the separate parts, which are very simple, and so constructed, that any part, if accidentally damaged or lost, can be at once replaced without trouble, and at small cost, a stock being always kept on hand for that purpose. Sole manufacturer, C. E. Crawley, 17 Gracechurch Street, London, E.C. Sole agent for Wilson's new Patent Oil Press, and Wilson's new Patent Cotton Press.

Any further information may be obtained in the Exhibition building, where attendance will be given daily, between the hours of ten and four.

[82]

DAVIS, JAMES, *Ulverstone.*—Iron pyrites (sulphur ore) from the Millom Mining Company (Limited), Millom, Cumberland.

[83]

DAWES, W. H. & G., *Denby Iron Works, Derby.*—Coal and ironstone.

[84]

DENBY, W., 3 *Denby Place, Sidmouth.*—Mosaic table composed of siliceous pebbles found at Sidmouth.

[85]

DENMAN, LORD, *Stoney, Middleton.*—Grit stone from the district.

[86]

DEVON AND COURTENAY CLAY COMPANY, *Newton Abbott.*—Pipe, potters', and china clays.

[87]

DEVON GREAT CONSOLS MINE, *Tavistock.*—Copper ores.

[88]

DEVONSHIRE, DUKE OF.—Slate in block and manufactured, from Burlington Quarries, Ulverstone, Lancashire.

[89]

DOVE, D., *Nutshill Quarries, Glasgow.*—Grindstones.

[90]

DOWLAIS IRON COMPANY, *Dowlais, Merthyr Tydvil.*—Samples of manufactured iron.

[91]

DUNCAN, FALCONER, & WHITTON, *Carmyllie Quarries, by Arbroath.*—A step; plate landing and pavement slabs.

[92]

DYBALL, T., *Kirton Lindsay,* for SIR CULLING EARDLEY.—Iron ore.

[93]

EAST CORNWALL ARSENIC COMPANY, 9 *Parade, Plymouth.*—Arsenical mundic, unrefined arsenic, refined arsenic, and lump arsenic.

Samples of Arsenical mundic, Unrefined arsenic, Pure white arsenic, finely ground, Pure white lump arsenic, from the works of the company at Hornbarrow, showing the different stages of the manufacture.

Inquiries, &c., may be made of the Secretary, MR. JOSEPH SEALE, 9 Parade, Plymouth; or of the London Agents, MESSRS. JOHN B. DRAYTON & Co., 30 Great St. Helens.

[94]

EASTWOOD & SONS, *Derby.*—Samples of iron.

[95]

EBBW VALE COMPANY, & PONTYPOOL IRON COMPANY, *Ebbw Vale, Newport, Monmouthshire.*—Minerals, tin-plates, and iron manufactures.

[96]

EDDY, JAMES RAY, *Carleton, Skipton, Yorkshire.*—Lead ores, with vein stone.

[97]

EDWARDS, WOOD, & GREENWOOD, *Tame Valley Colliery, Tamworth.*—Iron pyrites and fire clay.

[98]

ELLAM, JONES, & CO., *Maskeaton Mills, Derby.*—Emery, and oxide of iron paint, made from the ore, expressly for iron work.

[99]

ELLIS & EVERARD, *Markfield Granite Quarries, Leicestershire.*—Paving setts—broken for macadamizing; specimens for building, &c.

[100]

EVANS & ASKIN, *Birmingham.*—Nickel, cobalt, and German silver.

[101]

FARNLEY IRON COMPANY, *Farnley, near Leeds.*—Samples of coal, ironstone, pig, boiler-plate, tyres, angle-iron, rivets, and fire-clay goods.

No. 1, 2. Farnley ironstone, raw and calcined.	No. 12. Puddled iron.
3, 4. Farnley coal and coke (better bed).	13. Samples of railway tyre bars.
5. Limestone.	14. „ ditto, bent cold.
6, 7, 8, 9. Samples of Farnley pig metal.	15. „ boiler plate.
10. Blast furnace dross.	16. „ angle iron.
11. Refined metal.	17. „ bar and rivet iron.

[102]

FAYLE & CO., 31 *George Street, Hanover Square.*—Blue clay for the manufacture of earthenware.

[103]

FINNIE, ARCHIBALD, & SON, *Kilmarnock.*—Steam and house coal exported at Troon and Ardrossan, Ayrshire, Scotland.

[104]

FIRTH, BARBER, & CO., *Oak's Colliery, near Barnsley.*—Specimen of Barnsley seam, steam and house coal.

The specimens exhibited of the BARNSLEY BED OF COAL from the Oaks Colliery, show the full thickness of the seam, and its divisions into hard and soft coal. The pits from which it is produced are 860 feet in depth. The hard coal is upon the Indian Council and French Admiralty lists, and is well adapted for iron making, steel converting, and for locomotive and marine engines; the soft is valuable for domestic purposes. Agents in London at King's Cross Station, Messrs. Beale and Walker.

[105]

FITZGERALD, RICHARD, Clerk, *Clare View, Tarbert, Co. Kerry.*—Peat from Aughrim, near Tarbert, Co. Kerry.

[106]

FORSTER, G. B., *Cowpen Colliery, Blyth.*—Model of coal pit, with cages and apparatus.

[107]

FORSTER, R., *Gateshead.*—Grindstones.

[108]

FOWLER, W., & CO., *Sheepbridge Iron Works, Chesterfield.*—Coal, and ironstone of which armour-plate iron is made.

[109]

FRANKLIN, F., *Galway.*—Polished marble.

[110]

FREEMAN, W. & J., *Millbank Street, Westminster, and Penrhyn, Cornwall.*—Granites and stones. (*See page* 10.)

[111]

FRYAR, MARK, *School of Mines, Glasgow.*—Plans and drawings relating to mining.

FREEMAN, W. & J., *Millbank Street, Westminster, and Penrhyn, Cornwall.*—Granites and stones.

[Obtained a Medal and Certificate at the Exhibition of 1851.]

W. & J. Freeman exhibit specimens of granites from the Cornwall and other quarries; building stones from the oolite of Portland, used at the British Museum, and numerous other edifices; stones from the Bath and Painswick quarries; magnesian limestone from Huddlestone, used in the erection of York Minster, and other churches; sandstone from Hare Hill, and other quarries in Yorkshire; flag and landing stones from the same locality, used extensively for the London footways and buildings; millstone grit, for bridge and dock works.

The works supplied by Messrs. Freeman include the docks of Keyham, Chatham, Deptford, Jarrow, Commercial, East and West India, Birkenhead, Liverpool, Hull, &c.; the harbours of refuge at Alderney, Dover, and Portland; bridges over the Thames and Medway; lighthouses at Beachey Head, Bishop's Rock, Guernsey, and the Basses in the East Indies; the plinth and lodges in front of the British Museum, and the monoliths in the King's Library of that building; the plinth at the Royal Exchange, and the steps and landings for the terraces at the Crystal Palace; and the obelisk from the Exhibition of 1851, since erected in Chelsea College.

The polished granites in the obelisk at Scutari, and the pedestal for the statue of Carlo Alberto at Turin, containing stones upwards of twenty feet in length; the pedestal for the statue of Richard Cœur de Lion, in front of the Houses of Parliament, each by Baron Marochetti; and the monoliths for the mausoleum erected to the memory of her late Royal Highness the Duchess of Kent, from the design of Mr. Humbert, were executed at the polishing works connected with their quarries at Penrhyn.

[112]

GAMMIÉ, GEORGE, *Shotover House, Oxfordshire.*—Native Oxford ochre.

[113]

GARDNER, ROBERT, *Sansaw, Shrewsbury.*—Grinshill building stone; copper ore; barytes.

[114]

GARLAND, T., *Fairfield, Redruth.*—Arsenic.

[115]

GENERAL MINING COMPANY FOR IRELAND, *Westmoreland Street, Dublin.*—Zinc ores, spelter, fire-clays, and ochres, from Silvermines, Tipperary.

[116]

GEOLOGICAL SURVEY OF THE UNITED KINGDOM, *Geological Survey Office, 28 Jermyn Street.* —Published maps and sections of England, Scotland, and Ireland, 1-inch and 6-inch scales.

[117]

GIBBS & CANNING, *Tamworth.*—Glazed stoneware sewerage pipes, fire-bricks, and terracotta.

[118]

GILBERTSON, W., & Co., *Swansea.*—Tin-plates.

[119]

GILKES, WILSON, PEASE, & Co., *Middlesborough.*—Samples of pig iron and test bars; samples of iron ores.

[120]

GODDARD, EDWIN, for EDWARD BLAKE, *Newton Abbott.*—Tobacco-pipe clay; potters' clay; papermakers' clay; china clay.

[121]

GOLDSWORTHY, THOMAS, & SONS, *Hulme, Manchester.*—Emery, emery and glass cloths and papers, whetstones, and polishing-stones; knife-cleaning machine.

[122]

GOVERNOR AND COMPANY OF COPPER MINERS IN ENGLAND (*Cwm Avon Works Glamorganshire*, W. P. STRUVE, Esq., Manager of the Works), Offices, 10 *New Broad Street Mews, London, E.C.*—Coal; iron mine, iron, copper, yellow metal, tin-plates, chemicals, &c.

[*Obtained Certificate of Honour for tin-plates, and Prize Medal for railway iron, at the Great Exhibition of 1851; and Grande Médaille d'Honneur for railway iron at the Paris Exhibition, 1855.*]

Copper.

Ingot, cake, wire bar, sheets, ships' sheathing, copper rails for gunpowder magazines, bolts and strips rolled thin, to show the malleability of the metal.

Trade mark.

Yellow Metal (alloy of copper and zinc).

Sheets, ships' sheathing, and rails.
Rails, bolts, and composition nails.

Trade mark.

Iron produced and manufactured at Cwm Avon Works.
Pig iron for rails.

These rails and sections of rails are exhibited in Class V.
Various sections of rails made at these works.
One large bridge rail, 90 feet long, 58 lbs. per yard.
One flanch rail, 63 feet long, 3¾ lbs. per yard.

———

Fish-plates, merchant bars.

Sheet iron, known as Canada plates, made from iron specially prepared.

Trade mark.

Samples of wrought iron tested.

Chemicals.

Miscible naphtha (Pyroxilic spirit), as supplied to the Board of Inland Revenue for making "Methylated spirit."
Copperas, limesalt black or brown, and white or gray.

Minerals and specimens of Iron, illustrating the process of preparing Iron for Tin-plates.

Argillaceous iron ore (known as "Welsh mine").
Cold blast pig iron, made expressly for plates.
"Stamps" refined with charcoal only.
Bars.
Black or tole plate, in the form of a book, rolled to exhibit the malleability of the metal, weight ½ oz. per square foot.
Block tin specially refined for use.

Tin-plates (iron superficially alloyed with tin).

Trade Marks.

E C C. or V S.	1 C. DLDxx. III.	1xxxxx. Dxxxxx.	DBD.
C A.	1 C.		
B I.	1 C.		

Terne Plates (iron superficially alloyed with tin and lead).

Trade Marks.

E C C. or V S.	D.
C A.	1 C.
B I.	1 C.

A mechanical contrivance for facilitating calculations, invented and patented by Mr. Robert Dunlop, one of the Company's agents at Cwm Avon.

No. 1 embraces calculations from $\frac{1}{32}$ of an unit to upwards of 90,000 in multiplication; and from $\frac{1}{32}$ of a penny to twenty shillings in division.

No. 2 embraces calculations of whole numbers or decimals for multiplication or division for any sum containing from one to nine figures.

[123]

GOWANS, JAMES, *Rockville, Merchiston Park, Edinburgh.*—Boring machines, wedge, and galvanic apparatus for blasting.

The following are exhibited:—
1. Machine for boring holes in stone, upon the drill principle.
2. Machine for the same purpose, upon the ram principle.
3. Expanding wedge used in place of the pinch.
4. Galvanic battery and apparatus used for blasting at Redhall quarry, Edinburgh, and also at the Exhibitor's Railway Works, and elsewhere.

[124]

GRAHAM, ABRAHAM, Stone Merchant, *Huddersfield.*—Building stones and hard paving-stones.

[125]

GRANVILLE, THE EARL, *Shelton, Staffordshire.*—Minerals and pig iron.

[126]

GRAY, JAMES, M.D., *Glasgow.*—Modification of Davy's lamp.

[127]

GREAVES, JOHN W., *Portmadoc, North Wales.*—Roofing slates.

[128]

GREAVES & KIRSHAW, *Warwick, and South Wharf, Paddington.*—Hydraulic lias, lime, and cement; smooth polished lias stone.

[129]

GREENWELL, G. C., *Radstock.*—Sections of and specimens from Somersetshire coal-field,

[130]

GREENWELL, G. C., for WESTBURY IRON COMPANY (Limited), *Westbury, Wilts.*—Section of ironstone and furnace products.

[131]

GREGORY, J. R., 25 *Golden Square, London.*—Minerals, fossils, and rocks; Devonian fossil fishes from Scotland.

[132]

HALIFAX CORPORATION, *Halifax, Yorkshire.*—Building and other stones, coals, and ironstone in and near Halifax.

Specimens of building and other stones; also of coals, clays, iron pyrites, and fossiliferous remains found in the town and parish of Halifax.

[133]

HALL, J. & T., *Derby.*—Marble and spar vases and ornaments, and mosaic works in marble.

[134]

HALL, JOHN, & Co., *Stourbridge.*—Stourbridge fire-clays, gas-retorts, furnace bricks, melting-pots, crucibles, &c.

[135]

HALLIDAY, THOMAS C., *Greetham, Rutland.*—Clipham stone; blocks all sizes, will stand all weather.

[136]

HAMPSHIRE, J. K., *Whittington Collieries, Chesterfield.*—Safety apparatus, for raising and lowering persons in shafts.

[137]

HAMPSHIRE, MATTHEW, & Co., Stone Merchants, *Spring Street, Huddersfield.*—Building stones.

[138]

HANCOCK, ROBERT, *Polberro, St. Agnes, Cornwall.*—Pulverized ore dressing machine.

[139]

HARPER & MOORES, *Lower Delph Clay Works, Stourbridge.*—Glasshouse-pot clay, retorts, fire-bricks, lumps, &c.

[140]

HARRIS, JOSIAH, *Newton Abbott, Devonshire.*—Ores of iron, tin, lead, copper, blende, manganese, bismuth, and antimony.

[141]

HARRISON, AINSLIE, & Co., *Newland Furnace, Ulverstone.*—Lindal Moor hematite, puddling ores, and Lorn pig iron.

[142]

HARRY, G., *Swansea.*—Copper, silver, iron, zinc, and nickel ores, and metals.

[143]

HAWKSWORTH, WILLIAM, & Co., *Linlithgow.*—Cast steel, engravers' steel plates, patent steel rifle barrels, and tubing.

[144]

HEATH, EVANS, & Co., *Aberdare.*—Steam coal.

[145]

HEAVEN, W. H., *Lundy Island, Clovelly, North Devon.*—Specimens of Lundy Island granite.

[146]

HEGINBOTHAM, PETER, & SON, *Shallcross Mills, Whaley Bridge, near Stockport.*—Sulphate of barytes, unbleached, bleached, and unmanufactured.

[147]

HENDERSON, G. W. M., *Fordell, Fifeshire.*—Carved block of sandstone.

[148]

HENDERSON, JAMES, C.E., *Truro, Cornwall.*—A plan and section of a Cornish mine.

[149]

HENGISTBURY IRON MINING COMPANY, *Christchurch, Hants.*—Iron ore.

[150]

HENSON, ROBERT, 113A *Strand.*—Ornaments in marble and minerals.

[151]

HEWLETT, ALFRED, for the EARL OF CRAWFORD AND BALCARRES, *Haigh Colliery, Wigan.*—Cannel coal for making gas.

[152]

HIGGS, SAMUEL, & SON, *Penzance.*—Model of tin-dressing floors, safety-lamp, and specimens of tin and copper ores.

[153]

HILL, FREDERICK, *Helston.*—Specimens of ores, metals, minerals, clay, marl, elvans, and stone produced in Helston mining district.

[154]

HIRD, DAWSON, & HARDY, *Low Moor Iron Works.*—Samples of iron in various stages of manufacture.

[155]

HOLLAND, SAMUEL, & Co., *Portmadoc, Carnarvonshire.*—Roofing slates.

[156]

HOLMES, JOHN, *Bolton Wood Quarries, Bradford.*—Monument sculptured by Francis Stake & Co., Bradford, Yorkshire.

[157]

HOLROYD, JAMES, & SONS, Stone Merchants, *Brighouse.*—Flag and building stones.

[158]

HOOPER & MOORE, *Stourbridge.*—Fire bricks, &c.

[159]

HOWARD, HON. JAMES, 1 *Whitehall Place, London.*—Forest of Dean stone, coal, iron ore, clay, and pottery.

[160]

HOWARD, RAVENHILL, & Co., *King and Queen Iron Works, Rotherhithe, London.*—Patent bridge links rolled entire.

Improved Patent Links; for suspension and Girder bridges, roofs, and other purposes; the bar and heads rolled into form at one heat, avoiding the uncertain process of welding.

[161]

HOWARD, THOMAS, C.E., *Bristol.*—Samples of the building and other stones in the Bristol district.

[162]

HOWIE, JOHN, *Hurtford Colliery, Kilmarnock, Scotland ;* Shipping ports, *Troon and Ardrossan.*—Three pieces of coal.

[163]

HUNT, JOHN, *Porthleven, Helston, Cornwall.*—Model of patent ore-separator ; portable gold-washer ; phosphate of lead.

[164]

HYNAM, JOHN, 7 *Prince's Square, Finsbury.*—Purified dried fullers' earth, for blanket and cloth manufacturers and silk dyers.

[165]

IBBERSON, JOHN, Stone Merchant, *Lockwood, Huddersfield.*—Building stones.

[166]

IRVING, GEORGE VERE, Esq., *Newton, Lanarkshire, N. B.*—Minerals of the Leadhills district, Lanarkshire, Scotland.

[167]

JACKARD, E., & Co., *Ipswich.*—Coprolites from green sand and crag.

[168]

JENKINS, W. H., & Co., *Victoria Place, Truro, Cornwall.*—Ochre, umber or bistre brown, used for painting, paper-staining, paper-making, &c. ; fluorspar, flux for smelting, making fluoric acid, &c. ; felspar glaze for earthenware, porcelain, &c.

[169]

JENNINGS & Co., *Swansea.*—Arsenical ore, arsenic unrefined, refined crystals, powdered and lump arsenic.

[170]

JENNINGS, WILLIAM, *Victoria Street, Hereford.*—Specimen of "Three Elms Quarry" stone, near Hereford.

[171]

JOHNSON, MATTHEY, & Co., 78 & 79 *Hatton Garden, London.*—Platinum, and preparations of the precious metals.

[172]

JOHNSON, W. W. & R., & SONS, *Limehouse, London.*—Wetterstedt's patent metal for roofing and other purposes.

[173]

JONES & CHARLTON, *Duckinfield, Manchester.*—Self-extinguishing detector and patent safety-lamp.

[174]

JONES, DANIEL, *Bradford-on-Avon, Wilts.*—Stone from Bath Farleigh Downs.

[175]

JONES, DAVID, *Hay, South Wales.*—Specimen of gray sandstone from Pontvain Quarries, near Hay, South Wales.

[176]

JONES, DUNNING, & Co., *Middlesborough.*—Pig iron.

[177]

JONES, I., *Swansea.*—Flat chain.

[178]

JONES, W., *Port Tennant, Swansea.*—Fuel for steam purposes.

[179]

JORDAN, H. K., 2 *Clifton Wood Terrace, Clifton, Bristol.*—Minerals.

[180]

JORDAN, JAMES B., *Museum of Practical Geology, London, S.W.*—Models of mineral forms, constructed of cardboard.

[181]

JORDAN, THOMAS B., 15 *Union Grove, Clapham.*—Improved miners' theodolite, and model of Holm Bush Mine.

[182]

JORDAN, WILLIAM HATH, 14 *Langham Street, Regent Street.*—Model of pit-frame and safety-cage.

[183]

JULEFF, JOHN, *Fore Street, and Pednandrea, Redruth, Cornwall.*—Cornish assay crucibles for copper, silver, &c.; goldsmiths' and metallurgists' crucibles.

[184]

KAY, WILLIAM, *Hayhill, Ochiltree, Ayrshire.*—A pair of curling-stones.

[185]

KAYE, GEORGE, Stone Merchant, *Ryecroft Edge, Huddersfield.*—Building stones.

[186]

KELL, RICHARD, & Co., *Newcastle-on-Tyne.*—Grindstone manufactured at Gateshead Fell, suitable for all purposes.

[187]

KING BROTHERS, *Stourbridge Fire-Clay Works, Stourbridge.*—Clay retorts; bricks; section of clay as it is found in the mine.

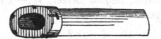

The exhibitors are proprietors of Stourbridge clay, and manufacturers of glass-house pots, retorts, and crucibles, and every description of fire-bricks and other fire-clay goods. Their branch works are at 38 Lichfield Street, Birmingham. Applications for tenders and tracings must be addressed Stourbridge.

[188]

KINSMAN, REV. R. B., M.A., *Tintagel, Camelford.*—Specimens of roofing slates from the Tintagel slate quarries.

[189]

KIRKSTALL FORGE COMPANY, *Leeds.*—Samples of manufactured malleable iron, showing excellence of quality.

[190]

KNIGHT, F. WINN, *Exmoor, South Molton.*—Spathic and other iron ores from Exmoor Forest. (*See page* 16.)

[191]

KNOWLES, ANDREW, *Highbank, Pendlebury.*—Safety-cage for coal-mines—Owen's patent.

[192]

LAMACRAFT, W., *Newton Abbott.*—Clays.

[193]

LAW LIFE ASSURANCE SOCIETY, *Fleet Street, E.C.*—Green and black marble, from Connemara, Co. Galway.

KNIGHT, F. WINN, *Exmoor, South Molton.*—Spathic and other iron ores from Exmoor Forest.

Sample of red, brown, and spathic iron ores, and clay ironstone from some of the principal veins in Exmoor Forest.

No. 1 is from the Hangley Cleve lode; No. 2, Topdeer Park lode; No. 3, Double lode; No. 4, Ebenezer Rogers lode; No. 5, Roman lode; No. 6, Cornham Ford lode; No. 7, Woolcombe lode; No. 8, Huel Eliza lode; No. 9, Picked Stones lode; No. 10, Blue lode; No. 11, Hoar Oak lode; No. 12, North Forest clay ironstone.

These iron ores are found in large veins in the Devonian slates of West Somerset, in the high lands which rise about ten miles west of Taunton, and run westward into the sea near Ilfracombe and Combmartin.

Old workings, supposed to be Roman, are found on the outcrops of these veins along the whole length of the Brendon and Exmoor Hills.

These ores are precisely similar to those which have long been worked in the Siegen district of Western Germany.

The value of this class of ores in producing steel and iron of the finest qualities was not known in England until the Great Exhibition of 1851, when their prototypes were brought over and shown by the continental iron-masters as the ores from which their finest steel and iron were made.

The old workings of West Somerset were then reopened; and their ores are now worked in considerable quantities in the Brendon Hills by the Ebbw-Vale Company, who have laid down a mineral railway from the mines to the port of Watchet.

These veins contain in some places more than 20 feet in width of solid iron ore.

The ores may be seen in great strength in the Exmoor Hills; but until a railway is constructed to one of the adjoining ports they cannot be worked at a profit.

The lands between the Exmoor mines and the sea are in the hands of few and friendly parties, and no Act of Parliament would be necessary for the formation of such a line.

These ores are very rich, and contain no trace of sulphur or phosphoric acid, or any other deleterious matter.

Address of Exhibitor—

Fred. Winn Knight, Esq., M.P., Simonsbath Lodge, Exmoor, Somerset.

[194]

LAYCOCK, JOSEPH, *Newcastle.*—Cast-iron jointed prop, used in the working of pillars in coal-mines.

[195]

LEE MOOR PORCELAIN CLAY COMPANY, *Plympton, Devon.*—Porcelain clay or kaolin, fire, and architectural goornite bricks.

[196]

LEESWOOD GREEN COLLIERY COMPANY, *near Mold.*—Section of the celebrated Leeswood Green cannel coal.

[197]

LEETCH, JAMES, 68 *Margaret Street, Regent Street, W.*—Preparation of fluorspar on cloth, paper, &c., for grinding and polishing.

[198]

LEISS, FREDERIC, 30 *Southampton, Street, Strand.*—A collection of articles manufactured of the mineral mica—patent.

[199]

LEVER, ELLIS, *West Gorton Works, Manchester.*—Flexible tubing, fly-door, and brattice, used in ventilating mines.

ELLIS LEVER is the inventor and sole manufacturer of the flexible tubing, for the ventilation of shafts and exploring drifts in mines. He makes also, in every width, improved brattice and door-cloth, for air-courses and stoppings in the workings of fiery mines. The brattice cloth of his manufacture, was used in restoring the ventilation in the shaft of the unfortunate Hartley Colliery.

[200]

LEVICK & SIMPSON, *Newport.*—Iron ores.

[201]

LILLESHALL COMPANY, *Shiffnal, Shropshire.*—Minerals, castings, malleable specimens.

[202]

LIVINGSTONE, ALEX. S., *Llanelly, Carmarthenshire.*—Patent fuel for marine, locomotive, smelting, and domestic purposes, for all climates.

[203]

LIZARD SERPENTINE COMPANY (Limited), 20 *Surrey Street, Strand, London.*—Various works and specimens of serpentine.

[204]

LLANGOLLEN SLAB AND SLATE COMPANY (Limited), 4 *South Wharf Road, Paddington.*—Enamelled slate work; large slate slab.

[205]

LLETTY SHENKIN COAL COMPANY, *Cardiff and Aberdare, South Wales.*—Thomas's Merthyr smokeless Welsh coal.

[206]

LOMAS, JOHN, & SONS, *Marble Works, Bakewell, Derbyshire,* and S. BIRLEY, *Ashford.*— Eight specimens of the Derbyshire marbles.

[207]

LONDONDERRY, MARCHIONESS OF, *Seaham Hall.*—Three blocks of Pensher sandstone, and model of Seaham harbour and town.

[208]

LONGMAID, WILLIAM, *Arthur Lodge, St. Philip's Road, Dalston.*—Specimens of iron alloyed with gold and platinum.

[209]

LOWES & ROBINSON, *Stanhope, Darlington.*—Case of minerals; section of Weardale strata.

[210]

LOWRY, J. W., 45 *Robert Street, Hampstead Road, N.W.*—Engravings of fossils for the Geological Survey of United Kingdom.

[211]

LUCAS & BARRATT, *Stockton-on-Tees.*—Pig iron.

[212]

LUMBY, JOHN, *Stamford.*—Ironstone, gray and black; pyrites, coal, fire and terra-cotta clay.

[213]

LUND HILL COAL COMPANY, *Lund Hill Colliery, Barnsley.*—Specimens of Barnsley seam, steam, and house coal.

This specimen shows the thickness of the "BARNSLEY" SEAM OF COAL as worked at this colliery; it is divided into soft or house coal, and hard or steam coal. The soft is very suitable for domestic purposes, and for making gas, of which it yields a large amount, and of brilliant quality; the hard portion is very valuable as a steam coal, both locomotive and marine, and is upon the English Admiralty list. The pits are 660 feet in depth. Agent, Mr. Thorneycroft, King's Cross Station.

[214]

MACDONALD, ALEXANDER, *Polished Granite Works, Aberdeen.*—Specimens of granite used in building, decoration, memorials, and general purposes.

[*Obtained the Prize Medal in Class XXVII., in 1851; and the Silver Medal in Paris, in Class XIV., in 1855.*]

No. 1. Polished red granite jointed Doric column, showing the closeness of, and flush surfaces of joints, when built in pieces by the exhibitor's patent process. In this way constructive and decorative erections of any size or form are made.

No. 2. Polished red granite pedestals for busts, vases, groups, &c.

No. 3. Gray granite Gothic headstone memorial, showing "fine axing," and the contrast between axed and polished surfaces.

No. 4. Polished red and blue granite Gothic baptismal font.

No. 5. Polished blue granite tomb,—specimen of cemetery memorial, which will retain its colour and polish under all atmospheric changes. Made of same granite as the sarcophagus executed for H.R.H. the Duchess of Kent's tomb at Frogmore.

No. 6. Polished red granite chimney-piece for public rooms. The polish and colour cannot be destroyed by smoke, or in any other way. Two large slabs, vases, and circular shafts, showing the material, red and blue.

No. 7. (Placed in court between Mineral and Agricultural Departments.) Polished red granite public drinking-fountain in operation. By experience granite is found effectually to withstand the action of water and frost, and not to contract any stain from vegetation.

[215]

MAGNUS, L. S., *Chatham, and 3 Adelaide Place, London Bridge.*—Coals, and products; Magnus's patent coke; iron ores.

Among the articles exhibited by Magnus and Son will be found a sample of SOUTH BRANCEPETH GAS COAL.

These coals give an extraordinary quantity of gas of high illuminating power, and a coke of superior quality: they are shipped at West Hartlepool.

These coals and coke, as well as other descriptions of coal, can be obtained of Messrs. Simon Magnus and Son, who will undertake the delivery in any part of England or at any port.

Price lists of every description of coal may be obtained by application to the exhibitors, at their offices, 3 Adelaide Place, London Bridge.

[216]

MARGAM TIN-PLATE COMPANY, *Taibach, Glamorgan.*—Tin and terne plates; sheet, bar, and iron, best charcoal quality.

The Margam Tin-Plate Company also exhibit a portfolio of various sizes of their NF brand of tin-plates.

[217]

MARLBOROUGH, DUKE OF, *Blenheim Palace, Woodstock.*—Specimens of iron ore from Fawler Mines, Charlbury, Oxfordshire.

[218]

MARSHALL, E. S., 31 *John Street, Tottenham Court Road.*—Gold and silver leaf—illustrative of the malleability of metals.

[219]

MARTIN, E., & SON, *St. Austell.*—China clay and China stone.

[220]

MARTIN, REBECCA, *Higher Blowing House, St. Austell.*—Specimens of china clay and china stone.

The specimens exhibited are the finest qualities of china clay and china stone used in the manufacture of earthen- | ware and porcelain; and also of the purest and best clay (kaolin) for bleaching and general purposes.

[221]

MATTHEWS, J., *Royston.*—Coprolites.

[222]

M'CALL, ROBERT, *near Limerick.*—Fine magnetic iron ore, similar to that of Sweden, America, &c.

[223]

MEESON & CO., *Grays, Essex; George Yard.*—Manufactured and unmanufactured products of Grays Chalk Pits.

[224]

MEIK, THOMAS, *Sunderland.*—Model of the mode of shipping coals.

[225]

MERSEY STEEL AND IRON COMPANY, *Liverpool.*—Cranks, shafts, and other forgings.

[226]

MICHELL, R. R., & CO., *Marazion, Cornwall.*—Model of tin-smelting furnace. Moulds, tools, kettles, &c.

[227]

MICHELL, SARAH, *St. Austell, Cornwall.*—Decomposed granite or clay, washed and unwashed; also washed and prepared for market.

[228]

MICKLETHWAIT, RICHARD, *Ardsley House, Barnsley, Yorkshire.*—Three grindstones from the Old Oaks Quarry, Barnsley.

[229]

MITCHELL, WM. BRIGHTMORE, Mineral Surveyor, 16 *Broom Hill, Sheffield.*—Coals; building, fire, and grinding stones; ironstones; minerals of South Yorkshire.

[230]

MITCHELL, WM. BRIGHTMORE, Mineral Surveyor, *Sheffield.*—Ores and other minerals of the High Peak district of Derbyshire.

[231]

MONA MINE COMPANY, *Amlwch, Anglesey, North Wales.*—Specimens of the produce of copper mining and smelting.

[232]

MONK BRIDGE IRON COMPANY, *Leeds.*—Yorkshire iron and minerals; patent combined cast steel and iron tyres.

[233].

MONTEIRO, L. A., 51 *Manchester Street, Manchester Square, W.*—A many-coloured stalagmite.

[234]

MOORE & MANBY, 3 *Billiter Square, London, and Dudley.*—Specimens of iron for engineers and others.

Trade mark.

Trade mark.

Descriptions of manufactured iron of best qualities supplied by MOORE & MANBY:—

Flat bars from ⅜ to 12 inches wide.

Round bars from ¼ to 8 inches diameter.

Square bars from ⅛ to 5 inches.

Half round, feather and square edge to 6 inches wide.

Bevelled, octagon, hexagon, oval, moulding, and every other description of fancy iron.

Best, best best, and treble best rivet iron, plating bars, &c.

Hoop and strip iron from ⅛ to 10 inches wide.

Sheets—single, double, and lattin.

Roofing sheets—corrugated and galvanized iron.

Nail sheets and hoops, nail rods and flat slit rods.

Boiler plates—best, best best, and treble best; all sizes.

Gasometer and tank plates; all sizes.

Ship, bridge, girder, and flitch plates; all sizes.

Ribbed and chequered foot plates; all sizes.

Canada and tin-plates, coke and charcoal sheets, &c.

Angle, equal and unequal sided, and double angle.

Tee, equal and unequal, and double tee.

Sash bars and trough iron of various sections.

Rolled girder, joist, and beam iron; all sizes.

Bulb, bulb angle, bulb tee, and deck beam iron.

Fencing and telegraph wire, black and galvanized.

Contractors, permanent, bridge, and tram rails.

Locomotive, coach, carriage, and waggon tyres.

Locomotive and other fire bars of various sections.

Railway axles, forgings, and use iron of all descriptions.

Railway spikes, fish plates, bolts, &c.

Railway iron work and stores of every description.

Best Yorkshire iron supplied of the various brands.

Hot and cold blast melting and forge pig iron.

Rolls turned for irregular sizes according to agreement.

All information as to prices, &c., can be obtained at 3 Billiter Square, or Dudley.

[235]

MORCOM, J., *St. Austell, Cornwall.*—Manganese and iron ores.

[236]

MORE, F., *Linley Hall, Shropshire.*—Model of lead field. Ancient lead, spades, &c.

[237]

MOREWOOD & ROGERS, *Stratford, Essex.*—Sheets of iron and other metals for roofing, &c.

The space allotted to MESSRS. MOREWOOD & ROGERS is covered by a shed of their new PATENT CONTINUOUS GALVANIZED ROOFING SHEETS. This roofing, which combines lightness, strength, and durability, can be applied at less cost than common asphalted felts, and further recommends itself by the ease with which it can be applied by unskilled labour; by the rapidity with which buildings can be covered with it; and by the important fact that sheets can be made of any required length.

The great and special advantages which such a material possesses are obvious: any labourer on a farm, or in a factory, who can use a hammer will be quite capable of applying this material, no greater skill being needed than if the building had to be covered with canvas or felt.

Corrugated sheets of Patent Continuous Metal for roofing or upright work can be supplied of any lengths up to twenty feet, without additional charge for extra length by the exhibitors, from whom licences may be obtained for working their patent. Applications should be addressed to MOREWOOD & ROGERS, Dowgate Dock, Upper Thames Street, E.C.

[238]

MORGAN, RICHARD, & SONS, *Llanelly, Carmarthenshire.*—Anthracite malting coal.

[239]

MOSER & SONS, *Southwark, London.*—Sections of rolled iron.

[240]

Moulded Peat Charcoal Company, *Fenchurch Street, London.*—Charcoal; foundry blacking; iron and tin-plate specimens; peat products.

[241]

Muckleston, E., the Rev., *Stoke Cobham.*—Stone from Whitesbourne quarry, Shropshire.

[242]

Murphy, John, *Penzance.*—Inkstand; pair of vases; figure of Apollo; pair of Indian vases, &c.

[243]

Murray, Adam, 24 *New Street, Spring Gardens.*—Anthracite from Broadmoor and Landshipping.

[244]

Murray, Thomas, *Chester-le-Street, Durham.*—Working model of underground steam-engine.

[245]

Museum of Practical Geology, *Jermyn Street.*—Model of Holmbush Mine, constructed by T. B. Jordan, Clapham.

[246]

Mylne, R. W., 21 *Whitehall Place.*—Map,—tertiary and cretaceous districts,—France, England, &c., with contoured seas.

[247]

Newall, D. H. & J., *Granite Works, Dalbeattie.*—One monument and one fountain..

MONUMENT AND FOUNTAIN IN GRAY GRANITE.

[248]

Newcastle, Duke of, K.G., *Shireoak Colliery, Worksop.*—Steam coal; ironstone in permian strata; views of colliery.

[249]

NICHOLLS, JOHN, *Trekenning House, near St. Columb.*—Copper and lead ores from Freton Mine; slates from Penpethy Quarry, near Delabole; porphyry from quarries near Newquay.

a. China stone from Trerice, St. Dennis, Cornwall.

b. Manganese and iron ores from two lodes in Trerice, St. Dennis.

c. Copper and lead ores from Trelow, in St. Issey, Cornwall.

d. Three specimens of porphyry from quarries near Newquay, Cornwall; obtainable in blocks of enormous size.

e. A block of trap stone, of exceedingly durable quality, from a quarry on the manor of Cannalidgey, in Cornwall.

f. Hard Cornish slates, from Penpethy quarry (adjoining Delabole) near Camelford—"Princesses" and "Duchesses."

All the above are obtainable in large quantities from quarries and mines on land belonging to one exhibitor who is ready to receive applications for setts or licences.

[250]

NICHOLSON, MARSHALL, *Whittington Collieries, near Chesterfield.*—Stone curb, and arching for coal-shaft bottom.

[251]

NIXON, TAYLOR, & CORY, *Cardiff.*—Navigation steam coal supplied to H. M. yacht, Warrior, and Black Prince; and sections.

This coal is shipped at Cardiff, Newport, Swansea, Briton-Ferry, and Liverpool; it is wrought solely from the celebrated "Upper Four Feet Seam," in the Aberdare Valley, which is the best steam coal in the world: and is shipped by Nixon, Taylor, & Cory, of Cardiff.

It is used on board Her Majesty's yacht; the frigates Warrior and Black Prince; and by the Cunard Line; the West India Royal Mail Company; the Peninsula and Oriental Company; the Hamburg and New York Company; the Liverpool and Montreal Ocean Steam Ship Company; the London and St. Petersburg Company, &c., &c. Reference can be made to any of these companies.

One pound of this coal has been found to evaporate more than 10 lbs. of water. It burns freely without smoke; is perfectly clear from iron pyrites, clod, shale, or other impurities; so much so that the engineer of the Atlantic Royal Mail Company's steamer, Prince Albert, during her passage of eight days from Galway to St. John's, Newfoundland, had only to clean the boiler fires twice.

This coal possesses a further advantage, viz., that the small that may be caused by breakage in transit will coke, or adhere sufficiently, so as when thrown into the furnace, to prevent its falling through the fire bars to waste. This quality of the Aberdare Steam Coal is unusual and most valuable.

Section of seams or beds of coal in the Aberdare Valley and Merthyr districts, in the order in which they occur in the section of the coal fields. The nine seams are mixed indiscriminately, and shipped and sold by other colliery proprietors under one name, as if of uniform quality.

	Thickness.
	ft. in.
1. Graig coal	2 6
2. Gothloon coal	4 0
3. Yard coal	2 9
4. "Upper four feet coal " . . .	6 0
5. Six feet	4 0
6. Red coal	2 9
7. "Nine feet coal "	10 6
8. Dirty coal	4 0
9. Seven feet coal	7 0
Total thickness of coal . .	43 6

After the "Upper four feet" coal, the seam called the "Nine feet" is the best steam coal in the above section: the whole of the others are very inferior in quality.

[252]

NORTH GUNBARROW CHINA CLAY COMPANY, *Newton Abbott.*—Porcelain clay, &c.

[253]

NORTHUMBERLAND, DUKE OF, K.G., *Alnwick Castle.*—Five pieces of freestone, from Alnwick, Denwick, Rothbury, Harlow, Hill and Thorngrafton Quarries, all in the county of Northumberland.

[254]

NOWELL & ROBSON, Stone Merchants, *Summerleys, Idle, near Leeds.*—York landings, paving and block stone.

[255]

OAKES & CO., *Alfreton, Derby.*—Coal and iron.

[256]

OKEY, S. F., & Co., *Castleford Iron Works.*—Coal; coke; bricks; Lincolnshire iron ore and iron.

[257]

ORD & MADDISON, *Darlington.*—Specimens of lime, limestone, paving stone, road stone, ironstone, building stone, millstone, and marble.

[258]

OXLAND, ROBERT, 42 *Park Street, Plymouth.*—Model of the furnace used in dressing tin ores containing wolfram; a series of natural and artificial compounds of tungsten.

[259]

PACKARD, EDWARD, & Co., *Ipswich.*—Specimens of coprolites, fossil bones, and super-phosphates of lime manufactured therefrom.

The following are exhibited :—

Coprolites (so called) and fossil bones and remains of animals, from the Upper Green Sand, washed and reduced to a fine powder ready for treating with acid, containing 60 per cent. of phosphates. Price 50s. to 55s. per ton.

The Suffolk Craig variety of mineral phosphates, which form the cheapest source of phosphates yet dis-covered, similarly prepared, yielding 56 per cent. of phosphates. Price 45s. to 50s. per ton, according to the demand.

Super- or bi-phosphate of lime, manufactured from the above, well adapted to root cultivation. Price 80s. to 90s. per ton, according to quality.

[260]

PALMER, C. M., *Newcastle-on-Tyne.*—Specimens of coke.

[261]

PARK-END COAL COMPANY, *New Fancy Pit.*—Park-end high delf and smith coal.

[262]

PARKINSON, JOHN, 81 *Cheapside.*—Devonshire minerals from the parishes of Ashburton and Ilsington, Dartmoor; Bethell's anthracite coke; Dr. Smith's patent peat fuel, fire igniters, deodorizing pastilles, pipes, &c.

[263]

PARKSIDE MINING COMPANY, *Whitehaven.*—Hematite iron ore, with section showing stratification of ore and superincumbent strata.

[264]

PATENT METALLIC FUSE COMPANY, *Wadebridge.*—Metallic safety-fuses for blasting; waterproof, and will not "hang fire."

These fuses are adapted for all blasting purposes: they are more economical, surer in action, and afford greater protection to life and limb than the pervious fuses, which often explode uncertainly, and spoil by damp. They were tested in blasting operations before the Miners' Association of Cornwall and Devon at the Royal Cornwall Polytechnic Exhibition, and obtained the Society's prize medal.

[265]

PATENT PLUMBAGO CRUCIBLE COMPANY, *Battersea Works, S.W.*—Crucibles for melting brass, steel, and other metals; portable furnaces and other requisites for assayers and dentists. (*See page 23.*)

[266]

PAULL, JOSEPH M., *Alston, Cumberland.*—Ores of iron, lead, copper, and zinc, as extracted; improved cage for the use of miners.

[267]

PEAKE, SAMUEL, *Berwig Quarry, Minera, Wrexham, and Whitsburn Quarry, Salop.*—Stone for building and paving purposes.

[268]

PEARCE, WILLIAM, *Truro.*—Candelabrum in serpentine, &c.; granite pedestal, made from part of the boulder used for the Duke of Wellington's sarcophagus.

[269]

PEARCE, W., Jun., *Boscawen Bridge, Truro.*—Inlaid serpentine and steatite tables, column mausoleum, and dolphin tazza.

PATENT PLUMBAGO CRUCIBLE COMPANY, *Battersea Works, S.W.*—Crucibles for melting brass steel, and other metals; portable furnaces and other requisites for assayers and dentists.

No. 1.

Furnace for melting Metals.

No. 3.

Muffle.

No. 4.

Assay, or Annealing Crucible.

No. 2.

Muffle Furnace, for Assayers, Dentists, Enamellers, etc.

No. 5.

No. 6.

Scorifier.

No. 7.

Patent Plumbago Crucible and Cover, for melting Gold, Silver, Brass, etc., These melt on an average forty pourings. and are made of any shape and size, to hold from 1 to 1,000 lbs.

No. 8.

Patent Plumbago Crucible, Cover, and Muffle for melting silver, as used in the various Royal Mints.

Patent Plumbago Crucible for melting Steel, malleable Iron, etc.

No. 9.

London Clay Crucible, round or square, for refining Gold.

No. 10.

Roasting Dishes.

No. 11.

Skittle pots, for purifying Jewellers' sweep.

The above may be seen at the Company's Stand, in Class I.
Price Lists and Testimonials free on application to the works as above.

[270]

PEARSON, EMMA MARIA, 11 *The South Quay, Great Yarmouth.*—Amber, jet, agates, jasper, chalcedony, and petrifactions.

The pebbles, &c., exhibited in this case were found on the beach between Great Yarmouth and Caistor. They are found, especially jasper, in great profusion on that coast, and several fine collections exist in the town of Yarmouth. The specimens exhibited are merely average examples of what may be gathered every low tide.

[271]

PEARSON, WILLIAM, *Heddon Quarry, Northumberland.*—Freestone suitable for building purposes, such as docks, piers, bridges, houses, &c.

[272]

PEASE, J. & J. W., *Darlington.*—Model of Upleatham ironstone mines; Cleveland iron, ironstone, limestone, &c.

[273]

PERRENS & HARRISON, *Stourbridge.*—Stourbridge clays, burnt and in the raw state; retort and fire-bricks, ordinary make.

[274]

PHILLIPS & DARLINGTON, 26 *Gresham Street, London.*—Patent fuel, from partially coked or torrefied coal.

[275]

PHILLPOTTS, I., *Newport, Monmouthshire.*—Risca black vein steam coal; miners' tools.

[276]

PHIPPARD, THOMAS, *The Priory, Wareham, Dorset.*—Pottery clays, and sands for glass from Carey, Wareham, Dorset.

[277]

PIKE, W. & J., *Wareham, Dorsetshire.*—Clays for fine pottery, &c.

[278]

PIRNIE COAL COMPANY, *Leven, Fife, N. B.*—Cannel coal, suitable for oil or gas manufacture.

The cannel coal, found in the Pirnie colliery, near Leven, Fifeshire, is highly suitable for the manufacture of oil and gas. The brown part is 8 inches, and the black 16 inches in thickness. By chemical analysis the former is found to yield 78·75, and the latter 59·17 gallons of crude oil per ton. The first named gives 13,500 cubic feet of 28-candle gas; and the average yield of the whole seam is 10,800 cubic feet of 20-candle gas per ton. Full particulars may be learned by applying to the agent at the colliery.

[279]

POLGLAZE & VICTOR, *Wadebridge.*—Metallic safety fuses.

[280]

POLKINGHORNE, W., *Tywardreath, Cornwall.*—A synopsis of the Cornwall ticketings for copper ores, from 1800 to 1860.

[281]

PORT NANT GRANITE COMPANY, *Pwlheli, North Wales.*—Piers, and permanently-rough pavement for bridges.

[282]

PORTER, WILLIAM, 21 *Pitt Street, Old Kent Road.*—Porter's patent millstones. Advantages: durability and perfection of joints.

[283]

POTTER, ADDISON, *Newcastle-on-Tyne.*—Fire-clay gas-retorts, blast-furnace lumps, and fire-bricks.

[284]

POWELL, THOMAS, & SONS, *Cardiff.*—Coal, and plan of colliery.

[285]

POWELL, WILLIAM JOHN, *Tisbury, Wilts.*—Specimens of coralline, flint, &c., from the oolites of Tisbury, Wilts.

[286]

PRICE, DR. DAVID SIMPSON, F.C.S., &c., Consulting Chemist, &c., 26 *Great George Street, Westminster.*—Iron, iron ores, &c.

Specimens illustrating the manufacture of iron in the most important districts of Great Britain, with analyses of the specimens.

[287]

PULLING, ROBERT & WELLINGTON, 10 *New Broad Street Mews, London.*—Iron nuts, screws, spikes, bolts, &c., for railways.

Iron.	R. & W. P. *Crown Iron.*
Hexameter, square, and rose-head bolts with round and square necks with screws.	Hoop, sheet, bar.
Holster-pin and box for rolling-mills.	Boiler plate.
Wood screws, dog-head spikes.	Nail rod.
Hook and drawing spikes.	
Iron rivets.	

[288]

PURIFIED FUEL COMPANY (Limited), 16 *George Street, Mansion House.*—Block fuel obtained from coal, from which a portion of the liquid distillates have been eliminated, and which possesses a higher evaporative power than ordinary coal.

[289]

QUEENSGATE WHITING COMPANY, *Beverley, Yorkshire.*—Fine, hard, and soft Paris white.

[290]

QUILLIAM, THOMAS, *Castletown, Isle of Man.*—Specimens, finished and unfinished, of Manx marble and stone.

[291]

RAMSAY, G. H., & SONS, *Derwenthaugh, Newcastle-on-Tyne.*—Fire-clay retorts, for gas-making; fire-clay goods of various descriptions; cannel coal, coke, and coking coal.

[292]

RAY, JOHN, Esq., *Kilburne Colliery, Derby.*—Specimens of Kilburne coal and ironstone.

[293]

RAY, J., *Ulverstone.*—Slates.

[294]

RAYNES, LUPTON, & Co., *Liverpool.*—Sets and channel stone for paving streets, from Penmaenmawr, Carnarvonshire.

[295]

READWIN, T. A., F.G.S., *Stretford, Manchester.*—British gold ores.

[296]

REDRUTH LOCAL COMMITTEE, *Redruth, Cornwall.*—Mineral produce of West Cornwall.

[297]

REID, P. S., *Felton Colliery, Chester-le-Street.*—Improved underground ventilating furnace.

[298]

RENWICK & NICHOLSON, *Newcastle-on-Tyne.*—Coal and coke from Broom's colliery.

[299]

RHIWBRYFDIC SLATE COMPANY, *Portmadoc.*—Roofing slates.

[300]

RHORYDD SLATE COMPANY, *Portmadoc.*—Roofing slates and ridges.

[301]

RHOS COLLIERY COMPANY, *Llanelly.*—Anthracite coal.

[302]

ROBINSON & SON, *Stanhope, Darlington.*—Section of Weardale strata.

[303]

ROBINSON, WALTER, & Co., *Gospel Oak Works, Tipton.*—Sheet iron, black, tinned, galvanized; tinned, and the tinned galvanized, both flat and corrugated.

[304]

ROBSON, ROBERT, *New Town Hall, Newcastle-on-Tyne.*—Specimens of freestone, Wideopen, Kenton, and Brunton Quarries.

[305]

ROGERS & RAWLINGS, *Bradford-on-Avon.*—Font in Bath stone, and model of part of the interior of Bemerton church.

Carved font, in Bath stone, from the Bethell Quarries, Bradford-on-Avon, Wilts. Designed by O. F. Hansom, Esq., Clifton, and executed by Mr. W. Farmer.

This font is exhibited by the proprietors of the quarries, for the purpose of illustrating the capabilities of the stone, for carving and building purposes.

[306]

ROGERS, E., *Abercarn.*—Iron ores, plans and description.

[307]

ROGERS, P., *Swansea.*—Enamelled slates and marble.

[308]

ROSS OF MULL GRANITE COMPANY, 35 *Parliament Street, S.W.*—Polished red granite, marble; also rough blocks.

The specimens exhibited are from the island of Mull, Argyllshire, where there are four magnificent quarries which the Granite Company have just opened out, and are now working extensively. Blocks of the largest size, suitable for docks and other similar works, can be procured with ease and at small expense. The specimens of red granite are equal to the best marbles for all polishing purposes, and are admirably adapted for works of an ornamental character. Information may be obtained by application to Captain Copeland, the company's agent, at the above address.

[309]

RUDDOCK, SAMUEL, 22 *Bloomfield Terrace, Pimlico, S.W.*—Statuette of St. Agnes.

[310]

RUSSELL, JOHN, *Newport, Monmouthshire.*—Steam, coking, and household coal, from Tyr Nicholas Colliery, Cwm Tylery.

The analysis of these coals will be seen on reference to a list reporting upon all coals in use for Her Majesty's Navy, as ordered by the House of Commons on the 30th June, 1858; and they will be found to stand highest for their evaporative power, and the very small quantity of clinker. These coals are now being used under contract for the navy at Portsmouth, Plymouth, and foreign stations.

The results of the experiments are as follows :—

1. The New Black Vein.

Water evaporated for each 1 lb. of coal consumed, calculated from 100 degrees constant temperature of feed-water, 9·56 cubic feet.

Water evaporated per hour, calculated from the same temperature, 50·67 cubic feet.

Per centage of clinker, 0·79.

Per centage of ash, 5·75.

They possess a high per centage of carbon with only a trace of sulphur; white ash, and are perfectly free from anything injurious to bars or boilers. Burning brightly and getting up steam easily, they possess this advantage, that the small will get up steam as well as the large, and it makes an excellent coke for locomotives and all other smelting purposes.

These coals are shipped at the exhibitor's wharf, or in the very commodious docks at Newport, Monmouth, where vessels of the largest class can load at all times with perfect safety.

2. The New Rock Vein.

Analysis by Dr. Percy :—

Carbon	..	..	..	89·81
Hydrogen	..	..	..	5·19
Oxygen and Nitrogen	..	..		5·00
				100·00

In the middle of this coal a live frog was found at the depth of 600 feet, March 10th, 1862.

[311]

SALT CHAMBER OF COMMERCE, *Northwich.*—Rock salt, marine salt, and other manufactured salt of various countries.

[312]

SALTER, J. W., 28 *Jermyn Street.*—Geological map, coloured on a new principle.

[313]

SANDERS, WM., 21 *Richmond Terrace, Clifton, Bristol.*—Coals and iron ores.

[314]

SCARTH, W. T., *Raby Castle, Darlington.*—Freestones, limestones, basalt, ironstones, lead ore, flag and slate from Teesdale district.

[315]

SCHLESINGER, JOSEPH, *George Street, Birmingham.*—Turkish emery on cloth and paper, and specimen of its manufacture.

[316]

SCHNEIDER, HANNAY, & Co., *Barrow in Furness.*—Model of blast furnaces for smelting hematite iron ore.

[317]

SCHULL BAY COPPER MINING COMPANY, 33 *Great Winchester Street, London, E.C.*—Malachite, and other copper ores.

[318]

SCOTTISH IRONMASTERS: Baird, W., & Co.; Merry and Cunningham; Dixon, W. S.; Houldsworth & Co.; Wilson's trustees; Addie, R.; Wilsons & Co.; Dunlop & Co.; and others, *Glasgow.*—Ironstones from which Scotch pig-iron is made; also specimens of pig iron.

[319]

SEAFIELD, EARL OF, *Cullen House.*—Serpentine, steatite, graphic-granite, asbestos, from Portsoy; Cairngorm crystals from Strathspey.

[320]

SECCOMBE, JAMES, *Pendowry, Liskeard, Cornwall.*—Crystallized oxide of copper.

[321]

SEWELL, EDWARD, *Fulneck, Leeds.*—Topographical and sectional model of Tong ironstone field, west of Leeds, Yorkshire.

[322]

SHELTON BAR IRON COMPANY, *Stoke, Staffordshire.*—Samples of boiler-plate and manufactured iron.

[323]

SHEPHERD, T., *Bath.*—Crossway stone.

[324]

SHEPHERD & EVANS, *Aberdare.*—Smokeless Curnamman Merthyr steam coal.

[325]

SHIELD & DINNING, *Langley Lead Works, Haydon Bridge.*—Lead ores, lead smelting and refining.

Messrs. SHIELD DINNING exhibit specimens of lead ore, lead, litharge, silver, &c., for the purpose of illustrating the process of lead-ore smelting and refining.

[326]

SIM, W., *Granite Works, Glasgow.*—Specimens of granite from the rough block, to the highest class of polished work. (*Nave.*)

[327]

SIMON, LOUIS, *Springfield Works, Nottingham.*—Bronze powder, varnish, and printing-ink.

[328]

SIMPSON, OCT. N., *Little Casterton Freestone Quarry, near Stamford.*—A perfect oolite, which will stand any weather.

[329]

SLATER, D., *Whitley.*—Rough jet.

[330]

SMAILE, R., & Co., *Newcastle-on-Tyne.*—Pressed crucibles.

These crucibles are composed of Stourbridge clay and other pure materials, and are manufactured by pressure in moulds, whereby equality in size, thickness, and solidity are insured. They are made in sizes holding from 15 to 400 lbs. weight; will bear the fusion of copper and other metals, high degrees of heat, and sudden changes of temperature, and can be used for several consecutive days. They are now in regular use in the principal foundries in the north of England, and other parts of the country.

[331]

SMITH, E. J., *Gateshead.*—Stones.

[332]

SMITH, RICHARD, *The Priory, Dudley.*—Minerals; hot and cold blast pig iron, and manufactured iron.

[333]

SMITH, SYDNEY, Marble Turner, *Ashford, near Bakewell.*—Three black Derbyshire marble vases, with handles.

[334]

SOPWITH, THOMAS, 43 *Cleveland Square, W.*—Illustrations of lead-mining from the Allenhead mines.

[335]

SOWERBY & PHILLIPS, *Newcastle-upon-Tyne.*—Waldridge Wallsend Hutton seam, gas, and smiths' coal. (*See page* 29.)

[336]

SPARK, H. K., *Darlington.*—Coal, coke, ironstone, iron, and fire-bricks.

[337]

SPARKS, W., *Crewkerne, Somerset.*—Stone and limestone.

[338]

SQUIRES, C., & SONS, *Stourbridge.*—Model of glass-house furnace, &c.

[339]

STAINIER & SON, *Silverdale, Newcastle, Staffordshire.*—Bars for ships' knees.

[340]

STARK, J. C., *Torquay.*—Devonshire marbles.

[341]

STICK, H., & Co., *Swansea.*—Tin-plates and iron.

[342]

STICKLEY, J., *Cross Street, Hatton Garden.*—Leaf gold, and other beaten metals.

[343]

STRATON & CARGILL, *Arbroath, Forfarshire.*—Polished and dressed pavement.

[344]

SUNDERLAND LOCAL COMMITTEE, 13 *Bridge Street.*—Model of docks and harbour entrance.

SOWERBY & PHILLIPS, *Newcastle-upon-Tyne.*—Waldridge Wallsend Hutton seam, gas, and smiths' coal.

The exhibitors being the proprietors of the Waldridge Wallsend Colliery, desire to call the attention of gas companies and exporters to their Waldridge Wallsend Hutton Seam gas coals, acknowledged to be the best portion of the Hutton Seam in the county of Durham.

It possesses the necessary elements for obtaining the largest yield of gas of high illuminative power, and also produces a first-class coke; it is used and much approved of by nearly all the principal European gas companies.

By the system of screening adopted by MESSRS. SOWERBY & PHILLIPS, the gas coals designed for shipment to foreign ports are freed from all impurities; thus increasing the yield of gas, and completely avoiding a great waste in quantity, and consequent loss in freight, particularly where overland carriage from the port of delivery is necessary.

The Waldridge small coal will be found to be the best now in use for smiths' purposes, being perfectly free from sulphur. It is extensively used, and its qualities are well known in nearly all the continental ports.

———

SOWERBY & PHILLIPS are in a position to ship coals at Shields and Sunderland Dock (at which ports vessels of the largest size are always afloat while loading), and possessing by their own railway, a direct communication with the line of the North-Eastern Railway Company, they can when necessary despatch coals by the most direct and speedy route to all parts of the country.

———

"Newcastle-on-Tyne, 14th Feb., 1862.
"Messrs. Sowerby & Phillips,
  Proprietors of Waldridge Colliery.

"Dear Sirs—It is with the greatest pleasure that we testify to the excellent quality of your Waldridge Wallsend coals for coke and gas, from the great satisfaction which they have given at numerous gas works, during our uninterrupted export of several thousand tons per annum, since 1845.

"We are,
"Yours very truly,
"C. F. JACKSON & Co.,
"Coal exporters, Exchange Buildings,
"Newcastle-on-Tyne."

(Copy.)   "Lyons, 21st Feb. 1862.

"I, the undersigned Augustus Genin, administrative Engineer of the Gas Companies whose head establishment is at Lyons, have pleasure in certifying that for fourteen years, I have always preferred for the supply of our works situated in France, Spain, and Italy, the Waldridge Coal, and that I have obtained from it the best product.

"This coal which was at first pointed out to me by my friend, Mr. J. B. Stears, has been supplied in preference by the house of C. F. Jackson and Co., of Newcastle.

"Accept my friendly wishes,
(Signed)   "AUG. GENIN."

———

(Copy.)   "Lyons, 6th March, 1862.

"I certify that having for a dozen years employed in those of our gas works situated so as to be supplied by English collieries, the coals from the Waldridge pit, I acknowledge that according to their yield of gas and coke, they are equal in quality to the best gas coal from the Newcastle basin.

(Signed)   "EMELIE VAUTIER,

"Engineer of the Gas Companies of the towns of Besancon, Bourg, Dole, Metz, Reims, Angers, Limoges, Clermont, Ferrand, le Puy, Montauban, Perpignan, Agen, Alais, Valence, Grenoble, Venice, Trieste, Padoue, Vincenne, Trevise Verone, Florence, Malaga.

"Viewed for authentication of the signature placed above,

Lyons, 7th March, 1862.

(Signed)   "The Mayor of the Second District,
"PUVONOY."

(Seal.)

[345]

SUTTON & ASH, *Snow Hill, Birmingham.*—Patterns of sections of rolled iron. (*See page* 30.)

[346]

SWANSEA LOCAL COMMITTEE.—Copper, silver, iron, zinc, and nickel ores, and metals.

[347]

SWEETLAND, TUTTLE, & Co., 55 *Old Broad Street, and Britonferry, Wales.*—Specimens illustrative of the manufacture of copper.

[348]

TASKER, THOMAS, *Billinge Hill Quarry, near Wigan.*—Scythe-stones, and grindstones for steel grinding.

SUTTON & ASH, *Snow Hill, Birmingham.*—Patterns of sections of rolled iron.

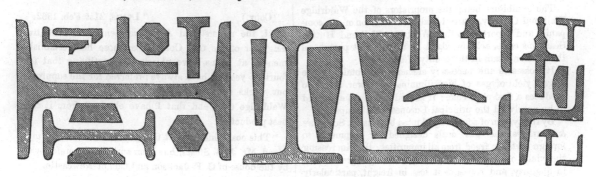

The following sections are rolled in many sizes and various thicknesses :—

Rolled iron girders, for fire-proof buildings, to 18 in.; flat bars, to 12 in.; rounds to 8 in.; square, to 5 in.

Bevelled, half-round, oval, octagon, hexagon, fire, sash, and patent shoe bars; angle, tee, rivet, cable, and boat guard iron. Fencing and drawn wire. Single, double, and latten sheets. Boiler-plates, hoops, strip, nail rods, galvanized iron, and corrugated sheets for roofing.

Ship and boat knees, cart-arm moulds, and all kinds of hammered iron.

Lowmoor plates, bars, angle, rivet iron, &c., &c.

Every description of melting iron, hot and cold air, in Staffordshire, Shropshire, Welsh, and Scotch, of the most approved brands.

Sheets of sections will be sent on application.

[349]

TAVISTOCK COMMITTEE, *Tavistock* (J. Matthews, Secretary).—Copper, tin, lead, iron, and other ores; building stones, clay, &c.

[350]

TAYLOR, BROTHERS, & Co., *Leeds.*—Tyres, cranks, axles, &c.

[351]

TAYLOR, H., *Coal Trade Office, Newcastle-on-Tyne.*—Plans and sections of coal-fields of Durham and Northumberland.

[352]

TEAGUE, MARTIN, *St. Paul, Penzance.*—Obelisk from the same granite block as the monument to "Dolly Pentreath."

[353]

TERRET, J., *Coleford, Gloucestershire.*—Brick tiles and pipes.

[354]

THOMAS, HENRY, *Lisburne Mines, Cardiganshire.*—Lead ores.

[355]

THOMAS, HENRY.—Silver-lead ore from Glogfach mine, Cardiganshire.

[356]

THOMAS, HENRY—Silver-lead ore from Log-y-las mine, Cardiganshire.

[357]

THOMAS, HENRY.—Silver-lead ore from Frongoch mine, Cardiganshire.

[358]

THOMAS, HENRY.—Silver-lead ore from Cwmystwith mine, Cardiganshire.

[359]

THOMAS, HENRY.—Silver-lead ore from Cefn-cwm-brwyns mines, Cardiganshire.

[360]

THOMAS, HENRY, *Lisburne Mines, Cardiganshire.*—Silver-lead ore from Goginan mines, Cardiganshire.

[361]

THOMAS, HENRY.—Silver-lead ore from Cwm-Erfin mine, Cardiganshire.

[362]

THOMAS, HENRY.—Silver-lead ore from East Darren mine, Cardiganshire.

[363]

THOMAS, HENRY.—Silver-lead ore from Nanty mine, Montgomeryshire.

[364]

THOMPSON, HATTON, & Co., *Bilston.*—Iron and tin-plate articles.

[365]

THOMPSON, WILLIAM, 11 *Elmer Street, Grantham.*—Specimens of Ancaster freestone and rag.

The specimens exhibited are produced on the original "Estate for general building purposes."

[366]

THWAITES, J., *Bristol.*—Specimens illustrative of Mitford's new method of cutting precious stones.

[367]

TOMLINSON, ABEL, *Bakewell, Derbyshire.*—Oblong inlaid marble table; oblong marble specimen table, geometric design.

[368]

TONKIN, J., *Pool, Cornwall.*—Section of tin and copper lode Dolcoath, in stone.

[369]

TOWNSHEND, WOOD, & Co., *Swansea.*—Railroad, bar, and sheet iron, tin, terne, and black plates.

[370]

TRASK, CHARLES, *Norton-sub-Hamdon, near Ilminster, Somerset.*—Specimens of ham stone.

[371]

TRICKETT, GEORGE, 21 *Cannon Street, E.C.*—A burnt iron column.

[372]

TRICKETT, SAMUEL, Stone Merchant, *Isle of Dogs.*—Specimens of stones, granite, and marble, for building, paving, and monumental purposes.

[373]

TRICKETT & HOLDSWORTH, Quarry Owners, *Horsforth, near Leeds.*—Bramley Fall stone, for docks, bridges, and basement course of large buildings.

[374]

TROTTER, THOMAS, & Co., *Winnalls Hill, near Coleford.*—Lathe-turned columns. Ashlar, sawn and planed, from Brixdale quarries. Coal.

[375]

TRUSCOTT, CHARLES, & Co., *St. Austell, Cornwall.*—China clays, china stones, and bleaching clays.

[376]

TUFFLEY, MRS. ESTHER, *Avening, near Stroud, Gloucestershire.*—Model of a staircase in Painswick stone.

[377]

TURNBULL, M., Jun., *Tranwell, Morpeth, Northumberland.*—Specimen of freestone.

[378]

TURNER, CASSONS, & CO., *Portmadoc.*—Slates and slate slabs.

[379]

TURNER, JAMES, 1 *Hall Bank, Buxton.*—Derbyshire marble vases and tables, inlaid with various marbles.

[380]

TYLER, JAMES WILLIAM, 4 *Wood Street, Westminster.*—Improvements in the manufacture and laying of pure thick zinc.

Improvements in the manufacture and laying of sheet zinc, as applied to building and roofing purposes. Models of construction, and method adopted to allow of free contraction and expansion, without confining the sheets with nails or solder. Specimens showing the pure and malleable quality of the metal used, from the purest mines of the Vielle Montagne Company.

[381]

TYM, JOHN, *Castleton, Derbyshire.*—Specimens of fluorspar, or blue-john vases, candlesticks, chalices, thermometer, &c.

[382]

USWORTH COLLIERY, OWNERS OF, *Usworth, Washington, Durham.*—Freestone.

[383]

VIGRA AND CLOGAN MINING COMPANY.—Gold, and gold ores from Merionethshire.

[384]

VINT, GEORGE, & BROTHERS, *Idle, near Leeds.*—Obelisk in stone from their Gazeby quarries.

[385]

VOSS, JAMES, *Woodyhide, Corfe Castle, Dorset.*—Purbeck marble.

[386]

WAGSTAFFE & CO., *Fremator Quarries, near Tavistock.*—Pedestal of granite, 4 feet 8 inches by 16 inches.

[387]

WALCOTT, GEORGE, 24 *Abchurch Lane.*—Gas retort bed, full-size set of six.

[388]

WALKER, DAVID, Quarrier and Pavement Merchant, *Turin Hill Quarries, near Arbroath.*—Arbroath pavement, or flagstones manufactured.

[389]

WARING, CHARLES HENRY, *Neath.*—Miners' safety-lamps, which cannot be opened without first extinguishing the light.

[390]

WARNER, A., 31 *Threadneedle Street.*—Iron treated by chemical processes.

[391]

WARNERS, LUCAS, & BARRETT, *Norton Furnaces, near Stockton-on-Tees.*—Samples of pig iron and railway chairs.

[392]

WATSON, HENRY, *Newcastle-on-Tyne.*—Safety-lamps used in the coal-mines of Northumberland and Durham.

[393]

WAYNE, T., *Aberdare.*—Iron.

[394]

WEARDALE IRON COMPANY, *Tudhoe and Tow Law Iron Works, Ferryhill, Durham.*—Iron, steel, and their minerals.

[Obtained a First Class Medal at Paris, 1855.]

Specimens of spathose iron ore, brown ore, coke, pig iron, bar iron, boiler-plates, hoops, wire, cast steel, axles, wheel tyres, cranks, &c. &c.

The whole of the specimens of pig, or cast iron, exhibited by the WEARDALE IRON COMPANY, as well as all their specimens of malleable iron and steel, have been produced from the ores of Weardale, a valley of the county of Durham, extending westward from near the city of that name, to the springs which form the sources of the river Wear at the western boundary of the county, where it is separated from Cumberland by the mountain ridge which looks down on Nenthead and Alston Moor.

The geological formation is the carboniferous, or mountain limestone, series, and the space through which the ore is found, in separated veins and masses, extends about fifteen miles from east to west; its average width being, from north to south, about six or seven miles.

The ores are all of the same general nature, consisting of the spathose, or sparry carbonate of iron, sometimes in a highly crystalline condition, and exhibiting distinctly the usual forms of its crystallization, and sometimes more compactly aggregated, and exhibiting those forms less visibly. It is certain that the ores of Weardale have been all deposited in this state of sparry carbonates; but in very many places, where the superjacent and contiguous rocks and soil have been much shattered and dislocated by the passage and intersection of veins or troubles, the carbonic acid has been expelled from the ore, and has been replaced by oxygen through atmospheric access, and the ore has passed into the state of a *brown hematite, i. e.,* into hydrated peroxide of iron. Generally in such cases it has become impoverished by the infiltration of earthy matter suspended in the percolating water, and by being permeated by the altered ore, which, when first altered, is brought into a pasty or semifluid state.

The specimens of highly crystalline and silvery white iron, exhibiting large lamellar plates or planes of great brightness (labelled A) are called "*silvery steel pig iron.*" It is in fact and strictly *steel*, and steel, too, of great purity; although from its hardness, and also from its tendency to cleave in the direction of the larger plans, rendering it not malleable, it cannot be used as steel.

At the moment when, after reduction from the state of ore to that of metal, it first passes into fusion, it is entirely malleable and is good steel, but afterwards changes with great rapidity, the greater in proportion to the degree of heat, and passes into the state of the specimens exhibited.

The ores are smelted at Tow Law, in Weardale, with coke made from the coal found there, which is of great purity, and this silvery steel pig iron is produced from the finest of these crystallized ores, selected for the purpose, specimens of which are exhibited and labelled correspondingly with the letter A.

This peculiar variety of iron is the same as is made and used in Germany from the same kind of ore, and known there by the name of *spiegel eisen*, or *specular* iron, and the chemical analysis, which will be found below, of one of these English specimens, and of the German one, which will be found along with them, and which is labelled for the purpose of distinction B (but which is stated to have been smelted not with *coke*, but with *charcoal*), will serve to show that they are, notwithstanding, as identical in composition, as they obviously are in their external characters.

The other specimens of cast iron exhibited, are of *grey pig iron*, as generally made from the same Weardale ore, but taken as it comes from the mine, or of its ordinary or average quality. It consists, however, chiefly of a more compact or less highly crystalline variety of the sparry carbonate, of which some specimens are shown and labelled C, the same letter being also used to mark these specimens of iron. These specimens comprise the qualities usually distinguished as No. 1 and No. 3, as used for foundry purposes, *i. e.*, for re-melting and re-casting into various forms of what are called cast metal goods. They also comprise the quality called No. 4; which is also partially used for the same purpose, but which is chiefly employed in the "*puddling*" process; *i. e.*, for conversion into malleable iron and steel. For these purposes it is not quite equal, but yet *not greatly inferior* to the *silvery steel pig iron*; and the whole of the iron produced from these Weardale ores possesses a peculiar fitness for making steel of a superior quality, in a degree very far beyond the produce of ores, of much greater richness, as regards the quantity of iron they contain. In fact they may be called, and considered as distinctively *steel ores.*

A portion of the brown hematite produced as before described, is also used in mixture with the sparry carbonates; and specimens thereof, labelled respectively with the letters D and E, exhibit the variable degree, and the manner, in which that ore is mingled with the rocks and earths, adjacent to the place of its deposit.

The specimens of bar iron marked "Tudhoe" are made from the produce of the brown ores. The tensile strength is about 25 tons per square inch. The bar iron and boiler plates marked "Weardale" made from pure spathose pig iron—are remarkable for ductility, and possess a tensile strength of about 28 tons per square inch.

The specimens of cast steel are made from Weardale spathose iron by the atmospheric process.

(Copy) "Assay Office and Laboratories,
29 Gt. St. Helens, Bishopsgate Street Within.
London, 25th Nov., 1861.
"Sample marked No. 2, Weardale 'Spiegel Eisen' sent by Charles Attwood Esq., contains :—

Iron	...	...	...	...	99·510
Manganese	...	...	...	...	none
Carbon	...	...	...	...	0·065
Sulphur	...	...	...	...	none
Phosphorus	...	...	...	...	a trace
Silica	...	...	...	...	0·140
Loss	...	...	...	...	0·285
					100·000

(Signed) "MITCHELL & RICKARD."

(Copy) "Assay Office and Laboratories,
29 Gt. St. Helens, Bishopsgate Street Within.
London 25th Nov., 1861.
"Sample marked No. 1, 'German Spiegel Eisen' sent by Charles Attwood Esq., contains:—

Iron	...	...	...	...	98·655
Manganese	...	...	...	...	none
Carbon	...	...	...	...	0·210
Sulphur	...	...	...	...	a trace
Phosphorus	...	...	...	...	ditto
Silica	...	...	...	...	1·062
Loss	...	...	...	...	0·073
					100·000

(Signed) "MITCHELL & RICKARD."

[395]

WELSH SLATE COMPANY, *Portmadoc, Carnarvon.*—Slates and building slates from quarries near Festiniog, Carnarvon.

[396]

WESCOMB, C., *Exeter.*—Tables of chalcedonies, jaspers, agates, petrified woods, &c., lead ores, spiral fluted nails.

[397]

WESTON & PRICE, *West Bromwich, near Birmingham.*—Bar-iron and railway fastenings.

[398]

WHARNCLIFFE SILKSTONE COLLIERY, *Sheffield.*—Coal.

[399]

WHEELER, PHILIP, & Co., *St. Austell, Cornwall.*—China clay (kaolin) and stone for porcelain and earthenware, and for bleachers and paper manufacturers; sulfate d'alumine, &c.

[400]

WHITELAW, JOHN, Manager, *Preston Grange Colliery, Prestonpans, Scotland.*—Model of miners' safety-cage, also applicable to hoists.

[401]

WHITEWAY & Co., *Kingsteignton, near Newton Abbott, Devonshire.*—Specimens of tobacco-pipe and potters' clays raised by them, with some manufactured articles from such clays.

[402]

WICKLOW COPPER MINE COMPANY (Limited), 43 *Dame Street, Dublin.*—Iron pyrites, rich in sulphur.

[403]

WILLIAMS, RICHARD, & Co., *Portmadoc, Carnarvonshire.*—Slate ridges for finishing roofs, superseding lead, tiles, &c.

[404]

WILLIAMSON, CLEMENT, *Plas-yn-Morfu, Holywell, Flintshire.*—Ores of lead and zinc in the rough, and dressed.

[405]

WILLIAMSON, JOHN, *Kerridge, Macclesfield.*—Building stones, steps, landings, &c.; granite paving setts.

[406]

WILLIAMSON, ROBERT, 18 *Lothian Road, Camberwell.*—Working model, to illustrate Williamson's improved system of ventilating collieries.

[407]

WILSON, GEORGE BESLY, *Forest Hall, Newcastle-on-Tyne.*—Specimen of freestone for building.

[408]

WILSON, JOHN, *Grantham.*—Carved font in Ancaster stone.

[409]

WILSON, SIR T. M., *Charlton House.*—Founders', and other sands.

[410]

WIMSHURST'S PATENT METAL FOIL COMPANY, 20 *Cannon Street, E.C.*—Sheet of cut lead, one mile long.

[411]

WOLSTON, RICHARD WALTER, *Brixham, Devonshire.*—Wolston's Torbay iron paints and composition for coating materials under water.

These paints are applicable to general purposes, and resist in a remarkable degree the action of the atmosphere, and sulphureous and other gases, as well as aqueous influences.

In the year 1853, a trial was authorized by the Admiralty in Woolwich, Devonport, and Keyham dockyards. The trial in Woolwich dockyard was on a caisson, and on a large surface of iron-roofing; in Devonport yard, on the iron and wood work of a crane erected on the sea wall at the anchor wharf, and various other surfaces of wood and iron. The trials have been officially reported on, and were so satisfactory as to result in an extensive use of the paint, not only in all the dockyards, but also in the Royal Arsenal and War Departments for corrugated iron roofs and buildings, and especially for painting the wood and iron huts at Shorncliff, Colchester, and Curragh camps; as also the huts previously covered with coal tar, at Pembroke dock and on Woolwich common.

The caisson in Woolwich dockyard, painted nine years since, is in a perfectly sound condition, both under water and between wind and water.

On the important question of *expense*, the official report stated, that 62lbs. of "Wolston's Torbay Iron Paint effectually cover as much surface as 112 lbs. of either white or red lead."

These paints have been found to stop corrosion even after it has set in to a considerable extent; and are therefore particularly valuable for the preservation of corrugated and other iron roofing; of which the following are remarkable instances :—

1. In the year 1859, two of the iron roofs over the slips Nos. 8 and 9 in Pembroke dockyard were found on inspection to be so corroded, as in the opinion of the authorities to need entire renewal; but in lieu of this, trial was made of Wolston's Torbay Iron Paint, and two coats were applied. The result has been most satisfactory, renewal now being unnecessary, and a very considerable outlay being thereby saved to the department.

2. In the year 1853, the corrugated iron roofing over the forges and mills at the Aberdare Iron Works were painted with two coats of Wolston's Torbay Iron Paint. They have had one coat since, and on examination by the engineer of the works in January, 1862, the roofs were found to be in good condition; no corrosion having taken place, notwithstanding a constant discharge of steam and gases passing over them for a period of nine years.

3. In the spring of 1852, the iron pillars supporting the Fish-market at Brixham were painted with Wolston's Black Torbay Paint. The paint is now (January, 1862), after ten years' exposure, in a perfectly sound condition, and has effectually protected the iron from corrosion, although the building is situated close to the sea, and subject to all the damp vapours of the harbour and sea coast.

The base of these paints being iron, they are free from those properties which in lead paints are so prejudicial to health and destructive to iron. The numerous testimonials received by the manufacturer prove the paints to possess the following valuable properties :—

1. They effectually protect iron from corrosion, and stop corrosion even after it has set in.

2. The body and covering qualities are so good, that three coats are equal to four of *lead* paints.

3. For priming wood (and for finishing, where the colour suits), the second coat bears out nearly equal to the third of lead paint. The same effect will be found when applied to stucco or compo fronts.

4. As a *stainer* the colouring properties are so intense, that where ochre, umber, and other stainers are used, half the quantity of Torbay brown accomplishes the work with better effect and less trouble.

5. The black, for general purposes, retains its lustre longer than other black paints, and is peculiarly valuable for ship painting, as it does not fade or get rusty by the action of sea water, and owing to its covering properties is really cheaper than the ordinary black paint of much less price.

6. These paints resist intense heat, and stand well on galvanized iron, and on materials previously coated with coal tar, where all other paints fail. They also resist the effects of sulphuretted hydrogen, without loss of colour, and likewise repel the action of acids longer than other paints.

These paints are extensively used by numerous railway, harbour, and gas companies, breweries, ship-owners, iron and wood ship-building and engineering establishments.

Specimens showing the condition of the paint on iron and wood, after various periods of exposure, are exhibited.

[412]

WOMBWELL MAIN COAL COMPANY, *near Barnsley.*—Wombwell coal and section; Froddingham iron-mine, and section.

[413]

WOOD & DAGLISH, *Hetton Colliery, Durham.*—Mode of working and ventilating coalmines, and conveyance of coals underground.

[414]

WOOD, THOMAS, & Co., *Cliff Wood and Spinkwell Quarries, Bradford.*—Ashlar Stone.

[415]

WOODHOUSE & JEFFCOCK, *Derby.*—From Shipley Collieries, Derby. Top and bottom, hard and soft coals.

[416]

WOODHOUSE & JEFFCOCK, *Derby.*—From Victoria Colliery, Warwickshire. Slate, rider, ell, and two-yard coals, and ironstones.

[417]

WOODHOUSE & JEFFCOCK, *Derby.*—From Cinderhill Colliery, Nottingham. Top and bottom hard and soft coals.

[418]

WOODHOUSE & JEFFCOCK, *Derby.*—From Granville Colliery, Derbyshire. Main and little coals.

[419]

WOODHOUSE & JEFFCOCK, *Derby.*—From Wyken Colliery, Warwickshire. Slate, ell, rider, and two-yard coals.

[420]

WOODHOUSE & JEFFCOCK, *Derby.*—From Moira Colliery, Leicestershire. Main coal.

[421]

WOODHOUSE & JEFFCOCK, *Derby.*—From Oakerthorpe and Highfield Collieries, Derby-shire. Bottom, hard, and furnace coals and ironstones.

[422]

WOODHOUSE & JEFFCOCK, *Derby.*—From Baddesley Collieries, Warwickshire. Rider and two-yard coals.

[423]

WOODHOUSE & JEFFCOCK, *Derby.*—From Gresley Colliery, Derbyshire. Main and little coals.

[424]

WOODHOUSE & JEFFCOCK, *Derby.*—Swanwick Colliery, Derbyshire. Top, hard, and Dunsild coals.

[425]

WOODHOUSE & JEFFCOCK, Civil and Mining Engineers, *Derby.*—A model of the Shipley Colliery, in the county of Derby, and specimens of coals and ironstones.

[426]

WOODRUFF, T., 4 *Quadrant, Buxton.*—Derbyshire tables, vases, &c.

[427]

WOODWARD BROTHERS, *Ruabon.*—Building stone, grinding stones, and scythe stones.

[428]

WRIGHT, JAMES, & SON, *John Street Polished Granite Works, Aberdeen.*—Red Peterhead polished granite hexagon vase, &c.

[429]

YNISCEDWYN IRON COMPANY, *Swansea.*—Foundry and forge anthracite pig iron, refined metal, anthracite coal, and iron ores.

[430]

YNISCEDWYN BRICK AND PIPE COMPANY, *Swansea.*—Sewerage pipes, vases, tazzi, chimney pots, and faced bricks.

[431]

YSTALYFERA IRON COMPANY, *Swansea.*—Anthracite pig, refined metal, bars, angles, rivets , boiler plate, tin, terne and Canada plates, cut nails.

---

[432]

GADLY'S IRON COMPANY, *Aberdare, Glamorganshire.*—Iron, and steam coal.

[433]

LAYCOCK, J., & Co., *Seghill Colliery, Newcastle-on-Tyne.*—"Carr's Hartley" steam coal.

CLASS II.

CHEMICAL SUBSTANCES AND PRODUCTS, AND PHARMACEUTICAL PROCESSES.

SUB-CLASS A.—*Chemical Products.*

[458]

ADAMS, J., *Victoria Park, Sheffield.*—Chemicals.

[459]

ALLEN, FREDERICK, Manufacturing Chemist, *Bow Common.*—Aniline and fine chemicals.

[460]

ALBRIGHT & WILSON, *Oldbury.*—Phosphorus, amorphous phosphorus, chlorate of potash, precipitated sulphur, and Crew's chloride of zinc.

[461]

ALLHUSEN, C., & SONS, *Newcastle-on-Tyne.*—Refined alkali, soda ash, crystal of soda, bicarbonate of soda, and bleaching powder.

[462]

ANDREW, FREDERICK WILLIAM, 3 *Neville Terrace, Queen's Elm, Brompton.*—Designs, manufactures; colla-ceramica, and petroconine, for repairing antiquities.

[463]

AVRIL, JOHN, 12 *Castle Street, Holborn.*—Insect-killing powder and patent apparatus.

[464]

BAILEY, JOHN, *Shooters Hill, Longton, Staffordshire.*—Colours for porcelain, earthenware, and glass.

The exhibitor, whose business has been established for more than forty years, manufactures all descriptions of colours used in making porcelain, earthenware, and glass. He also supplies all kinds of potters' materials, and will forward samples, price lists, &c., to any address, upon application.

[465]

BAILEY, WILLIAM, & SON, *Horseley Fields Chemical Works, Wolverhampton.*—Chemicals

[466]

BAKER, EDWARD, & SONS, *Birmingham.*—Nonpareil paste and liquid blacking; pure black-lead—powder and block.

[467]

BAKER, FRANCIS B., *Hampton Court.*—Crystals of sulphate of magnesia, copper, and alum.

[468]

BALKWELL & Co., *Plymouth.*—Metallic arsenic, lump arsenic, crystallized arsenic, and ground white arsenic.

[469]

BARNES, JAMES B., 1 *Trevor Terrace, Knightsbridge.*—A series of volatile organic acids and their ethers, &c.

[470]

BARRELL, JAMES, 26 *Upper Eaton Street, Pimlico.*—Crystal plate powder, for all description of electro-plated or silver goods.

[471]

BARTLETT, BROTHERS, & Co., *Devonshire Wharf, Camden Town, N.W.*—Silicates and aluminates of soda and potash, fused and in solution; with specimens of insoluble glass (for the induration of stone, or the manufacture of artificial stone) resulting from the combination of the above alkaline solutions of silica and alumina without heat. Also specimens of artificial pumice, Bath, and Caen stone, manufactured from the waste dust or chippings of the said stone, combined with the above insoluble glass.

[472]

BELL & BLACK, 15 *Bow Lane, Cheapside, London.*—Patent wax vesta wire, fusees, and congreve matches.

[473]

BELL, I. L., *Newcastle-on-Tyne.*—Aluminate of soda. Oxichloride of lead.

[474]

BERGER, S., & Co., *Bromley-by-Bow.*—Rice starch.

[475]

BETTS, ALFRED, 41 *North Bar Street, Banbury.*—Boot and harness blackings; polishing paste for metals; inks.

[476]

BLAYDON CHEMICAL COMPANY, *Newcastle-upon-Tyne.*—Chemical manures, and materials used in their manufacture.

[477]

BLINKHORN, SHUTTLEWORTH, & Co., *Spalding.*—Patent composition for removing fur and other incrustations from steam boilers.

[478]

BLUNDELL, SPENCE, & Co., *Hull and London.*—Varnishes, colours, paints, oils, oil-seeds, oil-cake, and chemicals.

[*Obtained the Prize Medal, London, 1851; and the First Class Medal, Paris, 1855.*]

The appended list shows the various articles manufactured by BLUNDELL, SPENCE, & Co.:—

Colours, paints, oils, and varnishes of every description.

Blundell's patent dryer, a cheap and powerful dryer for painters, floor-cloth makers, &c.

Blundell's improved marine composition for the prevention of corrosion and fouling on iron ships' bottoms—deep flesh-colour, and ready mixed for use.

Varnishes for house and coach-painters, japanners, &c.

Church varnish for interiors, quick and hard setting.

Blundell's oak stain, a new soluble brown; ½ lb. to 1 lb. dissolved in one gallon of water makes a stain of great depth and beauty, much used in interiors.

Blundell's cooling oil, for preventing heated bearings, especially adapted for the shafts of screw steamers.

Stucco paint of all shades, chiefly intended to imitate stone.

Colza oil, superior double refined for burning.

Blundell's pale drying oil for painting in zinc white, and other light colours. It dries rapidly, and can be used without spirits of turpentine.

Green composition for wooden ships, used extensively on the bottoms of fishing smacks, and is found superior to any other material. It sets in four hours.

Emerald, Schweinfurt, or Paris greens, as supplied to the Continent, East Indies, China, &c.

Boiled and refined linseed oil, &c.

Blundell's linseed cakes, branded ⑧, extensively consumed throughout England and Scotland.

Resident Agent, New York, E. Hill, 180 Front Street.

Resident Agent, Melbourne, R. A. Fitch, 74 Flinders' Lane East.

[479]

BOLTON & BARNITT, 146 *Holborn Bars.*—Chemical products.

[480]

BORWICK, G., *Little Moorfields.*—Baking powder.

[481]

BOUCK, JOHN T., & Co., 32 *Dickenson Street, Manchester.*—Sulphate of copper, nitrate of lead, sulphur, salts of ammonia, and tar products.

482]

BOWDITCH, REV. W. R., *Wakefield.*—Purification of gas from sulphur. Safety-lamps : one for oil ; two for gas.

[483]

BOWER, J., *Leeds.*—Chemical products.

[484]

BRAMWELL & Co., *Newcastle-on-Tyne.*—Prussiate of potash.

[485]

BRAY & THOMPSON, *Heybrook Alum Works, Chatterley, near Tunstall, Staffordshire.*—Alum.

[486]

BRODIE, B. C., F.R.S., *Oxford.*—Graphite, chemically disintegrated and purified.

[487]

BROOMHALL, JOHN, *London.*—White and blue crystal and powder starch, from rice, wheat, potatoes, and sago.

[488]

BRYANT & MAY, *London.*—Safety-matches, which ignite only on the box ; and other chemical lights.

[489]

BUCKLEY, J. (THE TRUSTEES OF THE LATE), *Manchester.*—Sample of copperas, or sulphate of iron.

[490]

BUSH, WILLIAM JOHN, 30 *Liverpool Street, E.C.*—Essences and essential oils.

[491]

CAHN, DAVID, 12 *North Buildings, Finsbury Circus.*—Blocks for printers, especially for copper-plate and lithographic.

[492]

CALLEY, SAMUEL, *Brixham, Devon.*—Torbay iron ores and metallic paints. Manufactured iron ochres.

Patent composition for ships, metal sheathing, iron ships, iron, wood, and other surfaces; and also the celebrated Torbay iron ore, metallic paints, and mineral ochres. Prices and testimonials may be had on application at the works.

[493]

CARR, T., *Birkenhead.*—Soluble super-phosphate.

[494]

CATTELL, DR., *Euston Square.*—Purified gutta percha ; varnishes ; lacquers ; metallized, stained, and enamelled surfaces ; carbons ; inks.

[495]

CHANCE, BROTHERS, & Co., *Alkali Works, near Birmingham.*—Soda, salts of ammonia, copperas, acids, and artificial manures.

[496]

CHICK, GEORGE BREILLAT, *Bristol.*—Indigo stone blue for laundry use, and patent black-lead for stoves.

G. BREILLAT CHICK's newly invented patent cylindrical black-lead. The patent consists in incorporating black-lead, by a peculiar process, with certain oils of a polishing nature ; thus giving additional brilliancy to the lead, and preventing rust upon the surface of the grate ; all other kinds of black-lead yet used are defective in this respect. A grate polished with Chick's patent lead in the spring, when fires are left off, will remain a brilliant jet, and quite free from rust, till the autumn. The patent lead is so packed as to be used without causing the slightest dust, and a servant may black-lead every grate in the house without soiling her hands.

This most useful invention is protected by her Majesty's Royal Letters Patent, granted the 6th of August, 1858.

G. BREILLAT CHICK's newly discovered laundry blue, prepared from pure indigo. This beautiful blue, from its peculiar chemical combination, not only gives to all descriptions of linen, lace, muslin, and every variety of fine fabrics a clearness and whiteness equal to new, but neutralizes the effects of all acid or alkali, of an injurious character left after washing.

It is made into thumb, lion, and Queen's shapes.

Purchasers should see that the trade mark, " G. B. C.," is on each fig or cake. Without this none are genuine.

[497]

CHURCH, ARTHUR HERBERT, B.A., F.C.S., Analytical Chemist, 170 *Great Portland Street, W.*—Rare chemical products.

[498]

COLLINGS, H. A., 48 *Whiskin Street, London.*—Jewellers' rouge ; block lead ; steel protector ; urn and polishing powders.

[499]

COLMAN, J. & J., 26 *Cannon Street, London, E.C.*—Mustard, starch, and blue.

STARCH is made from a variety of cereals, but that which is most approved is manufactured from wheat and from rice.

WHEAT.—The process of manufacturing from wheat is as follows :—The wheat is coarsely ground, and put into a vessel, which is filled with water. After a certain number of days a natural fermentation commences ; in its progress the starch is liberated from the gluten, albumen, &c. When the fermentation has entirely ceased, the starch is obtained by sundry washings and deposits, and is perfectly pure and white ; it is then, in a liquid state, put into long narrow boxes, and, after having had the greater portion of the moisture drained from it, is broken into pieces of about six inches square, papered, and then placed in a stove or kiln, and subjected to the needful degree of heat, for thoroughly drying it. In the act of drying it forms into the crystals in which starch is usually seen. When it is required of a blue tinge a quantity of smalt is introduced.

RICE.—The first process of making starch from rice is different to that adopted with wheat. It requires to be immersed in a caustic alkaline solution, which has the same effect on rice that the fermentation has on wheat, *i. e.*, it causes the disintegration of the particles. The rice then undergoes a levigating process, and is washed and deposited in the same manner as wheat. Several patents have been taken out for the manufacture of rice starch, of which two are by the exhibitors.

FIRE-PROOF STARCH.—J. & J. Colman have for some time been directing their attention to the manufacture of a starch to render fabrics stiffened therewith, non-inflammable. They have at length succeeded in their attempt, and are now making an article, under Letters Patent, which fully answers the desired end.

SATIN GLAZE STARCH.—This starch, though used in a very fluid state, is unusually strong, and more economical than the common starch ; it does not require boiling, and as the clearness, colour, and glaze which it imparts to laces and the finer fabrics of linen are permanent, it is strongly recommended.

PATENT WHITE STARCH.—This starch is manufactured on the same principle as the satin glaze starch, the only difference being that the former is blue and the latter white.

Specimens.

Wheat starch.	Bengal rice.
Patent rice starch.	Madras rice.
Patent white starch.	Gluten.
Satin glaze starch.	Rice fibre.
Patent fire-proof starch.	

INDIGO BLUE.—In preparing indigo for laundry purposes, it is moistened with water and ground as fine as possible between horizontal stones, then mixed with starch, and levigated till it acquires a sufficient consistency to be converted into "figs," commonly known as "thumb blue," or into cakes called "tittle."

Specimens.

Thumb blue.	Tittle.	Pure indigo.

[500]

Condy, Henry Ballmann, *Battersea.*—Condy's patent fluid or natural disinfectant, and other hygienic preparations.

[501]

Cowan & Sons, *Hammersmith Bridge Works, Barnes.*—Bones; animal charcoal. Patented improvements for revivifying animal charcoal.

[502]

Cox & Gould, *Chicksand Street, Whitechapel.*—Acetic acid, as manufactured from wood, and its products.

[503]

Crisp, Edwards, M.D., *Chelsea.*—Specimens of the bile of five hundred animals, and forty photographs from nature.

[504]

Davis, A., 30 *Union Street, Bishopsgate.*—Polishing paste.

[505]

Davy, Macmurdo, & Co., 100 *Upper Thames Street; Works, Horney Lane, Bermondsey.*—Mercurial preparations; photographic and other chemicals.

[506]

Dawson, Daniel, *Miln's Bridge, Huddersfield.*—Benzole, nitro-benzole, aniline, and magenta powder.

[507]

De La Rue, W., F.R.S., & Müller, H., *Bunhill Row, E.C.*—Rare chemicals.

[508]

Doubleday, Henry, *Coggeshall, Essex.*—Dextrine, for giving a superior finish or lustre to textile fabrics.

Dextrine, for giving a superior lustre or finish to textile fabrics. It boils to a clear white solution, similar to white gum-arabic, and is to be used in the same way as starch, either in lieu of, or in combination with it. It forms an excellent size for the use of paper and other manufacturers. Price 34*l.* per ton.

[509]

Dunell, R. G., *Ratcliff Highway.*—Artists', decorators', paper-stainers', painters', and export colours.

[510]

Dunn, Arthur, *Dalston, N.E.*—Marking-ink pencils, &c.

[511]

Dunn, Heathfield, & Co., *Princes Square, Finsbury, London.*—Chemical and pharmaceutical products, and photographic chemicals.

[512]

Emery, Francis, & Son, *Cobridge, Staffordshire.*—Specimens of porcelain, glass, and earthenware colours on china tablets.

[513]

Eschwege, H., *Mincing Lane.*—Potable wood spirit and naphtha.

[514]

Evans, Thomas, 18 *Newland Street, Pimlico.*—Blacking; brush and sponge compositions for harness; saddle polish.

[515]

Everett & Co., 51 *Fetter Lane, London.*—Blacking, and varnish for boots.

[516]

FENN, JAMES, 4 *North Terrace, South Street, Grosvenor Square, W.*—Blacking, varnish, waterproof dubbing, furniture oil and cream, &c.

[517]

FLEMING, A. B., & Co., *Chemical Works, Leith, Edinburgh.*—Vegetable carbon; the deepest and most intense black yet invented.

[518]

FOOT, C., & Co., *Battersea.*—Acetic, nitric, and other acids—commercial and pure, and dyes.

[519]

FOULKES & WALLWORTH, *Birkenhead.*—Cement of great tenacity for glass, wood, &c.; nursery and toilet powder.

[520]

GASKELL, DEACON, & Co., *Widnes Dock, Warrington.*—Bleaching powder, alkalies, and colours.

[521]

GILES & BARRINGER, *Hackney Wick, London, N.E.*—Starch, bleached spices, harness and metal polish, and blacking.

[522]

GREATOREX, FREDERICK, 281 *King's Road, Chelsea.*—Liquid blacking.

[523]

GRIMWADE, RIDLEY, & Co., 31 *Great St. Helens, London;* 69 *St. Clements, Ipswich.*—Anti-corrosive paint for preserving all kinds of external wood, iron, plaster, stucco, and brickwork.

This improved anti-corrosive paint is superior to every other description for the preservation of out-buildings, whether of wood, plaster, or brick. It is admirably adapted for preventing the decay of old stone and brick buildings, and is strongly recommended to noblemen, public companies, emigrants, and all connected with our colonies. Testimonials may be obtained on application.

[524]

HAAS & Co., *Leeds.*—Dyes.

[525]

HALLETT, GEORGE, & Co., 52 *Broadwall, Blackfriars.*—Antimony, and preparations therefrom, including antimony paint.

[526]

HARE, JOHN, & Co., *Temple Gate, Bristol.*—White-lead, Brunswick greens, chrome yellows, &c.

The exhibitors are manufacturers of white-lead, painters' colours of every description, varnishes of the finest quality, purified, quick, and hard drying linseed oil; also importers of olive and other oils. Established 1782.

[527]

HAWORTH & BROOKE, 33 *Lower King Street, Manchester.*—Refined indigo; sulphate of indigo carmine of indigo; oxides of tin.

[528]

HIRST, BROOKE, & TOMLINSON, *Leeds.*—Acetic acid and acetates; naphtha; chemicals; pharmaceutical preparations; varnishes, &c.

[529]

HOLLIDAY, READ, *Chemical and Lamp Works, Huddersfield ;* 128 *Holborn Hill, London.*— Tar products—benzole, aniline, &c.

The exhibitor is the patentee of the "Self-generating gas-lamp;" and a distiller and rectifier of the following coal oils and other products :—

Tar.
Naphtha.
Naphthaline.
Naphthalamine.
Para-naphthaline.
Nitro-naphthaline.
Benzole.
Nitro benzole.
Aniline.

Aniline colours.
Creosote.
Pitch.
Paraffine.
Carbolic acid.
Picric acid.
Ammonia liquor.
Sulphate of ammonia.
Arsenic acid.

Warehouses at 128 Holborn Hill, and 3 Leather Lane London ; and 28 Rue d'Enghien, Paris. Branch works at Sheffield, Bradford, Oldham, and Blackburn.

[530]

HOPKIN & WILLIAMS, 5 *New Cavendish Street, W.* — Chemical and pharmaceutical products.

[531]

HORNER, JAMES B., Merchant, *Lincoln.*—Prize manures. Silver cups awarded, 1858.

[532]

HOWARDS & SONS, *Stratford, Essex.*—Quinine ; other cinchona alkaloids ; cinchona barks ; fine medicinal and manufacturing chemicals.

[533]

HULLE, JACOB, *Lombard Road, Battersea.*—Quinine, cinchonine, strychnine, brucine, morphine, &c., and their salts.

[534]

HUMFREY, YOOLE, & Co., *Southwark.*—Paraffine coal and its products ; paraffine candles, and oils.

[535]

HURLET AND CAMPSIE ALUM COMPANY, *Glasgow.*—Alum, red and yellow prussiates.

[536]

HUSKISSON, WILLIAM, & SONS, 77 *Swinton Street, W.C.*—Chemical products.

[537]

HUTCHINSON & EARLE, *Widnes Docks, near Warrington, and Liverpool.*—Specimens illustrating the process of alkali manufacture.

1. Pyrites.
2. Nitrate of soda.
3. Vitriol (sulphuric acid).
4. Common salt.
5. Salt-cake.
6. Slack (small coal).
7. Limestone.
8. Black-ash.
9. Black-ash liquor.
10. Salts.
11. Soda ash, unground.
12. Soda ash, ground.
13. Refined soda ash unground.
14. Refined soda ash, ground.
15. Soda crystals.
16. Crystals in process of conversion into bicarbonate of soda.
17. Caustic soda.
18. Bicarbonate of soda, unground.
19. Bicarbonate of soda, ground.

[538]

HYNAM, J., *Princes Square, Finsbury.*—Matches, vestas, and fusees.

[539]

James, Edward, *Sutton Road, Plymouth.*—Starches, blues, black-leads; some useful products employed in their manufacture.

[540]

Jarrow Chemical Company, *South Shields.*—Soda, alkali, bicarbonate of soda, Epsom salts, bleaching powder, &c.

[*Honourable Mention in the Report of the Exhibition of* 1851, *for bicarbonate of soda and massive specimen of crystal soda.*]

Manufactories at Tyne Docks, near South Shields; Friars Goose, near Gateshead; Willington Quay, near Newcastle-on-Tyne.

Specimens of Chemical Products.

Articles exhibited, with Price List for April, 1862.— These prices include casks, and delivery free on board in the Tyne, or to rail at South Shields, Newcastle, or Gateshead. The prices are also quoted in French weights and money, 25 francs being calculated as equal to 1*l.* sterling, and 1000 kilogrammes as equal to 2205 lbs. avoirdupois.

1. Crystal Soda. Price 4*l.* 10*s.* per ton, or 110*fr.* 75*c.* per 1000 kilos.

2. Best White Refined Alkali, 40 to 52 per cent. Price 2¼*d.* per cent. per cwt. : *e. g.* 40 per cent. costs 7*l.* 10*s.* per ton; equivalent to 62·78 deg. Descroisilles, at 184*fr.* 60*c.* per 1000 kilos. 52 per cent. costs 9*l.* 15*s.* per ton; equivalent to 81·62 deg. Descroisilles, at 240*fr.* per 1000 kilos. This quality is obtained by evaporating and calcining the solution from which the crystal soda is made.

3. D. P. Alkali, 54 to 56 per cent. Price 2½*d.* per cent. per cwt. : *e. g.* 54 per cent. costs 11*l.* 5*s.* per ton; equivalent to 84·76 deg. Descroisilles, at 276*fr.* 90*c.* per 1000 kilos. This is of similar quality to the above, but of higher strength.

4. Pure Alkali, containing 58 per cent. of soda. Price 18*l.* per ton; equivalent to 91·03 deg. Descroisilles, at 443*fr.* 5*c.* per 1000 kilos. This alkali is very nearly chemically pure.

5. Caustic Soda, containing 72 to 75 per cent. soda. Price 3*d.*¼ per cent. per cwt. : *e. g.* 72 per cent. costs 18*l.* per ton; equivalent to 113 deg. Descroisilles, at 443*fr.* 5*c.* per 1000 kilos. This is the most concentrated form in which soda is produced. By using caustic soda, soapmakers can dispense with lime in preparing their leys. It is packed in casks, or in cylinders of sheet iron.

6. Soda Ash, or Unrefined Alkali, 50 to 53 per cent. Price 2¼*d.* per cent. per cwt. : *e. g.* 50 per cent. costs 8*l.* 17*s.* 1*d.* per ton; equivalent to 78·48 deg. Descroisilles, at 217*fr.* 95*c.* per 1000 kilos. This alkali contains a small quantity of insoluble matter.

7. Bicarbonate of Soda, ground and unground. Price 12*l.* per ton, or 295*fr.* 35*c.* per 1000 kilos.

8. Glauber's Salts, or Crystallized Sulphate of Soda. Price 6*l.* per ton, or 147*fr.* 65*c.* per 1000 kilos.

9. Sulphate of Soda, calcined. Price 5*l.* per ton, or 123*fr.* 7*c.* per 1000 kilos.

10. Refined Sulphate of Soda, calcined. Price 6*l.* per ton, or 147*fr.* 65*c.* per 1000 kilos. Prepared quite free from iron, for glassmaking.

11. Bleaching Powder. Price 10*l.* per ton, or 246*fr.* 14*c.* per 1000 kilos.

12. Oil of Vitriol, concentrated; made from sulphur. Price 7*l.* 10*s.* per ton; equivalent to 66 deg. Beaumé, at 184*fr.* 60*c.* per 1000 kilos. Carboys charged 3*s.* 6*d.* each.

13. Rough Epsom Salts, for agricultural purposes. Price 2*l.* 15*s.* per ton *in bulk*, or 67*fr.* 70*c.* per 1000 kilos.

14. Refined Epsom Salts. Price 8*l.* per ton, or 196*fr.* 90*c.* per 1000 kilos.

[541]

Johnson & Sons, *Basinghall Street, London.*—Lunar caustic, in various shapes; photographic and other chemicals, and preparations.

[542]

Johnson, W. W. & R., & Sons, *Limehouse, London.*—White-lead in different stages of manufacture and colours.

[543]

Johnstone, Robert, *Black Works, Agar Town, St. Pancras, London.*—Samples of vegetable and spirit blacks.

[544]

Jones, John Milton, *Gloucester.*—Composition for waterproofing, softening, and preserving leather; specific for foot-rot in sheep.

[545]

Jones, Orlando, & Co., Inventors, Patentees, and Manufacturers, *York Road, Battersea.*— Specimens of starch from rice.

[546]

Jones, W. J., Dyer to Her Majesty, 12 *Victoria Road, Belgravia.*—Chemical products, and their application in dyeing and cleaning.

[547]

Judson, Daniel, & Son, 10 *Scott's Yard, Bush Lane, City.*—Dyes and dye-stuffs.

[548]

Kane, William Joseph, *Chemical Works, Dublin.*—Chloride of lime, sulphate of soda, sulphuric acid, hydrochloric acid.

[549]

KINGSTON, SAMUEL, Auctioneer, Valuer, and Estate Agent, *Spalding.*—New paint, especially adapted for iron and external work.

[550]

KLABER, HERMAN, *Albion Place, London Wall, E.C.*—Wax vestas of all descriptions; flaming fusees; Vesuvians; English and foreign matches.

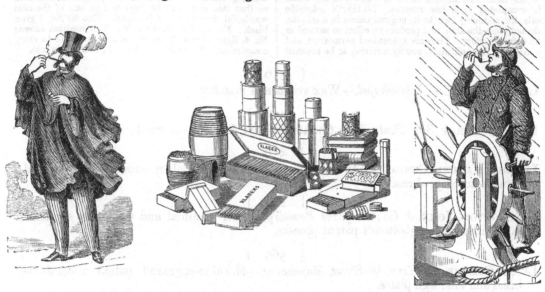

The exhibitor is the sole consignee of CRAY'S PATENT FLAMING FUSEE, the best cigar-light for open-air use, neither wind nor rain extinguishing the flame, and the ash not falling off to the injury of clothing.

Price-lists and samples will be forwarded to merchants and the trade on application.

The newly-invented Stearine Matches, are entirely free from unpleasant odour, and recommended to the attention of exporters.

[551]

KUKLA & CO., *Pentonville Road.*—Artificial salt.

[552]

LAMBERT, WILLIAM THOMAS, 9 *Tabernacle Row, Finsbury, London.*—Refined mercury, putty powders, chemically prepared black-lead, &c.

This black-lead, for domestic use and other polishing purposes, is a preparation of plumbago, with chemicals which facilitate its application and tend to preserve the material, to which it is applied. Its application is unattended by dust, waste, or injury to the hands. The putty powders are made from pure metals, chosen especially for the purpose. The specimens Nos. 8 and 9 show in a remarkable manner the importance of attention to this point. The samples of mercury have been thoroughly refined and purified by W. T. Lambert's process.

[553]

LANGDALE, E. F., 72 *Hatton Garden.*—Essential oils, fruit essences, hair dye, cantharidine.

[Honourable Mention, 1851.]

Specimens of the following manufactures of E. F. Langdale are exhibited :—
Ess. ol. almonds, free from prussic acid, 28*s.* per lb.
Ess. gin and ess. brandy, 8*s.* 6*d.* per lb.; 1 lb. added to 50 gals. of plain spirit, makes immediately a fine London gin or cognac brandy, without the use of a still. Wine, liqueur, and spirit flavouring of every description.
Fruit-essences.—Raspberry, strawberry, pine-apple, and every other sort, from 2*s.* 9*d.* to 12*s.* 6*d.* per lb.
English ol. mint, 32*s.* per lb.; English ol. lavender,

40*s.* per lb.; ol. jasmin, 30*s.* per oz.; ol. cloves, 3*s.* per lb., bergamot, 8*s.* to 14*s.* per lb.; lemon, 6*s.* to 12*s.* 6*d.* per lb.; and all others at market prices.
The Premium Hair-dye, requiring only one application, is instantaneous, harmless, and scentless in action, and may be had of any colour, at 3*s.* 9*d.* per case; cantharidine for reproducing and thickening hair, 3*s.* 9*d.* per case.
The *Lancet* report on E. F. Langdale's laboratory may be found in that journal of 10th Jan., 1857.

[554]

LANGLEY, WILLIAM, 3 *Salters' Hall Court, Cannon Street, City.*—Fine colours, and bronze powders.

Upon application, W. Langley will forward price lists and samples of the various leaf metals, gums, and colours made and imported by him.

[555]

LEATHART, CHARLES, 19A *High Street, Newington Butts.*—A perfumed oil for permanently dyeing the hair in one minute; atter brown for dyeing wool; also a green without arsenic.

MRS. C. G. LEATHART undertakes the restoration of ladies' hair to its original colour by a new process, which is completed in a few minutes. Leathart's colourific oils, for restoring hair to its original colour in a minute, do not stain the skin, and produce an effect so natural as to defy detection. They are guaranteed permanent, and are simple in application, merely requiring to be brushed through the hair. The hair does not so much as require washing either before or after the application of the oil, so that this may truly be regarded as one of the most wonderful discoveries of its kind. The oil No. 1 gives black; No. 2, dark brown; No. 3, a medium brown; No. 4, light brown of various shades, suitable to every complexion. Prices 5s. 6d., 10s. 6d., or 21s. per case.

[556]

LETCHFORD & CO., *Whitechapel.*—Wax vestas and matches.

[557]

LEWIS, JACOB, & SON, *Pontardawe Chemical Works, Swansea.*—Acetate of soda.

[558]

LONDON MANURE COMPANY, 116 *Fenchurch Street.*—Artificial manures; raw material·; manufactured products.

[559]

LONGBOTTOM, JOHN, & CO., *Belgrave Foundry, Leeds.*—Animal and vegetable substances carbonized by Longbottom's patent process.

[560]

LUCAS, GEORGE, 44 *Kennedy Street, Manchester.*—Machine-engraved, patent mineral-filled brass and zinc sign-plates.

[561]

MACKAY & CO., *Inverness.*—Permanent manure; chemical manufacture supplying all the elements extracted from soil by roots and cereals.

[562]

MANDER BROTHERS, *Wolverhampton, and* 363 *Oxford Street.*—Varnish and japan manufacturers: cabinet of varnishes and gums.

[*Obtained the Medal of Honour at the Paris Exhibition of* 1855.]

MANDER BROTHERS have recently introduced several important improvements into the manufacture of varnishes, by which they can secure their greater brilliancy, durability, and unvarying excellence.

By the careful observation of 60 years, and by giving their exclusive attention to the production of varnishes, they have thus succeeded in bringing them to a high state of perfection.

They can refer with satisfaction to the numerous contributions to this exhibition, which have been finished with their manufactures, particularly to the beautiful decorative and other works of the following well-known firms, to which special reference has been kindly permitted, viz :—

Carriage builders.—Messrs. R. & F. Offord, Wells Street, Oxford Street, London; Messrs. McNaught & Smith, Worcester.

Decorator.—Mr. T. Kershaw, 38 Baker Street, Portman Square, London.

Slate enameller.—Mr. G. E. Magnus, Pimlico slate-works, London.

Japanners.—Messrs. John Bettridge & Co., Royal Papier Mâché Works, Birmingham.

[563]

MARSHALL, JOHN, SON, & CO., *London and Leeds.*—Cudbear, orchill, indigo, carmine, lac dye, dye-woods, &c.

[564]

MASON, C. F. ALPHA, 13 *Walcot Place, Kennington Road, S.*—Blacking, exhibited on calf leather; other preparations for boots.

[565]

MAY & BAKER, *Garden and Phœnix Wharves, Battersea.*—Mercurial and other chemical products.

[566]

MELINCRYTHAN CHEMICAL COMPANY, *Neath.*—Acetates and other products derived from the dry distillation of wood.

[567]

METROPOLITAN ALUM COMPANY, *Bow Common.*—Alum.

[568]

MILLER, GEORGE, & CO., *Glasgow.*—Products of Boghead mineral, and suitable lamps; products of coal tar.

[569]

MOCKFORD & CO., 7 *Mincing Lane.*—Copperases, acids, caustic soda, Glauber salts, ochres, Venetian reds, chemicals, and colours.

[570]

MORSON, THOMAS, & SON, *Southampton Row and Hornsey Road.*—Chemical and pharmaceutical products.

[571]

MUSPRATT, BROTHERS, & HUNTLEY, *Liverpool, and Flint, North Wales.*—Products of the soda manufacture; chloride of lime; chlorate of potash; sulphate of alumina.

[572]

NAYLOR, WILLIAM, 4A *James Street, Oxford Street.*—Samples of varnish; a tried pattern of each varnish.

[573]

NEWMAN, J., *Soho Square.*—Pigments.

[574]

ODLING, ANSELM, 30 *Glasshouse Street; Vauxhall.*—Patent ammonia made by sulphuric acid, charcoal, and coal gas.

[575]

PALING, *Newark, Nottinghamshire.*—Starch; printers' flour; cattle food manures; and turnip fly preventative.

[576]

PARSONS, FLETCHER, & CO., *Bread Street.*—Italian wheaten starch and gluten; Indian rice starch.

[577]

PATENT NITRO-PHOSPHATE COMPANY, 109 *Fenchurch Street, E.C.*—Manures,—and materials used in their manufacture.

[578]

PATENT PLUMBAGO CRUCIBLE COMPANY, *Battersea Works, S.W.*—Samples of plumbago, black-lead, graphite, &c., in the natural and manufactured state.

[579]

PEACOCK & BUCHAN, *Southampton.*—Compositions for ships' bottoms; specimens of iron, wood, and copper; specimens of barnacles, &c., taken from ships coated with copper, zinc, and red-lead. (*See page* 48.)

[580]

PEGG, HARPER, & CO., *Derby.*—Painters' colours; plaster of Paris; barytes; mineral white, and emery.

[581]

PERKIN & SONS, *Greenford Green, Middlesex.*—Specimens illustrating the manufacture and application of W. H. Perkin's patent aniline purple.

[582]

PINCOFFS & CO., *Manchester.*—Patent commercial alizarine and garancine.

[583]

POTTER, W. H., 23 *Clapham Road Place, Surrey.*—Manures.

[584]

REA, JAMES, 115 *Wardour Street.*—Shellacs, resins, and varnishes.

PEACOCK & BUCHAN, *Southampton.*—Compositions for ships' bottoms; specimens of iron, wood, and copper; specimens of barnacles, &c., taken from ships coated with copper, zinc, and red lead.

PEACOCK & BUCHAN'S Improved Compositions for Ships' Bottoms, &c., are the best preservatives known against corrosion and fouling on iron and other ships. They give additional speed, and shortly after immersion become slimy like the back of a fish.

The "Atrato," "Himalaya," "Simla," "Shannon," "Nubia," "Delta," "Ceylon," "Pera," and other fast steamers, have always used the No. 2 Composition from the commencement of their career, and still continue to use it with unimpaired speed.

The Spanish Government, after trying experiments with every known composition for two years, have decided on using these compositions for the Spanish navy, and have recently ordered ten tons to be sent to their naval dockyards.

The following gratifying communications have been lately received:—

"Swansea, July 10th, 1861.

"Messrs. Peacock & Buchan,

"Gentlemen,—Our two iron ships, 'Deerslayer' and 'La Serena,' have just returned from the West Coast of South America, the former having been absent from England eight months, and the latter eleven months on the voyage. Your composition has answered well, and effectually kept them from fouling. We shall continue to use it, believing it to be superior to any other coating we have tried.

"We are, Gentlemen, your obedient servants,

"Pro HENRY BATH & SON,

"CHAS. BATH.

"P.S. When the 'Deerslayer' arrived home last voyage from Chili, after an absence of seven and a half months, and was put in dock, we found her so clean that we believe she might have made a second Chili voyage without docking."

"HER MAJESTY'S SHIP 'DEFENCE.'—The great success that has attended Messrs. Peacock & Buchan's compositions for ships' bottoms is manifesting itself more and more every day, and the Lords Commissioners of the Admiralty, after proving its merits on the bottom of the iron troop-ship 'Himalaya,' for a series of years, in voyages to the West Indies, Cape of Good Hope, and Mediterranean, also on various other iron steamers in the navy, in competition with other compositions, have decided to apply it to Her Majesty's ship 'Defence,' and she is now being coated with it. We understand that one of the chief merits of this preparation is its entire freedom from any admixture of *copper*, so that no galvanic action can take place to the injury of the iron plates and rivets—a very essential point in the preservation of the future navy of England, as it is beginning to be generally acknowledged that ere many years pass away, our wooden walls must give place to 'iron sides,' and this paint will occupy the place of copper sheathing, as at present used in the navy, at a much reduced cost, whilst the saving in repairs to our iron fleet, will, in future years, reduce our Navy Estimates considerably, although no doubt it will cost an immense sum to organize an iron fleet in the outset."—*Hampshire Independent*, October 26, 1861.

Several compositions for ships' bottoms having been patented within the last few years containing copper (the patentees being doubtless in ignorance of the injurious effects of copper on iron), Messrs. Peacock & Buchan conceive it to be their duty to inform the public of the results of their experiments with preparations of copper commenced upwards of twenty-four years ago, and laid aside in 1847,* and herewith annex a letter from the Superintendent of the Peninsular and Oriental Company on this interesting subject, after examining the professional opinions of some of the first practical chemists of the day.

From J. R. Engledue, Esq., to Messrs. Peacock & Buchan.

"Peninsular and Oriental Company's Office,
Southampton, Oct. 12.

"Messrs. Peacock & Buchan,

"Dear Sirs,—I am much obliged for your (Mr. Peacock's) letter on the subject of galvanic action on the bottoms of iron ships, accompanied by the professional opinions of Dr. Noad, Dr. Normandy, and Dr. Medlock, against the use of copper preparations for coating. My own experience is quite in accordance with these gentlemen's views as well as your own: I remember that fearful results took place on the bottoms of the late steamers 'Pasha' and 'Madrid,' belonging to this company, by the use of Baron W——'s Copper Composition † after only six months' trial, and I have never allowed it to be again used on any of the company's ships, whereas our iron ships that have been using red lead and your composition since the year 1848, are as sound and good as the first day.

"I have lately had the 'Euxine' scraped bright for examination. Her bottom is perfect, not a plate defective; whereas I learn that three iron ships of about the same size and age as the 'Euxine,' which I am told have been using a preparation of copper on their bottoms, have lately either been condemned or require new bottoms; we have not shifted a plate, and scarcely a rivet, in any of the company's ships, except the 'Haddington,' which vessel also had Baron W——'s copper preparation on her for some time.

"I continue to hear very satisfactory results of the use of your composition on our iron fleet in India and Australia, which you will be pleased to know.

"I remain, Dear Sirs, your obedient servant,

(Signed)   "J. R. ENGLEDUE,

"Superintendent of the Peninsular and
Oriental Company."

* See Pamphlet.   † Oxide of copper with naphtha.

For information, &c., application should be made to the manufacturers direct, Southampton; to Alfred Brett & Co., 150 Leadenhall Street, London; to Mr. Peter Cato, Drury Buildings, Water Street, Liverpool; or to Messrs. McSymon & Potter, Sailmakers, Glasgow.

[585]

RECKITT, J., & SONS, *Hull.*—Starches, blues, and black-leads.

[586]

REEVES & SONS, 113 *Cheapside.*—Fine pigments.

[587]

RICHARDSON, BROTHERS, & Co., 17 *St. Helen's Place, London.*—Refined saltpetre.

[588]

ROBERTS, DALE, & Co., *Manchester.*—Oxalic acid; caustic soda; chemical products; pigments; aniline colours; toilet soaps.

[589]

ROOTH, JOHN SAMPSON, *Chesterfield.*—Naphtha; acetic acid; acetates of lime and lead; iron liquor; charcoal, &c.

[590]

ROSE, WILLIAM A., 66 *Upper Thames Street, London.*—Colours, varnishes, &c.

White-lead, dry and ground in oil, red-lead, litharge, white zinc, powdered and ground; various colours for house-painters', ship-builders', and railway companies' use, anti-oxide for iron bridges, &c. Varnishes for coachmakers' and builders' use; oils for painting purposes, lubricating and burning; greases for railway carriages, waggons, hot necks, wire ropes, &c.; cotton waste for cleaning machinery; tar, pitch, and rosin.

[591]

ROWNEY, GEORGE, & Co., 51 *Rathbone Place.*—Fine pigments.

[592]

RUMNEY, ROBERT, *Ardwick Chemical Works, Manchester.*—Illustrations of new chemical used in dyeing and calico printing; introduced since the Great Exhibition of 1851.

[593]

RUMNEY, ROBERT, *Ardwick Chemical Works, Manchester.*—Silicates of soda and potash; uric acid and compounds.

[594]

RUMSEY, WILLIAM S., 3 *Clapham Rise, Surrey, S.*—Chemical productions for polishing all kinds of metals.

[595]

SAVORY & MOORE, 143 *New Bond Street.*—Chemicals.

[596]

SCOTT, WENTWORTH L., *Westbourne Park, London.*—Fabrics dyed with patent dianthine and aniline green; various mordants.

[597]

SHAND, GEORGE, Chemist, *Stirling.*—Specimens of tar, and chemical products derived from animal, mineral, and vegetable substances.

[598]

SHANKS, JAMES, *St. Helen's, Lancashire.*—A cycle of processes for the manufacture of chlorine.

[599]

SIDEBOTTOM, ALFRED, *Camberwell.*—Painting executed with an aqueous chemical vehicle that will resist water and atmospheric influences; chemical letter-copying fluid; hæmatoxylin, lakes, &c.

[600]

SIMPSON, MAULE, & NICHOLSON, 1 & 2 *Kennington Road.*—Chemical products from coal tar; benzole, nitro-benzole, aniline, mauve, magenta, &c.

[601]

SMITH, BENJAMIN, & SONS, *Spitalfields.*—Archil, cudbear, and patent orchelline; lichens from whence produced; dyed specimens.

[602]

SMITH, T. L., & Co., *St. James' Road, Holloway.*—Starch.

[603]

SMITH, T. W., *Lower Street, Islington.*—Magenta, lake, and other pigments.

[604]

SMITH, T. & H., *London and Edinburgh.*—Products from opium, alöin, caffeine, &c.

[605]

SPENCE, PETER, *Pendleton Alum Works, Manchester, and Goole Alum Works, Goole.*—Alum, and raw and calcined shale.

[606]

SPRINGFIELD STARCH COMPANY, 104 *Upper Thames Street, London, E.C.*—Starch and British gums.

[607]

STANFORD, EDWARD CHARLES CORTIS, *Worthing, Sussex.*—New products obtained by the destructive distillation of seaweeds.

[608]

STENHOUSE, J., F.R.S., &c., *Rodney Street, Islington.*—Rare chemicals.

[609]

STIFF & FRY, *Redcliff Street, Bristol.*—Starch, and other products from rice and wheat.

[610]

STRUVE & Co., *Royal German Spa, Brighton.*—Artificial mineral waters.

STRUVE and Co. prepare the waters of Selters, Tachingen, Vichy, Geilnau, Carlsbad, Ems, Adelheidsquelle, Obersalzbrunnen, Püllna, Seidschütz, Friedrichshall, Marienbad, Eger, Kissingen, Spa, and Pyrmont.

These waters are identical in their composition with those of the natural springs, and the chalybeates contain the full amount of carbonate of iron, in which respect they are superior to the imported ones.

[611]

SYMONS, THOMAS, Manufacturing Chemist, *Derby.*—Oil of vitriol; sulphate of ammonia; colcothar.

[612]

TUDOR, SAMUEL & WILLIAM, *London, and Lead Works, Hull.*—Carbonate of lead; white-lead of commerce.

[613]

VERSMANN, FREDERICK, 7 *Bury Court, St. Mary Axe, London.*—Wolfram ores; colours; tungstate of soda; ladies' antiflammable life-preserver. (*See page* 51.)

[614]

VINCENT, CHARLES W., 2 *Greyhound Court, Milford Lane, W.C.*—Varnishes for making black and coloured printing-inks.

[615]

WALKER ALKALI COMPANY, *Newcastle.*—Hyposulphite of soda; patent resin size; soda crystals; sulphate of zinc; alkali.

[616]

WALLIS, GEORGE & THOMAS, 64 *Long Acre, London, W.C.*—Resins; oils; extracts; varnishes, &c.

[617]

WARD, F. O., *Hertford Street, Mayfair.*—Series illustrating new process for extracting alkali from natural alkaliferous silicates.

VERSMANN, FREDERICK, 7 *Bury Court, St. Mary Axe, London.*—Wolfram ores; colours; tungstate of soda; ladies' antiflammable life-preserver.

THE LADIES' LIFE PRESERVER FROM FIRE.—Ladies' dresses and other textiles steeped in a solution of this compound are rendered non-inflammable, without injury to texture, colour, or appearance.

Manufacturers and Licencees.— Briggs & Co., Great Peter Street, Westminster.

Wholesale Agents.— Johnson & Sons, 18A Basinghall Street, City.

[618]

WARD, F. O., *Hertford Street, Mayfair.*—Series illustrating new process for separating the animal and vegetable ingredients of mixed rags.

[619]

WARD, JOHN, & Co., 452 *Garscube Road, Glasgow.*—Kelp, and its products.

[620]

WHAITE, H., 24 *Bridge Street, Manchester.*—Composition for painting flags.

[621]

WHITE, JOHN & JAMES, *Shawfield Works, Glasgow.*—Bichromate of potash.

[622]

WHITWORTH, GEORGE, & Co., *Jamaica Row, Bermondsey.*—Concentrated fish manure for wheat, oats, barley, &c. (*See page* 52.)

[623]

WILKINSON, HEYWOODS, & CLARK, *Battle Bridge, London, N.*—Varnishes; japan; colours, dry and ground; oxidized oils, &c.

Varnishes, japan, and gold size, for coach-makers' use.

Copals and oak varnishes for the use of decorators and painters, especially lucca oil-varnish, for white work and delicate woods.

Oxidized oil—showing its application to linen and paper.

Complete samples of gum resins, copals, animis, damas, &c., &c.

General assortment of colours adapted for coach painters, artists, decorators, house, ship, and sign painters, paper stainers, and colourers. Attention is especially directed to the greens of Messrs. W. H. & Co., on their wall board, warranted thoroughly permanent in oil.

WHITWORTH, GEORGE, & Co., *Jamaica Row, Bermondsey.*—Concentrated fish manure for wheat, oats, barley, &c.

Messrs. WHITWORTH & Co. recommend their fish manure with every confidence to the attention of farmers and agriculturists, as being the fertilizer calculated to produce a healthy, sound, and heavy crop.

	£	s.	d.	
Concentrated fish manure, for wheat, oats, barley, turnips, &c.	6	0	0	per ton.

This manure contains all the properties of Peruvian guano, and it is considerably richer in soluble phosphate, the want of which often causes guano to fail. These properties make it fit to be classed as one of the best general manures known. The quantity required is four to five cwt. per acre, either drilled or sown broad-cast.

	£	s.	d.	
Fish—Superphosphate of lime ...	6	0	0	,,
Fur waste	5	0	0	,,
Grass manure	3	3	0	,,
Nitro salt (containing particles of nitrate of soda)	2	2	0	,,

The manures are sent in bags at the above prices free to any railway station or wharf in London. A single trial will prove the efficiency and economy of the manures.

Analysis of Whitworth's Fish Manure, made by Messrs. Way & Evans.

"15 Welbeck Street, W., July 27, 1861.

"DEAR SIRS,—We beg to hand you the analysis of the fish manure made by you. It contains a fair quantity of phosphates, and more than the usual amount of nitrogenous matter.

"We are, dear Sirs, yours truly,

"WAY & EVANS."

(Copy—ANALYSIS.)

Moisture	13·01
Organic matter, combined water and salts of ammonia	44·30
Sand	6·42
Biphosphate of lime	9·10
Neutral soluble phosphate	14·20
Insoluble phosphate of lime	8·71
Anhydrous sulphate of lime	22·43
Alkaline salts, &c.	1·08
	100·00

Nitrogen	... ...	6·89
Ammonia	... ...	7·76

Second Part.

Analysis of Whitworth's Fish Superphosphate of Lime made by Dr. Letheby.

"41, Finsbury Square, January 18th, 1852.

"DEAR SIR,—I have to report that the sample of fish superphosphate manure which you left with me for analysis has the following composition :—

Moisture	22·06
Organic matter	11·85
Free sulp. acid	13·73
Soluble phosphate of lime	23·14
Insoluble phosphate	4·04
Sulphate of lime	23·14
Sand and oxide of iron	2·04
	100·00

Ammonia	...	3·70

"According to the usual mode of computing the value of this manure it would be worth 11*l*. 10*s*. per ton.

"I remain yours truly,

"Mr. Whitworth."　　"HENRY LETHEBY.

[624]

WILSHERE & RABBETH, *Great Western Road, Paddington; 2 Alexander Terrace, Ledbury Road, Bayswater.*—Samples of varnishes and colours.

Extra pale body-varnish for coach-makers	26/ per gallon.	Pale furniture varnish	18/ per gallon.
Ditto, ditto, for decorators	22/ 24/ 26/ ,,	Scarlet lake	20/ 24/ per lb.
Pale carriage varnish	14/ 16/ 18/ ,,	Crimson ditto	20/ 24/ ,,
Pale copal ditto	16/ 18/ ,,	Purple ditto	20/ ,,
Pale oak ditto	10/ 12/ ,,	Pure chromes	1/6 ,,

[625]

WILSON & FLETCHER, *Jubilee Street, Mile End, London, E.*—Aniline and aniline colours; emerald green and other pigments.

[626]

WILSON, JOHN, & SONS, *Hurlet, near Glasgow.*—Alum; alum-cake; gelatine and pearl hardening, made from bones.

WOTHERSPOON, WILLIAM, *Glenfield Starch Works, Paisley.*—Glenfield patent starch, manufactured entirely from sago flour, and used in the Royal laundry. (*See page 54.*)

PARTIAL VIEW OF THE BLEACHING DEPARTMENT

OF THE

GLENFIELD STARCH WORKS, PAISLEY.

[627]

WINSOR & NEWTON, 38 *Rathbone Place, W., and North London Colour Works, Kentish Town, N.W*—Fine colours.

[628]

WOOD, E., *Port-hill, Stoke-on-Trent.*—Borax, boracic acid, and china glaze.

[629]

WOOD & BEDFORD, *Leeds.*—Orchil and cudbear.

[630]

WOTHERSPOON, WILLIAM, *Glenfield Starch Works, Paisley.*—Glenfield patent starch, manufactured entirely from sago flour, and used in the Royal laundry. *(See page 53.)*

Although it is little more than twenty years since these extensive works were established, they are now by far the largest of the kind in the country, covering nearly two acres of ground, and giving employment to upwards of 250 persons.

The works are situated to the extreme south-west of the important manufacturing town of Paisley, and near to the foot of "Gleniffer Braes," rendered famous by the poet Tannahill.

These works are a series of brick, iron, and glass buildings, admirably planned, and having every appliance which science can afford for lightening the labour and improving the health of the workpeople.

A complete network of railways intersects the various departments, on which trucks are constantly employed transporting the materials used in the manufacture of the Glenfield Starch from one department to another, and the perfect system of supervision and division of labour are much to be admired, making the various operations fit into each other like tenons into mortices.

The raw material from which the well-known Glenfield Starch is made is East India sago, which is imported by the manufacturers themselves. It may be remarked here, that other starches are made principally from wheat, and on that account the "Glenfield Starch," independent of its superior qualities, has a claim upon public favour, as not interfering with the staple food of the people.

There are two stoves for drying the starch, one of which is the largest in Scotland, or perhaps in the world.

After the starch is dried it is again returned by rail to the bottom of the hoist, whence it is raised to the upper floors to be packed in the small parcels, which, in their blue and green wrappers, are so well known throughout the country, and of which upwards of thirty millions are sold annually. The packing department alone is divided into four branches, the workers in which are females; each department is spacious, clean, and perfectly ventilated.

The warehouse and its arrangements form a model of perfection. This part of the works, which is of recent erection run s through the centre of the premises; it is fireproof, and forms a bulwark between the two principal buildings, to prevent the spread of fire. At the end of this building, and opposite the gateway, are the loading platform and machinery, at which half a dozen lorries can be entirely loaded within an hour.

From the high position which the Glenfield Starch has acquired, it is not to be wondered at that the proprietors have often had to raise actions in the Court of Chancery in England, and the Court of Session in Scotland, to suppress spurious imitations of their manufacture; but it is to be hoped that a Bill on Trade Marks, which would greatly mitigate this crying commercial evil, will soon become the law of the land.

Purchasers may obtain the genuine Glenfield Starch, by observing the trade mark, as well as the name of W. Wotherspoon, the manufacturer, which is on each packet.

[631]

WRIGHT, FRANCIS, & Co., 11 *Old Fish Street, Doctors' Commons, E.C.*—Pharmaceutical preparations and chemicals.

[632]

YOUNG, JAMES, *Bathgate.*—Specimens of paraffine made from different kinds of coal.

[634]

YOUNG, J. W., *Neath, Glamorganshire.*—Paint and paint pigments.

SUB-CLASS B.—*Medical and Pharmaceutical Processes.*

[644]

ALLEN & HANBURYS, *Plough Court, Lombard Street, London.*—Drugs and pharmaceutical preparations.

[645]

BASS, JAMES, 81 *Hatton Garden, London.*—Pharmaceutical products.

Intended to facilitate the exhibition of the following medicines :—

WHITE POPPY.—A preparation from the dried capsules, to be mixed with simple syrup, for making pure syrup of white poppy extemporaneously.

SENNA.—Aromatic syrup. An agreeable and efficacious form for administering senna, especially to children.

RHUBARB.—Aromatic syrup. An improved substitute for the syrup of rhubarb commonly used.

BALSAM OF TOLU.—Concentrated syrup.

GINGER.—Concentrated syrup.

ORANGE PEEL.—Concentrated syrup. These concentrated syrups, diluted to the prescribed extent with simple syrup, will form syrups of the articles specified of full strength and great purity.

TARAXACUM.—A clear liquid preparation of this medicine, obtained from the fresh root by a direct process, and without artificial heat.

CUBEBS.—A concentrated solution of the resinous extract, and essential oil of cubebs.

ERGOT OF RYE.—A solution of the active principles of this substance, four fluid drams equal in strength to one dram of ergot.

PERUVIAN BARK.—Fluid preparations from the pale yellow and red varieties of cinchona, being convenient substitutes for the ordinary decoction and infusion of bark.

[646]

BASTICK, WILLIAM, *Brook Street, London.*—Medicaments prepared by improved processes, which insure their uniform therapeutic activity.

[647]

BROWN, THOMAS BELLISSON, 103 *Icknield Street, Birmingham.*—Cantharidine blistering tissue ; tissue dressing ; transparent plaster ; cantharidine horse blister.

[648]

BULLOCK & REYNOLDS, 3 *Hanover Street, Hanover Square, London.*—Chemical and pharmaceutical products.

[649]

CURTIS & CO., Manufacturing Chemists, 48 *Baker Street, London, W.*—Pharmaceutical preparations, and new inhaler.

[650]

DARBY & GOSDEN, 140 *Leadenhall Street, London.*—Pharmaceutical products.

[651]

DENOUAL, JULES, 1 *Walpole Street, New Cross, S.E.*—Nauseous and alterable drugs enclosed in soluble gelatine. (*See page* 56.)

[652]

DICKINSON, WILLIAM, Chemist, *Cambridge Street, and Queen's Gardens, London.*—An improved series of medicinal preparations.

[653]

DUNCAN, FLOCKHART, & CO., *Edinburgh.*—Chloroform prepared from pure alcohol; chloroform prepared from methylated alcohol; chloric ether.

[654]

GARDNER, J., M.D., 23 *Montague Street.*—Pharmaceutical chemicals.

[655]

HOLLAND, WILLIAM, *Market Deeping.*—Essential oils; vegetable extracts; dried plants and roots.

[656]

HOOPER, WILLIAM, 7 *Pall Mall East, S.W.*—Chemical and pharmaceutical preparations.

DENOUAL, JULES, 1 *Walpole Street, New Cross, S.E.*—Nauseous and alterable drugs enclosed in soluble gelatine.

DENOUAL'S SUPERIOR CAPSULES. — These beautifully finished capsules are made with the most genuine drugs; they are inclosed in a perfectly soluble envelope composed of gelatine, gum, and sugar, and their great superiority has brought them in great demand with the druggists and the public. Such a capsule, combining the greatest qualities with cheapness, has long been a desideratum.

Their shape facilitates their ingress, and the gentle solubility of the envelope allows the dissolution to take place in the stomach without the unpleasant effects produced by common capsules. Capsules should dissolve in the stomach; for if they do not, the drugs they contain cannot be absorbed by the system, and will, consequently, produce no effect. They are put up in boxes of 36 each, with directions for use, and each box is guaranteed by the seal and signature of "DENOUAL."

A superior extra large capaiba capsule, containing 20 minims (three forming a dose), the finest sold in Paris ... ... per box 2*s.* 6*d.*

Best, ordinary size, green label, four forming a dose ... . per box 1*s.* 6*d.*

A very superior capsule, of cubeb-oil and copaiba ... ... per box 2*s.* 6*d.*
Copaiba, pepsine, and bismuth ... 3*s.* 0*d.*

Cod-liver oil, castor oil, turpentine, Norwegian tar, ether, chloroform, and all kinds of capsules.

The attention of the medical faculty, and of all those who have to prescribe, is particularly called to Denoual's Compound Capsules of Iodidum Ferri, containing one grain of iodide of iron and four of cod-liver oil.

Also Denoual's Oleidum Pearls, highly recommended by eminent medical men for diseases of the chest, phthisis, severe coughs, chlorosis, debility, and many other diseases

Price 2*s.* 6*d.* per box of 36. Directions for use in each box.

[657]

LAMACRAFT & Co., 6 *Upper Rathbone Place.*—Court plaster, medical plasters, &c.

[658]

LAURENCE, W. H., 163 *Sloane Street, S.W.*—Cod-liver oil.

[659]

LE MAOUT, *Princes Street, Soho.*—Gelatine capsules enclosing nauseous drugs.

[660]

MACFARLAN, J. F., & Co., *North Bridge, Edinburgh.*—Chemical preparations from opium, green-heart bark, galls, and methylated spirit.

[Obtained the Prize Medal in 1851.]

Chemical preparations from opium, green-heart bark, galls, and methylated spirit; morphia and its salts; codeine and salts; beberine and its sulphate; tannin, gallic acid, and ink; chloroform, ether, and hyponitrous ether.

[661]

MAJOR, JOSEPH, V.S., 5 *Park Lane, Piccadilly, W.*—Medicine chests, and horse and cattle medicines.

[662]

MOFFAT, GEORGE DICKSON, *Dundas Street, Edinburgh.*—Pure medicinal cod-liver oil.

[663]

MURRAY, SIR JAMES, M.D., *Anatomy Office, Temple Street, Dublin.*—Specimens of fluid magnesia, camphor, and aërated extract of bark.

Sir James Murray exhibits his aërated magnesia and camphor, and specimens of bitters, barks, and resins digested in these fluids; with printed descriptions.

[664]

PHARMACEUTICAL SOCIETY OF GREAT BRITAIN, COMMITTEE OF THE, 17 *Bloomsbury Square.*—Systematic collection of drugs and preparations used in medicine.

[665]

RANSOM, WILLIAM, Manufacturing Chemist and Distiller of essential oils, *Hitchin.*—Medicinal extracts and English essential oils.

[666]

SQUIRE, PETER, 277 *Oxford Street, London.*—Chemical and pharmaceutical products.

[667]

TUSTIAN, JOHN, *Milcombe, near Banbury, Oxon.*—Rosæ gallica ; conf. rosæ ; conf. rosæ canin ext. hyoscyami.

[668]

USHER, R., *Bodicot, near Banbury.*—Rhubarb and other medicinal herbs.

[669]

WATERS, ROBERT, 2 *Martin's Lane, Cannon Street, London.*—Quinine wine : the finest tonic known to science.

QUININE WINE.—A preparation in which the sulphate of quinine is held in solution, without the aid of sulphuric or mineral acids. The intense bitterness of the sulphate is neutralized, whilst its medicinal properties are enhanced by the peculiar process used. Dr. Hassall says, "it is a useful and excellent preparation." Dr. Andrews, E. Cousens, Esq., M.R.C.S., and the medical profession generally, as well as the press, speak of it in the highest terms of approbation.

[670]

WATTS, JOHN, & Co., 107 *Edgeware Road, London.*—Pharmaceutical preparations ; extracts, fluid and solid, &c.

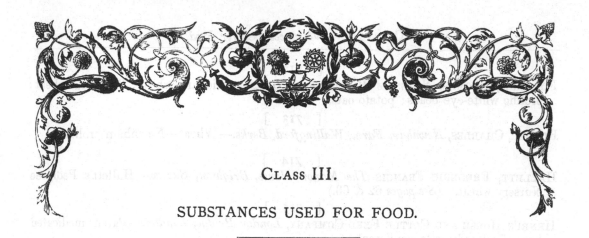

CLASS III.

SUBSTANCES USED FOR FOOD.

SUB-CLASS A.—*Agricultural Produce.*

[700]

ADKINS, THOMAS K., *Wallingford.*—English flour manufactured by Callard's patent process.

[701]

ASPREY, JAMES, *Sandleford, Newbury, Berks.*—White trump wheats; chevalier barley; black Tartar oats.

[702]

BAKERS, WHITE, & MORGAN, *Hibernian Chambers, London.*—British and foreign hops.

[703]

BARRY, DYKES, & Co., *Type Street, Finsbury, and Grand Surrey Docks, London.*—Chicory, cocoa, and mustard.

[704]

BROWN & POLSON, *Paisley, and 23 Ironmonger Lane, London.*—Patent corn flour and patent starch.

[705]

BUTLER & McCULLOCH, *Covent Garden Market, London.*—Dried medical plants, flowers, roots, and seeds.

[706]

CAHILL, MICHAEL, Land Agent, *Ballyconra, Kilkenny.*—Wheat, oats, and wool.

[707]

CARTER, JAMES, & Co., 238 *High Holborn.*—Samples of seeds, flowers and floral designs.

[708]

CHAMBERS, W. E., *Corn Market, Mark Lane.*—Cereals.

[709]

CHITTY, EDWARD, *Guildford.*—Wheaten flour (best whites).

[710]

CHRISTIE, WILLIAM, *Steam Flour Mills, Chelsea.*—Wheat manufactured into flour, showing its produce.

[711]

DAVIS, EDWARD JOHN, *Globe Wharf, Mile End Road, London.*—Compressed hay and other forage.

This is a new mode of packing hay and other forage for transport, without using iron hoops or other bands. Its advantages are, a great reduction of bulk, and facility of making packages of small weight. It can be supplied in compact cakes weighing as little as 20 lbs., while the cubic space occupied is only about one-third of the measurement of hay compressed in the ordinary manner, as supplied for consumption by horses and cattle on board ship, or for purposes of war. It has this further advantage, that the hay or other forage may be combined with oats or other grain in any proportions that may be desired. In this combined form it was supplied for the use of the Cavalry, Royal Artillery, and Military Train of her Majesty's army during the late war in China, and gave great satisfaction; the proportions used being 12 lbs. of hay and 10 lbs. of oats packed together in one cake, so that each cake contained one day's food for one horse.

[712]

FORDHAM, THOMAS, *Snelsmore Hill, Newbury, Berks.*—Chidham wheat; Talavera wheat; prolific white-eye beans; potato oats.

[713]

FULLER, CHARLES, *Newnham Farm, Wallingford, Berks.*—Wheat—Newnham prolific.

[714]

HALLETT, FREDERIC FRANCIS, *The Manor House, Brighton, Sussex.*—Hallett's Pedigree Nursery wheat. (*See pages* 62 & 63.)

[715]

HENRI'S HORSE AND CATTLE FEED COMPANY, *London Bridge, London.*—Patent medicated horse feed and cattle condiments.

[716]

IRWIN, ELIZABETH, *Ballymore, Boyle, Ireland.*—Black oats.

[717]

KIRK & SWALES, *New Wortley, near Leeds.*—Grain, flour, and malt.

[718]

KITCHIN, JOSEPH, *Dunsdale, Westerham, Kent.*—Pocket of Golding hops.

[719]

LIVERPOOL COMMITTEE OF THE INTERNATIONAL EXHIBITION OF 1862, *Liverpool.*—Imports and their appliances.

[720]

MACKEAN, WILLIAM, *St. Mirren's, Paisley.*—Corn flour and starches.

[721]

PACK, THOMAS HENRY, *Ditton Court, Maidstone, Kent.*—Pocket of hops.

[722]

PAINE, CAROLINE, *Farnham, Surrey.*—Pocket of best Farnham hops; one small case of ditto.

[723]

PALING, W. & E., *Newark, Nottinghamshire.*—Cattle food and cattle condiment.

[724]

POLSON, WILLIAM, & Co., *Paisley.*—Patent Indian corn flour; starch from rice, Indian corn, and sago flour.

[725]

RAYNBIRD, CALDECOTT, & BAWTREE, Seed Merchants, *Basingstoke.*—Specimens of seed-corn and seeds.

[*Obtained Prize Medal,* 1851.]

[726]

ROBINSON, BELLVILLE, & Co., 64 *Red Lion Street, Holborn, London.*—Patent barley and patent groats.

The exhibitors are manufacturers of patent barley, patent groats, pearl barley, oatmeal, groats, &c.

[727]

SIMPSON, ALEXANDER, *Steam Mills, Snow Hill, Birmingham.*—Condimental food for cattle, for rearing and feeding.

[728]

STEVENS, RICHARD, *Collyweston, Northamptonshire.*—Wheat, barley, beans, and oats.

[729]

STRANGE, WILLIAM, *Banbury, Oxon.*—Wheat and beans.

[730]

STYLES, THOMAS, 148 *Upper Thames Street.*—Ashby's groats for making gruel in a few minutes.

[731]

SUTTON & SONS, *Royal Berks Seed Establishment, Reading.*—Collection of seeds and specimens of grasses, &c.

[Obtained Medal and Certificate at the Great Exhibition of 1851.]

The exhibitors are seedsmen by appointment to Her Majesty the Queen, and his late Royal Highness the Prince Consort: also to the Government Gardens of India, and the Royal Agricultural Society of the Cape of Good Hope.

One thousand of the most distinct and popular varieties of seeds with English and botanical names. Dried specimens of one hundred species and varieties of grasses grown separately in one plot of ground, by S. & Sons. Specimen clusters of twelve distinct sorts of African imphee or sorghum. A collection of cones and seed pods gathered from useful and ornamental trees in various parts of the world. Exact representations of various argricultural roots, which have taken distinguished prizes at the principal agricultural meetings of England, during the autumn of 1861.

Among the one thousand sorts of seeds exhibited will be found one hundred and twelve sorts of grasses, one hundred and twelve sorts of various farm seeds, two hundred and twenty-four sorts of kitchen garden seeds, four hundred and fifty sorts of flower seeds, several varieties of cotton, and numerous sorts of fruits both English and Foreign.

The illustrations contain, among many others—

Sutton's champion Swedish turnip.

Skirving's Liverpool Swedish turnip.

Hardy purple-topped Swedish turnip.

Sutton's greentop yellow turnip.

Sutton's purple-topped yellow turnip.

Lincolnshire red-topped turnip.

Sutton's imperial green globe turnip.

Yellow globe mangel-wurzel.

Elvetham long red mangel-wurzel.

[732]

TAUNTON, WILLIAM, *Redlynch, Salisbury.*—Corn and seeds.

The exhibitor can supply agricultural seeds and seed corn of the finest qualities, such as the samples exhibited.

Prices may be learned on application at No. 97 Seed Market, Mark Lane.

[733]

TAYLOR, JOHN, & SONS, *Bishop's Stortford, Herts.*—White, coloured, amber, and brown malt.

[734]

THORLEY, JOSEPH, *Newgate Street, City.*—Thorley's food for cattle—a condiment; Thorley's feeding meal—corn substitute.

[735]

WEBB, RICHARD, *Culham House, Calcot, Reading.*—Mummy Talavera wheat; varieties of cob-nuts and filberts grown at Calcot.

[736]

WELLSMAN, JOHN, *Moulton, Newmarket.*—Pale malt chevalier barley; oats; barley grown from oats.

[737]

WOOLLOTON, C., & SONS, 246 *Borough.*—British and foreign hops.

[738]

WRENCH, JACOB, & SONS, *London Bridge, E.C.*—Favourite English cereals, &c., of the London Corn Market.

[739]

WRIGHT, ISAAC, & SON, *Great Bentley, Essex.*—Grass, ferns, and agricultural seeds.

The exhibitors have been engaged for the last thirty years in the collection of the British grass seeds, with a view to the permanent improvement of pastures. References are permitted to a number of gentlemen who have obtained fine pastures by the use of the seed supplied by Wright and Son. Price 3s. per acre. They also supply agricultural seeds of every description.

HALLETT, FREDERIC FRANCIS, *The Manor House, Brighton, Sussex.*—Hallett's Pedigree Nursery wheat.

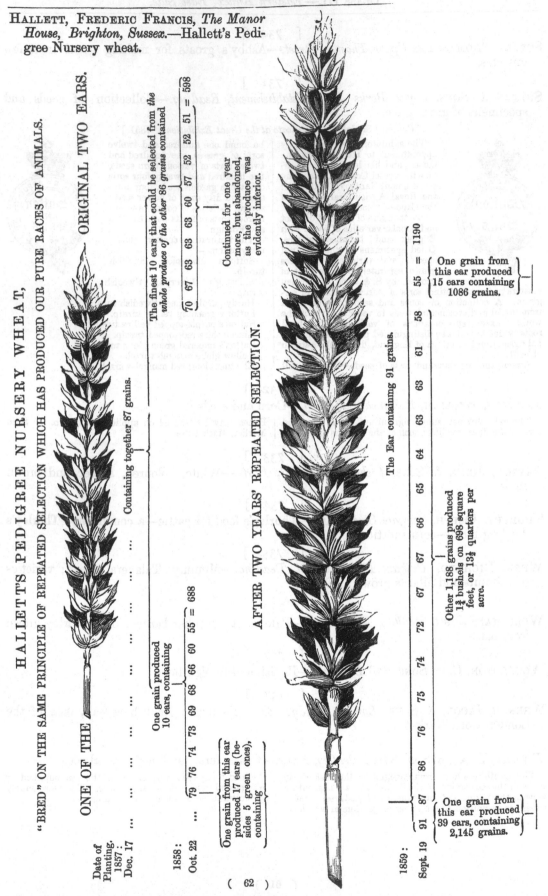

HALLETT'S PEDIGREE NURSERY WHEAT,

"BRED" ON THE SAME PRINCIPLE OF REPEATED SELECTION WHICH HAS PRODUCED OUR PURE RACES OF ANIMALS.

ORIGINAL TWO EARS.

Containing together 87 grains.

The finest 10 ears that could be selected from the *whole produce of the other 86 grains contained*

70 67 63 63 63 60 57 52 52 51 = 598

Continued for one year more, but abandoned, as the produce was evidently inferior.

ONE OF THE

AFTER TWO YEARS' REPEATED SELECTION.

The Ear containing 91 grains.

91 87 86 76 75 74 72 67 67 66 65 64 63 63 61 58 55 = 1190

One grain from this ear produced 15 ears containing 1086 grains.

Other 1,188 grains produced 1¾ bushels on 698 square feet, or 13½ quarters per acre.

One grain from this ear produced 39 ears, containing 2,145 grains.

Date of Planting. 1857 : Dec. 17

1858 : Oct. 22 ... 79 76 74 73 69 68 66 66 60 55 = 688

One grain produced 10 ears, containing

One grain from this ear produced 17 ears (besides 5 green ones), containing

1859 : Sept. 19

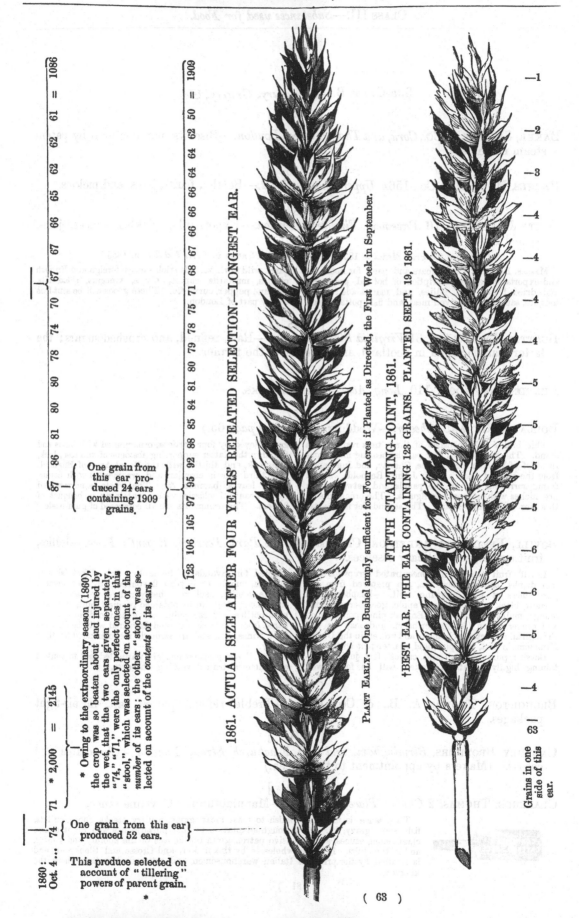

1860 :
Oct. 4 ...

71 * 2,000 = 2145

74

74

87 86 81 80 81 80 78 78 74 70 67 67 66 66 65 66 62 62 61 = 1086

† 123 106 105 97 95 92 88 85 84 81 80 79 78 75 71 68 67 66 66 64 64 62 50 = 1909

One grain from this ear produced 24 ears containing 1909 grains.

* Owing to the extraordinary season (1860), the crop was so beaten about and injured by the wet, that the two ears given separately, "74," "71," were the only perfect ones in this "stool," which was selected on account of the *number* of its ears; the other "stool" was selected on account of the *contents* of its ears,

One grain from this ear produced 52 ears.

This produce selected on account of "tillering" powers of parent grain.

1861. ACTUAL SIZE AFTER FOUR YEARS' REPEATED SELECTION.—LONGEST EAR.

PLANT EARLY.—One Bushel amply sufficient for Four Acres if Planted as Directed, the First Week in September.

FIFTH STARTING-POINT, 1861.

†BEST EAR. THE EAR CONTAINING 123 GRAINS. PLANTED SEPT. 19, 1861.

—1
—2
—3
—4
—4
—4
—4
—5
—5
—5
—5
—6
—6
—5
—4

63

Grains in one side of this ear.

SUB-CLASS B.—*Drysaltery, Grocery, &c.*

[752]

BAKER, SIMPSON, & Co., *Cork, and Thames Street, London.*—Biscuits manufactured by patent steam machinery.

[753]

BARNES, MORGAN, & Co., 156A *Upper Thames Street.*—Bottled fruits, jams, and pickles.

[754]

BATTY & Co., 15 & 16 *Pavement, Finsbury, London.*—Export oils, pickles, sauces, jams, bottled fruits, &c.

[*Obtained the Prize Medal in 1851, and Honourable Mention in Paris Exhibition, 1855.*]

MESSRS. BATTY & Co. prepare and pack for home use and exportation, every description of bottled, preserved, and dried fruits, jams, jellies, and marmalade : hams, tongues, sausages, preserved meat, and fish, potted meats, game, wild fowl, &c.; also rich sauces, foreign and English pickles, mustards, capers, olives, vinegars, salad oils, cayenne pepper, curry, &c. These goods will be sent free to any part of London.

[755]

BEATTIE, JOHN, & Co., 31 *Virginia Street, Glasgow.*—Raw, refined, and crushed sugars; the latter manufactured in Scotland, and superior to the former.

[756]

BEXFIELD & WOOD, 110 *Long Acre.*—Wedding cake.

[757]

BOLLAND, RICHARD, *Chester.*—Wedding cake. (*See page 65.*)

This brides' cake consists of three tiers resting on a stand. The first tier has four panels, bearing medallions in relief of Wisdom, Providence, Charity, and Innocence, from the original designs of Sir Joshua Reynolds. A rich frame surrounds each of the panels, and between them are niches with appropriate figures. The ornaments on this tier are chiefly Gothic. The second tier is an octagon relieved by four porticos, ornamented with busts and Cupids; the latter supporting festoons of the rose, leek, shamrock, and thistle, with the Royal Arms of England. The third tier is circular, and ornamented with cornucopiæ leaves, banners, &c. The whole is surmounted by a beautiful classic vase, containing a bouquet of flowers. The ornaments are all composed of gum paste.

[758]

BOVILL, FREDERICK ANDERSON, Chemist, &c., 24 *Park Terrace, Regent's Park.*—Jellies, fruit syrups, and culinary essences.

Bovill's fruit essences, or concentrated syrups for summer drinks, balls, parties, &c., are prepared from the pure juices of the following fruits :—Raspberry, red currant, black currant, cherry, apple, gooseberry, Seville orange, mulberry, foreign and English pine-apples, lemon and ginger lemon (for ginger-beer), and may be had through all respectable chemists and grocers in the United Kingdom, in pint, half-pint, and quarter-pint bottles. Bovill's pure ox feet and calves' feet jellies, containing highly nutritious matter, will be found well adapted for invalids or the table. Calves' feet (sherry, madeira, punch, or noyeau flavouring), ox feet (lemon, orange, vanilla, raspberry, cherry, red currant, strawberry and pine-apple), can be obtained in quart, pint, and half-pint bottles. These jellies are guaranteed to be prepared from the fresh feet, and flavoured with the choicest fruit, wines, &c., and are strongly recommended for their purity.

The above preparations are warranted not to contain any chemical fruit-flavouring whatever.

[759]

BROUGHTON, THOMAS A. B., & Co., *Bristol.*—Treble refined patent salt, in air-tight packages.

[760]

CADBURY BROTHERS, *Birmingham, and* 148 *Fenchurch Street, London.*—Chocolates and cocoas. (Makers by appointment to the Queen.)

[761]

CLARENCE, THOMAS, 2 *Church Place, Piccadilly.*—Manufacturer. Cayenne sauce.

This sauce is used as a relish to roast meat, game, poultry, steaks, chops, cutlets, fish, soup, gravy, &c. Its thorough adaptation to this purpose has won for it a first class among sauces, and extensive patronage in the houses of the nobility and gentry, and in the clubs. It is sold wholesale by the maker, and Crosse and Blackwell, and is retailed by the principal Italian warehousemen and sauce dealers throughout the kingdom.

BOLLAND'S WEDDING CAKE.

[762]

CLERIHEW, WILLIAM, *Richmond Hill, Aberdeen* (late of *Ceylon*, Coffee Planter).—Drawing illustrative of his patent process of curing coffee.

[763]

CLYDE SUGAR REFINERS' ASSOCIATION, *Greenock.*—Samples of sugar-refining produce.

[764]

COCKS, CHARLES, *Reading.*—The celebrated Reading sauce; pickles, and other sauces.

[765]

COLLIER & SON, *Steam Mills,* 10, *Foster Street, Bishopsgate.*—Cocoa, chocolate, chicory, and coffee—roasted, raw, and dressed.

Messrs. Collier & Sons are coffee, cocoa, and chicory roasters (by their patent enamelled cylinders), chicory importers, and chocolate manufacturers.

The following specimens or samples are exhibited :—
Raw coffee as imported.
Raw coffee cleaned and dressed.
Roasted coffee not cleaned.
Roasted coffee cleaned and dressed.
Refuse taken from coffee in dressing.
Raw Dutch and German chicories as imported by J. C. & Son.
Dutch chicory, roasted by J. C. & Son's patent process.
Ditto nibs,   ditto.
Ditto granulated, ditto.
Ditto ground,   ditto.

Raw Trinidad cocoa-nuts.
Roasted Trinidad cocoa-nuts.
Roasted Trinidad cocoa-nibs.
Roasted Trinidad cocoa flaked.
Cocoas and chocolates of various descriptions, and Patent Chocolate Powder.
Collier & Son's chocolate powder is prepared from the finest Trinidad cocoa-nibs. The popularity which this article has attained, justifies the manufacturers in recommending it as one of the most agreeable, and at the same time nutritious, beverages that can be taken. It may be purchased from any grocer in London or the country, and wholesale at the steam mills of the manufacturers. Established in 1812.

[766]

COLMAN, J. & J., 26 *Cannon Street, London, E.C.*—Mustard, starch, and blue.

The mustard of commerce is manufactured from mustard seed, of which there are two descriptions—black, usually termed "brown," and white. It is grown in Kent, Essex, Lincolnshire, Cambridgeshire, Yorkshire, and Holland. The mode of preparation is as follows :—The seeds are crushed between rollers, pounded in mortars, and dressed through sieves of varied fineness; and, according to the fineness of the farina, it is designated "genuine," "double superfine," "superfine," "fine," "aromatic," or "seconds." Oil is extracted from the refuse or bran, and is used for burning, dressing cloth, &c., &c.; the cake is used as manure, and is by many preferred to other descriptions of cake, as it is thought materially to lessen the ravages of the fly and wireworm on turnips.

The mustard which is most confidently recommended, is made from the finest qualities of seed; it is pure flour of mustard, and is called "genuine," and "brown;" the "double superfine" is made from the same description of seed, mixed with a very slight proportion of the best wheaten flour, and is by many preferred to the "genuine" on account of its more delicate flavour.

In the manufacture of other descriptions than "genuine" and "brown," flour, tinted or stained by finely powdered turmeric root, is used.

Specimens.

Genuine mustard.   Double superfine mustard.
Flour of brown seed.   Flour of white seed.
Brown seed.   White seed.

[767]

COPLAND & CO., 30 *Bury Street, St. Mary Axe, London.*—Preserved meats, fruits, vegetables, &c.

[768]

COXSHALL, JOHN, *Waltham Abbey, Essex.*—Gingerbread in various forms.

[769]

CROSSE & BLACKWELL, *Soho Square, London.*—Pickles, sauces, jams, fruits, and preserved provisions for all climates.

[770]

DAKIN & CO., 1 *St. Paul's Churchyard.*—Collection of teas.

[771]

DAWSON & MORRIS, 96 *Fenchurch Street, E.C.*—Isinglass.

[772]

DEWAR, THOMAS, *Newcastle-on-Tyne.*—Mustard, and process of manufacture.

[773]

DODSON, HENRY, 98 *Blackman Street, London.*—Improved patent unfermented bread; unfermented nursery biscuits; biscuit powder; cakes.

[774]

DORGUIN, ERNEST, 9 *Baker Street, Portman Square.*—Cho-ca, chocolate, and bonbons.

[775]

DUNCAN, A. M'E., & CO., *Gorey, Jersey.*—Preserved animal and vegetable substances.

[776]

DUNN & HEWETT, *Pentonville Hill, London, N.*—Lichen Icelandicus, or Iceland moss, and other cocoas. (*See page* 68.)

[777]

DU PARCQ, C., *Jersey.*—Manufactured cocoas in powder; Jersey cider in bottles.

[778]

ELDER, ALEXANDER, *Edinburgh.*—Royal Holyrood sauce.

[779]

FADEUILHE, V. B., 29 & 30 *Botolph Lane.*—Patent dry milk in powder.

[780]

FAHRMBACHER, M., 4 *Sion Square, E.*—Artificial confectionery.

[781]

FARMER, J., & CO., *Edgeware Road, W.*—Cocoa; cocoa fat refined.

[782]

FORTNUM, MASON, & CO., 180 *Piccadilly.*—Collection of preserved fruits.

[783]

FRY, JOSEPH STORRS, & SONS, 11 & 12 *Union Street, Bristol, and* 252 *City Road, London.*—Series illustrating the manufacture of chocolate and cocoa. (*See page* 69.)

[784]

GAMBLE, POWER, & CO., 78 *Fenchurch Street, London, and Cork.*—Preserved provisions, in hermetically closed tin cases.

[*Obtained the Prize Medal of* 1851.]

The exhibitors preserve meats, vegetables, &c., of all kinds, in hermetically closed cases; and warrant them to keep sound for years, and to be fit at any time for immediate use.

Price lists and testimonials from various celebrated authorities, naval and military commanders, Arctic voyagers, &c., can be had on application.

[785]

GARRARD, JOHN T., *Needham Market, Suffolk.*—Fine sugar-cured, smoked Suffolk hams; breakfast bacon, chaps, and ox-tongues.

[786]

HARRISON, R. & J., *Jack Lane Mills, Leeds.*—Pure Durham mustard, mustard seeds, and prepared chicory.

[787]

HART, J., *St. Mary Axe, City.*—Isinglass.

DUNN & HEWETT, *Pentonville Hill, London, N.*—Lichen Icelandicus, or Iceland moss, and other cocoas.

The "Iceland Moss Cocoa" is strongly recommended by medical men, on account of its nutritious properties, in cases of debility, indigestion, and pulmonary disease. It may be obtained of most grocers, price one shilling and fourpence per pound.

Dunn's Essence of Coffee will keep good in any climate. A cup of coffee can be made from the essence in one minute.

Dr. Hassall, Dr. Normandy, and others have testified to the genuineness of the manufactures of this firm.

[788]

HASSALL, A. H., M.D., 74 *Wimpole Street.*—Specimens illustrating the adulteration of food.

[789]

HAY, 6 *North Audley Street, Grosvenor Square.*—Improved Dutch rusks for invalids of weak digestion, and infants.

These rusks are recommended by some of the first members of the medical profession as a light and nutritious article of food, free from all tendency to acidity. To beef-tea (in which they immediately form, as it were, a uni-form jelly) they are a most useful adjunct; while with tea or coffee they are an agreeable light repast for the invalid or convalescent. They can be forwarded in tin cases, price 5s. or 6s., to any part of the country.

FRY, JOSEPH STORRS, & SONS, 11 & 12 *Union Street, Bristol, and* 252 *City Road, London.*—Series illustrating the manufacture of chocolate and cocoa. [*Obtained Prize Medals at the Exhibitions—London,* 1851; *New York,* 1853; *Paris,* 1855.]

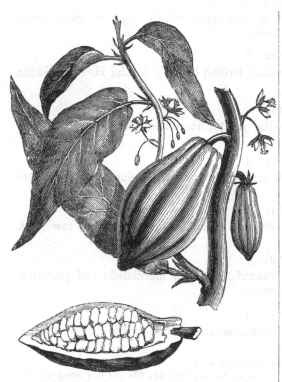

LEAF, FLOWER, AND FRUIT, OF THE THEOBROMA CACAO, WITH POD OPENED.

I. *Botanical Specimens.*

1. Branches of the cocoa tree (Theobroma Cacao).
2. Leaves   do.
3. Flowers   do.
4. Pod cut open showing the fruit which forms the cocoa of commerce.
5. Section of the wood of the cocoa tree polished, and other botanical illustrations.

II. *Specimens of Raw Cocoa as Imported from various Countries.*

1. Cocoa from Caraccas.
2. Do.   Guayaquil.
3. Do.   Para.
4. Do.   Bahia.
5. Do.   Trinidad, very fine quality.
6. Do.   do.   fine red.
7. Do.   do.   light red.
8. Do.   do.   fair gray.
9. Do.   Grenada, very fine.
10. Do.   do.   fair red.
11. Do.   Dominica.
12. Do.   St. Domingo.
13. Do.   Jamaica.
14. Do.   Africa, and other varieties.

III. *Illustrations of the Stages of Manufacture.*

1. Roasted cocoa (Caracca).
2. Do.   (Trinidad).
3. Do.   (Grenada).
4. The husk or "shell;" chiefly used in Ireland.
5. Cocoa nibs; the kernel of the nut bruised and separated from the husk. In this form cocoa is extensively used, and when boiled, these nibs produce a clear and fine-flavoured Cocoa.
6. Cocoa nibs ground; used as above, but more easily prepared for the table.
7. Pure chocolate, made solely from the cocoa nibs.
8. Pure chocolate, combined with sugar to produce cake chocolates and confectionery chocolates.
9. Do.   do.   flavoured with vanilla.
10. Mexican vanilla, fine quality.
11. Bourbon   do.   do.
12. Common vanilla.
13. Chocolate in powder, rendered perfectly soluble in boiling water.
14. Soluble cocoa.
15. Cocoa with the oil extracted.
16. The oil of the cocoa nut (or cocoa butter).
17. Chocolate cast in moulds of various shapes.
18. Iceland moss, for combining with cocoa, and other illustrative articles.

IV. *Chocolate and Cocoa as sold by J. S. Fry and Sons.*

1. Fry's cake chocolate, first quality, not sweetened, known as "Churchman's."
2. Fry's cake chocolate, and other descriptions of similar character.
3. Fry's cake chocolate, first quality, with sugar, known as "Victoria Chocolate."
4. Fry's cake chocolate, first quality, with sugar and vanilla, known as "Prince Albert Chocolate."
5. Fry's cake chocolate, other descriptions of sweetened.
6. Fry's chocolate confectionery in great variety, including sticks, drops, &c., packed in elegant boxes.
7. Fry's chocolate creams, a delicious sweetmeat.
8. Fry's soluble chocolate.
9. Fry's chocolate or cocoa paste.
10. Fry's chocolate in powder, in canisters.
11. Fry's homœopathic cocoa.
12. Fry's Iceland moss cocoa.
13. Fry's rock cocoa.
14. Fry's flake cocoa.
15. Fry's soluble cocoa, in packets.
16. Fry's pearl cocoa, and other varieties of chocolate and cocoa.

[790]

HEXTER, H., *Eccles on Street, Pimlico.*—Currie powder, vanilla, and essences.

[791]

HILL & JONES, *Jewry Street, Aldgate.*—Biscuits, lozenges, comfits, jujubes, boiled sweets liquorice, fruit syrups, preserved peel, and jams.

[792]

HOWARD & CO., *Scott Street Mills, Hull.*—Howard's British laundry starch ; Howard's Indian confection flour.

[793]

HUNTLEY & PALMER, *Reading and London.*—Biscuits for home and foreign trade ; wedding and other cakes.

[794]

JAMES, JOSEPH ELLIS, *Birnam, Scotland.*—Birnam imperial sauce; volunteer sauce ; Garibaldi sauce. (*See page 71.*)

[795]

JONES, RICHARD, & F. H. TREVITHICK, 30 *Botolph Lane, London.*—Azotized raw meat, poultry, &c.

[796]

KEILLER, JAMES, & SON, Confectioners, and manufacturers of marmalade and preserves, *Dundee.*—Confections, marmalade, and preserves.

[797]

LANGDALE, E. F., 72 *Hatton Garden.*—Ol. almonds free from prussic acid : culinary essences, syrups, liqueurs.

[Honourable Mention, Exhibition 1851.]

Concentrated fruit-syrup essences, prepared to keep in all climates, ¼ pts. 7s., pts. 12s. per doz.; culinary essences for flavouring, 8s. per lb. ; dried herbs for flavouring soups 2s. to 6s. per doz. ; essence of almonds free from prussic acid 8s. 6d. per lb. ; compounds for flavouring liqueurs, curaçoa, marasquino, &c., 20s. per lb. ; compounds for flavouring American drinks, mint-julep, bull's milk, cocktail, &c. 24s. per doz. ; essence of gin and brandy, for making London gin and cognac brandy without use of a still 8s. per lb.; 1 lb. to 50 gals. of plain spirit.

[798]

LEBAIGUE, HONORÉ, 9 *Langham Street, London, W.*—Confectionery, gum-paste figures, and fancy goods.

[799]

LEWIS, J. R., 16 *Gould Square, City.*—Liquorice root and extract.

[800]

LIVERPOOL PRESERVED PROVISION COMPANY, *Liverpool.*—Provisions preserved in hermetically sealed packages.

The articles exhibited will retain their flavour and freshness in any climate for several years. By this process the sailor or passenger at sea, the soldier in the trenches, the yachtsman on his cruise, or the sportsman on the Moors, can be supplied with fresh fish, soups, entrées, fowls, joints, vegetables, game, and fruit ready cooked, and capable of preparation for use when required.

[801]

McCALL & STEPHEN, *Adelphi Biscuit Factory, Glasgow.*—Plain and fancy biscuits—machine made.

[802]

McCALL, JOHN, & CO., 137 *Houndsditch, London.*—Preserved provisions.

[803]

McCLELLAND, GEORGE, *Wigtown, N.B.*—Preserved potato, and extract of Irish hops.

[804]

McCRAW, EDWARD CHARLES, *Winsford, Cheshire.*—Patent steam-made salt.

JAMES, JOSEPH ELLIS, *Birnam, Scotland.*—Birnam imperial sauce; volunteer sauce; Garibaldi sauce.

These sauces are compounded of the purest and choicest ingredients, and do not depend upon cayenne pepper for their piquancy. The use of them is in the highest degree consistent with health, as they materially assist the process of digestion, while stimulating the appetite. They impart a piquant flavour to cold meats, and an agreeable relish to made dishes, ragouts, hashes, soups, and stews, and are eminently desirable as fish sauces. They may be purchased in London of W. James, Coliseum Hotel, Portland Road.

IMPERIAL
SAUCE

GARIBALDI
SAUCE

VOLUNTEER
SAUCE

[805]

MACKAY, J., 121 *George Street, Edinburgh.*—Quintessences from spices and herbs, and other culinary preparations.

[806]

MACKIE, JOHN WYSE, 108 *Princes Street, Edinburgh.*—Rusks and biscuits.

[807]

MAKEPEACE, SAMUEL, *Merton, Surrey, S.*—Preserved herbs, vegetables, and herbaceous mixtures, flavouring essences, &c.

[808]

MARSHALL & SON, *Tavistock House, Covent Garden.*—Lazenby's Harvey's sauce ; Dr. Witney's condiments ; pickles, sauces, &c.

[809]

MARSHALL, T. W., 2 *Richmond Terrace, Grosvenor Street, Camberwell, S.*—Crystallized liqueurs and creams.

[810]

MARTINEAU, DAVID, & SONS, Sugar Refiners, *London.*—Illustrations of sugar refining.

[811]

MOORE, E. D., & Co., *Wood's Eaves, Newport, Salop.*—Concentrated milk : its combination with cocoa and chocolate. Concentrated wort.

[Obtained the Prize Medal at the Exhibition of 1851.]

The prize medal was awarded to Moore's Patent Concentrated Milk, and its combinations with chocolate and cocoa, for their novelty, utility, and economy. They are prepared for use by the addition of boiling water only.

These preparations are extensively used by voyagers and invalids, and are found by all to be delicious and nutritive beverages. Joseph B. Bull & Co. are the sole preservers under E. D. Moore's patent. The farm and works are situated at Wood's Eaves, Staffordshire ; the office and warehouse at Littleworth.

PRICES.

E. D. Moore's Patent Concentrated Milk (half-pint equal to 7 half-pints in liquid)	Half-pints per doz.	15s.
E. D. Moore's chocolate and milk .	doz. lbs. .	24s.
Do. do. . .	doz. ½lbs. .	12s.
E. D. Moore's cocoa and milk. . .	doz. lbs. .	20s.
Do. do. . .	doz. ½lbs. .	10s.

[812]

MORTON, JOHN THOMAS, 104, 105, 106 *Leadenhall Street, London ; Clayhills, Aberdeen, Scotland.*—Preserved provisions and jams.

[813]

MYZOULE, J. H., 72 *Southampton Street, Pentonville Road, N.*—Confectionery.

[814]

NELSON, DALE, & CO., *Bucklersbury, London ;* (Works) *Warwick.*—Brazil and patent isinglass ; gelatine ; gelatine lozenges.

[815]

PARSONS, FLETCHER, & CO., *Bread Street.*—John's nutritious corn flour ; Cowpe's dietetic and homœopathic cocoas.

[816]

PARTRIDGE, EDWARD, 22 *Leadenhall Street.*—Pickles, sauces, preserved fruits, preserved meats, &c., for exportation.

[817]

PEEK, FREAN, & CO., *Works, Dockhead, London, S.E. ; City Offices,* 37 *Mark Lane.*—A variety of steam-made biscuits, &c.

In the manufacture of these biscuits, the latest improvements of practical science are combined with purity of ingredients, producing biscuits of intrinsic excellence, and agreeable appearance.

Peek, Frean, & Co. having paid special attention to the qualities most suitable for the colonies and other countries as well as for the home trade, can with confidence recommend their selection to shippers. The advantages of their position enable them to execute all orders with promptitude. Goods for exportation are packed in air-tight tins to insure their arrival in good condition.

In all parts of the United Kingdom their biscuits may be obtained from respectable grocers and others.

[818]

PHILLIPS & Co., 8 *King William Street, City.*—Collection of teas.

[819]

RECKITT & Co., *Hull.*—Machine and fancy biscuits manufactured by steam-power.

[820]

ROBB, ALEXANDER, 79 *St. Martin's Lane, London, W.C.*—Infants' and invalids' food; wedding and other cakes, and biscuits.

[821]

SCHOOLING & Co., 14 *Great Garden Street, Whitechapel, London.*—Genuine confectionery in penny packets, &c.

[822]

SCOTT, WENTWORTH LASCELLES, *Westbourne Park, London, W.*—Table showing various articles of food and drink, and their adulterants.

[823]

SHACKLE, MARIA & RICHARD WILLIAM, 10 *Sussex Terrace, Camden Town.*—Ornamental confectionery in great variety. (*See page* 74.)

[824]

SMITH, SUTTIE, & Co., *Arbroath.*—A glass case containing lozenges, confections, jujubes, marmalade, orange and lemon peel, jams, and jellies.

[825]

SMITH, GEORGE, & Co., 23 *Little Portland Street, London.*—Isinglass, gelatine, and extract of calves' feet.

[826]

SPRATT, JAMES, 118 *Camden Road Villas, London.*—Patent dog-cakes, suitable for cats, poultry, and pigs.

827]

STANES, J., 4 *Cullum Street, City.*—Coffee branches in various stages of growth.

[828]

THOMAS, E., *Ealing Lane, Brentford.*—Flowers in sugar.

[829]

TURNER, G. & R. H., 111 *High Street, Borough.*—Wedding cakes.

[830]

VICKERS, JAMES, 23 *Little Britain.*—Specimens of isinglass in the rough and manufactured state.

[831]

WARE, G. R., Manufacturer and Importer of French Confectionery, 11 *Marchmont Street, London.*—French chocolate and bonbons.

[832]

WARRINER, G., Instructor of Cookery to the Army, *Aldershott.*—Preparations to facilitate cookery in all its branches.

[833]

WEBSTER, JOSEPH MUNDAY, 58 *Pall Mall.*—Webster's "Royal Old English Sauce," for venison, fish, &c.

[834]

WESTON & WESTALL, 115 *Lower Thames Street.*—Refined salt.

SHACKLE, MARIA & RICHARD WILLIAM, 10 *Sussex Terrace, Camden Town.*—Ornamental confectionery in great variety.

MARIA & WILLIAM SHACKLE have at all times a large variety of ornaments, for brides', Savoy, twelfth, christening, and birthday cakes. Besides the specimens exhibited a further variety may be seen at their show-room. They supply and ornament bridecakes, conducting their business on ready-money principles.

WOTHERSPOON, ROBERT, & CO., *Glasgow and London.*—Wotherspoon's Victoria lozenges, uncoloured, in packets; general confectionery and marmalade.

The above engraving represents the manufactory of WOTHERSPOON'S VICTORIA LOZENGES, which are quite a novelty, a vast improvement in every respect upon the old-fashioned lozenges, and can only be produced in perfection by their Patent Steam Machinery.

To give purchasers a guarantee of their genuineness, and to prevent the possibility of having a spurious article palmed upon them, they send out these lozenges in packets only, [which are labelled "Wotherspoon's Victoria Lozenges, and bear their full name and address. These packets are retailed at 1*d.*, 2*d.*, 4*d.*, 8*d.*, and 1*s.* 4*d.* each, respectively, which are the same prices as are charged for the ordinary inferior kinds, and are therefore beyond dispute entitled to universal preference,—a position which the demand for them proves they are rapidly attaining.

These lozenges are flavoured with peppermint, cinnamon, lemon, rose, musk, lavender, ginger, clove, and a variety of the purest essences, and are entirely free from all colouring matter.

The particular points of superiority of the Victoria Lozenges over the old-fashioned kinds are numerous, but the following are sufficient to be instanced here, viz:—

Their perfect cleanliness : being manufactured by self-acting steam machinery, they are entirely free from working of the hands, which is inseparable from, and so objectionable in, the old process.

Their improved shape : being quite smooth on the surface, and having no sharp edges like the ordinary lozenges, they have a much more pleasant feeling in the mouth.

Their purity : being manufactured from the finest sugar by a process which will not admit of adulteration, they can be used with perfect confidence.

Their safeness : being free from all colouring matter, they are uninjurious to the most delicate.

Their delicacy : being flavoured with the finest essential oils and essences only, they impart a most delightful taste to the mouth and fragrance to the breath.

Their guaranteed genuineness : the manufacturer's name being on every packet, purchasers are assured of the genuineness of the article.

Their moderate price : being retailed at the same price as the ordinary kinds, they are beyond dispute the cheapest confections made.

In short, they are injurious to none, beneficial to most, delicious to all, and are admired alike by adults and juveniles.

They may be obtained from grocers, druggists, confectioners, &c., and wholesale from the makers, Robert Wotherspoon & Co., Manufacturers of Scotch Marmalade and General Confectionery, 36 to 48 Dunlop Street, Glasgow, and 66 Queen Street, City, London.

[835]

WIGNALL, R. H., 98 *London Road, Liverpool.*—Royal original Everton toffee; improved original cocoa-nut ice.

The manufacture of the famous "Everton toffee" has been established one hundred and eight years. During this time it has been favoured with extended and exalted patronage; and in our own day has been supplied to Her Majesty and the Royal family, H. R. H. the Duke of Cambridge, the Right Hon. the Earl Russell, and other distinguished consumers. The original formula for this popular sweetmeat has never been copied, and remains in the family of Molly Bushell, the first maker, of whom R. H. Wignall is a grandson. It is supplied in tin cases, which are kept ready packed, of various sizes, and despatched on receipt of draft or money order.

[836]

WOOD, GODFREY, 15 *Commercial Street, Leeds.*—Ornamental brides' cakes; ornamental christening cakes.

Subjoined is a price list of articles exhibited and manufactured by Godfrey Wood :—
Ornamental brides' cakes (as exhibited), 10 to 100 guineas.
Wedges for distribution (as exhibited), 3s. per lb.
Christening cakes, 2 to 10 guineas.
Yorkshire game pies for presents, 3 to 10 guineas.

The above articles can be sent (packed in cases) to any part of the United Kingdom, and will be guaranteed perfect on delivery.
The exhibitor contracts for wedding breakfasts.
Orders sent by post will be punctually attended to.

[837]

WOTHERSPOON, JAMES, & Co., *Glasgow and London.*—Lozenges and comfits made by machinery; Scotch marmalade, jams, &c.

[838]

WOTHERSPOON, ROBERT, & Co., *Glasgow and London.*—Wotherspoon's Victoria lozenges, uncoloured, in packets; general confectionery and marmalade. (*See page* 75).

[839]

WRIGHT, FRANK, *Kensington.*—Essences for summer beverages, made from fresh fruit only, unfermented, free from chemicals.

SUB-CLASS C.—*Wines, Spirits, Beer, and other Drinks, and Tobacco.*

[851]

ARCHER, JOHN ALEXANDER, *Broadway, Westminster.*—Tobacco; cavendish, negro-head, and roll.

[852]

BAKER, F., *Virginia Mills, Stockport.*—Manufactured tobacco and cigars.

[853]

BASS, RATCLIFF, & GRETTON, *Burton-on-Trent.*—East India pale ale; No. 3 Australian ale; strong ale.

The ales of BASS, RATCLIFF, & GRETTON may be obtained in butts (108 gallons), hogsheads (54 gallons), barrels (36 gallons), and kilderkins (18 gallons), from the brewery, Burton-on-Trent; from their stores, of which a list is subjoined; in cask, as well as in bottle, wholesale from all respectable wine and beer merchants; and retail, on draught, and in bottle from the licensed victuallers.

London...	3 Wharf, City Basin, E.C.
Liverpool	28 James Street.
Manchester	34 Corporation Street.
Dublin	66 Middle Abbey Street.
Cork	10 Lavitt's Quay.
Belfast	10 Hill Street.
Glasgow	43 Dunlop Street.
Newcastle-on-Tyne ...	Trafalgar Goods Station.
Birmingham	Newhall Street.
Stoke	Company's Wharf.
Wolverhampton	Market Street.
Bristol	Tontine Warehouses, Quay Head.
Nottingham	1 Long Row.
Derby	Corn Market.
Devon & Cornwall ...	42 Union Street, Plymouth.
Shrewsbury	Wyle Cop.

[854]

BIGGS, AMBROSE, *Birmingham.*—Manufactured tobacco.

[855]

BOLLMANN, CONDY, & CO., 48 *Halfmoon Street, Bishopsgate, London.*— Malt vinegar; patent concentrated pure malt vinegar.

[856]

DYER, WILLIAM, *Littlehampton.*—British champagne, closely resembling foreign. Ingredients wholesome. Cost, only 2*s.* 4*d.* a gallon.

[857]

ENGLAND, GEO. JOS., *Dudley.*—Various descriptions of malt.

[858]

EVANS & STAFFORD, *Leicester.*—Stilton, Leicester, and Derby cheese; cigars and tobacco. (*See page* 78.)

[859]

FOWLER, J., & CO., *Prestonpans, N.B.*—Beer and India pale ale.

[860]

FRYER, DANIEL, *Epney, Stonehouse, Gloucestershire.*—Cider and perry.

[861]

GARRETT, NEWSON, *Aldeburgh, Suffolk.*—Patent crystallized malt.

[862]

GOODES, GEORGE & SAMUEL, 51 *Newgate Street, London.*—Cigars, tobaccos, and snuffs.

[863]

HEATLEY, JAMES, *Alnwick, Northumberland.*—Manufactured tobaccos.

[864]

HICKS, JOSH. R., *East Bergholt, Suffolk.*—English wines. Dr. Hassall's report, with prices, will be forwarded on application.

[865]

HILTON, ABRAHAM, *Barnard Castle.*—Rum shrub.

[866]

HOOPER, WILLIAM, 7 *Pall Mall East, S.W.*—Artificial mineral waters.

[867]

HUGGINS, EDWARD STAMFORD, 2 *Albert Street, Derby.*—Liqueur orange brandy.

[868]

HYAMS, MICHAEL, Manufacturer, *Bath Street, London.*—Collection, with models, illustrating improvements in the manufacture of cigars.

[869]

JONAS, E., BROTHERS, 78 *High Holborn.*—Cigars and tobacco.

EVANS & STAFFORD, *Leicester.*—Stilton, Leicester, and Derby cheese; cigars and tobacco.

CHEESE STORES

The following makes of cheese, of which samples are exhibited, are selected from the finest dairies :—

Cream Stilton.	North Wilts.	Ditto Ox tongues.	Cheddar loaf.
Leicester.	American.	Cheshire.	Choice bacon.
Ditto, Toasters.	Choice Leicester hams.	Cheddar.	Ditto lard in tins.
Derby.			

CIGAR MAKERS ROOM Nº 1

CIGAR MAKERS ROOM Nº 2

The exhibitors are manufacturers of the following Cigars :—

Cabanas.	La fragancias.	Perfections.	Carallos.
Regalias.	Sevillanas.	La Floritas.	Prince Consorts.
Trabacas.	Pruebas.	La Conchas.	Manillas.
Lopez.	Recompenzas.	Pillous.	Pilots.
Kings.	Woodvilles.	Sultanas.	Bengals.
Queens.	La favorites.	Emperors.	Mexicans.
Regents.	Dy J. Patrons.	La Jarnas.	Cigarros.
Imperials.	Cubas.	Unions.	Esmeraldas.
Eminentes.	Partagas.		
Principes.	Salvadoras.	They exhibit a case containing Havana, Giron and	
Yaras.	Eldorados.	Esmeralda, Regalia, Trabuca and Great Easterns.	

[870]

KENT, W. & S., & SONS, *Upton-on-Severn.*—French brandy and vinegar; British vinegar, cider, perry, cordials, and brandy.

Messrs. Kent exhibit the following home and foreign produce :—
Table and pickling vinegar, Nos. 18 and 24.
Choice cider and perry.

British brandy, and liqueur cordials.
Grande champagne cognac brandy, vintages 1851, 1855, and 1858.
First quality of French wine vinegar.

[871]

MART & Co., 130 *Oxford Street, W.;* (Wholesale) *Three Crown Square, Borough.*—Wines, preserved fruits, &c.

[872]

PITT & Co., 28 *Wharf Road, City Road, London.*—Pitt's patent tonic (aërated quinine) water.

This Aerated Water is the result of extensive chemical research, and has been submitted to several London physicians, from whom it has met with unqualified approval. It is considered by the proprietor to be of sufficient importance to patent, that being the only means by which the public can be protected against fraudulent imitations, and it is now offered under the most flattering testimonials. Its properties are antacid, cooling, and refreshing, combined with all the advantages of Soda Water; it gives strength to the stomach and tone to the whole nervous system, and is especially adapted to persons feeling depressed from mental or bodily excitement, imparting strength to those who suffer from nervous irritation, indigestion, or loss of appetite.

TESTIMONIAL FROM DR. HASSALL.
"Chemical and Microscopical Laboratory,
74 Wimpole Street, Cavendish Square, W.
19th December, 1860.
"I have carefully analyzed PITT'S TONIC WATER. The idea of combining a tonic like quinine with an aerated water is a good one, and the practical difficulties in the way of carrying it out have been entirely overcome in this preparation.

"It is a pleasant, refreshing tonic, and invigorating beverage, strengthening to the digestive organs, and calculated to promote appetite; it is also an excellent restorative to the stomach weakened by any excess or indulgence.

"From its composition and properties, PITT'S TONIC WATER ought to a great extent to supersede the use of soda and other aerated waters."

"ARTHUR HILL. HASSALL, M.D., Lond."

Author of the Lancet Sanitary Commission; author of " Food and its Adulterations," "Adulterations Detected," and other works.

The tonic water may be obtained of Messrs. Veillard & Co., Eastern Area of the Exhibition. Numerous medical testimonials may be had on application.

[873]

RICHARDSON, SANDERS, & Co., *Hope Brewery, near Notting-hill Gate.*—A new description of beer.

[874]

SALT, THOMAS, & Co., *Burton-upon-Trent.*—Pale and Burton ales for home consumption and exportation.

[875]

SHARMAN, ALFRED, *Walham Green.*—Salugenic beverage (a new drink), made from fruit of the carob tree.

[876]

SILICATED CARBON FILTER COMPANY, *Bolingbroke Gardens, Battersea, S.W.*—Filtered liquors.

[877]

TAYLOR, HUMPHREY, & Co., *Shawfield Street, Chelsea.*—English liqueurs, cordials, and flavoured spirits.

Taylor, Humphrey, & Co. exhibit specimens of liqueurs, compounds, and spirits, manufactured and distilled by them, comprising the following, viz :—

Maraschino.
Curacoa, sweet.
Curacoa, dry.
Crème de Noyau.
Crème de the.
Crème de vanille.
Crème de fleur d'orange.

Crème de parfait amour.
Crème de rose.
Crème d'abricot.
Anisette.
Ratafia.
Extrait d'absinthe.
Peppermint cordial.

Cinnamon cordial.
Cherry brandy.
Orange brandy.
Ginger Brandy.
Apricot brandy.
Raspberry brandy.
Orange bitters.
Milk punch.

Curaçoa punch.
Apricot punch.
Essence of punch.
Chartreuse.
Pine-apple shrub.
Green ginger liqueur.
Aniseed cordial.
Cloves cordial.

British brandy of very superior quality. Plain spirit, absolutely pure, manufactured by a new process.

[878]

WALKER, ALFRED & WILLIAM, 3 *Peartree Street, Goswell Street, London.*—Exhibition ginger and British-made wines.

[879]

WILLS, W. D. & H. O., & SONS, *Bristol.*—Best bird's-eye, roll, and other choice tobaccos.

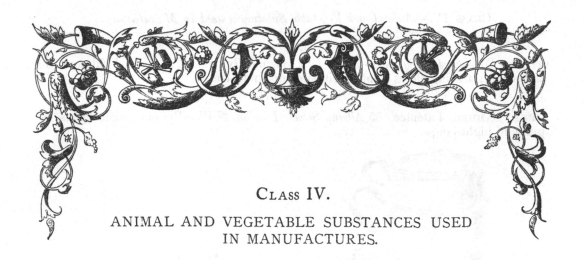

CLASS IV.

ANIMAL AND VEGETABLE SUBSTANCES USED IN MANUFACTURES.

SUB-CLASS A.—*Oils, Fats, and Wax, and their Products.*

[910]

BARCLAY & SON, 170 *Regent Street, London.*—Bleached wax; candles of various materials, night-lights, &c.

[911]

BAUWENS, FELIX LIEVIN, *Oil Works, 15 St. Anne Street, Westminster, S.W.*—Candles, soap, and oils.

[912]

BRECKNELL, TURNER, & SONS, *Haymarket, London.*—Shade for candlesticks.

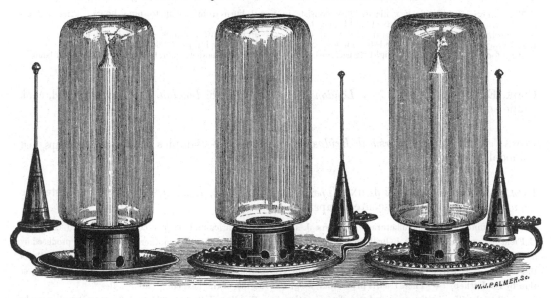

W.J.PALMER,Sc.

The accompanying illustrations represent designs for shades for hand-candlesticks. The improved shade, A, is made with the upper part curved inwards, in order to prevent the too sudden ingress of air, which by deflecting the flame causes the grease or wax to melt too quickly; the consequence of which is, that an uncovered candle, or even one protected by the ordinary shade, cannot be carried about without running over, and frequent spilling of the grease. The shape of the upper part of the registered shade shields the candle to a great extent from the downward flow of air, and it may be carried along rapidly without the annoyance arising from guttering, and the liability of dropping the grease or wax about the room, or upon the dress. The lower part of the shade is fitted into the band or ring of metal, B, the size of which is made to fit the socket of the candlestick, to which the shade is to be applied. This improvement in the shade adds also to its ornamental appearance.

[913]

CANTRILL, THOMAS, & Co., *River Terrace, York Road, King's Cross, London.*—Railway and other greases.

[914]

CATTELL, DR., *Euston Square.*—Oils; fats, chemically treated; proofed fabrics; silk hat bodies; tracing cloths; anti-attrition compounds, &c.

[915]

CLARKE, SAMUEL, Patentee, 55 *Albany Street, London, N.W.*—Pyramid night-lights, and pyramid night-lamps.

The superiority of Clarke's night-lights, consists in their giving double the amount of light and heat of other night-lights, and in their burning without flickering. They are admirably adapted for nursery use, for heating food, water, &c., and for lighting passages. They will burn from seven to nine hours. Price 7½d. per box.

Pyramid night-lamps, 1s. and 1s. 6d. each.
Hot-water lamps, 2s. 6d. and 3s. 6d. each.
Pyramid food-warmer, 6s. each, complete with lamp, 7s.

[916]

COOK, EDWARD C., & CO., *East London Soap Works, Bow, London.*—Yellow, mottled, curd, and soft soaps.

[917]

COWAN & SONS, *Hammersmith Bridge Works, Barnes.*—Samples of household soaps, and model of works.

[918]

FIELD, JOHN, CHARLES, & JOHN, *Upper Marsh, Lambeth, London.*—Paraffine, and stearine candles; sealing-wax; scented soaps.

The paraffine candles manufactured by Messrs. J. and C. Field far surpass all others, in brilliancy of appearance and in illuminating power. In support of this statement, the following is extracted from a report of an examination by Dr. Letheby :—" These results prove, that, weight for weight, the illuminating power of paraffine is rather more than 22 per cent. greater than that of spermaceti, about 40 per cent. greater than wax, 46 per cent. greater than stearic, and 58 per cent greater than composite. Or, to estimate it in another way, the light produced by 98 lbs. of paraffine candles is equal to that of 120 lbs. of spermaceti, or 138 lbs. of wax, or 144 lbs. of stearic, or 155 lbs. of the best composite candles."

Field's celebrated " United Service Soap " may be purchased in the form of tablets, price 4d. and 6d. each.

[919]

GIBBS, D. & W., *City Soap Works, London.*—Specimens of manufactures; hard, soft, and scented soaps.

[Prize Medal at the Great Exhibition, 1851.]

Specimens of composite, household, extra pales, yellow, and marine soaps, for general use; curds and mottled for manufacturers; curds, palm, and patent soaps for the silk trade; the $\frac{B S}{M}$ soft soap (free from smell), as supplied to H. M. Government, the Mail Steam Packet Companies, and the principal London hospitals; Naples tablets, old brown Windsors, honey, and various new kinds of toilet soaps.

[920]

GOSSAGE, WILLIAM, & SONS, *Warrington.*—Specimens of soap and of silicate of soda.

[921]

HALE, WARREN S., & SONS, 71 *Queen Street, London.*—Stearic acid; British sperm and composite candles.

[922]

HEMANS, MRS. H. W., 4 *St. James's Terrace, Clarendon Road.*—Wax flowers.

[923]

KNEVETT & AUSTIN, 22 *Mortimer Street, Regent Street, W.*—Flowers in wax and new materials.

[924]

KNIGHT, J., & SONS, *Soap and Candle Works, Old Gravel Lane, E.*—Primrose soap.

[925]

LAMBERT, ELIZABETH B., *Spring Villa, Tunbridge.*—A Kentish bank near Tunbridge, in July, modelled in wax.

[926]

LANGTON, BICKNELL, & SONS, *Newington Butts, S.*—Sperm oil and spermaceti, in various stages of manufacture.

[927]

LUMSDEN, ISABELLA, 8 *Trevor Terrace, Rutland Gate.*—Bouquet of wax flowers, in frame.

[928]

MACKEAN, WILLIAM, *St. Mirren's, Paisley.*—Household and toilet soaps.

[929]

MAKEPEACE, ELIZA, *Merton, Surrey, S.*—Wax flowers, orchidæ, anatomized leaves, and innocuous wax.

[930]

MARSHALL, J. & W., *Selby.*—Oils and refined oils.

[931]

MEECH, H. J., 3 *North Place, Kennington Road, Lambeth.*—Wax figures.

[932]

MINTORN, J., 106 *New Bond Street.*—Models of flowers in wax, and materials used.

[933]

MINTORN, JOHN HAYNES, 33 *Soho Square.*—Models of flowers in wax.

[934]

MITTON, THOMAS, *Old Square, Blackburn.*—Improved dip and mould tallow candles.

[935]

NEIGHBOUR, GEORGE, & SONS, 127 *Holborn.*—Specimens of oil for manufacturing and machine purposes.

[936]

OGLEBY, C., & Co, 58 *Paradise Street, Lambeth.*—Refined spermaceti, paraffine, and stearic acid; with candles made of them.

[937]

PENFOLD & MARTIN, *Tenison Street, Lambeth.*—Tubular candles.

[938]

PIERSON, J., 66 *Mortimer Street, W.*—Flowers modelled in wax.

[939]

PRICE'S PATENT CANDLE COMPANY (Limited), *Belmont, Vauxhall, London.*—Series of specimens illustrating improvements in the manufacture of candles, oils, night-lights, and glycerine.

[940]

RICH, W., 14 *Great Russell Street, Bloomsbury.*—Wax figures and flowers.

[941]

ROBIN & HOUSTON, *Paisley.*—Soap.

[942]

ROSE, WILLIAM ANDERSON, 69 *Upper Thames Street, London.*—Railway grease and other lubricating compounds ; paints, varnishes, oils, &c.

Railway carriage grease for fast trains ; grease for hot climates ; lubricating greases for contractors' use and mining purposes ; oils for lubricating, burning, and painting purposes.

Varnishes for coachmakers' and builders' use ; white lead, red lead, litharge, white zinc and colours for house-painters', ship-builders', and railway companies' use ; anti-oxide for iron bridges ; cotton waste for cleaning machinery ; tar, pitch, and rosin.

[943]

ROWE, T. B., & Co., *Thames Soap Works, Brentford, W.*—Specimens of various soaps for domestic and manufacturing purposes.

The exhibitors are the manufacturers of the "Brentford mottled," "Imperial pale" toilet and other soaps for domestic purposes, and also of

1. White or curd soap.
2. Refined ditto.
3. White oil ditto.
4. Strong brown ditto.
5. Pure oil soap.
6. Red palm ditto.
7. Scouring ditto.
8. Strong mottled ditto ;

and every variety used in the processes of manufacture by bleachers, clothiers, dyers, lace manufacturers and dressers, fine-paper makers, spinners, silk-throwsters, &c.

[944]

SENTIS, JULES, *Abercorn Street, Paisley.*—Stearine ; oil and soap manufactured entirely from grease recovered from soap-suds.

[945]

SHIPLEY, MISS JANE, Teacher of Wax Modelling, 34 *Carter Street, Greenheys, Manchester.*—Flowers modelled in wax.

[946]

SYMONS, MRS., 9 *Devonshire Terrace, Notting Hill Gate.*—Wax flowers.

[947]

TAYLOR, WILLIAM, & Co., *Leith.*—Soaps, stearic acid, stearic acid and composite candles.

[948]

TREWOLLA, MRS. RICHARD, *Halesowen.*—Group of wax flowers.

[949]

TUCKER, F., & Co., *Kensington.*—Wax, sperm, stearine, composition, and bleached tallow candles. Specimens of decorated candles, and bleached tallow.

FRANCIS TUCKER & Co. are wax chandlers, candle manufacturers and oil merchants to the Queen, and His Royal Highness the Prince of Wales, 61 High Street, Kensington, and 18 South Molton Street, Grosvenor Square, London, W.

They are patentees of wax candles with platted wicks. Their manufactory at Kensington was established in 1730.

Price lists of candles, oil, soap, &c., sent on application.

[950]

WEST OF ENGLAND SOAP COMPANY (Limited), *Plymouth.*—Manufacturers', toilet, and domestic soaps ; paraffine and composite candles. (*See page* 85.)

[951]

WILKINS, PRISCILLA, 49 *St. Paul's Road, Kennington, S.*—Wax flowers and fruit.

[952]

WILLIS, MARGARET H., *Marshside, Lower Edmonton.*—Wax flowers, ornamental leather work, and wax for making flowers.

[953]

WILLIAMS, JOHN, & SON, *Clerkenwell, London.*—Hard, soft, and fancy soaps, with illustrative and descriptive processes.

WEST OF ENGLAND SOAP COMPANY (Limited), *Plymouth.* — Manufacturers', toilet, and domestic soaps ; paraffine and composite candles.

The following specimens are exhibited.

MANUFACTURERS' SOAPS.

SILK THROWSTERS.

No.
1. White oil soap, for China silks.
2. Special mild curd soap, for Japan ditto.
3. Palm soap, for Bengal ditto.

SILK SPINNERS (WASTE).

4. Pure curd soap, for China and Japan silks.

SILK DYERS.

5. Best white soap, for fancy colours.
6. Brown oil soap, for blacks.
7. Brown oil soap, for boiling off.

WOOLLEN MANUFACTURERS.

No.
8. White West of England soap, for milling and mellowing woollen cloths.
9. Brown ditto, for scouring.
10. Brown soap, for scouring sale yarn (Scotch market).
11. Brown soap, for worsted spinners and manufacturers.

CALICO MANUFACTURERS.

12. Vienna soap, for sizing calicos.
13. Feeding soap, for bed-ticks, Nankeens, &c.

CALICO PRINTERS.

14. Best oil soap, for madder reds, pinks, and madder purples.

15. White oil soap, for paper-makers.

Specimens of Silk, Woollens, and Calicos, in the manufacture of which the above Soaps have been used, are shown in the Company's Case.

TOILET SOAPS.

No.
16. Transparent glycerine soap, being soap in its purest form, in pillars, shaving sticks, and tablets.
17. Finest toilet soap, variously and highly perfumed.
18. Treble-scented brown Windsor, in bars and slides.
19. Great improved brown Windsor, ditto.
20. Musk brown Windsor, ditto.
21. Extra brown Windsor, ditto.
22. Pure honey, ditto.
23. Otto of rose, ditto.
24. Lavender, ditto.

No.
25. Elder-flower soap, in bars and slides.
26. Turtle oil, ditto.
27. Pure glycerine, ditto.
28. Floating glycerine, ditto.
29. Almond and glycerine, ditto.
30. Sunflower oil, ditto
31. Brown Windsor, in wrappers, made up to suit all home and foreign markets.
32. Soft soap.

DOMESTIC SOAPS.

No.
33. West of England, best household soap.
34. Fine curd, for general use.

No.
35. Mottled, for scouring.
36. Marine, for salt-water purposes.

CANDLES.

No.
37. West of England kohinoor.
38. West of England sperm.
39. West of England opaline.
40. West of England Ceylon wax.
41. West of England wax.

No.
42. West of England composite.
43. West of England carriage lights.
44. West of England ship lights.
45. West of England chamber candles.
46. West of England tapers.

47. **West of England Night Lights.**

SUB-CLASS B.—*Other Animal Substances used in Manufactures.*

[965]

AZÉMAR, J. C., *The Waldrons, Croydon.*—Specimens of ivory turning.

The centre piece is an allegorical work representing the Temple of Industry with all its attributes. As peace is essential to the pursuits and progress of sciences and arts, the emblems of war appear as if cast out from the precincts of the edifice. Religion being essential to the success and stability of all enterprises, the cross crowns the whole.

Temple of Industry, 80 guineas.

African black-wood cup, deep square pattern, 20 guineas; large ivory cup, 15 guineas; flat cup, with rhinoceros horn base, 15 guineas; cocus-wood cup, bound with ivory rings, and stem of African black-wood, pattern of superposed lozenges, 5 guineas.

African black-wood cup, bound with ivory rings, 8 guineas.

Oval ivory frame, deep open-work, 10 guineas.

Freemason's gavel, forming a hand-bell, 5 guineas.

Hollow, scented, wood ball, with light ivory stand, price 3 guineas.

Box top, deep open-work, with thistle centre, 4 guineas.

Ditto, with lighter work, 4 guineas.

Price of the case complete, 170 guineas.

Application may be made to Mr. H. Dixon, turner, 29 Gracechurch Street, City.

[966]

BARNES. S. & T., 3 *Shouldham Street, W.*—Ivory, wood, and bone, hair, tooth, and nail brushes.

[967]

BARRY BROTHERS, *Meriton's Wharf, London, S.E.*—English sheepskins, showing an improved growth of wool.

[968]

BERENDT & LEVY, *Leeds.*—Samples of low wools.

[969]

BERTHOLD & PHILLIPS, 31 *Gloucester Terrace, New Road, Commercial Road East.*—Tortoise-shell combs.

[970]

BILLINGTON, MISSES, *Lord Street, Southport.*—Group of shell flowers.

[971]

BUXTON, WILLIAM, *Limetree Lodge, Rotherhithe.*—Wools grown in the United Kingdom.

[972]

CANTOR & Co., 6 *Houndsditch, London, N.E.*—Turkey sponges.

[973]

COPE, R., & SONS, *Uttoxeter.*—Cabinet-makers' glue.

[974]

COX, J. & G., *Gorgie Mills, Edinburgh.*—Gelatine and glue.

[975]

DARNEY, JOHN, & SONS, Glue Manufacturers, *Kinghorn, Scotland, and Drury Lane, London.*—Scotch glue, sizing, &c.

[976]

DOBSON, JOHN, Comb-maker, *Joseph Street, Leeds Road, Bradford.*— Buffalo-horn, and tortoiseshell combs.

[977]

DORRIEN, CHARLES, *Ashdean, near Chichester.*—Merino wools grown in Sussex.

[978]

DUTTON, T. R., 19 *Holywell Row, Shoreditch.*—Wood and ivory carvings and turnings.

[979]

FENTUM, MARTIN, 85 *New Bond Street, W., and 8 Hemmings Row, W.C.*—Works in ivory and hard woods.

[980]

FISHER, WILLIAM, & SONS, *Orchard Place, Sheffield.*—Umbrella, matchet, and knife handles of pressed horn.

[981]

FOX, THOMAS BARKER, 37 *St. John Street, Devizes.*—Wiltshire Southdown fleeces, hog and ewe.

[982]

GLASS, G. M., *Brandon Street, Walworth.*—Gelatine.

[983]

GREEN, JOHN, 7 *Sherborne Street, Islington.*—Sheet gelatine used for tracing, wrappers for confectionery, and valentines.

[984]

GURDON-REBOW. *Wyvenhoe Park, Colchester.*—Sheep's wool.

[985]

HASTILOW, CHARLES, 3 *Queen Street, Worship Street, E.C.*—Chessmen, draughtsmen, billiard and bagatelle balls, and fancy goods.

[986]

HEINRICH, J., *Lower Kennington Lane.*—Tortoiseshell combs.

[987]

HITCH, MARK, *Eversham, Worcestershire.*—Imitation tortoiseshell combs, which resist the action of damp atmosphere.

[988]

JACOB, BERNARD, 68, *Leadenhall Street, City, London.*—Shells and shell-work in all branches.

[989]

JAQUES, JOHN, & SONS, 102 *Hatton Garden, London.*—Fancy ivory goods.

[990]

JEWESBURY, H. W., & Co., 1 & 2 *Mincing Lane, E.C.*—Varieties of cochineal.

[991]

JOHNSON, PETER, Amateur Turner, *Wigan.*—Specimens of concentric turning in wood and ivory.

[992]

JOWITT & SONS, *Leeds.*—Wools.

[993]

LAMMLER, G., 2 *South Street, Finsbury.*—Carving in ivory.

[994]

LUBLINSKI, ROBERT, 183 & 185 *City Road.*—Carved ivory and other fancy handles for umbrellas, parasols, &c.

[995]

MANNINGS, GEORGE, *Wedhampton, near Devizes.*—Teg and ewe fleeces of South Down wool from Wilts.

[996]

MARLBOROUGH, DUKE OF, *Blenheim, Oxon.*—Oxfordshire Down wool, and blankets manufactured therefrom.

[997]

MASON, G., *Yateley, Hants.*—British silk and flax.

[998]

MILLER, HENRY, 4 *St. Edmund's Place, Bury St. Edmund's.*—Specimens of spiral turning by a patent lathe.

[999]

MOORE, WILLIAM SAL., 47 *Perceval Street, E.C.*—Ivory, bone, and wood, hair, tooth, nail, and shaving-brushes.   (Illustrated process.)

Ivory hair, hat, and cloth brushes.
Ivory tooth, nail, and shaving brushes.
Ivory hand-glasses, and glove stretchers.
Ivory powder boxes, and shaving rollers.
Ivory paper-knives, tooth and nail rollers.
Ivory turnery and fancy goods.
Bone tooth and nail-brushes.
Bone shaving and fancy brushes of every description.
Wood hair-brushes veneered with ivory.
Satin rosewood and ebony hair-brushes of all varieties and qualities.

A very superior bone tooth-brush made for exportation, stamped and unstamped, and packed in boxes, always in stock.

Specimens of every stage of the manufacture of the above articles can be seen in detail in exhibitor's case.

W. S. Moore invites the special attention of merchants, shippers, perfumers, dressing houses, and all wholesale factors to the superior and extensive stock always on hand for selection.

[1000]

NIMMO, THOMAS, & Co., *Rivald's Green Works, Linlithgow, N. B.*—Superior glues and gelatine.

[1001]

NUPPNAU, EDMUND, 27 *Norfolk Street, Strand.*—Vases, cups, &c., turned in ivory.

[1002]

OLLEY, THOMAS GEORGE, 98 *Bolsover Street, London, W.*—General turnery and work by compound action lathe.

[1003]

PLAYNE, CHARLES, *Nailsworth, Stroud, Gloucestershire.*—Ornamental turning in ivory.

[1004]

PROCKTER & BEVINGTON, 124 *Grange Road, Bermondsey.*—London-made glues.

[1005]

PUCKRIDGE, F., 56 & 57 *Kingsland Road.*—Goldbeaters' skin.

[1006]

RICHARDSON, E. & J., *Newcastle-on-Tyne.*—Glues and gelatines.

[1007]

ROYAL AGRICULTURAL SOCIETY OF ENGLAND, 12 *Hanover Square, W.*—Wool.

[1008]

RYLEY, E. C., *Great Prescot Street, E.*—Specimens of amateur turnings in turnery and hard wood.

[1009]

SALOMONS, A., Amateur, *Old Change, E.C.*—Articles in ivory (turned).

[1010]

SAMUEL, M., 7 *East Smithfield.*—Shells, matting, canes, &c.

[1011]

SANDS, T. C., *Mortimer Street, Leeds.*—Burry wool cleaned by machinery.

[1012]

SASSÉ, P. C., 53 *Wynyatt Street, Clerkenwell.*—Looking-glasses, paper-knives, card-cases, chessmen, &c., in ivory.

[1013]

SISSON, JOHN, & SON, *Kendal.*—Mane, clipping, dressing, and small-tooth comb manufacturers.

"The horn comb manufacture is of considerable antiquity in this town, having been in existence more than a century, and is carried on with great spirit at the present time by Messrs. John Sisson and Son. This establishment has been in the same family since 1794, Joseph Sisson having founded it in that year. The firm maintains a high reputation for the production of a particular description of combs for horses, outrivalling, perhaps, every other house in the trade throughout the kingdom in that article. London, Edinburgh, and Glasgow are the chief marts. Most of the combs are for domestic consumption; but some of the wholesale houses in London export Messrs. Sisson's produce. The manufacture is stimulated by a steam-engine and machinery of modern construction."—*Nicholson's Annals of Kendal.*

[1014]

STAIGHT BROTHERS, 35 *Charles Street, Hatton Garden.*—Specimens of patent coral: ivory combs, pianofore keys, &c.

[Obtained Prize Medal at the Exhibition of 1851.]

CORAL SUPERSEDED BY THE PATENT CORALLINE.

Specimens of Patent Coralline may be seen in Section I., Class IV. It is highly esteemed for jewelry purposes, and is also adapted for ornamenting works of art. Particulars can be had by applying to the patentees and sole manufacturers.

Messrs. Staight Brothers are also ivory merchants, and cut ivory for veneering in the spiral form; one length of veneer cut by them was exhibited in the London Exhibition of 1851, measuring 55 feet long; and being without a joint, obtained the prize medal. Messrs. Staight Brothers also manufacture ivory into combs, pianoforte keys, knife handles, chessmen, billiard balls, &c., &c.

[1015]

STEWART, ROWELL-STEWART, & CO., *Aberdeen Comb Works, Aberdeen, and* 13 *Grocers' Hall Court, Poultry, London.*—Horn, tortoiseshell, and india-rubber combs.

The Aberdeen Comb Works are the largest in the world, covering upwards of two acres of ground, and employing 700 hands; but in 1854, when ladies' back combs were very much in fashion, these works employed 1100 hands.

The following extracts are from "Chambers's Edinburgh Journal," No. 396, 2nd August, 1851, and may be interesting to the general public. Since that time, however, there has been, along with many improvements, a great increase in the power of production.

"We come now to treat of the grand era in the comb trade—of the time when it was destined, like the great staple manufactures of our country, to undergo a revolution. The introduction into the trade of machinery and steam power, with, as a collateral result, the division of labour, is at once suggestive of an important stride in the march of progress. About the year 1828 Mr. Lynn invented a machine of a singularly ingenious design and construction, having for its principal object that of cutting two combs out of one plate of horn or tortoiseshell; and two years afterwards Messrs. Stewart & Co. commenced the manufacture in Aberdeen. To the first of these circumstances the trade was indebted for the successful idea of a machine, which effected at the same time a saving of half the material, and an increase of produce almost inconceivable. To the latter it is still more indebted for the first application of steam-power to the machinery; and, what we think of infinitely greater importance, the introduction of those true principles in the philosophy of production so logically contended for by Adam Smith, a philosophy which, in its legitimate application, has the invariable effect of elevating alike the character of the produce and the producers.

"There are two chief divisions in the second article, horn; namely, buffalo and ox horns, both of which are imported from various parts of the globe. Buffalo-horn is, however, for the most part used in the manufacture of knife-handles, and such-like articles in the cutlery trade. In comb-making it is chiefly used for dressing-combs; and, generally speaking, all combs of a deep black colour are formed of this material. The best buffalo-horns are obtained from the East Indies, and incomparably the finest are those of the Indian buffalo from Siam. We were shown a beautiful specimen of Siamese horns, which, from their extraordinary dimensions, had been preserved and polished. One of them measured 5 feet from tip to base, 18½ inches in circumference at the widest part, and weighed 14 lbs. Some conception may be formed of the extraordinary size of an animal which can support such a weight on the frontal-bone, if we recollect that a good specimen of an English ox-horn weighs only 1 lb.

"After taking a look at the steam-engine, which is of fifty horse power, and we were informed the largest of the horizontal kind in Scotland, we proceeded to the first stage of the manufacture, where the horns are cut into assorted sizes by means of a circular saw. A horn is twice cut transversely, and afterwards, if a large one,

longitudinally. The tips or extremities of the horn here cut off are sent to Sheffield, where they are converted into table-knife and umbrella handles; and in this operation 16,000 horns are cut up in a week. Instead of being divided in this manner, the hoofs in their first stage are, after being boiled for a certain time, to render the fibre soft, cut into two pieces; or rather the sole is stamped out by means of vertical punching-machines of the same irregular conformation. The specimens of elaborate and skilful ornamentation displayed here, especially on ladies' braid-combs, were truly admirable; and one pattern in particular was shown us wherein there was a species of chain, formed of beautifully-stained horn, interwoven with the head of the comb, which, although we examined minutely, and knew there must have been a joint in each alternate link, we nevertheless failed to discover it.

"The aggregate number produced of all these different sorts of combs averages upwards of 1200 gross weekly or about 9,000,000 annually; a quantity that, if laid together lengthways, would extend about 700 miles. The annual consumption of ox-horns is about 730,000, being considerably more than half the imports for 1850; the annual consumption of hoofs amounts to 4,000,000; the consumption of tortoiseshell and buffalo-horn, although not so large, is correspondingly valuable: even the waste, composed of horn-shavings and parings of hoof, which, from its nitrogenized composition, becomes a valuable material in the manufacture of prussiate of potash, amounts to 350 tons in the year.

"There are so many beautiful instances of the division of labour here exhibited, that the task of selecting is not easy. But let us take for an example the cheapest article in the trade; namely, the side-combs, sold retail at 1*d.* per pair—an article that, in its progress from the hoof to the comb—finished, carded, and labelled 'German shell' undergoes eleven distinct operations. This comb, then, which twenty years ago was sold to the trade at 3*s.* 6*d.* per dozen, can now be purchased in the same way for *two shillings and sixpence per gross!* thus effecting a reduction in price of about 1600 per cent.

"As a curious illustration of the value of labour, we give the following comparative estimate of the produce of the three materials:—

	Value. £		Value. £	Increase per cent.
1 cwt. of shell,	200	produces combs,	275	37½
1 ton horns,	56	„ „	150	168
1 ton hoofs,	12	„ „	36	200

Regarded in this aspect, in the relation of labour to material, we find that hoofs—intrinsically the least valuable of the three materials—become, with the application of labour, the *most valuable*—that is, proportionably: and the converse holds good in the case of tortoiseshell. The important relation labour bears to the produce may be estimated from the fact, that this establishment pays a larger sum of weekly wages than is now paid for the important business of cotton-spinning in Aberdeen."

[1016]

TUCKER, EDWARD, & Co., *Belfast, Ireland.*—Bleachers' starch, specially adapted for linens.

[1017]

TUCKER, H., *Fleet Lane, Farringdon Street, E.C.*—Goldbeaters' moulds; and skin for scientific and other purposes.

[1018]

VOTIERI, J., 24 *Upper Park Street, Islington.*—Carvings in shell and stone.

[1019]

WRIGHT, FREEMAN, *Needham Market, Suffolk.*—Imperial and crown glues, made from pieces of hides and skins of cattle.

[1020]

YOUNG, B., & Co., *Spa Road, Bermondsey, S.E.*—Size, glue, and gelatine.

SUB-CLASS C.—*Vegetable Substances used in Manufactures, &c.*

[1033]

ADAMSON, R., Gardener, *Balcarres, Fifeshire.*—Baskets for fruits and cut flowers.

[1034]

AGAVA PATENT HAIR COMPANY, *Newlay, near Leeds.*—Fibre of the Agavé, raw and manufactured. (*See page* 92.)

[1035]

ALDRED, THOMAS, 126 *Oxford Street, London, W.*—Bows, arrows, and archery accoutrements; fishing-rods and tackle.

[*Obtained a Prize Medal at the Great Exhibition of 1851, and at New York.*]

Thomas Aldred has been appointed manufacturer of archery accoutrements and fishing tackle to the Emperor and Empress of the French, the Emperor of Brazil, and the Queen of Denmark. He imports Italian and Spanish yew; is the maker of the celebrated glued-up triangular rods; and manufactures bows, arrows, Thames rods, winches, lines, flies, &c. Catalogues may be obtained of these goods gratis. The prices of them are moderate, and they may be obtained in any quantity, wholesale, retail, or for exportation.

[1036]

ALLEN, M., 17 *Percy Street, Bedford Square.*—Models of plants, showing the blossoms, seed vessels, &c.

[1037]

ANDERSON, R., *Dunkeld, Perthshire.*—Salmon and trout flies.

[1038]

BAILEY, JOHN, Wholesale Manufacturer of Woodware, *King's Cliffe, Northamptonshire.*—Butter-prints, taps, spoons, spice-boxes, &c.; bread waiters, &c.

[1039]

BAZIN, GEORGE, 9 *Denmark Place, Wells Street, Hackney.*—Patent taper swan-quill floats and artificial bait.

[1040]

BELOE, WILLIAM LINTON, *Home Place, Coldstream, Berwickshire.*—Fishing-rods, reels, lines, flies, &c.

[1041]

BERNARD, J., 4 *Church Place, Piccadilly.*—Fishing-rods, tackle, flies, &c.

AGAVA PATENT HAIR COMPANY, *Newlay, near Leeds.*—Fibre of the Agavé, raw and manufactured.

THE AGAVE PLANT.

1. Raw fibre of the agavé.
2. Undyed agava, prepared for stuffing.
3. Dyed agava, prepared for stuffing.
4. Glass box, containing 8 lbs. of agava, under a pressure of 30 lbs.
5. Model mattress with springs of the usual depth, stuffed with agava.
6. The same without springs.
7. Cushion covered with seating, woven and stuffed with agava.
8. Agava prepared for weaving.

The merit claimed for this substance is, that it is a perfect substitute for horse-hair, a long-sought desideratum, and one of growing necessity; indeed, there are few articles for which a substitute has been more needed than horse-hair. The increasing demand for upholstery, mattresses, &c., arising out of the luxurious habits of the time, has so enhanced the price as to render its use in anything like purity almost an impossibility. All kinds of adulteration have been resorted to, and numerous substitutes have from time to time appeared; but not until the substance now exhibited was adapted and perfected, was success achieved. The superior advantages of the agava are these :—It is half the price of the hair generally used; is much cleaner; will more effectually resist moisture; will not become matted; retains its inherent strength and elasticity; and thus entirely removes all excuse for the adulteration of horse-hair with pig and cow-hair—materials which, notwithstanding they are known to be retentive of disease, vermin, and dirt, are now so generally used.

The agava fibre is extracted from the American aloe (*Agavé Americana*), a plant which grows wild in Mexico, and alone supplies this deficiency. It is a stemless plant, provided with large succulent spiny leaves, from the centre of which rises a flower-stalk of considerable height, bearing a magnificent head of large handsome flowers, sometimes as many as 4000 in number. In its native country the leaves are bruised and macerated in water, and afterwards beaten; their fibres are then separated and spun into a strong thread, from which rope, hammocks, fishing-nets, textile fabrics, and articles of clothing are made.

The ancient Mexicans employed it for the manufacture of paper, some of their curious MSS. being written on a material made from the fibre. The celebrated intoxicating beverage named *pulque* is also derived from this and other species of agavé, and from this beverage, again, a strong spirit, denominated *mezikal*, much resembling Scotch whisky, is distilled.

Attention has for some time been attracted to its applicability to various useful purposes; but it was not till a chemical process was discovered whereby its vegetable properties could be destroyed, that it was adopted as a stuffing material. By means of this process the fibre assumes a rounded form, and acquires a degree of strength, elasticity, and softness, previously unknown.

[1042]

BLACHE & Co., 21 *Wilson Street, Finsbury Square.*—Knife-cut veneers; walnut, rosewood, mahogany, and other woods.

[1043]

BLAKE, E., & Co., *Mill Street, Lambeth.*—Flax and Indian fibres, with woven fabrics from the same.

[1044]

BOLLANS, WILLIAM, Wood-Turner and Carver, *King's Cliffe, Northamptonshire.*—Wood turnings and carvings.

[1045]

BURLEY, ROBERT, & Co., *Glasgow.*—Patent steel-core and machine-made handles for hammers, picks, &c.

[1046]

CAMP, WILLIAM, 81 *Tottenham Court Road.*—Arm clubs, American pins and skittles, and other specimens of turning.

[1047]

CHEVALIER, BOWNESS, & SON, 12 *Bell Yard, Temple Bar, W.C.*—Fishing-rods and fishing-tackle.

The exhibitors have always in stock a large selection of superior salmon and trout rods, flies, &c., all of their own manufacture. They can supply complete cases, with sets of tackle, and flies, suitable for India, Canada, Norway, and all parts of the Continent. Their business has been established upwards of a century.

[1048]

CLARK, GEORGE F. H., & Co., *Camomile Street.*—Prepared resinous gums for varnish and hat manufacturing.

[1049]

CLARK & Co., 79 *Cannon Street West, London.*—India-rubber fabrics and felt. (*See page* 94.)

[1050]

CLARKE, JOHN ROBERT, 26 *Trafalgar Street, Walworth.*—Mosaic Tunbridge ware, inlaid with woods in their natural state.

[1051]

CLARKSON, T. C., 56 *Stamford Street, Blackfriars.*—Articles made in cork.

[1052]

CLEMENCE, HENRY, 55 *Upper Stamford Street, Waterloo Road, London, S.*—Specimens of cork, and corks manufactured by hand labour.

Specimens of various descriptions of manufactured corks :—

	Per gross.				Per gross.	
	s. d.	s. d.			s. d.	s. d.
Long claret corks, white	from 3 6	to 7 0	Black daffy corks		„ 0 10	„ 1 3
Port and sherry corks, ditto	„ 3 6	„ 6 6	„ phial ditto		„ 0 4	„ 0 8
Ale and beer corks, ditto	„ 2 0	„ 3 0	Homœopathic corks of all sizes		„ 1 9	„ 2 0
„ „ black	„ 1 9	„ 2 6				
Soda-water corks	„ 2 0	„ 2 9	White and black shives, bungs, and taps supplied at the lowest current prices.			
Ginger-beer ditto	„ 0 8	„ 1 6				
White daffy ditto	„ 1 6	„ 2 6	Wholesale and retail. Merchants and shippers supplied. Established 1851.			
White phial ditto	„ 0 6	„ 1 0				

[1053]

COHEN, CHARLES, 18 *Bury Street, City, E.C.*—Sticks; canes for umbrellas and parasols.

[1054]

COLES, WILLIAM FLETCHER, 52 *Aldermanbury*, and 61 *Paul Street, Finsbury, E.C.*—Cork and its compounds.

The following specimens are exhibited :—

Cork of various kinds and thickness. Cole's patent cork linings for the uppers of boots and shoes. Hat bodies. Thin cork for the inner soles of boots and shoes.

Patent compound cork carpeting (Dunn's patent), plain or figured.

CLARK & Co., 79 *Cannon Street West, London.*—India-rubber fabrics and felt.

The following are exhibited, viz.—

1. Waterproof fabrics of all kinds; including single and double texture cloths and garments, Clark's patent ventilating waterproof garments, artificial card leather, blankets for calico printers, sheeting for waggon covers, and all kinds of vulcanized india-rubber fabrics.

2. Airproof fabrics; including Clark's patent airproof cushions, beds, and mattresses.

3. Vulcanized india-rubber for mechanical purposes; including valves, washers, packing, hose and tubing for steam, water, and gas purposes.

4. Vulcanized india-rubber thread for all descriptions of elastic web.

5. Vulcanized india-rubber in the sheet, in any length and 60 inches wide.

6. Manufactured rubber cut into sheets.

7. Artificial leather for bookbinding, paper-hangings, &c. &c.

8. Clark's patent india-rubber felt, for packing goods in bales and cases, for ship sheathing below copper, &c. &c. This new and valuable material is a combination of cotton wool, or fibres of flax, and india-rubber, forming a durable, cheap and waterproof fabric. The following interesting trial of its peculiar adaptation to the packing of goods was made at Lloyds', in London, and the following is a copy of a certificate signed by fifty members :—

"*Lloyds', London, 28th June.*

" *We, the undersigned members of Lloyds', certify that we have witnessed the following trial of ' Clark's Patent India-rubber Felt.' Two bales of gray shirtings, packed by Messrs. Southgate & Co., were immersed for four days in water, in which a sufficient quantity of bay salt was dissolved, and on opening the bales, the goods packed with the india-rubber felt were found perfectly dry, while the goods packed with tarpaulin were quite saturated with water. The result is so satisfactory to us, that we have great confidence in the india-rubber felt, as a substitute for oilcloth and tarpaulin, and we will in preference insure goods packed in this material.*

" *We must also remark that the experiment of the wooden case lined with the felt, and containing several articles, was as satisfactory as the above.*"

[1055]

COLLYER, ROBERT HANHAM, M.D., F.C.S., *Alpha Road, N.W.*—Paper materials, raw to completed states, with machinery.

[1056]

COSSENS, EDWARD JOSEPH, 15 *Little Queen Street, Holborn.*—The Normandy basket-seller, with baskets carved in elder pith.

[1057]

COSSER, ROBERT, Fancy, Enamelled, and Gold Basket Manufacturer, 13 *Stucley Terrace, Hampstead Road.*—Specimens of fancy basket-work.

The productions of Robert Cosser may be procured from Miner's fancy repository, Lowndes Street, Belgrave Square, or from the manufacturer. They comprise the newest designs in gold rustic flower-baskets, flower-stands, and ornamental stands of every description for the drawing-room. Ladies' work-baskets and work-tables lined with silk or satin, suitable for presentation, R. Cosser decorates basket-work after any design required.

[1058]

COTTON SUPPLY ASSOCIATION, *Manchester.*—Cotton samples, and cotton tree.

[1059]

COW, P. B., & Co., 46 *Cheapside, London.*—India-rubber waterproof fabrics, and vulcanized india-rubber goods.

[1060]

DAHMEN, M. A. J., *Park Villa, Peckham.*—Fibre and vegetable substances connected with textile fabrics and paper.

[1061]

DANKS, J., 56½ *Webber Row, S.*—New invented door mats, made of cocoa-nut fibre and wool.

[1062]

DEED, JOHN S., & SONS, 451 *Oxford Street, London.*—Cocoa mats, matting, and worsted hearth-rugs.

John Deed & Son are engaged in the several businesses of curriers, morocco, roan, skiver and calf leather dressers, manufacturers of sheep and lamb skin wool rugs, cocoa mats, matting, &c.

Examples of some of these manufactures will be found in Class XXVI., the following being exhibited in Class IV.

1. Specimens of mats for doors and entrance halls, made entirely from the fibre of the cocoa-nut.

2. Specimens of matting for churches, public buildings and offices, made entirely from the fibre of the cocoa-nut.

3. Specimens of mats made from cocoa-nut fibre, with coloured yarn or worsted borders in fancy designs.

4. Yarn and worsted hearth-rugs made to match carpets, tesselated pavements, &c.

[1063]

DIMSDALE, T. J., *Forest Lane, Forest Gate, Essex.*—Vegetable fibres for making paper, and paper made therefrom.

[1064]

DUFFIELD, JAMES, 12 *Great Chapel Street, Oxford Street, London, W.*—Embroidery and butter-stamp, pastry stand, gum-paste mould, and dairy utensil manufacturer.

[1065]

DUFFIELD, JOSEPH, 28 *Brandon Street, Walworth.*—Oval and round carved butter-stamps, and beaters with impressions.

[1066]

EVERARD, H. W., *Union Mills, Manchester.*—Vulcanized india-rubber brace, surgical, and other webs; braces, belts, &c.

[1067]

FARLOW, CHARLES, 191 *Strand.*—Improved fishing-rods and tackle; artificial bait; winches; swivels; split cane rods.

[1068]

FARRANT, RICHARD E., 16 *Queen's Row, Buckingham Gate.*—Carved bread, butter, and cheese plates; potato bowls.

[1069]

FAUNTLEROY, ROBERT, & Co., 99 & 100 *Bunhill Row, Finsbury, London, E.C.*—Foreign hard woods, dye-woods, fancy woods, &c.

Robert Fauntleroy & Co. exhibit a large model of the west front of the Royal Exchange, constructed of various specimens of hard and other woods, to the number of three hundred or more, together with the corozo nut or vegetable ivory, coquilla, cahoan and betel nuts. The whole grouped and arranged so as to display the intrinsic merit of each, for turnery and other purposes.

[1070]

FAUNTLEROY, ROBERT, & SONS, *Potter's Fields, Tooley Street, London.*—Foreign hard woods, ivory, and mother-o'-pearl shells.

[1071]

FORSTER, T., *Streatham, Surrey, S.*—Articles in vulcanite (ebonite), made from vulcanite india-rubber waste.

The exhibitor is the patentee of a mode of utilizing india-rubber waste.

The whole of the black articles exhibited are produced from waste india-rubber, and will be found not inferior to those manufactured from the best india-rubber, though the price is considerably less.

The coloured samples for dentists' use are made from the best materials that can be obtained.

[1072]

GATES, T. F., 31 *Lower Belgrave Street, Pimlico.*—Anatomized leaves.

[1073]

GIEHR, ROBERT, 4 *George's Row, City Road.*—Chairs and fancy baskets.

[1074]

GOGGIN, CORNELIUS, 13 *Nassau Street, Dublin.*—Irish bog-oak ornaments.

[1075]

GOUGH & BOYCE, 12 *Bush Lane, London.*—Kamptulicon—an improved elastic floor-cloth, warm, noiseless, and durable.

[1076]

GOULD, ALFRED, 268 *Oxford Street.*—Fishing-rods of cane, hickory, and other woods; eel traps, &c.

[1077]

GOWLAND & Co., 3 *Crooked Lane, London Bridge.*—Every description of fishing tackle.

[1078]

GUTTA PERCHA COMPANY, *Wharf Road, City Road.*—Articles in gutta percha. (*See page* 97.)

[1079]

HANCOCK, JAMES LYNE, 266 *Goswell Street, London, E.C.*—Vulcanized india-rubber for manufacturing, scientific, and domestic purposes.

[1080]

HAWE, J., 7 *Adelphi Terrace.*—Preserved natural flowers.

[1081]

HEEKS, MARGARET HANNAH, 61 *White Lion Street, Pentonville, N.*—Wicker baskets of every description, including a balloon car.

[1082]

HEINRICH, J., 36 *Lower Kennington Lane, S.*—Combs.

GUTTA PERCHA COMPANY, *Wharf Road, City Road.*—Articles in gutta percha.

[*Obtained the Council Medal at the Great Exhibition of 1851.*]

APPLICATIONS OF GUTTA PERCHA.

TUBING.

For conveyance of water.
Conveyance of chemicals.
Conveyance of liquid manure.
Watering gardens and streets.
Washing carriages, windows, &c.
Sprinkling water in maltings.
Suction pipes for fire-engines.
For ventilation.
Syphons.
Hearing apparatus for the deaf in churches and chapels.
Speaking-tubes in counting-houses, warehouses, shops, public institutions, on shipboard, and in mines.
Domestic telegraph in lieu of bells in private houses.
The medical man's midnight friend.
Speaking apparatus for omnibuses.
Railway conversation tubes.
Hogar pipes for mines.
Alarum tubes for ditto.
Union joints and elbow pipes.

DOMESTIC, &c.

Soles for boots and shoes.
Chamber service.
Window blind cord, clothes' line.
Lining for bonnets.
Wine coolers.
Foot baths.   House pails.
Noiseless curtain rings.
Ear trumpets, cornets.

FOR PUBLIC ESTABLISHMENTS.

Viz: Hospitals, Asylums, Workhouses, Schools, Prisons, &c.
Bowls and soap dishes.
Water jugs and basins.
Drinking-cups.   Fire buckets.
Chamber utensils.
Speaking tubes.
Night pans, bed ditto, bed slips.
Waterproof canvas.

ELECTRICAL, &c.

Covering for electric telegraph wires.
Insulating stools.
Battery cells.
Handles for discharging rods.
Electrotype moulds.
Galvanic batteries.

SURGICAL.

Splints.   Caustic holders.
Thin sheet for bandages and dressings.   Stethoscopes.
Ear trumpets.   Bed straps.
Bed pans and bed slips for invalids.   Pessaries.
Medical man's midnight friend.   Vagina tubes.
Male and female urinals.

CHEMICAL.

Carboys.   Stopcocks.
Vessels for acids, &c.
Syphons.   Lining for tanks.
Tubing for conveying oils, acids, alkalies, &c.
Flasks, bottles, jugs.   Acid pumps, pourers, and scoops.
Funnels.

FOR OFFICES, &c.

Inkstands.   Ink cups (in lieu of glass).   Pen trays.
Cash bowls.   Tubes for conveying messages.
Architects' and surveyors' plan cases.
Washing basins, &c.

MANUFACTURING.

Buckets.   Mill bands.
Pump buckets, valves, clacks, &c.   Washers.
Pumps for acids.   Oil cans.
Felt edging for paper makers.
Bosses for flax mills.   Flax holders.   Shuttle beds for looms.   Covers for rollers.
Bowls for goldsmiths.
Round bands and cord.
Breasts for water wheels.
Cutting boards for glove makers.

GUTTA PERCHA COMPANY—*continued.*

AGRICULTURAL.

Tubing for conveying liquid manure. Stable buckets.
Spreaders for liquid manure.
Lining for manure tanks.
Driving bands for thrashing-machines, &c.
Stuffing for horses' feet.
Probangs for cattle. Whips.
Dumb jockeys. Saddle brackets, anti-crib-biters.
Bridle and harness hooks.

MINING.

Hogar pipes. Miners' caps.
Speaking-tubes. Syphons.
Tubes for ventilation.
Pump buckets. Valves and clacks. Alarum tubes.

DECORATIVE, &c.

A variety of mouldings in imitation of carved oak, rosewood, &c., for the decoration of rooms, cabinet work, &c. Brackets.
Picture frames. Mirror frames.
Daguerreotype frames.
Mourning card frames.

FANCY ARTICLES.

Counter trays. Baskets.

Whips. Vases, shells.
Watch stands.
Ornamental inkstands.
Card, fruit, pin, and pen trays.
Bouquet holders. Paper weights. Bread trays.
Biscuit trays. Toilet trays. Vine trays. Cotton trays.
Pin cushions. Decanter stands. Snuff-boxes.
Tobacco boxes.

MISCELLANEOUS.

Fire buckets. Tap ferules.
Coloured material for amateur modelling.
Cricket, bouncing, and golf balls. Police staves.
Guards for fencing sticks.
Life-preservers. Paper for damp walls. Beds for paper cutting machine knives.
Fringe for mourning coaches.
Skates. Bottling boots.
Corrugated sheet for wine packing. Talbotype trays.
Official seals. Dolls.
Powder flasks.
Collodion baths and dippers.
Washers for carriage wheels. Ditto for cold water pipes.
Welting cord for ladies' dresses. String boxes.
Chessmen. (May be used for the game of draughts.)

Manufactured by the Gutta Percha Company, Patentees, and sold by their wholesale dealers in town and country.

[1083]

HINKS, JOSEPH, 64 *George Street, Birmingham.*—Hard and soft wood turnings.

[1084]

HODGES, R. E., 44 *Southampton Row, Russell Square.*—Patent india-rubber accumulators or springs.

These springs stretch to six times their normal length, and are made of any degree of strength, from 1 lb. up to 500 horse-power. They are used in printing, agricultural, sawing, and other machinery. They are also used for mineral-boring and well-sinking, working at very high velocities instead of counter-balancing weights; and for giving the softness of a spring to rotary machinery, drums, cylinders, &c. They are largely used for door springs. They can be adapted for driving machinery, boats, light locomotives, &c.; for cable and towing springs; and for preventing jerk, jolt, jar, shock, and vibration generally.

[1085]

HOLLINGSWORTH & WILLOUGHBY, 2 & 3 *Wenlock Road, N.*—Veneers cut by their patent knife machinery.

[1086]

HOOPER, WILLIAM, 7 *Pall Mall East, S.W.*—Vulcanite and vulcanized india-rubber goods.

[1087]

HORSEY, JAMES, 36A *Belvidere Road, London, S.*—Articles in india-rubber, plain and coloured, for personal use, &c.

[1088]

HOWARD, J., *London Road, Luton, Beds.*—Blocks for shaping ladies' hats and bonnets.

[1089]

HYAMS, MICHAEL, *Bath Street, London.*—Prepared thistle-down—proposed substitute for silk, and for other useful purposes.

[1090]

HYDE, EDMUND, *Kingston-on-Thames.*—Barsham's patented cocoa-nut fibre brushes, and mats made therewith.

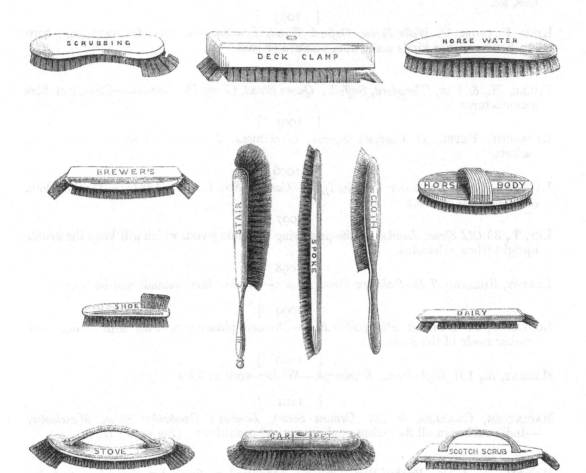

Brushes of every description are made of Barsham's Patent Cocoa Fibre. Their great durability, and very moderate price, have established their value and caused an increasing demand for them. To prevent imposition by brushes of an inferior quality being sold for Barsham's, every patent fibre brush is stamped

"*J. Barsham's Patent, Kingston-on-Thames.*"

Mats made of Barsham's Patent Cocoa Fibre are also in great repute.

[1091]

JAMES, JOHN, Jun., 1 *Cleveland Terrace, Bath.*—New models of basket-work.

[1092]

JONES & Co., 111 *Jermyn Street, London, S.W.*—Salmon and trout rods, reels, lines, flies, &c.

[1093]

KING, FRANCIS, 56 *Wells Street, Oxford Street, London.*—Brooms for sweeping; horse brushes and other kinds made from piassava or bass.

[1094]

KOLLE, H., & SON, *Glemsford, Suffolk; Queen Street, Cheapside, London.*—Cocoa-nut fibre manufactures.

[1095]

LATARCHE, PETER, 18 *Coldbath Square, Clerkenwell, London.* — Wickered flasks and baskets.

[1096]

LEATHER-CLOTH COMPANY (Limited), 56 *Cannon Street West, London.*—Leather-cloth. (*See pages* 102 & 103.)

[1097]

LEE, T., 33 *Old Street, London.*—Life-preserving swimming-vest, which will keep the wearer upright when exhausted.

[1098]

LENTON, RICHARD, 7 *Bartholomew Street, Exeter.*—Wicker flower-stands and bird-cages.

[1099]

LUDBROOK, S., *Bancroft Place, Mile End.*—Dressed piassava or bass, with brooms and brushes made of the same.

[1100]

MACKAY, A., 107 *High Street, Edinburgh.*—Wicker-work articles.

[1101]

MACINTOSH, CHARLES, & Co., *Cannon Street, London; Cambridge Street, Manchester.*—India-rubber in all its various applications and conditions. (*See page* 101.)

[1102]

McNEILL, F., & Co., *Bunhill Row, London.*—Asphalted roofing, ship sheathing, and dry hair felts; compound vulcanized rubber for steam joints; kamptulicon.

[1103]

MADDEN, SUSANNA, 56 *Long Lane, West Smithfield, E.C.*—Skittles; skittle and round balls.

[1104]

MASON, G., Esq., *Yately, Hants.*—Specimens of flax and silk cultivated at Yately, Hants.

[1105]

MEYERS, B., *Mill Lane, Tooley Street.*—Canes, sticks, whips, &c.

[1106]

MORLEY, JOHN, 12 *Carrington Street, Nottingham.*—Artificial salmon and trout flies.

[1107]

MORRIS, CHARLES, 4 *Mountnod Square, Lewisham Road, Greenwich.*—Combs, comb-making tools, and fancy baskets.

[1108]

NOBLE, G. & J. A., 4 *George Yard, Lombard Street.*—Textile fibres.

MACINTOSH, CHARLES, & CO., *Cannon Street, London; Cambridge Street, Manchester.*—India-rubber, in various applications and conditions.

[*Council Medal awarded at the Great Exhibition of 1851.*]

Charles Macintosh & Co. are the patentees of the vulcanized india-rubber, and manufacturers in general of caoutchouc articles.

The following is a summary of the articles exhibited :—

IN CLASS IV.

India-rubber, raw, and in progressive stages of manufacture; varnishes; waterproof and air-proof fabrics; elastic thread and general india-rubber manufactures.

CLASS V.

Railway buffers and bearing springs; carriage blocks and springs; wheel tires; locomotive hose, &c.

CLASS VIII.

Mechanical articles for stationary and marine engines; joint rings; printers' blankets; artificial card leather; hose; tubing, &c., &c.

CLASS XVII.

Surgical and hospital instruments and apparatus; vulcanite dental rubber, chemical articles, &c.

CLASS XXVII.

Waterproof clothing; military, naval, travelling, sporting, and veterinary.

CLASS XXIX.

Educational appliances; inflated globes, maps, raised types for the blind, elastic bands, and other stationery.

For details see C. M. & Co.'s illustrated descriptive handbook of their manufactures.

3 Cannon Street West, London, E.C.; and Cambridge Street, Oxford Street, Manchester (C. M. & Co.'s only establishments).

LEATHER-CLOTH COMPANY (Limited), 56 *Cannon Street, West, London.*—Leather-cloth.

The Leather-Cloth Company (Limited) are the sole Manufacturers of Crockett's Leather-Cloth. Patent printed and gilded Leather-Cloth. Embossed Leather-Cloth.

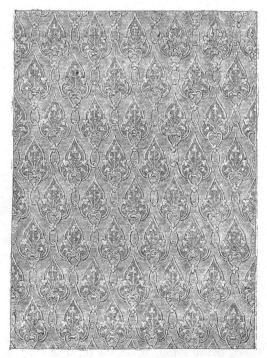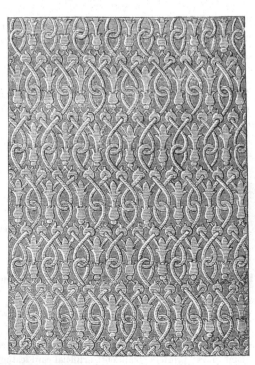

PATTERNS OF LEATHER-CLOTH FOR WALL HANGINGS, &c.

LEATHER-CLOTH COMPANY (Limited)—*continued.*

 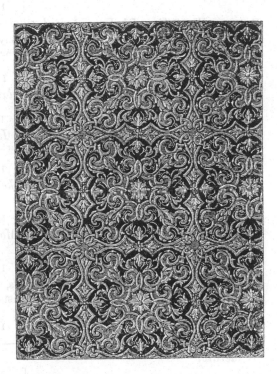

PATTERNS OF LEATHER-CLOTH FOR WALL HANGINGS, &c.

These articles are extensively used both in this country and on the Continent for wall hangings, table covers, the seats of chairs and couches, for lining carriages, for fancy bags, hassocks, and numerous other purposes.

Many of the patterns of printed, gilded, and embossed Leather-Cloth combine all the beauty of gilded leathers with far greater durability and at about one tenth of the cost.

A. Lorsont, Managing Director. Warehouses—56 Cannon Street West, London. 104 Boulevard de Sebastopol, Paris. Works—West Ham, Essex.

[1109]

NORTH BRITISH RUBBER COMPANY (Limited), *Edinburgh.*—India-rubber manufactures; boots, shoes, and rubber for mechanical purposes.

[1110]

OLIVER, WILLIAM, & SONS, 120 *Bunhill Row, Finsbury, London.*—Specimens of fine mahogany and rare foreign woods.

[1111]

PACKER, ROBERT LEWIS, 38 *Union Street, Lambeth Walk, London, S.*—Improved glove stretchers, and powder-boxes.

[1112]

PARKES, ALEXANDER, *Birmingham.*—Patent Parkesine of various colours; hard elastic, transparent, opaque, and waterproof.

[1113]

PEACH, J., & SONS, *Derby.*—Derby silk lines; registered improved salmon line.

[1114]

PETERS, W., & SON, 71 *Long Acre, London.*—Fishing-tackle.

W. Peters & Son manufacture flies and baits of every description upon the most approved principles. They hold the appointment of fishing-rod and tackle makers to Her Majesty the Queen.

[1115]

PILLINER, S. A., 4 *Hatfield Place, Blackfriars.*—Anatomized leaves.

[1116]

PLUMMER, STEPHEN, 84 *Church Road, Islington.*—Models of St. Paul's and Salisbury cathedrals.

[1117]

RAYNBIRD, HUGH, Land Agent, *Basingstoke.*—Specimens of timber, bark, hoops, &c., from Hampshire woods and coppices.

[1118]

RECKITT & Co., *Eureka Works, Hulme, Manchester.*—Patent American leather-cloth, and table baize.

[1119]

ROBERTSON, ALEXANDER, *Holloway Mills, London, N.*—Patent barrel-package of wood, improved substitute for tinned-iron canisters.

[1120]

ROUTLEDGE, T., *Eynsham Mills, Oxford.*—Esparto, or alfa, and half-stuff for paper manufacture.

[1121]

SCOTTISH VULCANITE COMPANY (Limited), *Edinburgh.*—Patent vulcanite (hard rubber and gutta-percha) combs, whalebone substitute, &c.

This company manufactures patent vulcanite combs, knife-handles, whalebone substitute, &c. This is a new and important manufacture of india-rubber and gutta percha, destined to be of permanent and almost universal adaptation. It at once supplants all the appliances of whalebone. Being infinitely more durable, and susceptible of a higher finish, it receives impressions as clear and sharp as the finest carved ivory, is capable of being worked out in an endless variety of elaborate designs, and has also the recommendation of great economy. This compound is fitted to take the place of the following substances, viz.: enamel, ivory, buckhorn, whalebone, &c. It not only makes a good substitute for these material but is also in reality superior in quality, in some respects, to the natural substances. The hardest compound resembles marble, that which is less hard ivory and buckhorn, and that which is still softer buffalo-horn and whalebone. In general it possesses more durable properties than any of these, except marble, and is even more substantial in some respects; because in all degrees of hardness, it has a great degree of toughness or tenacity, and the property of retaining the shape into which it has been moulded and heated.

[1122]

SCOTT, WENTWORTH LASCELLES, *Westbourne Park, Bayswater, W.*—Specimens of cotton, in "fasciculæ," showing length of staple.

[1123]

Seithen, Anton Bruno, 1 *Wharf Road, City Road.*—New manufacture of corks, and apparatus for grinding corks in lieu of cutting.

[1124]

Shepherd, Briggs, & Co., *Portobello Mills, Wakefield.*—Cocoa-fibre and Manilla mats and mattings.

[1125]

Silver, S. W., & Co., 66 and 67 *Cornhill.*—Articles in india-rubber and ebonite.

[1126]

Simmonds, Peter Lund, 8 *Winchester Street, Pimlico.*—Collection of nuts, seeds, fibres, &c., scientifically named, and their applications.

[1127]

Skilbeck, J., *Upper Thames Street.*—Woods and articles used in dyeing.

[1128]

Smee, William, & Sons, 6 *Finsbury Pavement, London.*—Specimens of woods used in the manufacture of household furniture.

[1129]

Smith, Thomas, & Sons, *Herstmonceux, Hurst Green, Sussex.*—Basket manufactures.

[1130]

Smith, William & Andrew, *Mauchline, Ayrshire, and 61 Charlotte Street, Birmingham.*—Scottish fancy wood-work.

[1131]

Spill, George, & Co., *Hackney Wick, E.C., and 149 Cheapside, London, E.C., and 9 High Street, Bristol.*—Vegetable leather; leather-cloths; waterproof fabrics; and machinery band manufacturers. (*See page* 106.)

[1132]

Steinitz, Charles, *London Parquetry Works, Grove Lane, Camberwell, S.*—Collections of exotic furniture woods, and of diaphanic woods.

[1133]

Stevens, M., *Royal Mews, Pimlico.*—Anatomized leaves.

[1134]

Stevens, W., 14 *Great Russell Street, Bloomsbury.*—Wax figures and flowers.

[1135]

Swaab, S. L., Oculist, 9 *Hunter Street, Brunswick Square.*—Prepared India fibres, flax, hemp, and fibres converted in silk and cotton.

[1136]

Tayler, Harry, & Co., 19 *Gutter Lane, Cheapside, London;* Works, *Deptford Green.*—Kamptulicon for floors, knife-boards, lunatics' cells, and horse-boxes. (*See page* 107.)

[1137]

Taylor, Benjamin, 169 *St. John Street Road.*—Vegetable ivory turnings.

[1138]

Toplis, T. & J., *Ashby-de-la-Zouch.*—Flower stands, work-baskets, &c.

SPILL, GEORGE, & Co., *Hackney Wick, N.E.,* and 149 *Cheapside, London, E.C.,* and 9 *High Street, Bristol.*—Vegetable leather; leather-cloths; waterproof fabrics; and machinery-band manufacturers.

INTERIOR VIEW OF WORKS. ONE OF THE MACHINERY ROOMS.

Nos. 1 to 10.—Spill's enamelled vegetable leather for carriage hoods, knee boot aprons, railway carriage cushions, antigropelos, gaiters, and military accoutrements, made of any colour or substance.

Nos. 11 to 20.—Morocco vegetable leather for carriage ini ngs, furniture covering, children's and ladies' shoes reticules, bags, office table-covers, &c. &c., made of any colour or substance.

No. 21.—India-rubber waterproof mineralized overcoat, or pocket siphonia, of superfine India cloth, weighing only six ounces, warranted to withstand any degree of heat under 400 Fahrenheit.

No. 22.—India-rubber mineralized overcoat, light-coloured surface for tropical climates, non-attractive of heat, and very durable.

No. 23.—Waterproof vegetable leather military cape, scarlet, made without sewing, and the material without weaving; very durable, and warranted suitable for any climate. It can be made to any other regimental uniform colour.

No. 24.—Enamelled vegetable leather antigropelos for riding, with side "steel spring" fastenings.

No. 25.—Enamelled vegetable leather gaiters for walking, with side "steel spring" fastenings.

No. 26.—Enamelled vegetable leather gaiters for riflemen, with knee-cap cushion for rifle practice, with side "steel spring" fastenings.

No. 27.—Enamelled and morocco vegetable leather buskins for walking (in colours), with side fastenings of buttons or steel springs.

No. 28.—Patent improved machinery belting, made to any length in one piece, and in any width up to twelve inches, warranted not to be affected by heat, grease, or water, will not stretch or slip on the pulleys, and is exceedingly strong, every inch in width of No. 1 quality will sustain a weight of 2000 lbs.

No. 29.—Patent improved machinery belting, No. 2 quality, every inch in width will sustain a weight of 3000 lbs.

No. 30.—Patent improved machinery belting, No. 3 quality, every inch in width will sustain a weight of 4000 lbs.

Manufacturers of improved vulcanized and mineralized india-rubber garments and piece goods, patent machinery bands, enamelled and morocco vegetable leather, and leather-cloth fabrics, waterproof oil clothing, sou'-westers, aggon and rick covers, and vegetable leather gaiters. Japanners and embossers.

TAYLER, HARRY, & CO., 19 *Gutter Lane, Cheapside, London;* Works, *Deptford Green.*— Kamptulicon for floors, knife-boards, lunatics' cells, and horse-boxes.

PATTERN OF KAMPTULICON.

KAMPTULICON is a felted article composed of india-rubber, gutta-percha, and cork, and is applied to numerous purposes, such as covering floors, knife-boards, the cells of lunatics, horse-boxes, and for the packing of railway chairs, &c. For floors it is usually made plain and of a light-brown colour; but it may be coloured of any tint to suit the taste of customers, or ornamented with designs of Egyptian, Grecian, Etruscan, or mediæval character. The pattern is printed, leaving the original surface as much as possible exposed, by which it is rendered a medium warmth between carpet and oil-cloth. As a covering for knife-cleaners it possesses all the advantages of leather at about one-fourth of the cost. For lunatics' cells—the walls and floors being covered with kamptulicon, if from half an inch to one inch in thickness, the resiliency of the material prevents the inmates doing themselves any personal injury, while, from its being a non-conductor of heat, it conduces to the maintenance of an equable temperature. It is already adopted by the governors of Bethlehem Hospital, and of some other asylums. It is of great service for lining the boxes or covering the backs of the stalls of kicking horses. By deadening the sounds of the blows, it has a great tendency to cure this vicious habit, while, by its elasticity, it prevents injury to the horse itself. It is used in the royal stables, and in those of many noblemen and gentlemen. It also makes an admirable floor for riding-schools; preventing noise, lessening the shocks in the falls of riders, and saving the horses' feet the concussion of hard pavement.

[1139]

TRELOAR, THOMAS, 42 *Ludgate Hill, London.*—Mats, matting, rugs, brushes, hassocks, &c., of cocoa-nut fibre.   (*See page* 109.)

[1140]

TRESTRAIL, F. G., & Co., 19 & 20 *Walbrook.*—Kamptulicon, or india-rubber and cork floor-cloth.   (*See page* 110.)

[1141]

TUCK, J. H., & Co., 35 *Cannon Street, London, E. C.*—Patent elastic packing and rubber manufactures, for steam-engines and other mechanical purposes.

No. 2.

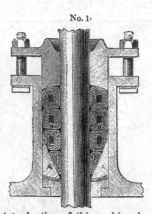

No. 1.

Before the introduction of this packing, hemp, plaited or otherwise, was the material most commonly employed in the stuffing-boxes of steam-engines, &c. This plaited hemp requires frequent removal, otherwise it becomes hard, injuring the rod or moving surface, and even when an excessive quantity of tallow or oil is used for lubrication it is extremely difficult to maintain a good vacuum. To meet these very serious objections the elastic core packing was invented. It consists of a roll of properly prepared canvas, having an elastic core or centre; this roll is cut and bent into rings, as shown in Diagram No. 2. The packing thus made when used in connection with the metallic cone or lining fitted into the bottom of the stuffing-box (the object of which is to bring the packing directly against the rod or rubbing surface), produces a better vacuum, reduces the friction, effects a great saving in oil and tallow, does not become hard, nor does it require drawing, but is gradually worn away.

Diagram No. 1 is a section of a stuffing-box packed with the patent packing A, and having a lining B inserted in the bottom.

Diagram No. 2 shows the patent packing and lining ready to be placed in the box.

[1142]

TURNBULL, T., *William Street, Portland Town, St. John's Wood.*—Specimens of wood sawn by an improved method.

[1143]

WALDEN, SAMUEL J., *Whitefriars, E.C., and Walham Green, Fulham, S.W.*—A variety of articles in wicker-work; baskets, tables, and chairs.

[1144]

WALKER & STEMBRIDGE, *Ducksfoot Lane, London.*—Gums and gum-resins of every description for manufacturing purposes.

Having devoted their attention for many years past exclusively to gums and gum resins, the exhibitors are prepared to supply (wholesale only) every description used in the arts and manufactures on the best market terms.

Shellac.

Sticklac.

Seedlac.

Animi.

Copal.

Damar.

Arabic.

Senegal.

Tragacanth.

Sandrac.

Mastic.

Benzoin, &c. &c.

Agents.—Manchester : Mr. Underwood, 20, Greenwood Street.

Paris : Mr. James Watt, 15 Rue de l'Exchequier.

Hamburg : Messrs. Steffensen & Co., 46 Brauer Strasse.

[1145]

WANSBOROUGH, JAMES, *Grove, Guildford Street, Southwark.*—Waterproof flocked cloth, and hard and soft india-rubber goods.

[1146]

WARNE, WILLIAM, & Co., 9 *Gresham Street West, London, E.C., and Tottenham.*—Manufactures of india-rubber.

TRELOAR, THOMAS, 42 *Ludgate Hill, London.*—Mats, matting, rugs, brushes, hassocks, &c. of cocoa-nut fibre.

[*Obtained Prize Medals*—*London*, 1851; *New York*, 1853; *Paris*, 1855; *Brussels*, 1856.]

1. Matting, plain and with figured borders, for covering halls, passages, waiting-rooms, offices, aisles of churches, and public buildings. It possesses extraordinary durability, and is not affected by damp or wet.

2. Mats and rugs for doorways, railway carriages, &c., plain and with figured worsted borders.

3. Mats and rugs in ornamental borders.

4. Pompeian mats, "Salve," and "Cave Canem."

5. Hassocks and kneelers for church use.

6. Brushes for household and stable use.

7. Mattresses filled with prepared cocoa-nut fibre, as a substitute for horse-hair.

DOOR MAT.

DOOR MATS.

TRESTRAIL, F. G., & Co., 19 & 20 *Walbrook*.—Kamptulicon, or india-rubber and cork floor-cloth.

PATTERN OF KAMPTULICON.

Thick, plain	5s.	
Ditto, printed	5s. 6d. to 5s. 9d.	Per
Thin, plain	4s.	square
Ditto, printed	4s. 6d. „ 4s. 9d.	yard.

F. G. Trestrail & Co.'s Patent Coloured Kamptulicon, manufactured with F Walton's patent india-rubber substitute, is impervious to wet, indestructible by damp or heat; soft, noiseless, and warm to the feet. It is far superior to any other material ever invented for the covering of floors, and is especially adapted for the aisles of churches, halls, public offices, railway stations, libraries, smoking, billiard, and bath rooms, &c., &c. It is made plain, coloured, or figured in imitation of carpets, mosaics, or other pavements; and also in a variety of different patterns expressly designed for the material.

This Kamptulicon differs from all others in this important particular, that it is coloured right through, and therefore instead of having the appearance which the ordinary material has when worn, it will, with occasional washing, preserve its colour to the last.

Warehouses—19 & 20 Walbrook, E.C. Manufactories—South London Works, Lambeth, and Chiswick.

[1147]

WELLS & HALL, 60 *Aldermanbury.*—Elastic braids and fabrics.

Messrs. Wells and Hall's patent vulcanized india-rubber cords, braids, and webs, are made from Charles Macintosh & Co.'s super thread, which is durable, and also permanently elastic.

[1148]

WEST HAM GUTTA-PERCHA COMPANY, 18 *West Street, Smithfield.*—Gutta-percha ; gutta-rubber ; telegraph wire.

[1149]

WHITEHEAD, THOMAS, 37 *Eastcheap, London.*—Straw envelopes for packing glass bottles.

[1150]

WILDEY & CO., 7 *Holland Street, Blackfriars Road, London.*—Mats, matting, &c., of cocoa-nut fibre.

[*Obtained Prize Medals—London,* 1851 ; *New York,* 1853 ; *and Paris,* 1855.]

The following preparations of the fibre of the outer husk of the cocoa-nut, and articles manufactured from the same are exhibited :—

1. In a curled state, to be used as stuffing for mattresses, chairs, sofas, carriages, &c., substitute for horsehair, wool, and other substances. Its peculiar qualities are durability, cleanliness, cheapness, and salubrity.

2. In a drawn state, to be used as a substitute for bristles in making brushes and brooms, both for household and stable purposes.

3. Fibre prepared for spinning.

4. Yarns and cordage spun from fibre by machinery in this country, and also by the Cingalese.

5. Floor-mattings, woven by hand and power-loom, as supplied to Her Majesty's Office of Works, for Palaces, Public Buildings, and Government Offices.

6. Plain and ornamental door and other mats.

7. Netting for sheep-folds.

8. Thatching cord.

9. Nosebags for horses.

10. Cider cloths.

11. Mats for oil pressing.

[1151]

WILSON, A. & G., 19 *Waterloo Place, Edinburgh.*—Variety of fishing-tackle, consisting of rods, reels, lines, gutwork, and flies.

[1152]

WRIGHT, C., 376 *Strand, W. C.*—Fishing-rod, tackle, and archery.

[1153]

WRIGHT, J., *Kelso, Scotland.*—Artificial flies and casting lines.

SUB-CLASS D.—*Perfumery.*

[1163]

ATKINSON, JAMES & EDWARD, 24 *Old Bond Street.*—Perfumery and articles for the toilet.

[1164]

BAYLEY & CO., 17 *Cockspur Street.*—Perfumed essences, oils, distilled waters, pomades, creams, and toilet soaps. (*See page* 112.)

[1165]

BENBOW & SON, 12 *Little Britain, E.C.*—Perfumery and toilet articles.

BAYLEY & Co., 17 *Cockspur Street.*—Perfumed essences, oils, distilled waters, pomades, creams, and toilet soaps.

Bayley & Co., perfumers to the Royal Family and foreign Courts, manufacture the following articles, samples of which are exhibited :—

ESS. BOUQUET.

From this perfume becoming the peculiar favourite of his Majesty George IV., arose many imperfect imitations of the article, which continue to be sold under the name of Bouquet du Roi, Esprit de Bouquet of George the Fourth, &c., but the ESSENCE BOUQUET, exclusively prepared by Bayley & Co., Cockspur Street, London, is the only article entitled to those appellations, and which possesses an unrivalled and distinct fragrance. This perfume has become a favourite in many foreign courts and cities.

Bayley & Co. make "the Ess." of one quality and one price only. The following perfumes are also of their peculiar preparation :—

BOUQUET DE LA REINE VICTORIA.

Jockey Club Bouquet.		Almond Blossoms	
Bridal	„	Esprit Victoria	
Army & Navy	„	„ Albert	
Balmoral	„	„ Unis	
Wellington	„	„ du Château	
Cavalry	„	„ Magnolia	
Court	„	„ de Fleurs	
Prince of Wales	„	„ Oriental	
Windsor	„	„ Verveine	
Esterhazy	„	„ Vetivert	
Kensington	„	„ Muguet	
Princess Alice	„	„ Réséda	
Princess Royal	„	„ Jasmin	
Empress	„	„ de Tubereuse	
L'Empereur	„	„ Fleur d'Orange	
Sweet Briar	„	„ Mousseline	
New Mown Hay	„	„ Violette Double	
Cuir de Russie	„	Esprit de Rose	
Spring Flowers		White Rose	
Forest Flowers		Provence Rose	
Summer Blossoms		Essence de Rose Mosscuse	

Essence of Maréchale
„ of Geranium
Essence Frangipane
„ Souveraine

Double Essence of the Wood Violet
Extrait Chypre

Extrait de Patchouli
Eau de Portugal
„ Miel
„ d'Hongrie de Montpellier
Eau Suave

Eau à Brûler
Bois de Santale
Lavender water, pts., ½ pts., ¼ pts.
Honey water

Otta of rose soap tablets		Rose soap tablets.	
Spermaceti	„	„	A l'Ambre Musque soap
Ess. Bouquet	„	„	tablets.
Almond	„	„	Glycerine soap tablets.
Violet	„	„	Brown Windsor „
Orange Flower	„	„	White „ „
Winter	„	„	Palm „
Indian	„	„	

Spermaceti shaving tablets
„ „ paste

Hanover shaving paste.
Italian cream.

Windsor soap, white
„ „ brown, highly perfumed
„ „ Sims's old

Hemet's Essence of Pearl for the teeth and gums.
„ Pearl Dentrifrice „ „

Cold cream, in pots
Pomade Divine, for bru'ses, &c.
Honey paste

PREPARATIONS FOR THE HAIR.

Aroma al Cariense
Marrow pomade oil
Wood violet

White rose.
Oriental
Bears' grease.

[1166]

BONUS, WILLIAM E., 9 *Charles Street, Manchester Square.*—Fruit essences; ancient and modern hair dyes; cantharidine.

[1167]

BREIDENBACH, F. H., 157B *New Bond Street.*—Perfumery.

[1168]

CLEAVER, F. S., 32 & 33 *Red Lion Street, Holborn, W.C.*—Fancy soap and perfumery.

[1169]

CONDY, BROTHERS, & CO., 15 *Garlick Hill, London.*—Essential oils and extracts; artificial flavourings and fruit essences.

[1170]

DELCROIX & SON, 39 *Great Castle Street, Regent Street, London.*—Perfumes, pomades, oils, cosmetics, sachets, and Eau-de-Cologne.

[1171]

EDE, R. B., & CO., 21 *Bow Lane, London, E.C.*—Perfumery and domestic requisites.

[1172]

EWEN, JAMES, 17 *Garlick Hill, London.*—Clarified fats for chemical, culinary, and perfumery purposes.

[1173]

GOSNELL, JOHN, & CO., 12 *Three King Court, Lombard Street.*—Perfumery and soaps; hair brushes and other kinds of brushes.

[1174]

HIRST, BROOKE, & TOMLINSON, *Leeds.*—Perfumed toilet soaps and perfumery.

[1175]

KEITH, GEORGE, 55 *Great Russell Street, Bloomsbury.*—British perfumery and freezing-powders for hot climates.

[1176]

LANGDALE, EDWARD FREDERICK, Distiller of Essential Oils, 72 *Hatton Garden.*—A collection of oils.

[1177]

LEWIS, JAMES, 6 *Bartlett's Buildings.*—Perfumes extracted by cold process; toilet and iodine soaps; marrow oil.

[1178]

LLOYD, W. A., 19 *Portland Road, Regent's Park.*—Aquarium.

[1179]

LOW, ROBERT, SON, & HAYDON, 330 *Strand.*—Fancy soaps, perfumery, ivory and inlaid hair-brushes; tortoiseshell, india-rubber, and ivory combs. (*See page* 114.)

[1180]

MOREAU, T., 88 *Regent Street.*—Rouge Végétal, Blanc de perles, Crême de l'Impératrice, Parfumerie en général. Wholesale and Retail.

The following are exhibited :—

Moreau's Crême de l'Impératrice renders the skin beautifully white, soft, and transparent. It removes sunburns, freckles, and other discolorations of the skin.

Ladies' Oriental Book of Beauty. A new discovery just registered for the Exhibition.

Bouquet International, a perfume for the handkerchief.

Rouge Végétal and Chinese leaf are beautiful and natural in colour, and harmless in their effects. Blanc Liquide for the arms and neck, gives a soft and velvet-like appearance resembling the bloom of youth. Blanc de Perle in powder, Poudre de Riz, and Noir de Ristori (which gives a brilliant lustre to the eyes).

Since 1781 Moreau has been the fournisseur of all the courts of Europe.

Every requisite for the toilet, may be procured at T. Moreau's, de Paris, Perfumer, 88 Regent Street, London. Established in Paris 1781.

Madame Moreau attends ladies either at her establishment or at their own residences.

LOW, ROBERT, SON, & HAYDON, 330 *Strand.*—Fancy soaps, perfumery, ivory and inlaid hair-brushes; tortoiseshell, india-rubber, and ivory combs.

Robert Low, Son, and Haydon are manufacturers of the choicest articles of perfumery, fancy soaps, hair-brushes, &c., some of which they here enumerate.

Low's highly perfumed brown Windsor soap obtains decided preference over all others in every part of the world.

Low's celebrated honey, glycerine, olive-oil, and other fancy soaps.

Low & Co. have always in stock a large assortment of hair-brushes in ivory, wood, and bone; also combs in tortoiseshell, buffalo horn, and india-rubber. Tooth and nail brushes of superior manufacture. The hair of these brushes is warranted not to come out.

Low's vanilla tooth-paste, a most valuable article for cleansing the teeth and gums, and sweetening the breath.

Low's Syrian liquid hair dye, instantaneous, permanent, and easy of application.

Low's cold cream for healing chapped skin at all seasons.

Low's celebrated well-established perfumes.

Jockey club.	Frangipanni.
Ess. bouquet.	Queen of Alps.
Empress.	Wood violet.
Fragrant.	New-mown hay.

Also new perfumes for the present year: World's fair, and Victory bouquet.

Low's celebrated preparations for the hair.

Manufactory: 330 Strand (opposite Somerset House), London.

[1181]

PEARS, A. & F., Inventors and Manufacturers, 91 *Great Russell Street, Bloomsbury, London.*— Genuine transparent soap.

This soap undergoes a process by which all the superfluous alkali is entirely removed. Its colour is acquired by age, its perfume has also been studied so as to make it most agreeable. It is made in square cakes, oval tablets, and balls for washing; and in round cakes and sticks for shaving.

Pears's Transparent Shaving Stick saves time and trouble to the shaver, and also renders the process of shaving more easy and cleanly than the old mode of using the shaving-dish.

Their case contains pieces of transparent soap manufactured by them 35, 20, 10, 5, and 3 years ago.

"The jury have tried transparent soap 25 years old, manufactured by A. & F. Pears, of which A. Pears was the inventor, and found it very good."—See Jurors' Report of the Great Exhibition of the Industry of all Nations, 1851.

[1182]

PERKS, SAMUEL, *Hitchin, Herts.*—Essential oil of lavender, &c.

[1183]

PHILLIPSON & Co., *Budge Row, Watling Street, St. Paul's.*—Fancy soaps; perfumery; brushes; combs; toilet preparations.

[1184]

PIESSE & LUBIN, 2 *New Bond Street.*—Sweet scents from flowers, and other perfumery. (*See page* 116.)

[1185]

PRICE, NAPOLEON, & JOHN LYON, 158 *New Bond Street, and* 3 *George Yard, Lombard Street.*—Choice perfumery, fancy and transparent soaps, and golden oil.

[1186]

RICHARDSON, J., 30 *Bishopsgate Street Without.*—Fancy soaps and perfumery.

[1187]

RIMMEL, EUGENE, Manufacturing Perfumer, 96 *Strand, London.*—Perfumery, perfumery materials, toilet soaps, and perfume vaporizer. (*See pages* 118 & 119.)

[1188]

ROBSON, J. M., 32 *Lawrence Lane, Cheapside.*—Fancy soaps and perfumery.

J. M. Robson imports the various essential oils and French extracts used in perfumery, and manufactures perfumes, fancy soaps, &c. He is the sole proprietor of the celebrated "Kalosgensis" sauce, and the inventor of the renowned "Rose of England" soap. He also keeps a stock of combs, brushes, and all other articles required for toilet use, of which price lists may be obtained by application. Sample cases from 10*l.* upwards, suitable for any part of the globe, are supplied on the shortest notice.

[1189]

SAUNDERS, JAMES TOUZEAU, 148 *Oxford Street.*—Specimens of various articles of perfumery, including several novel products. (*See page* 117.)

[1190]

THOMPSON, J., 6 *King Street, Holborn, W.C.*—Toilet soaps and distilled perfumes.

[1191]

VICKERS, SHORT, 12 & 13 *Boat Lane, Leeds, Yorkshire*—Perfumes, pomades, general perfumery, &c.

Perfumery, fancy soaps, sponges, and every other toilet requisite may be obtained, wholesale and retail, from

S. Vickers, at the "Acme of Fashion," established in the year 1804.

PIESSE & LUBIN, 2 *New Bond Street, London.*—Sweet scents from flowers, and other perfumery.

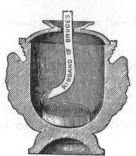

Section of Vase.

MESSRS. PIESSE & LUBIN are the inventors of several novelties for toilet use; manufacturers of perfumery; flower-farmers; distillers of the odours of plants, and importers of musk, ambergris, civet, and otto of roses. The following is a condensed list of their manufactures and preparations.

CONCENTRATED ESSENCES OF FLOWERS—Primitive Odours for perfuming the Handkerchief.

Magnolia	Wood Violet	Orange Blossom	Australian Wattle	Civet
White Rose	Volkameria	Orange of Portugal	Clematis	Ambergris
Cedrat	Limette	Sweet Pea	Wallflower	Lotus of Egypt
Sweet Daphne	Musk	Tuberose	Southernwood	Hoya-Bella
Sweet Briar	Mitcham Lavender	Clove Pink	Reseda	Kus-Kus, or Vitivert
Winter Green	Geranium	Acacia	Provence Rose	Patchouly
Neroly	Cedar Wood	Heliotrope	Mignonette	Water Lily
Bergamot	Forget-me-not	Lemon	Tea Rose	Fragrant Phlox
Meadow Queen	Moss Rose	Ambergris	Santal Wood	Narcissus
Hyacinth	Jonquil	Jessamine	White Lilac	Erica Odorata
Spring Violet	Lily of the Valley	Verbena Leaf	Syringa	Allamandra
Citronella	Lemon Thyme	Honeysuckle	Citron	Chypre

Sold in bottles, 2s. 6d., 5s., 10s., 20s., and 40s. each.

BOUQUETS AND NOSEGAYS—Mixed Odours for Scenting the Handkerchief.

Frangipanni, an Eternal Perfume	Baroness Rothschild's Bouquet	Albion Nosegay.
Piesse's Posy	The Cottage Flower	Royal Horticultural Garden Bouquet
Odoratissima	Wild Flowers	Jolly Dog
New Bond Street Nosegay	Box-his-Ears (sequel to Stolen Kisses)	Young Lubin
Bouquet Millefleurs	Rondeletia	Something New!
Her Majesty's Perfume	H.R.H. Prince of Wales' Perfume	Bouquet of all Nations
Empress Eugénie's Nosegay	(Smallest bottle of this essence is 20s.)	St. Valentine's Nosegay
Bouquet du Napoléon III.	The Flower of the Day	Mousselaine
Royal Hunt Bouquet	Early Spring Flowers	Bosphorus Bouquet from the Valley
Jockey Club Perfume	The Thorny Rose	of Sweet Waters
Yacht Club Nosegay	Maréchale	Buckingham Palace Perfume
Stolen Kisses—for 1861	Neptune, or the Naval Nosegay	Curious Essence
Zouave, this Nosegay contains "all	Flowers of Erin	Chinese Bouquet
the Perfumes of Arabia"	Kiss-me-Quick	Our Village Nosegay
Ess. Bouquet	Flowers of Scotland	Fleur de Mauve
Prince Arthur's Choice	Perfume of Paradise	

Sold in bottles, 2s. 6d., 5s., 10s., 20s., and 40s. each.

Purchasers taking an assortment of half a dozen will be charged at a reduced price. New perfumes every year.

The Sportsman's perfumes, three bottles in a box, 7s., consisting of Royal Hunt, the Tally Ho! and Yacht Club Bouquet. The Wedding perfumes, three bottles in a box, 7s., or three boxes, 20s., containing Orange Blossoms, Lily, and Violet. Scented Shells, from the Maldive Islands, 2s. 6d. per doz. Satchet Powders of dried flowers.

THE FOUNTAIN FINGER RING. Registered August 1st, 1860. The delight of all who have seen this little conceit is most gratifying to its inventor. It is at once useful and ornamental. By the least pressure, the wearer of the ring can cause a jet of perfume to arise from it at any time desired—thus every one can carry with them to a ball, concert, or public assembly, enough scent for the evening.

The rings can be filled with perfume with the greatest ease—thus: press the ball at the back of the ring nearly flat, pour scent into a cup, and dip the ring into it; the elasticity of the ball will then draw the perfume into the interior till full. Each ring will hold about half an ounce of the perfume.

Visitors to the sick will find a ring filled with Hungary Water, the antiseptic qualities of which are so valuable, to be of the greatest service, both to invalid and visitor.

RIBBON OF BRUGES, FOR FUMIGATION. — Draw out a piece of the ribbon, light it, blow out the flame, and as it smoulders a fragrant vapour will rise into the air. [Entered at Stationers' Hall.]

SCENTED GEMS.—Curiosity is excited to know how these gems are capable of yielding fragrance like a natural flower, and from what country they come. As they are moved about in the *petite boite* which contains them, they exhibit the beauty of the kaleidoscope, and exhale the most delightful odour.

Catalogues post free to all applicants.

Works by SEPTIMUS PIESSE, Analytical Chemist:—

1. "The Art of Perfumery, with the methods of obtaining the odours of plants." Crown 8vo, 60 wood engravings (third edition), 10s. 6d.

2. "Chemical, Natural, and Physical Magic." Crown 8vo, 30 wood engravings, 3s. 6d.

3. "The Laboratory of Chemical Wonders." Crown 8vo, illustrated, 5s. 6d.

Longman, Green, & Co., Paternoster Row; and of the Author, 2 New Bond Street, London.

Shippers and exporters are treated upon unusually liberal terms.

For shipping discounts—see export price list.

SAUNDERS, JAMES TOUZEAU, 148 *Oxford Street.*—Specimens of various articles of perfumery, including several novel products.

SAUNDERS'S FACE POWDER, OR BLOOM OF NINON, is a most delicate preparation for beautifying the complexion, free from anything which can injure the skin.

The FACE POWDER has a delicate roseate hue, and is preferable to all other preparations for preserving and clearing the complexion.

In the East Indies and other tropical countries it has been found of immense advantage in preserving the beauty of the complexion from the influence of climate.

Packets 6*d.* and 1*s.*, free for 8 or 16 stamps.
Boxes 2*s.* 6*d.*, free for 40 stamps.

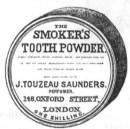

THE SMOKERS' TOOTH POWDER has been in use some years, and gives unqualified satisfaction; it prevents the discoloration of the teeth from smoking, and imparts fragrance to the breath.

Price 1*s.* per box.

SAUNDERS'S GUARDS' HAIR DYE, instantaneous in action, moderate in price, perfectly harmless to the hair or skin, and dyes a good black or brown.

Prices 2*s.* 6*d.*, 3*s.* 6*d.*, 5*s.*, 10*s.*

SAUNDERS'S QUILLAIA BARK HAIR WASH, a natural saponaceous wash, prepared from the bark of a tree (Quillaia Saponaria) found in South America.

This wash is useful as an astringent in case of weak hair, cleansing the skin of the head in a most surprising manner without the evil effect of soap on the hair.

Price 2*s.* 6*d.* and 4*s.* 6*d.*

SAUNDERS'S FLORAL PERFUMES are prepared with great care from every scent-giving plant or flower. Each perfume leaves upon the handkerchief a lasting odour of the flower from which it is distilled in all its freshness.

FASHIONABLE BOUQUETS.—Jockey Club, Frangipanni, Gerards' Bouquet, Prince of Wales, and every new favourite perfumes. Price from 2*s.*

SAUNDERS'S ENGLISH LAVENDER WATER, pure without the admixture of any other perfume, distilled from the finest Mitcham lavender flowers. Price 1*s.* 6*d.* to 7*s.*

SAUNDERS'S SHILLING PERFUMES, in great variety of perfumes, and of excellent quality.

SAUNDERS'S MEDICATED SOAPS, intended to be used under medical direction, supply a convenient and novel means of diffusing such medicaments as are usually prescribed for the external treatment of skin diseases.

The substances included in the various soaps are combined in due medical proportion with pure olive-oil soap, and form bland and agreeable preparations suited to the several cases in which they may be prescribed.

A list of soaps forwarded on application.

SAUNDERS'S CHEW STICK DENTIFRICE, prepared from the branches of a climbing shrub (Gouania Dominigensis) common in Jamaica and other West India Islands.

An eminent botanist thus describes it : "In powder it forms an excellent dentifrice; its aromatic bitter producing a healthy state of the gums; the mucilage it contains working up by the tooth-brush into a soap-like froth." Price 2*s.*

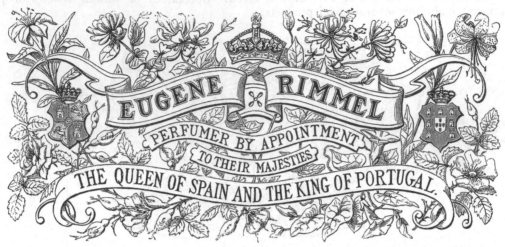

LEFT-HAND CASE.

TOILET SOAPS.	s.	d.
Pure Glycerine Soap .. .. .. per cake	0	6
The Queen's Soap, Russian, Spanish, and other fancy Soaps .. .. per cake	1	0
Honey, Mallow Flower, Aromatic Herbs, Windsor, Honeysuckle, and various other Toilet Soaps, in 1 lb. bars and in packets .. .. .. 1s. and 1		6

SHAVING SOAPS.	s.	d.
Cream of Almonds, Ambrosial Cream, Rose, and Pistachio Cream .. ..1s. and 1		6
The Officers' Shaving Soap, in metal tubes .. 1		0
Honey and Glycerine Shaving Soaps 0		6
Malaktikon, or Emollient Shaving Soap 1		0

RIMMEL's Distilled Violet Water, a new preparation for the toilet, is exhibited in a fountain designed by E. Rimmel and executed by Poitevin, illustrating the art of distillation. The public will thus be enabled to test and appreciate the delicious and refreshing fragrance of this article, which is sold in elegant Parian bottles at 3s. 6d.

CENTRE CASE.—FOUNTAIN OF RIMMEL'S DISTILLED VIOLET WATER.

RIGHT-HAND CASE.—PERFUMERY.

Rimmel's Magic Vines, and other Fruit Trees, containing scent; price from 7s. 6d.

Rimmel's Floral Trees and Bouquets; each flower exhales its natural fragrance; price from 5s.

Rimmel's New Perfumes, the Exhibition Bouquet, Victoria Bouquet, Prince of Wales Bouquet, Jockey-club, Wood Violet, Ess Bouquet, Solferino, Magenta, African Flowers, Rimmel's Bouquet, &c.; price from 2s. 6d.

Kwei-hwa, a Chinese Perfume, in a silk box, 3s. 6d.

Rimmel's Toilet Vinegar, Extract of Lavender Flowers, Verbena Water, and Eau de Cologne; price from 1s.

	s.	d.
Glycerine Cold Cream .. .. .. 1s.		0d.
Glycerine Paste for the hands 1		6
Rose-leaf Powder, for the toilet .. 1		0
Rimmel's Lotion, for the complexion .. 2		9

	s.	d.
The Queen's Pommade, in stoppered bottles .. 1s.		0d,
Glycerine Pommade 1		0
Parisian Cream, in glass vases with plated tops 2		6
Marrow Oil, in glass boxes with wooden tops . 2		6
Nutritive Cream, in cut glass bottles 5		0
Brillantine, for imparting gloss to the beard, &c. 3		6
Glycerine and Egg Wash 3		6
Detersive Pommade 2		6
Indelible cosmetiques 1		6
Royal Dentifrice 1		0
Coral Tooth Paste 1		0
Elixir for the teeth 2		6
Perfumed Almanacs 0		6
Illustrated Sachets .. .. .. 6d. and 1		0
Benzoline for removing spots 1		0

A perfumery museum, showing the principal apparatus and materials used in its manufacture, with their technical names and places of production, is exhibited by E. Rimmel in a separate glass case, at a short distance from his stand.

Descriptive Catalogues to be had on application.

RIMMEL'S PERFUME VAPORIZER is a newly-invented apparatus for diffusing the fragrance of flowers, and purifying the atmosphere in apartments, ball-rooms, theatres, &c. The various points of superiority it offers on pastilles, papers, ribbons, and other means in use hitherto may be thus briefly summed up.

I. It diffuses the perfume of any flower in all its freshness and purity.

II. The vapours produced are so delicate and refreshing that they cannot affect even the most nervous persons, and their elasticity is such that they spread over a vast area in a very short time.

III. This process entirely neutralizes the vitiated air generated in theatres, ball-rooms, and other assemblies. It also completely removes the smell of tobacco.

IV. It effectually purifies the air in dwelling-houses, and counteracts unpleasant and noxious effluvia arising from drains, gas, or any other cause.

V. It is invaluable for the sick chamber, substituting balmy and soothing vapours for a close atmosphere.

VI. It is strongly recommended to travellers, and will also be found very reviving at sea to fumigate close cabins, and alleviate the sufferings of sea-sickness, by producing a pleasant atmosphere.

VII. The perfumes used in this process possessing a watery basis are not liable to ignition.

VIII. This apparatus forms an elegant drawing-room ornament, and is sold at a very moderate price, which places it within the reach of all classes.

IX. It has been submitted to Dr. Letheby, Dr. Hassall, and other eminent authorities, who have all expressed the highest opinion of its merits in a sanitary point of view. It has also been very favourably noticed by the *Times, Morning Post, Star, Telegraph, Herald, Standard, Builder, Atlas, Technologist.* Extracts from these and other papers will be found in a more detailed prospectus.

X. It has been used on board of her Majesty's steam yacht, at the Lord Mayor's banquet, at her Majesty's Theatre, Covent Garden, Drury Lane, Princess's, Lyceum, Hanover Square Rooms, and other public and private entertainments, where it has always given the greatest satisfaction.

PRICES OF THE VAPORIZERS.

	£	s.
No. 1. Bronze	0	6
No. 1. Plated	0	12
No. 2. Bronze	0	16
No. 2. Plated	1	4
No. 3. Bronze	1	0
No. 3. Plated	1	12
No. 4. Bronze	1	12
No. 4. Plated	2	8
Elegant china, from	1	1
Fancy patterns, from	0	15

Marine Vaporizers with safety-lamps, as used on board of the Peninsular and Oriental Company's boats, £1 10s.

———

PERFUMES TO BE USED IN THE VAPORIZERS.

Ordinary compounds from 2s. 6d.
Best compounds from 3s. 6d.

The vaporizer can only be used with the compounds prepared specially for the purpose by E. Rimmel, as other perfumes would not produce the desired effect, and might cause accidents.

———

RIMMEL'S AROMATIC DISINFECTOR is a cheaper apparatus, working on the same principle as the vaporizer, but chiefly used for sanitary purposes. It has been adopted by the Royal College of Surgeons and the principal hospitals, and will be found exhibited in Class XVII. The price of it is 2s. 6d., including a bottle of aromatic disinfecting compound.

[1192]

WARRICK BROTHERS, *Garlick Hill, London, and Rue Fodéré, Nice.*—Essential oils, perfumes, pomades, &c.

[1193]

WHARRY, JAMES, *Chippenham, Wilts.*—Treble-distilled lavender water.

[1194]

WHITAKER & GROSSMITH, 120 *Fore Street, Cripplegate, E.C.*—Perfumery and toilet soaps.

[1195]

YARDLEY & STATHAM, 7 *Vine Street, Bloomsbury, London.*—Fancy soaps and perfumery.

LONDON: PRINTED BY WILLIAM CLOWES AND SONS, STAMFORD STREET AND CHARING CROSS.

SECTION II.

CLASS V.

RAILWAY PLANT, INCLUDING LOCOMOTIVE ENGINES AND CARRIAGES.

[1227]

ADAMS, W. B., *Holly Mount, London.*—Wheels, springs, and rail-joints.

[1228]

ALLAN, ALEXANDER, *Perth.*—Straight-link valve motion, pressure gauges, &c. (*See page* 2.)

[1229]

ANDERSTON FOUNDRY COMPANY, *Glasgow.*—Permanent way materials.

[1230]

ARMSTRONG, SIR W. G., & Co., *Elswick Engine Works, Newcastle-upon-Tyne.* — General traffic engine and tender, East Indian Railway Company.

[1231]

ASHBURY, JOHN, *Openshaw, Manchester.*—A saloon carriage. A goods waggon. Specimens of wheels and axles, axles, tires, and bar iron.

[1232]

AYTOUN, ROBERT, 3 *Fettes Row, Edinburgh.*—A railway break.

[1233]

BAIN, MʻNICOL, & YOUNG, *Edinburgh.*—Wrought-iron simultaneous-acting gates for railway level crossings, and wire fencing.

[1234]

BAINES, WILLIAM, & Co., *London Works, Smethwick, Birmingham; 35 Parliament Street, Westminster, S.W.; 76 Rue de la Victoire, Paris.*—Railway plant. (*See page* 3.)

ALLAN, ALEXANDER, *Perth*.—Improvements in the expansion valve gear of steam-engines. Straight-link valve motion.

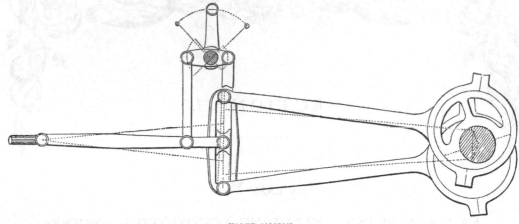

VALVE MOTION.

By this arrangement simultaneous movement is given to the eccentric rods and link, and to the valve rod, in opposite directions, by short levers placed on opposite sides of the reversing shaft, thereby obtaining a straight link.

This valve motion is easier of reversal, balance weights are dispensed with, and the sliding movement of the block is reduced. The only fixings required are the reversing shaft brackets.

Most accurate results as regards an equal distribution of the steam, can be obtained; while from the simplicity of the motion and from the link being straight (in place of curved), repairs are more economically executed.

Pressure gauges : indications by water rising in the gauge, compressing air within it.

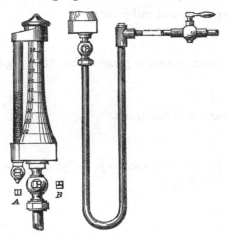

PRESSURE GAUGE.

Pressure, by this gauge, is indicated by the more or less compressed or expanded condition of an accurately measured quantity of atmospheric air contained within the body of the gauge, which is acted upon by cold water contained within a bent tube.

The pressure, acting on the water, forces it up the glass tube, and is indicated by the height to which the water rises, a graduated index being marked on the body of the gauge.

The principal features of this gauge are its simplicity, the durability of its parts, and facility of re-adjustment.

The air, which is the elastic spring, can be renewed at pleasure by simply turning the cocks to the positions shown (at A, B); and when the new atmosphere is thus admitted, the cocks must be reversed.

These gauges are capable of being made to suit any pressure from 1 to 5 lbs. for blast furnaces, and from 1 to 15, or up to 300 lbs. per square inch, for steam boilers or other purposes.

Compound buffer, springs with independent action, giving double resisting power in small space.

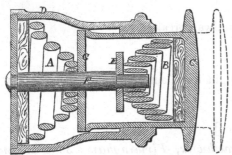

COMPOUND BUFFER.

The improvement in this buffer consists in obtaining increased resistance within a given space.

The springs used may be of steel or india rubber, as preferred, provided the principles of the arrangement be followed.

For the purpose of showing the action, the buffer is illustrated with its plunger at half stroke; the dotted lines show the position of the plunger when at rest.

It will be observed that while the first spring (A) is acting between the bottom of the cylinder (D) and the *inner* end of the plunger (C), the second spring (B) is acting between the *front* of the plunger (C) and the washer (E) fixed on the pillar (F).

There is thus a compound action; and it will be seen the resistance is exactly equal to the united power of the two springs, or, double that of an ordinary buffer.

BAINES, WILLIAM, & Co., *London Works, Smethwick, Birmingham; 35 Parliament Street, Westminster, S.W.; 76 Rue de la Victoire, Paris.*—Railway plant.

The following are exhibited:—

BAINES'S FURTHER IMPROVED SELF-CLEANING SWITCHES, offering the following advantages, viz.: The additional depth of the tongues enables the bottom flanches uncut to pass under the main rails, giving greater width of base and stability of switch. The sliding surfaces of the chairs, being placed obliquely to the seat of the main rails, and surrounded with inclined planes, enable the switch by its own action, to clear away any dirt or stones that may have lodged between the tongue and the main rail, thereby avoiding a prolific cause of accidents. This switch is perfected by an improved lever box, so arranged that nothing can impede the action of the weight. These switches gained the prize medals at the Exhibition of 1851.

BAINES AND WOODHOUSE'S PATENT CROSSING. The heart or V piece is forged solid, with its upper and lower face exactly alike, and is steeled for a length of twenty inches from the point; it takes the vertical bearing from the sides, by means of projections resting on corresponding seats in the chairs. By this arrangement the crossing can be turned over, when the upper surface is worn, and the same amount of wear can be had from the lower face.

The broad end is provided with side channels for the reception of the fish plates, to connect it with the main rails. The wing and check rails are all exact counterparts of each other. By transposing, and turning over the wing and check rails, the wearing surfaces can be renewed eight times. Thus effecting a very great economy.

By applying suitable fish plates, these crossings can be applied to any description of rail.

BAINES'S PATENT UNION-PLATE GIRDERS are specially adapted for ship building, architectural, or engineering purposes, their peculiar construction causing an absence of any strain upon the bolts or rivets.

BAINES'S PATENT TURNTABLE is constructed of his patent union plates, so arranged that when fixed, a lateral and vertical union is effected, making the skeleton frame of the revolving top as rigid as a single girder, and possessing great strength combined with lightness. These tables are well adapted for shipment abroad, as they can be packed in a small compass, and are of less weight than the ordinary table. Engine tables of any dimension can be constructed on this plan. These inventions are patented in England, America, and on the Continent.

[1235]

BARKER, D., 10 *Turret Grove, Clapham.*—Steam telegraphic and fog signal.

[1236]

BATESON, SAMUEL STEPHEN, 17 *Bolton Street, London.* — Patent feed-water heating apparatus, with internal perforated safety tube.

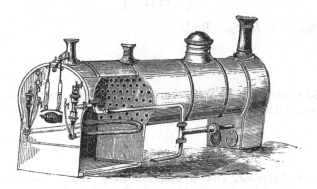

MODEL OF PATENT FEED-WATER HEATING APPARATUS, WITH INTERNAL PERFORATED SAFETY TUBE.

By this invention, the feed-water is forced through a tube or coil placed in the fire-box before it enters the boiler, thereby receiving an amount of heat nearly equal to the temperature of the water in the boiler. The effect of this is to prevent the generation of steam from being checked, as is the case when cold feed-water is injected.

By the use of the internal perforated safety-tube, there is no possibility of the water in the coil assuming the spheroidal condition. Each end of this internal tube is connected with the water space of the boiler, and in the event of any tendency of the water in the coil towards a spheroidal condition, a small jet of water is forced by the pressure of the water in the boiler through the perforation nearest the spot, which at once restores circulation and prevents any risk of injury to the coil.

Three locomotive engines on the London and North-Western Railway have been fitted with this apparatus, with eminently successful results in the rapid generation of steam, and economy of fuel.

In the working of the express engine, No. 248, during the six months ending 30th November, 1861, as compared with the preceding six months, the saving amounted to 27·9 per cent. upon the fuel consumed, the miles run being about 16,500 in both cases. These figures are taken from the official tables of the Company.

By the adoption of the principle involved in this invention, all boilers containing water in tubes are rendered perfectly safe in working.

[1237]

BAYLISS, SIMPSON, & JONES, 43 *Fish-street Hill.*—Iron hurdles, fencing, cable chains, anchors, screw-bolts, spikes, &c.

[1238]

BEYER, PEACOCK, & Co., *Gorton Foundry, Manchester.*—Locomotive express passenger engine and tender, designed for the South Eastern of Portugal Railway Company.

[1239]

BIDDELL, G. A., *Ipswich.*—Patent chilled railway crossings, as manufactured by Ransomes & Sims, Ipswich.

These chilled crossings have been extensively and successfully introduced during the last six years, both at home and abroad, several thousands being now in daily use.

Their simplicity, economy, and extreme durability, at once recommend them to the notice of railway engineers, and the result of a fair trial, as invariably, leads to their adoption.

RANSOMES & SIMS, Ipswich, the proprietors of the patent, will be pleased to furnish further particulars as to prices, &c., upon application.

[1240]

BROWN, G. & I. & Co., *Rotherham Iron Works.*—Patent solid iron tires, also patent solid steel-faced tires.

[1241]

BUTTERLEY IRON COMPANY, *Derby.*— Rail.

[1242]

CLARK, GEORGE, 30 *Craven Street, Strand.*— Gas signals for railways, tunnels, telegraphs, lighthouses, ships, and fire-alarms.

[1243]

COPLING, JOHN, ESQ., Inventor, *The Grove, Hackney, N.E.*—Railway signal (patented), single or double—guards' communication with drivers, and passengers' with guards.

The communication between guards and drivers is by means of the steam whistle, or a spring bell on the engine and guard's van; that from passengers to guards by a spring bell on the guard's van. The signals, in both cases, are worked by small wire ropes from a reel or drum; while, in the latter, side lines, with a flag or lantern, indicate to the guard the compartment from which the signal proceeds. On signal being made by a passenger in case of accident or other emergency, the guard can ascertain the cause of alarm without, or previous to, stopping the train, by safe and easy passage along the roofs of the carriages to the compartment, or private carriage, indicated by the signals. Odd carriages, not fitted with the apparatus, can be let into the train at junctions, &c., without interfering with the working, as the wire ropes will be suspended over them and kept level by the balance weights. Spare compartment-lines can be always kept in store in the guard's van, and affixed instantaneously by means of spring-hooks. This apparatus is simple, cheap, and not liable to get 'jammed,' or out of order. If desired, the upper line (guard's and driver's), which is free from the control of passengers, can be used alone,—without the lower or passengers' line. The signal rods are kept lower than the chimney or luggage gauge.

[1244]

CORLETT, HENRY LEE, *Inchicore, Dublin.* — Continuous rails; cellular brackets; joint chairs; carriage and waggon buffing springs.

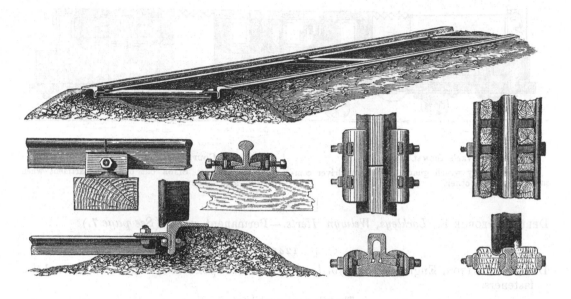

The illustration exhibits a new form of continuous, surface-supported rail, in two parts, bolted together, so as to 'break joint.' Timber sleepers, chairs, and fish plates are dispensed with; while at the same time durability of structure, improved drainage, a secure and level surface, and facility of executing repairs is attained, combined with economy in construction. This rail is laid down in the Western Annex of the Exhibition Building. The joint chairs for bridge and foot rails are of cast iron — wedge keys of wrought iron are introduced either above or below the flanges of the rail, and are set up, and held in position, by screw bolts and nuts. These chairs may be laid on sleepers or otherwise as desired. Cellular cast iron brackets, bolted longitudinally at either side of a double T rail, are also exhibited. The cells in brackets may be filled with compressed timber, asphalt, concrete, broken stones, ballast, or other similar material. This arrangement is particularly applicable to street railways.

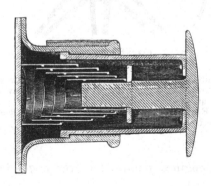

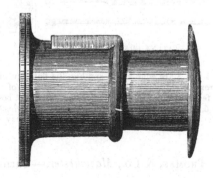

The BUFFING SPRINGS are Spencer and Corlett's combined Patent, and are manufactured by John Spencer & Sons, Newcastle-on-Tyne.

The illustration represents a waggon buffer in elevation and section; the casing, plunger, and head are all of wrought iron, combining great strength with lightness. The plunger is without a central bolt, and cannot under any circumstances fall out, being retained in position in the casing by corresponding projections. The spring is an improved volute, provided with ribs, whereby additional strength and diminished friction is attained.

[1245]

DAVIDSON, JOHN, *Leek, Staffordshire.*—System of communication between passengers, engine-driver, and guard on railways.

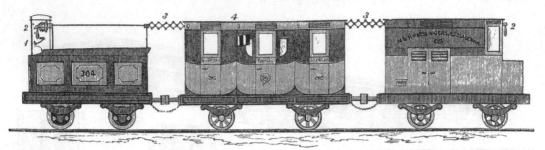

1. ENGINE-DRIVER'S SIGNAL BELL.
2. HANDLES by which guard and engine-driver communicate to each other.

3. SIGNAL COUPLINGS.
4. THE ALARM SLIDE out of a compartment when rung by a passenger.

[1246]

DERING, GEORGE E., *Lockleys, Welwyn, Herts.*—Permanent way. (*See page* 7.)

[1247]

DIXON & CLAYTON, Engineers, *Bradford.*—Patent rolled spoke iron; railway wheels and tire fasteners.

The following are exhibited:—

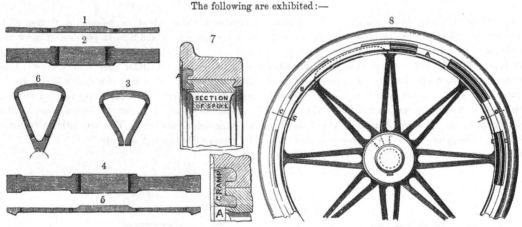

PATENT CRAMP FASTENINGS.

PATENT ROLLED SPOKES.

Figs. 1 and 3 are edge views, and Fig. 2 a flat view of a bar of iron suitable for making the spokes, and part of the rim, or felloe of a railway carriage wheel, with cast boss. Figs. 4, 5, and 6 spokes to form solid wrought iron wheel.

Fig 7. A, represents cramp fastening, which may be continuous as at B, Fig. 8; and tire and rim set down all round, or at intervals, as at C C; or it may be in segments of any convenient length, as at D D

E E are grooves in the edge of wheel rim and tire to receive the cramp fastener.

[1248]

DUNN, THOMAS, & Co., *Manchester.*—Turntables, engines, pumps, &c. (*See pages* 8 *and* 9.)

[1249]

EDINGTON, THOMAS, & SONS, *Phœnix Iron Works, Glasgow.*—Railway chairs and sleepers.

[1250]

ENGLAND, GEORGE, & Co., *Hatcham Iron Works, London, S.E.*—Locomotive engine with tender; also traversing screw-jack for railway purposes.

DERING, GEORGE E. *Lockleys, Welwyn, Herts.*—Improved permanent way of railways.

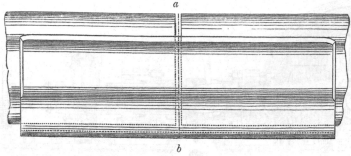

a

b

SIDE VIEW OF CLIP-JOINT.

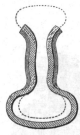 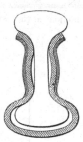 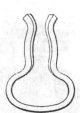

SECTION OF SPRING CLIP WHEN NOT EXPANDED BY THE RAILS.

SECTION THROUGH DOTTED LINE *a b.*

END VIEW OF SPRING CLIP.

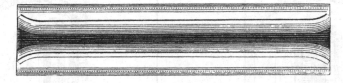

PLAN OF SPRING CLIP.

1. SPRING CLIP FISH-JOINTS, of tempered steel; affording the advantage of increased strength and smoothness at the joint, by reason of the powerful and uniform pressure of the Spring Clip. Any wear or loosening that may at any time occur is immediately repaired by the inherent tendency of the Spring Clip to collapse. Safety, simplicity, and economy are likewise insured by the absence of bolts, nuts, &c., and of the necessity for constant attention and labour which they entail,—one single piece of metal taking the place of the ten or fourteen separate parts which constitute the ordinary 'fish-joint.' The Figures show the adaptation of the Spring Clip to rails of the double-headed section, and it is applicable to other forms with equal advantage.

2. SPRING KEYS, of tempered steel; the most important advantages of which consist in the firmness with which they hold the rails, and that whilst possessing every qualification of the wooden key, without its defects, they are calculated to last at least ten times as long. The Spring Key never becomes loosened by vibration, owing to its unfailing tendency to expand, and is totally unaffected by hygrometric changes. It may be used either with intermediate or joint chairs; and forms, with the latter, a rail-joint equal to the ordinary 'fish,' at less than one-half the cost.

3. SPRING TRENAILS, of tempered steel; which possess like advantages with the spring keys, in point of efficiency and durability, over both wooden trenails and iron spikes. Owing to its permanent tendency to expand, the Spring Trenail cannot be loosened by vibration, although extracted readily, and without injury, when needful. It is not affected by weather, and cannot be broken by the tangential strain exerted at curves, or otherwise.

Examples are shown of rails united at the ends by 'hard-soldering' or 'brazing.' Brazed Joints are exhibited which have been severely tested by sledge-hammering,—the result of such treatment, when carried far enough, being to break the iron into fragments without the joint yielding. A pair of joints of this description are exhibited which have recently been taken out of the up main line of the Great Northern Railway, where they have carried the whole traffic for nearly four years, without renewal or deterioration. Eighty-six thousand locomotive engines, and nearly four million wheels of rolling stock, have passed over these joints, which are as sound and perfect as when first made.

Agents for the Patentee—WILLIAM L. GILPIN & Co., 10 St. Swithin's Lane, City, London.

DUNN, THOMAS, & CO., *Windsor Bridge Iron Works, Manchester.*—Turntables, traversers, cranes, engines, boilers, pumps; hydraulic machinery.

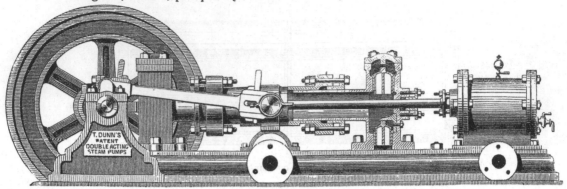

PATENT DOUBLE ACTION STEAM PUMP.

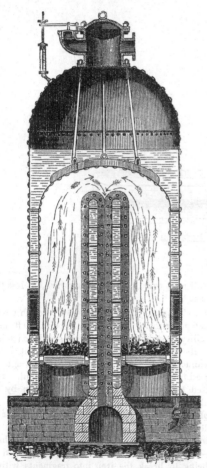

PATENT DOUBLE DOWN DRAUGHT SMOKE BURNING BOILER.

PATENT SINGLE DOWN DRAUGHT BOILER.

MACHINERY IN FULL OPERATION.

Wrought Iron Geared Locomotive Engine Traverser, 20ft. long.—Dunn's Combined Patents.

Wrought Iron Frame, deep-wheel easy running Traverser, 15ft long.—Dunn's Combined Patents.

Wrought Iron Plate, deep-wheel easy running Traverser, 15ft. long.—Dunn's Combined Patents.

Wrought Iron Surface Turntables, 15ft. diam.—Dunn's Patent.

Cast Iron Solid Ring Turntables, centre and curb for a 13ft. diam., as used in H. M. Dockyards.—Dunn's Patent.

Cast Iron Turntable, as used in Her Majesty's Store Houses, 8ft. diam. Patent self-foundation.

Double Action Steam Pump.—Dunn's Patent.

Improved Hydraulic, self-balanced Cross-head Wheel Forcing Machine.

Dunn's Improved Hand Pumps for Railway Stations, Tanks, Agriculture, Ships, &c.

Roof over part of these goods, in yard.—Dunn's Patent.

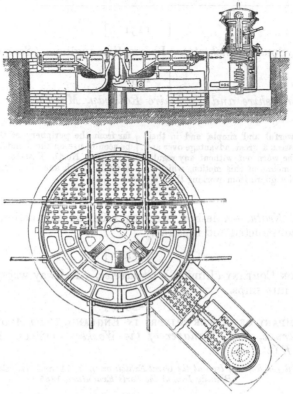

DUNN'S PATENT SOLID RING TURNTABLE, AS USED IN H. M.'S DOCKYARDS.

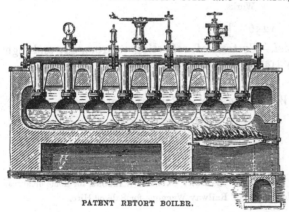

PATENT RETORT BOILER.

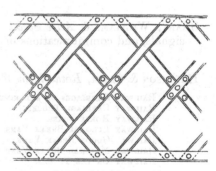

PATENT BRIDGE WORK.

MODELS.	DRAWINGS.

MODELS.

Wrought Iron Geared Locomotive Engine Traverser. —Dunn's Combined Patents.

Wrought Iron Frame deep-wheel easy running Traverser.—Dunn's Combined Patents.

Wrought Iron Traverser for heavy Carriages.—Dunn's Combined Patents.

Wrought Iron Traverser for Carriages. — Dunn's Combined Patents.

Cast Iron Solid Ring Turntable, as used in Her Majesty's Dockyards.—Dunn's Patent.

Wrought Iron Engine Beam Turntable.—Dunn's Patent.

Wedge beam Turntable.—Dunn's Patent.

Safety Carriage and Break for High Speeds.—Dunn's Patent.

Wrought Iron Lattice, and Steel Lattice, Basket Bridge Work.

Ditto, made from Rail Bars and Ribbed Iron.

DRAWINGS.

Dunn's Improved Hydraulic Self-balanced Cross-head Wheel-forcing Machine.

Dunn's Patent Wrought Iron Engine beam Turntable.

Traverser, sheet A.—Dunn's Combined Patents.

Traverser, sheet B.—Dunn's Combined Patents.

Traverser, sheet B.—Dunn's Combined Patents.

Dunn's Improved Hydraulic Machine for testing Cables, Beams, and Anchors.

Dunn's Patent Retort Steam Boiler.

Rose's Patent Multitubular Steam Boiler.

Dunn's Patent Double-action Steam Pumps.

Dunn's Improved Steam Travelling Cranes.

Improved Steam Wharf Cranes.

Improved 30 tons Travelling Crane.

Dunn's Patent Double-draught Smoke-burning Boiler.

Improved Horizontal Engine, as working Machinery at the Crystal Palace.

Dunn's Patent Cast Iron Solid Ring Turntable, as used in Her Majesty's Dockyards.

[1251]

FAIRBAIRN, W., & SONS, *Manchester.*—Locomotive engine. (*See page* 11.)

[1252]

FAY, CHARLES, *Lancashire and Yorkshire Railway, Manchester.* — Continuous railway-carriage breaks.

These breaks are powerful and simple, and in their self-adjusting motion possess a great advantage over any other. The blocks will be worn out without any regulation being required, by means of this motion, which at the same time prevents the guard from working them too far from the periphery of the wheels. The breaks may be seen in use on the London and North Western, Great Northern, North Eastern, Lancashire and Yorkshire, West Midland, and other railways.

[1253]

GARDNER, SANKEY, *Neath.*—Axle-box, securing efficient connection with spring. Truck-buffer, cheaply constructed and repaired.

[1254]

GLOUCESTER WAGGON COMPANY (Limited), *Gloucester.*—Railway waggon with iron body, for discharging coal into ships.

[1255]

GOVERNOR AND COMPANY OF COPPER MINERS IN ENGLAND, *Cwm Avon Works, Glamorganshire ; W. P. Struvé, Esq., Manager of the Works.* — Offices : 10 *New Broad Street Mews, London, E.C.*

Obtained Prize Medal for Railway Iron at the Great Exhibition of 1851, *and Grande Médaille d'Honneur for Railway Iron at the Paris Exhibition,* 1855.

ONE BRIDGE RAIL, 90 feet long, 58 lbs. per yard.
ONE FLANCH RAIL, 63 feet long, 3¾ lbs. per yard.

Various sections of RAILS made at the Cwm Avon Works.

[1256]

GRANT, WILLIAM, 6 *Alice Street, Liverpool.*—System of reflecting mirrors, day and night signals, and communications on railway trains, to prevent accidents.

[1257]

HARRISON & CAMM, *Rotherham Waggon Works, Masbro'.*—Railway coal waggon.

RAILWAY CARRIAGES OF ALL KINDS.
RAILWAY CARRIAGE TRUCKS.
RAILWAY HORSE-BOXES.
RAILWAY LUGGAGE BREAK VANS.
RAILWAY GOODS BREAK VANS.
RAILWAY MINERAL BREAK VANS.
RAILWAY CATTLE TRUCKS.

RAILWAY SHEEP TRUCKS.
RAILWAY GOODS WAGGONS, OF ALL SORTS.
RAILWAY TIMBER TRUCKS.
RAILWAY COAL WAGGONS.
RAILWAY COKE WAGGONS.
RAILWAY MINERAL WAGGONS.
RAILWAY BALLAST WAGGONS, for home or abroad.

Harrison & Camm manufacture every description of Railway Waggons for Sale or Hire.

[1258]

HATTERSLEY, WILLIAM, 135 *St. George Street, E.*—Passengers' signal for railway carriages, for ready communication with drivers, guards, &c.

[1259]

HENSON, WILLIAM FREDERICK, Civil Engineer, 15 *New Cavendish Street, Portland Place, London.*—Railway buffers and bearing springs.

The advantages of these buffer springs are, that they are combined in their action, whereby a greater power of resistance is offered with a less weight of steel than any other spring now in use, and at a less cost.

The grooved steel bearing springs possess great strength, combined with durability and cheapness in their manufacture, doing entirely away with spring fittings.

They are made by Bradley & Co., who are manufacturers of every description of railway buffers and bearing springs, Rishworth patent springs, and Price & Hawkins' patent fish plates.

For further information, application should be made to Bradley & Co., Broomhall, Sheffield.

[1260]

HOY, J., 6 *Pickering Place, W.*—Railway signal.

FAIRBAIRN, WM., & SONS, *Manchester.* — Goods locomotive constructed for the Midland Company.

The boiler is composed entirely of thick-edged plates, and is double-riveted throughout. No angle iron is used, the barrel of the boiler being flanged where attached to the tube plate. The principal dimensions are :— Boiler barrel—11 ft. 6 in. long, 4 ft. 3 in. diameter. Fire box—4 ft. 9 in. long, 4 ft. 3 in. wide. Plates of barrel—⅞ in. thick. Plates of fire-box—⅜ in. thick. Copper fire-box—4 ft. 0¼ in. long, 3 ft. 7 in. wide, and 4 ft. 8½ in. deep. Brass tubes—2 ft. diameter, 180 in number. Heating surface—1160 square feet. Cylinders —16 in. diameter, 24 in. stroke. Wheels—5 ft. 2 in. diameter, six coupled with outside cranks. Crank axle has four bearings—the inside 6¾ in. diameter, and the outside 6 in.; the leading and trailing axles have outside bearings 6 in. diameter. The Engine is designed by Mr. Kirtley, Engineer to the Midland Railway Company.

[1261]

HUGHES, HENRY, *Falcon Works, Loughborough.* — Models of plant used by railway contractors.

1. Drawing of a locomotive engine for contractors and mineral railways, and all purposes where a light engine is required to ascend steep gradients and turn sharp curves £450 0

2. Model of end-tipping waggon to hold three cubic yards. These waggons are made of stout elm, and put together with the best ironwork . 13 0

3. Model of side-tipping waggon . . . 14 0

4. Model of land cart for earth work . . 4 10

5. Model of contractors' or builders' travelling crane to lift three tons. 55 0

6. Model of strong dobbin cart with patent wheels £8 5

7. Model of an improved horse power which occupies a very small space and gives out the full power of the horse 18 0

8. Hughes's patent combined iron and wood wheel, which possesses great strength and durability, and entirely obviates the decay and shrinkage of wooden wheels.

9. Model of a stout 2-horse cart for contractors' purposes 17 0

[1262]

ISCA FOUNDRY COMPANY, *Newport, Monmouthshire.*—Switches &c. (*See page* 13.)

[1263]

KINGSTON, WM. H., A.B., Trin. Coll. Dublin, *Bandon.*—Means of verbal communication on railway trains.

Extract from the Report of Colonel Yolland, R.E., to the President of the Board of Trade :—
'I have examined Mr. W. H. KINGSTON's plans for "verbal communication between the passengers and guards, and guards and engine drivers on railway trains," and I have the honour to report that it is generally described in Mr. Kingston's circular, dated 25th June, 1859. The tubing is intended to be placed on the tops of the carriages, an arrangement being proposed for getting over the difficulties likely to arise from the inequality in the height of the carriages which usually form a train, as well as for the wriggling motion of the carriages when travelling.

'The subject has evidently been fully considered by Mr. KINGSTON, and the arrangements proposed are very ingenious.'

[1264]

KITCHIN, RICHARD, *Warrington.*—Weighing machinery, cranes, and railway plant. Full-size models and drawings.

The following machines are exhibited :—

1. A SIX-TABLED ENGINE-WEIGHING MACHINE, as used by the chief locomotive engineers of Great Britain, but with the addition of Hind's patent steelyard, by which it is rendered the most complete compound weighing machine extant. No locomotive engine stables can be considered complete without this machine.

2. A HIND'S PATENT WEIGHING CRANE of twenty tons' power, for an overhead travelling crane. This machine will raise an article, and, while holding it suspended, will indicate its weight. This will be found a most useful apparatus in foundries, boiler works, &c.

3. A variety of WEIGHING MACHINE STEELYARDS, and their fittings.

4. Drawings of TURNTABLES, CRANES, and WEIGHING MACHINERY.

[1265]

LANAUR, L., 4 *South Street, Finsbury.*—Axle-boxes and bearings.

[1266]

LILLESHALL COMPANY, *Shiffnal, Shropshire.*—Colliery locomotive. (*See page* 14.)

[1267]

LITTLE, CHARLES, 71 *Little Horton Lane, Bradford.*—Safety coupling for railway waggons.

By means of this invention, much of the danger to human life attending the ordinary method of coupling, is obviated; and time is saved in the marshalling of railway trains. Some of these couplings are in use by the Midland, Great Northern, Manchester, Sheffield and Lincolnshire, and other railway companies.

[1268]

LLOYDS, FOSTERS, & Co., *Old Park Iron Works, Wednesbury.*—Wheels and axles, turntables, cranes, tires ; samples of iron.

Manufacturers of all kinds of Railway Plant; including bridges of wrought and cast iron, turntables, wheels and axles, switches and crossings. Also, of the very best descriptions of tires, axles, boiler plates, and bar iron.

ISCA FOUNDRY COMPANY, *Newport, Monmouthshire.*—Switches, crossings, chairs, lever-boxes, axle-boxes, chilled and dobbin wheels. Lithographs &c.

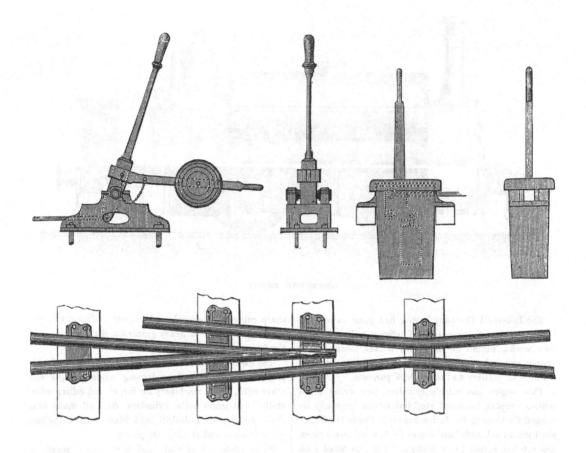

1. A SET OF PATENT BESSEMER CAST STEEL SWITCHES AND STOCK RAIL.

   PATENT WELDED CAST STEEL CROSSING.

1a. DOUBLE HEAD SWITCHES, AND SOLID WELDED CROSSING. (ORDINARY RAILS.)

2. SET OF PARSONS'S PATENT SWITCHES.

3. CARR'S PATENT FILLED CROSSING.

4. WELDED CROSSING. (VIGNOLE'S RAIL.)

5. GRAND RUSSIAN CROSSING WITH SOLID POINT.

6. SOLID WROUGHT STEELED CROSSING POINT.

7. CARR'S PATENT OBTUSE OR OVER CROSSING.

8. OBTUSE CROSSING OR ANGLE. (VIGNOLE'S RAIL.)

9. CONTRACTOR'S CAST IRON CROSSING.

10. VARIOUS CONNECTING RODS FOR SWITCHES.

11. OVERGROUND LEVER BOX AND WEIGHT.

12. UNDERGROUND LEVER BOX WITH WEIGHT ENCLOSED.

13. EXCENTRIQUE LEVER BOX AND WEIGHT.

14. EXCENTRIQUE LEVER BOX AND WEIGHT.

15. DISC SIGNAL CAPSTAN LEVER BOX.

16. QUADRANT SIGNAL LEVER BOX.

17. SELF-ACTING DISC SIGNAL LEVER BOX.

18. CRANK BOX.

19. SAYERS'S PATENT CHAIR.

20. MARSH'S BRACKET CHAIR.

21. FENTON'S PATENT JOINT CHAIR.

22. WROUGHT-IRON DOBBIN CART WHEELS AND AXLES.

23. CONTRACTOR'S CHILLED WHEELS.

24. CARR'S PATENT AXLE-BOX.

25. SPECIMENS OF WAGGON AXLE-BOXES.

26. WOOD'S PATENT TURNTABLE, WITH WROUGHT-IRON ROLLER PATH.

27. CARRIAGE OR WAGGON TURNTABLE, WITH CAST-IRON ROLLER PATH.

28. ENGINE BALANCE TURNTABLES, WITH OR WITHOUT GEARING.

29. PILLAR, HOSE, AND SWING WATER CRANES.

Lilleshall Company, *Shiffnal, Shropshire.*—An extra strong colliery or contractors' locomotive for curves and heavy gradients.

LOCOMOTIVE ENGINE.

The Lilleshall Company having had great experience in the working of locomotives of different makers in their own works, submit for exhibition a tank locomotive of simple and substantial construction, proved to be most suitable for colliery and contractors' purposes.

This engine has outside cylinders, four wrought-iron wheels coupled, hardened steel-link motion, expressly arranged for keeping the boiler unusually low in the frame, steel piston rod, slide bars, copper fire box and steam pipes, brass tubes, patent brass fittings. It is also fitted with the Lilleshall Company's patent compensating buffers, which adapt themselves to take an equal strain round sharp curves. The whole is built extra strong, to resist the wear and tear of heavy gradients, sharp curves, and the frequent inequalities of colliery roads.

The exhibitors are manufacturers of all kinds of high pressure expansive and condensing engines, sugar and other mills, heavy machinery for forges and rolling mills, chilled and grain rolls, cylinders, &c.; all made from their well-known Lilleshall cold blast iron, of the best workmanship, and at moderate prices.

Some specimens of coals and argillaceous ironstones, from which Lilleshall pigs are made, may be seen in Class I.

[1269]

London and North-Western Railway Company, *Works at Crewe.*—Locomotive engine and tender. (*See page* 14.)

[1270]

——————— Apparatus for supplying water to tenders whilst in motion. (*See page* 15.)

[1271]

——————— Wrought iron chair. Duplex safety valve. (*See page* 15.)

[1272]

McConnell, James, *West Houghton, Bolton-le-Moors.* — Self-acting railway signal for day and night.

[1273]

Macintosh, Charles, & Co., 3 *Cannon Street West, London; and Cambridge Street Manchester.*—Vulcanised rubber buffers, bearing springs, &c.

[1274]

Manning, Wardle, & Co., *Boyne Engine Works, Hunslet, Leeds.* (*See page* 16.)

LONDON AND NORTH-WESTERN RAILWAY COMPANY, *Works, Crewe.* — Locomotive engine and tender.

LOCOMOTIVE ENGINE. — Designed and built by Mr. Ramsbottom, Locomotive Superintendent, Crewe, and exhibited as a specimen of a first-class passenger engine. It is fitted with patent pistons duplex safety valves and lubricators, and is adapted for burning coal with great economy. An engine of this class ran the American express on the 7th January 1862, a distance of 130½ miles without stopping, at an average speed of 54 miles per hour. The tender attached is fitted with Mr. Ramsbottom's apparatus for taking up water whilst running.

LONDON AND NORTH-WESTERN RAILWAY COMPANY, *Works, Crewe.*

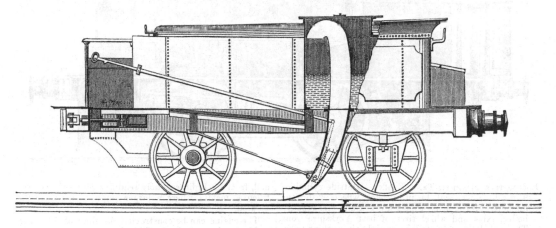

APPARATUS FOR SUPPLYING WATER TO TENDERS WHILST IN MOTION.

This is the invention of Mr. Ramsbottom, Locomotive Superintendent, Crewe. The plan has been in daily operation on the Chester and Holyhead Railway since it was first adopted in the winter of 1859—60. By it various quantities of water, from 1,200 gallons downwards, can be picked up at speeds ranging from 22 to 50 miles or upwards per hour. In the running of the Irish mails, the arrangement has the effect of reducing the dead weight of the tender about 6 tons, equal to the weight of a loaded carriage.

LONDON AND NORTH-WESTERN RAILWAY COMPANY, *Works, Crewe.*—Wrought iron chair for permanent-way. Duplex safety valve.

WROUGHT IRON CHAIR.—Invented by Mr. Ramsbottom, Locomotive Superintendent, Crewe. The above are made from rolled bars of suitable section, and are cut off whilst hot to the requisite breadth. The rails are held by the sides and shoulders, and are not in contact with the bottom of the chair. The lower head of the rail is consequently not indented, so that when inverted the second head is as good as the first. Both chairs and keys are of wrought iron, so that they do not get loose nor break. The chair has also a very broad base in proportion to its weight, and is therefore not so easily crushed into the sleeper. It is, moreover, applicable to a variety of sections of rails, by merely altering the form of the keys or filling-in pieces.

DUPLEX SAFETY VALVE.—This arrangement of safety valve, the invention of Mr. Ramsbottom, Crewe Works, is intended to prevent the boiler to which it is applied from being subjected to any pressure in excess of that to which it is adjusted. If any weight is put upon the lever, it has the effect of reducing the pressure, instead of increasing it, as in the ordinary arrangement. It requires no spring balance, and gives a much wider opening for a given excess of pressure, than the ordinary valve.

[1275]

MORRIS, E., 8 *Albert Square, Clapham Road, S.*—Patent iron wedge for securing railway rails in their chairs.

[1276]

MOULTON & CO., *Bradford, Wilts.*—Buffers.

MANNING, WARDLE, & CO., *Boyne Engine Works, Hunslet, Leeds.* — Locomotive tank engine, for contractors, collieries, &c.

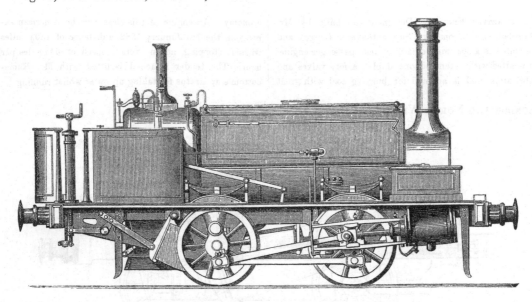

LOCOMOTIVE MINERAL TANK ENGINE.—Outside cylinders 9 in. diameter, and 14 in. stroke; wheels 2 ft. 9 in. diameter, all coupled; copper fire-box and brass tubes; boiler, axles, and wheel tires, of best Yorkshire iron. The tank holds 250 gallons; weight, in working trim, 10¼ tons.

This little tank engine was designed expressly for the mineral traffic at iron works, collieries, &c., and will go round any curve where an ordinary railway waggon will pass.

It is also admirably adapted for contractors' purposes; the wheels being small, it will ascend steep gradients, and, from its lightness, may readily be worked over contractors' metals, where a larger engine could not safely be used.

The engine can be constructed for lines of 3 ft. gauge and upwards, and the buffers placed to suit any special mineral or ballast waggons.

PHOTOGRAPHS.—The frames contain photographs of some of the many classes of engines, boilers, and other machinery, made by the same firm.

For prices and further particulars apply to MANNING, WARDLE, & CO.

[1277]

MURPHY, JAMES, *Railway Works, Newport, Monmouthshire.*—Pair of dovetailed-tire railway wheels, and safety bolt and nuts.

[1278]

NEATH ABBEY IRON COMPANY, *Neath.*—Locomotive engine for collieries, mine works, and quarries.

[1279]

NEILD & CO., *Dallam Iron Works, Warrington, Lancashire.*—Railway wheels, axles, tires, and bar iron, &c.

[1280]

NEILSON & CO., *Hyde Park Locomotive Works, Glasgow.* — Eight-feet wheel express engine.

[1281]

NETHERSOLE, W. E., *Swansea.* — Model of the frame of a railway-waggon, showing exhibitor's side-chain arrangement.

[1282]

NETHERSOLE, W. E., *Swansea.*—Model of improvements in draw gear and end-tipping waggon flaps.

[1283]

NEWALL, JAMES, *Bury, Lancashire*—Continuous railway breaks; signal, and patent gas apparatus for lighting railway trains.

[1284]

ORDISH & LE FEUVRE, 18 *Great George Street, Westminster.* — Ordish's patent elastic chairs and sleepers.

This system of fastening rails in chairs or sleepers is by utilizing the elastic power of the cast iron in the disposition of the metal for this purpose. The cast iron keys and jaws are provided with obtuse angled ratchets, and when driven in, the jaws are sprung asunder, thus holding and securing the rail, as the keys cannot come out, or be driven out, without springing the jaws further asunder.

[1285]

OWEN, WILLIAM, *Phœnix Works, Rotherham.* — Wrought engine and carriage wheels, stamped; patent axles and solid tires.

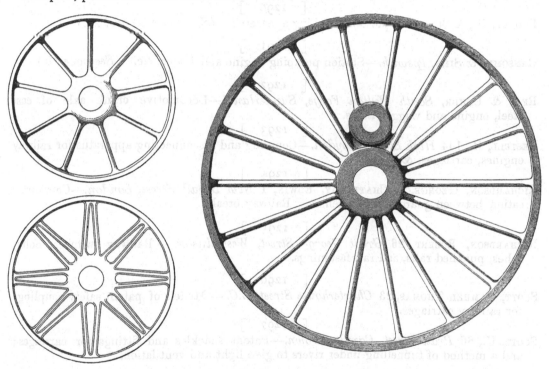

ARBEL'S PATENT STAMPED SOLID WROUGHT-IRON WHEELS.

OWEN'S PATENT AXLES.

PATENT ROLLED.
SOLID WELDLESS TIRES.

PATENT STAMPED SOLID WROUGHT-IRON WHEELS. — The peculiar recommendation of these wheels is their process of manufacture, being made in one piece under an immensely powerful hammer, by which perfect solidity is insured; so much so, that on cutting any of these wheels to pieces in a lathe, no trace of welding can be discovered.

OWEN'S PATENT ROLLED SOLID WELDLESS TIRES. — These tires possess the following advantages :—

They are made from a solid mass into a circular form, so that no alteration of the structure takes place in bending.

The whole surface of the tire, when at a welding heat, is subject to the action of the hammer — thus perfect soundness is necessarily obtained. After hammering, they are again heated to a welding heat, and afterwards rolled by patent machinery into a perfectly true ring, and thus welding is avoided, rendering breakage by that process impossible.

They are finished by the patent rolling process perfectly true to any dimensions required, and turning and boring is therefore unnecessary, and the external skin of the iron is preserved for wear, whilst the quality of the iron is greatly improved.

[1286]

PARSONS, P. M., 9 *Arthur Street West, London Bridge, London, E.C.* — Patent railway switch.

[1287]

PATENT SHAFT AND AXLETREE COMPANY, *Brunswick Iron Works, Wednesbury.*—Wheels, axles, tire iron, tire fastening. Models &c.

[1288]

PERMANENT WAY COMPANY, 26 *Great George Street, Westminster.*—Rail joints for railways; preserved timber for sleepers.

[1289]

PERRY, H. J., Jun., 3 *Greenwich Rd., Greenwich.*—Working model of atmospheric railway.

[1290]

POOLEY, H., & SON, *Liverpool.*—Weighing apparatus. (*See pages* 20 *and* 21.)

[1291]

RANSOMES & SIMS, *Ipswich.*—Station pumping engine and boiler, &c. (*See page* 19.)

[1292]

REAY & USHER, *South Hylton Forge, Sunderland.*—Locomotive crank axle of cast steel, engine and waggon axles.

[1293]

RESTELL, R., 144 *High Street, Croydon.*—Coupling and disconnecting apparatus for railway engines, carriages, &c.

[1294]

RICHARDSON, GEORGE, & CHATTAWAY, EDWIN, 1 *New Broad Street, London.*—Communication between guard and engineman. Railway break.

[1295]

RICHARDSON, ROBERT, 26 *Great George Street, Westminster.* — Railway switches, bolts, fishes, punched rails, and rail fastenings.

[1296]

SCOTT, SAMUEL THOMAS, 23 *Charterhouse Street, E.C.* — Models of patent safety couplings for railway carriages.

[1297]

SCOTT, U., 66 *Pratt Street, Camden Town.*—Patent shackles and fittings for carriages; and a method of tunnelling under rivers to give light and ventilation.

[1298]

SEATON, W., 44 *Albemarle Street.*—Safety saddle-rail. (*See page* 22.)

[1299]

SHARP, STEWART, & Co., *Atlas Works, Manchester.*—Goods engine fed by two of Giffard's injectors, and fire box arranged for burning coal.

[1300]

SIMONS, W., & Co., *London Works, Renfrew.*—Railway chairs, sleepers, and foundry castings.

[1301]

SPENCER & SONS, *Newcastle-on-Tyne.*—Cast-steel tires, spring buffers, &c. (*See page* 23.)

[1302]

STEVENS & SON, *Darlington Works, Southwark.*—Semaphore signals. (*See page* 24.)

[1303]

STRAFFORD, CAPT. P. P., *St. James's Square.*—Self-acting railway signal.

[1304]

TIZARD, WILLIAM LITTELL, C.E., 12 *Mark Lane, London.*—A bolt and nut fastening washer for railway fish-plates, &c.

RANSOMES & SIMS, *Ipswich.*—Combined steam station pumping engine and boiler; compressed railway fastenings; sundry castings.

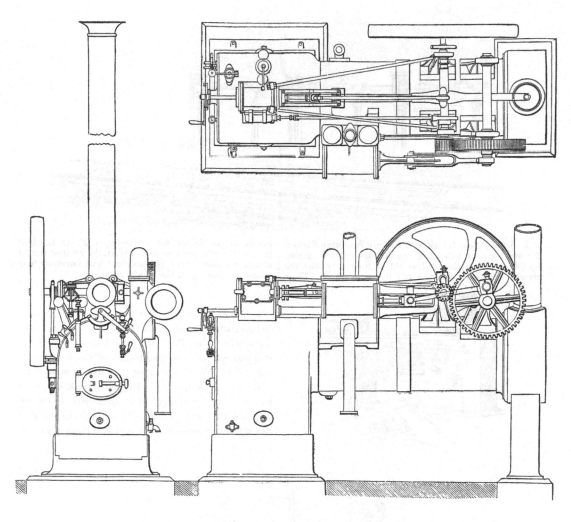

A THREE-HORSE POWER PUMPING ENGINE.

This engine is especially applicable for filling water-tanks at railway stations, and such similar purposes. The pump and the engine are both mounted upon a multitubular (locomotive shape) boiler, which rests upon two cast iron pedestals, and requires no further fixing. This engine is capable of raising in 10 hours between 4,000 and 5,000 cubic feet of water, from a depth of 15 feet under the ground where it stands to a height not exceeding 100 feet above the same level. But the engine is equally adapted for any other work for which engines of its power are used.

The pump is lined with brass; the bucket packed with leather; the valves of india-rubber seated on brass, and so arranged that they can easily be taken out, two at a time, for inspection or renewal.

Weight — about 45 cwt.

Measurement — about 140 cubic feet.

A SERIES OF COMPRESSED KEYS AND TRENAILS, for Railway and other purposes.

These fastenings are used very extensively on the English lines of railway, and have been almost entirely adopted in laying the railways in India and Australia. They are made of the best wood, which is cut out much larger than the finished size, is very carefully desiccated, and then compressed by machinery. These fastenings fit the chairs and rails accurately; they last longer than uncompressed fastenings; and when driven into their places in the permanent way, they return, under the influence of rain and moisture, towards their original size before compression, thus completely filling the hole into which they are driven, and holding the rail firmly in its place in the chair, and the chair firmly to the sleeper. The permanent way of the Great Northern Railway is laid with these compressed keys and trenails, and the speed and comfort attained on it prove them to be an admirable fastening. RANSOMES & SIMS have a patent for injecting these fastenings with the vapour of creosote, which adds but little to the expense and greatly increases their durability.

A SAMPLE SERIES OF CASTINGS MADE BY UNSKILLED LABOUR — intended to show the perfection in production which may be obtained by the use of Patented Moulding Machinery.

POOLEY, HENRY, & SON, *Liverpool*; *London House* — HENRY POOLEY, SON, & CO., 89 *Fleet Street, E.C.*—Railway, commercial, and mining weighing apparatus.

Obtained the Prize Medal in 1851.

1. Set of Patent LOCOMOTIVE ENGINE WEIGHING TABLES, for weighing and balancing, or adjusting, engines; giving by one operation the total weight of the engine, and the weight imposed upon the rail by each wheel. Their use is to enable the superintending engineer to adjust the springs of engines so as to obtain the greatest amount of tractive power that is consistent with immunity from danger of running off the line at curves.

2. 'PILE' WEIGHING MACHINE. For rolling mills, and specially for rails. Its use will be obvious to any ironmaster. The 'Piles' are formed upon a small truck, standing upon the weighing portion of the frame next to the 'Piler.' When the amount of iron to make a rail of the required weight is piled upon the truck, it is pushed forward to the workman, to be transferred to the furnace, and thence to the rolls. By this machine, the great loss attending guess work in such operations is avoided.

3. MACHINES FOR ROLLING MILLS AND FORGES — The rollers facilitate the loading and unloading of heavy rails and forgings, the locked and enclosed pent-house preserves it from weather and pilferage when exposed on the public quay or yard. Two sets of labourers can be employed at once — one loading and the other unloading, avoiding all loss of time in waiting between the loads.

POOLEY, HENRY, & SON, *Liverpool*; *London House* — HENRY POOLEY, SON, & CO., 89 *Fleet Street, E.C.* — Weighing apparatus — *continued.*

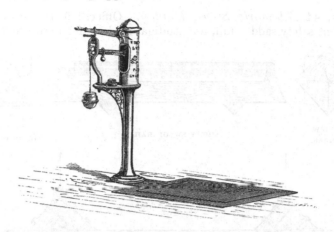

4. THE RAILWAY GOODS WAREHOUSE MACHINE.— First introduced by Messrs. Pooley at the Liverpool and Manchester Railway, 1835; and as the specimen exhibited shows, now greatly improved in design, construction, and exactitude. It is 'dormant' except when put in gear by the man in charge. The weighing-table forms part of the floor, and encumbers no space. Its accuracy is equal to the best scale-beam, whilst labour and cost are economised at least 50 per cent. It is only by means of these machines that the heavy merchandise traffic of railways could be despatched with adequate speed; it has, therefore, become the machine of the goods trade generally, not only for railways, but for general commerce.

5. THE PARCELS OFFICE MACHINE. — combining the instantaneous self-acting indications of the spring dial machine with the strength and convenience of the platform weighing machine. The load to be weighed is simply deposited on the platform, when, on drawing down the lever by the suspended handle, the weight is seen at once upon the face of the dial.

6. The PARCELS OFFICE MACHINE, for use upon the counter; very exact in its indications, but not so speedy as the dial. The low price, combined with great correctness, are its recommendations.

7. PASSENGERS' LUGGAGE MACHINE.—No railway station can be complete, and no company safe from fraud, without one.

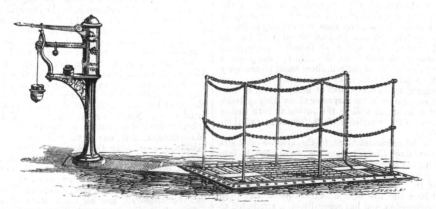

8. THE CART WEIGH-BRIDGE is of the smallest size, but it exhibits the principle and construction adopted for vehicles of every form, and of every capacity used upon highways. The present example is specially fitted for farmers' use; and, being cheap, very easily erected, without any masonry, and requiring no mechanic to erect or remove it from place to place, is well adapted for the farm.

SEATON, WILLIAM, 44 *Albemarle Street, London.* Offices: 5 *Parliament Street, Westminster.*—Patent safety saddle rail, longitudinal timbers for permanent way.

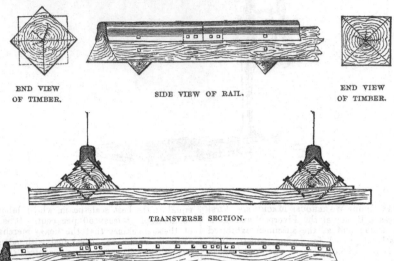

END VIEW
OF TIMBER.

SIDE VIEW OF RAIL.

END VIEW
OF TIMBER.

TRANSVERSE SECTION.

PERMANENT WAY, CONTINUOUS BEARINGS.

The advantages attending the adoption of this system of Permanent Way, may be shortly stated as follows:—

1. The Rail is supported throughout by a solid and continuous bearing of timber, having a bearing surface on the ballast of 17 inches in width.

2. The Rail has a firm surface bearing on the timber of 288 inches per lineal yard, and the pressure being at right angles to the flange, it has a tendency to compress the fibre of the timber, rendering it firmer and harder in proportion to the pressure applied.

3. The Rail and Sleeper are liable to no decay or injury from rain and wet; the form, being pyramidal, has no surface on which water can lodge.

4. The mode of joining and supporting the ends of the rails, by an under saddle-plate, is found to be thoroughly effective, and much less costly than any existing mode.

5. This system requires no iron or other chairs; no keys, fish plates, pins, spikes, or trenails; or separate pieces, liable to become loose; its only fastenings are a few bolts on each side of the rail, there being no tendency in the rail to quit its seat, or to work loose.

6. In regard to economy, both of first cost and of subsequent maintenance, it far surpasses all existing systems, a fact which is due, as well to the saving of both timber and iron, as to the simplicity of its fastenings and their immovability.

7. To the latter circumstance, to the absence of chairs or other fastenings likely, by accident, to be brought into contact with the flanges of the wheels, and to the perfect mode of joining the rails, we may look for an entire absence of that class of dangerous accidents, so common of late, where the tire of a wheel has given way from a violent concussion, and has generally been attributed to unknown causes.

The merits of the system, then, may be summed up as follows:—

1. SIMPLICITY OF CONSTRUCTION.
2. SAFETY.
3. ECONOMY IN FIRST COST.
4. ECONOMY IN MAINTENANCE.

This system has been laid down for trial on several of the Main Lines of Railway, and has been amended from time to time, as improvements have suggested themselves. The portion of Railway which may be referred to as embodying the whole of these improvements, as affording the best illustration of the superiority of the system, and to which the foregoing description is in all respects applicable, is part of the Down Line of the Great Western Railway near Kensal Green, which was laid in May, 1858, over which the whole of the large traffic of that Company out of London has, in the interval, passed. At the present time (May, 1861), the ballast on that line has not been disturbed for twelve months; the grass is growing between the rails; and the fastenings are as firm and immovable as on the day when it was laid down.

Attention is requested to the testimony of ROBERT BENDON DOCKRAY, ESQ., one of the earliest and most experienced Engineers in the kingdom, and who had, for eighteen years, under ROBERT STEPHENSON, ESQ., the charge of the Maintenance of the Permanent Way of the London and North Western Railway. This extract is taken from a Report made by Mr. DOCKRAY to CAPTAIN MARTENDALE, R.E., one of the Government Inspectors of Railways, soon after the Road was first laid; and the note which follows it, and which has recently been obtained from Mr. DOCKRAY, records his opinion of the system after it has been under trial for three years.

Plans and Models of this system of Permanent Way can be inspected, and all information obtained at the Offices, 5 Parliament Street, Westminster; where, if desired, contracts will be entered into for its Construction and Maintenance for 7, 14, or 21 years, at rates varying from ten to fifteen per cent. lower than those of other systems.

SPENCER, JOHN, & SONS, 124 *Fenchurch Street, London, and 5 Westgate Street, Newcastle-on-Tyne.*—Cast steel tires, volute spring buffers, springs, steel and files.

JOHN SPENCER & SONS manufacture CAST STEEL, BLISTER STEEL, and SPRING STEEL, FILES of every description, HAMMERED CAST STEEL TIRES, RAILWAY CARRIAGE and ENGINE SPRINGS, and PATENT VOLUTE SPRING BUFFERS for stations, and rolling stock of all kinds.

1. CAST STEEL INGOT FOR A TIRE.
2. DITTO HAMMERED DITTO.
3. DITTO FINISHED ROLLED TIRE.
4. HAMMERED CAST STEEL SLIDE BAR.
5. DITTO PISTON ROD.
6. VOLUTE SPRING BUFFER, WITH WROUGHT IRON CASING.
7. DITTO, WITH WROUGHT IRON PLUNGER.
8. DITTO, WITH CAST IRON CASING.
9. SPENCER & CORLETT'S COMBINED PATENT VOLUTE BUFFER, WITH WROUGHT IRON PLUNGER AND HEAD.
10. DITTO, ENTIRELY WROUGHT IRON.
11. ALLEN'S PATENT COMPOUND VOLUTE BUFFER.
12. SPENCER'S PATENT RIBBED-END ENGINE BEARING SPRING.
13. CARRIAGE BUFFER SPRING.
14. CARRIAGE BEARING SPRING.
15. SPENCER'S PATENT RIBBED-END WAGGON BEARING SPRING.
16. SPECIMENS OF ADAMS' PATENT VARYING LOAD ABUTMENT SPRING.
17. SPECIMENS OF PLAIN AND RIBBED VOLUTE SPRINGS, AND OF THE ROLLED STEEL EMPLOYED IN THEIR MANUFACTURE.

The volute spring obtained the only prize awarded for springs at the Hyde Park International Exhibition of 1851. The construction has since been improved by adding ribs rolled upon the strip of steel to strengthen it, and keep the coils apart from each other, giving the spring more freedom of action.

Stevens & Son, *Darlington Works, Southwark, S.E.*—Patent iron semaphore railway signals, and compensating signal wire apparatus.

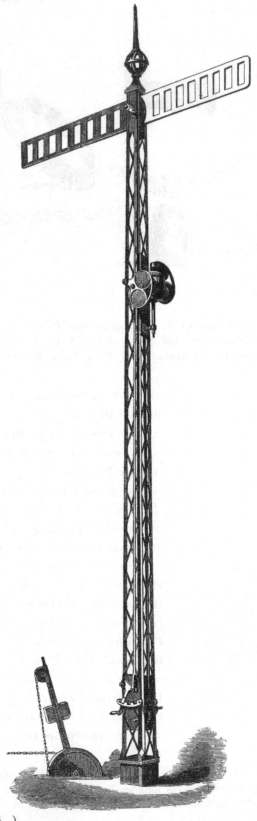

1. Patent Wrought Iron Semaphore Railway Station Signal, fitted with 8″ lens lamp for oil, and improved apparatus, complete, ready for work.

One of these signals, when fixed at a railway station, acts for both the 'up' and 'down' lines for day and night. It is the most durable and effectual signal in use, and, although recently invented, it is already adopted on many of the lines in the United Kingdom, also in India, Australia, &c.; and, being made of open iron work, it is not acted on by the most violent gales; while the strong cast iron base renders it most secure and impervious to the decay to which timber signals are liable.

2. Patent Wrought Iron Distant Signal, fitted with Brydone's patent candle signal lamp and apparatus complete, ready for work.

These signals are fixed, in many instances, 1800 yards from the railway stations, and are worked at that distance with the greatest facility, and with no more difficulty or uncertainty of action, than at 100 yards from the station.

3. Patent Cast Iron Distant Signal.

These are made in cast iron, where only short signals are required; they possess all the advantages of the wrought iron signal, for heights not exceeding 20 ft., and are less in price.

4. Patent Compensating Pullover Lever, with ratchet weight and chain complete.

These levers are for working the auxiliary signals at a distance from the station, the lever being fixed at the station, or on the junction platform. The advantage of these over the ordinary levers is, that by means of the ratchet balance weight and rack fitted to the lever, the expansion or contraction of the wire through the variation of temperature is compensated.

5. The Patent Compensator.

These are placed at intermediate points between the lever and the signal, in cases where wires of extraordinary length are required.

6. Patent Point Indicator.

These are fixed at railway stations, &c., where there are frequently a number of diverging lines. It is desirable to have an efficient apparatus to show when the shifting rail or point is open or shut. These indicators show most distinctly, by day or night, the state of the points. By day the disc divides, and by night the lamp placed at the back of, or between the disc or discs, shows a red, green, or white light.

[1305]

TRUSS, THOMAS S., C.E., 53 *Gracechurch Street, London.*—Patent cushioned railway chair, and packing.

SIDE ELEVATION.

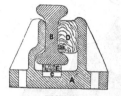

SECTION.

T. S. TRUSS, C.E., 53 Gracechurch Street, London, is the patentee and manufacturer of chemically prepared woollen packing for pipe joints, tanks, bed-plates, railway chairs, &c., &c.

By the application of this packing to railway chairs, the chair and the under head of the rail are entirely protected from friction.

The working head will last much longer, as the violent concussion caused by the train is absorbed by the packing, and thus the nature of the metal is preserved.

This packing is made of any required size, form, or thickness, and is rendered almost indestructible by the process through which it is passed.

Reference to the sectional drawing of railway chair: A, chair; B, rail; C, patent packing; D, wood key; E, wood seating.

[1306]

VICKERS, ARCHIBALD, *Bristol.*— Method of opening, shutting, and fastening four gates simultaneously, applicable to railway crossings.

The object of this invention is the prevention of accidents and loss of life at railway crossings. Some short time ago an accident occurred on the Midland railway, at Bristol, by which three young girls endangered their lives. At the adjourned inquest the gate-keeper deposed, that on the morning of the fatal occurrence he found it impossible to close the gates, owing to the violence of the wind, which, in spite of his efforts, blew them open again after each attempt. Finding the gates unclosed, the unfortunate girls attempted to cross the line, and one of them was killed. The small model exhibited was designed and constructed during the adjournment of the inquest.

[1307]

WALKER, WILLIAM, 3 *Atholl Lane, Edinburgh.*—Invoice box and ticket-keeper, can be supplied by the inventor at £9 per 100.

[1308]

WESTON & GRICE, *Stour Valley Iron Works, West Bromwich.* — Bar iron and railway fastenings.

[1309]

WISE, F., 22 *Buckingham Street, Adelphi.*—Ramié's railway chairs. (*See page* 26.)

[1310]

WRIGHT, JOSEPH, & SONS, *Saltley Works, Birmingham.*—First-class carriage, constructed for the Egyptian Railway.

This carriage is constructed with framework of Moulmein teak, and panels of papier maché. It is provided with an upper roof, as a protection from the sun; the lower roof being fitted with lamps and movable ventilators. The interior is trimmed with light drab morocco; the windows in both doors and quarters are made to fall, and are fitted with Venetian sun-shades and spring curtains. Ventilators, to be opened or closed at pleasure, are provided above the windows. The under-frame is of wrought iron of the most approved construction; and the carriage is mounted on solid wrought iron wheels, having Beattie's patent tire fastening. The axle-boxes are also Beattie's patent, to work either with oil or grease.

[1311]

WRIGHT, PETER, *Constitution Hill Works, Dudley.*—Patent railway wheels and railway axles.

[1312]

WRIGHT, T., *George Yard, Lombard Street.*—Permanent way.

[1313]

FAIRBAIRN, GEORGE, *Manchester.*—Working model of a tank locomotive engine.

Wise, Francis, C.E., 22 *Buckingham Street, Adelphi, London.*—Railway chairs without wedge or bolt (Ramié's patent).

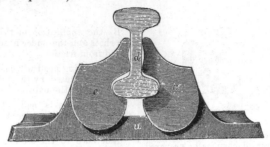

ELEVATION.

This chair (Ramié's patent) secures the rail without the employment of any wedge, bolt, or key, is self-tightening, and affords the greatest possible facility for placing, reversing, and removing the rail. Its action is such as to firmly secure and support the rail, maintaining its under head entirely uninjured, so that when reversed, instead of being 'chair-worn' into notches, as is the case where ordinary chairs are used, it is equal for working purposes to a new rail. It also avoids the evil which arises when a chair is so constructed that the under sides of the upper head of the rail rest upon rigid abutments, which, under traffic, act as anvils, between which, and passing wheels, the rail head is speedily hammered out.

It is well known that chairs of the ordinary kind, in which the rails are secured by wooden keys, are a constant source of trouble, annoyance, and expense, owing to the continual expansions and shrinkings of the keys, which, notwithstanding constant care and the employment of men along the line continually 'tightening up,' are rarely, if ever, in a condition to hold the rails in the chairs with anything like firmness and solidity; and in very numerous instances (perhaps one-half), are so loose as to fall out with the slightest push, and are without any effect whatever in holding the rails.

The natural consequences of this defect, under traffic, are to cause a continual hammering of the rails upon the chairs, which very quickly produces deep indentations in the rails, and thereby converts their under-tables into a kind of rack or succession of notches, which, upon the inversion of the rails, constitute a sort of corduroy road, and act as an efficient agent in the rapid destruction of the rolling stock, and of annoyance to the passengers over it. In some cases more serious results occur—see case of Taylor *v.* Manchester, Sheffield, and Lincolnshire Railway ('Times,' March 20, 1862), where £400 damages were awarded to plaintiff for injuries received, owing to the carriage in which she was travelling, and other carriages of the train, leaving the line. In this case it was clearly shown in evidence, that the keys whereby the rails ought to have been *held* were scattered along the line, and the accident

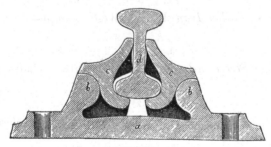

SECTION.

was attributed to that cause. This is but one (and a comparatively unimportant instance) among a great number of cases in which accidents—some of them attended with terrible results to life and property—have arisen from carriages leaving the line; and is merely put forward as being so clearly and unmistakably traceable to the imperfection of the ordinary method of attempting to secure the rails in their chairs.

Ramié's chair consists of three parts, as will be seen on reference to the accompanying elevation and section. The main casting, *a*, is secured to the sleeper by spikes or trenails, in the ordinary manner, and is formed with curved abutments *b*, upon which rest the 'tumble' jaws *c*, which carry the rail *d*. The weight of the rail, pressing upon the lower parts of the jaws, causes their upper parts to close upon its web with a force which is amply sufficient to maintain it securely in position when not under traffic. On a train passing over, the amount of grip or force with which the rail is held is increased directly in proportion to the passing weight.

Should the sleeper carrying the chair become beaten down into the ballast below its proper level, attention is at once drawn to the fact by the upper parts of the jaws standing slightly away from the web of the rail. Although by this it is at once apparent that the sleeper requires packing in order to keep the rail up to exactly its proper level, it does not in any way deteriorate from the security of the hold or grip upon it under traffic, as when the weight of a carriage comes upon the part, the rail is sprung downward until the jaws smoothly take their bearing, and grip firmly and solidly upon its web.

Ample play is allowed between the several parts of the chair, so that the accuracy of ordinary casting is quite sufficient to insure its efficiency.

Several years' experience of the working of this chair under the heaviest traffic, shows that it acts in the most perfect manner, and never allows chattering to occur between itself and the rail.

Further particulars may be obtained on application to Mr. Francis Wise, C.E., as above.

Class VI.

CARRIAGES NOT CONNECTED WITH RAIL OR TRAMROADS.

[1338]

ALDEBERT, ISAAC, 57 *Long Acre.*—A barouche landau. (*See page* 28.)

[1339]

ANDREWS, ARTHUR, 14 *Above Bar, Southampton.*—Light and elegant 'Eugénie' park phaeton.

[1340]

ANGUS, HENRY, *Newcastle-on-Tyne.* — A useful light one-horse double-seated brougham, with segmental front and improved break.

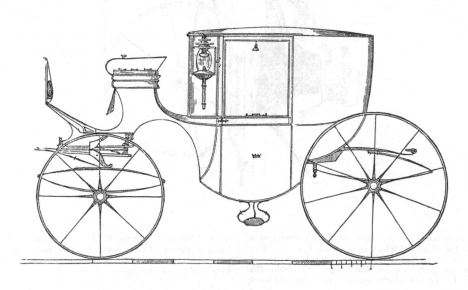

ANGUS'S LIGHT BROUGHAM.

ALDEBERT, ISAAC, 57 *Long Acre.*—A barouche landau, constructed with steel instead of iron, and fitted with Aldebert's patent noiseless springs.

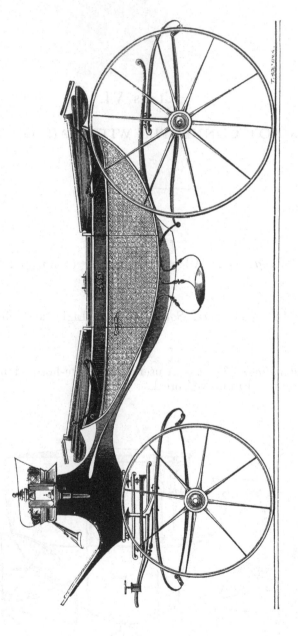

A BAROUCHE LANDAU.

A BAROUCHE LANDAU, constructed with steel instead of iron, the object of which is to obtain lightness, while the size of the carriage is not decreased. The wheels are made of foreign timber, so hard, tough, and durable that they can be worked much lighter than those ordinarily used. The body is hung on Aldebert's Patent Noiseless Springs, giving all the ease, comfort and quietude of C and under springs, without the additional expense or weight of perch carriages.

[1341]

BENNION & HEALEY, *Liverpool.*—Paragon brougham, circular front, roomy for four inside, high wheels, weight 8 cwt.

The peculiarity of the Paragon Brougham consists in the increased accommodation it affords, with greatly reduced weight and draught; and, in the employment for the springs and ironwork, of the Bessemer steel and iron, which combine increased toughness and endurance, with reduced weight, and unusual lightness of appearance. There is also an opera board, which is readily convertible into a luggage platform when required.

[1342]

BLACK, H., & SON, *Berners Street, Oxford Street.*—A light C and under-spring coach.

[1343]

BOOKER & SONS, 13 *and* 14 *Mount Street, Grosvenor Square.*—A 'sociable.'

[1344]

BOYALL, RICHARD JOHN, *Grantham Carriage Manufactory.* — Park or road phaeton. *(See page 30.)*

[1345]

BRABY, JAMES, & SON, *Newington Causeway, Southwark.*—A spring waggon, with improved patent wheels and break.

[1346]

BRIGGS, GEORGE, & Co., 45 *Wigmore Street.*—A carriage.

[1347]

BURNETT, EDMUND, *Ashford, Kent.*— Gorilla cart, to form either cart or sleigh.

[1348]

BURTON, HENRY LESNEY, 12 *Nowell's Buildings, Liverpool Road, Islington, N. London.* Perambulators and propellers.

[1349]

CAMPBELL, FREDERICK, Coach-Builder, *Dumfries.* — Sporting cart of Scotch elm varnished, adapted for dogs, luggage, or game.

[1350]

CAMPBELL, ROBERT FELIX, 8 *Brook Street, Gloucester Place, Hyde Park.*—Apparatus for the prevention of accident to carters, &c.

[1351]

CASE, C. J., 36 *Jamaica Street, Commercial Road East.*—Small model of an omnibus, made entirely of brass and steel.

[1352]

CHANTLER, JOHN DALE, *Ardwick Coach Works, Manchester.* — Light four-wheeled carriage.

[1353]

CLARKE BROTHERS, *Shiffnal, Salop.*—Patent tubular iron carriage shafts.

CLARKE BROTHERS are the patentees and manufacturers of tubular iron carriage shafts. CLARKE & TIMMINS, 10 Soho Square, W., their London agents, will supply all information as to prices &c.

BOYALL, RICHARD JOHN, *Grantham Carriage Manufactory.*—Handsome park or road phaeton, hung upon inverted double C-springs, remarkably easy and very light.

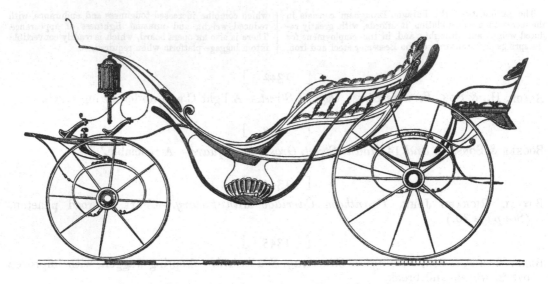

A LADIES' DRIVING PHAETON, OR FOR A POSTILION TO RIDE AND DRIVE.

A Ladies' Driving Phaeton for park or road, with a seat for servant behind. It is hung on improved inverted double C-springs and leather braces, whereby it is rendered remarkably easy. It is fitted with Collinge's patent axles, is elegantly painted and trimmed, and finished with plated furniture and richly-ornamented lamps, in the highest style of decoration.

Besides the specimen carriage exhibited, the most noticeable manufactures of this Exhibitor are, his Medium Brougham, which is at once roomy, comfortable, and light; his Grantham Sociable; and a variety of waggonettes and wicker carriages. In all of these, the designs and workmanship are of superior character.

References are permitted to many of the leading members of the aristocracy.

[1354]

COCKSHOOT, JOSEPH, Manufacturer, *New Bridge Street, Manchester.*—Medium-sized brougham, of lightest possible construction, combined with elegance and utility.

[1355]

COLE, W., Coach-Builder, *Kensington.*—Brougham, C- and under-springs.

A C- and under-spring Brougham, the body suspended with long leather braces, rendering it remarkably easy; supported by a crane-necked perch carriage, and patent axles. This carriage is made with every attention to elegance, lightness, and durability, combined with all improvements. It is lined with rich blue silk, the body painted a fine claret, relieved with silver mountings, the carriage and wheels scarlet. It has handsome silver lamps, improved patent concealed hinges to the doors, and an improved step, combining the double advantage of an inside step for an invalid, if required. It is fitted with patent ventilators.

[1356]

COOK & HOLDWAY, 12 *Mount Street, Grosvenor Square.*—A sociable landau with an improved registered head.

[1357]

COOPER, BLACKFORD, & SON, 140 *Long Acre.*—Specimens of carriage laces and fancy trimmings.

[1358]

CORBEN & SON, 30 *Great Queen Street.*—A dioropha. (*See page* 32.)

[1359]

COUSINS, EDWARD, *Alfred Street, High Street, Oxford.*—A pony carriage.

[1360]

CROSS, T. W., & Co., *Hunslet Road, Leeds.*—Bath chair and perambulators.

[1361]

DART & SON, 12 *Bedford Street, Covent Garden.*—Coach lace.

[1362]

DAVIES & SON, 15 *Wigmore Street.*—Sociable landau, the panels partly in imitation of turned open sticks, concealed self-acting steps.

[1363]

DAVIES & SONS, Coach-Builders, *Northampton.*—A trotting phaeton.

DAVIES & SONS exhibit a Trotting Phaeton with movable hind seat, built and finished with materials and workmanship of the best description; Collinge's patent axles, with wrought boxes, and tough steel tires to wheels. Price 60 guineas.

A large assortment of Carriages is always on sale and building to order, at the manufactory.
Carriages built for exportation.
DAVIES & SONS' Dog-Carts are patronised by 500 gentlemen.

[1364]

EDWARDS, SON, & CHAMBERLAYNE, 21 *Newman Street, W.*—A light fashionable four-wheel carriage, painted and lined green.

[1365]

ELL, GEORGE, & Co., *Euston Works, Euston Road, London.*—An improved van. (*See page 33.*)

[1366]

EVANS, JAMES, *Tarlton Street, Liverpool.*—An improved two-wheeled Hansom cab, secured by Royal Letters Patent.

EVANS'S IMPROVED PATENT CAB.

The improvement consists in the application of metallic springs to the shafts near their junction with the front of the vehicle, thus securing a combined action which removes the unpleasant motion common to vehicles of the ordinary construction. Uneven and irregular roads produce no effect on the cab, as by means of the patent springs and joint an equal balance is preserved between the vehicle and the shafts. The draught is considerably lightened (a very important feature). The patent includes several other improvements, an important one being the reduction of the weight of the vehicle, which weighs only 6 cwt.

The price of Evans's Improved Patent Cab varies according to style and finish. Prices will be forwarded on application.

CORBEN & SONS, 30 *Great Queen Street, London.*—A dioropha carriage.

A DIOROPHA CARRIAGE.

The above engraving represents the Dioropha as a close carriage. The upper half is movable by the aid of a balance weight, cord, and pulleys attached to the coach-house ceiling; a folding or phaeton head is then fixed on, which, with the addition of folding flaps, makes it a perfect open carriage, as shown below. It is hung on CORBEN & Sons' improved inverted C-springs and leather braces, which render it as easy as a carriage on the ordinary C-springs and heavy perch, and do not increase its weight beyond the usual elliptic spring carriages. This kind of spring has been found particularly advantageous, and can be applied to almost any carriage.

THE DIOROPHA AS AN OPEN CARRIAGE.

ELL, GEORGE, & Co., *Euston Works, Euston Road, London*, Wheelwrights.—An improved van for general purposes; also models of heavy vehicles.

The following is a price list of the manufactures of these exhibitors :—

Improved Pillar and Standard Van, to carry 50 cwt., the same as the one exhibited	£40	0	to £46	0		
Ditto, to carry 4 tons	48	0	„ 54	0		
Builder's Van, to carry 2 to 5 tons	40	0	„ 60	0		
Corn or Flour Van, to carry 2 to 4 tons	40	0	„ 55	0		
Railway or Carrier's Van, to carry 2 to 6 tons	36	0	„ 60	0		
Furniture Van, complete	45	0	„ 55	0		
Stone Truck, to carry 6 to 10 tons	35	0	„ 50	0		
Timber Carriage, to carry 3 to 8 tons	28	0	„ 45	0		
Two-horse Brick Cart, to carry 1000 bricks	20	0	„ 30	0		
One-horse Brick Cart, to carry 500 to 600 bricks	13	0	„ 21	0		
Standard Dobbin Cart	9	10	„ 10	10		

Plank-sided ditto . . . . . . £9 0 to £10 0
Agricultural Cart . . . . . . 11 0 „ 18 0
Agricultural Waggon . . . . 30 0 „ 40 0
Light Spring Cart. . . . . . 16 0 „ 26 0
Joiner's Cart, to carry 30 cwt. . . 26 0 „ 28 0
Saw Mills Cart. . . . . . . 27 0 „ 30 0
Corn or Wine Cart . . . . . 24 0 „ 28 0
Improved Crank-Axle Stone or Slate Cart . . . . 32 0 „ 36 0
Improved dray. . . . . . . 25 0 „ 45 0
Improved Mortar Cart, with iron body 25 0 „ 32 0
Improved Cattle Conveyance . . . 34 0 „ 40 0
Builders' Hand-Carts. . . . . 5 10 „ 10 0

Trollies, Earth Waggons and Whims; Brickmakers', Excavators', Gardeners', and every description of Barrows; also, Ladders, Trestles, and Steps.

[1367]

FELTON, W. J. & C., 2 *Halkin Place, Belgrave Square.*—New brougham 'shofle,' comfort and lightness of brougham and cab united.

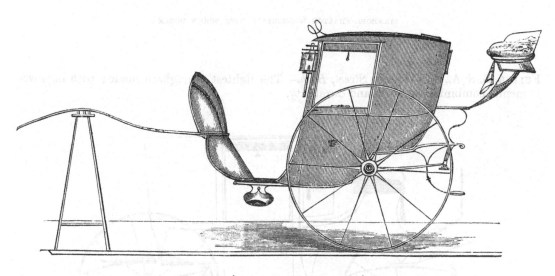

FELTON'S NEW BROUGHAM 'SHOFLE.'

[1368]

FINDLATER, WILLIAM, Coach-Builder, *Gas Street, Broad Street, Birmingham.*—Light brougham for one or two horses.

[1369]

FULLER, J., & SONS, *College Street, Bristol.*—Stanhope phaeton waggonette. (*See page* 34)

[1370]

FULLER, S. & A., *Kingsmead Street, Bath.*—Brougham weighing under 7 cwt. (*See page* 34.)

FULLER, J., & SONS, *Limekiln Lane, and College Street, Bristol.*—Stanhope phaeton waggonette, with screw break.

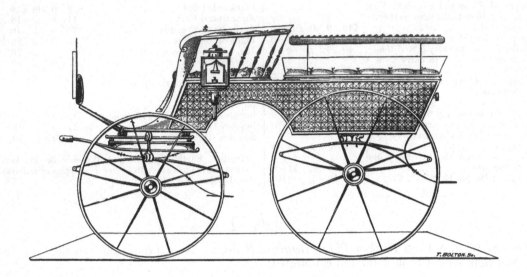

STANHOPE PHAETON WAGGONETTE WITH SCREW BREAK.

FULLER, S. & A., *Kingsmead Street, Bath.*—The lightest brougham made; with improvements, combining strength and durability.

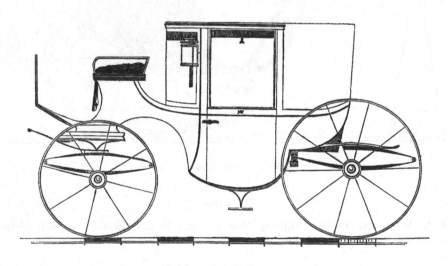

A FASHIONABLE BROUGHAM.

A fashionable Brougham upon the lightest possible construction consistent with strength and durability, and including all the latest improvements.

The weight of this Brougham is under 7 cwt. S. & A. FULLER, Manufactory and Show Rooms, Kingsmead Street, Bath.

[1371]

GITTINS, RICHARD, 28 *New Street, Dorset Square, London.*—New axletree, applicable to all kinds of carriages.

[1372]

GLOVER, JAMES THOMAS, *East Gate, Warwick.*—A light waggonette.

[1373]

HADLEY, C., 37 *Lower Hurst Street, Birmingham.*—Omnibuses, cabs, &c. (*See page* 36.)

[1374]

HALE, S. W., Manufacturer, 27 *Park Lane, Piccadilly.*—Elcho sociable, adapted for one or two horses.

[1375]

HALL & SONS, 98 *Long Acre.*—Barouche on elliptic springs. (*See page* 37.)

[1376]

HARVEY, JOSEPH, *Heron House, Richmond, Surrey.*—A patent two-wheel closed carriage.

[1377]

HAWKINS, JOSEPH, *Hatfield Street, Blackfriars Road.*—Arms, axletrees, and spring for all common road vehicles.

[1378]

HAZELDINE, GEORGE, 5 *Lant Street, Borough.*—Patent road van.

[1379]

HIGGINSON, CHARLES, Jun., 65 *George Street, Portman Square, W.*— Carriage heraldry.

[1380]

HIGGINSON, CHARLES, Sen., 15 *Henrietta Street, Manchester Square, W.*—Heraldic painting for carriages.

[1381]

HOLMES, H. & A., *Derby, Lichfield, and London.*—Park sociable, with improved landau head, upon C- and under-springs.

[1382]

HOLROYD, NOBLE, & COLLIER, *Halifax.*—Patent machine-made wheels; imitation wicker panelling; carved wood mouldings.

[1383]

HOOPER & Co., 28 *Haymarket.*—Sefton landau &c. (*See pages* 38 *to* 40.)

[1384]

HORSLEY, CHARLES, & SON, *Beccles.*—A light brougham.

[1385]

HOULGATE, FREDERICK, Carriage and Harness Manufacturer, *Scarborough.*—Handsome full-sized circular-fronted brougham. (*See page* 36.)

[1386]

HOWITT, W. J., 25 *Denmark Place, Soho; and* 52 *Parker Street, Drury Lane.*—C-springs and coach-smith work.

[1387]

HUTLEY, FREDERICK, 11 *Long Acre.*—New patterns in carriage laces.

[1388]

HUTTON, JOHN, & SONS, *Dublin.*—A round-fronted brougham; a very light Irish car.

HADLEY, CHARLES, *Lower Hurst Street, Birmingham.*—Single, double, and triple-bodied omnibuses, cabs, broughams, carts, waggons, hearses, &c.

Adding a forebody, A 1, in front of the present omnibus, brougham, and hearse bodies, A; lowering it to within a foot of the ground, avoiding steps; with hinged bottom; immured crank, and dwarf axles; also forming other separate bodies or recesses, B, or C, under, alongside, or upon it, all enclosed, and readily accessible by females.

Double-bodied saloon broughams and cars, seating six.

Brougham hearses, for mourners, bearers, and coffin.

Circular-fronted cab Broughams.

Circular wide-bodied Hansom cabs, seating three inside.

Double-bodied Hansom cabs, for three in and two out.

Double-bodied Hansom dog-cart cabs; seat five.

Hansom-cab hearses, for mourners, bearers, and coffin.

Widened beast transit carts, to carry two oxen abreast.

Watering carts, low and deep, on springs; high wheels.

Single bodied carts, lower and wider; higher wheels.

Lorry waggons, low, wide; openings to load each side.

Double bodied brewery, carrier, and other waggons.

Double bodied farmers' traps; for stock and produce.

Double-bodied carriers', brewery, and other carts.

Double-bodied scavengering and watering carts; to convey refuse, and water the roads simultaneously.

Scavenger and night-soil carts, to separate the liquid from the solid portions, to utilise labour, time, and cost.

Traction, transit, fire and power waggons, with portable engine combined, to propel itself: applicable also to other uses.

Boat waggons; for removing night soil from towns.

Vibrating flanged wheels, for rail, groove, or flat ways.

Metallic 'stepped,' and doubly secured axle box naves.

Stepped axles, to secure the nave at either end.

Cranked, dwarf, jib, sinuous, and expanding axles.

Breaks, to gain power down, for aiding up hill.

HOULGATE, FREDERICK, Carriage and Harness Manufacturer, *Scarborough.* — Handsome full-sized circular-fronted brougham.

HANDSOME FULL-SIZED CIRCULAR-FRONTED BROUGHAM.

The weight of the Brougham exhibited is 8½ cwt., price £135. Pole and splinter bar for a pair of horses, 5 guineas extra.

F. HOULGATE manufactures miniature broughams, ranging in price from £100 upwards. He can also supply on the most reasonable terms, and at the shortest notice, Hinge Yorkshire Sporting Carts, Malvern Dog-Carts, and carriages of every description.

[1389]

IVALL & LARGE, 56 *South Audley Street, and* 125 *Piccadilly.*—Four-in-hand coach, with patent drag.

[1390]

JONES, WALTER, 70 *Upper Seymour Street, London, N.W.*—Paintings for carriage decoration.

[1391]

KESTERTON, E., 94 *Long Acre.*—The 'Amempton' carriage. (*See page* 41.)

[1392]

KINROSS, WILLIAM, Coach-Builder, *Stirling.*—A two-wheel buggy, with hood to cover two persons.

[1393]

LARKINS, STEPHEN N., 6 *Limekiln Street, Dover.*—Propelling bathing machine.

HALL & SONS, *97 and* 98 *Long Acre.*—A barouche on elliptic springs, unusually easy and noiseless, and with self-acting body steps.

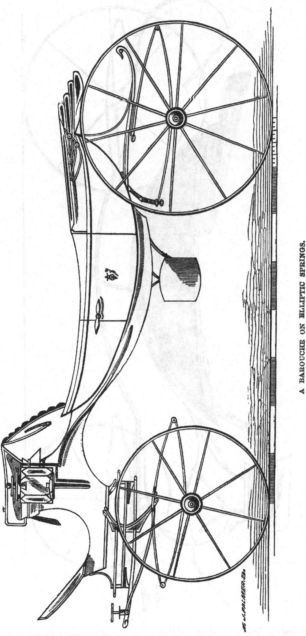

A BAROUCHE ON ELLIPTIC SPRINGS.

The above is a very handsome shaped Barouche, with self-acting steps, and hung on unusually easy springs. It embraces the twofold advantage of being a most roomy and commodious carriage, whilst at the same time it is so light in draught and construction, that the usual 'Brougham-sized horses' are more than equal to it. The painting is a rich lake picked out with carmine. The mountings are of silver. The lining is blue cloth and morocco, trimmed with handsome broad silk lace.

HOOPER & Co., 28 *Haymarket, London, S.W.*—A light 'Sefton' landau, with improved flat-falling head ; an improved light 'Craven' barouche, on C- and under-springs.

Obtained Prize Medal of the Great Exhibition of 1851.

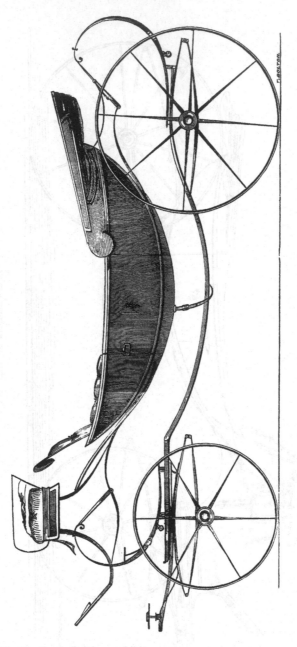

A LIGHT AND ELEGANT 'CRAVEN' BAROUCHE.

HOOPER & Co., Her Majesty's Coach-Builders, exhibit 1. A light and elegant 'Craven' Barouche, hung on under — and C-springs, with a perch of improved construction, made of wood so connected with iron by rivets and hammered edges, as to act on the principle of a tube with a wooden centre, combining lightness with greatly increased strength and safety. By the construction of the body, the commodious folding steps are so placed as not to be seen above the panels, thus enabling the latter to be made of a more than usually light and elegant form. It is also suspended at such a distance from the ground as to protect the occupants from the dust of the road. By the improvements introduced in the general construction of the individual and combined parts, and by the use of very tough steel instead of iron, where practicable, the utmost strength with the minimum of weight is obtained. The carriage is an example of the most recent improvements and combinations to effect elegance, lightness, and ease.

HOOPER & Co.—*continued.*

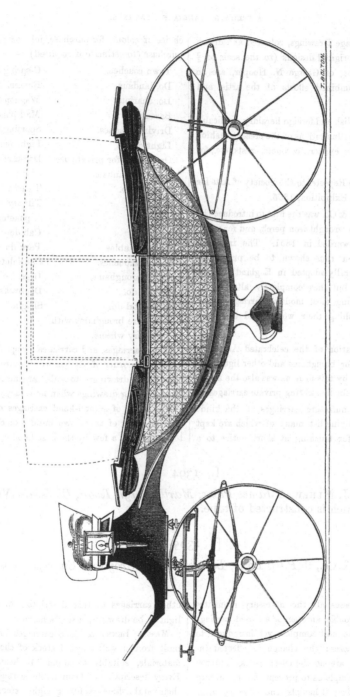

A LIGHT PAIR-HORSE 'SEFTON' LANDAU.

T. BOLTON.

2. A light pair-horse 'Sefton' Landau.—The improvements are on the same principle as those of the barouche before described; a careful combination of details giving the utmost strength and durability, combined with lightness.

The drawbacks hitherto considered inseparable from landaus (weight, and partial opening of the head), are overcome in the carriage exhibited. The head, by a very simple and efficient method of construction, is made to open as flat as a barouche, thus forming, when open, almost as airy a carriage, and when closed, a comfortable family carriage, equally adapted for London or country use. Being furnished with covered steps as a brougham, it can be used with or without a footman.

HOOPER & CO.—*continued.*

A SERIES OF CARRIAGE DRAWINGS.

3. A series of Carriage Drawings, coloured and drawn by J. Gilfoy from the original designs (to the scale of $\frac{3}{4}$ of an inch to the foot) of George N. Hooper, are an illustration of the combined efforts of the artist and practical constructor.

4. Medallions of English and foreign heraldry, applicable to dress carriages; also illustrations of the present fashion of grouping monograms, cyphers, coronets, crests, &c., for small carriages.

Mr. HOOPER was the Reporter to the Society of Arts for Carriages at the Paris Exhibition, 1855.

The firm of HOOPER & CO. was the first to introduce the C-spring Brougham on wrought iron perch, and for which a Prize Medal was awarded in 1851. The improved system of construction thus shown to be practicable has not only been generally adopted in England and the continent of Europe, but has completely altered the principle of constructing most modern carriages since 1851, greatly diminishing their weight and cost, and increasing their ease.

The regular importation of the celebrated American light hickory wheels for broughams and other light carriages, was first begun by this firm, as was also the application of photography for illustrating private carriages.

HOOPER & CO. manufacture carriages of the kinds named in the accompanying list, many of which are kept in an advanced state for finishing at short notice to a choice of colour, for purchase, job, or job with option to purchase (to estimate if required):—

Town coaches.	C-spring broughams.
Do. landaus.	Brakes.
Do. chariots.	Waggonettes.
Barouches.	Mail phaetons.
Driving coaches.	Sporting do.
Light do.	Light road do.
Omnibuses for private use.	Dog-cart do.
Barouche landaus.	Stanhope do.
Sociable do.	T carts.
Sefton do.	Tilbury and Spider
Elcho do.	phaetons.
Sociables.	Cab do.
Pony sociables.	Park do.
Light barouches.	Cabriolets.
Single broughams.	Gigs.
Double do.	Dog-carts.
Segmental do.	Sleighs.
Miniature broughams with hickory wheels.	

Dress carriages, and carriages for special purposes, are built to the order of persons who require them. In these cases small drawings 'to scale' are made, and also full-sized working drawings when necessary.

The stock of second-hand carriages consists of sound modern ones of their own build (some but little used), together with a few by the best London builders.

[1394]

LA ROCHE, J., & J. MEHEW, 5 *James Place, Marlborough Road, Chelsea.*—Velocipede, the iron work of which is constructed of tube.

[1395]

LENNY, CHARLES, & CO., 9 *Park Lane, London; and Croydon, Surrey.*—Landau sociable for one horse.

This carriage possesses all the necessary requisites for forming, without trouble, an open or a closed carriage at pleasure, and is the most complete yet introduced to combine the two purposes; the change is effected instantaneously. There are no detached parts, and the arrangements are so simple as to prevent the possibility of its getting out of order; it has also the advantage over other carriages of this description in being sufficiently light to be drawn by a single horse.

Messrs. LENNY & Co.'s carriages for exportation are built from a well selected stock of thoroughly seasoned materials, suitable to stand the heat of any climate. Every description of fashionable carriages can be seen at their establishments, filling eight extensive show-rooms.

[1396]

McDOUGALL, ARCHIBALD, & SON, 36 *Rupert Street, London, W.*—A one-horse van.

[1397]

McNAUGHT & SMITH, 9 *Tything, Worcester.*—Waggonette. (*See page 42.*)

KESTERTON, EDWIN, 93 *and* 94 *Long Acre, W.C.*—The 'Amempton,' forming a complete open and close carriage.

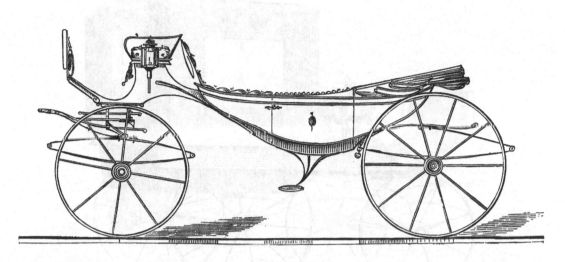

THE 'AMEMPTON' OPEN.

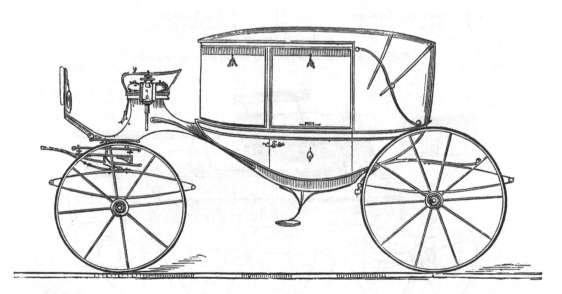

THE 'AMEMPTON' CLOSED.

EDWIN KESTERTON is the inventor and builder of the 'Amempton' and the 'Amempton Sociable,' each forming a complete open and close carriage, well adapted for home or colonial use, being roomy, light, and airy. Sociable Driving Phaeton, forming the light Driving Phaeton and the useful Waggonette. Patent two-wheeled Dog-Cart, light, strong, and giving entire freedom from the action of the horse. The exhibitor also manufactures Sociable Landaus, Broughams in various sizes, Cabriolet and Park Phaetons, &c. &c.

McNAUGHT & SMITH, *Tything, Worcester.*—A waggonette, with movable head and other improvements.

A REVERSIBLE WAGGONETTE CAPABLE OF FIVE DISTINCT FORMATIONS.

1. The open Family Carriage.
2. Close ditto.

3. Mail Phaeton.
4. Dog-Cart Phaeton.

5. Break or Luggage Cart.

Waggonette with enclosure, carrying four or six persons inside.

WAGGONETTE WITH OPEN SEATS.

Waggonette with open seats, and representing concealed folding-step as it appears drawn out over the front wheel; and self-acting folding-step behind.

For further particulars see McNAUGHT & SMITH's *own 'illustrated catalogue,' which may be obtained upon application.*

[1398]

MACNEE, JAMES, & Co., Coachmakers, 106 *Princes Street, Edinburgh.*—Improved landau Clarence carriage.

[1399]

MANN, J. H., *Twickenham, S.W.*—Park phaeton, with improved fore carriage.

[1400]

MASON, W. H., *Carriage Works, Kingsland Basin, and Clapton.* (*See page* 44.)

[1401]

MILFORD, THOMAS, & SON, *West of England Wheel Works, Thorverton, Devon.*—A pair-horse spring waggon, for town and road purposes.

[1402]

MULLINER, FRANCIS, *Northampton.*—A Fitzroy phaeton, constructed with malleable steel instead of iron; wheels of hickory.

[1403]

MULLINER, HENRY, *Leamington.*—Four-wheel dog cart, folds open and forms waggonette, head drops on.

[1404]

NEWHAM, EDWARD, *Market Harboro'.*—Light sociable phaeton, with seats and dash removable.

[1405]

NEWNHAM & SON, *Bath.*—Light Bath landau waggonette, with folding leather head and improved arrangement of interior seats.

The Bath Landau Waggonette with folding leather head, which can be instantly opened or closed; having glasses at the sides, front, and back, which drop into body; and an improved arrangement of interior seats. This carriage combines all the comfort of the sociable landau with the light draught of the waggonette. Its novelty consists in the application of concealed head joints, which draw the hoop sticks inwards, and prevent their protruding beyond the wings when lowered.

[1406]

NEWTON, JOHN, 10 *Werrington Street, London, N.W.*—Folding double-seated perambulator, with improved sheathed wheels.

[1407]

NURSE & CO., 200 *Regent Street, W.*—Sociable landau. (*See pages* 46 *and* 47.)

[1408]

OFFORD, R. & J., 79 *Wells Street, W.*—Carriages. (*See page* 45.)

[1409]

PARKER, F., 75 *Regent Street, Cambridge.*—Registered family cart with improved springs, free from knee motion.

[1410]

PARSONS, G., *Martock, Somerset.*—Wheels for common roads. (*See page* 48.)

MASON, W. H., *Carriage Works, Kingsland Basin, and Clapton.*—Waggonette, carrying ten persons, on improved principles, forming an open break, or exceedingly light omnibus.

WAGGONETTE FOR CARRYING TEN PERSONS.

W. H. MASON manufactures superior light Sociables; Broughams; Mail, and Driving Phaetons; Waggonettes forming perfect Stanhope Phaetons; Dog-Carts; road and town Buggies.

The carriage exhibited is strongly recommended for its extreme lightness of draught and luxurious roominess. It forms a convenient carriage for winter or summer use, can be successfully worked with one horse, and is made of various sizes.

OFFORD, R. & J., 79 *Wells Street, Oxford Street, London.*—Carriages.

OFFORD'S EXHIBITION LANDAU, OPEN.

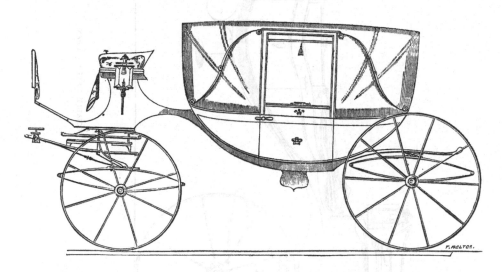

OFFORD'S EXHIBITION LANDAU, CLOSED.

These engravings represent a carriage manufactured by Messrs. OFFORD, of Wells Street, Oxford Street, London. It forms, as shown in the drawings, a perfect summer and winter carriage in one; and having no loose parts, can be opened or closed at any time in a few minutes. Several novel, striking, and commendable points deserve notice. The wheels are manufactured partly of wrought iron and partly of wood, combining lightness with increased strength. Upon the front wheels are exhibited Offord's New Patent India-rubber Tire, affixed to them without flanges in a manner ensuring continuous adhesion. These tires are productive of much comfort to the riders, and effect a considerable saving in the wear of the vehicle. The under-carriage is formed of iron-work of a new and improved design, remarkable for its lightness, elegance, and durability.

The window frames are specimens of another patented invention. They are made of soft india-rubber, united with the well-known solid material called vulcanite, or hard india-rubber. They are noiseless, elegant in appearance, and, from the nature of the hard material, are calculated to last as long as any carriage.

The interior is made more cheerful than usual by increasing the size of the windows, and by the addition of an extra one in the back, by means of which ventilation can be obtained without draught; or a current of air circulated quite through the vehicle when desired.

The silk and trimmings are new and elegant in design, and tastefully arranged. When opened, the head or upper part falls very flat, and presents none of the usual unsightly projections. The steps are made to open and close with the door by a new and very simple method, for which Letters Patent have been obtained.

NURSE & Co., 200 *Regent Street*, W.—Sociable landau on elliptic springs.

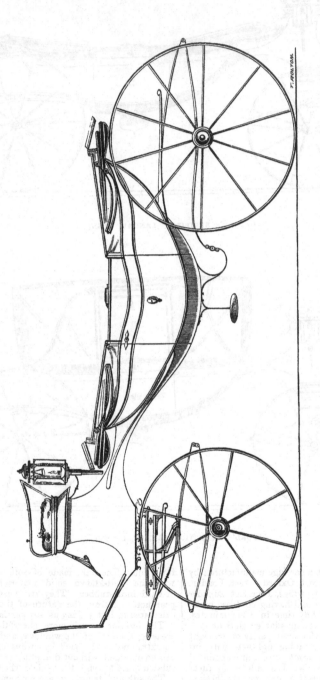

SOCIABLE LANDAU, OPEN.

NURSE & CO.—*continued.*

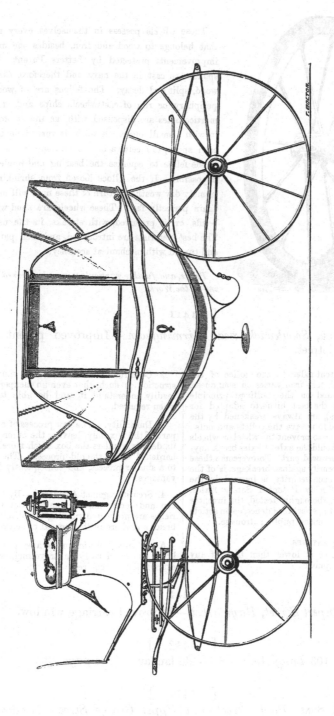

SOCIABLE LANDAU, CLOSED.

PARSONS, GEORGE, *Martock, Somerset.*—Patent wheels, specially adapted for common roads in tropical or other climates.

These wheels possess in themselves every advantage that belongs to wood and iron, besides the mechanical improvements protected by Letters Patent. The iron spokes are cast in the nave, and therefore, cannot, like wood, split and decay. The felloes are of wood, as the periphery or rim of cast wheels chips and cracks; but mortice-holes are dispensed with, as the spokes are let in with a small auger. A worm is turned on the end of each spoke, to receive a boss or nut, which is screwed up to the felloe to equalise the bearing and render fretting impossible. If the felloes loosen from shrinking in extremely dry weather, a turn of these nuts will make them again perfectly fast. These wheels are shod with whole bonds, or, if preferred, with streaks, in the usual way. The best materials are introduced, and each part is fitted and turned with mechanical accuracy.

Prices, and further particulars may be learned by applying at the Works.

[1411]

PARTRIDGE, EBENEZER, *Smethwick, near Birmingham.*—Improved patent (Collinge and Mail) carriage axletrees.

In presenting these patent axles to the notice of the public it may be observed, that iron varies in soundness, and that axles constructed on the Collinge principle (though admitted to be the best hitherto adapted for general carriage purposes), are always weakened by the shouldering down required to receive the collets and nuts; and by having no protection to prevent the wheel or wheels of carriages running off, should the axle or axles break anywhere in the journal, or screwed part. To overcome these defects, and to obtain a security against breakage, is of the utmost importance, and such security is insured by the use of E. Partridge's Trebly Patented Safety Axles. The principle having been thoroughly tested, the inventor offers them to the trade with perfect confidence, as an article which must extensively command public patronage.

ADVANTAGES.

1. Cheapness.—The price is lower than that of any other axle now before the public.

2. Construction.—The extreme simplicity of its construction is such, that even an inexperienced person will readily understand it, and be able to adjust the parts when required.

3. Durability.—The new process of hardening the inner part of the box only, leaving the outer part of it to retain its density, renders the box much more durable than those hardened under the old process. The axles are subject to a similar process, and consequently have the same advantages.

4. Security against Accidents.—By the use of an inner cap and screw pin, any one or all of the wheels of carriages are prevented running off, should the axle or axles break anywhere in the journal or screwed part.

5. A direct lubricator is also provided for giving a little oil occasionally in travelling, when no particular examination is required.

[1412]

PATERSON, T., 15 *Rupert Street, Haymarket.*—Improved carriage window.

[1413]

PEARCE & COUNTZE, 103 *Long Acre.*—Sociable landau. (*See page* 49.)

[1414]

PETERS, THOMAS, & SONS, *Park Street and Upper George Street, London, W.*—A park barouche and a brougham.

[1415]

REAY & USHER, *South Hylton Iron Works, Sunderland.*—Axle block forgings, finished under the forge hammer.

PEARCE & COUNTZE, 103 *Long Acre.*—Sociable landau, and materials used by them in carriage building.

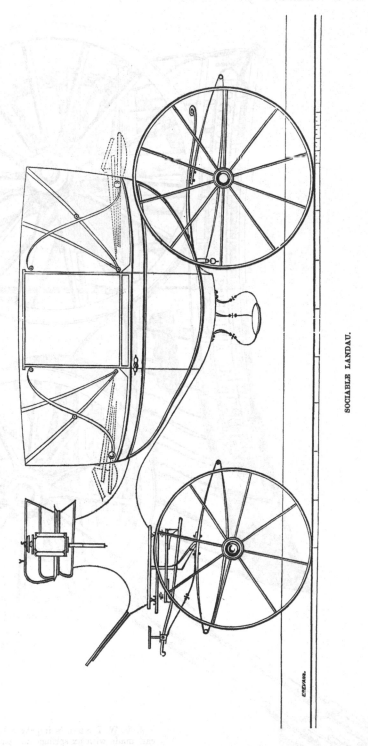

SOCIABLE LANDAU.

A SOCIABLE LANDAU, as finished from the body and carriage-maker's benches, before the painting and lining is commenced, exhibiting the wood, ironwork, and quality of workmanship employed in the construction of the exhibitors' carriages.

Specimens of the materials used by the exhibitors in carriage building.

Every description of fashionable carriage may be seen finished, and in the various stages of progress, at Pearce and Countze's manufactory.

CLASS VI.

[1416]

RENDALL, JOHN & WILLIAM, *High Street, Stoke Newington, N.*—Improved coal van to carry two tons.

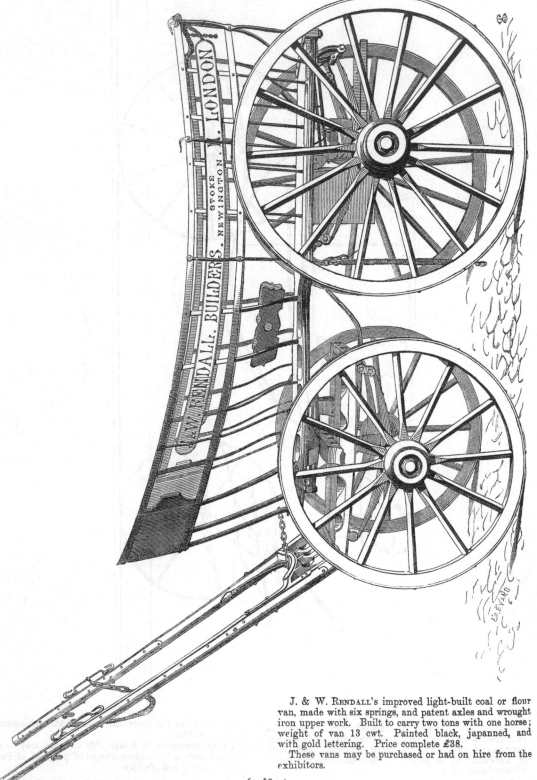

J. & W. RENDALL's improved light-built coal or flour van, made with six springs, and patent axles and wrought iron upper work. Built to carry two tons with one horse; weight of van 13 cwt. Painted black, japanned, and with gold lettering. Price complete £38.

These vans may be purchased or had on hire from the exhibitors.

[1417]

RIDGES, JOHN EDWARD, *The Tudor Coach Manufactory, Cleveland Road, Wolverhampton.*— A miniature landau carriage to open and close.

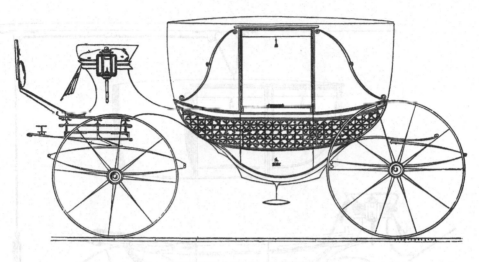

SOCIABLE LANDAU, CLOSED.

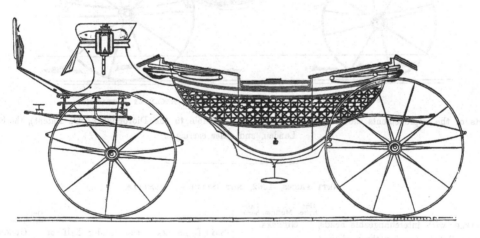

SOCIABLE LANDAU, OPEN.

The above engravings represent a Sociable Landau Carriage, weighing only 9½ cwt. This is the only carriage that will make a perfect open and close carriage without requiring even a wrench to change it. It has large front windows, and is perfect either open or closed. It is quite manageable for one horse, and is also an elegant vehicle for two horses. In either form it is highly suitable for ladies' use. Carriages of all descriptions, finished fit for immediate use, may be purchased from the exhibitor or had on hire. Drawings and estimates will be sent on application.

[1418]

RIGBY & ROBINSON, 7 *Park Lane, Piccadilly.*—Elcho barouche landau. (*See page 53.*)

[1419]

ROBINS, WILLIAM, 1 *Church Street, Lambeth.*—Patent revolving break for omnibuses and other vehicles.

[1420]

ROCK & SON, *Hastings.*—A Dioropha with patented improvements.

Obtained Prize Medals at the Exhibition of 1851, and at Paris in 1855.

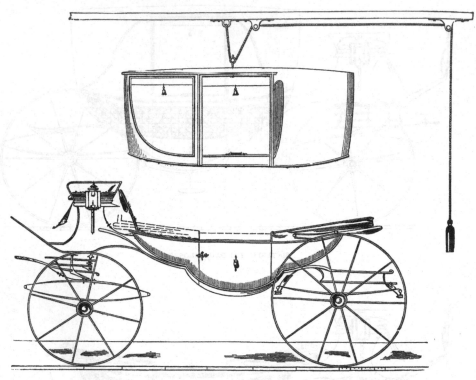

SOCIABLE WITH INTERCHANGEABLE HEADS, AS RECENTLY PATENTED.

Parts of the improvements shown in this carriage are applicable to the Dioropha (Rock's Patent), the Sociable Landau, and other carriages.

NETT PRICES, 1862, NOT INCLUDING RUMBLES.

	One-horse.	Medium.	Pair-horse.
	GUINEAS.		
SOCIABLE, with interchangeable heads, as exhibited, forming three distinct carriages, viz., coach, landau, and barouche	210	235	265
DIOROPHA, with coach head, interchangeable with barouche head and flap, as exhibited in 1851 and 1855, forming either coach or barouche . .	180	200	225
DIOROPHA, with landau head, interchangeable with barouche head and flap, forming landau or barouche .	190	210	240

	One-horse.	Medium.	Pair-horse.
	GUINEAS.		
SEMI-DIOROPHA — the hinder half of landau head a fixture; the front half to remove, and change for barouche flap	170	190	215
SOCIABLE LANDAU, with patented improvements	160	180	200
SINGLE BROUGHAM . . .	120	130	140
DOUBLE BROUGHAM, with round or elliptic front	130	140	150
OPEN SOCIABLE, with half-head .	130	145	160
Ditto, enclosed, with glass doors, &c. .	145	165	180
WAGGONETTE	65	75	90

DOG-CARTS, from 30 guineas; VICTORIA PHAETONS, 55 to 90 guineas. PONY PHAETONS, from 25 guineas.

RIGBY & ROBINSON, 7 *Park Lane, Piccadilly, London.*—Elcho barouche landau, combining elegance, extreme lightness, and ease.

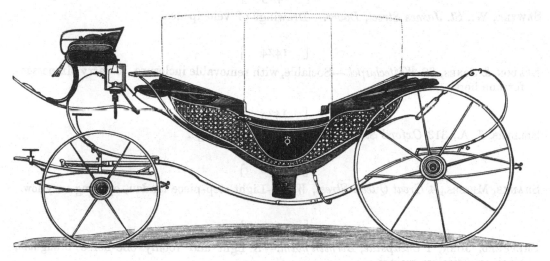

THE ELCHO BAROUCHE LANDAU, INVENTED AND MANUFACTURED BY RIGBY AND ROBINSON.

[1421]

ROGERS, I., *North Audley Street, London.*—Sociable landau, for one horse; a perfect open or close carriage.

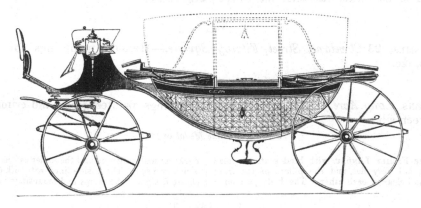

MINIATURE SOCIABLE LANDAU.

The above engraving represents a Miniature Sociable Landau—the only kind of perfect open and close carriage combined, without movable parts, and within the management of one horse, ever introduced. Its design in either form is elegant: when closed, it is cheerful, having a large front window. Others, finished and building, with every description of fashionable carriage, are always kept in stock, fit for immediate use.

[1422]

 Rogers, William, & Co., *Bristol.*—The ' Clifton Waggonette.' (*See pages 56 and 57.*)

[1423]

Sawyer, W., *St. James Street, Dover.*—Drawings of velocipedes.

[1424]

Seadon & Jones, 60 *Whitechapel.*—Sociable, with removable inclosure, to carry six persons, for one horse.

[1425]

Sellers, J. A., 313 *Oxford Street, W.*—Model of a carriage.

[1426]

Shanks, Messrs., 4 *Great Queen Street, W.C.*—Light step-piece landau, opening very low.

[1427]

Shepherd, John, 1 *Cheapside, Birmingham.*—A light and roomy one-horse brougham, hung on noiseless springs.

[1428]

Sherwin, Joseph, *Tabernacle Walk, Finsbury.*—Omnibus, chaise, carriage, and cart axle-trees. Waggon and cart arms manufacturer.

[1429]

Shillibeer, George, 40 *City Road.*—Patent ' vis-à-vis ' omnibus, inside seats separated, outside seats reached from interior.

The Patent vis-à-vis Omnibus comprises advantages which the omnibuses in use do not possess, both as regards the public and proprietors; viz., free ingress to and egress from the interior. The passengers are not 'packed,' as in the present omnibuses; the outside seats are easily and safely reached, and females can ride outside with propriety. A capôte, easily raised, covers two-thirds of the outside passengers in wet weather, thereby greatly benefiting the proprietors. The carriage weighs one ton.

[1430]

Short, James, 23 *Cleveland Street, Fitzroy Square.*—Heraldic mountings for carriages, harness, &c.

[1431]

Silk & Sons, *Long Acre.*—Landau on horizontal springs, painted green and crimson, lined with green silk.

Obtained a Prize Medal in 1851.

A full size Family Landau, with hind seat, the head arranged to fall very flat, and the pillars of the front window to fall clear of each other. The body painted a dark transparent green, and the under carriage and wheels a rich crimson. The inside lined with silk of a neat and chaste design. The lamps and mountings in brass.

[1432]

Simpson, Hortensius C., *Shrewsbury.*—Car, carrying from one to six passengers, with extra luggage accommodation.

[1433]

Smith, John Bennett, *Green Street, Bath.*—Silver-mounted perambulator.

[1434]

STAREY, T. R., *Nottingham.*—Light landau (The Granville), with flat fall of head, new highly elastic springs, and silent wheels, with chain tires.

Obtained first-class Medal at the Paris Exhibition, 1855.

The 'GRANVILLE LANDAU,' clipper shape, forms a perfect winter and summer carriage for a pair of light riding horses. The head by a new arrangement falls quite flat, lower than has hitherto been accomplished. It is strengthened throughout with mild steel instead of iron, whereby a maximum of strength with a minimum of weight is obtained. It is hung on highly elastic springs with india rubber bearings, and Messrs. Apperley & Co.'s 'patent silent wheels,' with flexible iron tires bedded in india rubber, a combination that ensures greater durability to the wheels and carriage, as well as a remarkably soft, easy, and noiseless motion. The doorway is unusually wide, to suit the present style of dress; and while having a very large window on each side, the occupants are not unduly exposed to view. The colour of this carriage was suggested by Owen Jones, Esq.

[1435]

STARTIN & MACKENZIE, *Benacre Street, Birmingham.*—Headed phaeton, with driving-seat, suitable for export.

[1436]

STEVENSON & ELLIOT, 177, 179, and 181 *King Street, Melbourne, Australia;* Branch Factory, *Stirling, Scotland.*—Light phaeton, with movable side glasses.

BAROUCHE with movable side glasses. This carriage is similar to those built by the exhibitors in Melbourne. It combines lightness with great strength. The construction of the wheels is somewhat unusual, the felloes being bent by steam and having only two pieces in each wheel. They are so constructed as to be capable of resisting shocks of a severe nature, and are in every respect very durable. This is a desirable carriage for this country, India, or Australia; it is fitted for one horse or a pair of ponies; built of the best materials, extra and highly varnished. The broughams manufactured by the exhibitors are equally light. Communications to be addressed to our Branch Factory, Stirling, N.B.

[1437]

STOCKEN, FREDERICK, *5a Halkin Street, Belgrave Square, S.W.*—Carriage.

[1438]

STRICKLAND, HENRY, 9 *Macclesfield Street, Soho.*—Specimens of heraldic carriage painting.

[1439]

THOMSON, GEORGE, *Stirling, Scotland.*—Light waggonette, to close or open at pleasure, with reversible seat and set of harness.

[1440]

THOMSON, WILLIAM, *Perth.*—Registered four-wheel dog-cart.

THE PERTH FOUR-WHEEL DOG-CART, Registered, highly approved for lightness, elegance, and convenience. Price 50 guineas delivered. WAGGONETTE, easily convertible into a four-wheel Dog-Cart. Price 65 guineas delivered.

[1441]

THORN, W. & F., 19 *Great Portland Street, W.*—Summer and winter carriage. (*See page* 58.)

[1442]

THORNTON, E. M., 6 *Brooke Street, Holborn.*—Patent rein-clip; for the reins when out of hand.

ROGERS, WILLIAM, & Co., *Bristol.*—The 'Clifton Waggonette.'

AS A WAGGONETTE.

AS A DOG-CART.

The REGISTERED CLIFTON WAGGONETTE, one of the most elegant and novel carriages of this class yet introduced. It is capable of being transformed into four distinct equipages, viz. a Waggonette, Driving Phaeton, Dog-Cart, and Close Carriage. The elegant sweeps of the iron work and novel method of transforming it from a single to a pair-horse carriage; the lightness and ease of the springs, the peculiarity of ascent to the body, and the power of raising or lowering the canopy top, recommend it as one of the most useful carriages of its class. A large assort-

ROGERS, WILLIAM, & Co.—*continued.*   The 'Clifton Waggonette.'

AS A CLOSE CARRIAGE.

AS A DRIVING PHAETON.

ment of well seasoned carriages for exportation are always kept in stock, viz.: Clarences, Broughams, Sociable Landaus, Amempton Barouches, Victoria Phaetons, Waggonettes, Queen's Pattern, Albert and Napoleon Phaetons, 2 and 4-wheel Dog-Carts, Pony and Basket Phaetons. Any of the foregoing may be jobbed with option to purchase, or let for a consecutive number of years, on most liberal terms.

THORN, W. & F., 19 *Great Portland Street, W.*—Improved summer and winter carriage, perfect open or closed.

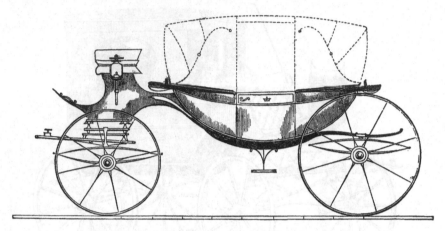

SUMMER AND WINTER CARRIAGE.

W. and F. THORN build to order every description of carriage, and have always on hand a large stock, which may be jobbed with option of purchase. They have had great experience in building carriages for exportation, and are the inventors of the patent equi-motive springs.

[1443]

THRUPP & MABERLY, Coach-Builders, 269 *Oxford Street, London.*—A light elliptic spring coach.

Exhibitors in the Great Exhibition, 1851; obtained Prize Medal at the Paris Exhibition, 1855.

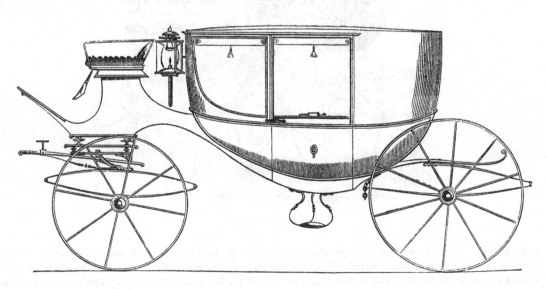

AN ELLIPTIC SPRING COACH.

[1444]

TURRILL, HENRY LEWIS, 67 *South Audley Street, and 22 and 23 Long Acre.*—Carriage.

H. L. TURRILL (late Robson & Co.), is Coach-Builder by appointment to Her Majesty, original builder of the 'Shofle, or Gentleman's Hansom,' and manufacturer of every description of carriage. Carriages are let on hire by the exhibitor for any period, with option of purchase.

[1445]

VEZEY, R. & E., *Long Acre, Bath.*—Brougham, with patent concealed step, noiseless springs, on india-rubber bearings.

Coach-Builders to Her Majesty, and patentees of springs on india-rubber bearings, and concealed descending brougham steps.

[1446]

WARD, JOHN, *Leicester Square.*—Invalid pleasure-ground chairs, and children's perambulators, to be drawn by hand or animal.

[1447]

WATERS, G. & SON, 5 *George Street, and 72 North End, Croydon.*—One open carriage.

[1448]

WATKINS & HORNSBY, *Duke Street, Birmingham.*—Patent carriage axles.

[1449]

WHITTINGHAM, THOMAS, & WILKIN, 136 *Long Acre, W.C.*—Carriage laces.

[1450]

WICKSTEED, FREDERICK, 18 *Upper St. Martin's Lane.*—Carriage drawings.

[1451]

WINDOVER, C. S., *Huntingdon.*—Carriage adapted for the four seasons. (*See pages* 60 *and* 61.)

[1452]

WOODALL & SON, *Orchard Street, London, W.*—A superior side-light coach, with improvements in ventilation.

[1453]

WOODBOURNE, JAMES, *Park Ironworks, Kingsley, near Alton, Hampshire.*—Improved model cart for general purposes.

[1454]

WYBURN & Co., 121 *Long Acre.*—A landau and a brougham. (*See pages* 62 *and* 63.)

Windover, Charles Sandford, *Huntingdon.*—A carriage adapted for the four seasons, forming a barouche, sociable, coach, and landau.

WINDOVER'S REGISTERED TESSATEMPORA,

The only Convertible Carriage adapted for the four Seasons.

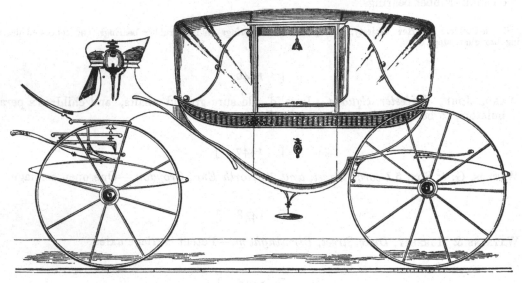

SPRING.

Price from 160 to 200 Guineas.

SUMMER.

Price from 115 to 135 Guineas.

WINDOVER, CHARLES SANDFORD—*continued.*

WINDOVER'S REGISTERED TESSATEMPORA,

The only Convertible Carriage adapted for the four Seasons.

AUTUMN.

Price from 135 to 160 Guineas.

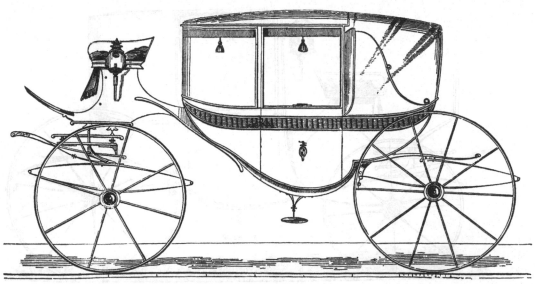

WINTER.

Price from 165 to 205 Guineas.

WYBURN & Co., Her Majesty's Coach-Makers, 121 *Long Acre, W.C.*—A phaeton, a landau, and a brougham.

A PHAETON.

A BROUGHAM.

WYBURN & Co.—*continued.*

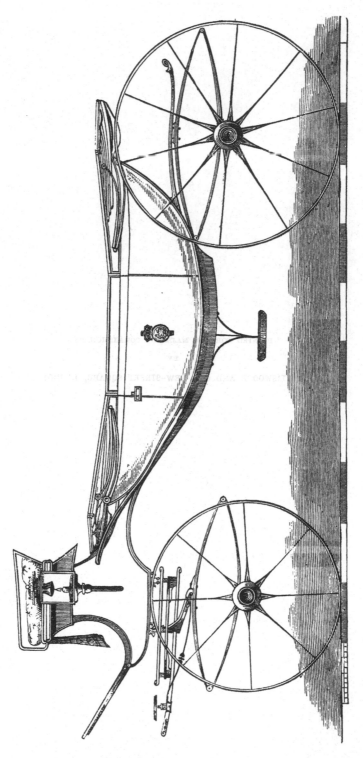

A LANDAU.

PRINTED FOR HER MAJESTY'S COMMISSIONERS
BY
SPOTTISWOODE AND CO., NEW-STREET SQUARE, LONDON

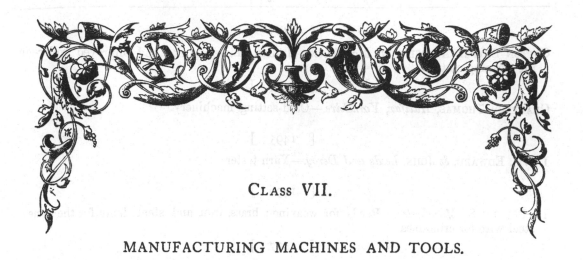

CLASS VII.

MANUFACTURING MACHINES AND TOOLS.

SUB-CLASS A.—*Machinery employed in Spinning and Weaving.*

[1486]

ANDERSTON FOUNDRY COMPANY, *Glasgow.*—Looms for weaving checked and fancy goods.

[1487]

APPERLY, JAMES, & Co., *Dudbridge, Stroud.*—Oiling, feeding, carding, condensing, grinding mills.

[1488]

BOOTH, JOSEPH, *Rock Street, Bury, Lancashire;* THOMAS WILLIAM CHAMBERS, 96 *Georgiana Street, Bury, Lancashire;* JAMES CHAMBERS, 100 *Georgiana Street, Bury, Lancashire.*—Loom for weaving; improved letting off, taking up, picking, and shedding motions, and reed-holder.

UNDERPICK LOOM, 36 in. reed space, can be made any width required, with letting off, picking, taking up, shedding, fast and loose reed motions, all new: with plain tappets and temple, suitable for calicoes, twills, domestics, &c. which by a simple arrangement are suitable also for ginghams, checks, plaids, &c. This loom may be worked safely at 250 picks per minute. Any of these motions can be applied to other looms. Prices and lithographs may be obtained free, by application to the makers, Tuer & Hall, Bury, near Manchester.

[1489]

CLARKE, I. P., *Leicester* —Reels, spools, and mill bobbins for cotton, silk, and linen thread.

[1490]

CLARKE, T. A. W., *Leicester.*—Machine for covering india-rubber rings.

[1491]

COMBE, JAMES, & Co., *Belfast.*—Flax machinery.

[1492]

COOK & HACKING, *California Iron Works, Bury, Lancashire.*—Self-acting heald-knitting machine.

This newly patented heald-knitting machine is entirely self-acting, simple in construction, requires little attention, and knits healds of any description with great rapidity. It may be seen in the machinery department of the Exhibition, or at Cook & Hacking's works, Bury, Lancashire.

[1493]

COTTON SUPPLY ASSOCIATION, *Manchester.*—Indian native churka, and improved roller-gin for cleaning cotton.

[1494]

CRABTREE, THOMAS, *Halifax, Yorkshire.*—Card-setting machinery.

[1495]

DAVIS, EDWARD, & JOHN, *Leeds and Derby.*—Yarn tester.

[1496]

DE BERGUE, S., *Manchester.*—Reeds for weaving; brass, iron, and steel dents for the same; steel wire for crinolines.

[1497]

DICKINSON, WILLIAM, & SONS, *Phœnix Iron Works, Blackburn.*—Loom for fancy weaving; fast loom, Taylor's patent; power loom with dobby, &c. (*See page* 3.)

[1498]

DIXON, JOHN, & SONS, *Steeton-in-Craven, Leeds*—Bobbins, rollers, keys, tree-nails, drawer knobs, and boxes.

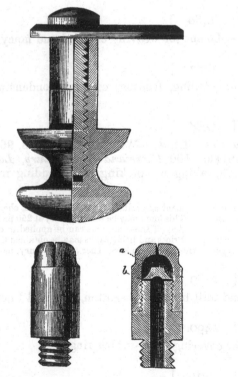
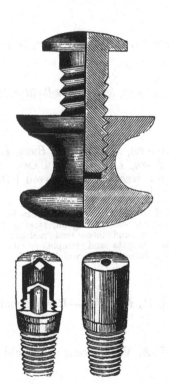
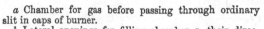

a Chamber for gas before passing through ordinary slit in caps of burner.

b Lateral openings for filling chamber *a*, their direction being at an angle with, or across the opening in cap of burner.

The following articles manufactured by Dixon & Sons are also exhibited :—

Turned pill and powder boxes and cases.
Heywood & Dixon's patent knobs and handles.
Oldfield & Dixon's patent gas burners.

[1499]

DOBSON & BARLOW, *Bolton, Lancashire.*—Machinery for opening and cleaning, preparing and spinning cotton. (*See page* 4.)

DICKINSON, WILLIAM, & SONS, *Phœnix Iron Works, Blackburn.*—Loom for fancy weaving ; fast loom, Taylor's patent ; power loom with dobby.

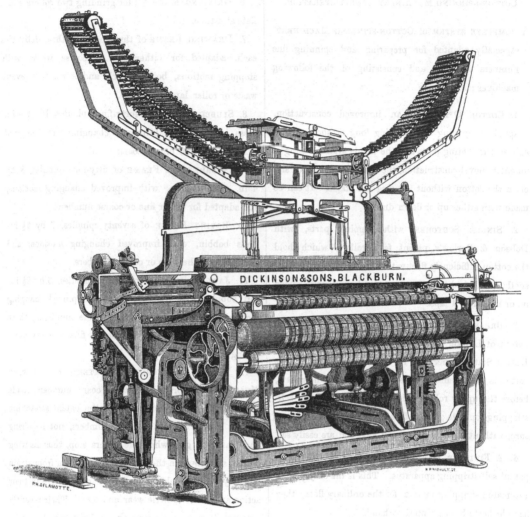

POWER LOOM.

The Exhibitors are patentees of improvements in looms ; the inventors of the Blackburn system, or over-pick motion for looms ; and original makers of sizing and warping machines.

MACHINERY EXHIBITED IN OPERATION.

POWER LOOM with multiple shuttle box and dobby combined, for weaving checks, ginghams, diapers, velvets, linsey-wolseys, &c.

POWER LOOM with Taylor's patent, for weaving plain and fancy twills, spots, satin checks, figured scarfs, &c. &c.

DOBBY LOOM with Taylor's patent, for weaving India scarfs.

MACHINERY EXHIBITED NOT IN OPERATION.

IMPROVED SIZING MACHINE (see Testimonials).

MACHINERY NOT EXHIBITED (see Drawings).

WINDING MACHINES for throstles or cops ; warping machines for improved sizing machine, winding-on or beaming machines, &c. &c.

Spinning and weaving mills are estimated for, completed, and put to work by this firm on the most improved system.

DOBSON & BARLOW, *Bolton, Lancashire.*—Machinery for opening and cleaning, preparing and spinning cotton.

COTTON-SPINNING MACHINERY IN FULL OPERATION.

A COMPLETE SYSTEM of COTTON-SPINNING MACHINERY, especially adapted for preparing and spinning fine numbers of yarn, and consisting of the following machines; viz. :—

1. COTTON OPENER of an improved construction, adapted for opening and cleaning long or short stapled cotton, the feeding parts and the inside gratings being of an entire novel construction, the object being to open and clean the cotton without injuring the staple. It can be made with either up or down drafts.

2. SINGLE SCUTCHER with lapping parts, with Dobson & Barlow's patent feed rollers, which hold the cotton sufficiently firm without crushing the seeds or shells. The feed table is so arranged as either to spread on or to double 4 laps.

3. BREAKER CARDING ENGINE. This is a compound patent of Dobson & Barlow, and Geo. Wellman of the United States, America. Its chief features are that the cotton is well opened and cleaned by the working rollers, before the upper roller will allow it to pass to the self-stripping top flats; these flats can be taken out at pleasure, without the use of a screw key, and are easily set.

4. A FINISHER CARDING ENGINE, with Wellman's patent self-stripping apparatus. This is the only perfect automaton stripping motion for the ordinary flats : they have hitherto been stripped by hand.

5. ASHWORTH'S PATENT LAP MACHINE, for making laps for the finisher carding engine, and combing machine, from slivers produced by the breaker card; the plates at each end of the lap revolve with it, and prevent the edges from being felted together as hitherto.

6. GRINDING MACHINE for grinding two rollers and a flats at a time.

7. DRAWING FRAME of three heads of three deliveries each, adapted for either fine or coarse work, with stopping motions, both at back and front, to prevent waste or roller lap.

8. SLUBBING FRAME of forty four spindles, 10 by 5 in. press bobbin, with improved changing motions, and adapted for either fine or coarse numbers.

9. INTERMEDIATE FRAME of fifty-four spindles, 8 by 4 in. press bobbin, with improved changing motions, and adapted for either fine or coarse numbers.

10. ROVING FRAME of seventy spindles, 7 by 3½ in. press bobbin, with improved changing motions, and adapted for either fine or coarse numbers.

11. JACK FRAME of eighty-eight spindles, 5 by 2½ in. soft bobbin, with tapering motion, improved changing motions, and especially adapted for fine numbers ; these frames are made so as to produce a fifty or sixty hank roving if desired.

12. DOBSON & BARLOW'S PATENT SELF-ACTING MULE, adapted for either fine or coarse numbers, made with double speed when required ; the patent stretching motion is well adapted for fine numbers, not requiring any change of gear when the rollers stop, thus insuring perfect steadiness in the carriage and freedom from strain on the yarn. The changing motions are positive in their action and liable to little wear and tear. Their recently patented improved winding on quadrant (which is perfectly self-acting and independent of the workman) ensures the yarn being properly wound on at the nose of the cop.

These mules are made to drive from either above or below, or with or without driving apparatus, and are adapted to spin any number from 1s. to 150s.

[1500]

DUGDALE, JOHN, & SONS, *Soho Foundry, Blackburn.*—Loom for twilled cloth ; loom for plain cloth ; cop winding machine.

COP WINDING MACHINE.

FAST REED POWER LOOM, for weaving heavy twilled cloth.

LOOSE REED POWER LOOM for weaving light fine cloth, with Dugdale's patent shedding motion attached, which enables the loom to run quicker, steadier, and with more ease to the yarn and healds, than the ordinary make of looms.

DUGDALE'S PATENT CONTRACTING COLLARS for roving machines, mule and throstle spindles, which by a novel arrangement for contracting the collar, when it or the spindle becomes worn, enables the spindle to run quicker with less vibration.

[1501]

FAIRBAIRN, P. & SONS, *Leeds.*—Rope spinning machinery.

[1502]

FERRABEE, JAMES, & CO., *Stroud, Gloucestershire ; and 75 and 76, High Holborn, London.*—Machines for forming bats of fleece, fulling and shearing woollen cloth.

WOOLLEN MACHINERY EXHIBITED.

FERRABEE'S PATENT MACHINE FOR FORMING BATS OF FLEECE in connexion with scribbling or carding machines, and for other purposes.

A PERPETUAL SHEARING MACHINE, adapted for finishing the cutting of fine woollen cloths.

FERRABEE'S PATENT FULLING MACHINE, adapted for woollen goods which vary in bulk and character ; intended to obviate wrinkling in the process of fulling.

[1503]

GATENBY & PASS, *Manchester.*—Reeds, dents, dent wire, &c., for weaving textile fabrics, also crinoline wire.

[1504]

GORDON, J., 3 *Billiter Square, London.*—Roller gin for cleaning cotton, worked by the foot.

[1505]

HALEY, JONAS, & SONS, *Cloth Hall Machine Works, Dewsbury.*—Improved rag grinding machine.

[1506]

HARDING, T. R., *Leeds.*—Card clothing, &c.

[1507]

HARDING, T. R., *Leeds.*—General mill furnishings.

[1508]

HARRISON, J., & SONS, *Blackburn, Lancashire.*—Looms and other weaving machinery for cotton, linen, &c. (*See pages* 6 *and* 7.)

[1509]

HATTERSLEY, GEORGE, & SON, *North Brook Works, Keighley, near Bradford.*—Two looms for weaving fancy goods.

[1510]

HENDERSON & CO., *Durham.*—Power-loom for weaving Brussels and velvet carpets.

[1511]

HETHERINGTON, JOHN, & SONS, *Vulcan Works, Manchester.*—Cotton cleaning, preparing, combing, and spinning machinery. (*See pages* 8 *and* 9.)

[1512]

HEWKIN, HENRY, *Oldham, Lancashire.*—Model of the Oldham Building and Manufacturing Company's cotton-mills.

[1513]

HIGGINS, WILLIAM, & SONS, *Manchester.*—Patent cotton machinery, automatic carding engines, drawing, roving frames, and throstle.

[1514]

HINE, RICHARD E., & CO., *Manchester.*—Machine for spinning, doubling, and twisting silk and other threads, wet or dry.

[1515]

HODGSON, GEORGE, *Bradford.*—Looms, with improvements up to the present time. (*See page* 11.)

Harrison, J., & Sons, *Blackburn, Lancashire.*—Looms and other weaving machinery for cotton, linen, &c.

Winding Machine for winding cotton yarn from the cop on to spools or bobbins, commonly called "warper's bobbins," for the purpose of warping or beaming.

This machine has an arrangement on one side for winding cotton or linen yarn from "Throstle bobbins" on to warpers' bobbins, and can be made of any number of spindles.

The spindles are arranged in such a manner that they are always kept on a level with each other.

The motion for shaping the bobbin is a very simple eccentric or "heart," by means of which the bobbin can be filled up in any form.

Warping Machine, on Knowles & Blackburn's patent, to wind the yarn from the warpers' bobbins on to beams for the sizing or dressing machine.

This machine is made on an entirely new principle; the rollers run on centres instead of on bearers as heretofore, thereby greatly diminishing the tension on the yarn and in a very great measure obviating breakages, the production being increased in the same ratio as the breakages are lessened. It is also supplied with a letting-back motion, whereby when a thread is broken, the motion of the beam or roller is reversed, so that the thread may easily be found and reunited. There is also a self-acting measuring and stopping motion by means of which the machine is immediately stopped when the required length of yarn is wound on the beam.

The drum or cylinder on which the beam revolves, is made in such a manner that it may be expanded or contracted according to the width of beam required to be used. Among other improved appliances is Messrs. Knowles & Blackburn's patent expanding and contracting comb.

This improved machine is capable of working more delicate yarn, and yarns of lower qualities, than other machines of the kind, and will in this respect effect a considerable saving. It is also very applicable to silk.

Sizing Machine, commonly called **Slasher,** for sizing or dressing, and afterwards drying the warp preparatory to being woven.

In this machine the yarn is brought from the warpers' beams through the boiling size, and over drying cylinders, after which it is wound on the weavers' beam. The use of the heald and reed is dispensed with, thus facilitating the management of the machine, and causing a saving of between 40 and 50 per cent. in the cost of labour. There is an arrangement for working the machine by friction, and for preventing any tension being put upon the yarn whilst in a wet state. Its elasticity is thus retained, and breakages in weaving almost altogether prevented,

causing considerable increase in the production. By this arrangement coarse and fine yarns can be sized with equal facility, as also yarns of medium and low qualities.

There are syphon boxes for the purpose of condensing the steam as it comes from the drying cylinders; or they can be connected with the size box by means of steam pipes, and the exhaust steam from the cylinders introduced into the size box for the purpose of boiling the size. In this manner *no* steam is wasted. Safety valves, to regulate the pressure of the steam previous to its passing into the drying cylinders, and also a safety valve to "blow off" should the pressure of steam accidentally get too high. There is an arrangement for letting out any water that may accumulate in the cylinders.

The cylinders themselves are made on an improved principle, with an aperture or manhole in the end of each, covered by movable plates, which can easily be removed to allow the cylinders to be cleaned out or repaired, and can with equal facility be replaced. The joints of these plates are perfectly steam-tight, and the manner of their application rather adds to than detracts from the strength of the cylinders.

Another arrangement of very great importance is that by means of which, simultaneously with the stoppage of the machine (at any time), the steam is shut off from the cylinders. The machine is also fitted up with Messrs. Knowles & Blackburn's patent expanding and contracting comb or rathe.

The production is about 100,000 yards of warp per week, or sufficient to supply at least 300 looms.

The machine can be made to dress warps suitable for any width of cloth.

Loom for weaving calicoes, shirtings, and printing cloths, also cambrics, jacconets, &c.; with self-acting temple to keep the cloth stretched to its full width whilst being woven. Self-acting positive taking-up motion for receiving or rolling-up the cloth.

The taking-up roller in this loom is composed of sheet-iron covered with composition. This roller always presents a perfectly level surface to the cloth, being on this account much superior to the ordinary wooden roller covered with emery, the disadvantage of which is, that it changes with the temperature—in damp weather becoming swollen, and in dry weather "warped" or crooked, causing great irregularity in the cloth.

This loom is also supplied with the weft stopping-motion, causing an instantaneous stoppage of the loom when the weft or shoot breaks or is absent. Metallic picking motion for propelling the shuttle. The advantages of this picking motion are greater durability and precision.

HARRISON, J. & SONS, *continued.*

Patent treading motion, by means of which a saving of upwards of 25 per cent. in wear and tear of "healds" or "heddles" is effected, and which conserves in a superior degree the "nap" or "cover" of the cloth.

This loom is on the loose-reed principle, and capable of being worked at a speed of 350 to 400 "picks" per minute, being double the usual speed. It can also be arranged to weave twilled and fancy cloths.

LOOM for weaving heavy domestics' twilled goods, and strong drills and tweeds.

This loom is on the fast-reed principle. It combines all the advantages of the above loom, together with modifications and arrangements necessary for weaving strong goods. It has a cast-iron taking-up roller, fluted and chased, and a patent break; also an improved appliance for preventing strain on the warp threads when the weft is being "beaten up."

LOOM to weave linens.

This loom combines many important improvements. It is supplied with patent self-acting positive letting-off motion, which delivers the warp as required by the taking-up motion for the cloth, which motion is also positive. These two motions work in concert, and with such precision, that the warp is delivered from the yarn beam with the same regularity when the beam is almost empty as when it is full.

The taking-up roller of this loom is covered with patent surfacing instead of emery. It is also supplied with the weft-stopping motion and other important appliances.

The yarns woven in this loom are spun by Messrs. Johnston and Carlisle, of Belfast.

In all these looms the cranks are made of one piece of iron, and bent by graduated pressure. The fibre of the iron by this process remains undisturbed, and renders the crank much stronger than when welded in the usual manner. The bend of the crank, which has heretofore been the weakest part, is now as strong as any other part of it.

Besides the above machines, J. Harrison and Sons, are makers of :—

KNITTING MACHINES on an improved principle, for knitting healds or heddles by power, by means of which a superior quality of heald is produced, with none of the irregularity which occurs in hand-made healds. Another important advantage in this machine is a saving of 50 per cent. in the cost of production.

FOLDING OR PLAITING AND MEASURING MACHINES by power, for measuring the cloth and laying it in folds after it comes from the loom, and previous to being put in bales or bundles.

This machine folds and measures the cloth with the greatest regularity and precision, and effects a very important saving in this department.

CLOTH PRESSES to press the cloth after it has been put into bundles.

DRUM WINDING MACHINES to wind cotton or linen yarns from the hank on to the warper's bobbins or spools.

WARPING MACHINES, specially adapted for linen yarns, with weighting motion, presser, &c.

DRESSING MACHINES on the Scotch principle, to dress and dry linen yarn, preparatory to being woven; with circular or sweep brushes and fans, steam chests, and organ pipes for drying the yarn.

SPOOLING OR PIRNING MACHINES, to wind linen and cotton yarns from the hank or from the bobbin, on to pirns or spools for the shuttle.

LOOMS on an improved principle, to weave fustians, beverteens, &c. with Woodcroft's patent section tappets, positive taking-up motion, self-acting temples, and other improvements.

WINDING MACHINES, suitable for winding yarns for fustian warps, on the best principle.

WARPING MACHINES, for fustian yarns.

SIZING MACHINE, specially adapted for sizing or dressing fustian warps, combining the systems of the sizing machine, and the dressing machine on the Scotch principle, with all their advantages.

LOOMS to weave worsted goods, plain and ancy.

WARPING MACHINES, specially adapted for silk.

LOOMS to weave silk on the newest and most approved principle, with spring reed, &c.

J. Harrison & Sons also supply every accessory connected with the weaving of cotton, linen, &c. &c.

HETHERINGTON, JOHN, & SONS, *Vulcan Works, Manchester.* — Cotton cleaning, preparing, combing, and spinning machinery.

MACHINES EXHIBITED.

A SINGLE SCUTCHING MACHINE, suitable for making laps for 40-in. cards, with enlarged fan, and improved grid bars and beater, specially adapted for surats or leafy cotton, previously opened in their improved opening machine with four porcupine beaters.

THREE CARDING ENGINES, each 40 in. on the wire :—

1. CARDING ENGINE, with main cylinder 45 in. diameter, doffer 22 in. diameter, 6 working rollers, 6 clearers with draw-box and coiling motion ; also 3 takers-in, with self-acting apparatus for working the lower one at variable speeds, so that when at the maximum speed it strips or cleans the main cylinder. This card is also fitted with improved doffing knife, and also under casings for saving the fly.

2. CARDING ENGINE, with main cylinder 40¼ in. diameter, doffer 18 in., with rollers, clearers, and taker-in, combined with patent self-stripping flats, arranged to be cleaned by a brush. The frequency with which the flats are cleaned may be varied to suit the class of cotton used. It has also an improved doffing knife.

CARDING ENGINE.

3. CARDING ENGINE, with main cylinder, doffer, and taker-in. In this card, two new and important improvements (Rivett's patent) are exhibited—First ; a novel method of making the rollers revolve with a peculiar advancing and receding movement, which facilitates their being stripped by a stationary comb or knife. Second : an improved means of stripping or cleaning the main cylinder, which by the arrangement of mechanism is self-acting, and brought into operation at certain intervals as desired. The main cylinder has a reverse motion applied to it, and a revolving brush is brought into contact with it, which thoroughly brushes out the wire on the cylinder.

These three cards have been selected to show some of the most recent improvements, but any of these motions is either applied singly to the usual card, or otherwise modified and combined with each other.

DRAWING FRAME of 1 head or 4 deliveries, with coiling motion and stop motion at the back, and also at the front for stopping the machine when either the feeding or delivery is deranged.

These machines are made with any required number of heads, and deliveries in each head.

SLUBBING FRAME of 60 spindles, 10 by 5 in. bobbin, with 3 lines of rollers, and single centrifugal presser.
ROVING FRAME of 120 spindles, 7 by 3¼ in. bobbin, with 3 lines of rollers, and single centrifugal presser.

These machines are made with increased length of spindle, and reversed bottom rail, so as to reduce the height of the frame ; and they have cones of increased size, with an arrangement adapted for extra long strap.

HETHERINGTON, JOHN, & SONS, *continued.*

These may be taken as examples of this class of machines—they are made with various numbers of spindles, and the size of bobbin is adapted to suit various numbers of yarns, and ranges from 11 by 5 in. to 4 by 2¼ in. accordingly.

THROSTLE FRAME of 232 spindles, 2 in. lift; the spindles are driven by bands from 2 tin cylinders, but are made to be driven by bands f.om single tin cylinder; or when required, by list or tapes from single tin cylinder and carrying pulleys.

These frames are made with various numbers of spindles and sizes of bobbins, and the doubling frame usually employed for making sewing thread or other doubled yarn is a modification of this machine.

SELF-ACTING MULE, Hetherington & Robertson's patent, with 400 spindles, but they may be made with any number of spindles up to 1,200 in each mule.

This mule is shown in the annexed woodcut, and is distinguished by its simplicity of construction, less power required for driving it, regularity of twist in the yarn,

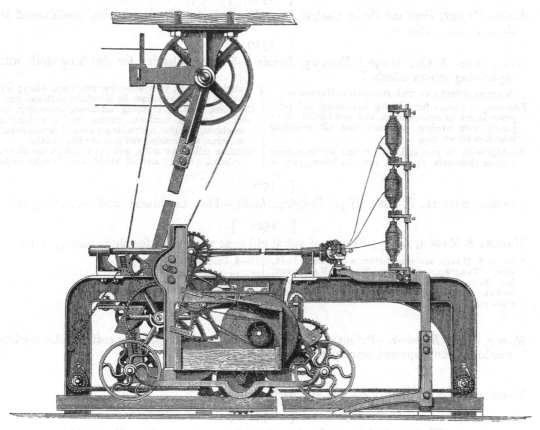

SELF-ACTING MULE.

increased production, and economy in working. The power is transmitted direct to the tin cylinder, and from thence force for propelling the various motions is distributed. The rim band, as used in other mules, and through which the greater proportion of the power required has to pass, is dispensed with, and a considerable expense from its constant wear and tear saved; the irregularity of twist from variation in tension, produced by different temperatures and other causes on these bands, is avoided; and as in this improved mule the drawing rollers are directly geared with the tin cylinders, an uniform twist may be relied on, and the yarn can be limited to the minimum requisite amount of twist, and an equal production secured with slower speed of spindle, and consequently a saving of power; or by driving a spindle at the usual speed a proportionably increased production of yarn of superior quality, from its uniformity of twist, is obtained. The backing-off (*i.e.* the

stopping and reversing of the spindles) is effected through a simple friction pulley on the tin cylinder shaft, instead of being conveyed through the rim band, and is consequently more rapidly and accurately performed. The rack pinion for moving the carriage outwards is driven from the gearing to the roller, and a positive proportionate speed obtained. The quadrant for regulating the winding-on is geared with the carriage motion, and the bands dispensed with. The cam shaft is driven by a positive motion, securing accuracy in all the changes of the mule. Great simplicity of construction is secured by having all the motions in the immediate vicinity of where they are required.

COTTON COMBING MACHINE of six heads and drawing head on Heilman's principle, with Hetherington's improvements.

Prices may be had on application to the Works.

[1516].

IRVIN & SELLERS, *Preston.*—Box-wood logs, cuttings, shuttles, bobbins, pickers.

The following specimens of BOXWOOD, &c. raw and manufactured, are exhibited :—

Boxwood in log; box-wood in cuttings, for engravers' blocks; rules, shuttle-blocks, joiners' tools and handles; bosses for flax-spinners; shuttles of boxwood and fruit-tree; bobbins and skewers of various kinds, used in the spinning and manufacture of cotton, wool, and silk.

Samples and price lists may be obtained by application stating the nature of the goods required. Foreign orders can be supplied on short notice with the best goods at moderate prices.

[1517]

JACKSON & GRAHAM, *Oxford Street, London.*—Jacquard carpet loom worked by steam power.

[1518]

JAMES, HENRY, *Portland Place, Coalpit Lane, Nottingham.*—Braid and whip machines of all descriptions. (*See page 12.*)

[1519]

KERR, JOHN, & Co., *Douglas Foundry, Dundee.*—5-roller calendar for finishing cloth, with equalizing screws attached.

MANUFACTURERS OF THE FOLLOWING MACHINES :—

Preparing machines for spinning flax, hemp, and jute; power looms for linen, canvas, and jute fabrics; Cox's patent weft winding machines, and all preparing machines for weaving.

Washing mills for yarn and cloth, rollers for immersing yarn in chemicals, squeezers, pans for boiling yarn or cloth by steam, yarn softening machines, bluing machines, and all machines for bleaching cloth and yarn. Calenders, beetles, cropping machines, measuring machines, rolling machines, drying machines, damping machines, mangles, hydraulic presses and pumps, and all machines for finishing and packing linen cloth, Grinding mills, saw mills, and gas works; water wheels, turbines, pumps; and all kinds of millwright work.

[1520]

LAWSON, SAMUEL, & SONS, *Hope Foundry, Leeds.*—Flax machinery and self acting tools.

[1521]

MACLEA & MARCH, *Leeds.*—System of spiral gill cone preparings for short hosiery wool.

MACLEA & MARCH are manufacturers of HEMP, FLAX, TOW, WORSTED, and SILK WASTE MACHINERY, &c. They exhibit in this class a system of IMPROVED SPIRAL GILL CONE preparings for short hosiery wool, viz. :—

1—8. Spindle cone drawing.
1—16. Spindle cone finisher.
1—54. Spindle cone roving.

[1522]

MASON, JOHN, *Rochdale.*—Patent machinery for preparing and spinning cotton; also woollen machinery on improved principles. (*See page 13.*)

[1523]

MORRISON, T. & G., *Paisley.*—Jacquard machine.

[1524]

NIGHTINGALE, W. & C., *Old Street, London, E.C.*—Patent horsehair-curling machine.

[1525]

OLDHAM, JOHN C., *Heywood, near Manchester.*—Variety of power loom shuttles.

[1526]

PARKER & SONS, CHARLES, *Dundee.*—Power looms and preparing machines for weaving flax, hemp, and jute.

[1527]

PERRY, JOHN, *Shipley, Field Mills, near Bradford.*—Machinery for preparing and combing wool; circular combs, gills, and fallers.

[1528]

PLATT, BROTHERS, & Co., *Hartford Iron Works, Oldham.*—Machinery for preparing and spinning cotton and wool. (*See pages 14 to 31.*)

[1529]

REYNOLDS, BROTHERS, *Belfast.*—Hackling machines.

HODGSON, GEORGE, *Bradford.*—Looms, with improvements up to the present time.

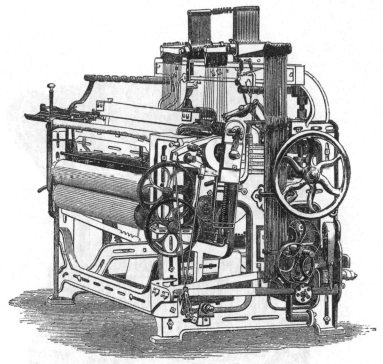

POWER LOOM.

The Exhibitor manufactures every description of power looms, with orleans, cobourg, shalloon, satin, serge, lena, jacquard, serge-de-berry, and every other kind of gearing, for weaving with plain, drop, rising,

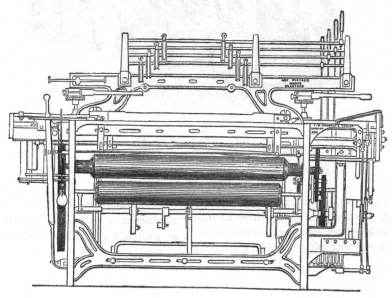

POWER LOOM.

sliding, or revolving circular boxes, for making cotton, worsted, alpaca, mohair, fancy or woollen cloths.

He is also the maker of a patent motion which works in connexion with the boxes, to weave plain, diamond twill square, or satin, and change at pleasure from one to the other without stopping the loom. This motion is particularly well adapted for weaving the German twill plaids, and has the advantage of a Jacquard machine up to 12 or 20 healds. It is simply worked by two plain tappits, whereby greater speed and better cloth are required.

JAMES, HENRY, *Portland Place, Coalpit Lane, Nottingham.*—Braid and whip machines of all
descriptions.

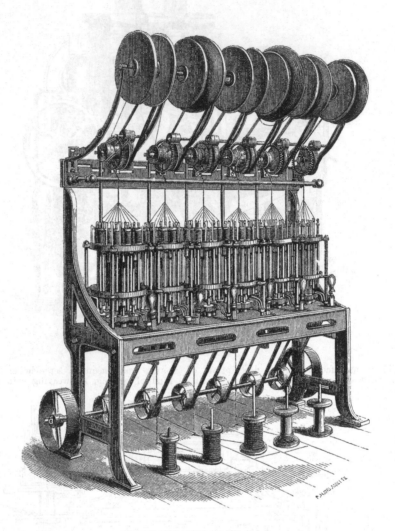

BRAID AND WHIP MACHINE.

BRAIDING MACHINES for covering elastic web, braids, crinoline steel, boot-laces, &c. which perform all these varieties of work without requiring alteration, in one-third less time, and with half the loss by wear of any other machines of the kind. The exhibitor also manufactures a new machine for braiding whips, which is capable of braiding a six-feet whip in five minutes.

[1530]

ROBINSON, J. & R., & Co., 30 *Milk Street, Cheapside, London.*—A Spitalfields silk-velvet loom
at work.

[1531]

ROWAN, JOHN, & SONS, 152 *York Street, Belfast.*—Machine for scutching flax, and other
fibrous substances.

MASON, JOHN, *Rochdale.*—Patent machinery for preparing and spinning cotton; also woollen machinery on improved principles.

MASON'S PATENT SLUBBING AND ROVING FRAMES.

The object of these improvements is, to secure larger production, greater durability, and at less cost.

This is accomplished, first, by continuing the collar (which is firmly fixed to the lifting rail) through the pinion wheel, up the inside of the bobbin, nearly to the top, where the bearing for the spindle is formed as shown at *a* in figs. 1 and 2.

In order to reduce the friction, the collar is made with a recess or hollow chamber inside, so that the spindle only fits at each end. The bobbin at its upper end runs on the spindle as usual, and friction upon the outside of the collar is prevented by its being made to pass at its lower end upon a flange, which projects upon the top of the pinion wheel. The bobbin entirely covers the collar, protects the bearing from injury by dust or other matter, and thus less oil is required.

The top of the flyer is left clear for piecing-up and doffing.

These advantages are much more manifest after the machines have been some time at work, when instead of having to reduce the speed, it is generally increased a little.

Although the bobbin barrel is about $\frac{3}{16}$ in. larger in diameter, it is not found to be a disadvantage; the frame starts better upon the empty bobbin, and a trifling addition to the diameter, when full, will hold the same length of slubbing or roving.

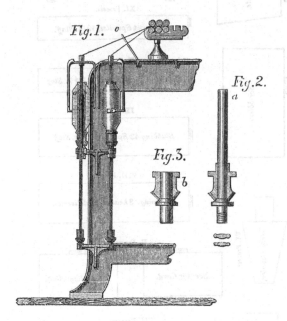

SECTION AND ELEVATION OF PATENT COLLAR, &c.
IN ROVING MACHINE.

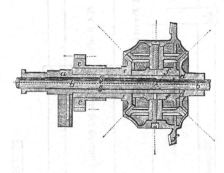

Fig. 4.
DIFFERENTIAL OR JACK MOTION.

The difference between this mode and the best arrangements of collar previously used, is shown at figs. 2 and 3, the bearing for the spindle in the one being at *a*, and in the other at *b*, a difference equal to the length of the lift (say 10 or 12 in. slubbing and 6 or 7 in. roving frames) in favour of this patent. From the increased steadiness of the spindles, there is less wear and tear by friction.

The second improvement is in the separating plates (as shown at *c* in fig. 1) which prevent the ends from becoming entangled, and thus reduce the waste. They are placed between the rollers and the spindles, with convenience for removing at pleasure, to facilitate doffing and cleaning.

The third improvement is in the application of a long boss to the differential or jack motion, as shown at *a a* in fig. 4. The main shaft of the said motion is at *b*, supported by the boss *a*, which according to the usual arrangement terminates at the pinion *c;* according to this improvement, however, it is extended to the point *d*, and the wheels *c, e, f* are mounted thereon instead of upon the shaft as is usual. To reduce the friction upon the driving shaft, the long boss is recessed or chambered out, as shown at *g*.

The advantages of the above arrangement are, steadiness in working, reduced friction, and greater durability. This is attained by the increased length and diameter of the bearings, which are of cast-iron. The motions of the box and shaft being opposed to each other, by the introduction of the long boss, the rubbing surfaces are separated, as will be seen from the annexed sketch.

The arrangement as regards the other parts of the apparatus is that in ordinary use.

The following machines are manufactured by John Mason, at the Globe Works, Rochdale:—

Openers, scutchers, lap machines, fans, single and double carding engines, grinding machines with cement and emery cylinders and rollers, drawing frames, &c.; Mason & Co.'s patent slubbing and roving frames; throstles with band and list wharves, winding machines, reels, &c.; teazers; woollen, worsted, and silk carding engines; self-acting cotton and woollen mules and power looms; patent condensor or endless carding engine.

PLATT, BROTHERS, & CO., *Hartford Iron Works Oldham.*—Machinery for preparing and spinning cotton and wool.

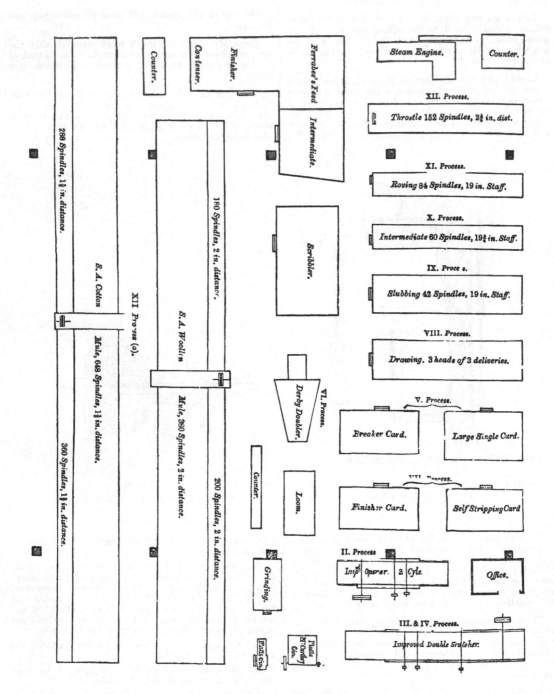

PLAN OF MACHINERY EXHIBITED.

PLATT, BROTHERS, & CO., *continued.*

THE CHURKA GIN.

This machine is composed of two rollers, the lower of which is covered with hard wood, and is 1½ in. diameter in its working part; the upper one is of steel, ½ in. diameter, with a finely fluted surface. They work in contact, and are coupled by gearing, so that their two circumferences travel at the same rate. It will clean all kinds of staple from hard seeds, one of its rollers being so small that the smallest seed cannot be taken in by the rollers.

When the fibre is separated from the seeds it is passed through the rollers and delivered, whilst the seeds as they are released drop through the grid in front of the rollers.

PLATT & RICHARDSON'S PATENT CHURKA GIN (exhibited)

The novelties and improvements introduced, consist in holding the rollers in contact, supplying them with seed cotton by a self feeder, and in preventing them from lapping. They operate as follows—The cotton containing the seed is spread on an endless travelling lattice, which conveys it to a series of three spiked rollers, the first of which revolving over the lattice and its circumference travelling at the same speed, holds the cotton; the second which travels much faster fills the spikes with cotton; whilst the third moves at an intermediate speed to the other two, its object being to prevent the second roller from carrying pieces of cotton on its surface.

The next operation is to strip the second roller and

COTTON GIN—HAND.

convey the cotton to the Churka rollers. This is done by a comb having a circular vibratory motion, given to it through an elastic connecting rod to prevent breakage in case of obstruction. After this operation of the rollers and comb, the fibres are loosened from the seeds, and are in the most favourable condition for being passed through the wood and steel Churka rollers. The steel roller is held in contact with the wooden roller by a weight and levers bearing upon its journals. A knife is fixed in a frame over the top of the steel roller to keep

it clear. This frame also carries a roller covered with leather, which runs in contact with the wooden roller: this knife and the roller prevent the steel roller from being wrapped with cotton, and can be lifted out of the way together.

The bottom or wooden roller is kept from wrapping by a fluted roller revolving under it on the delivering side, and driven by one of the other rollers.

This gin will separate from hard seed about 600 lbs. of clean cotton weekly.

PLATT, BROTHERS, & CO., *continued.*

THE MACARTHY GIN.

This is a machine for separating cotton fibre from its seed.

The original Macarthy gin consists of a roller covered with leather about 5 in. diameter, having a number of small grooves cut in spirals in its surface, making about one hundred and ten revolutions per minute. On the face of this roller is a thin steel plate acting against it with a slight pressure ; it is also furnished with a wire grid, upon which the seed with its fibre attached is pushed by hand against the face of the roller, which, by means of the

spiral grooves, and the adhesive nature of the leather surface, draws the fibre under the steel plate until the seeds come in contact with its edge. Whilst the fibres are thus held the seeds are pushed off by the edge of a bar which has a vertical vibratory motion, so as to pass the edge of the plate where the seed is held, and thus separate it from the fibre, which is carried forward and delivered by a fluted roller placed in front, and which revolves in the same direction as the Macarthy roller.

It is important to make the spaces of the grid to the size of the seed the machine is cleaning, for

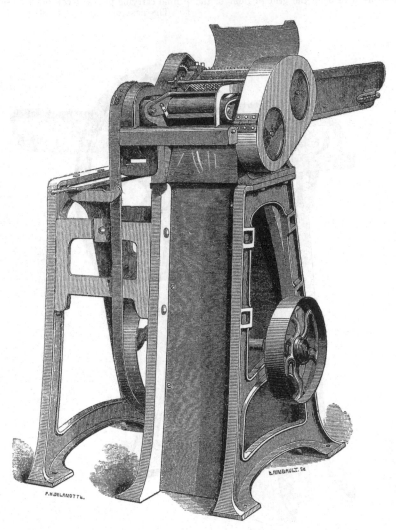

COTTON GIN—STEAM POWER.

if too coarse the seeds will pass through before they are cleaned, and if too fine they will accumulate.

ONE DOUBLE-ACTING MACARTHY GIN.

Platt & Richardson's patent (exhibited).

The novelties and improvements introduced are, in feeding the machine with seed cotton, which is placed on an endless travelling lattice, and conveyed by it to a series of rollers, the last of which is furnished with spikes, and travels at an increased speed, so as to separate

the tufts in detail from the sheet spread on the lattice. From this spiked roller, the tufts are transmitted to the Macarthy roller by a comb having a circular vibratory motion given to it, through an elastic connecting rod, by which breakage from obstruction is prevented ; also in the introduction of two bars with vertical vibratory motion, moving alternately from a double crank (Platt & Richardson's patent balance), for the purpose of pushing the seeds from the fibre whilst held by the steel plate.

These improvements cause an immense saving of labour, as hitherto each machine required an attendant,

PLATT, BROTHERS, & CO., *continued.*

and now one attendant can superintend several machines, whilst each machine will clean more than double the quantity.

This machine will clean all kinds of cotton, but it is especially adapted to such as contain soft and woolly seeds.

A machine 24 in. wide will separate from hard seed about 1000 lbs. of clean cotton weekly.

MIXING (FIRST PROCESS).—Selecting the bales and mixing the cotton is the first process in the cotton manufacture. It is done as follows—A selection of bales of cotton suitable to the class of yarn required is made, and their contents spread out in layers of each so as to form a stack called a "mixing," from the sides of which the cotton is taken vertically to supply the opener.

MACARTHY GIN. (PLATT AND RICHARDSON'S PATENT.)

THE COTTON OPENING (SECOND PROCESS). — This process is to open out the fibres of the cotton after it has been pressed in bales, and to extract the sand, dried leaf and other impurities imported with it, and it is important to do this without entangling or injuring the fibre. The machines used for this purpose are of various kinds, to suit the requirements of the trade.

NEW OPENING AND COTTON-CLEANING MACHINE.

The machine illustrated is recommended and used for cottons of short and middling staple. It comprises an endless lattice, upon which cotton is spread, and an iron roller with ribs on its surface, which together convey the cotton to a pair of fluted feed rollers, and is delivered by them to the first of a series of four cylinders which is furnished with twelve rows of teeth, the second, third, and fourth having only four rows of teeth. These cylinders revolve in the same direction, in journals or bearings supported by a horizontal framing, at a speed of about 1,000 revolutions per minute. These cylinders are all cased on the upper side with sheet iron, the first part of the under side of each is cased by angular bars with spaces betwixt them, forming a circular grid, which allows the dirt disengaged by the action of the cylinders to pass through to the floor.

PLATT, BROTHERS, & CO., *continued.*

The remaining part of the under casing is made from a perforated sheet of metal, which allows the dust to escape through whilst the cotton is passing over it.

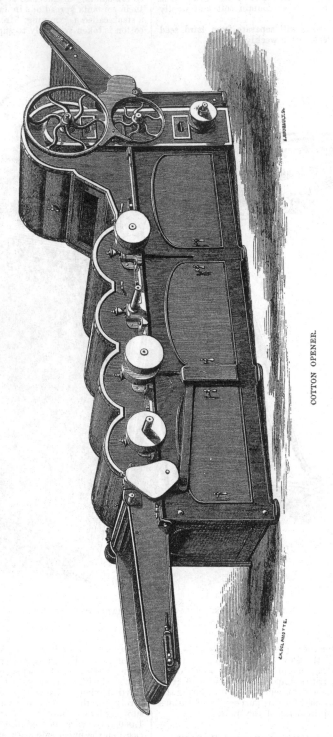

COTTON OPENER.

The first of these cylinders strikes the cotton from the feed rollers, passes it over the circular grid and perforated plate, and delivers it to the second cylinder, whence it receives a blow equal to the combined velocity of the two cylinders, and passes it on to the third and fourth cylinders, so that this action is repeated three times with the four cylinders, each making a deposit through its respective grid and perforated plate, and as the light fibre only

PLATT, BROTHERS, & CO., *continued.*

offers any resistance to the quick blows of these cylinders, it is impossible it should receive any injury in this operation.

The last or fourth cylinder drives the cotton over a straight grid to the back of two wire cylinders, when it is collected and afterwards stripped by two iron rollers, which carry it to a delivery lattice in front of the machine, so that throughout the whole passage of the cotton from the feed rollers to the wire cylinders, there is a continual deposit of impurities.

The two wire cylinders are exhausted by a fan which collects the dust within the casing of the machine and forces it into any place provided for its reception, by this means keeping the rooms where these machines are in operation perfectly free from dust.

These machines are also constructed with one or two cylinders, and with lap machines attached, so as to prepare laps to be afterwards fed up the scutcher.

ONE OPENING MACHINE, with two cylinders and lap attached (exhibited).

REMARKS.—The machinery previously used for opening and cleaning cotton having been found incapable of taking out the dried leaves and other impurities contained in the cotton imported to this country during the last few years, without materially damaging the cotton fibre, has called for the introduction of this machine to the trade, and it is found to be admirably suited to the purpose.

SCUTCHING AND LAPPING (THIRD PROCESS).—The machines are supplied with cotton from the opener in a uniform fleece by two methods; one by dividing a feeding lattice into a number of equal parts, and spreading uniformly upon each part a given weight of cotton to present to the feed rollers. The other is by driving the lattice and feed rollers of the scutcher at varying speeds in proportion to the thickness of cotton supplied, which speed is regulated by the rise and fall of the top feed roller multiplied by levers, so as to guide a strap communicating motion to the lattice and feed rollers, from a cone pulley revolving at a uniform rate to a second cone pulley. These pulleys are on parallel vertical axes attached to the sides of the feeder; thus when the feed roller rises, its speed is diminished, when it falls it is increased, and an almost uniform supply of cotton is presented to the first cleaning cylinder, which is furnished with twelve rows of teeth, that in revolving strike the cotton and pass it over a circular grid to a revolving beater with three blades, which then passes it over a second circular grid and a straight grid to a pair of wire dust cylinders that are exhausted by a fan. The cotton is then stripped from these dust cylinders by a pair of iron rollers, and passed through a second set of two pairs of feed rollers which revolve more quickly than the first, thereby delivering a thinner fleece to the second beater, which again passes it over a circular and straight grid to two other wire dust cylinders which are stripped by rollers as before. This latter pair of cylinders and rollers travel at three times the speed of the feeder, so that they deliver a fleece one-third the thickness first supplied to the machine.

The next operation is to form the cotton into a large roll or lap. This is done by the lap machine attached to the scutcher, forming together one machine.

The rollers which strip the last dust cylinder, deliver the fleece to a set of four callender rollers placed over each other, so that the cotton in passing through them receives three compressions, which form the fleece into a kind of felt; three of these callenders have their surfaces kept clean by bars of iron covered with flannel which are pressed in contact with them. The cotton then passes over one of two large fluted rollers which revolves in the same direction, and under a smaller plain roller which is above the fluted roller, and receives its motion from it by contact through the fleece; this small roller also cleans the second callender roller, by running against it in a contrary direction with a slight pressure (it also breaks the fleece when the lap is formed).

The fleece is now wound upon an iron tube slightly taper, that is placed in the channel between the two fluted rollers, and driven by contact with them, having gudgeons at each end, on which it receives pressure from two friction pulleys revolving in racks placed vertically and gearing into pinions upon a shaft across the machine. This shaft again communicates by gearing to a break pulley which has a slight pressure given to it by a lever that can be released by the foot of the attendant; by this means as each successive layer is wound upon the rollers, the break slips and allows it to rise. One of these fluted rollers has a worm on its axis geared into a wheel with such a number of teeth, that one revolution of it will indicate the length of fleece to form the lap required. On the same axis as this wheel is a tappet which stops the feeding motion and callenders, by pulling the support from under the hand levers that carry the end of the driving shaft. When the wheel drops out of gear, the two fluted rollers carrying the lap continue to revolve and break the fleece, the foot lever releases the break, the racks are lifted by a hand wheel, and the lap is taken out and stripped by dropping the small end of the tube upon a block of caoutchouc placed conveniently on the floor, and which, by its elasticity, causes the tube to rebound from the lap, when the attendant seizes it, lifts it out of the lap and again places it upon the machine, lowers the rack and friction pulleys by the hand wheel, lifts up the gear levers, and the process again commences.

THE SECOND SCUTCHING AND LAPPING (FOURTH PROCESS).—The machine used in this process is similar to the one previously described except in its feeder part, which is so arranged that three of the laps made by the first machine can be placed upon it so as to be uncoiled by the traverse of the lattice, and which is done as follows —Through the centre of the laps rods are inserted, the laps are then placed upon the lattice with the rods in slits or guides made in the framing to receive them, and thus keep them parallel; the laps are then uncoiled and spread upon the surface of the lattice in three layers on the top of each other, so as to present to the feed rollers a uniform fleece equal in thickness to that fed upon the first scutcher.

The machine is then set in motion, the cotton is passed through the feed rollers, and the remainder of the operation is precisely the same as in the machine previously described. By thus doubling the laps the fibres are more thoroughly mixed, and the fleece is made more uniform in thickness; and as the fleece must be uniform in its length and breadth as well, it is absolutely necessary that the beater should produce one uniform current of air, and thus waft the fibre over the straight grid direct to the wire cylinder. It is whilst the cotton is thus floating that the heavier impurities, loosened by the beaters, drop out and fall through the grids into the dust boxes. The laps formed by this machine are then taken to the breaker carding engines.

ONE NEW SCUTCHING AND LAPPING MACHINE (exhibited).

Novelties consist of an improved section of machine, by which more uniform currents are obtained, and better felted laps produced; in cylinders with teeth in combination with a beater with knives instead of having a beater only; in producing a uniform fleece by varying the speed of the feed roller; and lastly, in the covering and casing them with steel made by the Bessemer process, and in the application of Lord's patent feeder.

REMARKS.—By successive stages through a long series of years to the present time, the difficulties which originally presented themselves to the adoption of this class of machinery for cotton cleaning have been overcome, and the cotton can now be perfectly cleaned without injury to the staple or fibre, laps produced with fleeces uniform in length, breadth, and thickness, and so felted that they uncoil at the carding engine without any derangement of the felted fleece.

PLATT, BROTHERS, & CO., *continued.*

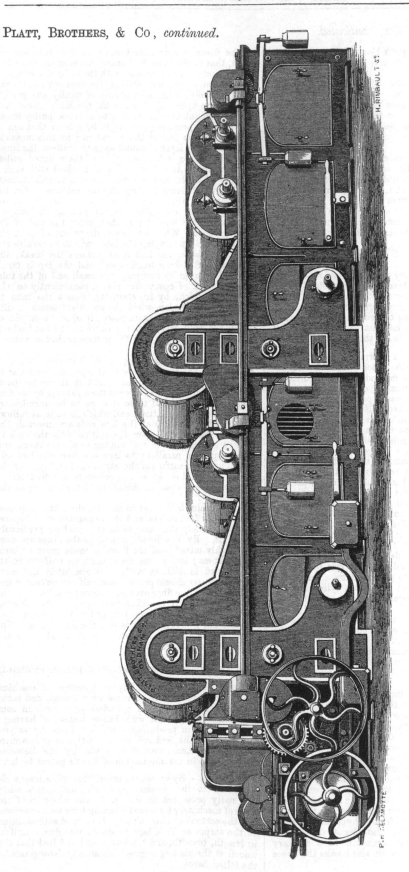

DOUBLE SCUTCHER.

PLATT, BROTHERS, & CO., *continued.*

THE FIRST CARDING ENGINE (FIFTH PROCESS) continues the operation from the lap machine to the drawing frame by a kind of combing process. For low coarse yarns one only is used (single carding) to change the lap fleece into a sliver, but for finer yarns, and for coarse yarns made from the best description of cotton,

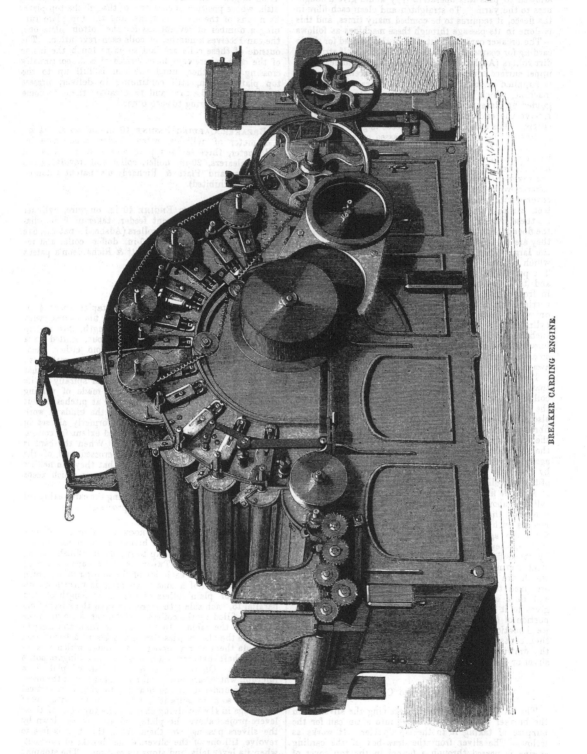

BREAKER CARDING ENGINE.

two cards are used, one acting as a "breaker" and the other as a "finisher," the laps to supply the latter being formed from slivers delivered by the breaker on a machine called the Lap Doubler. By this process the felted fleece delivered by the lap machine, with its fibres crossed in all directions, is combed and straightened, and the light

PLATT, BROTHERS, & CO., *continued.*

impurities still adhering to it, such as short fibre and the moss-like covering of the seeds, are taken out, for if allowed to pass this operation they would give a roughness to the yarn. To straighten and clean each fibre in the fleece, it requires to be combed many times, and this is done in its passage through these machines as follows—The breaker card illustrated (may also be used for single carding for coarse yarns) has a taker-in, three self-stripping dirt rollers (Adshead's patent), and the remainder of its upper surface covered with rollers and clearers. This card is supplied or fed by a lap of the fleece from the lap machine, which is placed on a roller at its feeding end, guided by plates at each side which have slots in them to receive the ends of the rods passing through the laps' centre. The unlapping of this fleece is governed by the motion of this roller; it is now passed over the plate to the feed roller which delivers it to the taker-in roller; at this point the combing or carding commences, whilst the fleece is held by the feed rollers travelling at a slow speed, the taker-in running much faster and having its surface covered with cards (see CARDS), a kind of wire brush covered with crooked teeth so fixed that the points of the teeth strike down into the fleece held by the feed rollers. As these fibres are combed the impurities separated fall to the floor; the taker-in passes the fibrous tufts of cotton as they are released by the feed roller on its under surface to the large cylinder which is also covered with cards and which revolves in an opposite direction to the taker-in. The points of its teeth incline in the direction of motion, and its surface travels much faster than that of the taker-in from which in passing it takes the fibrous tufts and carries them to the self-stripping dirt rollers, the cards on which have their hook point to face those of the cylinder, so as to hold in the interstices of their wires such impurities as they may receive, which are carried forward by their motion and stripped by a vibratory comb so as to form a roll on their upper surfaces, to be taken away at intervals.

These dirt rollers revolve with a very slow motion, so that they assist in stretching the fibres as well as in collecting the dirt. From the dirt rollers it passes under the first clearer to the first carding roller, whose hooks also face those of the cylinder, so as to straighten the fibres and divide any tuft remaining; this roller passes the fibres fixed in its teeth (by their antagonism with the cylinder) back to be stripped by the clearer, this again delivers them to the cylinder to be again divided by the same roller. This operation is repeated by each of the five rollers and clearers, till the tufts are all reduced to straight fibres, which pass on to the doffer (another cylinder about half the size of the main cylinder); the hooks of the doffer face the cylinder, its motion also recedes with it and travels at a much slower speed, the fibres are again stretched whilst they are left on its surface, they now pass on its under side to be stripped by the doffing comb, which is formed of thin plates of steel having fine straight teeth on their lower edge, which are hardened to prevent wearing; these plates are fixed to a channel-bar which is connected at each end to a crank running at a high speed, and which gives to it a vertical vibratory motion, so as to strip a portion of fleece from the face of the doffer by its downward motion and clear itself by its rising, the fleece is then contracted through a funnel and taken forward by the drawing rollers which deliver it in the form of a sliver or riband to the coiler and can.

COILER AND REVOLVING CAN MOTION.

This is a small machine for receiving the slivers from the breaker cards, and coiling it into a tin can for the purpose of taking it to the lap doubler. It works as follows—The sliver from the draw-box of the carding engine is passed through a funnel in the top cover of the machine, to a small pair of revolving rollers underneath, by which it is taken in and delivered through a tube and revolving plate to the can over which it is

placed: the top end of the tube is concentric and the lower end eccentric to its motion, *i.e.* the tube is placed at an angle. The can is situated below in a revolving dish, whose position is eccentric to that of the top plate; by means of these two motions, and the top plate running a number of revolutions for the bottom plate one, the can receives a number of coils each revolution. The outside of these coils are laid so as to touch the inside of the cans, where they form circles of coils continually crossing each other, until the can is full up to the top plate, which, still continuing to deliver, presses more sliver in the can, and thus causes them to come out without adhering to each other.

A BREAKER CARDING ENGINE 40 in. on wire, 40¼ in. diameter of cylinder, patent feeder, taker-in 9 in. diameter, three self-stripping rollers, four rollers and four clearers, 20-in. doffer, coiler and revolving can motions, and Platt & Richardson's patent balanced cranks (exhibited).

A BREAKER CARDING ENGINE 40 in. on wire, cylinder 40¼ in. diameter, patent feeder, taker-in 9 in. diameter, two self-stripping rollers (Adshead's patent), five rollers and four clearers, 20-in. doffer, coiler and revolving can motions, and Platt & Richardson's patent balanced cranks (exhibited).

CARDS.

These cards are made by fixing staples about ½ in. long, and ¼ in. wide, made of very fine wires, with a side bend in the middle of their length, into a strip of elastic cloth composed of caoutchouc, united to a number of layers of cloth, made from either linen, cotton, and wool, or a combination of these materials; these strips are about 1¼ in. wide, and in lengths that will cover the cylinder by being wound spirally on its surface. These staples or teeth are made of varying strengths of wire, and set in the cloth at pitches to suit the parts of the machine, as well as the kinds of work they are intended for; they were formerly all set in leather, which is still used to a limited extent for cotton, but still more generally for wool. When the card is wound tight on the cylinders the crossed end of the staples is pressed to its surface, so that they can neither rise nor fall, but have an elastic firmness which keeps them to the work.

A portion of the wire used in making the cards exhibited is from steel made by Bessemer's process.

THE LAP DOUBLER (SIXTH PROCESS) exhibited.—By this machine the slivers from the breaker card are formed into a fleece and coiled into a lap to supply the finisher card; it operates as follows—Two rows of tin cans containing slivers are placed on each side of the feeding table, which forms the section of a cone; this table is furnished with two pairs of plain rollers of the entire length of, and parallel to, each side; these rollers take the slivers from the cans filled by the coilers, and deliver them upon the surface of the table. In their course from the cans to the table the slivers pass through holes in a bar of iron to guide them over a curved plate, under which is a revolving shaft that carries a boss with three wings opposite each sliver. On the top edge of the curved plate is a fulcrum, which carries a small two-ended lever; the lower end hangs under it, and is heavier, to give it a vertical direction, so as to cause it to fall in contact with one of the wings in the revolving shaft. The top ends of these levers project above the plate, and are pressed down by the slivers passing over them, when the shaft is free to revolve, till one of the slivers either breaks or runs out, when its lever falls, and stops its motion. The stopping of this shaft puts in motion a cam that moves the strap upon the loose pulley, and stops the machine; the end of the sliver is again supplied, and the machine proceeds as

PLATT, BROTHERS, & CO., *continued.*

before. By this means missing slivers, or " singles," is entirely prevented, and the fleece is uniform.

Two slivers, one from each side, pass up the centre of the table, close to each other, from the apex of the cone, the others are supplied in equal divisions on each side, so as to fill the whole surface.

The lap machine is connected to the wide end of the table, and the first of its callenders that receives the sliver, travels at the same surface speed as the smaller rollers that supply the table from the cans, so that the slivers move in straight lines from one to the other, and are drawn over the table by mutual assistance, as the

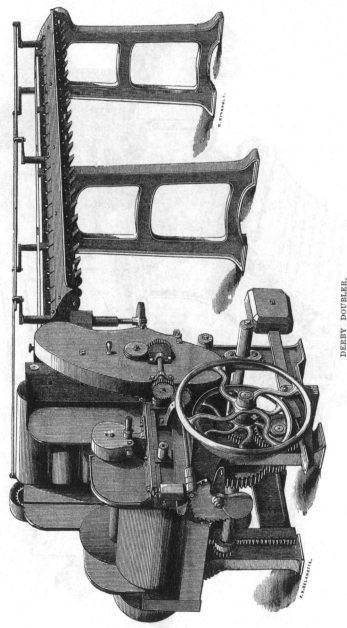

DERBY DOUBLER.

long ones up the centre would break if not assisted by those at the sides.

The machine that winds the lap is similar to that described for the scutcher, but the laps are wound upon wooden bobbins that are taken with them to the cards. These bobbins are weighted by Knowles' patent motion, consisting of an iron roller which presses on them

whilst their ends are formed against revolving washers, guided in their centres, without gudgeons, which facilitates the removal of the laps.

These machines are constructed to form laps, either one-third, one-half, or the full width of the lap required, as may be desired.

The novelties introduced are—improved stop motions

PLATT, BROTHERS, & CO., *continued.*

(Knowles' patent), revolving plates to lap ends, feeding table, and improved general construction of machine.

THE SECOND, OR FINISHER CARD (SEVENTH PROCESS).—The finisher card continues the operation of combing and cleaning commenced by the breaker.

In some cases for carding middling qualities, cards similar to the breaker cards before described, are used as finishers also ; so that if desirable both may be used for single carding.

For fine qualities those of the construction illustrated and exhibited are most generally used ; for fine qualities

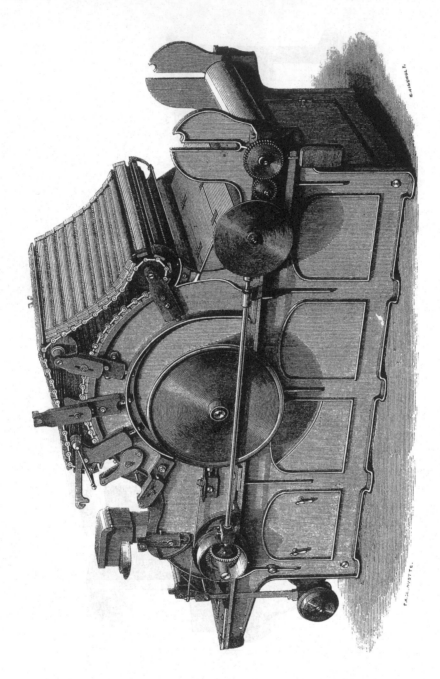

FINISHING CARDING ENGINE.

of still higher counts, this construction of carding engine is used for both breaker and finisher ; and for the finest qualities, it is used as a breaker card for cotton to be afterwards combed by the combing machine.

The finisher is supplied with laps formed by the lap doubler (before described) of 96 slivers from the breaker card, in order that the mixing of the cotton may be more thoroughly effected, and to ensure more perfect uniformity of the sliver.

These laps are placed between two rollers at the feeding end of the card which unlap the fleece and deliver it to the feeding roller ; they are guided at each

PLATT, BROTHERS, & CO., *continued*.

side by a plate to keep the fleece central with the cards. The taker-in roller combs it from the feed rollers, and carries it to the main cylinder, which is covered on a portion of its surface with a train of iron flats, covered with cards, and united at each end by links, so as to form an endless travelling lattice. This lattice is carried on shafts, having a slow motion, and supported by bearings in the general framing. Those flats in operation slide upon a curve that is adjustable to the cylinder.

The sliding portion of the flat is formed with a slight angle to the face, upon which the card is fixed, so that the point of contact with the cylinder will be near to the front or leading side of the card. Those flats not in operation slide on plain slips on each side of the cylinder to support them whilst the faces of the cards on their surface are ground true and sharp by a short disc of metal covered with emery, and running at a quick speed and at the same time traversing over the lengths of the strips of card on the flats, so as to form the points of wire to a true surface. The hooks of these cards face those of the cylinder, so that each flat combs the fibres as it passes on the face of the cylinder. The main cylinder and doffer are also made true by this method of grinding. The impurities separated are carried forward by the motion of this train, and are stripped off by a vibratory comb in front, when they fall into a box.

After passing the flats, the fleece is again combed and delivered as before described in the breaker carding engine.

Two FINISHER CARDING ENGINES, 40 in. on wire, cylinder 40¼ in. diameter, with patent feeder, taker-in 9 in. diameter, fifty revolving flats (Leigh's patent), eighteen of which are in action, and doffers 18 in. diameter, coilers and revolving can motions, guards to wheels, &c., and Platt & Richardson's patent balanced cranks (exhibited).

The novelties consist in the arrangement of the machine, so that the flats can be accurately ground whilst the card is working, and the other portions of the machine can be stripped and ground without being moved from their place, and in the application of a motion to stop the doffer when breakage of sliver or any other obstruction occurs.

REMARKS.—Until recently, the finisher cards were constructed without taker-in rollers, the main cylinder taking the fleece direct from the feeding roller, causing the fibres to fill the cards, and any impurities passing the feed rollers damaged the cards on this large surface. By using taker-in rollers, these evils are prevented, the fibres being delivered to the cylinder without pressure.

The original difficulties with the carding engine were to maintain true surfaces, on which the cards were fixed; (these being generally constructed of timber varying with every change of the atmosphere, had to be made true each time by grinding the full parts from the ends of the wires.) The cylinders and rollers were not carefully constructed so as to run with a steady motion.

The fixings for carrying the different journals were not capable of a fine adjustment, neither were they steady after being set. These defects prevented the cards working sufficiently near to each other without occasionally coming in contact, which destroyed the carding point. The above defects are now overcome by using iron instead of wood, and by the aid of machinery in the construction. The moving parts are capable of fine adjustment, and are as firm as the fixed ones when set. These improvements in construction cause less grinding and stripping to be required, as the finer and truer the points of the wire can be maintained, the clearer will be the card.

THE DRAWING FRAME (EIGHTH PROCESS).—By this process, the cotton already cleaned, carded, straightened, partially drawn and formed into ribands or slivers, is

doubled and further drawn by passing a number of those ends or slivers—say about six—over guides depressed by the weight of the sliver, through a series of four pairs of rollers, each pair travelling at a different speed; the difference in this case between the first and the fourth pair being about as one is to six, that is to say, that the circumference of the fourth roller travels through a space six times greater than the circumference of the first pair, and by so doing elongates or draws the sliver thus passed to six times the original length, and forming a single web, which is passed through a funnel to a pair of callender rollers, through which it passes to a coiling motion which deposits it in a revolving can, as described in the carding engine.

The sliver thus deposited being doubled six times and drawn six times is the same weight or thickness per yard as each of the slivers received by the back roller, and the object sought by this is to equalize the quality of the cotton and to make the slivers of uniform strength and texture by the combination. This process is repeated three times in this machine, and the amount of doubling and draft is equal in each case, say 216.

The guides depressed by the sliver in passing to the back rollers act as stop motions when the sliver breaks or runs out, by being thus released and coming in contact with a spider having a circular vibratory motion communicating to it through a catch box connected with a strap fork.

ONE DRAWING FRAME, with three heads of three deliveries each, four rows of rollers, the front row of steel made by the Bessemer process, and the back row fluted with coarse flutes, Leigh's top rollers to the front row, and coarse fluted top rollers to the back row, fitted with stop motions, coilers, and revolving can motions, and improved flats with endless traversing cloth, for cans 36 × 9 in. (exhibited).

The novelties introduced are in the use of rollers made from Bessemer's steel, Leigh's top rollers with revolving bosses, for front row; in an improved top clearer or flat which hangs upon hinges, and is provided with an endless cloth which clears the top rollers by travelling over them; its advantages are, a saving of power, labour, oil, and roller leather, it is much cleaner than the ordinary flat, there is less friction, and consequently less heat and electricity; the oil is less fluid, and the greasing of the pivots of the rollers is much better, facility of inspection is much greater, and the "flat waste" is never taken away by the sliver; in a stop motion for stopping the machine when the sliver breaks betwixt the front rollers and the callender, and which is driven from the same shaft and catch box as that used when the sliver breaks betwixt the can and the back roller.

THE SLUBBING, INTERMEDIATE, AND ROVING FRAMES (NINTH, TENTH, AND ELEVENTH PROCESS).—The single slivers of cotton delivered in the last operation of the drawing frame are now conveyed in their cans to the back of the slubbing frame. This frame is furnished with guides similar to those described in the drawing frame, over which the ends pass to a series of three pairs of rollers revolving at varying speeds, the speed of the first pair being to the speed of the last in the proportion of one to five, so that the sliver is again increased in length five times in passing through them. In front of these rollers are two rows of spindles which are furnished with flyers having two hollow legs, and upon these spindles, bobbins about 11 in. long are threaded. These spindles and bobbins are both made to revolve, but at varying rates, and from distinct and separate movements.

The cotton, now called slubbing, delivered by these rollers is partially spun or twisted by the revolutions of the spindles, passes through the hollow legs of the flyers, and is wound upon bobbins; two of these bobbins are then filled into the creel of the intermediate frame; the slubbings

Platt, Brothers, & Co., *continued.*

are then doubled by passing the ends of two of them through another series of three pairs of rollers, and joining, drawing, twisting, and winding them upon bobbins about 9 in. long, which are revolving upon spindles in front of the delivery rollers as before; two of these bobbins are then doubled in the creel of the roving frame, the process

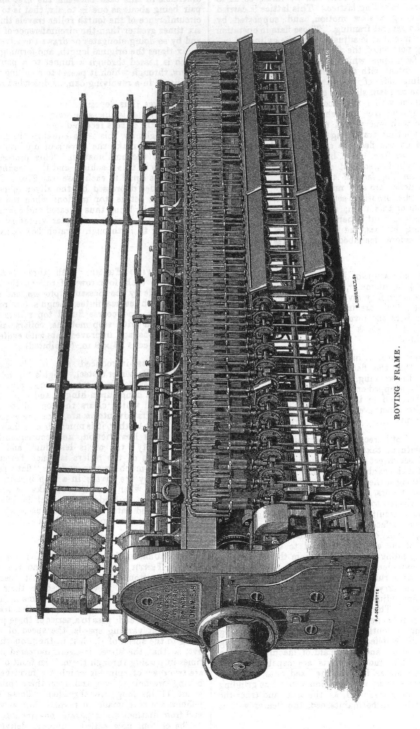

ROVING FRAME.

of drawing, twisting, and winding is again repeated, and the cotton, now called roving, is wound upon bobbins about 8 in. long, ready for being spun in the mule and throstle. The twisting of the cotton, after being delivered by the rollers of these machines, is effected by the revolutions of the spindles, the slubbing or roving is passed

PLATT, BROTHERS, & CO., *continued.*

through a hole on the top of, and down one of the legs of the flyer to its finger or presser, round which it is coiled, and delivered to the bobbin; this presser hangs loosely upon the flyer leg, but is parallel with and carried round by it at a uniform rate, causing a uniform pressure to be given to the bobbin through its weight and the resistance of the air in its circuit. As the bobbin is being wound, it is caused to traverse up and down the spindle against the finger, so as to equally distribute the roving. The winding of the roving upon the bobbin is regulated by increasing or diminishing its speed accordingly as the bobbin follows the flyer or the flyer follows the bobbin. Frames are made in both ways.

When the bobbin follows the flyer its speed must be increased as its diameter increases by winding, or the roving will be irregularly stretched or broken.

The speed of the front roller delivering the roving and the speed of the spindle which twists it is constant.

In these frames the bobbin follows the flyers, and the first motion communicating with the bobbin is at its greatest speed when the bobbin begins to wind, the speed gradually diminishing as the layers are wound on. This diminution of speed is effected by moving a strap upon two conical drums, one concave and the other convex, the speed of the concave drum is constant; these drums also communicate motion to a rail which, in its up and down motion, traverses the bobbin upon the spindle, and by this means regulates the speed of this traverse to suit the increased diameter of the bobbins.

The length of these drums is arranged to suit the diameter of the bobbins to be filled, so that when the strap has been traversed across the drum, the bobbin has attained its full dimensions when the frame knocks off.

The bobbins being now filled, are taken off and exchanged for empty ones; the end of the convex drum is raised so as to release the strap, which is wound back to the opposite end of the drum by means of a rack and pinion, and the frame is ready for starting again.

ONE SLUBBING FRAME of 42 spindles, three rows of rollers, with Leigh's top rollers to the front row, back rows (top and bottom) fluted with coarse flutes. Double centrifugal pressure for bobbins 10 in. lift by 5 in. diameter, fitted with stop motions, indicator and improved flat with endless traversing cloth (exhibited).

ONE INTERMEDIATE FRAME of 60 spindles, three rows of rollers with Leigh's top rollers to the front row, back rows (top and bottom) fluted with coarse flutes. Double centrifugal pressers, for bobbins 9 in. lift by 4½ in. diameter, with iron creels, indicator, and improved traversing top clearer or flat (exhibited).

ONE ROVING FRAME of 84 spindles, three rows of rollers with Leigh's top rollers to the front row, back rows (top and bottom) fluted with coarse flutes. Double centrifugal pressers for bobbins 7 in. lift by 3¼ in. diameter, with iron creels, indicator, and improved traversing top clearer or flat (exhibited)

The novelties introduced in these machines, are Leigh's front top rollers with revolving bosses and coarse fluted back rollers (top and bottom) for better holding the cotton; in an improved flat with its endless travelling cloth which hangs upon hinges as in the drawing frame; more complete casing-up of the working parts; more convenient arrangement of setting-on and knocking-off rods, and in more effective and economical lubricating arrangements.

THE THROSTLE (TWELFTH PROCESS).—These machines are generally used for spinning yarn, for making warps, and winding it upon small bobbins; they have also been sometimes arranged for spinning weft and winding it in the form of cops, but never with good practical results, and always at a cost of increased complication in the mechanism. They are used for spinning from 40s. downwards.

The creel for supporting the bobbins filled with rovings to be spun by the throstle, is placed on the top of the frame between two sets of three pairs of rollers, and which travel at varying rates, the variation in this instance between the first back roller, and the third or front, being about one into eight. Through one of these three pairs of rollers each roving is drawn and afterwards passed through an eylet or guide wire, which is fixed in a bar of wood (hinged to the beam for supporting the rollers), and whose position when at work is immediately over the centre of the revolving spindles which twist the yarn: one row of which is supported by rails, parallel with and perpendicular to the rollers on each side of the machine. The tops of these spindles are furnished with flyers, round one leg of which the thread is coiled and passed through another eylet at the bottom to a bobbin which is threaded upon the spindle, and upon which the yarn is wound. The lower rail or bar for supporting the spindles, is fixed, and the upper one is movable, and upon it the bobbin rests; this rail or bar has an up and down motion given to it by means of racks and pinions in communication with a heart cam; the bobbin is thus moved up and down the spindle past the eylet of the flyer, and the yarn is equally distributed upon it in winding.

The motion of the bobbin round the spindle is variable, and is obtained from the tension of the yarn whilst winding, and as the revolutions of the spindle and flyer cause the yarn to drag the bobbin after them, and the weight and friction of the bobbin upon the movable rail acts as a break, the yarn is wound tight on its surface.

ONE THROSTLE of 152 spindles, 2 in. lift, three rows of rollers, Leigh's front top rollers, middle and back rollers self-weighted, lifting rails, top and bottom oiling plates (exhibited).

The novelties introduced are as follows—Oiling plates for both bottom and top spindle rails the whole length of the machine, which can be lifted by racks so as to allow the attendant to oil the whole of the spindles without interruption; also in an improved iron creel plate fitted with steel pegs for the tin tubes of the roving bobbins to revolve upon, and in an arrangement by which yarns of one count may be spun on one side of the machine, whilst those of another count are being spun on the opposite side.

SELF-ACTING MULE FOR SPINNING COTTON.

TWELFTH PROCESS (a).—These machines are used for drawing and twisting into yarn the rovings as prepared by the machinery before described, and coiling or winding it upon spindles in the form of cops by automatic means. Like the common hand mule jenny, this machine may be divided into two principal parts, one part fixed, and comprising the creels for supporting the bobbins, the rollers for drawing or elongating the fibres, the frame-work or headstock containing the movements for effecting the changes required in the operation, and for communicating motion to the movable portion of the machine called the carriage, which supports the spindles and the drum for imparting motion to them, and which is made to traverse in and out from the rollers upon iron rails or slips as the yarn is being drawn out or wound upon the spindles. The average length of this traverse or draw is about 63 in.

As the fibres of the roving are being drawn and delivered by the rollers, the carriage is caused to move from the rollers until it arrives at the end of the stretch, when it stops; the rollers and drawing-out motions are disengaged, the twist motion is acting, the spindles continue to revolve, until the quantity of twist necessary to be put in the yarn has been given, the change is then

PLATT, BROTHERS, & CO., *continued.*

made from the twist to the backing off, by causing the direction of motion of the spindles to be reversed, and the yarn to be uncoiled a little, so as not to break by the depression of the faller wire upon it. The winding-on

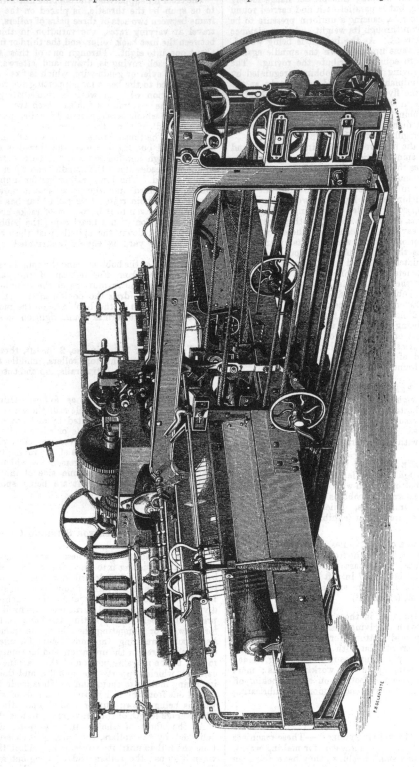

SELF-ACTING MULE FOR COTTON.

and taking-in changes are then made, the carriage advances to the rollers, the yarn is wound upon the spindles, and the operation is complete.

The different changes are effected by means of a cam

PLATT, BROTHERS, & CO., *continued.*

BACK OF SELF-ACTING MULE.

shaft in connexion with the long lever, which is acted upon by the traversing in and out of the carriage, the locking of the faller, and the revolution of the twist motion.

PLATT, BROTHERS, & CO., *continued.*

ONE SELF-ACTING MULE of 648 spindles 1⅞ in. distance, 16¼ in. spindles, three rows of rollers for two threads to each boss, Leigh's front top rollers, spindles driven by tin rollers, plate footsteps and bolsters, iron creels for single roving, back-shafts, and driven direct from the main-shaft (exhibited).

The novelties in this machine as exhibited and illustrated are :—

1. The introduction of foundation plates for supporting the headstock or principal framing of the fixed portion of the machine, the iron rails or slips upon which the carriage traverses, and the copping and taking-in motions, thus entirely preventing derangement of the working parts from deflection or bad floors.

2. The improved arrangement for driving the cam shaft by gearing instead of by friction, making the motion positive, and causing a more certain and noiseless action of the changes.

3. Simple and direct arrangement of rim or twist band (called double banding), by passing the band twice round the rim and all other twist pullies for driving the spindles, by which means we can work with slacker bands, have less strain in the bearings, less wear and tear of band, and a greater regularity of twist in the yarn.

4. A new taking-in motion, which is differential without having an eccentric band pulley or scroll. The circumference of the concentric pulley now used being equal to the length of the draw, stretch, or traverse of the carriage, gives to the band one uniform tension, ensuring greater delicacy of action in working, much greater durability, and less breakage of taking-in bands.
This motion being firmly fixed upon the foundation plate, and being connected directly with the carriage, has no tendency to lift it from the rails during its traverse.

5. Simple construction and arrangement of copping and faller locking motions with double copping plates, by which the copping rail may be taken out in any stage of the cop's progress, without disturbance of its working position.

6. The application of a governor or cop regulator for adjusting the winding-on motion to the formation of the cop, which is perfectly automatic throughout.

7. Improved construction of carriage or movable portion of the machine, and in the manner of connecting the square and the carriage together, combining greater accuracy, strength, and neatness ; and in the position and arrangement of the diagonal rods for strengthening the same ; and

Lastly. In the general construction and adaptation of the framing to form a casing to the working parts of the machine, the facilities for making changes when required for spinning varied numbers, in the introduction of a friction coupling, through which motion is transmitted to the taking-in motion, and which may be so adjusted as to slip in cases of obstruction to the free traverse of the carriage, thereby preventing breakages in the machine and banding.

GRINDING MACHINE used for grinding and sharpening the teeth of the cards on the rollers and flats of the carding engines.

WRAP DRUM AND SCALES, for measuring and weighing rovings.

WRAP REEL AND SCALES, for measuring and weighing yarns.

SET OF TACKLE, for nailing on cards.

ROLLER ENDING MACHINE.

MACHINE for forcing leathers on top rollers.

CASES containing samples of bottom and top rollers, spindles, flyers, and bobbins.

CASE showing cotton in its various stages of manufacture.

CASE, showing wool in its various stages of manufacture.

SET OF PHOTOGRAPHS of the machinery exhibited.

ONE POWER LOOM, 38 in. reed space, for weaving plain calicoes for shirtings, any kind of twills, fancy goods in cotton, or fine woollen, union cloth with cotton warp and woollen weft, fine linen goods or union cloths with cotton warp and linen weft (exhibited).

The novelties introduced are :—
An improved picking motion, which is worked from the first motion or crank shaft ; the picking shaft is provided with loose tongues, which are acted upon by cams every alternate revolution of the crank shaft ; this arrangement is exceedingly simple, is little liable to wear, and can easily be repaired in case of accident.
An improved surface taking-up roller, without glass or emery, and which is applicable for either light or strong goods.
An improved self-acting temple, and a new buffer or check-spring (instead of check-strap), to prevent breakage of cops or bobbins in the shuttles of the looms.

PREPARING WOOL.

Wool is prepared for the carding engine, first by shaking and having the dust extracted from it by a machine similar to that illustrated for opening cotton, and afterwards by oiling, for the purpose of softening it and preventing the short fibre from flying.
Dyed wool is also passed through a similar machine to be cleaned, and for the purpose of extracting the spent dye-wood from it, before being oiled.

CARDING WOOL.

Wool passes through a series of three machines in this process ; viz. the "scribbler," the "intermediate," or second card, and the finisher card and "condenser," and is supplied to the first on an endless travelling lattice, which has its surface divided into a number of equal parts, and upon each of which a given weight of wool is spread. This lattice carries it to a pair of feeding rollers, which draw in and deliver it to a taker-in roller that carries it on its upper surface under a guard roller to the breast cylinder ; this taker-in roller is formed of cast-iron, with a fine groove or thread cut on its surface, and into this groove is pressed the lower edge of a flat wire, whose upper edge is cut into teeth like those of a fine saw, so that it has the appearance of a roller with a number of fine saws placed at short intervals. The object of these teeth is to stretch the fibres until they are released by the feed rollers, when they pass forward in the spaces betwixt them, and any burrs or impurities that are carried on the surface of the wool are driven back by the guard roller which revolves at a quick speed, and has a number of ribs fixed longitudinally on its surface. This roller is so placed that its ribs clean from the points of the teeth any refuse, and deposit it in a box placed over the feed rollers to receive it. The taker-in roller is stripped by a clearer roller which passes the wool to a breast cylinder having two rollers and clearers revolving in suitable bearings over it to be broken up and prepared for the main cylinder. The breast cylinder is stripped by a clearer, which passes the wool to a large

PLATT, BROTHERS, & CO., *continued.*

cylinder, over which revolve four rollers and four clearers for carrying on the carding process, and a large roller, called a "fancy," covered with long toothed cards, and whose surface travels faster than that of the cylinder, the object of which roller is to prevent the cards of the cylinders from being clogged with grease and wool, to raise the wool to the surface, and to deliver it to the surface of the doffing cylinder, from which it is stripped

by a vibratory comb, as described in the cotton-carding process.

The fleece thus stripped from the doffer is formed into a sliver by being drawn through a revolving tube by means of a pair of rollers that are placed in front of the doffer, at one end of it, from which it falls upon an endless travelling cloth just over the floor, to be then conveyed to the second or intermediate card.

LOOM.

ONE SCRIBBLER ENGINE, 60 in. on wire, with common hand-spreading lattice feeder, one 12 in. patent burring roller, with guard and dirt box, working from patent feed rollers, with patent stripper, breast roller 27 in. diameter, with two rollers and two stripping rollers; cylinder 45½ in. diameter, with four stripping rollers, four rollers and fancy ; doffer 22 in. diameter ; roping apparatus, and floor creeper (exhibited).

ONE INTERMEDIATE, or SECOND CARDING ENGINE, 60 in. on wire, with Apperley & Co.'s patent diagonal feeding machine, patent taker-in 7 in. diameter, with

stripper 7 in. diameter ; cylinder 45½ in. diameter, four stripping rollers, four clearers, and fancy ; doffer 22 in. diameter (exhibited).

The sliver delivered by the scribbler card is formed into a fleece to supply this machine by an apparatus known as "Apperley's feeder," by taking the slivers from the travelling cloth, over the floor, and laying them in lines close to each other upon a number of endless travelling webs, which are driven by a shaft parallel to the feed rollers. These webs are of increased lengths, from side to side of the machine, so as to form an angle

PLATT, BROTHERS, & CO., *continued.*

to the feed rollers. By this means, the slivers pass obliquely, and a number of them are presented at the same time to the feed rollers, insuring greater uniformity of fleece ; which, after having passed the feed rollers, is acted upon by the first of 2 taker-in rollers, about 7 in. diameter. The first of these is covered with saw-like teeth (as described in the "scribbler"), and the second is covered with cards, and is by them passed to the main cylinder, which conveys it to the doffer, whilst the process of carding is going on with the rollers, clearers, and fancy, as before described in the "scribbler engine."

This doffer is also stripped by a comb, and delivered in a thin fleece to a travelling lattice supported on rollers, which are carried by levers in the form of a pair of compasses, and having one point fixed, and the other moving, with a small carriage on rails, which also carries two tin rollers. By this lattice and rollers the fleece is deposited in layers across the feeder lattice of the

FINISHER CARDING ENGINE,

which is placed at right angles to the intermediate, and which moves with a slow motion, so that each layer is placed a little behind the preceding one. This system of moving lattices is known as "Ferrabee's feeder," and its object is to lay the fibres so as to enter the feed rollers of the condenser card crosswise, to be again straightened and taken forward by two takers-in to the cylinder, rollers, and clearers, to be carded and passed to the doffer, upon the surface of which the fleece is now spread uniformly. The next operation is to strip it off in a number of small slivers or bands, and is called

CONDENSING.

There are a great variety of machines constructed for this purpose ; the one exhibited is of recent contrivance, and is patented and known as "Fairbairn's condensor." Its novelty consists in having small grooves cut round the doffer in equal divisions, and placing in each a thin blade of steel level with the point of the cards, where they receive the fleece from the cylinder. This steel blade follows the face of the cylinder at a short distance, therefore projects above the cards of the doffer so as effectually to divide the fibres that lay across this line, when they are taken by the side which has the firmest hold ; they now pass under the doffer and are stripped by a plain card roller placed in the front of the doffer, and which conveys them to two endless travelling sheets of leather. Each sheet is carried on two rollers and vibrates in contrary directions, so as to rub them into round felt slivers, or bands, to be wound on to two bobbins by surface contact with two rollers, so that each bobbin contains one-half of the threads delivered.

These bobbins are then filled into the creel of the self-acting mule.

Note.—For some qualities of woollen yarn the fleece is stripped from the doffer in bands, which are afterwards joined together by the piecing machine, and wound upon bobbins to be filled into the creel of the self-acting slubbing mule, to be partially spun and wound upon bobbins to supply the creel of the self-acting mule.

FINISHER CARDING ENGINE AND CONDENSOR, 42 in. on wire, Ferrabee & Co.'s patent bat feeding machine up to patent feed rollers and strippers, patent taker-in 7 in. diameter, with stripper 7 in. diameter, cylinder 45½ in. diameter, four stripping rollers, four rollers and fancy ; doffer 22 in. diameter, with Fairbairn's patent condensor, to deliver forty good threads and two waste ends (exhibited).

SELF-ACTING MULE FOR WOOL, 380 spindles, 2 in. distance, 18 in. spindles, to spin either upon the bare spindle, or upon wooden or tin spools, and from condensor or slubbing bobbins, either warp or weft yarn (exhibited).

The improvements and novelties introduced in connexion with the self-acting mule for cotton-spinning, are also introduced into this machine, in addition to which we have also introduced :—

A "double speed" or fast and slow motion of the spindles with two rims, the change being obtained by a traverse of the strap and two rims without the aid of either counter shaft or gearing, the rim out of action in each case being converted into a carrier pulley, enabling us to retain the double banding arrangement.

In a motion for giving out the necessary length of slubbing to be spun, which is so connected with the cam-shaft, as to give a simultaneous action of the delivering rollers and the drawing-out motion.

In a simple arrangement of a receding motion of the carriage during the twisting of the yarn, and which may be regulated to recede quickly or slowly, as the fineness of the yarn and the amount of twist may require.

In a simple arrangement for regulating the length of draw or traverse of the carriage, in accordance with the running up of the yarn and the recedence of the carriage.

The cards in use in these machines are made by—

For Cotton, Messrs. Joseph Sykes & Brothers, Lindley, near Huddersfield, and Mr. William Horsfall, Great Bridgewater Street, Manchester.

For Woollen, Messrs. R. & C. Goldthorpe, Cleckheaton, near Leeds.

The bobbins and skewers are supplied by Messrs. Lawrence, Wilson, & Sons, Cornholm Mills, near Todmorden, and the banding by Mr. Samuel Green, King Street, Oldham.

Messrs. Samuel Radcliffe & Sons, of Rochdale, and Messrs. Radcliffe Brothers, of Lower House and Wallshaw Mills, Oldham, are working the cotton machinery.

Messrs. H. & L. Newall, of Littleborough, near Manchester, and Messrs. the Executors of George Lawton & Sons, of Micklehurst, near Mosley, are working the woollen machinery.

The engine driving the machinery is made by Messrs. B. Hick and Son, Bolton.

Prices may be had on application to—

Messrs. PLATT, BROTHERS, & CO., in the Exhibition ; at their Works, in Oldham ; and at their Offices, St. Ann's Square, Manchester.

And from their Agents—

Russia	Messrs. DE JERSEY & CO., Manchester, St. Petersburg, and Moscow.
France, Belgium, Holland, Prussia, Bavaria, Italy and Savoy, Sweden and Denmark	Messrs. E. NATHAN & SINGTON, Manchester.
Saxony and Bohemia	Mr. W. W. DERHAM, Leipsig.
Vienna and Switzerland	Mr. F. E. SCHOCH, Vienna and Switzerland.
Spain	Mr. JAMES SYKES, Barcelona.

SHARP, STEWART, & CO., *Atlas Works, Manchester.*—Reel-winding machine, for silk, linen, or cotton sewing-thread.

[1533]

SMITH, WILLIAM, & BROTHERS, *Heywood, Lancashire.*—Woollen looms; jacquard damask loom, and half-woollen loom.

[1534]

SMITH & CO., *Stratford.*—Machine for separating cocoa fibres.

[1535]

STUART, JOHN & W., *Musselburgh, near Edinburgh.*—Patent fishing-net weaving loom.

[1536]

THOMPSON, JAMES, & CO., *Kendal.*—Card clothing.

[1537]

TUER & HALL, *Hope Foundry, Bury, near Manchester.*—Shearing machine, looms.

1. LOOM FOR WEAVING LINENS, strong fustians, nankeens, velvets, ticks, jacquard work, woollen goods, &c. This loom is applicable by a change of tapets at the end for weaving any kind of cloths, and may be worked at any required speed.

2. GINGHAM OR FANCY LOOM with rising and falling box, to weave one pick or more of each colour as may be desired, suitable for weaving ginghams, checks, plaids, drills, quiltings, light fustians, nankeens, heavy domestics, plain and twilled calicoes, &c.

3. CARPET LOOM to weave pile fabrics of any width required, invented in 1857. All the working motions are outside the loom, except the crank from which it is driven, by which free access to the working parts is obtained. The wire motion inserts 45 wires per minute at 2 picks per wire. This loom is also applicable to the weaving of Utrecht velvet for the lining of carriages, omnibuses, &c. One horse-power will turn 6 of these looms.

4. SHEARING OR CUTTING MACHINE with 2 revolving cylinders, 5 steel cutters on each cylinder for shearing fustian, velvet, and moleskin cloths, &c.; can be made on the same principle with 1 cylinder only at about two-thirds cost. The cylinders can be made of a larger or smaller diameter, with more or less cutters, as may be desired.

Prices and lithographs may be had on application to the makers.

[1538]

WALKER & HACKING, *Bury, Lancashire.*—Machinery for opening, scutching, preparing, and spinning cotton yarn.

[1539]

WARD, GEORGE, 77 *Darwen Street, Blackburn.*—Heald-knitting machine.

[1540]

WATKIN, S., *Bradford.*—Washers' rods for silk frames.

[1541]

WILSON, LAWRENCE, & SONS, *Cornholme Mills, Todmorden, near Manchester.*—Bobbins, tubes, spools, skewers, bosses, clearers, &c. (*See page* 34.)

[1542]

WHITESMITH, I., 29 *Govan Street, Glasgow.*—Power loom with six shuttles and twilling combined.

[1543]

WREN & HOPKINSON, *London Road, Manchester.*—Machinery for manufacturing cotton sewing-thread and spinning silk. (*See page* 35.)

WILSON, LAWRENCE, & SONS, *Cornholme Mills, Todmorden, near Manchester.*—Bobbins, tubes, spools, skewers, bosses, clearers, &c.

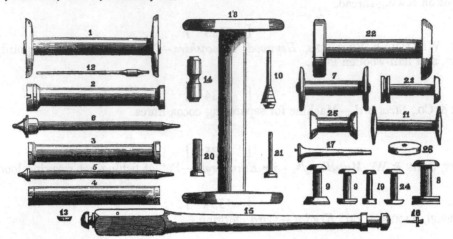

COTTON PREPARATION.

1. Slubbing soft bobbin.
1. Intermediate · ditto.
1. Roving   ditto.
    Ditto fine   ditto.

PATENT LONG COLLAR PRESS.
2. Slubbing tube, hooped.
2. Intermediate ditto.
2. Roving   ditto.

COMMON PRESS TUBES.
3. Slubbing, beaded ends.
3. Intermediate ditto.
3. Roving   ditto.
4. Slubbing, plain ends.
4. Intermediate ditto.
4. Roving   ditto.

SKEWERS FOR BOBBINS OR TUBES.
5. Slubbing, lancewood, or ash, and footed.
5. Intermediate ditto   ditto.
5. Roving   ditto   ditto.

SKEWERS FOR PATENT COLLAR TUBES.
6. Slubbing, ash and footed with box.
6. Intermediate ditto   ditto.
6. Roving   ditto   ditto.

WARPING AND SPINNING.
7. Warping bobbins.
7. Warping bobbins, feather edges.
7. Winding ditto.
8. Doubling ditto, for wet, solid.
    Ditto   ditto, for dry, jointed.
9. Throstle ditto, plain or painted.
9.   Ditto   ditto, Wilson's improved.
9.   Ditto   ditto, metal bushed.
10. Pin   ditto, soft or hard wood.
    Sally   ditto ditto   ditto.
    Weaver's   ditto.
11. Gasing   ditto.
    Bolling   ditto.

MISCELLANEOUS.
12. Twiner's skewers, lancewood.
    Reeler's cop ditto   ditto.
    Warping   ditto   ditto.
13. Spindle cop braids, box.
14. Throstle top clearers.
    Ditto under ditto.
    Mule top clearers.
    Ditto under ditto.
15. Picking sticks, turned.
    Ditto ditto, flat.
16. Carr's patent bobbin nails.

WOOLLEN PREPARATION.
Condenser bobbins.
1. Sliver   ditto.
7. Twister   ditto.
17. Warp   ditto.
17. Weft.   ditto.

WORSTED PREPARATION.
18. Drawing bobbins.
18. 1st finisher ditto.
18. 2d   ditto ditto.
18. Roving   ditto.
7. Warping   ditto.
19. Spinning ditto.
20. Spool   ditto.
21.   Do. shell ditto.

FLAX AND SILK PREPARATION.
22. Large headed bobbins.
    Small ditto ditto.
23. Spinning bobbins, large.
24.   Ditto   ditto   small.
    Box   ditto   ditto
25. Winding   ditto
26. Bosses and pulleys.
14. Clearers, and all other kinds.

Prices will be cheerfully forwarded for any of the above, or other descriptions of bobbins, &c. (plain or painted and varnished), on receipt of full particulars of size.

Orders (for home or export) will be carefully and promptly attended to. Address, Cornholme Mills, Todmorden, near Manchester.

The following superior advantages and facilities possessed by L. W. & Sons for producing the best possible article at the lowest remunerative cost, enable them to make it especially advantageous to purchasers intrusting them with their orders.

1. Forty years' practical experience in the business as bobbin manufacturers.

2. Upwards of 300 hands employed, all trained by the firm, being the largest establishment of the kind in the united kingdom.

3. Inventors and sole proprietors of patent machinery which produces superior workmanship, and guarantees the greatest uniformity in shape and size.

4. Immense stocks of prepared and well-seasoned timber always on hand, enabling them to execute orders to any extent at a very short notice.

The following eminent machinists have kindly permitted their names to appear as references:—Messrs. Platt Brothers & Co., Oldham; Parr, Curtis, & Co., Manchester; William Higgins & Sons, Manchester; Joseph Hetherington & Sons, Manchester; Walker & Hacking, Bury; John Mason, Rochdale; John Tatham, Rochdale; Wilson & Longbottom, Barnsley; Lord Brothers, Todmorden.

WREN & HOPKINSON, *London Road, Manchester.*—Machinery for manufacturing cotton sewing-thread and spinning silk.

MACHINES EXHIBITED IN OPERATION :—

MACHINES FOR THE MANUFACTURE OF THREAD AND SEWING COTTON.

THROSTLE, to draw and spin the cotton into yarn or thread. Patent revolving weights.

DOUBLING MACHINE, to twist two or more ends of yarn together, forming sewing cotton or thread.

CLEARING MACHINE, to remove irregularities in the thickness of the thread.

BOBBIN REEL, to wind the thread from the bobbins into hanks for dyeing or bleaching.

HANK-WINDING MACHINE, to wind thread from the hank upon bobbins after dyeing or bleaching.

BALLING AND SPOOLING MACHINE, to wind thread into balls or upon small wooden spools or bobbins ready for the retail market.

CUBING PRESS, to make up small bundles of thread for sale.

BUNDLING PRESS, to press and pack thread or yarn into large bundles for exportation.

PIRN-WINDING MACHINE, to fill small shuttle bobbins used in weaving.

MACHINES FOR THE MANUFACTURE OF SILK FOR WEAVING AND SEWING.

WINDING MACHINE, to wind raw silk upon bobbins from the hank as imported.

PATENT SIZING MACHINE, to assort the silk into various degrees of thickness.

CLEANING MACHINE, to remove irregularities in the thickness of the silk.

SPINNING MACHINE, to twist or spin a single thread of raw silk.

DOUBLING MACHINE, to wind and twist together several ends of silk, detecting the breaking of any one during the operation.

THROWING MACHINE, to twist or spin two or more threads of silk into one of greater strength.

SOFT SILK WINDING MACHINE, to wind silk from the hank after dyeing ready for weaving.

PATENT STRINGING MACHINE, to stretch dyed silk while immersed in steam, giving lustre to the surface.

Pair of NON-CONDENSING HIGH-PRESSURE STEAM ENGINES, diameter of cylinder 10 in. stroke 20 in.

STEAM GAUGE, Allen's patent.

HYDRAULIC PUMPS, to work a press for packing textile goods for shipment.

SELF-LUBRICATING PEDESTALS, Möhler's patent.

SET OF IMPROVED STOP VALVES, 2 in. to 8 in.

SHAFTS, WHEELS, PULLEYS, &c. for giving motion to the machinery.

SUB-CLASS B.—*Machines and Tools employed in Various Manufactures.*

[1551]

ALCOCK, JOHN, *Prescot.*—Lathe; upright tool with slide-rest on improved principle.

[1552]

ANNABLE & BLENCH, 28 *St. John Street, E.C.*—Patent horizontal printing machine for cheap and expeditious printing.

[1553]

ARMITAGE, M. & H., & Co., *Mousehole Forge, near Sheffield.*—Anvils, vices, hammers, &c.
Obtained the Prize Medal at the Exhibition of 1851.

[1554]

BARRETT, EXALL, & ANDREWES, *Reading, England.*—Patent aërated bread machinery (Dr. Dauglish's), and patent biscuit machinery.

[1555]

BERTRAM, GEORGE, *Sciennes Street, Edinburgh.*—80-inch paper making and cutting machine, fully mounted.

The exhibitor has had thirty years' experience in the manufacture of PAPER-MAKING MACHINERY. His manufactures include—

Every sort of useful CUTTING, WILLOWING, and DUSTING APPARATUS, for rags, waste, or straw.

REVOLVING DRUMS, for washing rags.

WASHING, BEATING, AND POACHING ENGINES.

PAPER-MAKING MACHINES of all widths, with single-sheet cutters attached, of a new and improved description, or cutters to cut from 6 to 8 reels at one time.

All kinds of SIZING AND DRYING MACHINES, detached or in connexion with paper-making machine, so as to make, size, dry, and cut the paper in one continuous unbroken web.

ROLLING, CALENDERING, AND GLAZING MACHINES, for writing papers, in the web, single sheets, or in copper plates.

NEW AND IMPROVED ANGULAR CONDENSING STEAM ENGINES of all sizes, very economical and useful for driving every kind of machinery.

HEAVY AND LIGHT GEARING of every description.

[1556]

BESLEY, ROBERT, & Co., *Fann Street, Aldersgate Street, London.*—Type-casting machine in operation.

[1557]

BEYER, PEACOCK, & Co., *Gorton Foundry, Manchester.*—Wheel lathe to turn and bore up to 7 feet diameter; and triple-headed slotting machine.

DOUBLE FACE-PLATE WHEEL LATHE, for turning railway wheels, and boring tyres up to 7 ft. diameter; adapted to turn two wheels at once upon their axle without torsion, or to turn two wheels, or bore two tyres respectively; or to turn a wheel, or bore a tyre upon one face-plate, whilst boring or bossing a wheel upon the other, with additional driving gear upon both head-stocks for boring or bossing wheels at quick speeds.

[1558]

BISSELL, WILLIAM, *Union Street, Wolverhampton.*—Flooring and bench cramp; machine for mortising wood; lifting jack. (*See page* 37.)

[1559]

BOARD, CHARLES, 7 & 8 *Barton, Bristol.*—Veneering press for veneering large or small flat surfaces.

[1560]

BRADBURY & Co., *Rhodes Bank Foundry, Oldham.*—Manufacturing and domestic sewing machines and binding guides.

[1561]

BRADLEY & CRAVEN, MESSRS., *Westgate Common Foundry, Wakefield.*—Patent plastic clay brick-making machine. (*See page* 38.)

BISSELL, WILLIAM, *Union Street, Wolverhampton.*—Flooring and bench cramp; machine for mortising wood; lifting jack.

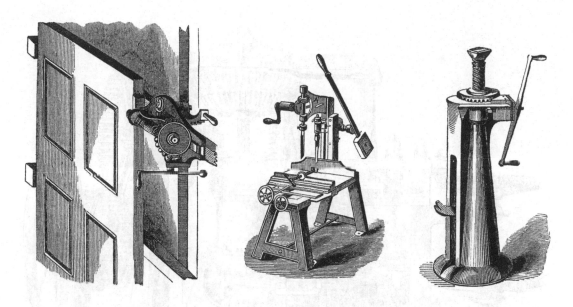

THE FOLLOWING ARE EXHIBITED :—

MORTISING MACHINE. This works upon a different and more powerful principle than any other mortising machine yet introduced. The power is obtained by an eccentric wheel working in a rack at the back of the slide, whereby greater leverage is obtained.

BISSELL'S PATENT FLOORING CRAMP. The force of this cramp is fully equal to 1 ton. It is adapted to joists from 2 to 4 in. and is the most expeditious and easy in working of any cramp in use.

BISSELL'S COMBINED LIFTING JACK.

[1562]

BRUNTON, J. D., *Barge Yard, Bucklersbury, E.C.*—Peat fuel, and machinery for preparation of same.

[1563]

BUCHTON, JOSHUA, & CO., *Well House Foundry, Leeds.*—Self acting engineers' tools.

[1564]

BUNNETT & CO., *Deptford, Kent.*—Brick-making machine.

[1565]

BURN, ROBERT, *Lochrin Engine Works, Edinburgh.*—Envelope and label dies.

[1566]

CARVER, WILLIAM, *Ducie Bridge Mill, and 5 Todd Street, Manchester.*—Sewing machines.

[1567]

CASSON, JOHN, *Wellington Street, Woolwich.*—Patent improved machines for dressing raisins, currants, and other dried fruits.

[1568]

CLARK, JOHN, 184 *Buchanan Street, Glasgow.*—Automaton ruling machine, hand machine for envelopes, Albion embossing press.

[1569]

CLARKE, T. A. W., *Leicester.*—Machine for covering elastic rings or threads by a new method.

[1570]

CLAYTON, HENRY, & CO., *Atlas Works, near Dorset Square, London, N.W.*—Patent brick, tile, and pipe machines. (*See page 39.*)

BRADLEY & CRAVEN, MESSRS., *Westgate Common Foundry, Wakefield.*—Patent plastic clay brick-making machine.

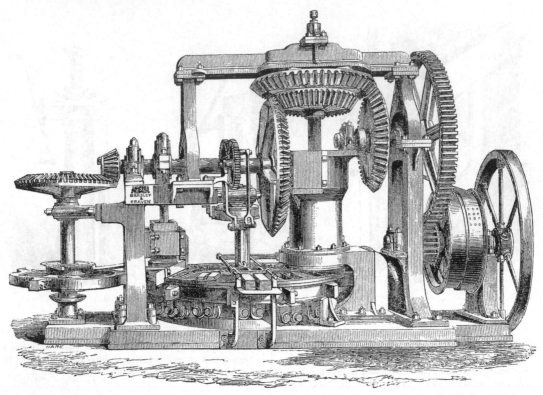

PATENT BRICK MOULDING AND PRESSING MACHINE.

The ground allotted to this firm by the Commissioners not being sufficient for the display of their dry clay and other powerful brick-making machines, the above engraving represents the only one their space will allow them to exhibit.

Any material capable of being manufactured into bricks, can be delivered to this machine in the state of dryness it leaves the earth, which, without the addition of any water, produces a superior pressed brick (with many clays), ready for immediate delivery to the kilns for burning. This is the case with several machines working the gault clay in Kent, which comes from the earth so dry, that when made by the machine, the bricks are immediately wheeled into the kilns. The clay on being dug from the earth is delivered to the machine, which grinds and works it into a close, dense, well amalgamated mass, and fills it into the moulds with great solidity.

The action of the machine is as follows:—One pair of the moulds (of which there are twelve in the face of the rotating table) receives the charge of clay at a time from the mill. During the moment that this operation is going on the table is stationary, and two other moulds that have been previously filled are being subjected to considerable pressure by pistons on the opposite side of the table to the mill, and two finished bricks that have been discharged by an inclined plane from the moulds, are delivered on to a creeper band by the action of the machine for removal to the kilns or sheds, perfect-pressed face bricks. Thus the only labour required is to supply the crude, fresh-dug clay to the mill, when the machine prepares, manufactures, and also delivers the bricks to the kiln men for burning.

This machine makes from 15,000 to 20,000 per day. Three of them are working at this rate for the Aylesford

Pottery Company, near Maidstone; and others in this neighbourhood, as well as in different parts of the country, are giving general satisfaction. To save any risk or disappointment to purchasers, the patentees invite manufacturers to test their own clays in the machines previous to incurring any outlay; and they will give every facility for doing so, the only charge being for carriage or freight of clay, when prepayment has not been made. The importance of such trials will be appreciated by practical men. This machine is on the same principle as their well-known dry clay machine, but is not so large nor so powerful.

The result of extensive practical experience, gained in working these machines in all kinds of earth, fully proves the great superiority of forming the clay in moulds, over forcing it through dies, and cutting it with a wire. The advantage lies in the greater truth in the form of the bricks, and also in making them without any water. A still more important advantage is, that the manufacturer is enabled to work any kind of clay, from sandy brick-earth, to the strongest clays mixed with breeze or ashes and sand, to reduce it, neither of which could be worked satisfactorily with a die or cut smooth with a wire. Furthermore, it enables the manufacture to be carried on through all seasons of the year.

Price of machine subject to 1s. per 1,000 royalty £250 0
Price of machine free from royalty, for export 500 0

Illustrated catalogues of dry clay and other machines for the manufacture of bricks and tiles, with references to those at work, may be had upon application.

CLAYTON, HENRY, & CO., *Atlas Works, near Dorset Square, London, N.W.*—Patent brick, tile, and pipe machines.

These are the champion prize machines of the Royal Agricultural Societies of England, Scotland, Ireland, France, Sardinia, Holland, Austria, Belgium, Hanover, &c.

They have obtained the—
First-class prize at the Great Exhibition of all Nations, London, 1851.
Gold medal of honour at the Universal Exposition, Paris, 1855.
Prize medal and diploma at the Great Exhibition, Amsterdam, 1853.
Gold medal prize at the Royal Exposition, Vienna, 1857.
First-class prize of the Royal Polytechnic Society, 1860.

HENRY CLAYTON & CO., inventors, patentees, and manufacturers of the UNIVERSAL BRICK-MAKING MACHINES, TILE-MAKING MACHINES, PRESSES, &c. have been patronized by H.R.H. (the late) Prince Consort, H.I.H. the Emperor of Russia, H.M. the Queen of Spain, H.I.H. the Emperor of France, H.M. the King of the Belgians, H.M. the King of Hanover, H.I.M. the Empress of Russia, and by Her Majesty's Government for home and colonial use, &c.

BRICK MACHINES of several sizes and of varied construction, according to the nature of the clay, adapted to the manufacture of solid, tubular, or perforated bricks, of any size or form to order, arranged for working either by steam, water, animal, or hand power.

DRAIN PIPE AND TILE MACHINES of various sizes and construction for the manufacture of agricultural drain pipes, sanitary tubes, roofing and paving tiles, and hollow goods of every description.

PRESSES for bricks and tiles, plain or ornamental.
CLAY MILLS, for washing, crushing, pugging, and screening.
MORTAR, LOAM, AND PEAT MILLS.
STEAM ENGINES, portable or stationary, of all sizes.

Detailed plans for an improved construction of kilns, drying rooms, and sheds.

Every description of sawing and constructive machinery for contractors' use, and machinery, tools, and utensils of every kind required in the brick, tile, or pottery manufacture.

The following are selected from a number of favourable notices of these machines:—

"They unquestionably bear evidence of great mechanical ingenuity, and are the most efficient apparatus yet before the public."—*Engineer.*

"Clayton's machines are simple, and judiciously arranged, combining rapidity of production and economy of manufacture."—*Practical Mechanic's Journal.*

"The problem solved."—*Artizan.*

"What the saw mill is to the timber, in our opinion, is Clayton's machine in the manufacture of bricks."—*Mining Journal.*

"In this machinery Mr. Clayton has proved his thorough knowledge of the mechanical means required, and of the material he has to deal with."—*Mechanic's Magazine.*

"Cheap and good bricks are now made by these machines;—a subject of national and universal importance."—*Builder.*

Machines may be inspected and clays tested at the manufactory. Descriptive catalogues sent free by post.

[1571]

COHEN, B. S., 9 *Magdalen Row, Great Russell Street.*—Pencil manufacturing.

[1572]

COLLEY, EDWARD E., 5 *West Cottages, West Street, Walworth.*—Working model of Hoe's printing machine.

[1573]

COLLIER, LUKE, *River Street, Rochdale.*—Confectioners' and biscuit bakers' machines; sugar mills, &c.

[1574]

CONISBEE, WILLIAM, 39 *and* 40 *Herbert's Buildings, Waterloo Road.*—A Main's patent printing machine, for bookwork and job printing. (See page 41.)

[1575]

COOK, D., & CO., *Glasgow.*—Patent steam riveting machine; bour pan, for evaporating sugar-cane juice. (See page 40.)

COOK, D., & Co., *Glasgow.*—Patent steam riveting machine; bour pan, for evaporating sugar-cane juice.

COOK'S PATENT STEAM PUNCHING, SHEARING, AND RIVETING MACHINE effects the three operations in one frame, or, when desired, it is made for riveting only.

It can punch thirty holes per minute in ordinary boiler plates. The action of the punch being instantaneous, and every stroke under the control of the keeper, insures both accuracy and speed. In riveting, ten holes can be closed up in one minute, and the work much superior to that effected by hand. Steam pressure required, 25 lbs. per square inch. Prices, designs, and testimonials to be had on application.

BOUR'S PATENT EVAPORATING PAN for the concentration of all liquids.

Having become the proprietors of the patent for this pan, D. Cook & Co. have introduced a considerable number of them into the various sugar-growing countries, and from the superior mode of construction which they have adopted, can recommend them with every confidence.

This pan consists of ten hollow discs of copper, about 3 ft. diameter, mounted on an axis 10 ft. long, and of a form which allows the exhaust steam, under a pressure of 2 lbs. per square inch, to communicate freely with all the discs, and, at the same time, carrying off the water of condensation at the other end. This revolver turns at a speed of twenty revolutions per minute, in a semi-cylindrical pan of copper, supplied with liquor from the battery, and will cook 12 cwt. of sugar per hour, from 20° Beaume, the temperature not exceeding 170° Fahrenheit.

Prices, designs, &c. may be had on application.

[1576]

COOKE, S., & SONS, *York.*—Amateurs' turning lathes and tools; circle-dividing and wheel-cutting engines.

[1577]

CORYTON, JOHN, 89 *Chancery Lane.*—Patent type-composer.

[1578]

COWAN, THOMAS WILLIAM, *Kent Iron Works, Greenwich.*—Patent air compressed machine-hammer for general forging. (*See page* 42.)

[1579]

COX & SON, 28 *and* 29 *Southampton Street, Strand, and Belvidere Road, Lambeth.*—Wood carving machine.

[1580]

CRAIG & SONS, 62 *Argyle Street, and* 68, *Glassford Street, Glasgow.*—Perforating machine; numbering or paging machine.

[1581]

CRAWHALL & CAMPBELL, *Glasgow.*—Horizontal boring machine, with adjustable bar, self-acting, and slide tables.

The exhibitors are manufacturers of machines and tools of every description for mechanical engineers. They exhibit a SELF-ACTING HORIZONTAL BORING MACHINE, having the boring bar adjustable in height by self-acting motion; with a bracket for supporting the end of the bar when necessary, and a strong bed with adjustable slide tables for carrying the work to be operated upon.

[1582]

DAY & SON, *Gate Street, Lincoln's Inn Fields.*—Lithographic and copper-plate presses.

[1583]

DEANE & DAVIES, 19 *Blackfriars Street, Manchester.*—Sewing machines, presses, gas apparatus, hand stamps.

[1584]

DE BERGUE, CHARLES, & Co., *Manchester, and* 9 *Dowgate Hill, London.*—Punching, shearing, riveting, and rivet-making machines, and steam hammers.

CONISBEE, WILLIAM, 39 *and* 40 *Herbert's Buildings, Waterloo Road.*—A Main's patent printing machine, for bookwork and job printing.

This well-known machine is simple in principle, substantial in construction, and economical in working. It will be observed in the accompanying illustration that the table is connected to the cylinder by direct gearing, thus causing them to move always in unison, producing a perfect impression; the gearing being cut by steam machinery, the arrangements for rapid feeding have been found superior to any other; each sheet being laid to elevating and depressing marks placed at the front of the feed table, giving facilities for certain and rapid feeding, and obviating to a great extent the use of the pointing arrangements which are provided for best register work.

The arrangement for inking will be found equal to the

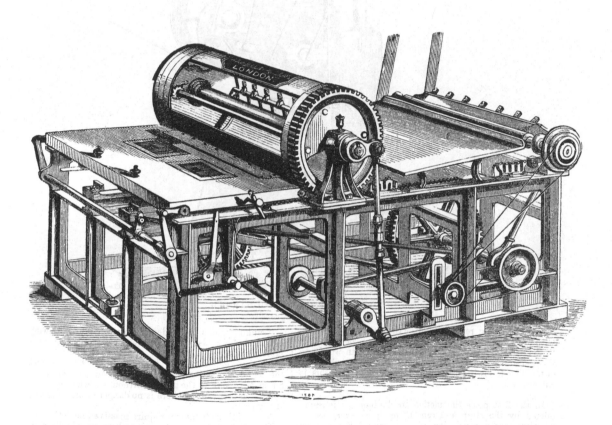

MAIN'S PATENT PRINTING MACHINE.

most expensive machine, and the whole mechanism is well and carefully made of the best material, by the best London workmen.

The "Main's" machine is manufactured solely by the proprietor at the Atlas Works, Herbert's Buildings, Southwark, S.; and sold by him, and by the sole agents, Harrild & Sons, Farringdon Street, London, E.C.

Sizes and prices:—

Fast jobbing, to print 19 by 14 in. with fly wheel		£70
Demy, to print 22 by 17 in. with fly wheel .		100
Double crown, 30 by 22 in. ditto . .		130
Double demy, 36 by 23 in.		150
Double royal, 40 by 28 in.		180
Newspaper size, 50 by 38 in.		220

Fly wheel for hand power, £12 for double demy, and £14 for the other sizes.

COWAN, THOMAS WILLIAM, *Kent Iron Works, Greenwich.*—Patent air compressed machine-hammer for general forging.

PATENT AIR COMPRESSED HAMMER.

The engraving represents an AIR HAMMER of 8 cwt. without compression. These hammers can be regulated to the utmost nicety for giving a blow equal in weight to the fraction of an ounce, and increasing it to about 45 cwt. The following is a description of the way this is effected:—

In the first place the motion to the hammer is transmitted by the strap and cam through a lever, which is raised every revolution of the cam. For a very light blow the cock at the bottom, which is usually open, is at this moment shut, causing the air to be compressed at the bottom of the cylinder, at the downward stroke of the hammer; this blow can be regulated by partially opening the cock. When a blow is required equal to the weight of the hammer itself, all the cocks are open. When a heavier blow is required the air is compressed by regulating the upper cock, which communicates with several chambers.

These hammers are recommended for general smiths' work, as they are very easily managed by any boy. The hammer block can be suspended at any part of the stroke, and the speed may be regulated the same as steam hammers.

Small hammers in sets of twos and fours driven from one shaft, and having conical speed pulleys, are very useful where rapidity of workmanship is required, as they are capable of giving about 360 blows per minute.

Trunk hammers are made on this principle for drawing out steel, &c.; also movable cylinder hammers for the same purpose.

All these hammers have very heavy anvil blocks, and the main frame being fixed on these blocks it is impossible for them to sink in the ground without the whole machine going together; hence there is no danger of breaking any of the parts.

None of these hammers require massive foundations.

There is no expense in having to keep up a boiler with high-pressure steam for these hammers, as they are driven by a strap from the usual main shafting, and there is very little foundation required. They are perfectly under the control of the hammer-man, and very soon pay for themselves. Price, from £65 upwards.

WINTON & COWAN'S PATENT HIGH AND LOW PRESSURE DOUBLE CYLINDER HAMMERS are recommended for large forgings. These hammers are made to any size required, the smaller ones having single frames, as the drawing above, and the larger ones, for iron manufactories, double frames, they being best adapted for manufacturing iron and steel.

These hammers effect a great saving in steam, as the steam which is used in raising the hammer, after it has done its work in the small cylinder, is allowed to enter the large cylinder, and give the blow. Price, from £100 upwards.

[1585]

DONKIN, B., & CO., *Near Grange Road, Bermondsey.*—Paper-making machine and paper-cutting machine. (*See page* 47.)

[1586]

DOULTON & CO., *Lambeth Pottery, S.*—Potter's wheel worked by steam, showing the process of manufacture.

[1587]

DUPPA, T. D., *Longville, Westanstow, Shropshire.*—Vice bench, for carpenters, coopers, &c.

[1588]

EASSIE, WILLIAM, & SONS, *Gloucester.*—A machine for manufacturing round mouldings in wood.

[1589]

EASTERBROOK & ALLCARD, *Sheffield.*—Engineering and railway tools, machines, tacks, crabs, ratchet-braces, spanners, screwing tackle, &c.

Taps, diameter in inches	1/16	3/32	1/8	5/32	3/16	7/32	1/4	9/32	5/16	3/8	7/16
Working taps each	1/7	1/7	1/7	2/0	2/0	2/5	2/10	3/8	3/3	4/0	4/10
Master taps, each	2/5	2/5	2/5	2/10	3/3	3/7	4/0	4/5	4/10	5/7	6/10
Machine taps each		2/0	2/5	2/5	3/0	3/3	3/7	4/5	5/3	6/5	

Taps, diameter in inches	1	1 1/8	1 1/4	1 3/8	1 1/2	1 5/8	1 3/4	1 7/8	2	2 1/4	2 1/2	2 3/4	3
Working taps each	5/7	6/5	7/7	8/10	10/5	12/0	13/7	16/0	18/6	23/3	28/10	35/3	42/5
Master taps each	8/0	9/7	11/3	12/10	14/5	16/9	19/3	21/7	24/0	28/10	35/3	41/7	49/7
Machine taps each	7/7	8/5	10/0	11/7	14/0	17/7	20/0	24/0	27/3	34/5	43/3	53/0	62/6

Whitworth's pattern screw stock, with dies and taps, taper, second, plug, and master, 3/8, 1/16, 1/2, 3/4, complete with tap wrenches and screw tools in polished case. Price £6 0 0

E. & A.'s registered 3-die screw stock, with dies and taps, taper and plug, 3/8, 1/16, 1/2, 3/4, and tap wrenches, in polished case £3 16 0

Screw stock No. 153, with dies and taps, 1 1/4, 1 3/8, 1 1/2. Price £5 12 0

Iron gas stock taps and dies, 1/4, 3/8, 1/2. . . . 2 0 0

Easterbrook & Allcard's patent universal ratchet brace. No. 1304, 12 in. 18/0; 16 in. 21/0; 20 in. 24/0.

E. & A.'s registered double action, No. 4235, 15 in. 36/0.

Ratchet braces, 012, 14 in. 18/0; 16 in. 19/6; 18 in. 21/0; 20 in. 22/6.

Ratchet braces, 14 in. No. 010, 13/0; 011, 15/0; 013, 30/0; 014, 52/6; 016, 16/0; 017, 19/6; 018, 16/0; 041, 22/6; 042, 36/0; 043, 13/0.

Extra strong black ditto, No. 040, 30 in. 45/0; 20 in. 30/0.

Cast-steel drills to fit the above, assorted 1/4 to 1 in. 10/0 per dozen.

Spanners, 12 in. No. 020, 8/6; 021, 6/6; 022, 8/0; 024, 9/3; 025, 8/6; 026, 4/9; 027, 4/9; 028, 2/7; 029, 2/7; 030, 4/0; 101, 3/3.

Screw wrenches, 10 in. No. 093, 3/0; 0214, 5/3; 0216, 6/3; 0217, 7/3; 0220, 9/6; 0221, 6/6.

Sash cramps 18 in. 11/0 pair; floor cramps, 18/0 each.

Pulley blocks, 2 and 3 sheaves, 4 1/2 in. dia. 45/0 per pair. Snatch block to match, 26/0.

Fluted rimers, set of 12 assorted 3/8 to 1 1/8 . . £3 10 0

Foot lathe, double gear, 6 in. centres, 5 ft. bed, hand-rest, face-plate, and centres complete £25 0 0

Compound slide rest for ditto 6 0 0

Traversing jack, 2 1/4 in. screw 8 16 0

Tripod jack, 2 1/4 in. screw 5 4 0

Bottle jack, wrought-iron case, 8 tons . . . 4 0 0

Ditto, cast-iron case, 3 tons 1 13 0

Haley's jack, 4 tons, £4; 6 tons, £4 16s.; 8 tons 5 12 0

Improved ratchet jack, 9 tons 5 10 0

E.& A.'s patent wrought-iron parallel vice, 5 in. 1 10 0

E. & A.'s registered adjustable vice, 5d. lb.

Portable vice bench, with 5 in. bright vice . 3 0 0

Hammers, chipping, cast-steel, 1/4 lb.

Boiler-makers' hammers, same price as chipping.

Sledge hammers above 8 lb. 4 3/4 d. lb.

Chisels, chipping, CS. 10d. lb.

Bench drilling machine, 30 in. £7 0 0

Drilling stoop adjustable, 50/0; No. 053, 32/0.

Scotch iron brace, with 36 bits, 25/0.

Permanent way cramp, 5 1/2 d. lb.

Boiler-punching bear, £5; cramp, 10/0.

Sets of miniature tools, made up from £1 to 40 0 0

[1590]

EASTWOOD, CHARLES, *Virginia Place, Leeds.*—Cutting and measuring machine, for cutting purposes in the brush trade.

[1591]

EASTWOOD, JAMES, & SONS, *Railway Iron Works, Derby.*—Steam hammer, samples of iron.

[1592]

EFFERTZ, PETER, 71 *Coupland Street, Manchester.*—Brick machine; drain-pipe machine; model of brick machine; drawings. (*See pages* 44 *to* 46.)

[1593]

EVANS, JOHN, & SON, 104 *Wardour Street.*—Amateurs' ornamental turning lathe, tools, and apparatus, and specimens of turning by amateurs.

EFFERTZ, PETER, 71 *Coupland Street, Manchester.*—Brick machine; drain-pipe machine; model of brick machine; drawings.

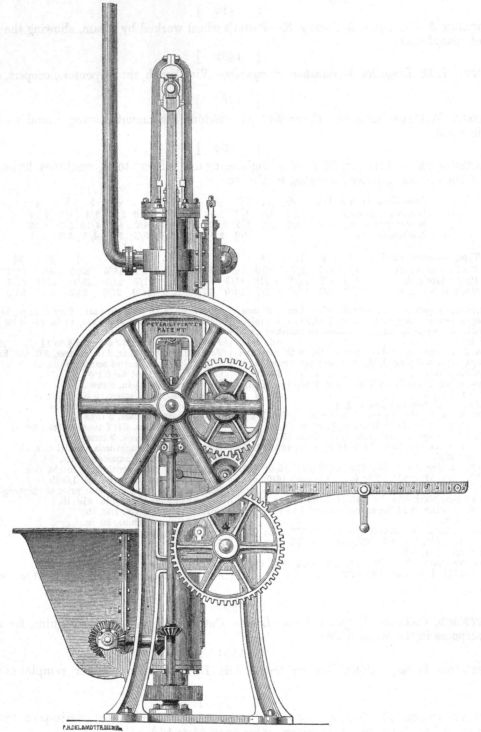

1.—BRICK-MAKING MACHINE.

The model of this BRICK-MAKING MACHINE is an elaborate and ingenious piece of mechanism, well worth the attention of visitors to the Exhibition.   It represents a machine calculated to make 75,000 highly finished bricks per day.

It cleans, mixes, and presses the rough clay into moulds the required size of the bricks, which are conveyed from the machine to the drying places or kilns on peculiarly constructed waggons, a model of which is shown with the

EFFERTZ, PETER, *continued.*

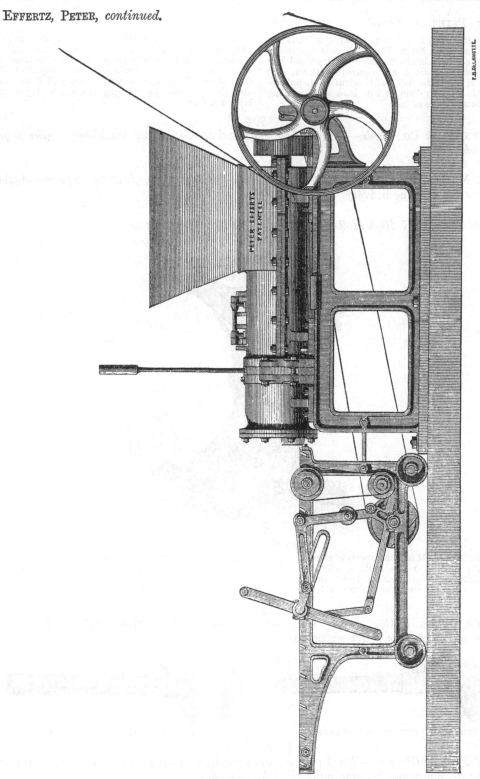

2.—PATENT BRICK AND TILE MAKING MACHINE.

model machine. The inventor has provided machines of four different sizes, constructed to produce respectively 25,000, 30,000, 50,000, 75,000 and upwards per diem. The illustration No. 1 represents a view of the smallest, or No. 1 machine, producing 25,000 bricks per day, a full-sized working machine is in the Exhibition.

EFFERTZ, PETER, *continued.*

The PATENT BRICK AND TILE MAKING MACHINE is shown in illustration No. 2. This machine is constructed to produce common bricks, roofing tiles, drain and floor tiles, and similar articles at any required length and size; it cleans and mixes the clay, and is calculated to produce 30,000 drain tiles per diem, of 2 in. bore, and, according to size or length more or less.

In the Exhibition, besides the above-named models, and full-sized working machine, are highly finished drawings on a large scale, representing several of the brick machines; the waggons used to convey the bricks from the machine to the drying places and kilns; and also the apparatus for carrying the rough clay to the brick machine.

[1594]

FAIRBAIRN, P., & Co., *Leeds.*—Engineering tools and rope-spinning machinery. (*See pages 48 to 54.*)

[1595]

FENTUM, MARTIN, 85 *New Bond Street, and* 8 *Hemmings Row, Leicester Square.*—Lathe and saw for working in ivory.

[1596]

FERRABEE, HENRY, 75 *High Holborn, London.*—The British sewing machine.

BRITISH SEWING MACHINE.

THE BRITISH SEWING MACHINE is specially adapted for family use. It makes a stitch which is exactly alike on both sides of the fabric, and it can execute perfect sewing at the rate of 5,000 stitches per minute. Price of a machine as illustrated and exhibited £10 0

Cabinet machines at various prices.

[1597]

FERRABEE, JAMES, & Co., *Stroud, Gloucestershire, and* 75 & 76 *High Holborn, London.*—Adjusting spanners or screw wrenches.

FERRABEE'S PATENT WEDGE SPANNER.   FERRABEE'S PROTECTED SPANNER.

[1598]

FORREST & BARR, *Glasgow.*—Wood planing and moulding machine, for ship builders, timber merchants, house and waggon builders. (*See page* 56.)

[1599]

FOX, BROTHERS, *Derby.*—Slide and screw-cutting lathe; vertical drilling machine; planing machine.

[1600]

GADD, WILLIAM, & SON, *Fishergate, Nottingham.*—Screwing machine upon a new principle.

DONKIN, B. & CO., *Near Grange Road, Bermondsey.*—Paper-making machine and paper-cutting machine.

Obtained the Council Medal in Class 6, in London, in 1851.

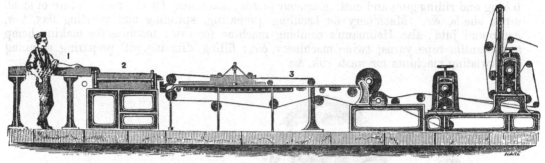

MACHINE WITH ENDLESS WIRE.

PAPER-MAKING MACHINE on the same principle as those erected by Mr. Donkin, of Bermondsey, at Frogmore, in Berks, in 1803, and at Twowaters, in Hertfordshire, in 1804, which were the first machines ever used for making endless paper. The machines for clearing the pulp, and for drying and cutting the paper, were subsequent inventions, and admit of great variation in their construction.

DRYING MACHINE.

1. Cast-iron sand catcher, coated with zinc.

2. Knot strainer, with brass plates, a parallel motion being given to this knotter, a uniform action over the whole plate is secured.

3. Machine with endless wire, 7 ft. 6 in. wide, 34 ft. long, with improved deckles, self-acting guide for the wire, and rider roll of perforated copper (Wilkes's patent).

4. Drying machine, consisting of 6 steam cylinders, 4 ft. diameter.

5. Two sets of smoothing presses.

6. Cutting machine of improved construction, for cutting the endless paper into sheets as it leaves the smoothing presses, without the intervention of reels.

Although the machines Nos. 3, 4, and 6, are drawn separately, they form one continuous machine; the pulp

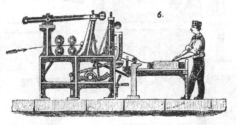

CUTTING MACHINE.

being supplied at one end, and the dry paper being delivered at the other, in sheets of the size required.

A machine of this description would make an endless sheet of paper about 20 miles long in 24 hours, which would cover about 17 acres, if kept continuously at work.

Fairbairn, P., & Co., *Wellington Foundry, Leeds; 36 Great George Street, Westminster, London,* Engineers' tools, including lathes; boring, drilling, slotting, planing and shaping machines; wheel-cutting, screwing, punching and shearing machines; plate-bending, forging and hot-iron machines, wood-cutting machinery, steam and travelling cranes; gauges, surface-plates and hand screwing apparatus. Special machinery for turning, boring and rifling guns and cutting armour plates; machinery for the manufacture of small arms, shells, &c. Machinery for hackling, preparing, spinning and twisting flax, tow, hemp and jute; also Heilmann's combing machine for tow; machine for making hemp and Manilla rope yarns; twine machinery, &c.; filling, dressing, gill preparing, spinning and twisting machines for waste silk, &c.

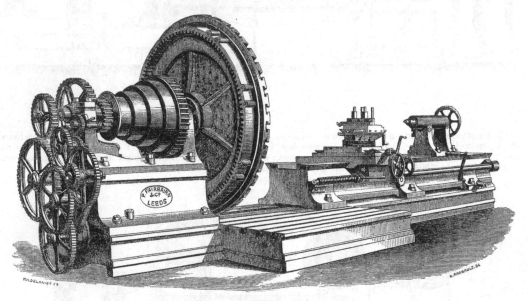

SELF-ACTING BREAK LATHE, 21 in. centres.

SELF-ACTING SLIDE BREAK LATHES, with treble-geared headstocks and large face plate. Strong base plate with sliding bed upon it moved by rack and pinion. Saddle with T-grooves for fixing work when boring, reversing motion for guide screw, self-acting patent surfacing motion, movable headstock, driving apparatus and screw keys.

These lathes are made with headstocks from 12 to 36 in. centres and upwards, and the diameter of work they will admit in the break from 4 ft. 5 ft. to 11 ft. respectively.

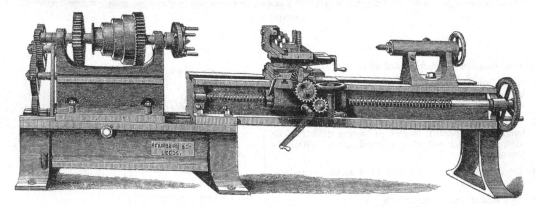

SLIDE AND SCREW CUTTING LATHE WITH MOVABLE GAP, 10 in. centres.

SELF-ACTING SLIDE LATHES, with doubled-geared headstocks, spindles up to 12 in. centres fitted in conical bearings of hardened steel; above that size in parallel bearings of gun-metal. The bed is fitted with guide screw (reversing motion, above 12 in. centres), self-acting patent surfacing motion, loose headstock, driving apparatus, face plates, Clement's driver, back stay and screw keys.

These lathes are made from 5-in. centres and upwards; screw cutting when required, or with movable gap as shown.

Also foot lathes, hand, surfacing, boring and special lathes, lathes for railway wheels, crank axles, guns, &c.

FAIRBAIRN, P., & CO., *continued.*

RADIAL DRILLING MACHINE, 1¾ in. spindle.

SELF-ACTING DOUBLE-GEARED RADIAL DRILLING AND BORING MACHINE, with the base plate arranged with bolt grooves, so that the work can be fixed to the top or the side as convenient.

Will bore up to 4 in. at a maximum radius of 36 in. The driving apparatus is self-contained, and the machine complete with drill chucks and screw keys. Diameter of steel spindle 1¾ in.

FAIRBAIRN, P., & CO., *continued.*

SELF-ACTING RADIAL DRILL, $2\frac{1}{4}$ in. spindle,

SELF-ACTING DOUBLE-GEARED RADIAL DRILLING AND BORING MACHINES, radial arm will revolve 320° and will move vertically up or down by power, at any position ; strong base plate with T-grooves for holding the work.   Driving apparatus, drill chucks and screw keys.

Diameter of steel spindle.	Will bore up to	Maximum radius.	Maximum height of spindle from base plate.	Feed of spindle.	Radial arm will move up or down.
$2\frac{1}{4}$ in.	6 in.	6 ft. 0 in.	6 ft. 0 in.	1 ft. 3 in.	3 ft. 0 in.
3 in.	12 in.	8 ft. 4 in.	7 ft. 6 in.	2 ft. 0 in.	4 ft. 4 in.

FAIRBAIRN, P., & CO., *continued.*

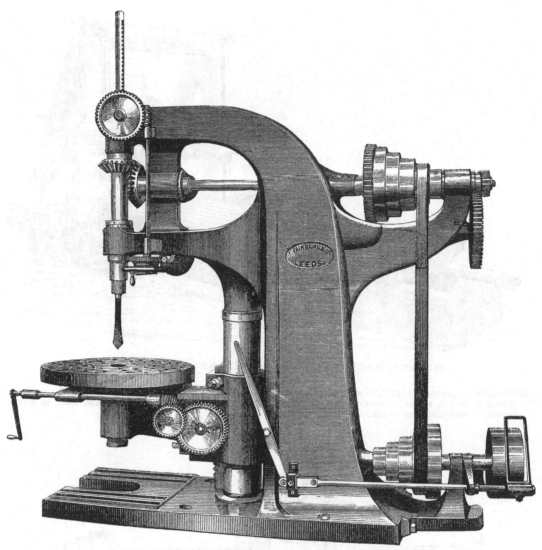

LARGE VERTICAL DRILLING MACHINE, 2¾ in. spindle.

SELF-ACTING VERTICAL DOUBLE-GEARED DRILLING AND BORING MACHINES, with the base plate and frame cast in one; revolving table so arranged on a radial arm as to leave the base plate clear for large work, and movable vertically by rack and pinion. Driving apparatus self-contained. Drill chucks and screw keys.

Diameter of steel spindle.	Will bore up to	Will take in diameter.	Feed of spindle.
1½ in.	1½ in.	2 ft. 0 in.	7 in.
1¾ in.	4 in.	2 ft. 8 in.	10 in.
2¼ in.	7 in.	3 ft. 10 in.	15 in.
2¾ in.	12 in.	5 ft. 0 in.	24 in.

FAIRBAIRN, P., & Co., *continued.*

SELF-ACTING PLANING MACHINE, to plane work 6 feet square, and 20 feet long.

SELF-ACTING PLANING MACHINE, to work either by screw or rack up to 3 ft. square; above that size, by rack and pinion. These machines are made to plane work from 1 ft. 6 in. square up to 14 ft. square, and to any length with 1 or more tool boxes or side tools as may be desired.

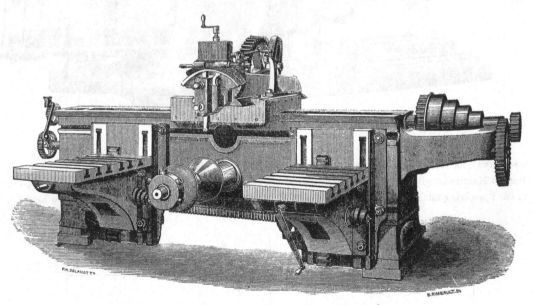

TREBLE-GEARED SHAPING MACHINE, 24 in. stroke.

SHAPING MACHINES, self-acting, in longitudinal, vertical, angular and circular motions, head-stock sliding upon the bed, and with quick return motion to tool, 2 front tables, movable vertically by rack and pinion, driving apparatus, and set of screw keys. These machines are made from 6 in. stroke upwards. Smaller machines are described elsewhere.

FAIRBAIRN, P., & Co., *continued*.

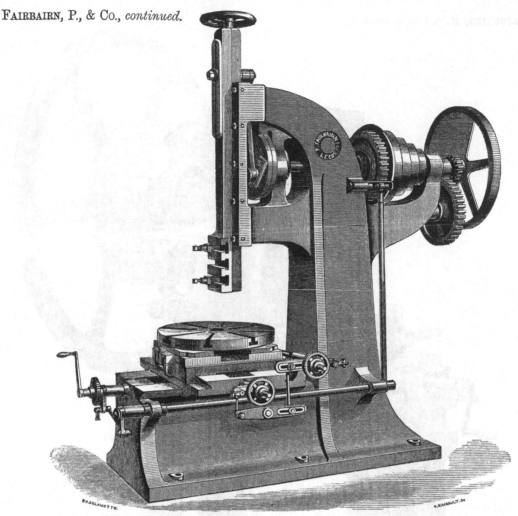

SLOTTING MACHINE, 9 in. stroke.

SELF-ACTING SLOTTING MACHINES with the upright and bed cast in one up to 20 in. stroke, above that size they are cast separately. Double-geared up to 20 in. stroke, and arranged for the gear to throw out for short and finishing strokes. Above 20 in. they are treble-geared; self-acting circular table capable of being inclined for key bed slotting, self-acting compound lower slides. Balanced vertical slide, arranged to hold one or more tools. Driving apparatus and screw keys. Smaller sizes described elsewhere.

Length of stroke.	Will admit in diameter.
6 in.	2 ft. 6 in.
9 in.	3 ft. 7 in.
12 in.	4 ft. 8 in.
16 in.	5 ft. 10 in.
20 in.	7 ft. 0 in.
25 in.	8 ft. 0 in.
30 in.	8 ft. 6 in.

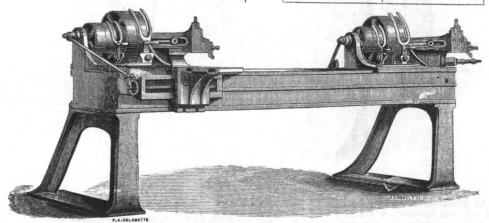

TWO SHAPING MACHINES, 3½ in. stroke upon one bed.

FAIRBAIRN, P., & CO., *continued.*

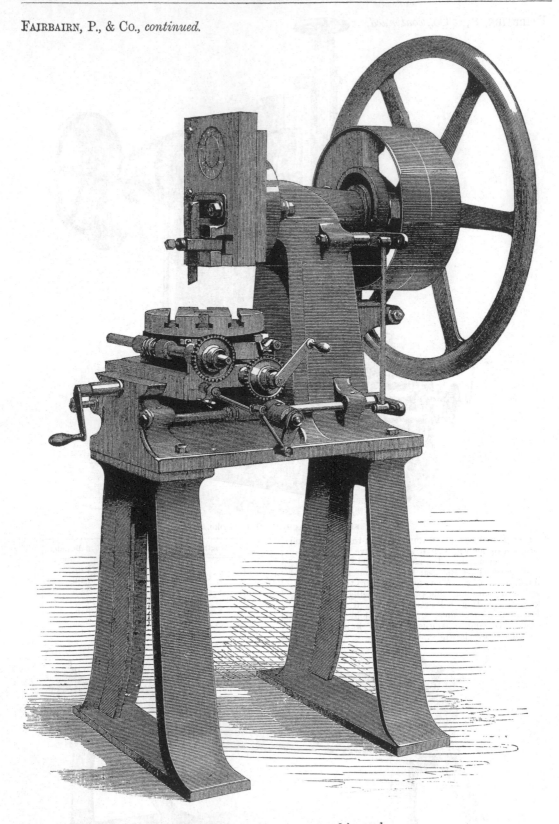

SELF-ACTING SLOTTING MACHINE, 3 in. stroke.

SMALL SINGLE-SPEED SLOTTING MACHINE, 3-in. stroke, placed upon standards or a fitter's bench, as desired. Capable of slotting flat work, 9 in. in length, and circular work, 9 in. diameter; both self-acting. Balanced slide, fly wheel, belt fork and screw keys.

FAIRBAIRN, P., & CO., *continued.*

SMALL SELF-ACTING SHAPING MACHINES 3½ in. stroke, with fast and loose pulleys for two speeds. Arranged for flat work with self-acting table adjustable vertically, and fitted with vice and screw keys; or for circular work with arbor and cones.

These machines can be had separately, to place upon a fitter's bench, or one, two, or three can be placed upon a bed, as may be desired.

A larger machine is also made 5 in. stroke on the same principle, but with gear. Nut-shaping apparatus extra.

These machines are very useful for shaping nuts, bolts, keys and a great variety of other work.

[1601]

GARRETT, B., 5 *Cumberland St. Camberwell.*—Imperial printing presses and bookbinding press.

[1602]

GARSIDE, HENRY, Maker, *Manchester;* GAIFFE'S Patent, YORK & CO., Proprietors, 2 *Royal Exchange Buildings, London.*—Electrograph engraving machine for engraving copper cylinders used in calico-printing, &c. (*See page 57.*)

[1603]

GEEVES, WILLIAM, *Caledonian Mills, New Wharf Road, Islington, N.*—Saw frame.

[1604]

GERISH, F. W., *East Road, City Road.*—A platen press, with rotary motive power.

[1605]

GHERLING, J., 15 *William Street North, Caledonian Road.*—Eylet machines, various tools, and steelyards.

[1606]

GIBBS, D. & W., *City Soap Works, London.*—Machinery for grinding and compressing soap.

Obtained Prize Medal at the Great Exhibition, 1851.

The following are exhibited by MESSRS. GIBBS :—

MILL, CANNON, AND MACHINERY FOR CRUSHING, GRINDING, AND COMPRESSING TOILET SOAPS. Soaps finished by this method, being mechanically as well as chemically combined, are rendered thoroughly pleasant and free in use, without any excess of alkali.

Steampan fitted with archimedean screw, soap frame, and patent foot-lever press for manufacturing composite household soap.

[1607]

GLASGOW, JOHN, *Trafford Street, Manchester.*—Screwing machine.

[1608]

GLEN & ROSS, *Greenhead Engine Works, Glasgow.*—Rigby's patent double-acting steamhammers, 2 cwt. and 5 cwt.

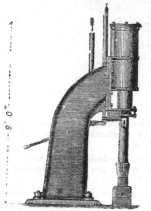 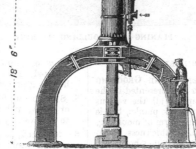 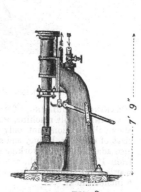

FIG. 2.   FIG. 1.   FIG. 3.

Fig. 1 is a representation of HAMMER FOR FORGING OR PUDDLING PURPOSES. They are made from 30 cwt. upwards. From the simplicity of their design and the substantial manner in which these hammers are constructed, they are easily kept in repair. The valve is so arranged that the under side of the hammer piston is never open to the atmosphere, whereby a great saving of steam is effected, especially when a large forging is under the hammer. To raise the hammer-piston, steam is admitted under it in the usual manner; but to accelerate its fall a communication is opened between the under and upper sides of piston. The upper side has an additional area equal to the cross section of piston rod, and the steam operating on this area produces a much sharper blow than can be obtained from the hammer falling by its own gravity only. This, with the height under frames, renders the hammer of great service in deep forging.

Fig. 2 shows a HAMMER FOR HEAVY SMITH-WORK. They are made from 5 cwt. up to 20 cwt. Full pressure of steam is admitted on the upper side of the hammer piston at pleasure, which gives great rapidity and power when required. They are very compact, and accessible to the workman on three sides.

Fig. 3 illustrates HAMMER FOR WORK OF A LIGHTER DESCRIPTION. They are made of 2 and 4 cwt. They differ from figure 2 only in having the anvil block and column cast in one piece.

Price, &c. may be learned by applying to the makers.

FORREST & BARR, *Glasgow.*—Patent safety derrick crane, for engineers, foundries, contractors, wharves, railways, quarries, and builders ; a planing and moulding machine.

The following machinery exhibited and manufactured by Forrest & Barr may be seen in operation in London, Glasgow, and many other towns in England, Scotland, and Ireland ; as also in America, Australia, and Russia, and many places on the Continent.

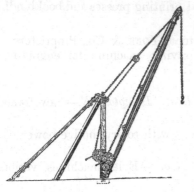

PATENT SAFETY DERRICK CRANE.

CRANES :—

1. PATENT SAFETY DERRICK CRANE as represented in accompanying illustration, driven either by hand or steam power, made of any size required, for lifting from 1 to 50 tons. These cranes are extensively used by house and ship builders, quarriers, saw millers, &c. ; and are highly valued for their convenience, and the security against the falling of the jib, which the patent arrangement affords.

2. FOUNDRY CRANES of all sizes, with improved gearing by which the suspended load can be moved, and set to the required position, with steadiness and precision.

3. WHARF CRANES of all sizes and of various descriptions, suitable for particular positions.

4. WAREHOUSE CRANE, and various other winches and hoisting apparatus required for the storing of goods.

5. PORTABLE STEAM WINCH for building purposes ; engine and boiler placed upon a carriage. This is a very compact and useful machine, and can be applied to a variety of purposes.

6. STEAM WINCH for ships' decks. This machine is extensively used for loading and discharging ships, &c.

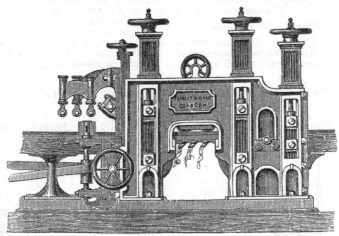

PLANING AND MOULDING MACHINE.

WOOD-WORKING MACHINERY :—

1 PLANING, MOULDING, TONGUEING, AND GROOVING MACHINE. This machine, which is represented in the above illustration, not only prepares all the various sizes of flooring, lining, and ship's deck planks, but is also adapted for the working of any form of moulding, from 12 in. in breadth, and 4 in. in thickness downwards. Ship builder's larger size, 24 inches broad and 6 inches thick.

2. MOULDING MACHINE, arranged for working mouldings only. This is a very beautiful, highly finished, and convenient machine. It has only been 3 years in use ; but during that time it has been much admired by all who have seen it in action. A number of samples of mouldings worked at the City Saw Mills, Glasgow, by it, are exhibited.

3. VERTICAL DIRECT-ACTING STEAM SAW FRAME, with single crank, for cutting square timber. This is a most substantial and compact machine. One has been in use over eight years, running at a high velocity, without requiring any repair, and has given the greatest satisfaction.

4. VERTICAL SAW FRAME for cutting square timbers, driven by belts and pulleys with crank shaft, either above or below the frame. These machines are constructed in a very substantial manner, and have been highly approved of.

5. SAW FRAME for cutting deals, driven from either above or below, and in which the deals are carried forward to the saws by rollers.

6. Combined machine for shipbuilders and others, comprising CIRCULAR SAW, BREAKING OR OPENING SAW, and SQUARING MACHINE.

7. CIRCULAR SAW TABLES of various other descriptions and sizes, with or without self-acting feed gear, and with improved guides.

————

Forrest & Barr also manufacture high-pressure condensing and compound steam engines, with steam-cased cylinder, variable expansive gearing, surface condenser, and other fuel-economising improvements ; dye-wood chipping and grinding mills ; grain mills ; sugar mills ; and shafting and gearing of every description.

Engravings of the foregoing machinery, and full particulars as regards price, &c. may be obtained on application.

GARSIDE, HENRY, Maker, *Manchester;* GAIFFE'S Patent, YORK & Co., Proprietors, 2 *Royal Exchange Buildings, London.*—Electrograph engraving machine for engraving copper cylinders used in calico-printing, &c.

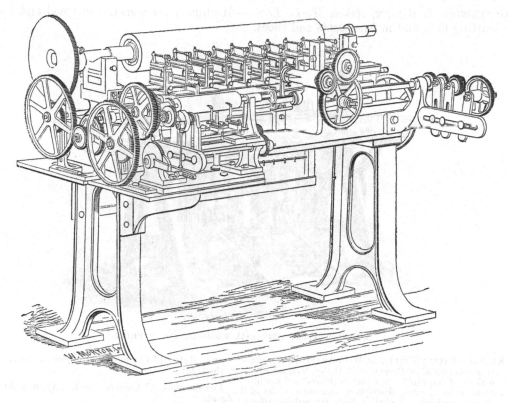

ELECTROGRAPH ENGRAVING MACHINE.

The machine represented in the accompanying illustration, is used for engraving the cylinders of copper or brass employed in the printing of woven fabrics and paper hangings.

A distinctive feature in this machine, apart from its general mechanical arrangement, is in the application of the subtle agency of voltaic electricity in communicating certain necessary movements, to important and delicate portions of the apparatus.

The cylinder to be engraved is first coated on its outer surface with a thin film of varnish, sufficiently resistant to the continuous action of the strongest acids. The required number of copies of the original design are then traced or scratched simultaneously by a series of diamond points, arranged on the machine parallel with the axis of the cylinder. The metallic surface is thereby exposed in certain parts, and a bath of nitric or other acid being afterwards used to etch or deepen the engraved portions, the operation is completed.

Each diamond point is in connexion with a small temporary magnet, and the entire series is so arranged *en rapport* with the original design, that intermittent voltaic currents are established, which result in the diamonds being withdrawn from action at proper intervals. The precise adaptation can be understood only by observation of the machine itself.

Amongst other special advantages of this apparatus, the facility with which engravings may be enlarged or diminished to any necessary extent, from the same original, is not the least important. Its capability of producing variety of result is very extensive.

[1609]

GRAFTON, HENRY, 80 *Chancery Lane.*—Cask and barrel machines, which cut and mould the wood hot; machine for making solvable paper tubes.

[1610]

GREENWOOD & BATLEY, *Albion Works, Leeds.*—Machinery for working in wood and metals, cutting files, and making boots and shoes.

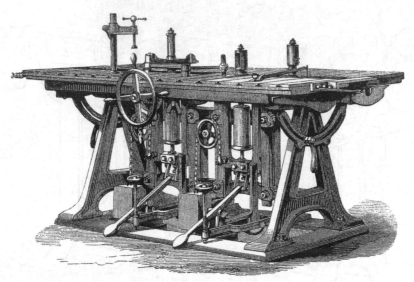

KINDER'S PATENT UNIVERSAL WOOD-SHAPING MACHINE.

KINDER'S PATENT UNIVERSAL WOOD-SHAPING MACHINE is manufactured by Greenwood & Batley, who are also makers of improved self-acting engineers' and machinists' tools of every description, and constructors of special machinery, including tools for making rifles and rifled artillery; moulding apparatus for shot and shell, &c.

London office, 20 Cannon Street. ARTHUR KINDER, Agent.

[1611]

GREIG, DAVID & JOHN, *Edinburgh.*—Paper-cutting machine, lithographic, copper-plate, and photographic presses; case of copper and steel plates.

[1612]

GUINNESS & CO., 42 *Cheapside, London, E.C.*—Patent shuttle sewing machine.

These machines are recommended for their simplicity, economy, and durability. Being moved by cranks from one shaft they are more easily worked, less noisy, and far less liable to be put out of order than any other machine; while they possess the additional advantage of enabling the operator to work either backwards or forwards.

Price, on tables, £10 each; in cabinets from £13, according to style and finish.

[1613]

HARRILD & SONS, 25 *Farringdon Street.*—Patent newspaper addressing machine, and other new printing materials.

[1614]

HARRISON, —, 16 *Bishopsgate Street Within.*—Magnetic printing press.

[1615]

HARRISON, C. W., *Lorrimore Road, Walworth.*—Electro-magnetic printing press.

[1616]

HARVEY, G. & A., *Albion Works, Glasgow.*—Machine tools.

[1617]

HAWKINS, JOHN, & CO., 16 *Station Street, Walsall,* and 38 *Lisle Street, London,* W.—Patent self-acting steam fly-press.

[1618]

HETHERINGTON, JOHN, & SONS, *Vulcan Works, Pollard Street, Manchester.*—Tools.

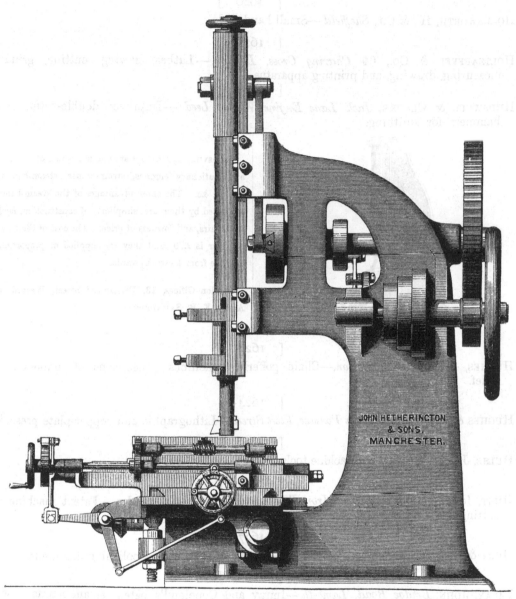

JOHN HETHERINGTON & SONS, MANCHESTER.

SLOTTING MACHINE.

THE FOLLOWING ARE EXHIBITED:—

12-IN. SELF-ACTING SLIDE AND SCREW-CUTTING LATHE, with fast and following back-geared headstocks, fitted upon planed cast-iron bed 12 ft. long, with case-hardened spindle in conical bearing; with carriage for screw-cutting or sliding, traversed by means of a regulator or guide screw, and rack for hand traverse, including change wheels; also compound slide rest and improved screwing stay. Each lathe is supplied with 2 face-plates, Clement's driver, tool rest, boring rest, screw keys, and the driving apparatus.

VERTICAL RADIAL DRILLING AND BORING MACHINE, with self-acting feed motion. The drill is adjustable on a radial arm, movable through an arc of 280° from a radius of 2 ft. 6 in. to 6 ft.; traverse of spindle 12 in.; vertical stroke of jib 2 ft.; capable of taking in an object 6 ft. from the floor; with holding-down bolts, screw keys, and the driving apparatus.

SLOTTING MACHINE WITH VARIABLE STROKE up to 14 in. self-acting longitudinal and transverse slides, and self-acting revolving worm table, and also adjustable table for giving the requisite taper to key beds. It is adapted for grooving wheels, also for paring and shaping objects externally and internally. Each machine is supplied with screw keys, and the driving apparatus.

COMPLETE SET OF HAND SCREWING TACKLE, from ¼ to 1½ in. Whitworth's thread.

Prices may be learned by application at the Works.

[1619]

HILL, PEARSON, *Bertram House, Hampstead, London.*—Post-office stamping machine, used in the English post-offices.

[1620]

HOLDSWORTH, H., & Co., *Sheffield.*—Small hand tools.

[1621]

HOLTZAPFFEL & Co., 64 *Charing Cross, London.*—Lathes, sawing cutting, grinding, measuring, drawing, and printing apparatus.

[1622]

HUDSWELL & CLARKE, *Jack Lane Engine Works, Leeds.*—Improved double-action steam hammers for smithing.

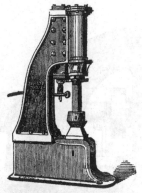

STEAM HAMMER.

HUDSWELL & CLARKE are manufacturers of locomotive and stationary engines, steam-cranes, steam-hammers, boilers, &c. The chief advantages of the steam-hammer exhibited by them are, simplicity of construction, facility for repairs, and lowness of price. The cost of their 4 cwt. hammer is £75, and they are supplied at proportionate expense from 1 cwt. upwards.

London Offices, 13, Parliament Street, Westminster. Agent, E. B. SAUNDERS.

[1623]

HUGHES, HESKETH, *Homerton.*—Chain goffering machines; specimens of embossing in relief.

[1624]

HUGHES & KIMBER, *Red Lion Passage, Fleet Street.*—Lithographic and copper-plate press, &c.

[1625]

HULSE, J. S., *Manchester.*—Machine tools. (*See pages 62 and 63.*)

[1626]

HUNT, JOHN, & Co., *Clay Hall Iron Works, Old Ford, Bow, London.*—Patent machine for cutting the teeth of wood or metal wheels.

[1627]

HUNT & ROSKELL, 156 *New Bond Street.*—Process of cutting and polishing diamonds.

[1628]

IMRAY, JOHN, *Bridge Road, Lambeth.*—Imray and Copeland's patent steam hammer, with hydraulic anvil and striker.

This hammer is worked by steam pressure, both for raising and dropping it. The valve is of the most simple kind, and the ports are arranged so as to give an elastic cushion at top and bottom, and thereby to save the piston and cylinder from damage, whatever be the force of the stroke, or the clearance between the hammer and anvil.

The hammer is fitted to the end of the piston rod with an intervening liquid cushion, which, without in the least affecting the intensity of the blow, saves the rod from being upset or otherwise damaged, the concussion being converted into a diffused fluid pressure between the hammer and rod.

The anvil is mounted on the ram of an hydraulic cylinder, fitted with a valve for regulating the ingress or egress of water, so that the anvil with blocks or work on it can be raised or lowered at pleasure; and a forging, with the necessary blocks or tools, occupying greater or less height, can be made to receive the blow at any required level. At the same time the shock is transmitted to the framing and foundation through a liquid cushion, which takes off the whole violence of the concussion, and thereby obviates the necessity for the great strength and solidity which are required for other hammers.

Anvils and strikers, constructed according to the hydraulic system, can be fitted to existing steam power or other hammers.

One of the patent hydraulic steam hammers can be seen in operation daily, at the works, 65, Bridge Road, Lambeth, London, where particulars can be obtained as to dimensions and prices.

[1629]

IRVIN & SELLERS, *Preston.*—Tools.

[1630]

JAQUES, JAMES, *Prescot.*—Spring dividers, and compasses of various sorts.

[1631]

JARRETT, GRIFFITH, 37 *Poultry, City, and* 66 *Regent Street.*—Patent endorsing, linen-marking, and embossing presses. (*See page* 64.)

[1632]

JOHNSON, J. R., & J. S. ATKINSON, 31 *Red Lion Square.*—Machinery for casting and finishing type.

[1633]

JONES, JONATHAN, 35 *Holywell Lane, Shoreditch.*—Machinery for turning lasts, boot trees, and all irregular forms.

[1634]

JONES, LAVINIA, *Bradford-on-Avon, Wilts.*—Miniature Albion printing press, cases of type, and furniture, with appliances.

MINIATURE ALBION PRINTING PRESS, cases of type, furniture, rules, and chases, arranged as a cabinet for private use.

Illustrative specimen printed sheets in Continental and Oriental spoken languages.

The exhibitor gives instruction to ladies in composition and press work, and takes orders for the above cabinets and presses. Communications received, and interviews arranged, at the private office of the exhibitor, 2 Bow Street, Covent Garden, W.C.

[1635]

JONES, WILLIAM, 246 *High Holborn.*—Embossing and screw stamping presses.

[1636]

KEILA, WEDDERSPOON, *Perth.*—Marmalade-cutting machine; cinnamon and cassia cutting machine.

[1637]

KEITH, WILLIAM, 11 *Three Crown Square.*—Improved sewing machine.

[1638]

KENDALL & GENT, *Salford, Manchester.*—Patent self-acting machine, for cutting tubes for engineers and boiler makers.

[1639]

KENNAN & SONS, *Fishamble Street, Dublin.*—Sculpturing machine; amateurs' lathes; specimens of mechanical sculpture and turnings.

THE FOLLOWING ARE EXHIBITED:—

MACHINE FOR COPYING WORKS OF ART, &c. from the round or flat, upon any scale, in ivory, wood, alabaster, &c. It is easily worked by one person. The movement for copying proportional straight lines is unique. The cutting is performed by a revolving tool, mounted on a bar with universal centre, and guided by a tracer applied to the original. It will copy the most intricate forms.

SPECIMENS OF MECHANICAL SCULPTURE, showing the powers of the machine.

ORNAMENTAL TURNING LATHE with improved slide rest and Kennan's universal geometric cutter; apparatus for cutting screws; improved chucks; cutter bars, &c.

SPECIMENS OF TURNINGS executed by Kennan's lathes.

AMATEURS' CIRCULAR SAWING MACHINE with parallel and angular gauges.

AMATEURS' VERTICAL DRILLING MACHINE.

[1640]

KERSHAW, J. & J., *Store Street Works, Manchester.*—Double stud lathe bench shaping machine.

[1641]

KIRKSTALL FORGE COMPANY, THE, *Leeds, and* 35 *Parliament Street, Westminster, S.W.*—Naylor's single or double action steam hammer. (*See page* 65.)

[1642]

LAMB, J., *Holborn Paper Mills, Newcastle, Staffordshire.*—Laying apparatus, attached to paper-cutting machine, felt not required. (*See page* 66.)

HULSE, J. S., *Manchester.*—Lathes ; planing, slotting, drilling, boring, screwing, wheel-cutting, punching, and shearing machines.

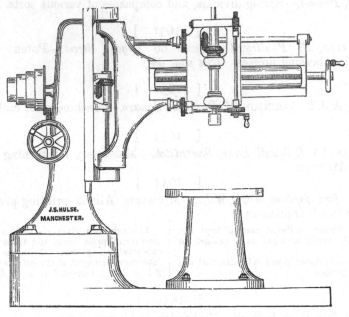

BORING MACHINE.

JOSEPH HULSE, who for seventeen years was with Messrs. Whitworth & Co. exhibits the following machine tools, viz.:—

    Slide and screw-cutting lathes, from 5 to 24 in. centres, of any length.
    Hand turning lathes, from 5 to 24 in. centres, of any length.
    Gap and break lathes, from 5 to 24 in. centres, of any length.
    Foot lathes, slide or hand.

Railway wheel turning lathes, for 4, 5, 6, 7 and 8 ft. wheels.
Headstocks, slide rests, universal chucks, to suit any lathes.
Planing machines, to plane from 1 ft. 6 in. to 10 ft. in width and height, and any length.
Brackets, for side planing.
Shaping machines, from 4 to 24 in. stroke of any length.
Bench shaping machines, from 2 to 5 in strokes.
Slotting machines, from 6 to 24 in. stroke.

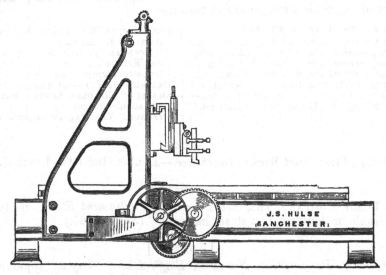

PLANING MACHINE.

HULSE, J. S., *continued.*

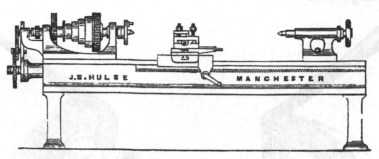

SCREW-CUTTING LATHE.

Bench slotting machines.
Vertical drilling machines, single and double geared.
Bench drilling machine, for hand or power.
Pillar drilling machines.
Radial drilling machines, single and double geared.
Horizontal radial drilling machine, ditto ditto.
Tube plate drilling machines.
Angle iron drilling machines.
Horizontal boring machines.

Portable boring apparatus.
Punching and shearing machines.
Bar-cutting machines.
Sawing machines for hot iron.
Sawing machines for boiler tubes.
Tube-cutting machines.
Plate-bending machines.
Wheel-cutting machines.
Cutter-forming machines.

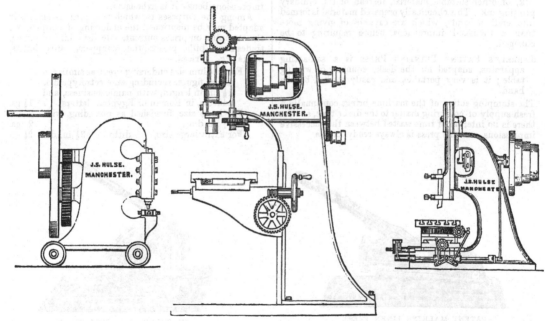

PUNCHING, DRILLING, AND SLOTTING MACHINES.

Nut-shaping machines.
Sawing machines, with circular saw for wood.
Ribbon saw.
Hydraulic mandril press.
Grindstone frames.
Portable vice benches.
Hand driving wheels.
Drill braces and frames.
Standard gauges.

Surface plates and straight edges.
Screwing machines, of any required range.
Screw stocks, dies and taps (Whitworth's threads and sizes).
Improved machine vices, suitable for shaping, planing, and drilling machines.
Cast-iron billiard tables.
Steam hammers.
Dynamometers.

JARRETT, GRIFFITH, **37** *Poultry, City, and* 66 *Regent Street.*—Patent endorsing, linen-marking, and embossing presses.

PATENT PRINTING PRESS. (Fig. 1.)

PATENT PRINTING PRESS. (Fig. 2.)

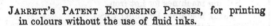

JARRETT'S PATENT ENDORSING PRESSES, for printing in colours without the use of fluid inks.

The very general objection to endorsement stamps, &c. where fluid ink is used, from the inconvenience incident upon the drying or caking of the ink, which renders the production of a correct or satisfactory impression so uncertain, has induced the patentee of the above press to substitute carbonic or other chemically prepared paper, silk, or other suitable material, instead of the ordinary printing ink. This chemically prepared material is formed into endless bands, which are capable of giving more than a thousand impressions before requiring to be changed.

JARRETT'S PATENT PRINTING PRESS is a self-acting apparatus, adapted for the desk, counter, or writing table; it is very portable, and easily worked by the hand.

The stamping action of the machine brings continually a fresh supply of colouring matter to the die or type, so that there is no interval of time wasted between the successive impressions, and the press is always ready for use.

In the press shown in Fig. 1, the die or type-holder is attached to the slide of the press by means of a taper-dovetailed key, so that it can be instantly removed for changing the dates, or for being replaced by another die.

In the press got up as Fig. 2, which is more particularly adapted for movable types, the dies or types are placed below with their face upwards, so as to be more easily changed. In this press the endless band being double the length of the press, will yield some thousands of impressions before it is exhausted.

Among the purposes to which this press is eminently adapted may be mentioned the endorsing of cheques, &c. the stamping on prices current, sale lists, bills of lading, tradesmen's bills, prescription wrappers, cards, letters, books, &c. Prices:—

For medium size endorsing press, as similar to either figure, including an electrotype die, 1½ in. in length, with name, business, and address, in Roman or Egyptian letters . . 1¼ gs.
For large size, furnished as preceding, with 2¾ in. die . . . . 2 gs.
For extra large size, ditto, 3½ in. die 2¼ gs.

PATENT MARKING LINEN PRESS.

IMPROVED EMBOSSING PRESS.

JARRETT'S PATENT SELF-INKING PRESS, for marking linen with indelible ink, applicable also for endorsing.

The above press entirely supersedes the pen, the hand-stamp, and the stencil-plate. It is portable, self-acting, and so easy in its application, that a child may by its means stamp a hundred pieces of linen in a few minutes.

The small size marking-linen press, including a prepared roller, a bottle of the best marking ink, together with an electro-plated die (not exceeding 1¼ in.), with the engraving of either a coronet, crest, or initials, or name, price 25s. complete.

JARRETT'S IMPROVED EMBOSSING PRESSES, for easily and effectively embossing coats of arms, crests, initials, residences, or business impressions, on note or letter-paper, envelopes, books, official documents, certificates, agreements, &c.

The dies in Jarrett's improved embossing presses are all of steel, polished and properly tempered. The counter-parts are of hardened copper.

Upwards of 30,000 different impressions taken by these presses may be inspected.

Price of press with engraved die complete, from . . . . 14s. 6d.

KIRKSTALL FORGE COMPANY, THE, *Leeds, and* 35 *Parliament Street, Westminster, S.W.*—Naylor's single or double action steam hammer.

*Obtained Prize Medal for railway wheels and axles, Class 5, Exhibition, 1851;
also Honourable Mention, iron and steel, Class 1.*

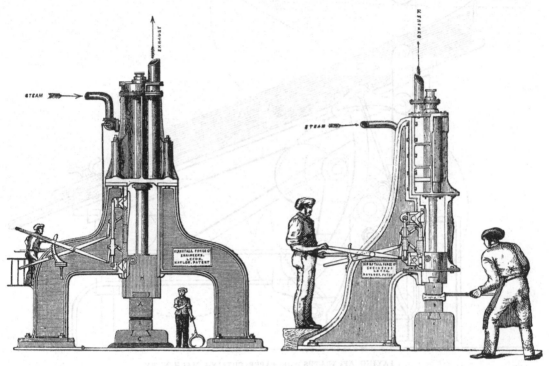

STEAM HAMMERS.

NAYLOR'S PATENT SINGLE OR DOUBLE ACTION STEAM HAMMERS.

The valuable improvements developed in these hammers, also the great advantages and capabilities which they possess over all others that have hitherto been invented, are the following :—

STEAM HAMMERS which have hitherto been constructed involve the same general principle of being lifted by steam pressure, and falling by gravity, the effect of the blow being dependent on the weight of the hammer, multiplied by the height of its fall.

The greater the distance it falls, consequently the greater the force of the blow, the slower is the speed of working. The great practical drawback to the more extended application of steam hammers, has been the impossibility of obtaining sufficient speed or quickness of stroke.

The advantages of the DOUBLE-ACTION STEAM HAMMER for forging are its being capable of working up to 200 strokes per minute (when required), which is from three to four times faster than any steam hammer hitherto constructed. The power can also be more than doubled, instantaneously, and as rapidly altered. The adjusting valve gearing also allows of instantly changing the length of stroke, and force of blow, by altering the position of the sliding wedges.

It is completely under the control of the hand gear, which is easy to work in any position. The rapidity of the stroke obtained by it is particularly advantageous for forgings requiring a great number of blows, by finishing the work at one heat, and saving both the fuel required for the second heat and the deterioration and waste of the iron.

This principle of hammer is also adapted for riveting wrought-iron bridges, girders, ship-building, &c.

LAMB, J., *Holborn Paper Mills, Newcastle, Staffordshire.*—Laying apparatus, attached to paper-cutting machine, felt not required.

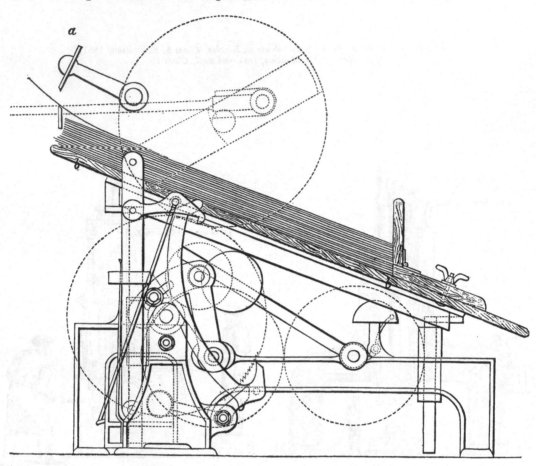

LAYING APPARATUS FOR PAPER CUTTING MACHINERY.

The above patented apparatus, applicable to machines for cutting paper, may fairly claim to be the completion of the paper-making machine. The object of it is to collect the paper in piles or heaps, and to dispense with the manual labour hitherto required for that purpose. *a* represents the knife, forming part of an ordinary paper-cutting machine; and *b* the patent laying table on which the paper is deposited. In proportion as the paper accumulates on the laying table, it is gradually lowered by means of self-acting mechanism.

When nearly a sufficient quantity of paper has been thus deposited on the laying table, a bell is rung by the machine to give warning; the machinery by which it is lowered soon after throws itself out of gear; the attendant then removes the piles or heaps of paper, the platform rises up to its original position, and the operation continues as before.

In most of the cutting machines a felt is generally required, on which the paper drops after being cut by the knife *a*, and an attendant is employed, in nearly every case, to catch the sheet or sheets of paper so cut, but by this apparatus the felt and the attendants are dispensed with, thereby not only effecting a considerable saving in wages, but avoiding the injury and waste resulting from finger marks.

This machine has been in successful operation at the Holborn Mills, Newcastle-under-Lyme, Staffordshire, for several years, and may be seen at work on application.

For further particulars apply to the Patentee, to MESSRS. HETHERINGTON & SONS, *Vulcan Foundry, Manchester, makers, or to* MR. WALTER IBBOTSON, *Agent for the same, 8 Dickinson Street, Manchester; and in the Exhibition Building, to* MR. S. MUIR, JUN., *of 40 Broad Street Buildings, London.*

The following amongst other testimonials is submitted :—

"*Hollins Paper Mills, Darwen, Lancashire.*
February 17, 1860.
MR. LAMB,

Sir,—We have worked the first of your laying machines upwards of three years, and the second nearly two years. We are quite satisfied with the working of them both, and we consider your machine a most useful auxiliary to the paper-cutting machine. In our opinion no cutting machine is complete unless your laying apparatus is attached thereto.

Yours truly,
C. POTTER & CO."

[1643]

LANCELOTT, JAMES, 4 *Clifton Terrace, Birmingham.*—Machine for making sheet-metal chains.

[1644]

LEE, H. C., 11 *Laurence-Pountney Lane, E.C.*—Knitting machine.

[1645]

LEGG, ROBERT, 14 *Owen's Row, Clerkenwell, London.*—Combined compressing and cutting machine for tobacco; Legg's 4-horse power steam engine.

[1646]

LEIGH, EVAN, & SON, *Manchester.*—Case of patent top rollers, with loose bosses; model of improved patent sailing and steam ship.

[1647]

LELY, AFFIFI, *Redditch.*—Machine for grooving sewing-machine needles; machine fo polishing the eyes of needles.

The following machines are exhibited.

PATENT GROOVING MACHINE, for grooving sewing machine needles.

COMBINED WEIGHING MACHINE AND PACKING PRESS for packing powders of any description, accomplishing 20 or more packets at one blow, or 10 cwt. in 10 hours.

BURNISHING MACHINE for polishing the eyes of needles. Various machines used in the manufacture of needles.

[1648]

LEWIS, JOSEPH, 51 *High Street, Bloomsbury, W.C.*—Patent machine for boring and fret cutting.

The machine exhibited is adapted for ladies' use.

It may be seen in action at Mr. Lewis' manufactory.

Price,

Without drill £7 0

With drill 8 0

The exhibitor is prepared to supply larger machines for the trade, at proportionately low prices and is a manufacturer of new inventions and machinery in general.

[1649]

LOCKETT, JOSEPH, SONS, & LEAKE, *Manchester.*—New and improved patent double-bar pentagraph engraving machine (Shield's patent).

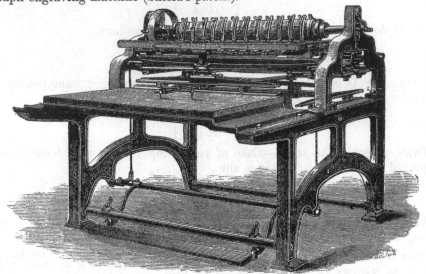

PATENT DOUBLE-BAR PENTAGRAPH ENGRAVING MACHINE.

By this machine the system of pentagraph engraving, now universal, receives further and more perfect development.

The advantages to the operator are :—

1. Any enlargement of sketch can be used from 2 to 10, thus allowing adaptation to the various peculiarities of design, and great economy of sketch-making and zinc-cutting in large designs.
2. Unerring fitting from both bars—durable working qualities—all bands and pulleys being dispensed with.
3. The sketch may be made to an approximate girth of roller ; thus enabling the zinc to be prepared before the rollers are applied.

4. Any angle to 3 inches may be given to cross-over lines, and fractional steppings at the side are greatly facilitated.
5. The pattern table is flat, a sensible relief to the workman.
6. The fittings for handkerchief rollers do not necessitate any change in the construction of the machine : the diamond bars are the same as for garment rollers.

IMPROVED PATENT DOUBLE-BAR PENTAGRAPH ENGRAVING MACHINE. (Rigby s Patent.)

This machine during the last five years has been widely adopted in Great Britain and on the Continent. By its use the system of pentagraph engraving has been mainly established.

[1650]

LYONS, MORRIS, 143 *Suffolk Street, Birmingham.*—Apparatus for depositing metals from new solutions, and for shaping a new plastic compound.

[1651]

MACLEA & MARCH, *Leeds.*—Double-wheel lathe, self-acting slithe lathe, planing, shaping, and slotting machines.

THE FOLLOWING MACHINES ARE EXHIBITED :—

DOUBLE-WHEEL LATHE, for turning up a pair of 6-ft. wheels on their axle.

7-in. CENTRE DOUBLE-GEARED SELF-ACTING SLIDE LATHE, 7 ft. bed.

DRILLING MACHINE, DOUBLE-GEARED, 12-in. traverse.

PLANING MACHINE, to plane 6 ft. long, 3 ft. wide, and 3 ft. high, self-acting in all cuts.

IMPROVED SHAPING MACHINE, 12-in. stroke on 4 ft. bed.

SLOTTING MACHINE, 6-in. stroke, to admit 2 ft. 6 in. diameter, with self-acting compound slides.

[1652]

McDOWALL, JOHN, & SONS, *Walkinshaw Foundry, Johnstone, Glasgow.* — Planing and moulding machine, and saw-bench, for wood-cutting.

[1653]

MACKENZIE, A., & CO., 32 *St. Enoch Square, Glasgow.*—New double-action cylinder sewing machine. (*See page* 69.)

MACKENZIE, ALEXANDER, & Co., 62 *North Frederick Street, Glasgow.*—New double action cylinder sewing machine, with specimens of work.

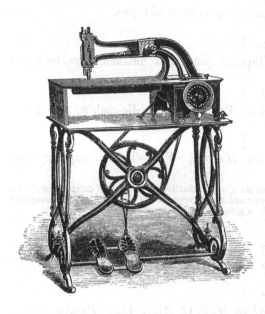

MACHINE, WITH TABLE, AND SUITED FOR ALL THE PURPOSES OF AN ORDINARY MACHINE.

MACHINE AS USED FOR CYLINDRICAL WORK, AS BOOT LEGS, TROUSERS, SLEEVES, &c.
PATENTED 7th FEBRUARY, 1862.

The continual demand for a machine capable of working in either direction at will, without the necessity of detaching and substituting other parts of machinery, which were in many cases laid aside, and always troublesome, led to the invention of this machine, where the same working parts operate in both actions.

This machine has a cylindrical arm, 15 in. long, and 2 in. diameter, enclosed in a brass tube; and at the will of the operator, can be made to sew either in the ordinary, or in the lateral direction from left to right, by simply turning the tube half-way round, and turning the presser (which is carried in a separate frame concentric with the needle) at right angles to its former direction.

[1654]

MCKERNAN, L, 98 *Cheapside.*—Sewing machines.

[1655]

MCQUEEN, BROTHERS, 184 *Tottenham Court Road.*—Process of plate printing.

[1656]

MARSHALL, THOMAS J., 80½ *Bishopsgate Without.*—Paper-making machines, patent pulp strainer, cutting machine, and watermarking rollers.

[1657]

MATHIESON, ALEXANDER, & SON, *Tool Works, East Campbell Street, Glasgow.*—Planes, mechanical, engineering, and edge tools.

[1658]

MIERS, W. J., 15 *Lamb's Conduit Passage.*—Machine for cutting geometrical forms.

[1659]

MILLER & RICHARD, *Edinburgh and London.*—Printed specimens of types.

[1660]

MILLS, J., late MILLS & ROBERTS, *Stockport.*—Tapered pins, and finished keys, by patent machinery.

[1661]

MILWARD, HENRY, & SONS, *Redditch, Worcestershire.*—Processes in needle-making machinery.

[1662]

MITCHEL, WILLIAM H., 16 *Newton Street, High Holborn, W.C.*—Type composing and distributing machines.

This machine has been largely and successfully used in America, as well as in some of the leading printing houses of England and Scotland. It may be applied to every description of plain book or newspaper work; and effects a large reduction in the cost of composition, in the wear and tear of type, and in the quantity of type required for a given amount of work.

[1663]

MORGAN & CO., *Paisley.*—Block-cutting machine.

[1664]

MORISON, JAMES, & CO., late J. & G. MORISON, *Abbey Mill, 15 Abbey Close, Paisley.*—Piano card-perforating machine.

[1665]

MORRALL, ABEL, *Studley Mills, London and Manchester.*—Needles and thimbles in process of manufacture.

[1666]

MORRISON, R., & CO., *Newcastle-on-Tyne.*—Steam hammer with piston and bar forged solid. (*See pages* 74 *and* 75.)

[1667]

M'QUEEN, BROTHERS, 184 *Tottenham Court Road.*—Copper-plate printing machine.

[1668]

MUIR, WILLIAM, & CO., *Britannia Works, Manchester.*—Machine tools. (*See pages* 71 *to* 73.)

[1669 .]

NAPIER, DAVID, & SON, 5 *Vine Street, and 51 York Road, Lambeth.*—Letter-press printing machine; machine for forming rifle bullets from cold lead by compression.

[1670]

NASMYTH, JAMES, & CO., *Bridgewater Foundry, Patricroft, near Manchester.*—Differential dividing, punching, and other machines; steam hammers, &c.

[1671]

NASH, R., *Ludgate Hill Passage, Birmingham.*—Presses, lathes, dies, tools, &c.

[1672]

NAYLOR, THOMAS, *Rainhill, near Prescot.*—All kinds of graving tools and broaches.

[1673]

NEILSON, WALTER MONTGOMORIE, *Hyde Park Foundry, Glasgow.*—Radial steam hammer, called a "steam smith."

MUIR, WILLIAM, & Co., *Britannia Works, Manchester*—Machine tools.

Prize Medal awarded at the Exhibition of 1851 ; Prize Medal awarded at the Paris Universal Exposition, 1855 ; Prize Medal awarded from the Society of Arts, 1855.

Fig. 1.—CENTRE DUPLEX LATHE.

Fig.1. MUIR'S SELF-ACTING 12-IN. CENTRE DUPLEX LATHE for sliding and screw cutting. Double-geared headstocks, with wrought-iron steeled mandrel, running in hardened cast-steel conical bearings, guide screw full length of bed, with patent self-acting screw bearers, bed 25 ft. long, with patent duplex slide, carriages with clamp nuts by eccentric, rack and pinion for quick return by hand, the slide rests have two releasing motions for drawing back to tool slides. There is also a new reversing motion for changing from right to left hand, screw cutting, without changing the wheels, of which there are 22. Clements and common drivers, face-plate, backstay, &c. &c. This lathe is specially adapted for cutting screws expeditiously, and for sliding shafting at once going over.

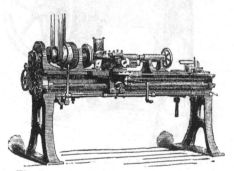

Fig. 2.—CENTRE DOUBLE-GEARED LATHE.

Fig. 2. MUIR'S PATENT 7-IN. CENTRE DOUBLE-GEARED LATHE, for sliding and screw-cutting, wrought-iron steeled mandrel running in hardened cast-steel conical bearings, guide screw full length of bed, which is 6 ft. long, reversing motion to slide and cut screws right or left hand without changing the wheels, of which there are 22, eccentric to lock the cone pulley, eccentric back-shaft, the slide carriage has clamp nut by eccentric, rack and pinion for quick return by hand, releasing motion to tool slide, clements and common drivers, 14-in. face plate, centre chuck backstay, hand rest, &c.

Fig. 3. MUIR'S PATENT 8-IN. FOOT LATHE, with 2 treadles, for screw cutting. Designed particularly for use on board steam vessels, for repairs afloat, or for the colonies, where labour is cheap, as with an assistant a workman will be able to do as much work as with a steam-power lathe of the same capacity.

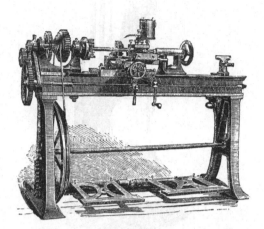

Fig. 3.—MUIR'S 8-IN. FOOT LATHE.

The same lathe is also fitted with four treadles for India, the wages of the natives being so low, it will in many places be found more economical than steam-power.

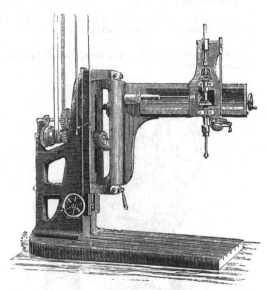

Fig. 4.—SELF-ACTING RADIAL DRILLING MACHINE.

Fig. 4. POWERFUL SELF-ACTING RADIAL DRILLING MACHINE, with vertical elevating slide radial arm, movable through an arc of 190°, to drill holes up to 10 in. diameter.

This machine is particularly adapted for drilling ends of boiler plates, large cylinders, and all work of a massive character, as it will take in an object 9 ft. high ; all holes within range of the machine can be drilled without removing the object.

MUIR, WILLIAM, & CO., *continued.*

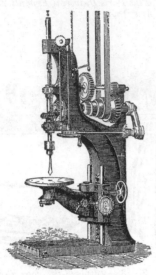

Fig. 5.—MUIR'S VERTICAL DOUBLE-GEARED DRILLING
MACHINE.

FIG. 5. SELF-ACTING VERTICAL DOUBLE-GEARED
DRILLING MACHINE, with circular revolving table on
a radial bracket, which can be raised or lowered on a
vertical slide by means of a worm wheel, so that when
the work is once fixed a hole can be drilled on any
part without moving it.

This drill is provided with a hardened steel locknut,
which entirely prevents any backlash in the spindle.

SMALL BENCH DRILLING MACHINE, to drill to ¾ in. by
hand or power.

Fig. 6.—MUIR'S SELF-ACTING UNIVERSAL SHAPING
MACHINE.

FIG. 6. SELF-ACTING UNIVERSAL SHAPING MACHINE, with
a variable stroke from ½ in. up to 6 in. Will plane an
object 2 ft. long, circular work of 12 in. diameter, and
can be changed to plane round, hollow, or flat surfaces,
without refixing the article operated upon.

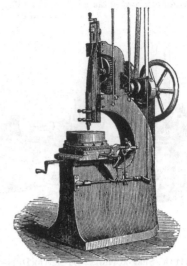

Fig. 7.—MUIR'S SELF-ACTING SLOTTING AND SHAPING
MACHINE.

FIG. 7. SELF-ACTING SLOTTING AND SHAPING MACHINE,
with a variable stroke up to 6 in. Will take in a wheel 3
ft. diameter, self-acting transverse and circular motions.

Fig. 8.—MUIR'S SMALL PLANING MACHINE.

FIG. 8. SMALL PLANING MACHINE, worked by hand or
power, with crank movement and elliptical wheels for
producing uniform motion in cutting, and treble speed
in return of the table.

Fig. 9.—MUIR'S PATENT GRINDSTONE APPARATUS.

MUIR, WILLIAM, & CO., *continued.*

Fig. 9. MUIR'S PATENT GRINDSTONE APPARATUS for grinding edge tools. The stones are regulated by means of a right and left hand screw, and a lateral motion is given to one of them by means of a cam, thus enabling the workmen to grind their tools with a degree of accuracy hitherto impossible, and also doing away with the great dust arising from turning-down stones, so injurious to the bearings of all machinery.

A prize medal was awarded for this machine at the Paris Exposition, 1855, and also by the Society of Arts during the same year.

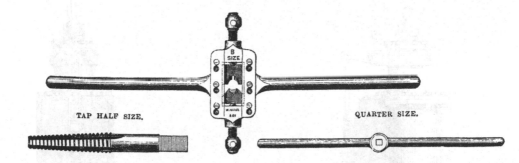

TAP HALF SIZE.   QUARTER SIZE.

Fig. 10.—MUIR'S SCREW STOCK, AND SCREWING TACKLE.

Fig. 10. A COMPLETE SET OF IMPROVED SCREWING TACKLE.

The dies are made so that one will serve as a guide, and the other as a cutter, which can be sharpened on a grindstone. The taps are fluted in a superior form for cutting; the cutting edge is a radial line through section of tap, which is found by experience to take about one-third less power than taps that have hitherto been in use; they are made to standard gauges.

The angle of the thread is 55° for all diameters, rounded both at the top and bottom.

SECTION OF INCH TAP.

Fig. 11.—MUIR'S PATENT COPYING PRESS.

Fig. 11. MUIR'S PATENT COPYING PRESS, with stand and drawers.

Fig. 12.—MUIR'S IMPROVED COPYING PRESS, QUARTO.

Fig. 12. MUIR'S IMPROVED COPYING PRESS, without stand. These presses are designed on the elliptic and screw principle; in quarto, foolscap, and folio sizes.

MORRISON, ROBERT, & CO., *Ouse Burn Engine Works, Newcastle-upon-Tyne.*—Steam hammer, with piston and bar forged solid.

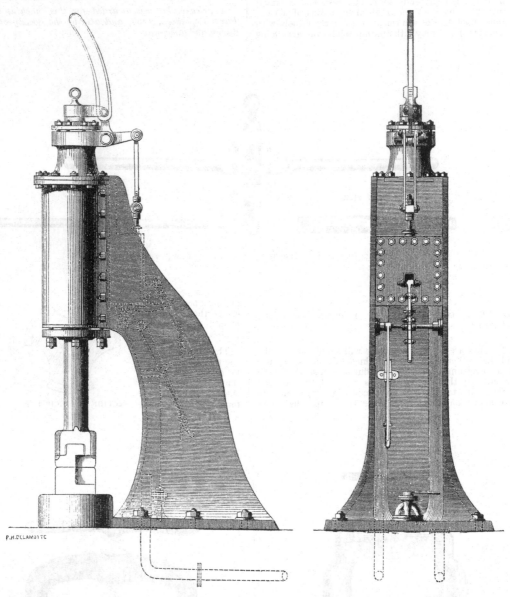

DOUBLE-ACTING STEAM FORGE HAMMER.

ROBERT MORRISON'S PATENT DOUBLE-ACTING STEAM FORGE HAMMER of 20 cwt. with hammer bar and piston forged solid together.

The hammer is in full operation.

The above engraving shows a front and side elevation of a 20 cwt. double-acting forge hammer, and fully details the whole of the gear connected therewith. The steam cylinder is firmly bolted to the single frame, which is made of a box form, the side of the box looking from the cylinder being omitted.

This frame also contains the steam chest, steam passages, steam and exhaust pipes, shown by the dotted lines in the engraving. The hammer bar, an engraving of which is given in the centre of the letter-press, is forged in one solid piece, with the piston and claw for holding the different faces required for various classes of work. Two small steel rings are inserted in grooves turned in the piston, and render it effectively steam tight without the

MORRISON, ROBERT, & CO., *continued.*

introduction of bolts, junk rings, or any additional parts calculated to destroy its solidity and simplicity. Above the piston the bar is planed flat on one side, a corresponding flat being left in the cylinder cover; this keeps the bar and the hammer face constantly in the same position relative to the anvil. On the top of the hammer bar there is a small roller which works in the slot of the slotted lever, shown attached to the cylinder cover; this lever, by means of a pair of links and a slide rod, gives motion to an ordinary box slide for admitting steam above and below the piston, and the slotted link is so shaped,

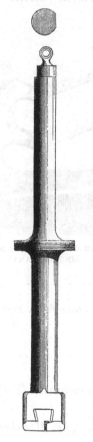

HAMMER BAR.

that equal spaces traversed by the bar at any portion of its stroke, produce correspondingly equal, though smaller motion in the slide. This slide once set requires no further alteration.

The larger of the handles is attached by links to a movable slide face, which can thereby be moved up and down by hand, and regulate not only the length of the stroke but also its height from the anvil, according as the piece to be forged may happen to be thick or thin.

The smaller of the handles is attached to an ordinary stop valve, and is used for either shutting the steam off entirely, or so far reducing its pressure as to strike light blows for swaying or other purposes. The steam pipe from the boiler is fixed to the underside of the frame, and runs up some distance inside the bellmouthed part of the steam passage cast in the frame, which forms a trap for collecting any water that may accumulate in the pipes; a small cock placed at the bottom forms a communication between this tap and the exhaust passage, and can be opened at any time to carry off the water. A similar cock is placed higher up, which opens a communication from the bottom of the cylinder to the exhaust, for the purpose of getting rid of condensed water.

If at any time it should be required to strike a single blow, all that is necessary is to open the stop valve, and raise the lever attached to the movable face, and as the bar rises, suddenly depress it, when a single blow of any degree of intensity can be given, according as the stop valve is more or less open.

The foundation, anvil, and bed plate require no particular description, inasmuch as their form and size must depend on the nature of the ground and situation in which the hammer is required to be placed.

The momentum of the bar in rising and its impact with the forging in its descent, regulate the action of the valve to the greatest nicety. After the delivery of the blow, no more steam is admitted, and as it requires scarcely an eighth of an inch opening of slide to raise the bar while working with heavy blows on hot iron, the full force of the falling of the bar without any check from the steam below is obtained at the commencement, the reduction in thickness of the forging consequent on the blow being sufficient to open the slide to admit the requisite amount of steam to lift the bar, the momentum of which being unchecked in its upward course, opens the slide considerably more in that position, and admits the steam freely on the top of the piston, so that in all cases a very firm and powerful blow is obtained.

[1674]

NEWBERY, RICHARD CHARLES, & CO., 4 & 5 *President Street West, Goswell Road, E.C.*—Patent enamelled cloth collars; machine for making the same.

MACHINE for the purpose of making ladies' and gentlemen's collars and cuffs from the patent enamelled | cloth, stitched, and with button-holes, &c. complete at one operation.

[1675]

NEWTON WILSON, & CO., 144 *High Holborn, London.*—Sewing machines; and patent carpet sweepers. (*See pages* 78 *and* 79.)

[1677]

OATES, JOSEPH PIMLOTT, *Erdington, Birmingham.*—Photograph of machine for making solid bricks, immediately fit for firing.

[1678]

PAGE, E., & CO., *Victoria Iron Works, Bedford.*—Brick and pipe machinery.

[1679]

PALMER, HENRY ROBINSON, 308 *Albany Road, London, S.*—Patent parallel-motion stamping, printing, endorsing, and paging machines.

ENDORSING AND STAMPING MACHINE,
DOUBLE FRAME.

ENDORSING AND STAMPING MACHINE,
SINGLE FRAME.

These machines as used by H. M. Government offices, are applicable for railway companies, bankers, and all firms using stamps for documents, tickets, &c.; also for clothiers, hotels, and any establishment requiring printing on textile fabrics with indelible or oil inks.

Prices, Double Frames.

No. 0	£1	6	0
No. 1	2	12	6

No. 2	£3	3	0
No. 3	3	15	0

Prices, Single Frames.

No. 0	£1	1	0
No. 1	1	15	0
No. 2	2	0	0
No. 3	2	7	6

[1680]

PARKER, W., & SONS, *Northampton.*—Boot and shoe making machine. (*See page* 77.)

[1681]

PATENT FILE MACHINE AND FILE MANUFACTURING COMPANY, THE, *Manchester.*—Self-acting machines for cutting files.

Two of F. Preston's patent machines for cutting files.

[1682]

PEARSON, WILLIAM, & CO., *Leeds.*—Cut-nail machine for headed nails; also various sewing machines.

[1683]

PERRY, THOMAS, & SON, *Highfields, Bilston.*—Chilled or case-hardened rolls for rolling metals.

[1684]

PETO, BRASSEY, & BETTS, *Birkenhead.*—Drilling machine and machine for punching holes at one operation.

PARKER, W., & SONS, *Northampton.*—Boot and shoe making machine.

MACHINE FOR ATTACHING THE SOLES OF BOOTS AND SHOES BY MEANS OF SCREWS, &c. Secured by Her Majesty's royal letters patent.

This simple and ingenious invention, from the facility and ease with which it can be worked, its portable construction, and the great saving of time and labour over the old system, cannot fail to recommend itself.

An inspection of it will fully convince the trade of its great advantages, not only as to economy, but in general usefulness.

Each machine is manufactured under the supervision of the exhibitors by experienced workmen, and is thoroughly tested before being sent out.

It occupies a very small space, can be transported with perfect ease, and worked in any situation commanding sufficient light.

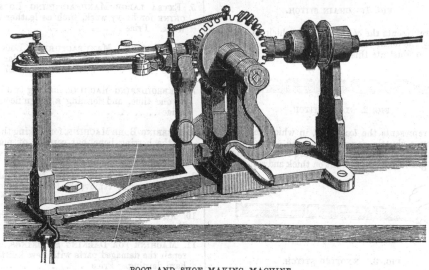

BOOT AND SHOE MAKING MACHINE.

The principal recommendations are—
Its entire simplicity—the use of it being acquired in a few hours by any boy of ordinary intelligence.
Its speed—being worked by hand, treadle, or steam power.
Its durability—being strong and fitted with mechanical precision, it remains in use for years without requiring repair of any kind.

The screw can be varied (by simply changing the die) to suit every description of work, from the lightest to the strongest.

Particulars can be obtained and the machine seen in constant use by applying to the exhibitors.

The proprietors are prepared to treat for the sale of this patent.

[1685]

PETTER & GALPIN, *Belle Sauvage Works, E.C.*—Printing machine. (*See pages* 80 *and* 81.)

[1686]

PINCHES, T. R., & Co., 27 *Oxendon Street, Haymarket, S.W.*—Medal press.

[1687]

PORTER & Co., *Carlisle.*—Patent lozenge and biscuit machine.

[1688]

POWIS, JAMES, & Co., *Victoria Works, Blackfriars Road, London.*—Sawing and wood-cutting machinery and steam engines. (*See pages* 82 *and* 83.)

[1689]

PRENTIS & GARDNER, *Steam Engine and Paper Machine Works, Maidstone, Kent.*—Patent knotter or paper strainer.

[1690]

PRESTON, FRANCIS, & Co., *Manchester.*—Letter-copying machines; embossing presses; bankers' stamping machines.

[1691]

REYNOLDS, J. G., 33 *Wharf Road, City Road, London.*—Machine for making tobacco-pipes.

Newton Wilson, & Co., 144 *High Holborn, London.*—Family and manufacturing sewing machines, and patent carpet sweepers. (Agents to the Grover and Baker Sewing Machine Company of Boston, U.S.A.)

The machines comprise the whole of the stitches at present known, of which the following are the leading:—

FIG. 1. CHAIN STITCH.

Fig. 1 represents the *chain stitch*, in which the loop is secured by the needle at the succeeding stitch, a part is left slack to illustrate this more clearly.

FIG. 2. LOCK STITCH.

Fig. 2 represents the *lock stitch*, in which the loop is secured by a shuttle carrying a second thread passing through it, the right extremity showing its appearance on hard and glazed fabrics; the left on thick and soft fabrics.

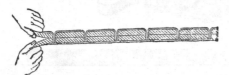

FIG. 3. KNOTTED STITCH.

Fig. 3 represents the *knotted stitch*, in which the loop is secured by an instrument carrying a second thread which enters the needle loop, leaving its own loop there, this second loop is then held open till the needle at its next descent enters it and the two loops are then drawn tight together. Fig. 3 shows the character of this stitch, and how securely even the last stitch is fastened by the action of the machine itself.

1. A large Manufacturing Machine of the last-described stitch, suitable for sewing artisan clothing, tents, &c. Price, complete, with stand . . £15 15
   In N. W. & Co.'s Catalogue, see Nos. 2 and 28.

2. Medium-size Machine, suitable for corsets, cloth caps, and upholstering. Price £13 13
   See Nos. 3 and 27.

3. Manufacturing Machine, Lock Stitch, for tailors, shoe makers, and stay makers. Price . . . £10 10
   See No. 5 in N. W. & Co.'s Catalogue.

FIG. 4. LOCK-STITCH MACHINE.

4. New style Manufacturing Lock-Stitch Machine, possessing the following features:—Great range of work, fine or coarse, simplicity, strength, durability, speed, quietness; adapted particularly to tailoring, boot making, and all trades where a great variety of work is done. Prices . . . . . £13 13s. and £14 14

5. Extra large Manufacturing Lock-Stitch Machine for heavy work, such as leather traces, harness work. Price £18 18

6. Small light Manufacturing Lock-Stitch Machine, new style, for shirt and collar work, &c. &c. Price . . . . £7 7s. and £9 9

7. Embroidering Machine, making two lines of sewing at one time, and forming a magnificent embroidery. Price £16 16

8. Herring-Bone Machine, for making the stitch known as the herring bone, but capable of plain work also. Price £20 0

9. Button-Hole Machine for tailors—all the button-hole apparatus removable for adaptation to general work. Price £25 0

10. Fine Button-Hole Machine for shirt work, &c. Price £20 0

11. Machine for Darning Stockings, constructed to repair the damaged parts with new knitting instead of hard darning. Price £15 15

Small ditto for family use £2 10

FIG. 5.
MACHINE FOR STITCHING SHOE SOLES ON TO THE UPPERS.

12. The Machine represented in fig. 5 for stitching shoe soles on to the uppers without the intervention of welts. The machine uses a waxed thread, making a perfectly flat seam, and completing the sewing of a pair of boots in three minutes.

This machine will be exhibited in operation in the Exhibition at a particular hour each day. The time may be ascertained from the attendants.

NEWTON WILSON, & CO., *continued.*

NEWTON WILSON, AND CO.'S FAMILY MACHINES.

FIG. 6. MEDIUM-SIZE KNOTTED-STITCH MACHINE.

FIG. 7. KNOTTED-STITCH MACHINE.

FIG. 8. CABINET AND MACHINE.

13. Fig. 6 represents a MEDIUM SIZE AND QUALITY KNOTTED-STITCH MACHINE, suitable for families and dressmakers, combining simplicity, speed, lightness, quietness, and great range of application, with the most perfect elasticity of stitch £11 11

14. LOCK-STITCH MACHINE, to fit the same stand as the last, and in same style £11 11

15. Fig. 7. Best style KNOTTED-STITCH FAMILY MACHINE,

on very elegant table and stand. Price, 16 to 20 guineas.

16. Fig. 8. Magnificent CABINET, in buhl and gold, with best machine.

17. New style FAMILY LOCK-STITCH MACHINE, highest finish.

18. New style FAMILY KNOTTED-STITCH MACHINE, highest finish.

FIG. 9. BOUDOIR MACHINE.

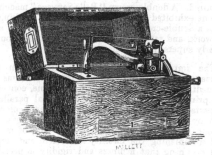

FIG. 10. MACHINE IN WORK-BOX.

19. Fig. 9. BOUDOIR MACHINE, knotted-stitch, runs both ways, machine detaches from driving arrangements by the act of closing. Price . . . £14 14

20. Fig. 10. MACHINE IN WORKBOX—a form perfectly portable and compact for travelling or exportation, knotted stitch. Price, £12 12s. to . . . £14 14

21. MACHINE FOR LIGHT WORK, chain stitch, including stand £5 5

22. Series of GUIDES and APPARATUS for facilitating the operations of the different machines, comprising hemming, felling, tucking, binding, cording, embroidering, and braiding above and below, or both at the same moment. The whole of these results being produced by machines Nos. 13, 15, 16, 18.

23. TABLEAU, showing the different operations of a single machine, knotted stitch. The whole of the work forming this magnificent tableau has been executed by machine No. 24 in Newton Wilson, and Co.'s catalogue (see fig. 6).

24. Three magnificent WAX MODELS of children, boy, girl, and infant, dressed in garments made entirely by these sewing machines, the infant seated in Messrs. Newton Wilson, & Co.'s patent chair.

25. Sample garments of different kinds, in show case, illustrating the application of the different machines.

26. PATENT CARPET SWEEPER, taking up the dust, &c. from carpets without damping, kneeling, or dusting. Applicable to all kinds of carpets. Price . 12s. 6d.

27. Carpet sweeper, with self-adjusting arrangement to brush. Price . . . . 15s. 0d.

28. Carpet sweeper, larger size ditto. Price . 20s. 0d.

29. Carpet sweeper, noiseless in action, for sick rooms. Price . . . . 18s. 0d.

30. Carpet sweeper, large size, ditto. Price . 24s. 0d.

The brushes can be renewed at the cost of ordinary brooms.

PETTER & GALPIN, *Belle Sauvage Works, E.C.*—Printing machine.

PETTER AND GALPIN'S DOUBLE PATENT NOISELESS "BELLE SAUVAGE" PRINTING MACHINE, as supplied to Her Majesty's Government, with S. BREMNER'S patented improvements, new design and registered framework; the simplest and best news, book, and general jobbing machine of the day, adapted to foot, hand, or steam power.

Messrs. Petter and Galpin's patent "BELLE SAUVAGE" machine has been entirely remodelled (see accompanying engraving), and the whole of Mr. Bremner's recently patented improvements added, for which new patterns have been made to an original design, registered according to act of parliament. The utmost attention has been paid to every detail, so as to render the "Belle Sauvage" machine simple and strong in construction, noiseless in working, and light and easy to turn by hand, combining all the facility of the hand-press with superior productive powers, both as regards speed and economy. It is unquestionably the most perfect, useful, and easily worked machine ever introduced for newspapers, book-work, and first-class jobbing, ruled headings, broadsides, &c.; and the facility with which it can be changed from one class of work to another, together with the little attention it requires beyond that of an ordinary hand-press, renders it the desideratum of the jobbing office.

The machine occupies but little space, is highly finished, and though sufficiently light to admit of its being erected in a press-room, is strong, powerful, and well-built; its mechanical arrangements and working parts are exceedingly simple, free from unnecessary noise and friction, and may be easily understood by an ordinary pressman; it requires little or no making ready, and can be driven with ease by hand, or by steam power, at the rate of from 1,000 to 2,000 impressions per hour, according to size of machine, capacity of layer-on, and class of work required. A double-crown "Belle Sauvage" machine, size of one exhibited Class 7 B, occupies less working space than a double-crown press, will do 6 times the amount of work, and save 75 per cent. in cost and labour, thus clearly superseding the hand-press.

The improved reciprocating motion given to the carriage, by means of compound levers or beams placed immediately in the centre of the machine, connected to the printing-table by well-fitted horizontal parallel rods, imparts to it a perfectly steady and even movement, thereby materially lessening the noise and heavy bodily labour which usually attend the working of machines, and diminishing the liability to accident and stoppage; and attaining that firmness and rigidity so necessary to good printing.

The cylinder being made to rest while the white sheet is taken and the printed one delivered (printed side upwards), ample time is afforded to lay the sheet correctly up to the register gauge; when if, by accident, the sheet (through its being too damp, sticking together, or having the corners turned down) has not been laid up in time, the layer-on has it in his power to stop the cylinder, without stopping the machine, and so prevent the blanket from being inked. By this means also he is enabled, before printing the sheet, to ink the forme two or more times at pleasure, in cases of posters, or other heavy solid formes, where more than an ordinary charge of colour is required. The grippers take the sheet while the cylinder is at rest (a point essential to good register), by means of a patent improved horizontal register gauge, which is capable of being adjusted to any given margin, whereby the necessity for changing the position of the forme is entirely obviated. The register gauge is carefully arranged in front of the laying-on board; attached to the gripper-bar, and closes with the grippers, thus securing perfect register without points. The usual pointing apparatus, however, can be attached to the machine.

TESTIMONIALS.

ON HER MAJESTY'S SERVICE.

"*Royal Laboratory, Woolwich Arsenal,*
"*4th April,* 1862.

"GENTLEMEN,—In reply to your letter of the 3rd February last, asking for my opinion of the merits of the Patent 'Belle Sauvage' printing machine, I beg to say that the two machines of the above description in use in this department, one of which has been at work for fifteen months, have given entire satisfaction. Their simplicity and economy, added to their power of producing large quantities of work correctly and expeditiously, render them very valuable.

"I am, gentlemen, your obedient servant,
(Signed) "E. W. BOXER,
"Superintendent Royal Laboratory.

"Messrs. Petter and Galpin, Belle Sauvage Works, Ludgate Hill, London, E.C."

FROM THE "DOVER EXPRESS," KENT.

"*Express Office, Dover, Feb. 18th,* 1862.
"GENTLEMEN,—I have much pleasure in informing you that the Improved Double Patent 'Belle Sauvage' Machine (Double Super-Royal) I have recently purchased of you answers in every respect, and to the fullest degree, my expectations. It works beautifully; runs with great ease; and may be said to work without noise, so little is there.

"The recent improvements, patented by Mr. BREMNER, and of which this machine affords an admirable specimen, are of the greatest service, especially to those who, like myself, have not been accustomed to machine work.

"I am, Gentlemen, yours truly,
(Signed) "JOSEPH T. FRIEND.

"Messrs. Petter and Galpin."

EXTRACTS FROM TESTIMONIALS.

"It is light and easy to turn by hand....A boy has no difficulty in working it at the rate of 900 to 1,000 per hour.....It prints well, with rapidity and ease, good register being easily obtained....We are perfectly satisfied with the admirable manner and ease with which she works....For commercial work, I cannot speak of it in terms too favourable....It works very easily; its register is perfect....I can strongly recommend it, too, for its simplicity of construction, non-liability to get out of order, and easy working....We work it by hand at about 1,000 per hour, but have worked it considerably above that rate....A lad of 15 or 16 years of age can turn it without difficulty....Just come out of the machine room (9.0 P.M.), leaving the "Belle Sauvage" (No. 5) working away to the tune of 1,000 an hour, apparently without any effort; and scarcely any noise can be heard; if you had to see it, I am sure you would be astounded.....Printers who are unacquainted with it in practice, can form no adequate idea of its usefulness."

Nearly 200 "Belle Sauvage" machines now in use.

The machine may be viewed in operation, testimonials seen, references given, illustrated prospectuses, with sizes, prices, and full particulars of the machine obtained, upon application to the proprietors and sole manufacturers, Messrs. Petter and Galpin, Printers, Engineers, and Machinists, Belle Sauvage Works, Ludgate Hill, London, E.C., and at the Great International Exhibition.

PETTER & GALPIN, *continued.*

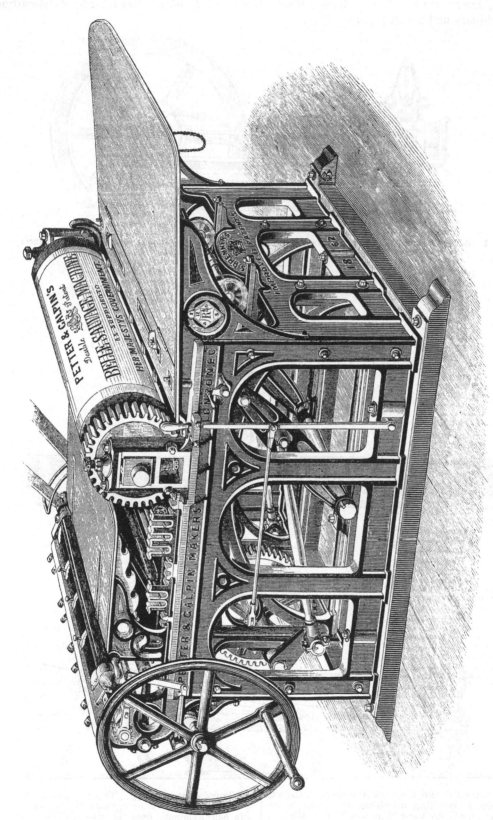

PETTER AND GALPIN'S DOUBLE PATENT " BELLE SAUVAGE " PRINTING MACHINE, WITH S. BREMNER'S PATENTED IMPROVEMENTS, NEW DESIGN AND REGISTERED FRAMEWORK.

POWIS, JAMES, & CO., *Victoria Works, Blackfriars Road, London.*—Sawing and wood-cutting machinery and steam engines.

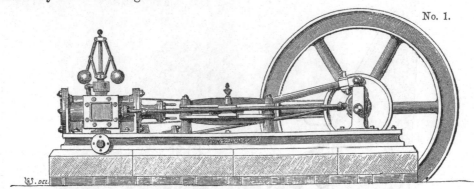

HORIZONTAL ENGINE.

No. 1. 6-horse power HORIZONTAL ENGINE, with all the recent improvements, combining compactness and extreme portability in case of removal.

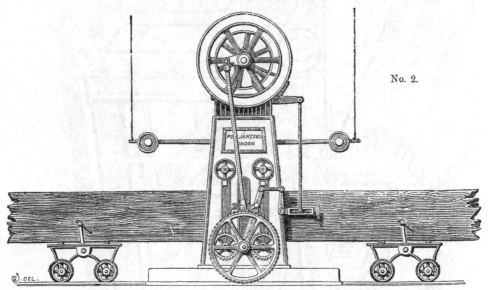

COMBINED TIMBER AND DEAL FRAME.

2. COMBINED TIMBER AND DEAL FRAME, for 24-in. logs, or two deals 24 by 7 in. | This is a small mill of itself, and one that no builder should be without.

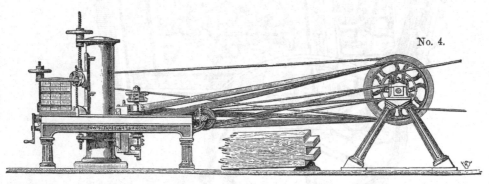

3. CONTRACTORS' AND BUILDERS' COMBINED MACHINE, for planing, moulding, and edging all four sides at one operation, any size under 12 in. wide by 6 in. thick.
4. Self-acting double or single TENONING MACHINE, for railway-carriage framing. This machine operates on four waggon soles at once, and completes four tenons in half a minute from the time the cutter strikes the wood.

POWIS, JAMES, & CO., *continued.*

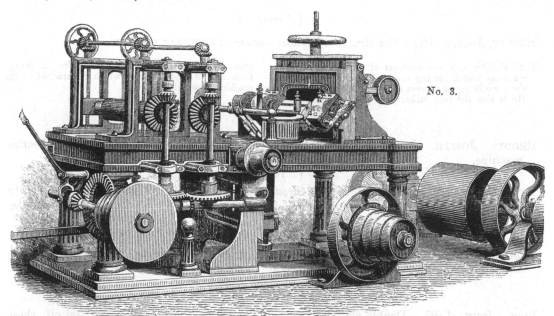

No. 3.

5. DOUBLE-DEAL FRAME, to cut 2 deals 14 by 4 in.

Most advantageous where mills are by the side of tidal rivers, as it is so constructed that no expensive or deep foundations are required.

6. COMBINED MOULDING, THICKNESSING, AND SQUARING-UP MACHINE, for carriage framing, door styles, &c.

7. IMPROVED BAND SAWING MACHINE, with Powis, James, & Co.'s patented adjustment for regulating the tension of saws, thereby preventing breakage.

8. SELF-ACTING CIRCULAR SAW BENCH, capable of breaking up logs 20 in. diameter, or cutting deals; made with or without bogies.

An excellent machine for colonists, where strength and portability are required.

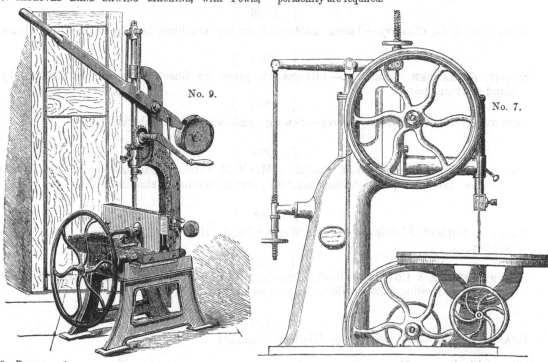

No. 9.

No. 7.

9. PATENT COMBINED MORTISING, TENONING, AND BORING MACHINE, for hard or soft wood.

This compact, strong, and useful tool is capable of doing the work of eight men. As a proof of its appreciation, three thousand are now in use.

10. MULTUM IN PARVO, or the general joiner.

This machine will saw, plough, groove, rebate, thickness, bore, cross-cut, and strike mouldings.

11. IMPROVED TENONING MACHINE, for joiners and the small work of railway-carriage builders.

12. STEAM MORTISING MACHINE, for joiners and railway shops.

The exhibitors are patentees and manufacturers of all kinds of wood-cutting and sawing machinery, portable and fixed steam engines, &c.

All letters and applications for drawings and prices should be addressed 26, Watling Street, E.C.

[1692]

RHODES, JOSEPH, *Hope Foundry, Morley, near Leeds.*—Patent rag machine.

The exhibitor is a manufacturer of woollen machinery, slubbing horses, piecing machines, tenterhook teazers, shake woolleys, rag shakers, shoddy and mungo machines. He is also the sole maker of a patent improvement applicable to mungo machines, for freeing the mungo from bits of cloth, one of which is attached to the machine now exhibited.

[1693]

RHODES, JOSEPH, *Grove Works, Wakefield.*—Steam hammer; punching and shearing machine.

[1694]

ROBERTS, RICHARD, & Co., 10 *Adam Street, Adelphi.*—Drawings of jacquard punching machine, and angle-iron punching machine.

[1695]

ROBINSON, THOMAS, & SON, *Rochdale.*—Sawing, planing, moulding, mortising, tenoning, and sharpening machines for working wood. (*See page* 88.)

[1966]

ROSS, JOHN, *Leith.*—Double-cylinder printing machine, with self-acting set-off sheet apparatus.

[1697]

RYDER, WILLIAM, *Bolton, Lancashire.*—Ryder's forging machine for rollers, spindles, bolts, studs, &c. ; patent machine for fluting rollers for cotton machinery.

[1698]

SALISBURY, S. C., *Coventry.*—Patent knot-stitch sewing machine, simple, durable, and cheap.

[1699]

SEGGIE, ALEXANDER, *Edinburgh.*—Lithographic press for finest work, can be wrought by hand or steam power.

[1700]

SERVICE, WILLIAM, *Mitcham, Surrey.*—Sewing machines with double-feed action.

[1701]

SHANKS & Co., 6 *Robert Street, Adelphi.*—Manifold drilling machine; mortising drilling machine; frictional guard drilling machine; steam hammer; shaping machine.

[1702]

SHARP & BULMER, *Middlesbro'.*—Little Wonder hand brick or tile machine; 5,000 bricks per day.

[1703]

SHARP, STEWART, & Co., *Atlas Works, Manchester.*—Workshop tools, wheel lathe, Giffard's injectors for feeding steam boilers. (*See pages* 86 *and* 87.)

[1704]

SHARRATT & NEWTH, *Clerkenwell.*—Glaziers' diamonds.

MACHINE FOR CUTTING OVALS.

BEAM COMPASS FOR CIRCLES.

SHARRATT & NEWTH, *continued.*

MACHINE FOR CUTTING CIRCLES.

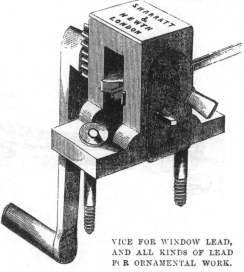

VICE FOR WINDOW LEAD,
AND ALL KINDS OF LEAD
FOR ORNAMENTAL WORK.

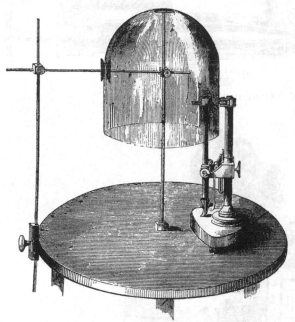

MACHINE FOR CUTTING ROUND AND SQUARE SHADES,
WITH TABLE AND FIXING APPARATUS.

MOULDS FOR CASTING LEAD.

GAUGE DIAMOND FOR SHADES.

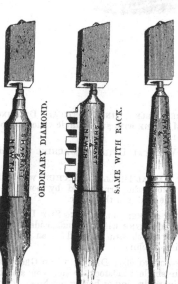

ORDINARY DIAMOND.

SAME WITH RACK.

DIAMOND MOUNTED IN SILVER WITH IVORY HANDLE.

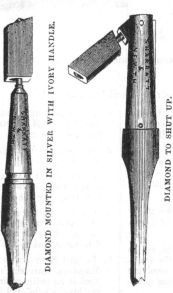

DIAMOND TO SHUT UP.

POCKET DIAMOND WITH PENCIL, LENGTH 4¼ INCHES.

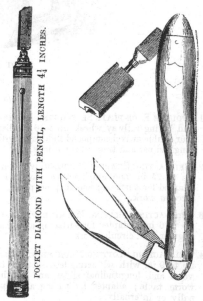

POCKET DIAMOND WITH PENKNIFE.

SHARP, STEWART, & Co., *Atlas Works, Manchester.*—Workshop tools, wheel lathe, Giffard's injectors for feeding steam boilers.

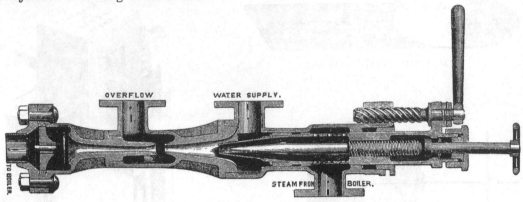

OVERFLOW   WATER SUPPLY.

TO BOILER.

STEAM FROM BOILER.

GIFFARD'S PATENT INJECTOR.

SLOT DRILLING AND GROOVING MACHINE.

Obtained the Council Medal in 1851.

1. DOUBLE FACE-PLATE SELF-ACTING LATHE, for turning and boring railway wheels up to 6 ft. diameter, with four double-swivel compound slide rests, with self-acting longitudinal and transverse motions.

2. SELF-ACTING SHAPING MACHINE, with variable stroke up to 25 in. and quick return motion for the tool; adapted for surfacing and shaping plane, angular, and circular work.

3. SELF-ACTING SHAPING MACHINE with variable stroke up to 6 in.; adapted for surfacing and shaping plane, angular, and circular work.

4. SELF-ACTING SLOTTING MACHINE with variable stroke up to 9 in. with self-acting feed motions to the transverse and longitudinal slides and to the revolving worm table; adapted for paring and shaping externally or internally.

5. DOUBLE-GEARED SELF-ACTING VERTICAL DRILLING AND BORING MACHINE with swivel table which can be raised by power.

6. SELF-ACTING VERTICAL DRILLING MACHINE with adjustable table.

7. INDEPENDENT RADIAL DRILLING MACHINE with self-acting feed motion, the arm raised by power and radiating from the centre of the pillar.

8. PATENT DOUBLE-GEARED SELF-ACTING SLOT DRILLING AND GROOVING MACHINE with two independent headstocks, either of which can be used as an ordinary drilling or boring machine.

9. PATENT SELF-ACTING SLOT DRILLING AND GROOVING MACHINE with single headstock, which can also be used as an ordinary drilling or boring machine. (This machine is exhibited in two sizes.)

SHARP, STEWART, & CO., *continued.*

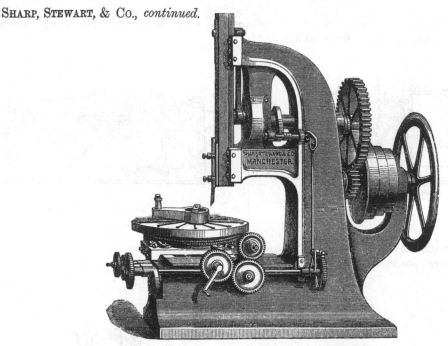

4. SELF-ACTING SLOTTING MACHINE.

11. MACHINE FOR PUNCHING HOLES.

10. PATENT DOUBLE-GEARED SLOT DRILLING AND BORING MACHINE, especially adapted for heavy, stationary, or marine work.

11. MACHINE FOR PUNCHING HOLES up to 1¼ diameter in 1¼ plates, with apparatus for disengaging the punch, and for shearing 1¼ plates; the shearing slide placed at an angle so as to cut off bars of any length.

12. SELLER'S PATENT SELF-ACTING BOLT AND NUT SCREWING MACHINE to screw up to 2 in. diameter. The bolt is withdrawn without stopping or reversing the machine.

13. SELLER'S PATENT SELF-ACTING BOLT AND NUT SCREWING MACHINE to screw up to 1 in. diameter.

14. PATENT SELF-ACTING MACHINE for winding cotton, linen, or silk sewing thread upon reels.

ROBINSON, THOMAS, & SON, *Rochdale.*—Sawing, planing, moulding, mortising, tenoning, and sharpening machines for working wood.

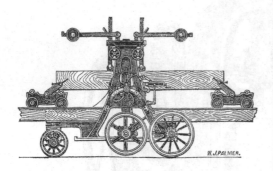

PATENT PORTABLE FRAME for sawing trees, logs, and deals into boards, planks, or scantlings.

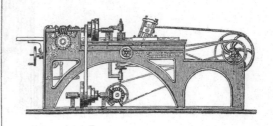

PATENT MOULDING AND PLANING MACHINE for planing floor and match boards, and working mouldings.

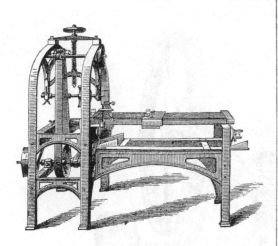

IMPROVED TENONING MACHINE for cutting single and double tenons, scribing sashes, trenching, and grooving.

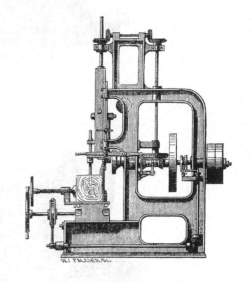

IMPROVED MORTISING MACHINE with self-acting variable descending and ascending motion to chisel.

LIST OF MACHINES MANUFACTURED BY T. ROBINSON & SON :—

Circular saw benches.	Moulding ditto.
Timber frames.	Squaring ditto.
Deal ditto.	Tenoning ditto.
Portable log and deal frames.	Veneer sawing machine.
	Endless band sawing do.
Planing machine.	Sweep ditto.

Mortising ditto.	Rifle-stock ditto.
Boring ditto.	Wood-turning lathes.
Wheel-spoke ditto.	

Complete sets of machinery designed and arranged for contractors, ship builders, carriage and waggon works, saw mills, planing and moulding mills, railway companies, Government arsensals and dock yards.

London Office, Unity Buildings, 8, Cannon Street, E.C

[1705]

SHEPHERD, HILL, & CO., *Union Foundry, Leeds.*—Machinery.

[1706]

SIEBE, AUGUSTUS, 5 *Denmark Street, Soho.*—Paper-knotting machine.

[1707]

SIEMENS, HALSKE, & CO., 3 *Great George Street, Westminster.*—India-rubber covering machines; submarine cable of new construction.

[1708]

SIMPSON, R. E. & CO., *Glasgow.*—Patent American single and double action shuttle sewing machines, with all the latest improvements.

[1709]

SINCLAIR, JOHN, 541 *Castle Hill, Edinburgh.*—Ornamented laid dandy roll for watermark on paper.

[1710]

SINIBALDI, MADAME C., *London.*—Chain machine, cranks, pistons, printing press, &c. (*See page* 90.)

[1711]

SMITH, ARCHIBALD, *Princes Street, Leicester Square, W.*—Patent machinery for making submarine cables and wire ropes.

[1712]

SMITH & COVENTRY, *Ordsal Lane, Manchester.*—Radial drill, improved screwing lathe, and other tools for cutting metals.

[1713]

SMITH & CO., *Marsh Gate Lane, Stratford.*—Patent machine for thrashing corn without injuring the straw.

[1714]

SMITH & HAWKES, *Eagle Foundry, Birmingham.*—Chilled cast rolls, tested iron, and bricks; testing machine, diagrams.

[1715]

SMITH, BEACOCK, & TANNETT, *Victoria Foundry, Leeds.*—Self-acting machine tools for shaping, slotting, turning, and rifling.

[1716]

SMITH, CHARLES, 30 *White Street*, late 140 *York Street, Hulme, Manchester.*—Grocer's patent soap-cutting machine.

[1717]

SMITH, EDWIN, *Cemetery Road, Sheffield.*—Pointing and carving machine for objects in the round and relievo.

SMITH'S PATENT POINTING AND CARVING MACHINE, for the use of sculptors and carvers in general, is capable of producing in marble, stone, wood, &c. statuary, busts, and other ornamental objects, in the round, and in relievo.

The combined advantages of this machine, are facility of execution and unerring accuracy.

[1718]

SMITH, JAMES, & CO., *Crown Court, Crown Street, Finsbury.*—The "English" continuous-motion shuttle sewing machine, simple, easy, durable, and cheap.

This machine, which is the first and only one on this principle, was invented and patented in this country. It comprises all the latest improvements; it is simple, durable, noiseless, and rapid; and its working may be easily learned.

Prices, on stand complete, £8, £10, £12.

[1719]

SMITH, JOHN, *Crorton, near Prescot.*—Nippers and pliers, and an assortment of Lancashire joint tools.

SINIBALDI, MADAME CELESTE, 5, *Albert Terrace, Notting Hill.*—Chain machine, cranks, pistons, axles, plates, printing press, bolts, arms, screws, brazed bath.

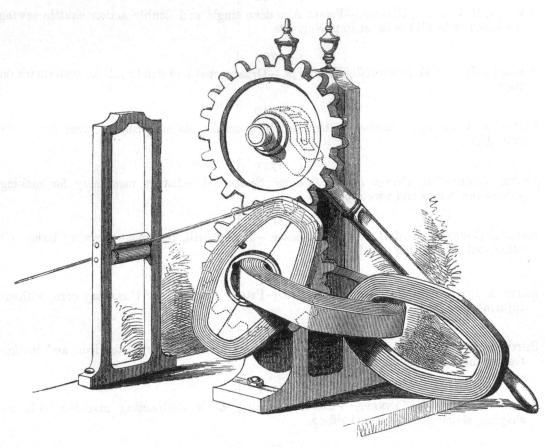

CHAIN MACHINE.

The accompanying woodcut shows this machine performing the double operation of making a link and joining it to one already formed. One end of the band of metal is held in the notch of the mould on which it is wound. The mould and the cog-wheel on which it is fixed are made in valves so as to open and admit and retain the links as they are formed. Two ribbons of metal may be used in making links by this process, one of iron and one of steel. This patent is the property of the Duke of Buccleugh.

[1720]

SMITH, JOHN & SON, 8 *Upper Fountain Place, City Road, E.C.*—Model moulds and rollers used in paper-making ; watermarks.

[1721]

SPELLER, WILLIAM, 14 & 15 *York Street, Blackfriars Road, S.E.*—Artesian-well boring tools and pump works.

[1722]

STEVENS' PATENT BREAD MACHINERY COMPANY, 10, *Old Jewry Chambers, London.*— Machinery for kneading dough, dispensing with the dirty hand-and-arm process. (*See page 93.*)

[1723]

STONE, JOSIAH, *Deptford, London.*—Machine for making 1000 cast-metal nails at one time.

[1724]

STOTHERT & PITT, *Bath.*—Machine for striking or scraping leather hides; model machine for rolling leather. (*See page 94.*)

[1725]

SWEET, A., 20 *St. James's Place, Hampstead Road, N.W.*—Case of challenge locks.

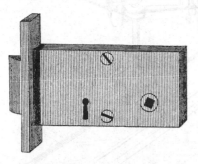 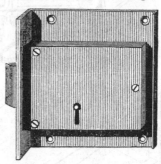

As these locks are new inventions, and as yet untested, the exhibitor refrains from commenting on their merits. At the same time he confidently challenges the inquiries and experiments of the most scientific locksmiths, and offers a premium of ten guineas to any person who shall open any one lock in less time than that specified upon it. Thus, the time given to open No. 3, which is a mortise letter lock, is from the opening to the close of the Exhibition. For No. 2 (a night-latch) the time allowed is 8 consecutive hours. The price of the former lock is £2, and of the latter 10s.

The exhibitor believes that in point of cheapness, security, and durability, these locks possess merits worthy of attention.

[1726]

THOMAS, W. F., & CO., 1 *Cheapside, and* 66 *Newgate Street, London.*—Sewing machines, and samples of work produced by them. (*See page 95.*)

[1727]

THOMPSON, ROBERT HENRY, *Her Majesty's Dockyard, Woolwich.*—Machine for joiners' purposes; horizontal sawing machine; tree-felling machine. (*See page 92.*)

[1728]

THWAITES & CARBUTT, *Vulcan Iron Works, Bradford, Yorkshire.*—Steam hammers and engineers' tools.

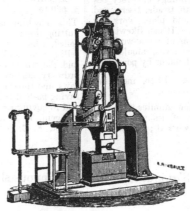

PATENT DOUBLE-ACTION SELF-ACTING STEAM HAMMER.

A 7-cwt. patent double-action self-acting STEAM HAMMER.
A 4-cwt. double-action SINGLE STANDARD HAMMER.
Pillar radial DRILLING MACHINE.
A 6-in. centre SLIDE AND SCREW CUTTING LATHE, with compound slide rest.

A ratchet DRILL-STAND MODEL.
Model of a very powerful PLANING MACHINE.
A 10-in. centre DOUBLE-GEARED SLIDE LATHE, with compound slide rest.

THOMPSON, ROBERT HENRY, *Her Majesty's Dockyard, Woolwich.*—Machine for joiner's purposes; horizontal sawing machine; tree-felling machine.

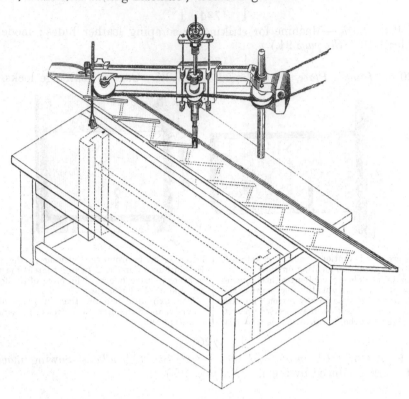

THOMPSON'S PATENT UNIVERSAL JOINER.

THOMPSON'S PATENT UNIVERSAL JOINER. Price, royalty included. £50 0

This is the most complete, simple, and inexpensive machine of its kind as yet invented. It prepares every description of joiner's work, including gothic heads, elliptic and all other curves; mouldings of every description; the strings of stairs, with treads, risers, and handrails; and also ornamental and plain work for cabinet-makers and coach-builders. It can likewise be used in masonry for the preparation of stone-work for windows, &c. whether straight or curved, moulded or plain. It can be worked either by hand or by power.

THOMPSON'S PATENT TREE-FELLER. Price, including royalty £40 0

This machine can be erected in a few minutes upon any land, irrespectively of the nature of its surface. It requires but little skill or power to work it; can be used in felling trees of any size. From the rapidity of its action, and simplicity of construction, will be found of great service to all timber dealers, and of especial value to colonists.

THOMPSON'S PATENT PORTABLE HORIZONTAL SAWING MACHINE. The price, including royalty, varies from £60 to £80 0

This machine is well fitted for siding trees, or for cutting them into planks, &c. on the spot where they are felled, whatever be the formation of the ground. The felling-gear can be adapted to form part of this machine, and the whole can be put together and worked by any ordinary labourer. From the very rapid motion of the saws, and the small degree of power necessary, an immense saving both of time and money will be found in converting and removing timber.

Agent, T. Meacham, 2 New London Street, Fenchurch Street, E. C.

[1729]

TIDCOMBE, GEORGE, & SON, *Watford, Herts.*—A continuous-sheet paper-cutting machine.

[1730]

ULLMER, F. & W., *Castle Street, Holborn.*—Patent cylindrical printing machine, patent diagonal paper-cutting machine.

[1731]

VERO, JAMES, *Atherstone.*—Machine to remove fur and wool from skins preparatory to leather making.

[1732]

VICARS, T. & T., & Co., *Wheatsheaf Foundry, Liverpool.*—Bread and biscuit machinery.

STEVENS' PATENT BREAD MACHINERY COMPANY, LIMITED, 10 *Old Jewry Chambers, London.*—Machinery for kneading dough, dispensing with the dirty hand-and-arm process.

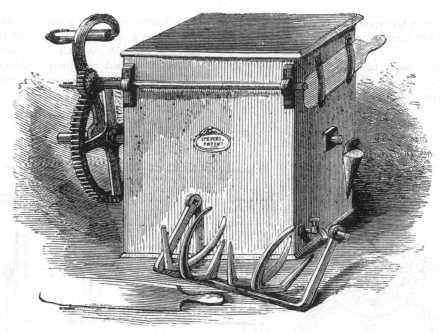

FAMILY BREAD MACHINE.

E. STEVENS' BREAD-MAKING MACHINES AND UNIVERSAL CAKE AND PUDDING MIXERS.

Patented in Great Britain, France, and Belgium.

This invaluable invention is alike suited for the use of private families and the largest public establishments. It has already been successfully adopted by Government, East India Council, several public institutions, bakers, as well as in private families. It ensures pure and superior clean bread, and repays its cost shortly. It is produced in sizes to mix from one quartern of flour to five sacks at one time, and is applicable for making every kind of bread.

Prices :

Family machines range from 35s. to £5 each, the former of which will mix at one time from 2 to 8 2-lb. loaves, while the latter will make from 15 to 30, with intermediate sizes. Machines suitable for public institutions, and bakers, range from £10 to £100 each; a trade machine capable of mixing 2 sacks of flour at one time, may be had as low as £30. Illustrated catalogues free of charge.

These machines are all made of the best materials, occupy but little space, are readily understood, and, owing to the simplicity of their construction, seldom or never get out of order.

The following are specimens of the numerous testimonials received:—

The Most Noble the Marquis of Sligo writes :—

"I have had your bread machine in use for the supply of my house for nearly four months, and I can most strongly recommend it; indeed, I have done so more than once to visitors, who left my house intending to procure one. It saves two-thirds of the labour of kneading, and enables any servant in the house to do the work; and I most strongly recommend it to every baking establishment on either a large or a small scale."

The Right Hon. Lord Camoys writes :—

"You are fully at liberty to say and publish, that I have one of your machines for making bread, and that I much approve of it."

Lieut.-Col. Colvill, Governor of the House of Correction, Cold Bath Fields, writes:—

"I am desired by the visiting judges to inform you that they are perfectly satisfied with the bread-making machinery which you have supplied to this establishment. The average consumption of flour daily here is ten sacks; the saving has been about 1s. 6d. per sack, or £4 7s. 6d. per week. The machine has been in constant use forty-six weeks, and in that time we have saved by the machine, £207. The bread is also much better; the cleanliness of the manufacture is admirable; it is a much healthier labour for the men, and the machine can be worked by any of the prisoners."

Deputy Commissary-General Robinson, of Aldershot Camp, writes:—

"Stevens' dough-making machine performs better in 20 minutes what occupies 45 by manual labour; and it has been proved to gain 12 lbs. of bread per sack of flour over what can be obtained by hand labour; the machine thus paying its own cost in a very short time."

Mr. M'Cash, master baker, of Stratford, London, writes:—

"I am perfectly satisfied with the whole operation of your dough-making machine. I believe the time is not far distant when the machine will be considered a necessity in all bakehouses, on account of its economy, and being alike a boon to master and man."

Mr. S. Shelton, Peterborough, writes:—

"I am a baker of thirty years' standing, and I confidently believe that no invention has ever given more benefits to the working man in any trade than your machine has to ours."

For further testimonials from noblemen, gentlemen, physicians, proprietors and heads of public establishments, master bakers, cooks, confectioners, &c. and opinions of the press, see the trade prospectus, which may be obtained by applying at the offices of the Company.

Stothert & Pitt, *Bath.*—Machine for striking or scraping leather hides. Model machine for rolling leather.

Cox's Patent Machine with Pitt's Patent Improvements for Striking or Scraping Hides.

The hide, being laid upon the lower wood roller, is gradually allowed to pass beneath the upper roller, which carries a sharp-edged spiral knife. The lower roller, being supported on springs, maintains a uniform but yielding pressure, and adapts itself to the varying thickness of the hide; the knife in the mean time scraping out the bloom most completely in the space of about three minutes. The machine is also used for rubbing down foreign shoulders, and for dressing offal.

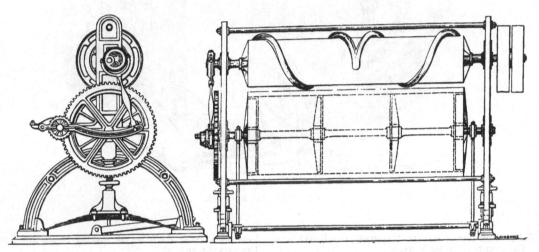

COX'S PATENT MACHINE FOR STRIKING OR SCRAPING HIDES.

Ripley's Patent Rolling Machine, improved and manufactured by Stothert and Pitt.

The hide being laid upon the table *A*, steam is admitted to the cylinder *B*, and propels the loaded roller box, *C*, alternately from end to end of the hide. The motion being entirely self-acting, the attendant has both hands free. The stroke can be lengthened or shortened at pleasure.

The machine will roll from 15 to 20 butts an hour.

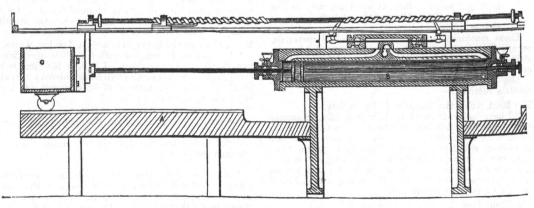

RIPLEY'S PATENT ROLLING MACHINE.

THOMAS, W. F., & Co., 1 *Cheapside, and* 66 *Newgate Street, London.*—Sewing machines, and samples of work produced by them.

DOUBLE-ACTION SEWING MACHINE.

1 *B.* Double-action machine for domestic purposes, shirt-making, and light work of every kind.

DOUBLE-ACTION SEWING MACHINE.

DOUBLE-ACTION SEWING MACHINE.

2 *B.* Double-action machine for the use of boot-makers, tailors, shirt-makers, &c.

2 *C.* Double-action machine for dress and mantle making.

[1733]

Victoria Sewing Machine Company, The, 97 *Cheapside.*—Sewing machines.

[1734]

Waterlow & Sons, *London.*—Railway ticket printing machine, to be worked by hand or power. (*See page* 97.)

[1735]

Watkins, Thomas, 89 *Bridge Street, Bradford.*—Porcelain guides, washers, steps, shuttle-eyes, &c., used in machinery.

[1736]

Watson, Henry, *High Bridge Works, Newcastle-on-Tyne.*—Improved knotter or pulp strainer, for paper makers.

A A shows the thickness of the margin.
B B B B B are ribs which stiffen the plate.

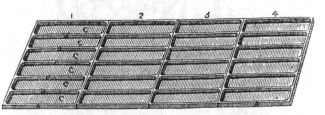

C C C C shows the slits cut between each rib.

KNOTTER BOTTOM IN FOUR PLATES, WITHOUT FRAME.

The Exhibitor manufactures
Brass and copper rolls for paper mills.
Jullion's patent pulp regulating elevator.
Gun-metal cocks, valves, water and steam guages, hydraulic rams, &c.
Large brass castings, brass and copper work for marine, locomotive, and other engines.

Safety lamp, the most improved in use in the north of England.
Armstrong's (Sir William) hydro-electric machines, for the production of electricity from steam.
Frames of brass or wood, with brass mountings, made to order.

[1737]

Weatherley, Henry, 54 *Theobald's Road.*—Confectioners' machines, &c., for hand or steam power.

[1738]

Weston & Horner, 80 *Whitecross Street, London.*—Patent self-feeding mortising machine.

This machine, which the inventor has had in private use for about twelve months, has been found to possess great advantages over those commonly employed, as will be readily apparent on examining it, and witnessing it in action. The bed, on which the material to be mortised is placed, is permanently fixed; the chisel-holder being raised or lowered, to suit its breadth. The "wedging" is obtained by inclining the chisel, which, being self-oiled at every incision, involves the least possible degree of force in the operation. The "feed," or shifting of the material, is effected by means of a semi-self-acting arrangement. The machine is worked by a treadle, and is adapted for mortising, boring, drilling, dovetailing, &c.

It will be sold at prices varying, according to finish and completeness, from £7 10s. to £15 0

[1739]

Whight & Mann, *Gipping Works, Ipswich.*—The "Excelsior" sewing machine.

This is a new and improved sewing machine, making the "double loop" or tight stitch." It is suitable for the use of families, manufacturers, dress, and mantle makers. The exhibitors keep in stock every requisite for working the sewing machine, such as needles, shuttles, bobbins, silks, cottons, &c.; and are also prepared to supply first-class lock-stitch machines, for heavy manufacturing, at reduced prices. Price lists and prospectuses may be obtained by application at the works, or at the London depôt, 122 Holborn Hill, E.C.

WATERLOW & SONS, *London.*—Railway ticket printing machines, may be worked by hand or power.

WATERLOW'S RAILWAY PASSENGER TICKET MACHINES.

These machines are manufactured of best materials in the best possible manner, and have been in use for several years at the offices of some of the principal railway companies in the United Kingdom, the British Colonies, and on the Continent, to whom we are permitted to refer, and whose experience forms the best guarantee of their speed, durability, and general efficiency. They are constructed with a fast and loose rigger, to work from a shaft, or may be driven by hand with perfect ease ; printing, perforating, numbering consecutively, either at one or both ends of the tickets at one operation, at the rate of 8,000 to 10,000 per hour.

The plain tickets are inserted in the tube A, pass along the plate F, and rise into the tube F, in numerical order.

When the machine is driven from a shaft, the printer has simply to insert the bundles of plain tickets in one tube, taking the printed ones out on the other side, the machine stopping of its own accord and ringing a bell when any derangement arises from an imperfect card or other cause.

List of some of the Railway Companies to whom these machines have been supplied.

Eastern Counties Railway, London.
Great Western Railway, London.
Midland Railway, Derby.
South Eastern Railway, London.
Great Northern Railway, London.
Buffalo and Lake Huron Railway, Canada.
Great Western Railway of Canada.
The Australian Railways.
Calcutta and South Eastern Railway.
Eastern Bengal Railway.
Vienna Railway.

The machines can be seen in operation, daily, at the Printing Offices, London Wall, where every explanation or information respecting them may be obtained.

UNPRINTED TICKETS, various colours and devices, perforated if required. A pattern card, containing 120 sorts, with prices of each attached, will be forwarded on application from any Railway Company, or their agent.

Great care is exercised in the preparation and examination of these tickets, and every defective one removed. The tickets are perforated singly, thereby ensuring the perforation falling in the centre, and the removal of any imperfect ticket—a great advantage over all others yet submitted.

TICKET CASES, made of teakwood or oak, and finished in the best manner, at prices varying according to the number of tubes required. A detailed price-list forwarded on application.

TICKET SCREW BOX, or tying-up machine.

TICKET NIPPERS, best steel, from 2/0 per pair.

WATERLOW & SONS, *continued.*

TICKET DATING PRESSES, with set of steel types and box for same, complete.

    Ribbon inking machine.
    Printing ribbon.
    Counting machines.
    Guard and dispatch cash boxes.

Station cash bags.
Season-ticket cases, &c.

PRINTED TICKETS, single or double journey, once or twice numbered, and perforated if required, striped, parti- or double-coloured, at prices varying according to quantities and description. Estimates furnished if desired.

[1740]

WHITFIELD, H., *Rainhill, near Prescot.*—Lancashire files.

[1741]

WHITMEE, JOHN, & Co., 70 *St. John Street, Clerkenwell, E.C.*—Mills ; weighing machines ; Tice's patent gas regulators ; Carley's patent elastic-stitch sewing machines.

[1742]

WHITWORTH, J., & Co., *Chorlton Street, Manchester.*—Machinery for cutting metals and timber. (*See pages 99 to 105.*)

[1743]

WILSON, W., *Campbellfield, Glasgow.*—Semi-dry pulverised clay brick-making machine.

[1744]

WOOD, J. & R. M., 89 *West Smithfield, E.C.*—Printing and stereotyping machinery, and type. (*See pages 106 and 107.*)

[1745]

WORSSAM, S., & Co., 304 *King's Road, Chelsea.*—Wood-working machines. (*See page 108.*)

[1746]

WRIGHT, JOHN, *Pathend, Kirkcaldy.*—Mould-making machines for producing printing surfaces, and specimens of typing, &c.

[1747]

WYLIE, ALLAN C. (Successor to JOHN CONDIE), 8 *Cannon Street, London.*—Two Condie's patent steam hammers. (*See page 109.*)

[1748]

YATES, W. S., *Stamford Street, North Street, Leeds.*—Machine to assort bristles into sizes for brush manufacturers.

The exhibitor is the sole inventor of machines for assorting bristles into their various lengths and sizes. One of these machines is exhibited. It consists of ten nippers, and separates the bristles into ¼ in. lengths with accuracy and rapidity, depositing them in suitable receptacles. With the exception of feeding, this machine is entirely self-acting.

Machines may be obtained from the maker of any size required, varying from one to ten nippers. They have already secured the approval of those who have employed them in dressing bristles, and have been adopted both by English and foreign manufacturers. The exhibitor will be happy to furnish further particulars as to prices, &c. on application.

[1749]

YOUNG, J. & T., *Ayr.*—Vertical saw frame, to cut battens from twenty-four inches broad, and from five inches thick.

YOUNG'S IMPROVED SAW FRAME to cut 2 battens at once, from 24 in. broad and from 5 in. thick, with

patent silent feed motion. Price at the Works, exclusive of saws and buckles, £130.

[1750]

YOUNG'S PATENT TYPE COMPOSING AND DISTRIBUTING MACHINE COMPANY (Limited), 77 *Fleet Street.*—Type-composing machine, and type composing and distributing machines. (*See pages 110 and 111.*)

WHITWORTH, JOSEPH, & Co., *Chorlton Street, Manchester.*—Machinery for cutting metals and timber.

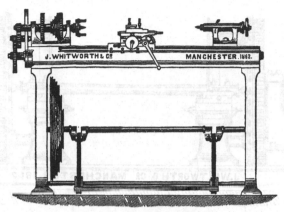

SELF-ACTING FOOT LATHE.

SELF-ACTING FOOT LATHES for the use of engineers, amateurs, philosophical-instrument makers.

They are suited for turning plain or ornamental work, for small sliding, screwing, and surfacing, and are made of the several sizes from 5 to 9 in. centres, and with any requisite length of bed. They are sometimes supplied with chucks of various descriptions and overhead motion.

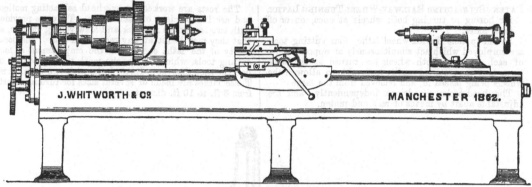

SELF-ACTING LATHE.

SELF-ACTING LATHES for sliding, screwing, surfacing, and boring.

They are furnished as single or with the patent duplex motion. The latter having two tools, will turn double the quantity of work that the single lathe can, and of a better quality. For sliding shafting, the duplex principle is invaluable, as the lateral strain put upon the shaft by the single tool is neutralised by the additional tool acting opposite to it.

The lathe exhibited is a medium size, with 10-in. centre and bed 10 ft. long. The sizes usually manufactured range from 6 in. to 36 in. in height of centre, and are made with any required length of bed.

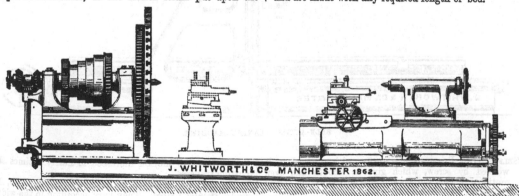

SELF-ACTING BREAK LATHE.

SELF-ACTING BREAK LATHES, with geared headstocks, large face plate, with wheel at the back, heavy foundation plate, planed and grooved on its upper surface, extending the entire length and breadth of the lathe.

The bed is movable on the foundation plate by rack and pinion, so as to form a break for admitting work of a large diameter.

These lathes are universal in their application, being suitable for sliding, screwing, surfacing, and boring, and for large and small work. They are made of several sizes, admitting work varying from 4 ft. to 10 ft. diameter, and of any length.

WHITWORTH, JOSEPH, & CO., *continued.*

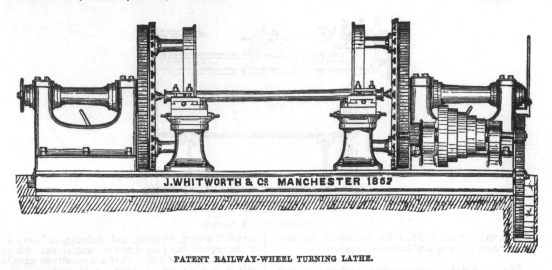

PATENT RAILWAY-WHEEL TURNING LATHE.

PATENT SELF-ACTING RAILWAY-WHEEL TURNING LATHE, for boring or turning both wheels at once, on or off their axles.

In the patent duplex wheel lathe, four cutting tools are employed which act simultaneously at opposite sides of each wheel. Both wheels are turned at the same time, and the rests are readily removable to allow of the wheels being placed in, and removed from the lathe.

The headstocks are driven independently, and are adjustable apart by means of rack and pinion.

The rests are worked by overhead self-acting motion, and are independent of each other. They are provided with two transverse motions and a swivelling motion, so that they may be set at any required angle. The peculiarity of the lathe consists in the employment of four cutting tools, which enables the work to be produced in half the time that the ordinary lathe with two tools takes. These lathes are made of several sizes for turning wheels from 3 ft. to 10 ft. diameter.

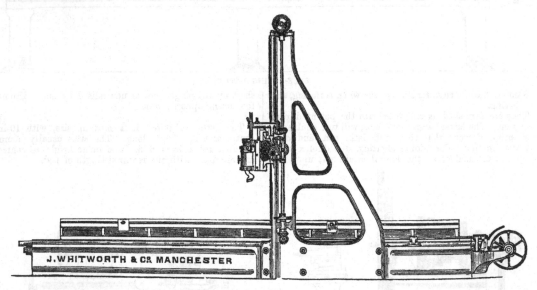

SELF-ACTING PLANING MACHINE.

SELF-ACTING PLANING MACHINES, with grooved table worked by screw, which gives a smooth and even motion ; driving pulleys and gear placed at the end of the bed.

These machines have self-acting motions for horizontal, vertical, and angular planing. Two of them are exhibited, namely, one E size, 22 ft. long, 5 ft. 6 in. wide, 5 ft. 6 in. high, provided with 2 reversing tools, each to plane both ways.

The other machine is smaller, being a C size, 9 ft. long, 3 ft. 6 in. wide, 3 ft. 6 in. high, and is provided with one tool holder and planes only one way, the table having its return traverse motion increased to nearly 3 times the cutting traverse.

The sizes manufactured are designated A, B, C, D, E, F, G, H, and I sizes. They vary according to the width and height of the object they will plane, namely from 18 in. by 18 in. up to 14 ft. by 14 ft. In length they are made to suit the work to be planed, and, as a general rule, will plane three-fourths of the length of their bed.

A CRANK PLANING MACHINE is also exhibited, in which the table is worked by a crank and lever, which imparts a uniform motion in cutting and a rapid one in returning.

WHITWORTH, JOSEPH, & CO., *continued.*

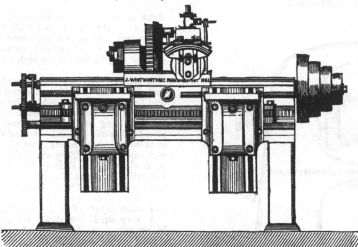

UNIVERSAL SHAPING MACHINE.

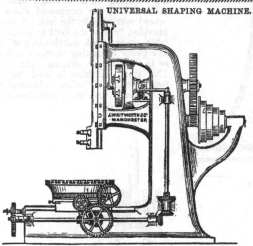

SELF-ACTING SLOTTING MACHINE.

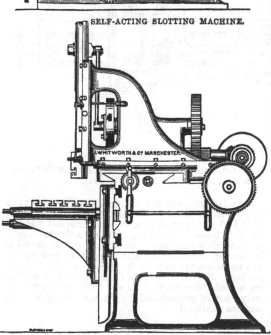

SELF-ACTING COMPOUND SLOTTING MACHINE.

PATENT UNIVERSAL SHAPING MACHINES, for shaping levers, cranks, connecting rods, and for work in general.

They have adjustable crank motion, self-acting motions for horizontal, verticular, angular, and circular work, and for internal curves. The tool-holder is provided with a segment wheel and worm, to which a self-acting motion is attached.

These machines have two independent tables for holding the work, which are adjustable vertically by means of screws and nuts, and longitudinally by means of racks and pinions. They are attached to the machine in front by bolts sliding in planed T grooves.

There are several sizes of these machines made, viz. A, B, C, D, E, F, G, H, I, and J sizes. The stroke of the tool varies from 5 in. in the A size to 42 in. in the J size, and the lengths of the bed from 3 ft. to 18 ft. respectively.

The two sizes exhibited are B and F.

The larger sizes, viz. C, D, E, F, G, H, I, and J sizes, are sometimes provided with two headstocks adjustable independently, and with independent self-acting motions. In this case three tables are usually provided, and the bed is made of any length, to suit the work to be done. This arrangement is suitable for planing or shaping both ends of connecting rods at the same time, or for general work.

SELF-ACTING SLOTTING AND SHAPING MACHINES, with independent upright framing, continuous vertical slide for tool holder, worked by a crank, table for holding the work fitted with transverse slides and a worm wheel for circular work. In some cases extra transverse slides are used for convenience of chucking and shaping work. Two sizes of this machine are exhibited, an A size, with 6 in. stroke, and a D size, with 18 in. stroke.

The small one exhibited has the extra transverse slides, and is placed on standards to bring it to a convenient height for the workman.

These machines are manufactured of several sizes, which are designated A, B, C, D, E, F, G, H, and I sizes.

The length of stroke varies from 6 in. in the A size, to 48 in. in the I size, and the diameter admitted from 2 ft. in the A to 10 ft. in the I size.

In the machines with stroke above 30 in. long the tool slide is worked by means of a screw with a quick return motion, and in the others by a crank.

SELF-ACTING COMPOUND SLOTTING AND SHAPING MACHINES, with one or more slotting headstocks mounted on the same slide-bed. The bed and tables of these machines are similar to those of the universal shaping machine. The work is fixed to the tables, which have vertical and longitudinal adjustments. The slotting headstocks and tools are made to slide along the bed, each tool having independent self-acting motions for plain and circular work. The several tools may be made to operate at different points of the same object at the same time, and as a general rule the work may be finished, before removal from the machine, with once fixing only.

WHITWORTH, JOSEPH, & CO., *continued.*

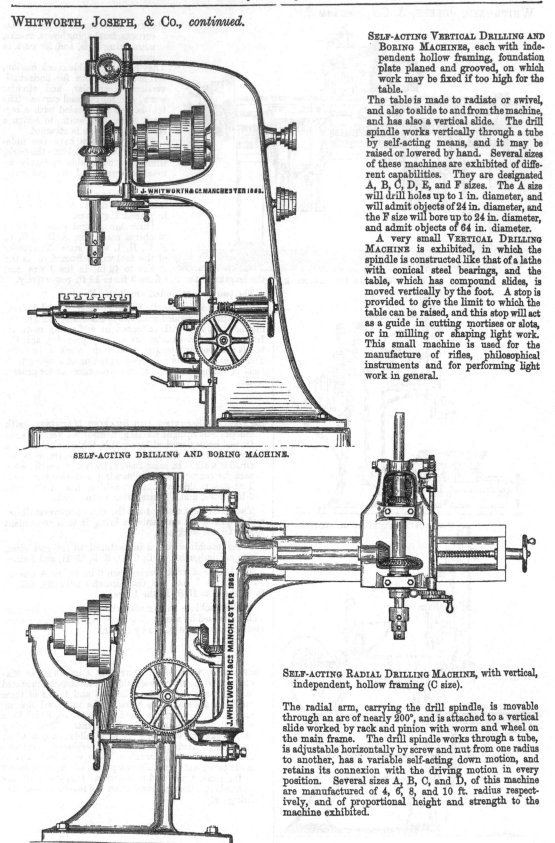

SELF-ACTING DRILLING AND BORING MACHINE.

SELF-ACTING RADIAL DRILLING AND BORING MACHINE.

SELF-ACTING VERTICAL DRILLING AND BORING MACHINES, each with independent hollow framing, foundation plate planed and grooved, on which work may be fixed if too high for the table.

The table is made to radiate or swivel, and also to slide to and from the machine, and has also a vertical slide. The drill spindle works vertically through a tube by self-acting means, and it may be raised or lowered by hand. Several sizes of these machines are exhibited of different capabilities. They are designated A, B, C, D, E, and F sizes. The A size will drill holes up to 1 in. diameter, and will admit objects of 24 in. diameter, and the F size will bore up to 24 in. diameter, and admit objects of 64 in. diameter.

A very small VERTICAL DRILLING MACHINE is exhibited, in which the spindle is constructed like that of a lathe with conical steel bearings, and the table, which has compound slides, is moved vertically by the foot. A stop is provided to give the limit to which the table can be raised, and this stop will act as a guide in cutting mortises or slots, or in milling or shaping light work. This small machine is used for the manufacture of rifles, philosophical instruments and for performing light work in general.

SELF-ACTING RADIAL DRILLING MACHINE, with vertical, independent, hollow framing (C size).

The radial arm, carrying the drill spindle, is movable through an arc of nearly 200°, and is attached to a vertical slide worked by rack and pinion with worm and wheel on the main frame. The drill spindle works through a tube, is adjustable horizontally by screw and nut from one radius to another, has a variable self-acting down motion, and retains its connexion with the driving motion in every position. Several sizes A, B, C, and D, of this machine are manufactured of 4, 6, 8, and 10 ft. radius respectively, and of proportional height and strength to the machine exhibited.

WHITWORTH, JOSEPH, & Co., *continued.*

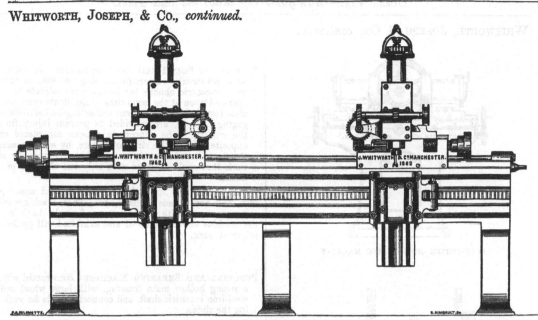

SLOT-DRILLING MACHINE.

IMPROVED SELF-ACTING SLOT-DRILLING MACHINE for cutting slots or mortises, and for general drilling, boring, and milling, constructed with horizontal slide-bed, grooved in front to receive the several tables used for holding the work, which are movable vertically and longitudinally. The drill spindle revolves in conical case-hardened steel bearings within a tube, which by preference is made octagonal in section, and is adjustable in its bearings, as in the spindle of a turning lathe.

The means for imparting both the rotary and the reciprocating actions, as well as the variable self-acting down motion to the cutter, are all contained within, and carried upon and with the main slide or headstock of the machine, which slides on the upper surface of the bed. These machines are made with one or several drilling headstocks and tables, and of various sizes, according to the class of work to be done by them. Attention is given to the construction of the drill spindle (applicable to drilling, milling, and boring machines in general) which is designed to prevent the cutting tool from having lateral play, and consequently will produce superior work, also to the means provided for adjusting the drill spindle whilst working, both transversely and longitudinally, and for regulating the traverse of the cutter.

Another description of slot-drilling machine is made suitable for cutting slots which do not run in the direction of the length of the object, as for example in the cross head of a steam engine. The main frame carrying the drill spindle is stationary. The crank motion is applied to one of the slides of the table carrying the work, and is so arranged as to give the traverse to the table at any angle according to the direction in which the mortise has to be cut. Compound slides are provided for adjusting the work, so that when several mortises are required to be cut, it is only necessary to fix or chuck once.

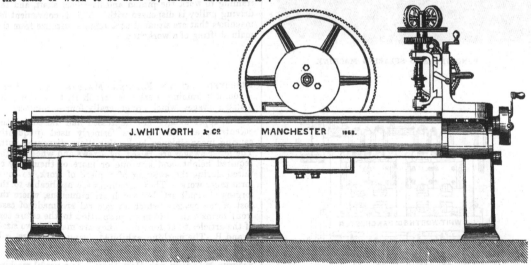

WHEEL-CUTTING MACHINE.

SELF-ACTING WHEEL-CUTTING MACHINE, to cut the teeth of spur, bevel, and screw wheels (in metal or wood).

Several sizes of this machine, C, D, and E, are made, the D size being exhibited. It will cut wheels up to 10 ft. diameter, and pinions of the smallest diameter required in engineer's work. Machines of a simple description and smaller sizes, A and B, are made for cutting the teeth of spur-wheels only, in which several wheels or pinions are placed side by side on an arbor, and the cutter is traversed through them by self-acting means, the dividing being accomplished by change-wheels and worm-wheel, as in the large machine exhibited.

WHITWORTH, JOSEPH, & CO., *continued.*

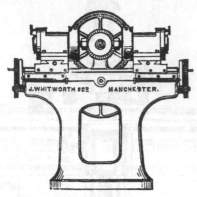

SELF-ACTING NUT-SHAPING MACHINE.

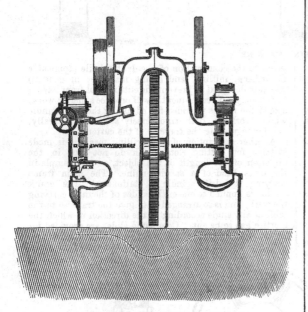

PUNCHING AND SHEARING MACHINE.

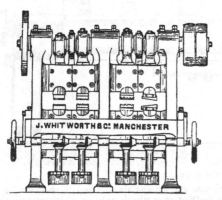

IMPROVED "RYDER'S" FORGING MACHINE.

SELF-ACTING BOLT HEAD AND NUT SHAPING MACHINE, with two circular cutters for shaping two sides at once, two concentric chucks for holding two objects to be operated upon at the same time. Duplicate compound slide rests, with independent self-acting and self-disengaging motions, are provided to prevent injury from the cutters. The concentric chucks are placed on opposite sides of the circular cutters, by which means the forces are balanced, and double the quantity of work that could be done with only one chuck is produced.

This machine is applicable for shaping and squaring nuts, bolt-heads, ends of shafts, &c. These machines are sometimes made with single cutter and single chuck, but the machine above explained and exhibited will produce the most work.

PUNCHING AND SHEARING MACHINES, constructed with a strong hollow main framing, with large wheel and steel-iron eccentric shaft, and connecting rods for working the slides.

Both the operations of punching and shearing may be regularly going on at the same time without interfering with each other. The large wheel is driven by a pinion placed on a shaft at the top of the machine, at each end of which is keyed a fly-wheel and at one end the driving pulley. An apparatus is provided for raising the punch without stopping the machine. Sometimes "traverse tables," for holding and moving the plates to be punched and sheared, are supplied, the correct division of the holes being secured by screw and change-wheels worked from a cam on the eccentric shaft of the machine. The several sizes of these machines are A, B, C, D, and E, which will punch respectively ½-in. to 2-in. holes through plates of a corresponding thickness, and will shear a similar thickness of plate.

BAR-CUTTING MACHINES are also manufactured of similar design to the above, or with only one slide.

In large machines for punching, shearing, or bar cutting, a steam engine of sufficient power is attached to the main framing to work the pinion and fly-wheel shaft, and the driving pulley is dispensed with. This is convenient for machines that are situated at a remote distance from the main shafting of a workshop.

IMPROVED RYDER'S FORGING MACHINE, with strong compact framing, steel eccentric shaft, for working the upper swages, which are generally four in number.

The upper swages are depressed and raised by the eccentrics, and the springs formerly used (frequently causing inconvenience by breaking) are dispensed with. The lower swages are raised by screws and wheels to the required height, and any one or more of them may be raised during the swaging of a piece of work, so as to form taper work. These machines are applicable to the forging of small articles which are numerous, where the cost of the swages (which in general are made of cast-iron) forms a small item in proportion to the entire cost of the articles to be forged. They are made of two sizes, A and B. The machine exhibited, being the B size, will admit a piece 6 inches square.

SAWING MACHINE FOR HOT IRON. Is a useful machine in connexion with the Ryder's forging machine.

It is arranged with a slide bed, on which the slide carrying the circular saw is movable longitudinally. The iron to be sawn is placed in an angular grooved bed, and the saw is drawn through it by means of a screw and nut.

WHITWORTH, JOSEPH, & CO., *continued.*

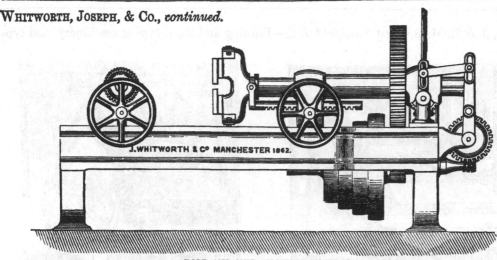

BOLT AND NUT SCREWING MACHINE.

IMPROVED SCREWING MACHINE for bolts and nuts, with hollow mandril, four radial dies, two on each side of the centre, complete set of dies and taps, with chucking apparatus for bolts and nuts.

The dies are cut by master taps, of double the depth of thread, larger in diameter than the working taps, so that the circle of the dies in contact, is the same size as the screw blank. A perfect guide is thus obtained, and a thread of correct pitch is formed at the commencement. The inner edges of the dies being filed off to an acute angle, they cut with ease without destroying the thread; and by the direction in which the dies are moved, their cutting power is preserved for the full depth of thread. Their action in cutting is similar to that of a chasing tool, which they resemble in form, and may in like manner be sharpened on a grindstone.

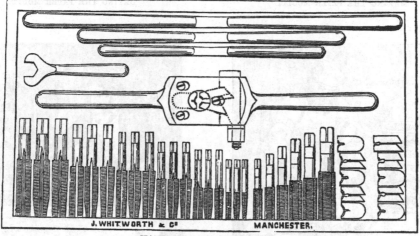

HAND SCREWING APPARATUS.

HAND SCREWING APPARATUS, with screw stocks, dies, taps, and tap wrenches, are furnished of all sizes to screw from $\frac{1}{16}$ in. to 3 in. diameter. The screw threads throughout are uniform in angle and shape.

INTERNAL AND EXTERNAL CYLINDRICAL GAUGES, being standards of size, are made exact to the measure of the realm and tested by the measuring machine. They are supplied in sets, in boxes, as exhibited.

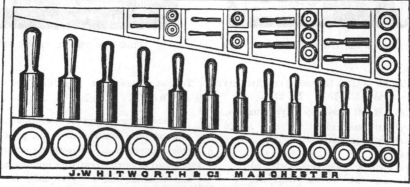

CYLINDRICAL GAUGES.

WOOD, J. & R. M., 89 *West Smithfield, E.C.*—Printing and stereotyping machinery, and type.

MAKING THE PAPER MOULD.

DRYING THE PAPER MOULD.

OPERATION OF CASTING A STEREOTYPE PLATE.

FINISHING THE PLATE FOR PRESS.

IMPROVED PAPIER MACHÉ STEREOTYPING APPARATUS, patented September 8th, 1860. No. 2180.

Prices of J. & R. M. Wood's stereotype foundries.

To cast a demy plate, and under	£20	0
To cast a quarto royal or folio crown	25	0
To cast a royal folio	35	0
To cast, type high, a newspaper, single and double columns, with Wood's patent core gauge	40	0

To cast a newspaper page, flat, size of the "Times" £135 0

The exhibitors are type founders, Columbian and Albion printing press manufacturers, and patentees of the improved papier maché stereotyping apparatus. They are prepared to execute orders for printing machinery, plant, and material, and every requisite for printing newspapers, books, and jobbing.

WOOD, J. & R. M., *continued.*

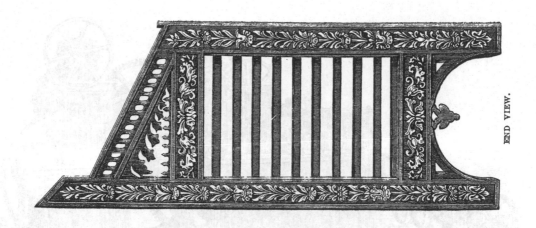

END VIEW.

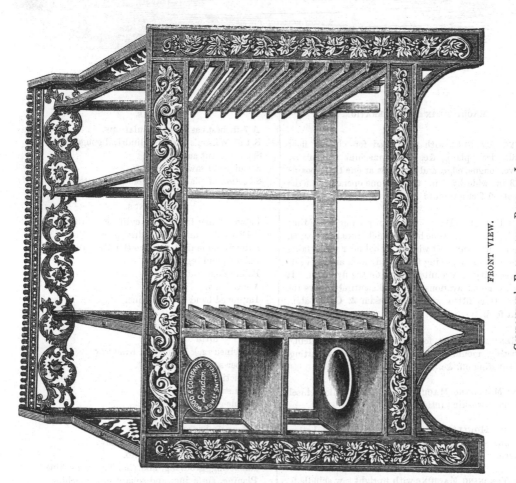

FRONT VIEW.

COMPOSITOR'S FRAME AND RACK IN IRON,

With galley shelf and enamelled iron basin for wetting type for distribution.

WORSSAM, SAMUEL, & CO., 304 *King's Road, Chelsea, S.W.*—Machines for sawing, planing, moulding, mortising, tenoning, grooving, rabbeting, &c.

ROLLER PLANING MACHINE.

MACHINES IN FULL OPERATION.

PLANING MACHINE with roller feed for planing floor boards, deck plank, &c. This machine will plane, groove, tongue, edge, and thickness at one time, boards of 13 in. wide by 6 in. thick at one operation, at the rate of 50 feet a minute.

PATENT PORTABLE DEAL FRAME, for sawing at one time two deals of 14 in. wide by 4 in. thick, into thin boards. This frame is supplied with an air cylinder which takes the weight of the swing frame and saws, and prevents the necessity of a counterbalance on the fly wheel. It requires no excavations, and works entirely above the floor. It is fitted with S. Worssam & Co.'s patent silent feed.

THE GENERAL JOINER, for sawing, grooving, tongueing, rabbeting, moulding, tenoning, boring, cross-cutting, and squaring off, &c. &c.

PATENT MORTISING MACHINE with square hollow chisel, and auger working inside.

IMPROVED MOULDING MACHINE with top and bottom and one side cutter, for working mouldings of any pattern.

SMALL TENONING MACHINE with upright saw spindle for cutting double tenons, and sliding plate fitted with spring stops.

TOOLS.

A 7-ft. best cast-steel circular saw.
Set of Wilson's patent cylindrical gouges.
Set of patent augers.
Set of band saw blades.
Selection of plane irons, tenoning cutters, &c.

DRAWINGS.

Patent timber frame with cylinder overhead.
   Ditto   ditto   to drive from below.
Patent double timber frame (for Russia).
Patent planking frame.
Patent double-deal frame.
Veneer saw.
Improved band saw machine.
Fret saw.
Large rack circular saw bench.
Small   ditto   ditto.
Combined rack bench and band saw.
Sleeper saw bench.
Self-acting saw bench.
Plain saw bench.
Double-edging bench.
Grooving and rabbeting bench.
Patent timber cross-cut saw.
Ransome's pendulum cross-cut saw.
Improved scantling   ditto.
Rabbeting, grooving, and chamfering machine.
Planing, surfacing, and squaring-up machine.
Whine's patent dovetailing machine.
Boring machine.

WYLIE, ALLAN C. (Successor to JOHN CONDIE), 8 *Cannon Street, London.*—Two Condie's patent steam hammers.

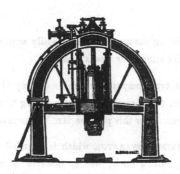

CONDIE'S PATENT DOUBLE-ACTING STEAM HAMMER.

The novelty and utility of this invention consists in the introduction of the steam into the hammer block, which also acts as the steam cylinder. By this simple and compact arrangement, the liability to breakage of the piston and piston-rod are avoided, these being the evils so much complained of in all other forms of steam hammers. These hammers are all constructed of the best materials and workmanship, and with Musgrave's improved arrangement of double-action and valve gearing. They are extensively used throughout Great Britain and Ireland; also, in Russia, Prussia, Austria, Belgium, Holland, Oldenburg, Bremen, Spain, Brazils, United States, Canada, East Indies, Australia, and in the arsenals and dockyards of the English, French, Russian, and Austrian Governments.

TESTIMONIALS.

From Messrs. J. Schultze & Co. Engineers, Oldenburg, Germany.

"JOHN CONDIE, ESQ.

"DEAR SIR,—In reply to your inquiry concerning the 30-cwt. steam hammer we received from you in 1852, we have much pleasure in stating that the same has been constantly working since that time for 5 puddling furnaces, and making forgings for heavy machinery. It has never been out of order, or required any repairs. In one word, we consider your hammer the most perfect machine of the kind that can be constructed.

"Believe us, dear sir, yours truly,
"J. SCHULTZE & Co.
"*Varel, Oldenburg, June 20, 1856.*"

From H. W. Harman, Esq. C.E., Marine Engineer, Northfleet, Kent.

"JOHN CONDIE, ESQ.

"DEAR SIR,—I beg to state that both the 10-cwt. and 30-cwt. patent steam hammers supplied to Mr. Pitcher, and erected here, have given me every satisfaction. In a time of pressure they enabled us to execute an immense amount of work in a very short period, their excellence consisting in the simplicity of the details, and their consequent non-liability to derangement. In our case boys work them, and this, coupled with a minimum of wear and tear, insures us two most economical and efficient machines.

"I am, dear sir, very truly yours,
"H. W. HARMAN, C.E.
"*Steam Factory,*
Northfleet Dockyard, Kent,
August 21, 1856."

From the Glasgow Iron Company, Glasgow.

Works. { Glasgow Iron Works, Glasgow.
{ Motherwell Iron Works, Motherwell.

"WE have had a 50-cwt. Condie's patent steam hammer at work for 5 years; a 40-cwt. hammer for 4 years; and

we have had other two 50-cwt. hammers erected and set to work within the last 6 months, all of which have given us the greatest satisfaction. We shingle puddled balls, scrap blooms, and slabs with them, and we find them most excellent machines for all these purposes. They are easily kept in repair, and at very little expense, and they are a great contrast, in these respects, to the metal helves we have had in use previously. We have another 50-cwt. hammer getting ready by Mr. Condie, which will supersede the last of our metal helves.

"ROBERT CASSELS.
"*23 St. Enoch Square, Glasgow,*
June 5, 1857."

From George Blaxland, Esq. H. M. Dockyard, Sheerness.

"JOHN CONDIE, ESQ.

"DEAR SIR,—In reply to your inquiry, the 50-cwt. steam hammer supplied and erected by you 12 months ago, has given most entire satisfaction, without any derangement of the parts. The foreman of the engine smithery department, having had many years' experience with steam hammers, affirms that yours is the most efficient he has ever used.

"I am, dear sir, yours truly,
"GEORGE BLAXLAND,
"*Chief Engineer.*
"*Steam-engine Factory, Sheerness,*
March 25, 1858."

From the North London Railway Company, Bow Road, London.

"JOHN CONDIE, ESQ.

"DEAR SIR,—I am happy to say the 10-cwt. steam hammer you made for this company about 2 years ago has given very great satisfaction. It has worked uniformly well since it was first started, and has required scarcely any repair. I consider it in every way a first-rate machine.

"Yours truly,
"W. ADAMS.
"*North London Railway,*
Locomotive Department, Bow Road Works,
November 22, 1859."

From Alex. Fulton, Esq., Glasgow Forge.

"JOHN CONDIE, ESQ.

"DEAR SIR,—In reply to your favour of the 2d instant, the 50-cwt. hammer erected in 1856, 7-ton hammer erected in 1857, and 4-ton hammer erected in 1858, have given me every satisfaction, and I consider them, from long experience with various hammers, the most perfect I have yet seen. The large working space between the framing of the 4-ton and 7-ton hammers, enables us to execute the heaviest class of forgings with great ease, and their general design have been much admired by all who have seen them. The 40-cwt. shingling hammer recently erected, will, I have no doubt, please equally well.

"I remain, dear sir, yours truly,
"ALEX. FULTON.
"*Glasgow Forge,*
Scotland Street, Glasgow,
October 8, 1859."

Prices, drawings, &c. on application to Allan C. Wylie, successor to John Condie, Unity Buildings, 8 Cannon Street, London, E.C., or to Messrs. John Musgrave & Sons, engineers, iron founders and boiler makers, Globe Iron Works, Bolton, Lancashire.

YOUNG'S PATENT TYPE COMPOSING AND DISTRIBUTING MACHINE COMPANY (Limited), 77 *Fleet Street.*—Type composing machine, and type composing and distributing machines.

For some time past the necessity of discovering a means for increasing the speed of setting-up types, and superseding the present slow hand method, has been strongly felt.

While printing from the composed types has, by the improvements in the steam press, been carried to a most advanced stage, setting-up by hand is not now done more quickly than it was 400 years ago by the earliest printers. In order to save a few minutes' time in printing, large sums are paid for improved steam presses, when much more time might be saved by the use of a well-devised composing machine.

The type-composing machine invented by the late Mr. James Hadden Young accomplishes this object completely, as a single example in reference to its use by a daily newspaper will show.

Let it be supposed that half an hour before the usual time of putting to press important news arrives, enough to extend over three columns of the paper, or say 45,000 types have to be set up in thirty minutes by hand : this would require the assistance of ninety compositors, each having a scrap of paper put into his hand to set up in such a manner that it may tally with his neighbour's piece, technically called "making even." With the machines, the work would be done in the same time by six players and twenty-two justifiers, therefore, only six pieces of copy, instead of ninety, would be required, and the system would besides, offer immense facilities for correction.

It must be remembered, too, that for this very work steam presses are waiting to throw off copies at the rate of 20,000 per hour, so that the saving of only five minutes would be a gain of 1,500 copies.

Mr. James H. Young's type-composing machine fulfils all the conditions necessary to make it of practical utility. It is simple, durable, not likely to get out of order, and causes no damage to the type. It is provided with separate keys for all the letters of a fount, to admit of each letter being set up in the order required by the compositor's copy, with a speed which is only limited by the skill of the player. In reference to this it will suffice to say that the present ordinary speed is at the rate of 12,000 to 15,000 types set up in an hour's time. The art of playing the machine can be easily acquired by any compositor after a few weeks' practice.

As the type-composing machine sets up the type in long lines, they require previous to going to press to be put in page. For this purpose Mr. Young invented his

JUSTIFYING APPARATUS, which is intended to replace the compositor's stick, which, however, it resembles. It is fixed to a frame, and is used as follows : — The compositor places the galley filled with the long lines set up at the composing machine. He slides one of these lines into the apparatus, divides it into the proper length, reads it, makes any obvious corrections, and having justified it, he moves a handle, by which the completed line is depressed, and room is made for a succeeding line. It is found that a compositor can justify at the rate of 4,000 types per hour, and the calculations of saving are founded on this rate, but it is probable that a rate of from 5,000 to 6,000 will be reached. So that if 12,000 types are set up by a player in an hour, three justifiers, as now, will nearly simultaneously have prepared that quantity for the reader.

Mr. James H. Young also invented a DISTRIBUTING MACHINE, which, besides collecting the different types of the same character together, sets them up in rows ready for the composing machine.

This operation is effected by means of distinguishing nicks cast or cut in the type.

In Mr. Young's distributing machine 71 per cent. of the types require only a single nick—a very shallow one —not larger than those now used to distinguish different founts ; 20 per cent. have two nicks, and the remaining that require more are, for the most part, thick-bodied types. The machine, attended by two lads, will prepare upwards of 18,000 types per hour, and, if desired, this quantity may be doubled simply by increasing its size.

It is calculated that by the use of Mr. Young's type composing and distributing machines 50 per cent. saved in the cost of composition.

Young's Patent Type-Composing and Distributing Machines Company (Limited),

WILLIAM YOUNG, Manager, 77, Fleet Street, London.

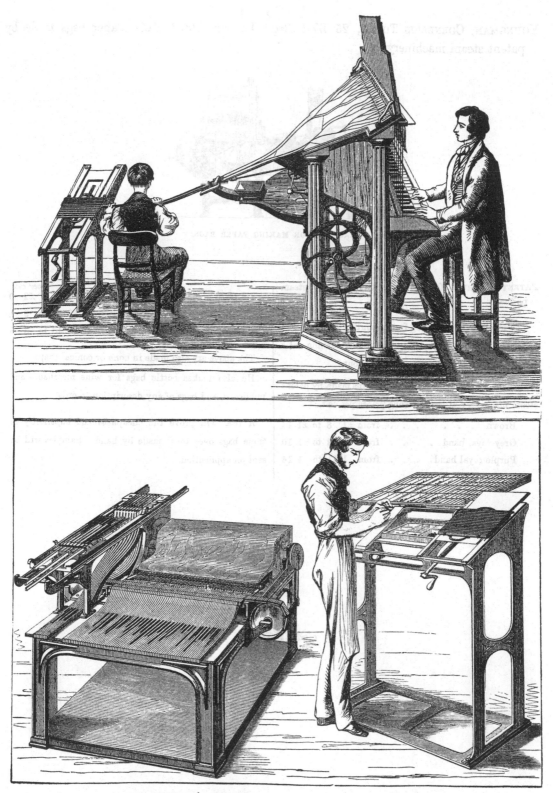

JAMES H. YOUNG'S TYPE COMPOSING AND DISTRIBUTING MACHINES.

[1751]

YOUNGMAN, CORNELIUS TIPPLE, 25 *West Street, Victoria Street, E.C.*—Paper bags made by patent steam machinery.

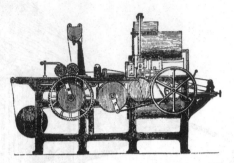

MACHINE FOR MAKING PAPER BAGS.

PATENT MACHINE for making paper bags by steam power.

The machine makes a bag from a continuous length of paper, folding, pasting, cutting, and finally turning out a perfect bag without the use of hand labour.

C. T. Youngman can supply paper bags made by patent steam machinery at the following prices per cwt. :—

Brown	from £1 8 to £1 14
Grey royal hand	from 1 4 to 1 10
Purple royal hand	from 1 11 to 1 14

	per cwt.
Glazed purple hand	£2 0
Tea	from £2 16 to 3 1
Small hand	from 2 16 to 3 6

The above are also made in cone or conical shape.

He also makes bottle bags for wine merchants and publicans, and bags of any description to order.

A trial will prove the cheapness and superiority of these bags over those made by hand. Samples will be sent on application.

LONDON

R. CLAY, SON, AND TAYLOR, PRINTERS,

BREAD STREET HILL.

Class VIII.

MACHINERY IN GENERAL.

[1780]

ADAMSON, DANIEL, & CO., *Newton Moor Iron Works, near Manchester.*—Twenty tons hydraulic lifting-jack ; and small patent steam boiler.

ADAMSON'S PATENT HYDRAULIC LIFTING-JACKS, of twenty, sixteen, ten, eight, and six tons power. These jacks are entirely self-contained, portable, power-ful, and light. One man using them, can lift as much as four men with the screw jack, and in much less time. They are made to lift from 4 to 50 tons weight.

[1781]

ADCOCK, JOHN, *Marlborough Road, Dalston, London.*—Carriage odometer, or improved distance indicator for wheel carriages.

[1782]

ALLEN, HARRISON, & CO., *Cambridge Street Mills, Manchester.*—Gun-metal fittings for marine, locomotive, and stationary engines.

[1783]

APPLEBY, BROTHERS, 69 *King William Street, City, London.*—Steam cranes ; vertical and horizontal engines ; wheels ; pumps. (*See page* 2.)

[1784]

ARMSTRONG, ROBERT, *North Woolwich, E.*—A vertical steam boiler, model, and drawing.

[1785]

ARMSTRONG, SIR W. G., & CO., *Elswick Engine Works, Newcastle-upon-Tyne.*—Models exhibiting Armstrong's hydraulic system.

[1786]

ASHTON, J. P., 2 *Upper Holland Street, Kensington.*—Steam engine and hoist.

[1787]

ASKEW, CHARLES, 27½ *Charles Street, Hampstead Road, N.W.*—Improved brewers' circular refrigerator, circulating boiler, and chimney cowl.

APPLEBY, BROTHERS, 69 *King William Street, City, London.*—Steam cranes; vertical and horizontal engines; wheels; pumps.

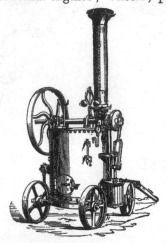

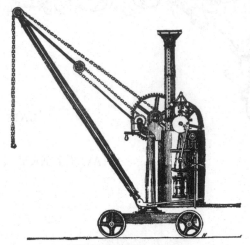

PORTABLE ENGINE.

PORTABLE STEAM CRANE.

C. J. APPLEBY'S IMPROVED PORTABLE ENGINE, complete as shown in the illustration, or mounted on circular tank, for fixing on boarded or other floors. Price, 3-horse power £75 0
    If without wheels or circular tank, £5 less.
C. J. APPLEBY'S IMPROVED PORTABLE STEAM CRANE, to swing in any direction, and lift 2 tons . . £185 0
    Larger sizes or wharf cranes to order.
C. J. APPLEBY'S HORIZONTAL ENGINE WITH MULTI-TUBULAR BOILER, the whole on feed water tank, and requiring neither brick-work for fixing, nor chimney stack. Price, 4-horse power £145 0
    With Cornish boiler £35 less.

CIRCULAR SAW BENCH, with planed iron top, fitted with 24-in. saw, improved adjustable fence, iron standards, fast and loose pulley, &c. Price £17 10
PORTABLE CRANE to swing in any direction, to lift 3 tons £45 0
WHARF AND WAREHOUSE CRANES of every description.
DOUBLE PURCHASE CRABS, with strap break, to lift 5 tons £6 15
LIFT AND FORCE PUMPS, with Appleby's patent indestructible clacks and oscillating valves . . . £2 2
YARD OR TANK PUMPS as above £1 5

[1788]

BAINES & DRAKE, *Glasgow.*—Engine and boiler mountings.

[1789]

BALFOUR, HENRY T., 16 *Adam Street, Strand, W.C.*—Quartz-crushing machine, and steam engine.

[1790]

BARNETT, S., 23, *Forston Street, Hoxton.*—Soda-water machinery. (*See page* 3.)

[1791]

BARRETT, EXALL, & ANDREWES, *Reading.*—30-horse power double-cylinder horizontal engine; high-pressure expansive condensing engine.

[1792]

BASTIER, JOHN URSIN, 19 *Manchester Buildings, Westminster.*—Patent chain-pump improved.

[1793]

BATE, J., & Co., 18 *Crescent, Birmingham.*—Improved bottle corking machine, machine for washing bottles.

[1794]

BAYMAN, HENRY, *Johnson Street, Old Gravel Lane, E.*—Double and single lifting jacks; a set of iron blocks; improved ship's hearth; and single winch.

BARNETT, SAMSON, 23 *Forston Street, Hoxton, N.*—Soda-water machinery.

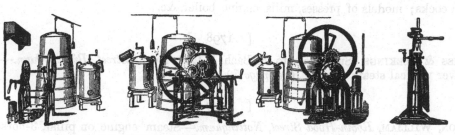

SODA-WATER MACHINES AS PLACED FOR USE.   BOTTLING MACHINE.

There are 3 sizes of the direct and beam action machines; their producing powers, and prices are :—

To make 200 dozen bottles per day	.	. .	£75 0
Ditto 160	ditto	ditto . .	. 70 0
Ditto 140	ditto	ditto . .	. 65 0

There are 2 sizes of the band-action machines :

To make 120 dozen bottles per day	.	. .	£55 0
Ditto 100	ditto	ditto . .	. 50 0

Many of these have been in constant use for 25 years, without requiring any repairs.

DOUBLE-ACTION MACHINES, for making lemonade and soda water at the same time, or for making either separately :—

To make 400 dozen bottles per day	.	. .	£150 0
Ditto 320	ditto	ditto	. . 130 0
Ditto 280	ditto	ditto	. . 120 0

A patent bottling apparatus is a very valuable addition to all the above machines, as it can be either used or not at pleasure, the usual nipple for the knee-bottling being on every machine. The advantage of the bottling machine is, that a person totally unacquainted with making aërated beverages, can, by this addition, immediately bottle it, as highly charged with gas as they please. Smaller machines are made for hotels and refreshment rooms, of the power of 60 dozen per day, £40; 40 dozen per day, £35; 30 dozen per day, £30. These machines are valuable, where the consumption is small, as the cost of the carriage is often more than making the

article itself, besides the advantage of having it always at hand, and always fresh. The improved bottling apparatus can be had separately.

The exhibitor, having had thirty years' experience in the manufacture of mineral-water machinery, and confining his attention to that and diving apparatus, every part has been the object of careful study ; and the requirements of those who use machinery where mechanical assistance cannot be obtained, have received due consideration.

The greatest purity is obtained when the condenser is lined with silver, and the plunger made of glass. The average cost of these additions is about £12, according to size.

Bottles, corks, wire, and all ingredients supplied.

Corks are usually packed in the same case, thus saving freight.

These improved soda-water machines are warranted superior to any hitherto manufactured, in solidity of construction, power, and simplicity. They are also admirably calculated for exportation, as they are packed in one case, without taking them to pieces, and can be set to work, and soda-water or lemonade made from them in half-an-hour after arrival. These machines are also used to manufacture ginger beer, orangeade, nectar, seidlitz, carrara, &c.

An illustrated pamphlet sent with each machine, containing full directions for use, and recipes for making soda water and all aërated beverages.

[1795]

BEAUMONT, FRANCIS WILLIAM, *Clapham.*—Self-acting steam boiler-feeding and general meter.

[1796]

BECK, J., 133A *Great Suffolk Street, Southwark.*—Valves for gas, water, and steam ; fire-cocks, &c.

[797]

BELLHOUSE, EDWARD T., & Co., *Eagle Foundry, Manchester.*—Steam engine ; hydraulic pumps and cocks ; models of presses, mills, engine boiler, &c.

[1798]

BELLISS & SEEKINGS (Successors to R. Bach & Co.), *Broad Street, Birmingham.*—2½-horse power vertical steam engine.   (*See page* 5.)

[1799]

BENSON, WILLIAM, *Robin-Hood Street, Nottingham.*—Steam engine on pillar, 3-horse power, new design.

[1800]

BLINKHORN, SHUTTLEWORTH, & Co., *Spalding, Lincolnshire.*—Patent fire engines, of great power, for service in the Industrial Department.   (*See page* 6.)

[1801]

BODMER, R. & L. R., 2 *Thavies Inn, Holborn, London.*—Safety valves, for steam boilers.

[1802]

BOTHAMS, JOHN C., *Salisbury.*—Water meter, measuring by capacity, continuous motion ; simple water meter ; high-pressure water tap, to check waste.

[1803]

BOWSER & CAMERON, *Glasgow.*—Five ton derrick crane.

[1804]

BRADFORD, THOMAS, *Manchester, and Fleet Street, London.*—Washing, wringing, drying, mangling, and knife-cleaning machinery ; drying closets ; churns.   (*See page* 7.)

[1805]

BRAY'S TRACTION ENGINE COMPANY (Limited), 12 *Pall Mall East, London.*—A traction engine for common roads.   (*See page* 8.)

[1806]

BRIDLE, HENRY, *Bridport, Dorset.*—Patent double-action refrigerator, for brewing and distilling purposes.   (*See page* 9.)

[1807]

BRIGGS & STARKEY, *Leeds and Liverpool.*—Washing, wringing, and mangling machines.   (*See page* 10.)

[1808]

BROUGHTON COPPER COMPANY, *Manchester.*—Copper rollers for printing ; copper and brass tubes for fire engines, and all descriptions of copper and brass work.

[1809]

BRYANT & COGAN, 55 *Broadmead, Bristol.*—Patent edge-laid leather mill-band.

[1810]

BUNNETT & Co., *Deptford, Kent.*—Concentric steam engine, working without fly wheel ; brick-making machine.

BELLISS & SEEKINGS (Successors to R. Bach & Co.), *Steam Engine and Boiler Works*, 13 *and* 14 *Broad Street, Islington, Birmingham.*—2¼-horse power vertical steam engine.

VERTICAL FIXED DIRECT-ACTING HIGH-PRESSURE STEAM ENGINE, with inverted cylinder 5 in. diameter, and 10-in. stroke. Nominal power, 2½ horses.

The cylinder is carried on a cast-iron standard, which is bolted to the bed or sole plate, 2 ft. 6 in. square. To this same sole plate, are also fixed the bearings for the crank shaft. The connecting rod and crank shaft are of the best hammered iron. The piston is fitted with steel spring, and metallic expanding packing. The cross-head is of wrought-iron, with adjustable gun-metal slide-blocks, capable of taking up the wear. The guides in

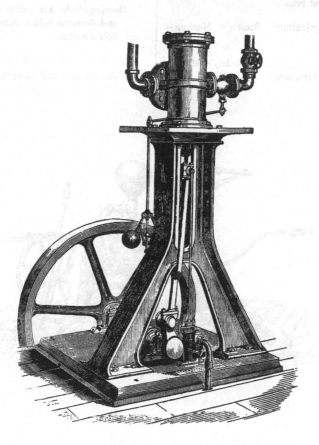

TWO AND A HALF HORSE-POWER VERTICAL STEAM ENGINE.

which they work, are truly bored in the standard itself. The pump is driven from the cross-head, and has consequently the same stroke as the piston. The governors act upon an equilibrium, or double-beat throttle-valve, through the intervention of only a single lever, and the comparative absence of resistance, renders their action peculiarly sensitive. The power is taken from the fly-wheel by means of a band. There are no extras required to render these engines complete and ready for fixing. From their simplicity of arrangement and construction they stand unrivalled in the number of purposes to which they can be applied, and their durability recommends them to the attention of all users of steam power for every stationary purpose, among which may be named, factory work, grinding, sawing, barn work, &c. for which they have been extensively adopted. The shape also is such as to pack into very little compass, being thus well adapted to the requirements of the exporter and colonial trader.

Prices complete as above :—

2½ horse power	. £34	. . with boiler	. £44
4 ditto	. . 64	. . ditto	. . 84
6 ditto	. . 90	. . ditto	. . 120
8 ditto	. . 112	. . ditto	. . 152
10 ditto	. . 130	. . ditto	. . 180
12 ditto	. . 144	. . ditto	. . 204

Higher powers in proportion.

With the boilers are supplied all the necessary fittings, glass water gauge, safety valve, check valve, fire doors, fire bars, beam, dead-plate, and damper. The boilers can be supplied with either cylindrical, Cornish, or multitubular to suit the locality.

BLINKHORN, SHUTTLEWORTH, & Co., *Spalding, Lincolnshire.*—Patent fire engines, of great power, for service in the Industrial Department.

PRIZES AWARDED :—

Manchester and Liverpool Agricultural Meeting, held at Bolton—Silver medal.

Yorkshire Agricultural Society's Meeting, held at Pontefract—First prize.

Peterborough Agricultural Society's Meeting—Second prize.

North Lincolnshire Agricultural Society's Meeting held at Brigg—First prize.

Manchester and Liverpool Agricultural Society's Meeting held at Ashton-under-Line—Silver Medal.

Agricultural Meeting held at Middleton—First prize.

Meeting of the Association of German Agriculturists and Foresters held at Schwerin, North Germany—Silver medal.

Exhibition of the Royal Cornwall Polytechnic Society held at Falmouth—Bronze medal.

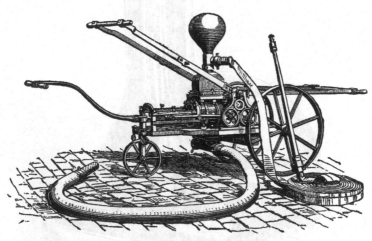

FIRE ENGINE.

PRIZE PATENT HORIZONTAL DOUBLE-ACTION FARM, MANSION, OR FACTORY FIRE ENGINE. This engine will discharge 100 gallons of water per minute, to an elevation of 100 ft. weather permitting, when worked by 8 men, is of very great power, exceedingly portable, made of the most durable materials, is not likely to get out of order, and effects a saving of 50 per cent. in labour for working. Price £30. It will throw a continuous stream of water with more force, and to a greater height than the engines generally in use. It possesses a double action, and being on the horizontal principle, is not likely to foul. When worked, the cylinder is always full of water, the air is excluded, and the flow of water is consequently freer, and more regular, than from the ordinary vertical barrels. Another important feature in its construction, is the formation of inlets to the valves, which can be opened in the course of a few seconds, and any obstruction which may take place may be removed almost instantaneously, thereby preventing the possibility of any serious delay at times when the services of an engine are imperatively necessary.

A complete set of fittings to the above, including suction and delivery hose, hose reel, branch pipe and nozzles, patent buckets, dam, &c. &c.

Size A, a smaller engine than the above, will discharge, when worked by 6 men, 70 gallons of water to an elevation of 70 feet, weather permitting.

A complete set of fittings to the above.

Full particulars, testimonials, reports of trials, and references, may be obtained of the inventors and manufacturers, Spalding, Lincolnshire.

BRADFORD, THOMAS, *Manchester, and Fleet Street, London.*—Washing ; wringing ; drying ; mangling ; knife cleaning machinery ; drying closets.

BRADFORD'S ORIGINAL COMBINED MACHINE, WITH IMPROVEMENTS PATENTED IN 1861 (see Nos. 2, 4, and 6),

Obtained every prize for which it competed (13 altogether) last year, 1861.

PATENT WASHING AND DRYING MACHINERY, LAUNDRY REQUISITES, &c.

Washing machines, original patent. Price,

No. 1 . . . . £3 10 0
No. 3 . . . . 5 10 0
No. 5 . . . . 6 10 0

Washing machines, improved patent, combined with wringing and mangling apparatus.

No. 2 . . . . £8 8 0

The most useful family machine.

No. 4 . . . . 12 12 0

Specially adapted for mansions, hotels, &c.

No. 6 . . . . 15 15 0

For large hotels or public institutions.

No. 8 . . . . 25 0 0

Fitted for steam or water power, and specially recommended for large public institutions, laundry contractors, and extensively adopted in larger sizes for various manufacturing purposes.

No. 10 . . . . £40 0 0

Similarly constructed to the above, but with two washing compartments and double-acting rollers.

Wringing machine or cottage mangle. Price,

No. 0 . . . . £2 12 6
No. 1 . . . . 3 3 0
No. 2 . . . . 4 4 0

Portable mangle with 3 rollers. Price,

No. 1 . . . . £5 5 0
No. 2 . . . . 6 6 0

This is a really useful and very convenient mangle.

Mangle, the original Baker or box, improved. Price £10 10 0
Ironing mangle, with heated cylinder . . 8 10 0
Drying machine, centrifugal 10 0 0
Ironing stove, with stand plate 2 10 0
Drying stove—Model.
Steam laundry—Plan of interior.
Churn—"New," "The Vortex."—Drawings. Made from 2 to 50 gallons.
The Guinea "Gem" Knife Cleaner.

BRADFORD'S NEW PATENT E. E. WASHING MACHINE.

The E. E. (Eccentric "Eclipse") machines will commend themselves to the favour of those who will take the trouble to examine them.

The "Boudoir" E.E. £1 10 0
The "Boudoir" E.E., combined . . . 4 4 0
The "Nursery" E.E. 2 10 0
The "Nursery" E.E., combined . . . 5 5 0
The "Cottager" E.E. 3 10 0
The "Cottager" E.E., combined . . . 6 6 0
The "Family" E.E. 5 10 0
The "Family" E.E., combined . . . 10 10 0
The "Contractor" E.E. 8 10 0
The "Contractor" E.E., combined . . 15 15 0
The "Contractor" E.E., combined, with Patent Reverse Gear 20 0 0

BRAY'S TRACTION ENGINE COMPANY (Limited), 12 *Pall Mall East, London.*—A traction engine for common roads.

BRAY'S TRACTION ENGINE.

This engine was built at the Company's factory, to the order of Her Majesty's Government, and is intended for permanent service in Woolwich Dockyard. Its construction embraces many improvements, and the introduction of several appliances of great importance, but the feathering principle of the wheels, which is the great distinctive feature of this Company's patent, is preserved intact. This principle consists in the circumference of the wheel having small apertures through which, by means of an eccentric, "blades" or teeth can be protruded or withdrawn as required, according to the nature of the ground over which the engine is travelling. In many cases the ordinary surface of the wheel is sufficient to gain the requisite amount of tractive power; the blades can then be thrown out at the top, or on that part of the wheel not coming in contact with the road; while, in the event of the ground being soft or slippery, or of the engine having to ascend a steep incline, the auxiliary power of the blades can be brought into action, and the additional bite or grip on the road obtained, as may be necessary to gain progress. This system does no damage whatever to, but, on the contrary, tends rather to improve the roads, as the breadth of wheel of the engine has much the same effect on their surface as a roller.

Power is transmitted to the driving wheels by means of pinions on the crank shaft, working in large rack wheels, which are fixed to the arms of each driving wheel near the peripheries. The engine having different gearing the speed or power may be altered as desirable. The engine exhibited is fitted with a drum which renders it available for driving any fixed or portable machinery of whatever nature; a derrick or steam crane, with which it can load its own waggons, &c.; and a capstan or cone barrel, whereby a rope, such as the fall rope of a tackle, may be hauled upon to any extent; so that, in addition to its tractive powers, it is applicable to all the purposes of a stationary or portable engine, which renders it particularly suitable for the service destined in Woolwich Dockyard.

The other special features of construction to be noticed in this engine are, the introduction of an improved steering gear, and of outside bearings for the driving wheels, which are mounted on springs on the inner and outer framings. By means of the first the engine is reduced to a state of the most perfect control, and can be guided with the greatest facility; and nearly all the jar, which involves extra wear and tear to the machinery, is obviated by the latter.

One of the Company's engines was employed in removing locomotive engines, the various castings, &c. for the large marine engines, and other heavy machinery, from different railway stations, manufactories, and the docks, into the Exhibition. The loads conveyed were sometimes as much as 45 tons, and reference can be made to the London and North Western Railway Company, Messrs. John Penn & Son, Messrs. Maudslays, Messrs. Humphreys & Tennant, and other eminent engineering firms, who employed the engine, as to its power and capabilities.

The above engraving represents an engine of still further improved construction, being built to the designs of the Company's engineer, Mr. D. K. Clark, C.E.

Further particulars respecting the engines, &c. may be obtained on application to Mr. S. H. Louttit, secretary to the Company, at the above address.

BRIDLE, HENRY, *Bridport, Dorset.*— Patent double-action refrigerator, for brewing and distilling purposes.

The cooling powers of this refrigerator surpass those of any hitherto in use: making unnecessary the employment of auxiliary coolers, and reducing boiling wort to nearly the same temperature as the water used for the purpose of reduction. By means of it the hottest weather ceases to be an obstacle to the production of a perfectly sound and brilliant article.

In introducing to the notice of brewers and distillers his double-acting refrigerator, the inventor feels justified in saying that he has succeeded in perfecting an apparatus which has long been sought after, calculated as it is to meet all the requirements of the trade.

The importance of securing a method by which wort can be cooled with rapidity in the hottest weather, need not be dwelt upon. A number of plans have been before proposed, but they have been accompanied by objections which have more or less interfered with their uniform success in working.

The improvements, however, which mark the invention now exhibited, are considered by practical brewers, who have inspected the apparatus, to obviate every difficulty which may have characterised those hitherto in use.

The advantages which attend its employment, may be briefly enumerated as consisting in the extraordinary cooling power which is rapidly attained, combined with the greatest cleanliness, strength, and simplicity, and united with the utmost economy of space and water, as well as cost.

Its mode of construction will be seen to guarantee its power. A series of flat pipes are arranged vertically in a case about 1 in. apart, through which the water passes in two streams, one over the other, in opposite directions, continuing through the whole length of the refrigerator, whereby a uniform temperature of the water in each pipe is preserved throughout its entire length and depth.

From the water traversing the pipes in the manner indicated, in such thin columns, and every particle of wort of necessity running round every pipe, and being by an obvious arrangement kept flowing in a continuous stream over the whole cooling surface of the refrigerator, it not only receives the cooling power of all the water employed, but the cooling influence of the atmosphere also.

If it should be objected by any who are accustomed to the various forms of refrigerators adopted, that the wort should not be exposed to the influence of the atmosphere, it may be stated that this apparatus admits of either method of working, allowing, if desired, the wort to traverse the inside of the pipes, subject to the external cooling influence of the water.

Absolute cleanliness is one of the great features in this refrigerator, as the pipes being perfectly flat and smooth, and standing edgeways, present but a slight surface for the deposit of sediment, especially as the wort is continually flowing around them in a rapid stream. What little may accumulate, can be cleaned off by passing a brush between the pipes; and as the bottom forms a hollow underneath every alternate pipe, it serves to empty the refrigerator (of wort when in use, or water when cleaning it) through openings in the side of the refrigerator into the draining pipe. As the whole of the wort and water is contained in the refrigerator, no wood cooler is required, which materially adds to its cleanliness: tinned copper, of which it is constructed, being easier to clean than wood-work. The cleansing of the inside of the pipes, may be still more easily effected, by fitting a movable cap to the end of every pipe, as in the one exhibited. This plan will, however, add to the cost of the refrigerator.

Its strength is such, from the pipes, although flat, being made in a series of small compartments, that they are able to bear almost any pressure of water that may be driven through them; and the sides of the refrigerator and the ends of the pipes being of brass, cast together and tinned, they cannot well be damaged.

Its simplicity is such that any workman, having a knowledge of a brewery, can use it; as all the inlets and outlets are connected by union joints, and the supply of wort and water regulated by stop cocks.

The various advantages which have been shown to belong to this refrigerator cannot fail to be appreciated, as valuable acquisitions to the manipulation in the art of brewing, especially when taking into consideration the economy of cost, which a reference to the price list will show.

The readiness with which this apparatus can be adapted to existing arrangements in any establishment is also a great recommendation. Indeed, the portability of those of a moderate size is such that they may be moved from one place to another with the utmost convenience, if necessary, only requiring a few inches' fall, dependent on the distance from the hop back.

The water, after its employment for cooling, may be made available for all purposes for which it may be needed. Its heat upon leaving the refrigerator is about 140°.

The results attending the use of this admirable apparatus have been most astonishing. The numerous firms adopting it, have found it so much to exceed their expectations, that they have spontaneously forwarded to the inventor their testimonials in its favour. These may be seen on application, together with price lists, and full particulars of its extraordinary cooling powers.

This refrigerator can be constructed without the copper bottom: in this case it is laid in a wood cooler, the passage of the wort, round the pipes, remaining the same. By adopting this plan, its power is slightly lessened, and for the purpose of cleansing, the apparatus must be lifted. Whichever arrangement is selected, the price will be the same.

LIST OF PRICES, &c.

The following is the guaranteed scale of sizes, together with the power of each for reducing the wort to 58° with water at 52°, accompanied by the cost, including royalty.

	Length.	Width.	Depth of Pipes.	Cooling Power per Hour.	Price.
No. 1	4 ft.	2 ft.	7½ in.	6 barrels	£ 50 0
No. 2	5 ft.	2 ft. 3 in.	7½ in.	8 barrels	63 0
No. 3	6 ft.	3 ft.	7½ in.	12 barrels	86 0
No. 4	7 ft.	3 ft.	7½ in.	14 barrels	95 0
No. 5	7 ft.	4 ft.	7½ in.	18 barrels	120 0
No. 6	7 ft.	4 ft. 4 in.	7½ in.	20 barrels	135 0
No. 7	8 ft.	4 ft. 4 in.	7½ in.	22 barrels	155 0
No. 8	9 ft.	4 ft. 4 in.	7½ in.	26 barrels	180 0
No. 9	10 ft.	4 ft. 4 in.	7½ in.	28 barrels	195 0
No. 10	8 ft.	4 ft. 4 in.	11 in.	32 barrels	220 0
No. 11	9 ft.	4 ft. 4 in.	11 in.	36 barrels	250 0
No. 12	10 ft.	4 ft. 4 in.	11 in.	40 barrels	275 0

The cost of those of larger dimensions can be obtained on application.

BRIGGS & STARKEY, *Leeds and Liverpool.*—Washing, wringing, and mangling machines.

Have obtained 47 first-class Prize Medals.

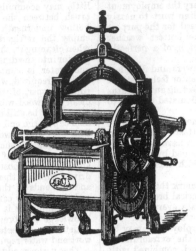

WASHING MACHINE.

THE patentees have had their machines tested in nearly all parts of the world, and have received the largest number of first-class prize medals and others, for improvements in their patent washing, wringing, and mangling machines, thus showing their superiority over all others at present in the market. A list of 10,000 references can be had, to persons who have their machines in regular use, on application to the manufactory. Intending purchasers can have machines sent to any part of the kingdom upon application, accompanied with a good reference, to the above address. Specimens may be seen and prices obtained in the Exhibition Buildings.

Prices varying from £1 16s. to £9.

[1811]

CAMERON, PAUL, *Glasgow.*—Steam pressure and vacuum gauges, improved self-acting lubricator.

[1812]

CARR, THOMAS, *New Ferry, near Birkenhead.*— Patent disintegrator mills ; patent fan blower.

CARR'S PATENT DISINTEGRATOR MILL, for disintegrating and mixing conglomerated phosphates, guano, chemicals, &c. Also for pulverising bone ash, boiled bone, chemical crystals, coal, and other unfibrous or brittle materials. Also for mixing purposes, such as converting brown sugars of various shades into one uniform sample. Price,

Without external wood casing £60
With external casing complete 64

This machine, which requires about 6 horse-power to drive it, is warranted thoroughly to break up, pulverise, and perfectly mix from 30 to 40 tons per day, of either hard and dry, or soft and damp conglomerated phosphate, guano, &c. without any inconvenience from becoming clogged or choked in the operation.

When applied to pulverise bone ash, or boiled bone, no mill driven by the same power can at all approach it in rapidity, as from 60 to 70 tons a day of these materials have been reduced by it to a powder, varying from dust up to the size of rice. For mixing purposes alone the machine has also given great satisfaction at sugar works, and other manufactories.

A small machine, capable of being worked by hand as well as by power, chiefly for mixing purposes, is also manufactured. Price, with iron casing, complete, £21.

Further information may be obtained either from the patentee, Thomas Carr, New Ferry, near Birkenhead, or from the manufacturers, Messrs. Richmond & Chandler, Salford, Manchester, either of whom will forward, on application, an illustrated circular, fully explaining the details and principles of the machine.

[1813]

CARRETT, MARSHALL, & CO., *Sun Foundry, Leeds.*—Compound direct-action condensing engine, double-action steam pump and fire engine, &c. (*See page* 11.)

CARRETT, MARSHALL, & CO., *Sun Foundry, Leeds.*— Compound direct-action condensing engine; double-action steam pump and fire engine; 3-horse engine.

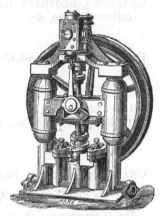

FIG. 1.   FIG. 3.   FIG. 2.

PATENT STEAM PUMPS to raise in one continuous stream from 3,500 to 100,000 gallons, from 50 to 100 feet, in 10 hours or upwards.

Fig. 1. Constructed for feeding stationary boilers, for 7,000 gallons in 10 hours, and sizes upwards.

Fig. 2. Constructed for feeding marine and locomotive boilers, for 7,000 gallons in 10 hours, and sizes upwards.

Fig. 3. For the above purposes, and also as a water lift, to raise up to 100,000 gallons in 10 hours. Fig. 3, as exhibited, is made double-action as a stationary fire engine, delivering in a perfectly continuous stream 10,000 gallons 125 ft. high per hour. All these modifications have inlet and outlet air-vessels.

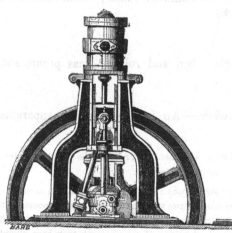

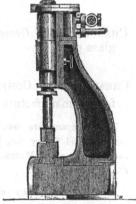

FIG. 7.   FIG. 6.   FIG. 8.

Fig. 6. A THREE-HORSE POWER PORTABLE ENGINE, upon an improved vertical-flued boiler of 4 horse-power, constructed of boiler plate welded and riveted, the interior being everywhere accessible by a man-hole for cleaning out. The boiler serves the purpose of foundation entirely.

Fig. 7. The same class of DIRECT-ACTION VERTICAL ENGINE, constructed from 1 to 25 horse-power. All the parts are direct-action and self-contained on one plate, and expansive motion for variable cut-off of steam is advantageously applied herein.

A PATENT HYDRAULIC RECIPROCATING ENGINE used as a motor, from 1 to 60 strokes per minute, for working bellows of organs, stone-sawing, and other reciprocating purposes. Two examples exhibited.

A NOMINAL 12-HORSE POWER HORIZONTAL COMPOUND ENGINE contained on one bed-plate and foundation, with one slide for varied degrees of expansion, working both

the high and low pressure cylinder, with the least possible distance for the port steam to traverse, with the freest openings into and out of cylinders, and with the working parts of one engine balancing the other. Condensor and air pump are double-acting in direct action. This engine has no dead centre, and will work at a maximum speed.

Fig. 8. PATENT SELF-ACTING 2-CWT. STEAM HAMMER, without levers, valve motion, or stuffing box; and for a variable stroke and intensity of blow.

A model of PATENT WATER CRANE worked by the pressure of Town's water or hydraulic pressure. This pillar crane will swing all round, and is of 3 powers, consuming water in proportion to weight raised only. These cranes are also made of 6 powers. For 3 powers the water is admitted on 1 and 2 and 2 + 1 = 3 area, and for 6 powers the areas are in the ratio of 1, 2, 3, from which any proportion up to 6 can be obtained.

[1814]

CATER, HENRY, 9 *Anchor Terrace, Southwark Bridge.*—Patent multitubular steam boiler.

[1815]

CHADBURN, BROTHERS, *Nursery, Sheffield.*—Patent metallic steam or water pressure gauges, tallow feeders, &c.

[1816]

CHALMERS, DAVID, 43 *Holmhead Street, Glasgow.*—Hot air engine.

[1817]

CHANDLER, JAMES, 10 *Mark Lane, London, E.C.*—Patent flat glass water gauges for steam boilers and other vessels. (*See page* 13.)

[1818]

CHANTRELL, GEORGE FREDERIC, 6 *Hatton Garden, Liverpool.*—Model of Chantrell's patent animal charcoal revivifying furnace for sugar refineries.

The 12-chamber size of this furnace is calculated to return from 60 to 70 tons per week.

The sizes of furnaces vary from 4 chambers to 48, or 12 rows of 4 chambers each.

These furnaces are in operation in all the leading refineries in the kingdom, and are cheap, durable, and efficient; effecting a saving in fuel of upwards of 50 per cent.

[1819]

CHAPLIN, A., *Glasgow.*—Carrying and traction engine for common roads ; steam crane used by the Commissioners ; ship's crane. (*See page* 14.)

[1820]

CHEDGEY, JOHN, *Grove, Southwark.*—Mangle, with glass bed and rollers, glass pump, and glass pipes.

[1821]

CHESHIRE SALT COMPANY (Limited), *Winsford, Cheshire.*—An improved steam apparatus for the manufacture of salt.

This Company are manufacturers of table, butter, common, and fishery salt, by a patent steam process, and proprietors of MESSRS. JUMP & HALL'S PATENT FIRE-FEEDERS.

MACHINERY IN OPERATION:—

JUMP & HALL'S PATENT STEAM-PAN, for the manufacture of fine or table salt.

STEAM-PANS ATTACHED TO THE BOILING PAN, for the purpose of making common or fishery salt.

JUMP AND HALL'S PATENT FIRE-FEEDER.

Samples of salt manufactured by the Cheshire Patent Salt Company (Limited).

These steam-pans and fire-feeders have been in successful operation for two years, and from the great economy in labour, wear and tear, fuel and heat, combined with their great simplicity of arrangement, they have proved themselves most valuable inventions.

The fire-feeders can be attached to any boiler, and can be seen daily at work at the Company's Works, at Winsford, Cheshire, where the whole of the salt-pans are erected on the patent steam principle.

[1822]

CLARK, D. K., 11 *Adam Street, Adelphi, London, W.C.*—Smoke-consumer, and feed-water heater.

[1823]

CLARK, JOSEPH LESTER, 2 *Sambrook Court, Basinghall Street, E.C.*—Patent fire bars for consuming smoke and economizing fuel.

CHANDLER, JAMES, 10 *Mark Lane, London, E.C.*—Patent flat glass water gauges for steam boilers and other vessels.

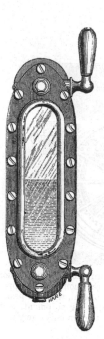

SINGLE GAUGE A.

UNIVERSAL GAUGE B.

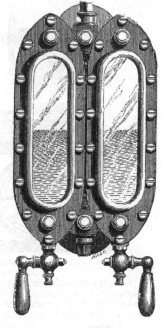

DOUBLE AND CHECK GAUGE C.

The chief advantages of these gauges are—strength, simplicity, durability, steadiness of water level, and perfection of sight.

GAUGE A. This gauge is intended more especially for new boilers.

GAUGE B. This gauge is intended to replace the common glass tube on existing boilers, or to be used in conjunction with the ordinary glass tube connexions for new boilers. These cases are made various lengths to suit various boilers, and can be attached, by simply removing the glass tube, and into the same stuffing boxes inserting the tubes attached at each end of the gauge. Then place the patent case between them, and finally screw the two tubes into it. Persons desirous of using this gauge with existing boilers, should send the following dimensions:—1st. Length of tube they are using. 2d. Diameter of ditto. 3d. Distance between the glands.

GAUGE C. The object of this gauge is, that the several indications should check each other; and should any accident occur to either of them, it can be shut off during repair, at the same time the opposite one can be used singly or as duplicate, or both can be shut off at the same time.

The exhibitor has always a large stock on hand.

The following are selected from a great number of testimonials to the value of these gauges:—

"*Metropolitan Board of Works, Engineers' Department, Spring Gardens, March 20, 1862.*

"Two of Mr. Chandler's patent flat glass water gauges have been fixed on two high-pressure engine boilers belonging to the Metropolitan Board of Works at St. George's Wharf, Deptford, for nearly two years. They have been exposed all the winter, and have withstood great pressure without leaking, cracking, or any defect whatever. The water-line can be readily seen by day and night, and they are well adapted for such boilers.

"J. W. BAZALGETTE, *Engineer.*"

"*Engine and Agricultural Implement Manufactory, Lynn, Feb. 28, 1861.*

"DEAR SIR,—I beg to say that the gauge I had of you answers the purpose very well. The size of the glass shows such a large column of water that the driver can see it some yards off. I have just had another, as you are aware, which I am now fixing on a new engine, and hope by harvest to want several others. I shall certainly continue to use them, as I think them far preferable to any other that I have seen, and would urge other makers to give them a trial; and if they do, I feel sure they would be equally pleased with them. Wishing you a large demand,

"I am, sir, your truly,
"JAMES CHANDLER." "R. S. BAKER, *Engineer.*"

"*Deptford, March 2, 1861.*

"DEAR SIR,—In answer to yours of the 25th ult., asking for a testimonial, I beg to say that the two gauges you put on the boilers here are the most perfect I have ever seen, they having withstood great pressure and much exposure. I have every confidence in them, and deem them worthy of the highest recommendation, and shall at all times be happy to speak in their favor.

"I remain, yours truly,
"R. A. RUMBLE, *Engineer.*"
"MR. J. CHANDLER."

"*Lambeth Waterworks, Kingston, July 8, 1861.*

"MR. CHANDLER.

"SIR,—In reply to your inquiry respecting your patent flat glass water gauges in use on two of these boilers here, I am happy to say they have given great satisfaction, and I am very pleased with them, as they have not been the least trouble since they started, which is nearly two years ago. I am sure any one who tries them will be highly pleased, especially those made according to your second patent.

"H. CARRUTHERS."

CHAPLIN, ALEXANDER, *Glasgow.*—Carrying and traction engine for common roads; steam crane used by the Commissioners; ship's crane.

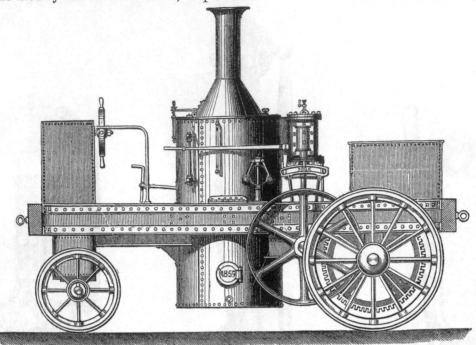

CHAPLIN'S PATENT PORTABLE STEAM ENGINES AND BOILERS.

From the strength, simplicity, and compactness of these engines, they are extensively used for general purposes, and also in situations where steam engines of the ordinary construction cannot be applied.

STATIONARY ENGINES (Fig. 1) require no building in, nor chimney stalk, and with the forced-combustion apparatus will burn inferior qualities of coal, wood, or peats. These engines are specially suited for shipment, and may be packed inside the boiler to economise freight.

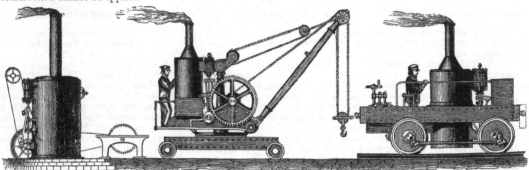

Fig 1. STATIONARY ENGINE.
From 1 to 30 Horse-power.

Fig. 2. PORTABLE STEAM CRANE.
30 Cwt. to 10 Tons.

Fig. 3. CONTRACTORS' LOCOMOTIVE.
6 to 27 Horse-power.

PORTABLE STEAM CRANES (Fig. 2) for wharf or railway, with wrought-iron carriages on wheels, sink motion, foot break, &c. all under the easy control of one man; the 4 and 5 horse power hoist and lower by steam, and twist by hand; the larger sizes hoist, lower, and turn round by steam.

CONTRACTORS' LOCOMOTIVES (Fig. 3) are adapted to work on rails or tramways of a gauge from 2 ft. upwards. They are complete and efficient locomotives, simple in construction, and the working parts easily got at for repair. They draw heavy loads at reduced speeds; for shipment these engines are usually sent in one package, ready for work on arrival.

ROADWAY OR TRACTION ENGINES (as illustrated above), are adapted for travelling over hilly or soft ground, for simply propelling themselves, or for taking behind them heavy loads at a speed, proportionate to the load, of from 2 to 10 miles an hour. Each engine is complete with coal and water tanks, &c. and under the control of one man.

CARRYING ENGINES adapted to carry loads up to 50 tons.

HOISTING ENGINES, on carriages of wood or iron, and iron wheels, with crab winch, &c. complete. The engine, break, &c. are under the easy control of one man.

HOISTING ENGINES, similar to above, but with pillar and jib, to swing about three-quarters round by hand.

LIGHT PORTABLE ENGINES, specially adapted for agricultural purposes, and for sawing, pumping, &c.; while, from their lightness and simplicity of construction, the 4 and 6 horse power are an easy load for one horse. The larger sizes are mounted on 4 broad roadway wheels, the front pair being made to swivel.

SHIP ENGINES specially suited for winding, cooking, distilling, &c. on board ships of every class; and for aiding the crew in performing the heavy work of the ship, such as heaving anchors, discharging cargo, hoisting heavy sails, &c. One fire serves both for the steam boiler and cooking and distilling apparatus, with a small consumption of fuel.

Prices and other particulars may be learned by applying at the Cranstonhill Hill Engine Works, Glasgow, or at the London depôt, Lambeth Wharf.

[1824]

CLAYTON, SHUTTLEWORTH, & CO., *Lincoln, and* 78 *Lombard Street, London.*—High pressure fixed and portable steam engines.

A 12-HORSE POWER HORIZONTAL FIXED STEAM ENGINE, manufactured by the exhibitors, with cylinder 11 in. in diameter, 16-in. stroke, governors, and all usual appendages, fitted on planed-up iron bed plate, complete.

Price, including Cornish boiler, 14 ft. long by 5 ft. diameter, made of Lowmoor and best Butterley plates £280 0

[1825]

CLOWES, FREDERIC J., 92 *Southwark Bridge Road, London.*—Patent metallic spring steam and vacuum gauges, and steam boiler fittings.

[1826]

COFFEY, JOHN A., *Finsbury.*—Pharmaceutical and other apparatus, stills, &c,

[1827]

COLQUHOUN & THOMSON, 1 *Laurence Pountney Hill, Cannon Street.*—Movable girder fire bars.

[1828]

COOMBE & CO., 30 *Mark Lane; Manufactory, Gower's Walk, London.*—French mill stones, flour machines, wire brushes, patent smut machines, general flour mill machinery.

(*Obtained medals at the Great Exhibition of* 1851, *for woven wire, &c.*)

COOMBE & Co. subjoin a list of the manufactures in which they are engaged.

General wire weavers, workers, and brush makers; builders of French millstones, and importers of French burrs; dealers in peak, Cologne, and grindstones.

Manufacturers of patent iron revolving flour-dressing machines, and Ashby's patent corn or smut machines. Improved steel machine wire; brushes of all kinds for machinery; weighing machines, scales, beams, steel mills, &c.; wood cylinders; iron millstone provers; mill chisels and picks; patent punched iron; hoisting chain; extra strong wire malt-kiln heads, and malt screens; separators for wheat, barley, oats; trucks, shovels, corn measures, sieves, brooms, &c.; leather straps, elevator webbing, tin buckets and rivets, gut, gutta-percha bands; waterproof cart covers, sacks, &c.; iron pulley blocks, screw jacks, &c.; improved patent bolting cloths.

They are also importers of Swiss and French silks.

FRENCH MILL STONE.

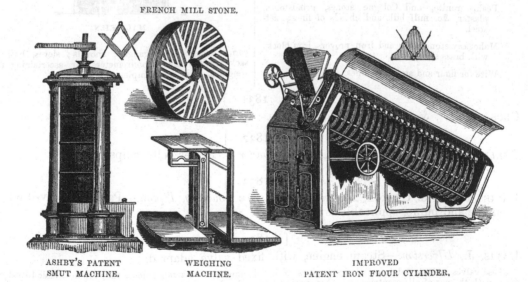

ASHBY'S PATENT SMUT MACHINE.   WEIGHING MACHINE.   IMPROVED PATENT IRON FLOUR CYLINDER.

[1829]

CORCORAN, BRYAN, & CO., *Mark Lane.*—Specimens of metallic cloth; model of malt kiln; silk flour-dressing machine, mill stones, &c. (*See page* 16.)

[1830]

COWAN, THOMAS WILLIAM, *Kent Iron Works, Greenwich.*—Patent nominal 16-horse power trunk engine. (*See page* 17.)

CORCORAN, BRYAN, & Co., *Mark Lane.*—Specimens of metal cloth; model of malt kiln; silk flour-dressing machine, mill stones, &c.

BRYAN CORCORAN, & Co. are the original makers of paper-machine wires, which they now weave to the width of 9 ft. They manufacture every sort of wire work, deckle straps, felts, dandy rolls, moulds, and every description of driving bands. Established 1805.

THE CASE OF SPECIMENS CONTAINS :—

Samples of wire-drawing in the various stages, from the bar of metal to the finest thread of wire.

3,000 yards of copper wire, (or nearly 1¾ miles) drawn out of an old penny-piece.

1,300 yards of brass wire, (nearly ¾ of a mile) weighing only 1 ounce.

1,000 yards of iron wire, (nearly ½ a mile) weighing only 1 ounce.

Samples of woven wire, from 1 to 28,800 holes in a square inch.

Fine and strong samples of various sorts; samples of Swiss silk, &c.

The largest millstone is 5 ft. 8 inches diameter in one solid block : a very rare specimen.

Millstones of various sizes, of the finest quality ever produced, for grinding wheat.

Peak, granite, and Cologne stones, grindstones, plaster, &c. mill bills and chisels of finest cast-steel.

Mahogany stone staffs and iron provers, iron blocks with brass sheaves.

Wire for flour and smut machines.

Silk dressing machines, elevators, and worms.

Separators for peas, wheat, &c.

Brushes of all sorts for machinery.

Corn measures of all description.

Sack chains, jiggers, punches, spanners, &c.

Swiss dressing-silk.

Blackmore's bolting cloths.

The exhibitors are also erectors of malt kilns on improved principles, as shown in model; makers of woven-wire kiln plates of any dimensions; malt and

MALT KILN.

corn screens; malt gauges; shovels; sieves, bushels, sack trucks, and chondrometers for ascertaining the weight of corn from sample.

[1831]

CROSS, T. W., & Co., *Leeds.*—Fire engines.

[1832]

CUNLIFFE, THOMAS, & SONS, *Ardwick, Manchester.*—Leather belts, skips, &c.

[1833]

DAVIES, JONAH & GEORGE, *Albion and Limerick Foundry, Tipton.*—Patent improved rotary engine and pump, applicable to all purposes.

[1834]

DAVIS, J., *Ulverston.*—Steam engine, with fixed valve adapted.

These valves are applied to oscillating engines to dispense with the use of all eccentrics and other gearing for

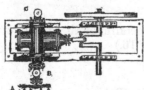

working the steam valve, or "reversing." From the

simplicity of arrangement, they are not liable to get out of order : the wear keeps the valve faces true.

The engine (plan of which is given) is adapted for marine, locomotive, and general purposes. The motion of the engine is reversed by simply moving the index or lever *A*, to the side or direction in which it is desired to move. *B* steam pipe, *C* exhaust pipe.

The model shows another arrangement of valve to effect the same purpose.

Price of engines, complete, from £7 to £10 per horse power, according to size. (Exhibited in Class VII B.)

COWAN, THOMAS WILLIAM, *Kent Iron Works, Greenwich.*—Patent nominal 6-horse power trunk engine.

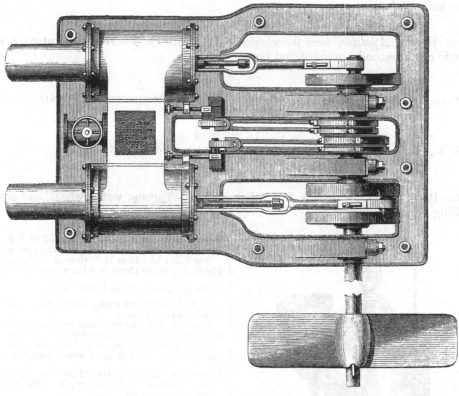

PATENT TRUNK ENGINE.

BURGH & COWAN'S PATENT TRUNK ENGINE of 16-horse power.

It is a well-known fact that the trunk engine is the most simple at present in use; but the immense friction of the trunks in their respective stuffing boxes, and their alternate exposure to the steam and atmosphere, render them highly destructive to steam and tallow.

T. W. Cowan, the sole manufacturer of Burgh & Cowan's patent engines, is desirous of introducing them to the public. The following are a few of the many advantages gained by the use of these improvements:—

1. The area is gained, hitherto lost in trunk engines, thereby a saving in space.

2. The immense stuffing boxes being entirely dispensed with, a great reduction in friction and packing is effected.

3. The trunks are never alternately exposed to the steam and atmosphere; also the moving or piston trunk is entirely frictionless, gaining a considerable saving in tallow and in steam.

4. The connecting rod is in the centre of the cylinder, and perfectly accessible to tighten and lubricate, which dispenses with the guides beyond the cylinder.

5. The guides being within the cylinder, and cast in the trunk, they never get loose, and are entirely out of harm's way.

6. In beam engines, this improvement entirely dispenses with the expensive and complicated parallel motion, thus rendering engines cheaper and simpler than those at present in use.

7. In high and low pressure engines, the high pressure is within the low pressure, while the areas of both are maintained. This is a great advantage over those at present in use.

8. The simplicity of the whole engine, together with the small space it occupies at any given horse-power, renders it highly advantageous, particularly for marine purposes.

9. In stationary engines the connecting rods are about six times the length of crank.

10. Marine engines made on the same principle as the above engraving are much lighter, take up less space, and are much cheaper to work than any other description of engines.

11. Vertical engines on this principle are particularly adapted for places where there is little room to spare; a 10-horse power engine only taking up the space of 1 foot 4 inches by 1 foot 8 inches.

12. In steam fire engines the pumps are connected by a rod, to the piston, through the bottom trunk, thereby taking up less space.

13. The high and low pressure engines are invaluable where fuel is expensive, as they save a great deal of steam that is altogether lost in other engines.

These engines being of the best materials and workmanship are found to be cheaper and work longer than any other engines.

[1835]

DAWSON, CHARLES S., *Thames Ditton, Surrey.*—Hydrostatic engine.

[1836]

DAWSON, JOHN, *Greenpark, Scotland.*—A machine for protecting the revenue derived from the manufacture of spirits.

[1837]

DEACON, HENRY, *Appleton, near Warrington.*—100 millions 4-wheeled counter; 3-wheeled electric clock, seconds, minutes, and hours.

[1838]

DINGWALL, WILLIAM, 4 *Idvies Street, Dundee.*—Patent water meter, with distributing valves placed in a movable diaphragm.

[1839]

DIXON, E., *Wolverhampton.*—Wrought-iron gas tubes and connexions.

[1840]

DONKIN, B., & CO., *near Grange Road, Bermondsey.*—Turbine water wheel, and gas valve. Drilling apparatus for mains.

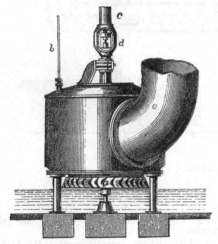

TURBINE WATER WHEEL.

TURBINE WATER WHEEL, suitable for a high fall of water.

a Revolving ring with buckets in a single casting.
b Shaft communicating with regulating valve tackle.
c Vertical shaft, transmitting the power.
d Bearing brass for supporting weight of revolving wheel, which bearing, being out of the water is readily accessible.
e Pipe for bringing water to casing.
This wheel, with a 40 ft. fall of water, would give a

power of 36 horses, or 33,000 lbs. lifted 1 ft. high per minute, and would make 150 revolutions per minute.

For falls under 15 or 16 ft. the casing *f* is unnecessary; the turbine being placed in a brick or wooden pit.

The advantages of the turbine are :—

A high speed, rendering the gearing comparatively simple and inexpensive.

Freedom from the inconvenience arising from floods, as the wheel will work immersed many feet under water.

An economy with regard to the useful effect, as compared with an ordinary water wheel, on any given fall.

IMPROVED VALVE FOR GAS MAINS, so constructed that there are no external working parts ; the one exhibited is for a main of 30 in. bore.

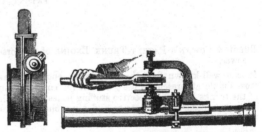

VALVE.   DRILLING APPARATUS.

UPWARD'S PATENT DRILLING APPARATUS.

This invention is calculated to prevent accidents, by furnishing the means of drilling holes in gas mains when laying service pipes, as the hole is both drilled and tapped, without allowing an escape of gas.

[1841]

DORWARD, WM. L., 15 *Camden Square, Camberwell.*—Rotary engines for ships' propellers, and other purposes.

[1842]

DUNCAN, THOMAS, 44 *West Derby Street, Liverpool.*—A water meter, from which power may be obtained for driving machinery.

[1843]

EADIE & SPENCER, *Glasgow.*—Iron tubes for boilers.

[1844]

EASTON, AMOS, & SONS, *Grove, Southwark.*—Patent centrifugal Appold pump, improved turbine, hydraulic ram, pumps, &c.

THE FOLLOWING MACHINERY IS EXHIBITED :—

An improved patent combined APPOLD'S CENTRIFUGAL PUMP AND STEAM ENGINE, for drainage of marsh lands or irrigation, and used also for graving dock, and other purposes. The machine exhibited is of 40-horse power nominal, and is driven by a pair of expansive condensing steam engines. It is capable of delivering 100 tons of water per minute at a mean lift of 6 ft. The principal advantages obtained by the arrangement are, compactness, economy, and the dispensing with the greater portion of the ordinary massive foundations, the machine being entirely self-contained.

Smaller patent APPOLD CENTRIFUGAL PUMPS, of improved construction, for general purposes. The construction is such, that the whole of the internal working parts, may be withdrawn, without disturbing the casing and framing.

Improved PATENT HYDRAULIC RAMS for supplying small towns, mansions, &c. with water, in sites where a small fall exists.

IMPROVED TURBINE on the "Tourneyron" principle, adapted for either high or moderate falls of water.

The arrangement adopted secures compactness, easy accessibility to the working parts, a greatly improved arrangement of regulating-gate for controlling the quantity of water, and an improved method of suspension.

PATENT REGULATING VALVE, for maintaining a constant and uniform steam pressure, with a varying pressure in the boiler, applicable to any situation, or any establishment, where both high and low pressure steam are required at the same time, from one boiler or one range of boilers.

Sundry smaller articles.

[1845]

EDWARDS, C. J., & SON, 32 *Great Sutton Street, London, E.C.*—Leather bands, leather hose, and fire buckets.

[1846]

EDWARDS, RICHARD, 12 *Fairfield Place, Bow, E.*—Models of machinery for pulverising mineral, vegetable, and animal substances.

[1847]

ENGLAND, G., & CO., *Hatcham Iron Works, Hatcham.*—Screw jack.

[1848]

EVERITT, A., & SONS, *Birmingham.*—Brass, copper, and iron articles.

TUBES IN BRASS, solid drawn, for locomotive, marine, and other boilers.
These tubes are always drawn taper, to give extra thickness at the end nearest the fire. The taper is inside, and given by drawing the tubes upon steel mandrels, the outer diameter of the tube is parallel.
The expanded ends are to show the quality and ductility of the metal.
Tubes in brass and copper of various dimensions for steam, gas, &c. all drawn solid.
Tubes in brass, ornamental, rope twist, &c. for chandeliers, coronas, &c.

BRASS, COPPER, STEEL, AND IRON WIRES.

Brass, iron, and copper wires, for weaving, drawn as fine as human hair. Brass wire for pins, drawn malleable

to allow the head to be formed upon it, and sufficiently stiff not to bend.
Steel wire for springs.
Iron wire for furniture springs.
Iron wire for weaving of various colours, such as white and blue. Iron wire galvanized with copper for furniture springs.
Iron wire tinned.
Iron wire galvanized with zinc, for electric telegraphs; hank of 1 cwt. Iron wire for ropes, best charcoal quality; hank of from 30 to 40 lbs.
Brass sheets of various descriptions thin rolled brass or latten.
All these articles are made at Messrs. Everitt's Works at Birmingham. Prices, and all information can be obtained there, and at their London offices, 33 Clement's Lane, E. C.

[1849]

FARROW & JACKSON, 18 *Great Tower Street, London, E.C.*—Machines, &c., used in the management of wines, spirits, oil, &c. (*See pages* 20 *and* 21.)

[1850]

FAWCETT, PRESTON, & CO., *Liverpool.*—Cane mill and engine; Aspinall's patent evaporating pan; vacuum apparatus; centrifugal machines. (*See pages* 22 *and* 23.)

FARROW & JACKSON, 18 *Great Tower Street, London, E.C.*—Machines, &c., used in the management of wines, spirits, oil, &c.

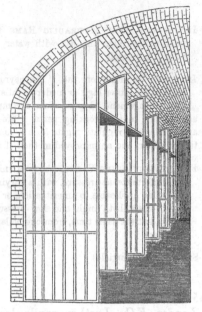

For Arched Vaults.

For Flat Ceilings.

WROUGHT-IRON WINE BINS.

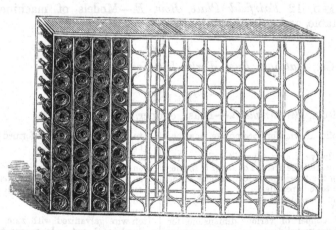

REGISTERED CELLULAR WINE BINS, WITH SEPARATE REST FOR EACH BOTTLE.

WROUGHT-IRON SCANTLING FOR CASKS.

FARROW & JACKSON, *continued.*

PATENT BOTTLING APPARATUS.

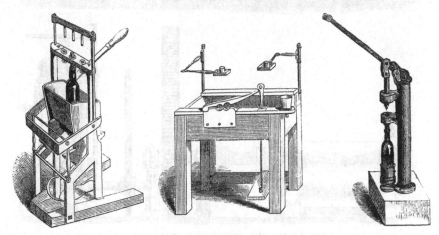

CORKING MACHINES.

BOTTLE-WASHING MACHINE.

PATENT CAPSULING MACHINE.

PATENT SPIRIT INDICATOR.

SAMPLE CASE.

TASTING SPITTOON.

FAWCETT, PRESTON, & CO., *Liverpool.*—Cane mill and engine; Aspinall's patent evaporating pan; vacuum apparatus; centrifugal machines.

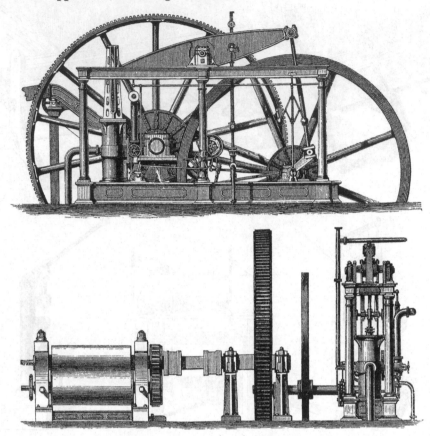

NON-CONDENSING ENGINE AND SUGAR-CANE MILL.

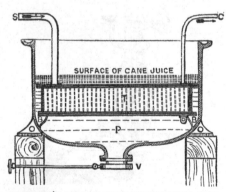

ASPINALL'S PATENT OPEN EVAPORATING PAN.

P—Represents the pan, which is filled with cane juice to about an inch or two above.

T—The tube box or steam chest which rests upon brackets, *B*, and is fitted with vertical tubes, open top and bottom, so that the cane juice has free passage through them; the steam which surrounds the tubes being admitted through

S—The steam pipe.

C—Is the pipe through which the condensed water escapes, thus arranged to avoid joints in the pan, and to make it simple and easy to cleanse, for by lifting out the tube box it leaves only the plain pan to wash out.

V—The discharge valve, worked by a hand-wheel and screw.

O—Outside brackets on which the pan rests, so that a few pieces of timber are all that are required for its support.

B—The inside brackets on which *T*, the tube box, rests.

MESSRS. FAWCETT, PRESTON, & CO., are also manufacturers of

STEAM ENGINES—land and marine, of every description, high or low pressure, combined or condensing, &c.

DREDGING MACHINERY.

BOILERS—of wrought-iron, Cornish, plain, flue, tubular, multitubular, &c.

MILLS—for rice, corn, shumac, mortar, sugar (horizontal or vertical, driven by cattle, wind, water, or steam-power, and on De Mornay's patent); also, rolling mills for iron.

SUGAR APPARATUS of every description.

CENTRIFUGAL MACHINES—(Patent).

VACUUM PANS and APPARATUS complete.

PANS—of all descriptions; wrought and cast iron, and copper patent evaporating pans: clarifiers, cisterns, and tanks of wrought and cast iron.

PRESSES—hydraulic and screw, worked by hand or steam, for oil, cotton, grapes, &c.

IRON ROOFS, GAS APPARATUS, COTTON GINS, &c.

ORDNANCE and AMMUNITION—smooth-bored or rifled cannon, in steel, brass, iron, and on Blakely's patent. Carriages, limbers, and ammunition waggons. Patent rifle-bullet presses.

Wrought-iron work, castings, and millwright work of all kinds.

Steam vessels, dredge boats, barges, &c. &c.

FAWCETT, PRESTON, & CO., *continued.*

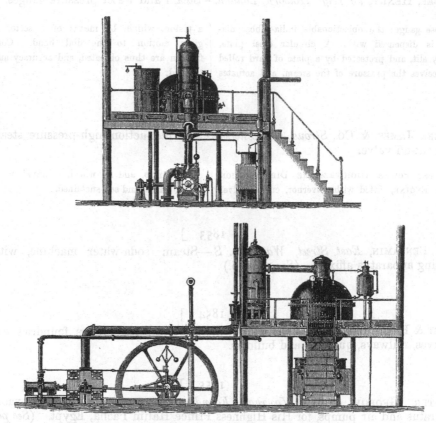

VACUUM APPARATUS FOR THE MANUFACTURE OF SUGAR.

FIG. 1.   FIG. 2.

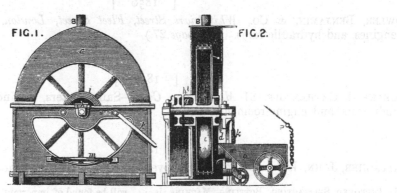

THOMAS' CENTRIFUGAL MACHINES.

Fig. 1 is an elevation of one of the improved machines; and Fig. 2 is a vertical cross sectional elevation of the same, and a side elevation of one of the trucks used to feed the apparatus.

A is the external casing, provided with B, opening to allow the moist air to escape ; C the revolving drum, for containing the charge to be operated upon ; D, a conical flanch surrounding the annular opening, through which the charge is inserted ; E, circular opening in the side of external casing, which is surrounded by an annular lip or flanch, which projects over the outer edge of the conical flanch D, to prevent any portion of the charge falling between the inner side of the outer cover and the outside of the revolving drum, by which arrangement the apparatus can be charged during the time it is in motion ; F, a loose pulley on axle of the revolving drum ; G, a fast pulley on ditto ; H, friction pulley on the axle of the machine, against which acts I, the friction strap ; K, the lever to actuate the break gear for stopping the machine ; L, division to prevent the syrup falling on the strap ; M, a transverse horizontal bar, upon which the truck rests during the time it is being emptied of its contents ; O, the improved truck for charging the machine, which is mounted on a pair of wheels, and a single one in front, the bearings of which turn on a vertical axis, to enable the carriage to be re-turned round with facility ; P, pulley block and tackle for elevating the rear of the truck.

[1851]

FERRABEE, HENRY, 75 *High Holborn, London.*—Steam and water pressure gauges.

In these gauges the objectionable india-rubber diaphragm is dispensed with. A circular steel plate, peculiarly slit, and protected by a plate of hard rolled brass, receives the pressure of the steam, and actuates a piston, which, by means of a sector and pinion, gives motion to the dial hand. Corrosion and friction are thus obviated, and accuracy and durability ensured.

[1852]

FERRABEE, JAMES, & CO., *Stroud, Gloucestershire.*—Direct-action high-pressure steam engine, with cut-off valve.

A 14-HORSE POWER HIGH-PRESSURE DIRECT-ACTION STEAM ENGINE, fitted with governor, cut-off valve, feed pump, and fly wheel, mounted in a substantial iron frame, and self-sustained.

[1853]

FLEET, BENJAMIN, *East Street, Walworth. S.*—Steam soda-water machine, with patent bottling apparatus affixed. (*See page* 25.)

[1854]

FORREST & BARR, *Glasgow*—Patent safety derrick crane, for engineers, foundries, contractors, wharves, railways, quarries, and builders.

[1855]

FORRESTER, GEORGE, & CO., *Vauxhall Foundry, Liverpool.*—Triple-effect vacuum pan apparatus and air pumps, for His Highness Prince Halim Pacha, Egypt. (*See page* 26.)

[1856]

FOWLER, BENJAMIN, & CO., *Whitefriars Street, Fleet Street, London.*—Force pumps, fire engines, and hydraulic rams. (*See page* 27.)

[1857]

FRIEAKE & GATHERCOLE, 81 *Mark Lane, City.*—Salinometers, engine counters, telegraph indicators, and engine-room fittings.

[1858]

GALLAGHER, JOHN, *Wolverhampton.*—Improved self-acting bottling machine.

This IMPROVED SELF-ACTING BOTTLING MACHINE is adapted to fill 6, 8, or 10 bottles simultaneously, and will be found of important service to bottlers of wine, spirits, and malt liquors.

[1859]

GALLOWAY, WILLIAM & JOHN, *Manchester.*—Models of land and marine boilers ; safety valve and lifting-jack.

[1860]

GERARDIN & WATSON, 43 *Poland Street, Oxford Street, London.*—Watson's patent beer engine, and tavern bar fittings.

FLEET, BENJAMIN, *East Street, Walworth, S.*—A steam soda-water machine, with patent bottling apparatus affixed.

This STEAM SODA-WATER MACHINE is an improved method of manufacturing and bottling soda-water, lemonade, ginger-beer, and all kinds of mineral waters and aërated drinks, by means of a patent screw bottling apparatus, which forces the cork into the bottle without the aid of a mallet, and being elevated by a treadle; and pressed firmly against a suitable packing ring, on the under side of the filling piece, the air is excluded, and the otherwise dangerous operation entirely prevented. A further improvement is the application of a small steam cylinder, combining in one machine, the apparatus for making the soda-water, and a steam engine for driving the same, which being connected to the same shaft, the fly wheels answer the purpose of both.

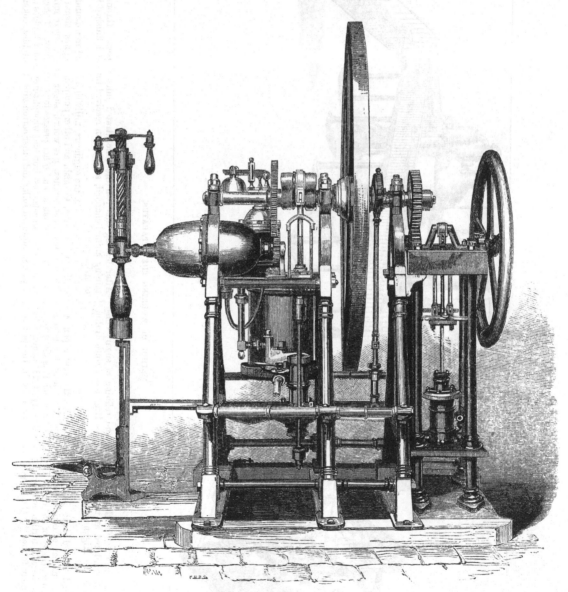

SODA-WATER MACHINE.

The machine produces 2,500 bottles per day (or over 200 doz.), and the principal features of the invention are the mechanical contrivances for the entire exclusion of all atmospheric air, and the ease with which it can be worked by non professional men.

The great success which has attended its working by the exhibitor, proves that from its solidity of construction, power, and completeness, it is a great acquisition to this increasing branch of trade.

FORRESTER, GEORGE, & Co., *Vauxhall Foundry, Liverpool.*—Triple effect vacuum pan apparatus and air pumps, for His Highness Prince Halim Pacha, Egypt.

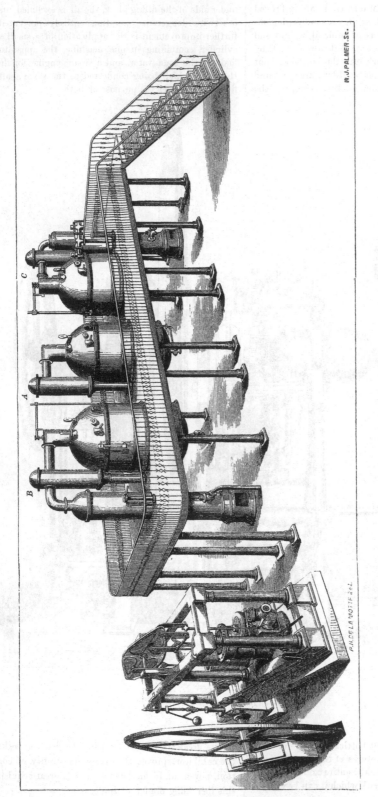

METHOD OF WORKING THE APPARATUS.

VACUUM PAN APPARATUS, for the manufacture and refining of sugar, constructed for and exhibited at the request of His Highness Prince Halim Pacha, brother to His Highness the Viceroy of Egypt.

The dessicated and filtered cane juice is received into the centre pan A, which contains a number of copper tubes surrounded by steam; the juice is here evaporated to a density of 15° to 18° Beaumé, after which it is discharged into the pan B, to be still further evaporated to a density of 28° Beaumé by means of the vapour formed by the evaporation of the juice in the pan A.

After leaving this pan the concentrated juice is passed through filters containing animal charcoal, and is then received into and finished in the vacuum or strike pan C.

G. Forrester & Co. are engineers, millwrights, and iron-founders, makers of stationary, marine, Cornish, and other pumping engines, steam boilers, steam dredging machines, cranes, gas works, sugar works, sugar mills, saw mills, corn mills, hydraulic and other presses for cotton, hay, oil, &c. ; centrifugal pumps, water wheels, cast-iron and wrought-iron girder bridges ; millwork and shafting, improved roller and other cotton gins, turntables, cokers, &c. &c.

FOWLER, BENJAMIN, & CO., *Whitefriars Street, Fleet Street, London.*—Force pumps, fire engines, and hydraulic rams.

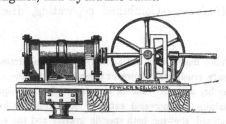

HORIZONTAL DOUBLE-ACTION PUMP.

1. FOWLER'S IMPROVED HORIZONTAL DOUBLE-ACTION PUMP (No. 143) for contractors' use, irrigation, and other purposes.

This pump is arranged either to lift large quantities of water from excavations, cuttings, docks, mines, &c., and delivering over embankments or through other channels; or it is suited for raising water, and forcing it through lengths of piping to any elevation required. Its merit consists in its extreme simplicity, and the ready means of working it direct from a portable or other steam engine. The pump discharges an equal quantity at both strokes of the plunger, and the valves are readily accessible.

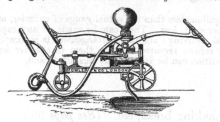

FIRE ENGINE.

2. FOWLER'S NEW AND IMPROVED FIRE ENGINE (No. 142), for towns, public buildings, mansions, manufactories, &c.

The principle of this pump is similar to the foregoing; it throws a large supply of water at both strokes of the plunger, the valves are readily accessible, it is fitted with an air vessel, and inlet and outlet unions, to connect hose piping, and with long handles at each end; it is mounted on a stout carriage, with wheels and drag handle. Its advantages are, great simplicity and few working parts, one barrel is made to do the work of two. It is well adapted for hot climates, as it will always be found ready for work after lying out of use.

DOUBLE-ACTION PUMP.

3. FOWLER'S IMPROVED DOUBLE-ACTION PUMP, mounted in frame with fly-wheel and handles (No. 138).

This pump is fitted with gun-metal plunger and brass bucket, and delivers the water at both up and down strokes in a constant stream. It is well adapted for supplying railway stations, public and private establishments, also for fixing on water lighters to supply ships with fresh water through hose pipes.

4. FOWLER'S IMPROVED DOUBLE-ACTION PUMP (No. 141), of a similar description to foregoing, mounted on base with pillar, forming air vessel and gear for driving by steam power.

This pump is adapted for manufactories, and other places where large quantities of water are required.

HOLMAN'S DOUBLE-ACTION PUMP.

5. HOLMAN'S PATENT DOUBLE-ACTION PUMP (No. 7), for steam power.

The advantage of this pump consists in the valve arrangements, being all contained in one chamber, readily accessible by the removal of a single plate.

DOUBLE-ACTION HAND FORCE PUMP.

6. FOWLER'S IMPROVED DOUBLE-ACTION HAND FORCE PUMP mounted on barrow (No. 46a).

This useful and powerful pump is well adapted for watering gardens, forcing water to a distance; for use as a small fire engine, and for a variety of useful purposes. It is thoroughly well fitted, and very economical in cost.

HYDRAULIC RAM.

7. The IMPROVED HYDRAULIC RAM (No. 60), for raising water for the supply of mansions, farms, &c. from a stream, brook, or spring, where a fall can be obtained. By this means water may be conveyed to a great distance and height.

This machine is entirely self-acting, and is capable of raising water 10 ft. high for every foot of fall obtained.

The figures in brackets relate to the numbers the various articles bear in B. Fowler & Co.'s general list.

[1861]

GIBBON, RICHARD, *Royal Brewery, Brentford, Middlesex.*—Combined separating, dressing, malt crushing machine.

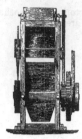

PATENT MALT-CRUSHING MACHINE.

This machine will thoroughly separate, dress, and evenly crush malt, however irregular in size the sample may be, without reducing any to powder, consequently obtaining an increased extract. An adjustable balance is affixed, showing both specific gravity and the extract obtainable per quarter.

A descriptive catalogue with prices, may be had of the patentee ; and also manuscript instructions for brewing India pale ale as brewed at Burton-on-Trent.

[1862]

GODWIN, RICHARD A., 151 *Newport Street, Lambeth.*—Flood pump, double-actioned ; retaining and other valves accessible by simply raising outlet valve.

The working parts being entirely at command, any "stoppage" that cannot be immediately remedied is impossible. As cheap and efficient water raisers they are unequalled ; one man with a 4-in. pump, 6-in. stroke, discharging 1,455 gallons of water per hour, being but 15 gallons less than the actual gauge of cylinder, or with 1 per cent. loss. They can be made of any capacity ; need no fixing ; and their arrangement is so simple that any repairs required to keep them in effective working condition can be done by any unskilled hand.

[1863]

GOODALL, H., *Derby.*—Machines for grinding and making bread, &c. (*See page* 90.)

[1864]

GOUGH & NICHOLS, *Back Quay Street Works, Manchester.*—Improved vertical portable engine, for contractors and others.

[1865]

GRAUTOFF, B. A., & CO., 4 *Lime Street Square, E.C.*—Steam and vacuum gauges and salinometers.

[1866]

GRAY, JOHN WILLIAM, & SON, 114 *Fenchurch Street, City, and* 1 *Margaret Street, Limehouse.*—Patent spherical steam engine.

[1867]

GREENING & CO., *Victoria Iron Works, Manchester.*—Fixed oscillating steam engine, with simplified surface valve.

[1868]

GREW, NATHANIEL, 8 *New Broad Street, City.*—Model of a locomotive engine for running on the ice ; scale one-eighth full size.

A model showing the general construction of an ICE LOCOMOTIVE sent to Russia last autumn, which has been successfully at work during the winter on the river Neva, between Cronstadt and St. Petersburg, conveying passengers and goods.

The full-size engine, of which a photograph is exhibited, was constructed by Messrs. Neilson & Co. of Glasgow, and bids fair to be a very useful agent, on the great rivers, and inland seas of Russia in transporting goods, &c. when the ordinary navigation is closed. The engine when in working trim weighs about 12 tons, has cylinders of 10 in. diameter by 22 in. stroke, with driving wheels of 5 ft. diameter ; these wheels are studded with steel spikes to obtain the necessary adhesion. The steering of the engine is accomplished, by shifting the front sledges in the required direction by means of an endless screw and worm wheel, working a pinion, gearing into a circular rack bolted to the sledge.

GREW, NATHANIEL, *continued.*

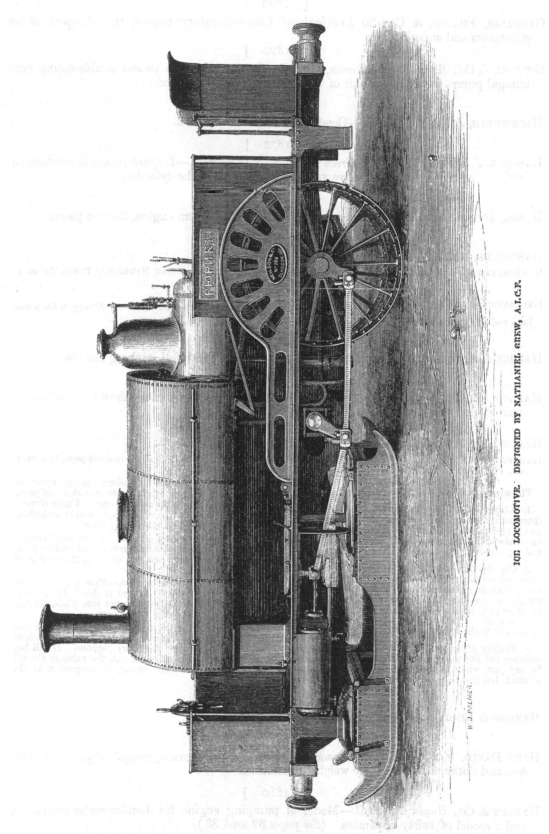

ICE LOCOMOTIVE. DESIGNED BY NATHANIEL GREW, A.I.C.E.

W. J. PALMER.

[1869]

GRIMALDI, FILIPPO, & Co., 30 *Bucklersbury, City.*—Rotatory boilers, the cheapest steam generators and superheaters.

[1870]

GWYNNE & Co., *Essex Street Wharves, Strand.*—Gwynne & Co.'s patent double-acting centrifugal pump, worked by a pair of their horizontal steam engines.

[1871]

HACKWORTH, J. W., *Darlington.*—Condensing engine and model.

[1872]

HANCOCK, J. & F., & Co., *Tipton Green Furnaces, Staffordshire.*—Improvements in condensing engines, by which a more effective vacuum is obtained in the cylinder.

[1873]

HANDS, JOHN, *Cardigan Street, Birmingham.*—Horizontal steam engine, 2-horse power.

[1874]

HANDYSIDE, ANDREW, & Co., *Derby.*—Brewing machinery.

HOP SEPARATOR—(Hodge's patent).

HYDRAULIC DOUBLE-ACTING PRESS, used for compressing hops and other materials.

PAIR OF DOUBLE-ACTING HYDRAULIC PUMPS, for working this press.

These machines are used at Messrs. Allsopp & Son's new brewery, Burton-on-Trent.

[1875]

HARGREAVES, WALMSLEY, *Crawshaw Booth, Manchester.*—Waterfall washing machine.

[1876]

HARLOW, R., *Stockport.*—Multitubular fire bridge and heat generator shown in section of steam boiler.

[1877]

HARRISON, JOSEPH, 8 *New Broad Street.*—Patent cast-iron boiler.

IMPROVED STEAM BOILER, system of Joseph Harrison, Jun., Philadelphia, United States.

The advantages claimed for this boiler are—

1. Adaptation.—It may be adapted to any form or use (particularly to mining or ordinary stationary purposes), and is available in places of difficult access, or where the materials, skill, and means could not be easily had for making boilers of the usual kind.

2. Capacity for sustaining pressure.—It will sustain with entire safety 2 or 3 times greater pressure than the boilers in general use, and from being the multiplication of a single unit, entire uniformity of strength in all its parts is secured, no matter how large the boiler may be made. It has been proved by hydraulic pressure of 500 lbs. to the square inch without injury.

3. Facility of repairs or renewal.—It has less than ordinary liability to get out of order. It can be renewed in any part when necessary, much more speedily, and at much less cost than boilers of the usual construction.

When repaired, it will, in all the renewed parts, be equally good as when new.

4. Explosion.—Serious explosions cannot occur in boilers of this construction, either from weakness of parts, too great pressure, or lowness of water. Under circumstances that would cause violent rupture in other boilers, every joint in this becomes a safety valve.

5. Facility of cleaning.—It can, by very simple means be kept free from injurious deposit, or incrustation in all its parts, with greater ease and certainty than boilers of the ordinary kind.

6. Facility of transportation.—However large the boiler may be, it can be carried in detail by a single man, and, if necessary, may be put into place, through an opening not more than 3 ft. square.

7. Economy of manufacture.—It can be made and kept in order at about one-half the cost of the boilers now generally used for stationary purposes. It will last equally long, and when worn out, the value of the old material will be much greater, in proportion to the original cost.

[1878]

HARRISS & RISSE, *New Oxford Street.*—Pressure gauges.

[1879]

HART, DAVID, *Whitechapel Road, London.*—Patent weighing crane, weighbridge for waggons, &c., and dormant and portable weighing machines.

[1880]

HARVEY & Co., *Hayle, Cornwall.*—Model of pumping engine for London water companies, and a model of safety apparatus. (*See pages* 32 *and* 33.)

[1881]

HAYES, EDWARD, *Watling Works, Stony Stratford.*—Portable steam engine; patent self-acting windlass for steam ploughing.

Obtained the Royal Agricultural Society's Silver Medal at Leeds, 1861.

HAYES' PATENT WINDLASS possesses the following advantages, which are peculiar to it alone, viz.—

The cultivator or plough can be instantly stopped by the anchormen at the headland, without stopping the engine, the engine continuing in motion as in thrashing or other work.

No signals are required; the work may be performed in foggy weather, or by moonlight, with perfect safety to the machinery.

One man can superintend both engine and windlass. A double-cylinder engine not required as the engine is not stopped.

No wheels are required to be put in or out of gear.

E. HAYES'S 8 and 10 HORSE ENGINES, designed and built extra strong for steam cultivation. Further particulars may be learned by reference to his catalogue.

OPINIONS OF THE PRESS.

In the notices of the Royal Agricultural Trials, Leeds Show, 1861.

Extract from THE TIMES, *July 17th*, 1861.—" Mr. Hayes, of Stony Stratford, exhibited a very clever windlass on the coiling principle."

THE ENGINEER, *July 19th*, 1861.—" The self-acting windlass of Mr. Edward Hayes, of Stony Stratford, was one of the important novelties in the show."

LEEDS MERCURY, *July 15th*, 1861.—"As a piece of mechanism this deserves as much attention as anything in the field."

BELL'S WEEKLY MESSENGER, *July 15th*, 1861.—" The construction of this machine was greatly admired."

MARK LANE EXPRESS, *July 15th*, 1861.—" Mr. Hayes, of Stony Stratford, has a novel form of windlass."

[1882]

HEPBURN & SONS, 25 *Long Lane, Bermondsey, London.*—Machine belts and leather.

[1883]

HERKLESS, WILLIAM, *Broad Close, Shuttle Street, Glasgow.*—Machine for grinding tanners' bark.

[1884]

HILL, JOHN, *Ashford, Kent.*—Improved flour dressing machine, with silent feed, revolving cylinder, and outside brush.

[1885]

HOLGATE, J., & CO., 33 *Dover Road, Southwark.*—Leather mill bands and hose pipes.

[1886]

HOLMES, F. H., *Northfleet, Kent.*—Magneto-electric machine and light; lighthouse regulators.

[1887]

HOPKINSON, J., & CO., *Huddersfield.*—Patent compound safety valve, steam engine indicator, mercurial steam and vacuum gauge, &c.

[1888]

HORTON, SON, & KENDRICK, *Southwark, London.*—Models of high pressure, marine and land steam-engine boilers.

[1889]

HOWORTH, JAMES, *Victoria Works, Farnworth, near Bolton.*—Patent self-acting Archimedian screw ventilators.

[1890]

HUGHES, J., & SONS, 91 *Dover Street, Borough.*—Millstones.

[1891]

HUMPHREYS & TENNANT, *Deptford Pier.*—Marine engine.

HARVEY & Co., *Hayle, Cornwall.*—Models of pumping engine for London water companies and of a safety-balance valve.

The accompanying engraving represents a SINGLE-ACTING CONDENSING ENGINE, on the Cornish principle, erected by Harvey & Co. engineers and founders, Hayle, Cornwall, and 12 Haymarket, London, for the East London Waterworks Company, at Lea Bridge. The cylinder is 100 in. diameter, and working stroke 11 ft. The pump is a plunger, 50 in. diameter, and 11 ft. stroke.

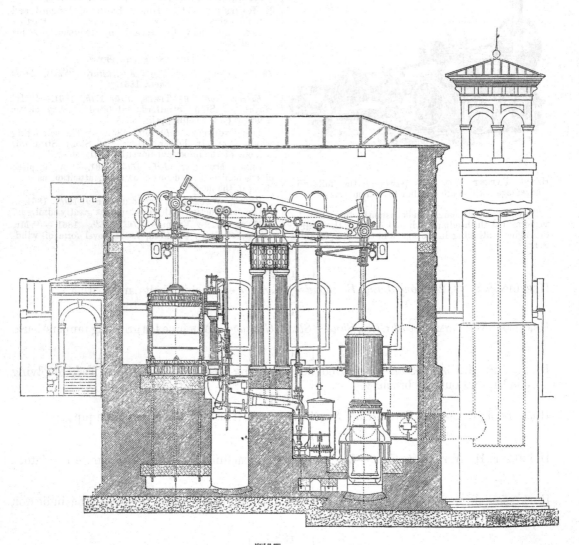

SCALE OF FEET

SINGLE ACTING PUMPING ENGINE.

This engine, when working full power, pumps about 9,000 gallons of water per minute, usually 140 ft. high, which water is conveyed into London by cast-iron pipes 36 in. diameter. The model exhibited by the above firm closely approximates to this engine. At the time of its erection in 1855, this was the largest machine for supplying water to towns ever constructed.

HARVEY & CO., *continued.*

In 1858, Harvey & Co. erected for the Southwark and Vauxhall Water Company at Battersea, a pumping engine, the cylinder of which is 112 in. diameter, weighing with its case 36 tons.

This engine, although the largest and most powerful ever built for such purposes, is of the most simple construction ; the steam valves are all on the equilibrium principle, and the arrangement of parts is such, that this colossal engine is as completely under control as those of the smallest size, and performs an enormous amount of work, without the slightest shock or noise.

The total quantity of water pumped for the supply of London daily amounts to 115,000,000 gallons. Of this large amount 79,000,000 of gallons are pumped by the single-acting engine, and considering that Harvey & Co. have erected nearly all the machinery for pumping the latter amount, and about 25 years ago first introduced into London this machine, of which the above-mentioned engines are examples, that firm has thought it advisable to exhibit a working model of a pumping engine, supplied with Harvey & West's valves, and complete in every respect. The pump of this model with the valve boxes are partially constructed of glass, thus allowing the action of the valves to be observed.

Like all great improvements, this class of engine has met with much opposition. Gradually but surely, however, it is superseding all others for supplying towns with water, and for all drainage purposes, and as now improved, it stands unrivalled for economy and durability. This is sufficiently proved by its adoption by the Southwark and Vauxhall, the Kent, the West Middlesex, the Grand Junction, and the East London Water Companies. The above companies now use this engine exclusively, and effect by so doing a very large saving of fuel. Some of these companies have worked their engines without intermission for twenty years, without requiring to stop for repairs.

The single-acting engine having been employed with such entire success for pumping water into London, it appears surprising that the same plan of engine is not to be employed for pumping the water out again in the form of sewage. An experiment is to be tried at Deptford with rotative engines, for pumping the sewage up from the low level sewers, thus going back to the plans adopted in London before the introduction of the single-acting engine, regardless of the experience of the most eminent water-works engineers in London. It is very desirable however, considering the immense interests at stake, that this question should be thoroughly investigated, before a farther outlay be decided on. For as the cost of working steam engines, and maintaining them in repair is a *daily charge,* a step in the wrong direction would entail enormous loss on the City of London, and the evil would be irremediable.

Harvey & Co. have had great experience in the manufacture of machinery for stamping and crushing ores. The space allotted however does not admit the introduction of models. This business has of late years become of great importance since the gold discoveries in Australia and California, and as future success in those countries, must depend on mining, suitable machinery for crushing and stamping, will daily become of greater importance.

The above firm have constructed pumping machinery expressly adapted for draining gold workings in Australia or other distant countries. Wrought-iron is substituted for cast-iron wherever practicable, thus at once decreasing cost of transport by reducing the weight ; and diminishing the risk of breakage to a minimum. This is even of more importance for the pumps than for the engine, as the weight of lifts is thereby so much lessened that the labour of fixing is reduced by about one-half.

The model exhibited is a type of these machines for draining mines, and by it the method of working may be readily understood even by those not intimately acquainted with the subject.

MODEL OF A SAFETY-BALANCE VALVE.

This is a model of a SAFETY-BALANCE VALVE, patented by W. Husband, and made by Harvey & Co. for pumping engines. It is fixed in and forms part of the main pipes for conveying the water ; by its action an engine always retains its load in case the main should burst. It performs the office of a stand pipe, and dispenses with this costly structure.

[1892]

HURRY, HENRY C., *Rookswood Villa, Worcester.*—An electro-magnetic motive engine.

[1893]

HUXHAMS & BROWN, *Exeter.*—Mill to grind bark for tanners; hydraulic lifts to raise ships and heavy weights, French burr millstone to grind wheat.

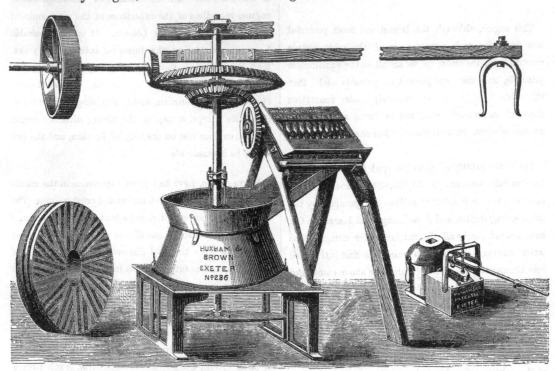

Obtained a Medal at the Paris Exhibition, 1855.

TANNERS' BARK MILL, grinding in the best manner for English tanneries, 25 cwt. a day by one horse, and about 5 tons a day by four-horse steam power. It separates the fibres of the bark thoroughly, therefore the tannin can be more easily extracted. It is not necessary to chop the long bark for this mill, as it shortens the bark, tears and grinds it at one operation.   Price £45 0

HYDRAULIC LIFT, raises or pushes heavy weights with far greater ease and much less cost than screws. Ships of from 200 to 400 tons can be gradually raised by it. One man and a boy have lifted with it a ship of 250 tons. It is useful for a variety of heavy work, and wherever a severe strain is to be slowly overcome.

Price £14 14

MILLSTONE of the best description of French burr for flour, not of the closest or hardest burr; will do the best miller's work for fine flour, except when the wheat is extremely hard; though full of fine pores which give cut to the last. Easier to dress well than closer stones.

[1894]

IMPERIAL IRON TUBE COMPANY, *Birmingham.*—Iron, brass, and copper tubes, and fittings for boilers, gas, steam, water, &c.

[1895]

IMRAY, JOHN, 65, *Bridge Road, Lambeth.*—Improved horizontal and vertical steam engines.

HORIZONTAL AND VERTICAL DIRECT-ACTING HIGH PRES-
SURE STEAM ENGINES, of the most simple and substantial construction.
        2 horse power vertical £35 0

5	horse power, vertical	. . . .	£80	0
7	ditto	horizontal	110	0
12	ditto	ditto	180	0
20	ditto	ditto	280	0

[1896]

INGLIS, A. & J., *Glasgow.*—Two drawings of the 200-horse power engines of H.M. Steam-ship "Chanticleer."

[1897]

KEY, JOHN, *White Bank, Kirkcaldy.*—Horizontal direct-acting screw engines of the collective power of 80 horses.

The exhibitor is a designer and manufacturer of hori-zontal direct-acting screw engines, of oscillating paddle wheel steam engines, boilers, &c. &c. Prices and other particulars may be learned by application.

[1898]

KING, C. B., 20 *Abingdon Street, Westminster.*—Design for traction engine and steam carriage.

[1899]

KING, J. CHARLES, 12 *Portland Road, Regent's Park, W.*—Tubular carriage axle, and wood washers.

[1900]

KIRKALDY, JOHN, & SONS, 166 *Wapping.*—Ship's portable fire engine.

[1901]

KNOWELDEN & CO., *Park Street, Southwark.*—Patent pumps, valves, hydraulic motive engines and cranes, safety valve, &c.

1. KNOWELDEN & EDWARDS'S PATENT DIAPHRAGM PUMPS.

The working barrel of these pumps, protected by the diaphragm, are uninjurable from grit, sewage matter, and all such causes, which act injuriously on the working barrels of all other pumps. The valves can be withdrawn, and any impediment to their free action, removed, and replaced in a minute. By the reversal of the handle the suction pipe becomes the outlet one, through which water may be forced at a great pressure; so that no inaction from choking, or injury to the barrel can take place in these patent pumps. These advantages render them invaluable for ships' use, mines, &c.

2. KNOWELDEN & EDWARDS'S PATENT DOUBLE ACTION DIAPHRAGM PUMP, a modification in make of the above described pump. For use of brewers, distillers, chemical works, &c.

3. KNOWELDEN & EDWARDS'S PATENT STEAM PUMPING ENGINE.

4. KNOWELDEN & EDWARDS'S PATENT 10-HORSE STEAM ENGINE.

5. KNOWELDEN'S PATENT SAFETY VALVE.

[1902]

LAIRD, JOHN, SONS, & CO., *Birkenhead.*—A pair of 40-horse power horizontal direct acting engines.

[1903]

LAMBERT, THOMAS, & SON, *Lambeth.*—Hydraulic press pumps; lift and force pumps; steam engine fittings in gun-metal. (*See page 36.*)

[1904]

LANSDALE, RICHARD, *Pendleton, Manchester.*—Patent compound rotary washing machine, with rollers for wringing or mangling. (*See page 37.*)

[1905]

LA ROCHE, PHILIP, 6 *Blacklands Terrace, King's Road, Chelsea.*—Improved beer engine; tapping cock, muller, and valves.

[1906]

LAWRENCE, H. M., & CO., *London Works, Sefton Street, Liverpool.*—Machine for making ice by steam. (*See page 40.*)

[1907]

LAWRENCE, JAMES, 5 *Formosa Terrace, Maida Hill, W.*—Patent refrigerator; mash-tun machinery; boiling and fermenting apparatus; plans and models.

LAWRENCE'S PATENT REFRIGERATOR, combining great cooling power, cheapness, and durability, with perfect cleanliness.
LAWRENCE'S PATENT REMOVABLE MASHING MACHINE, heat distributor, false bottom, sparger, &c.

LAWRENCE'S PATENT FERMENTING APPARATUS.
The above are in use at Burton-upon-Trent, and the chief towns of the United Kingdom; and also on the Continent and in the Colonies. Breweries erected and remodelled by contract.

[1908]

LEADBETTER, THOMAS, & CO., 13 *Gordon Street, Glasgow.*—Force pump; fire plug; hydraulic ram; water-closet.

[1909]

LEONI, S., *St. Paul's Street, N.*—Taps; steam cocks; machinery bearings; gas burners; ornamental wares of adamas, resisting heat, acids, wear, and friction.

[1910]

LILLESHALL COMPANY, *Shiffnal, Shropshire.*—Pair of blast engines. (*See pages 38 and 39*)

LAMBERT, THOMAS, & SON, *Lambeth.*—Hydraulic press pumps, lift and force pumps, steam engine fittings in gun metal.

HYDRAULIC PUMPS.
DEEP WELL PUMP.
LAMBERT'S PATENT REGULATING STEAM VALVE.
STEAM ENGINE FURNISHINGS, &c.

VAUCHER'S PATENT METAL BEARINGS.
STAND POSTS, SLUICE VALVES.
FIXED PUMPS AND SAFETY VALVES.

Society of Arts' Medal, 1847 ; Prize Medal, 1851 ; Bronze Medal, Amsterdam, 1854.

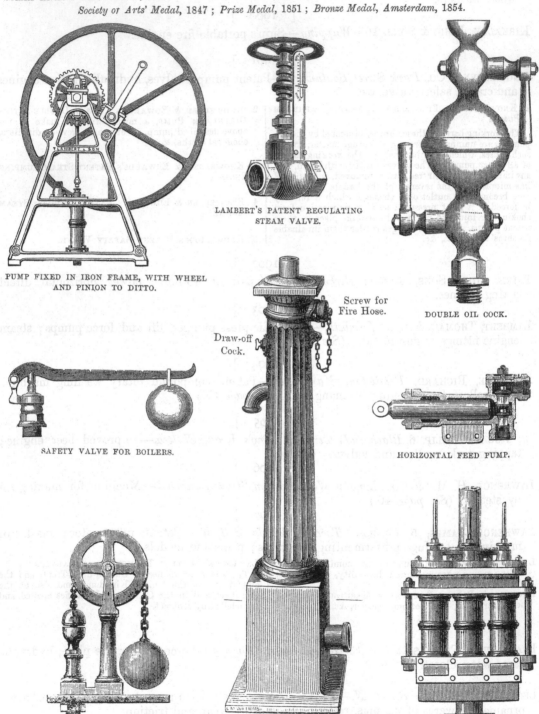

PUMP FIXED IN IRON FRAME, WITH WHEEL AND PINION TO DITTO.

LAMBERT'S PATENT REGULATING STEAM VALVE.

DOUBLE OIL COCK.

Screw for Fire Hose.

Draw-off Cock.

SAFETY VALVE FOR BOILERS.

HORIZONTAL FEED PUMP.

ALARM WHISTLE WITH STANDARD, CHAIN, AND BALANCE WEIGHT.

STREET OR YARD STAND-POST FOR FIRE OR ROAD-WATERING.

TREBLE-BARREL DEEP WELL PUMP.

Illustrated Catalogues may be obtained, post free, on application.

LANSDALE, RICHARD, *Pendleton, Manchester.*—Patent compound rotary washing machine, with rollers for wringing or mangling.

This invention, by its compound action, easy working, and complete efficiency, having won the unqualified praise of many eminent machinists, and approved itself to all purchasers unexceptionally, the patentee submits it to the public, assured that wherever its construction is understood, its merits will be admitted.

A barrel of 30 gallons' capacity is hung upon 4 centres, mechanically arranged to produce (without the least complication) a compound rotary motion, so that by the turning of a handle, the barrel containing the articles to be washed, revolves two ways—horizontally and vertically—at the same time. By this double movement the contents of the barrel are thoroughly agitated, and a washing process attained, upon a compound dash-wheel principle.

Over the inner surface of the wash-tub short wooden cones are studded about 6 in. asunder, which, when the machine is in motion, alternately rub the clothes and dash them through the water, repeating this operation with each revolution of the barrel, as the mechanical consequence of its compound rotation.

For easy working this machine is unsurpassed; its excellence in this particular being proved by one significant fact, viz. that much less power will turn the barrel when containing 20 gallons of water, than when containing only 2; strength equal only to that of a child is enough for working it. Combining, then, this vast advantage of light labour with the perfect cleansing treatment the clothes receive, in consequence of the peculiar mechanical action of the wash-tub, we have a result establishing beyond question the complete efficiency of the invention.

A five minutes' trial will fully demonstrate the general convenience of this machine. It simplifies washing, reducing it from an affair of skilled labour, to the trifling process of turning a handle. It is portable—the push of a hand or foot sufficing to wheel it about. The working parts are strong, simple, and cannot disarrange themselves; whilst of its compactness the best estimate may be formed by examining the following engraving.

WASHING MACHINE.

This machine having been designed expressly to meet every requirement of a well-appointed family laundry, the patentee requests a careful notice of its general arrangement. With a wash-tub of 30 gallons' capacity, are combined thick well-seasoned sycamore rollers, for wringing or mangling; which are so adapted that, without the use of any dripping boards, all drainage falls back into the open tub. To the rollers are affixed self-regulating weighted levers, which adjust themselves to every article passing through them, of whatever texture—thick or thin—with perfect nicety, and with a pressure such as greatly hastens the subsequent drying.

One other great convenience merits notice. By the simplest contrivance, both mangling and washing are worked with but one handle, which can be instantly adapted to either purpose. For well-combined working parts in little space, this machine will satisfy the most fastidious. Extreme measurement, 36 by 50 in.

Price £9 9

INSTRUCTIONS FOR WASHING.

Chip 1 pound of soap into 3 quarts of water, and boil into a ley. Half-fill the wash-tub with clothes and water, adding soap ley at discretion, and give each lot of clothes about a 4 minutes' wash. Next boil such articles as need it, and work them once again for 4 minutes, as a finish. Always, before the tub is turned, take care to screw the lid down tightly.

LILLESHALL COMPANY, *Shiffnal, Shropshire.*—Pair of blast engines.

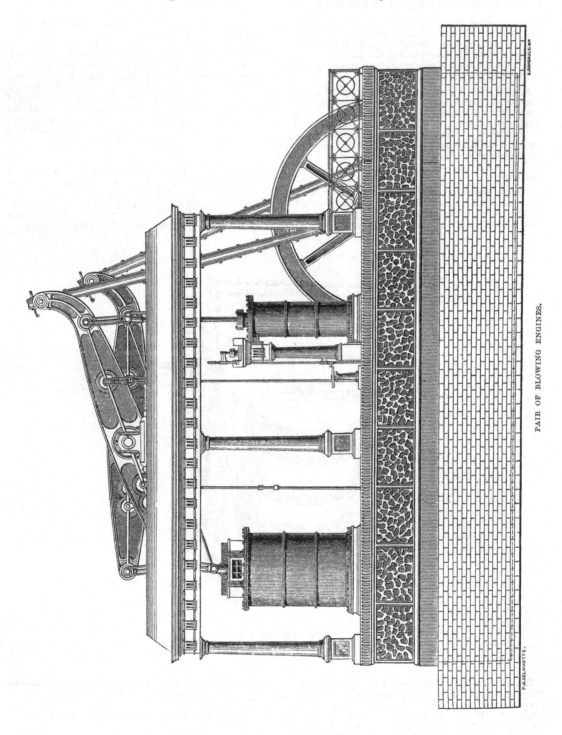

PAIR OF BLOWING ENGINES.

The small pair of BLOWING ENGINES sent for exhibition by the Lilleshall Company, are self-supporting, and fixed upon wrought-iron foundation, &c. for the convenience of exhibition. They are capable of blowing 2 cold-blast furnaces, are arranged to work together or separately with great economy, and are most simply and substantially built. A pair of large engines may be seen at the Works, of the same design, but fixed in a house, the beams resting upon a transverse entablature, supported by 4 massive columns; blowing 5 cold-blast furnaces.

The Company are manufacturers of all kinds of high-pressure expansive and condensing steam engines, and colliery and contractor's locomotives, especially adapted for heavy gradients and sharp curves, fitted with the Lilleshall Company's patent compensating buffers, which adjust,

LILLESHALL COMPANY, *continued.*

themselves to any angle, each buffer taking equal strain; and also Gifford's injector, most important improvements in this class of engine.

Blast engines fitted with 4 double-beat valves and expansive valve gear, working direct, of simple and durable construction (all the gearing and principal wearing parts being of hardened steel).

This Company also manufactures PUMPING ENGINES of all descriptions, acting direct or through a beam, working with high-pressure steam, cut off at any part of the

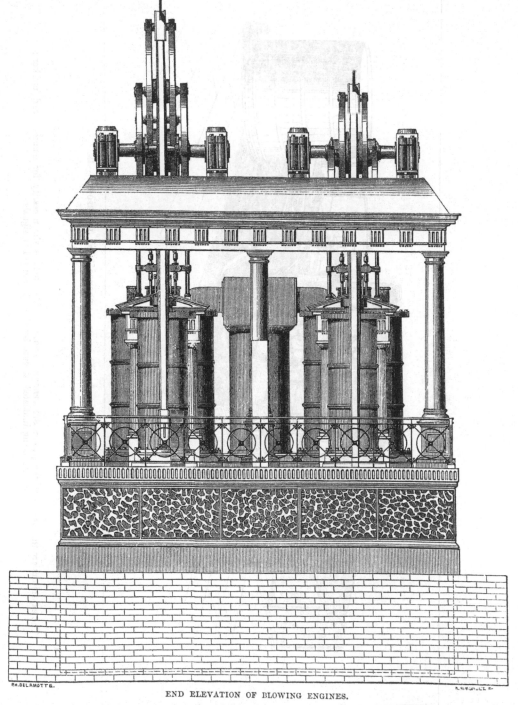

RH. DELAMOTTE.

R.N.W.UNULL

END ELEVATION OF BLOWING ENGINES.

stroke; condensing winding engines, horizontal, vertical, direct or beam, coupled or single, fitted with a new improved link motion (which gives the engine-man more perfect control); strong and massive steam-engines for rolling mills, sugar mills, saw mills; chilled and grain rolls; guides and other castings, where strength and durability is a desideratum. All these goods made from the Lilleshall cold-blast iron.

See specimens of coals, and argillaceous ironstones, from which their cold blast pigs are made, Class 1.

LAWRENCE, H. M., & Co., *Sandon Works, Sefton Street, Liverpool.*—Machine for making ice by steam.

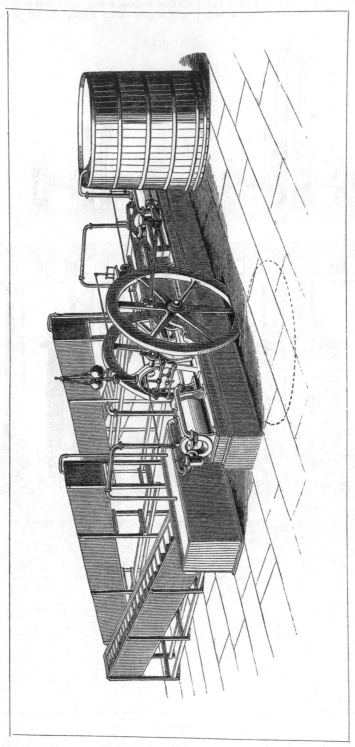

MAKING ICE BY STEAM. MESSRS. H. M. LAWRENCE & CO, LIVERPOOL.

The exhibitors are the makers of the patent SEMAPHORE TARGET, which obviates the necessity for mantelet and marker; the hits being indicated at once by the rise and fall of signals.

[1911]

LINDSAY, ROBERT BAIRD, *Laurel Bank, Paisley.*—Model to exhibit patent method of removing incrustations from marine and locomotive boilers.

[1912]

LLOYD & LLOYD, *Albion Tube Works, Birmingham.*—Wrought-iron tubes and fittings for gas, steam, water, &c.

Wrought-iron fittings in tees, elbows, crosses, &c.
Conducting pieces, various, all forged on the anvil.
Iron main cocks.
Taps, stocks, and dies for screwing.
Locomotive and other boiler-fittings in brass and gun metal.
Water gauges, whistles, &c.
Solid bottom stuffing box, gland steam cocks from ¼ in. bore upwards.

Patent lapwelded iron tubes, for locomotive, marine, and stationary boilers.

Wrought-iron butt-welded tubes, screwed and socketed, from ⅛ in. bore upwards.

Specimen of improved homogeneous metal tube, flattened and turned up at the ends, to show its perfect malleability.

[1913]

LLOYD, GEORGE, 70 *Great Guildford Street, Southwark.*—Patent noiseless centrifugal fan blowing machines, mine ventilators, &c.

Obtained the prize medal at the Great Exhibition, 1851; also the silver medal at the Paris Exhibition, 1855.

For melting iron and other metals; blowing smiths' forges; puddling furnaces; dessication; ventilating buildings, ships, sewers, wells, coalpits, and mines of every description; and forcing or exhausting hot or cold air at high or low pressure, for any purpose for which it may be required.

The machine exhibited (42-in.) will melt from 4 to 5 tons of iron per hour, or blow from 60 to 80 smiths' forges; or will deliver, for ventilation, 7,500 cubic feet of air per minute.

From the peculiar construction of this fan, it will do nearly double the amount of duty with the same amount of power as any other kind of fan, and from there being no back action on the blades by the air, it works entirely without humming noise.

Sizes made :—13, 16, 19, 22, 25, 30, 36, 42, 48 inches.

[1914]

LOUCH, JOHN, & CO., 69 *Fenchurch Street.*—Union joints and pipe fittings.

[1915]

LUMLEY & WATSON, 50 *Lower Shadwell, E.*—Steam crane, iron blocks, and crab winch.

[1916]

McCALLUM, DAVID, 1 *Octagon, Plymouth.*—Electro-magnetic engine.

[1917]

McFARLANE, WILLIAM, 39 *Stockwell Street, Glasgow.*—Patent cylinder mangle, washing and wringing machines.

[1918]

McGLASHAN & CO., *Drury Lane, W.C.*—Beer cooler.

[1919]

McGLASHAN & MERRYWEATHER, *Coal Yard, Drury Lane.*—Steam cocks; boiler fittings; plumbers' brass work; pumps; model refrigerator.

[1920]

MACINTOSH, CHARLES, & CO., *Cannon Street, London; and Cambridge Street, Manchester.*—Mechanical appliances of vulcanized rubber.

[1921]

McONIE, W., & A., *Scotland Street Engineer Works, Glasgow.*—30-horse power steam engine and sugar mill, with cane and megass carriers. *(See page 43.)*

[1922]

MACORD, R. H., 63 *Lower Thames Street.*—Machines, tools, and utensils used for bottling wine, spirits, beer, &c. (*See page* 44.)

[1923]

MANCHESTER WATER METER COMPANY, THE, *Tipping Street, Ardwick, Manchester.*— Water meters for general and trade purposes, steam boilers, &c. (*See page* 45.)

[1924]

MANLOVE, ALLIOTT, & CO., *Bloomsgrove Works, Nottingham.*—Engines, centrifugal sugar machines, washing and drying machinery.

PATENT CENTRIFUGAL SUGAR MACHINE, under driven, full operation.
PATENT CENTRIFUGAL SUGAR MACHINE, top driven, full operation.
PAIR DIRECT-ACTING STEAM ENGINES, for driving centrifugal sugar machines.

PATENT HAND-DRIVEN HYDRO EXTRACTOR, or wringing machine.

MODELS.
Patent improved washing machine, horizontal engine.
Patent hydro extractors with counter gearing.
Drawings of engines and centrifugal machinery.

[1925]

MARTIN, W. A., 55 *Great Sutton Street, E.C.*—Patent rocking furnace bars for land and marine purposes.

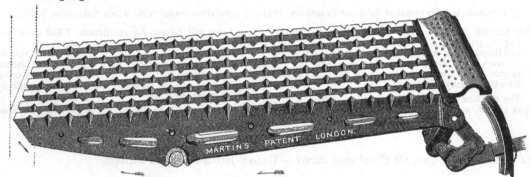

MARTIN'S PATENT ROCKING FURNACE BARS.

This invention is of great importance to mill owners, steam navigation companies, and all firms using steam power. The durability of the patent bars is extraordinary; they have surpassed every fire bar yet tried, being constructed on scientific principles, to insure strength and burning powers, and having been most rigorously tested. They are now standing the fiercest fires, under service, night and day, and are not in the least affected; the full effect of the furnace is maintained, and the highest steaming powers are produced. The fires are easily managed and effect a large saving in fuel and labour; the lever, with one touch, moves the bars, every one acting as a poker, instantly clears all the furnace, and removes clinkers, and all impurities.

The sea service bars, from their simplicity and efficiency, will be found most valuable to ocean steam ships. They do away with the laborious work in the stoke-hole; they cannot get foul or knocked out of their place; and with the dirtiest description of fuel, they will clean themselves, and maintain regular steam, and high speed.

References to large firms now adopting them and particulars, may be obtained by applying to the patentee.

Agents, Lankester & Son, Ironfounders, Southampton; Alston & Gourlay, British Iron Works, Glasgow.

[1926]

MAUDSLEY, SONS, & FIELD, *Lambeth.*—Marine engine.

[1927]

MAY, WALTER, & CO., *Birmingham.*—Double-cylinder steam engine and surface condenser; portable corn mill. (*See page* 46.)

[1928]

MERRYWEATHER & SON, 63 *Long Acre, London.*—Fire engines, hose, &c. (*See pages* 48, 49.)

[1929]

MICKELTHWATE, ARTHUR, *Sheffield.*—Patent metallic, hemp, and leather belting; metallic and leather boot and shoe soles.

[1930]

MIDDLETON, THOMAS, *Loman Street, Southwark.*—Murray's patent chain pump for sewerage drainage, or irrigation.

McOnie, W. & A., *Scotland Street Engineer Works, Glasgow.*—30-horse power steam engine and sugar mill.

THIRTY-HORSE POWER HIGH-PRESSURE STEAM ENGINE.

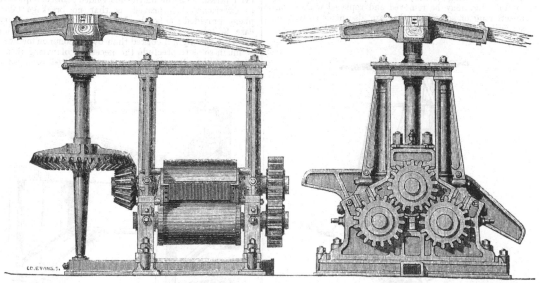

SUGAR-CANE MILL FOR WORKING BY CATTLE.

• A High-pressure Steam Engine and Sugar-cane Mill. | A Small Sugar-cane Mill, to be worked by cattle.

MACORD, R. H., **63** *Lower Thames Street*.—Machines, tools, and utensils used for bottling wine, spirits, beer, &c.

MACORD'S IMPROVED PATENT BOTTLING APPARATUS.

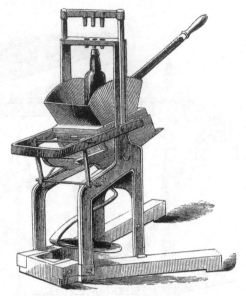

IMPROVED PATENT CORKING MACHINE.

This apparatus is far superior to the original "Masterman's Patent," and is the best in use. Its advantages are —1st: That being made entirely of metals, it is much more durable, and less likely to get out of order. 2dly: The cistern has two pinions, connected by a shaft, and two slides fixed to it, which are worked in the racks on the upright iron pillars by means of a lever handle, and thus raised or lowered with perfect ease, being kept level all the time (a great recommendation). It is also fitted with a tap at bottom, by means of which it may be thoroughly cleaned without removing. The syphons are fitted to the cistern with hinge joints; and, one pin forming the centre for all, they may be removed and replaced without any unscrewing or screwing; they are also made so that one set will answer both for pint and quart bottles.

The principle of this corking machine is, to force the cork into the bottle through a conical tube in contact with its mouth in such a position, as to form one continuous tube with its neck, and having the lower orifice so small, that the cork must be considerably compressed and compacted in passing through it. As the corks are impelled into the bottles by a lever, it must be evident, from the above principle, that all jarring against or even pressure on the bottles is avoided; the consequence, as experience has proved, is that no breakage takes place, provided the bottles be sound, and mere ordinary care be taken. Another advantage is, that the bottles can be much tighter corked than by the common method; so much so as to preclude the necessity of wiring them. The machine is portable, and constructed principally of iron.

MACORD'S BOTTLE-WASHING MACHINE.

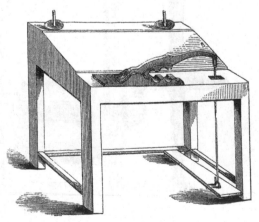

MACORD'S CORKING MACHINE.

This machine having been extensively used in the trade, is recommended with confidence as the best and quickest mode of cleaning bottles; it is simple, portable, and effective thoroughly cleaning all bottles without the aid of soot or grit; it requires no fixing, and is very durable.

This form of machine is in general use throughout the bottling trade; it is used with a leathern boot strapped on to the knee; and the bottle being held therein, the cork, after being sufficiently compressed by the machine, is driven into the bottle with a hard-wood driver.

MANCHESTER WATER METER COMPANY, THE (Limited), *Tipping Street, Ardwick, Manchester.*— Water meters for general and trade purposes, for steam boilers, works, warehouses, shops, offices, &c.

METER FOR GENERAL AND TRADE PURPOSES (see wood engraving).

These meters are constructed on the piston and cylinder principle, the piston having a reciprocating action. Their chief novelty consists in the use of a compound fluid motive valve to reverse the stroke of the piston, and change the direction of the effluent water; which object it effects, without concussion or stoppage in the flow. This has never before been accomplished in any high-pressure water meter with a single cylinder and piston, without the aid of springs or tumbling weights. The exterior of these meters consists of a strong case of cast-iron in three parts, bolted together. The lower portion forms the measuring cylinder, and is lined with brass, which is smoothly bored out. In this cylinder the piston works: it is packed with cupped leathers, similar to those used in hydraulic presses. The upper portion of the meter contains the compound valve and the wheelwork of the index. All the working parts are made of brass, and are therefore not liable to be affected by water. These meters have been practically and thoroughly tested for upwards of three years, and a large number of them are now used by water companies and others. They require no lubrication, and in accuracy and durability, they have far surpassed all other meters.

NEW PATENT STEAM-BOILER METER.

A meter to measure the water evaporated by steam boilers has long been a desideratum; but the necessity of using leather, india-rubber, and other flexible substances in packing the pistons of all positive measuring meters, has hitherto been the great difficulty. This difficulty has now been successfully removed in the Company's new boiler meter, which is constructed entirely of metal, on a principle that involves the smallest possible liability to become deranged, and that secures accuracy and efficiency in working. It is portable and convenient in form, and can easily and readily be attached and detached.

This meter can be placed at any distance from the boiler, or between the boiler and the pump. Its use will ensure the most accurate and reliable test of the best construction of boilers, fire-bars, and furnaces; and of the various kinds of steam economizers. It will also secure perfect tests of all descriptions of coal and other fuel, and of the work done by steam engines in proportion to the coal or other fuel consumed.

NEW PATENT OFFICE AND DOMESTIC METER.

The attention of water companies & the public generally is directed to the new water meter for private dwelling houses, offices, warehouses, shops, public houses, &c. The size of this meter is small, and the price is moderate. To water companies, who desire to economize their water by preventing waste, and to deal equally towards all their customers, this meter will prove of inestimable use; while to small consumers, for baths, stables, water closets, fountains, &c. it will afford the means of guaranteeing a supply of water at a fixed rate per 1,000 gallons, and remove any sense of injustice which may now be unavoidably experienced, in consequence of the charges for water being arbitrarily fixed, without any reference to the quantity used.

Water has hitherto been generally charged at a rate higher per 1,000 cubic feet than gas; but now that water meters can be had capable of measuring water as accurately as gas is measured, there is no longer any necessity to fix the charges for any class of consumers of water otherwise than by meter.

For further particulars, apply to the Manchester Water Meter Company, Limited, Tipping Street, Ardwick, Manchester.

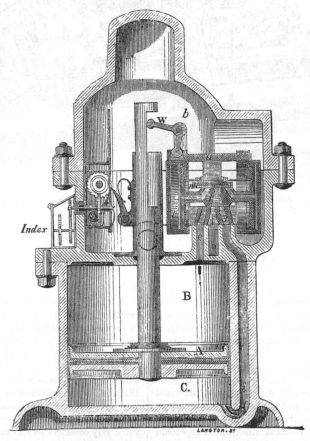

WATER METER FOR GENERAL AND TRADE PURPOSES.

MAY, WALTER, & Co., *Suffolk Works, Berkeley Street, Broad Street, Birmingham.*—An uniform power double-cylinder horizontal steam engine.

This engine is designed especially with a view to obtain uniform rotative power, and at the same time to carry the expansion of the steam to a great extent, for the purpose of ensuring the utmost amount of economy in fuel.

The cylinders are respectively 10 and 21 in. diameter, and the length of stroke is 2 ft. in each case; one external steam jacket, fed direct from the boiler, embraces both cylinders. The steam enters the small cylinder at the full pressure of the boiler, and when the piston has travelled half its stroke, is cut off and expanded through the remainder of the stroke. It is then exhausted into the wrought-iron reservoir, shown by dotted lines under the bedplate, this reservoir being jacketed with high-pressure steam from the boiler. Here the steam is stored up until the crank of the larger cylinder, which is at right-angles to that of the smaller one, has brought its piston to the end of the stroke, when the slide valve of the large cylinder opens and admits the steam from the above mentioned reservoir, and, as in the smaller cylinder, it is again cut off at half stroke, expanded through the remainder of the stroke, and exhausted into the condenser, which may either be a surface condenser, as in the case of the engine

W. MAY AND CO.'S DOUBLE-CYLINDER HORIZONTAL ENGINE.

exhibited, or an ordinary one, according to the circumstances of each particular case. The air pump, which is placed vertically, as being preferable to horizontally, is worked by a connecting rod from the end of the crosshead of the large cylinder.

The nearest approach that it is possible to obtain to perfectly uniform rotative power is arrived at by constructing from calculation, the indicator figure, that would be produced by each cylinder, and deducing therefrom the requisite proportions that should exist between their two diameters, and the points at which the steam should be cut off, in each respectively.

In connexion with the above described engine, is exhibited PERKIN'S PATENT SURFACE EVAPORATOR CONDENSER; the advantages of which may be summed up as follows, viz:—The supply of perfectly pure water to the boiler, which infallibly prevents all incrustation and priming. The more regular supply of water to the boiler. The condensers are cheap and very portable. Dirty or salt water is capable of being used for condensation; and existing high-pressure engines, may, by its use be converted at a moderate cost into condensing engines, and a very considerable increase of power obtained, without any additional consumption of fuel.

[1931]

ILLER & PIERCE, *Glasgow.*—Fire pump for ships.

[1932]

MIRRLEES & TAIT, *Glasgow.*—Steam engine and sugar mill in motion.

[1933]

MONCTON, E. H. C., *Wansford.*—Model of a steam generator.

[1934]

MOORE, EDWIN, *Depôt, 55 Upper Marylebone Street, W.*—Pressure gauge; all kinds of steam fittings.

[1935]

MORGAN, J. & Co., *Stafford Street, Birmingham.*—Block-cutting machine.

[1936]

MORRISON, R. & Co., *Newcastle-on-Tyne.*—High-pressure surface condensing expansive marine engine, cut-off variable.   (*See page* 50.)

[1937]

MURRAY, E., & COMPANY, 2 *Walbrook Buildings, City, London.*—Patent moving argand fire bars, patent metallic lubricant.

[1938]

NAPIER, D., & SON, 5 *Vine Street, & 51 York Road, Lambeth.*—Centrifugal machine for curing sugar; automaton mint weighing machine.

[1939]

NAPIER, ROBERT, & SONS, *Glasgow.*—Drawings of marine engines.

[1940]

NEAL, THOMAS, 45 *St. John Street, Smithfield.*—Patent grinding mills, for flour, ink, drugs, &c.

[1941]

NEEDHAM, JOHN, *School Brow, Warrington.*—Direct-acting horizontal steam engine, with adjustable eccentric for regulating valve.

[1942]

NEEDHAM & KITE, *Phœnix Iron Works, Vauxhall.*—Filter for semi-fluids.   (*See page* 51.)

[1943]

NEILL, E. B., 11 *Parliament Street, W.C.*—Ericson's caloric air engine, 2-horse power, no boiler, most safe and simple.

[1944]

NEWTON, KEATES, & Co., *Liverpool.*—Copper and brass articles for engineers, &c.

[1945]

NOBES & HUNTER, 16 *St. Andrew's Road, Borough, London.*—Leather for engineering and mechanical purposes, machine bands, hose, and buckets.

The exhibitors are curriers, and manufacturers of improved single and double leather bands for driving all kinds of machinery, copper-riveted leather and india-rubber hose-pipes for fire engines, steam and other purposes; leather fire buckets and leather for railways, engineering, mechanical, and ships' purposes; improved suction-hose, &c.

[1946]

NORMANDY, D. A., & Co., *London.*—Apparatus for obtaining aërated fresh water from sea water.

MERRYWEATHER & SON, 63, *Long Acre, London.*—Fire engines, hose, buckets, fire escapes, &c.

MERRYWEATHER AND SON'S PATENT STEAM FIRE ENGINE.

MERRYWEATHER & SON'S PATENT STEAM FIRE ENGINE, for service in any climate, is light, powerful, and compact; is mounted on a strong wrought-iron frame, with high wheels, and springs for rapid travelling; the pump, self-lubricating piston, and valves are of gun-metal, to work the foulest water without injury; the boiler is of steel, with copper tubes to generate steam quickly, and stand great pressure; and the pump will throw large or small bodies of water to great distances. The engine is fully equipped with suction and delivery hose, branch pipes, wrenches, tank, &c.

Prize Medal Great Exhibition, 1851, awarded to the " Prince Albert."
Prize Medal Paris Exhibition, 1855, awarded to " L'Empereur."

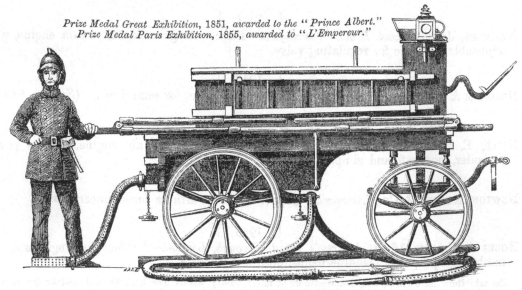

MERRYWEATHER AND SON'S IMPROVED LONDON BRIGADE FIRE ENGINE.

MERRYWEATHER & SON'S IMPROVED LONDON BRIGADE FIRE ENGINE, to be drawn by horses or men; with gun-metal pumps, pistons, and valves in separate valve-chambers; spherical copper air vessel, folding handles for 30 men, wrought-iron fore carriage, patent axles and springs, and delivery screws on both sides for 2 lines of hose. Fully equipped with suction-pipes, hose, branch-pipes, jet-spreaders, wrenches, &c.

MERRYWEATHER & SON, *continued.*

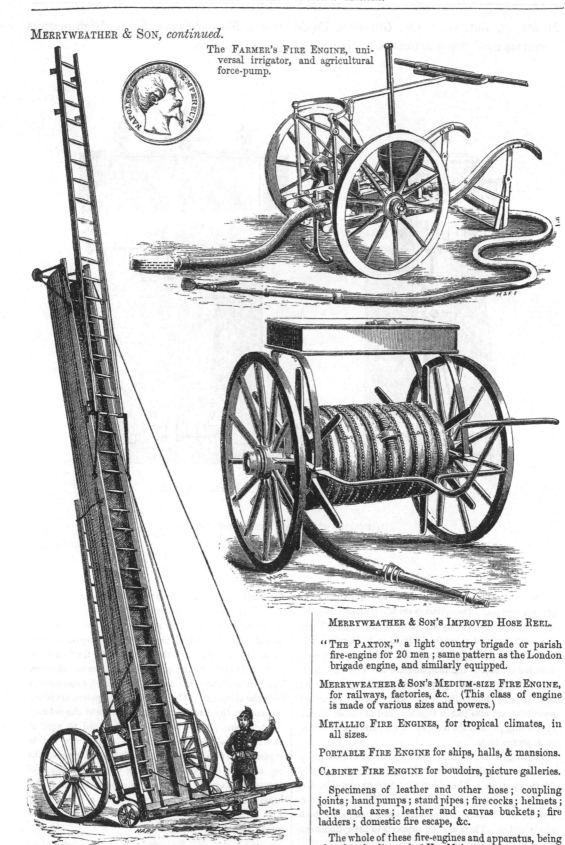

The FARMER'S FIRE ENGINE, universal irrigator, and agricultural force-pump.

MERRYWEATHER & SON'S IMPROVED FIRE ESCAPE, as used in London, Dublin, and many provincial and foreign towns, to reach 60 ft.

MERRYWEATHER & SON'S IMPROVED HOSE REEL.

"THE PAXTON," a light country brigade or parish fire-engine for 20 men; same pattern as the London brigade engine, and similarly equipped.

MERRYWEATHER & SON'S MEDIUM-SIZE FIRE ENGINE, for railways, factories, &c. (This class of engine is made of various sizes and powers.)

METALLIC FIRE ENGINES, for tropical climates, in all sizes.

PORTABLE FIRE ENGINE for ships, halls, & mansions.

CABINET FIRE ENGINE for boudoirs, picture galleries.

Specimens of leather and other hose; coupling joints; hand pumps; stand pipes; fire cocks; helmets; belts and axes; leather and canvas buckets; fire ladders; domestic fire escape, &c.

The whole of these fire-engines and apparatus, being placed at the disposal of Her Majesty's Commissioners, are stationed in various places, for the protection of the Exhibition building.

MORRISON, ROBERT, & CO., *Ouse-burn Engine Works, Newcastle-upon-Tyne.*—High-pressure surface condensing expansive marine engine, cut-off variable.

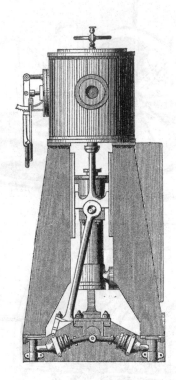

SIDE ELEVATION.

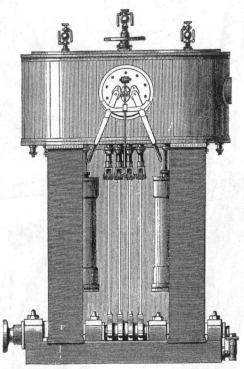

P.H.DELAMOTTE.

FRONT ELEVATION.

CONDENSING EXPANSIVE MARINE ENGINE.

The above engraving shows a side and front elevation of these engines. The cylinders are inverted, and with the slide chest are completely surrounded with steam. There is an ordinary and expansive slide to each cylinder, each worked direct by a single eccentric. The two main eccentrics are connected together, and are loose on the shaft, being retained in their proper positions, for going a-head or a-stern, by stops bolted to the shaft. The two expansive eccentrics are also joined together, and are loose on the connecting parts of the main eccentrics, and also provided with proper stops.

Starting and reversing the engines are effected by means of a small additional valve, introduced expressly for that purpose. There is an air pump on one side, and a cold water pump on the other side of the eccentric gear, worked direct from the piston; these pumps are both single acting. There are two supports to the cylinders on the starting side, but only one, which reaches nearly the whole length of the cylinders, on the other; in the latter are placed the tubes for the surface condenser. There is a door at the back by which to reach them, to replace or clean them. The remainder of the back support, the foundation plate, and the two front supports, all communicate together, and form a hot well to contain the distilled water pumped in by the air pump. The feed and bilge pumps are joined in front to the foundation plate, and are of the ordinary construction. These engines are intended to work at about 60 lbs. pressure above the atmosphere, and to expand from 6 to 8 times. There is a small wheel on the top of the slide chest, by turning which, the amount of expansion may be varied. This engine, though of 30-horse nominal power, stands on a space of 5 ft. 6 in. by 4 ft. The diameter of each cylinder is 18 in. and the stroke 18 in.

NEEDHAM & KITE, *Phœnix Iron Works, Vauxhall.*—Filter press for semi-fluids.

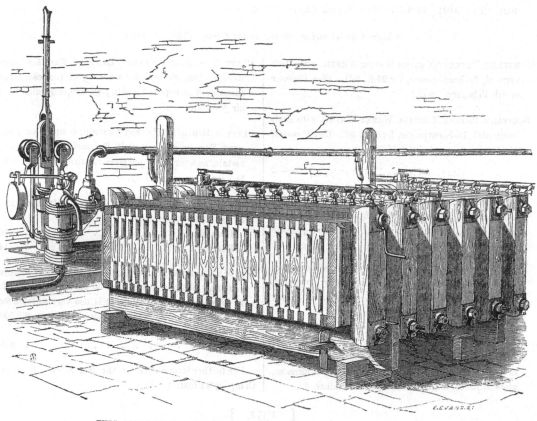

FULL SIZED FILTER PRESS, CONTAINING 600 FEET AREA OF DRAINAGE,
CAPABLE OF WORKING FROM 10 LBS. TO 100 LBS. PRESSURE UPON SQUARE INCH.

The PATENT FILTER PRESS for semi-fluids, manufactured by the patentees Needham & Kite, engineers, Phœnix Iron Works, Vauxhall, London, and Hanley, Staffordshire.

The patent filter press is used by the largest manufacturers of china and earthenware, for expressing the water from slip instead of boiling.

It is used by the largest brewers in London, Burton-on-Trent, and the United Kingdom, for rapidly clarifying drawings, and expressing beer from yeast and grounds; and also by the largest oil refiners and colour makers in the United Kingdom. It can be applied to all trades having large quantities of semi-fluids to deal with.

[1947]

NORTH BRITISH RUBBER COMPANY, *Edinburgh.*—India-rubber belting for machinery.

[1948]

NORTH MOOR FOUNDRY COMPANY, *Oldham.*—Turbines, fans, blast machines, steam turbine and fan, ship ventilators, steam engines, &c. (*See page 52.*)

[1949]

NORTON, L., 38 *Belle Sauvage Yard, Ludgate Hill.*—Model pumps; cloth tentering and wool-drying machine.

[1950]

ORKNEY, EARL OF, 3 *Ennismore Place, Hyde Park.*—Rotary engine.

NORTH MOOR FOUNDRY COMPANY, *Oldham.*—Turbines; fans; blast machines; steam turbine and fan; ship ventilators; steam engines, &c.

Obtained medal and certificates at the Paris Exhibition, 1855.

SCHIELE'S PATENT TURBINE WATER WHEEL, with shaft vertical, 30-horse power, for 25 ft. fall. (See *Engineer* of 8th February, 1862.)

SCHIELE'S PATENT TURBINE WATER WHEEL, with shaft horizontal, 15-horse power, for 50 ft. fall. (See *Practical Mechanics' Journal*, July, 1861.)

SCHIELE & WILLIAMS' PATENT VENTILATOR, for ships—steam-engine and fan combined—for ventilating the holds and cabins, cooling the engine-rooms and stoke-holes, and increasing the draught of the fires; will produce 600,000 cubic ft. of air per hour; space occupied, 3 ft. square.

HIGH-PRESSURE STEAM ENGINE, with expansion gear, variable to any extent, either by hand or by governor. Price (exclusive of expansion gear, which costs 20 per cent. extra),

    12-horse power engine £110
    8-horse power engine 70

PLATT & SCHIELE'S PATENT FAN for blowing smiths' fires, melting iron, blowing puddling and mill furnaces, glass furnaces, and for ventilating coal mines, &c.

PLATT & SCHIELE'S PATENT EXHAUST FAN, for drying wool, cotton, &c. and for a variety of purposes where exhaustion is required; will pass 300,000 cubic ft. of air per hour.

PLATT & SCHIELE'S PATENT COMPOUND OR HIGH-PRESSURE FAN (working model), for smelting and refining metals, and for other purposes requiring blast of 1 lb. to 2 lbs. pressure, and upwards.

SCHIELE'S PATENT PORTABLE SMITHS' HEARTH AND BLAST ENGINE combined.

    Blast engine £10
    Hearth 14

SCHIELE'S PATENT BLAST ENGINE, OR STEAM ENGINE AND FAN combined, will blow 30 smiths' fires, or melt 4½ tons of metal per hour. It is also very suitable for mine ventilation, as it will exhaust or produce 600,000 cubic ft. of air per hour.

    Price £70

For further information see the North Moor Foundry Company's illustrated lists.

[1951]

OXLEY, WILLIAM, & CO., *St. Mary's Churchyard, Parsonage, Manchester.*—Mill furnishings; lubricators; syphon boxes; air valves; strapping; sliver cans; gas works.

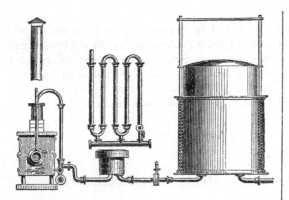

OXLEY'S IMPROVED PORTABLE GAS WORKS, suitable for mansions, railways, and small works, from £28.

MACHINES.

BOWDEN'S STEAM TRAP, or Syphon Box, for discharging condensed steam water.

    1. To carry 10 lbs. £1 10
    2. ditto 25 . . . . 1 15
    3. ditto 50 . . . . 2 10
    4. ditto 80 . . . . 4 0

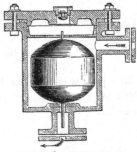

BOWDEN'S STEAM TRAP.

OXLEY'S SELF-ACTING LUBRICATORS, for oiling the journals of shafting while in motion. Price per dozen . £2 0 0

OXLEY'S SELF-ACTING LUBRICATOR.

Case containing specimen spindles, flyers, and various articles of mill furnishings.

[1952]

PARKIN, WILLIAM, 13½ *Lovell Street, Attercliffe Road, Sheffield.*—Metallic railway key; cast-steel piston head.

[1953]

The PATENT FRICTIONAL GEARING COMPANY, *Glasgow.*—Specimens and examples of the application of Robertson's patent frictional gearing.

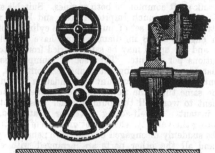

PATENT FRICTIONAL GEARING.

MACHINERY IN MOTION.

DOUBLE-CYLINDER STEAM WINCH, geared with wedge and grooved frictional wheels, with patent break movement, the steam cylinders 6 in. diameter, stroke 10 in. capable of lifting 2 tons.

STEAM ENGINE WITH CIRCULAR SAW for cutting hot iron, driven by wedge and grooved frictional wheels, steam cylinder, 7 in. diameter, stroke 10 in.; circular saw 3 ft. 6 in. diameter.

HOISTING AND TRAVERSING STEAM ENGINE for travelling cranes geared with frictional wedge and grooved bevel wheels, cylinder 5 in. diameter, stroke 10 inches.

SMALL STEAM ENGINE and MODEL of DAVISON'S PATENT

HOT AIR, geared with wedge and grooved frictional wheels.

SPECIMENS OF PATENT WEDGE AND GROOVED FRICTIONAL GEARING AND FASTENINGS.

Pair of spur wheels.

Pair of bevel wheels.

Pair of mitre wheels.

Pair of wedge and groove adjustable tyred plate wheels.

Wedge and grooved disc coupelings and keys.

Frictional screw motion.

Two models of hoists.

Rolled iron wedge and groove fastenings for iron structures ; girders, roofs, plate work, &c.

DRAWINGS.

Seven sectional drawings of patent wedge and grooved wheel surfaces.

Steam engine and main factory gearing, for increasing speed.

Incline hauling engine, with frictional gearing and patent break movement.

Steam engine and rolling mill geared with frictional wheels.

Steam engine with circular saw for cutting wood, geared with frictional wheels.

Single cylinder hoisting engine, geared with frictional wheels with patent break movement.

Steam engine and fan, driven by frictional wheels.

Frictional screwing rolls for straightening round bars and tubes.

Two drawings of warehouse hoists.

A drawing of shafts with wedge and grooved couplings.

Drawings illustrative of the action and application of frictional screws.

[1954]

PEEL, WILLIAMS, & PEEL, *Soho Iron Works, Manchester.*—Steam engine. (*See page 54.*)

[1955]

PENN, J., & SONS, *Greenwich.*—Marine engine.

[1956]

PERREAUX & CO., 5 *Jeffrey's Square, London, E.C.*—Patent India-rubber pump valves, and India-rubber as applied to mechanics.

[1957]

POTIER, WILLIAM, *Green Street, Wellington Street, Blackfriars Road.*—Gut wheel-bands.

[1958]

POTTS, JOHN, *Derby Lane, Burton-on-Trent.*—Working model of steam engine made of glass, showing the piston valves and other movements.

[1959]

PRELLER, C. A., 4 *Lant Street, London, S.E.*—Machine driving belts. (*See page 55.*)

[1960]

RANDOLPH, ELDER, & CO., *Glasgow.*—Drawing of Marine Engine.

1961]

RANSOMES & SIMS, *Ipswich.*—Portable double-cylinder steam engine, &c. (*See pages 56 and 57.*)

[1962]

RAVENHILL, SALKELD, & CO., *Glass House Fields, Ratcliff, and Orchard Wharf, Blackwall.*—Models of marine steam engines. (*See pages 58 and 59.*)

PEEL, WILLIAMS, & PEEL, *Soho Iron Works, Manchester.*—Steam engine, hydraulic press, and pumps for beet-root sugar works.

Fig. 1 represents a powerful HYDRAULIC PRESS, having a cylinder of 12 in. diameter, capable of exerting a pressure of 340 tons, with water at a pressure of 3 tons per square in. It is provided with extra large water ways, which facilitate expedition in running down the table. The columns for supporting the top of the press are of wrought-iron, and turned all over perfectly true. The recesses upon which the collars of the columns rest, are planed to one true surface, to insure a uniform bearing upon each corner of the framework.

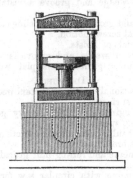

Fig. 1. HYDRAULIC PRESS.

Such presses are used for expressing the syrup or juice from beet-root, in the sugar manufactories of Southern Russia. The table has a channel along its four sides into which the syrup is collected. These presses are also applicable for a variety of other purposes; in some instances having the tables and under side of the top of the press planed true and smooth all over, for pressing paper, &c. They are also extensively used for packing cloth goods (or hay) tightly into small bales for exportation, &c.

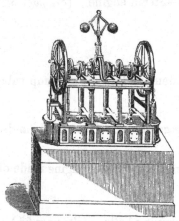

Fig. 2. HYDRAULIC-PRESS PUMPS.

Fig. 2 represents a set of HYDRAULIC-PRESS PUMPS worked by two independent steam engines, on the non-condensing principle of direct action, attached to the same framing, applicable not only to presses such as Fig. 1, but also to every description of hydraulic press.

This set consists of eight pumps, four being 1¾ in. diameter and four 1 in. diameter, all having a stroke of 3 in. Usually one of each size is used to each press, and the arrangement is such, that, by a self-acting apparatus, when a pressure of one ton to the square inch has been

reached, the larger pump ceases to act, and the final pressure is obtained by the use of the small pump alone.

The pumps receive motion from eccentrics fixed upon the crank shaft common to both engines. Suitable safety valves and also a much improved stop and let-off valve are attached to this set of pumps. The cylinders of the steam engines are 8 in. diameter and have a stroke of 16 in. and the speed may be safely varied from 80 to 100 revolutions per minute. This set of pumps possesses peculiar advantages, being entire and self-contained, consequently a very small amount of foundation is required. At the same time, the power in the cylinders is amply sufficient to work all the pumps under pressure at the same instant. A self-acting governor is attached, for regulating the velocity when a set or more of the pumps may be suddenly disengaged or otherwise; and the cylinders and all the other parts are arranged with every facility for taking to pieces to clean out, or repair. All the joinings at the junctions of the pipes, &c. are wholly metallic, no leather or other medium being used except round the working plungers. Attention is also directed to the very efficient mode adopted, for compensating the slackness occasioned by the wear of the slide blocks of the engine piston rods, as also the knuckle-joints of the pump rods.

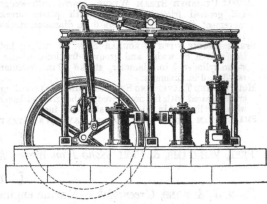

Fig. 3. VACUUM ENGINE.

Fig. 3 represents a steam engine, technically known as a "vacuum engine." It is nominally of 16-horse power, and is constructed on the non-condensing principle. It is fitted with two vacuum pumps 18 in. diameter, and 18 in. stroke, all the valves being of vulcanized india-rubber. The purpose for which this class of engine is employed, is for creating and maintaining a vacuum in the sugar boiling pans, thereby producing ebullition and vaporisation, at a much lower temperature than is usual in vessels used for such purposes, when subject to atmospheric pressure, by which means a superior quality and colour of sugar is produced. It is provided also with two additional pumps, one for supplying cold water to a cistern for general use, and the other for supplying water to the boiler. Either or both of these pumps may be used or dispensed with as circumstances require. The speed of this engine is 50 revolutions per minute, and power may be taken by a broad belt from the periphery of the fly wheel formed for that purpose, or by gearing from the fly wheel shaft. This engine and pumps are not only applicable to the work already referred to, but are also well adapted to sugar refineries, and may be made with the air pumps upon the double-acting principle, thus giving out twice the effect, in which case a cylinder proportionately larger will be necessary; and although the engine exhibited is arranged to work at a speed of 50 revolutions per minute, this may be varied at pleasure to a considerable extent.

PRELLER, C. A., 4 *Lant Street, London, S.E.*—Machine driving belts for transmitting power, made of leather, combining extraordinary strength with suppleness.

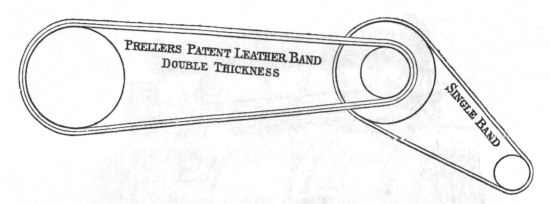

By experiments in the Woolwich Dockyard (made on October 24, 1855, and repeated on April 27, 1858), it has been ascertained, that Preller's leather is at least 50 per cent. stronger than tanned leather, and consequently far superior to all substitutes for leather.

Eminent engineers are of opinion that the thinness of a band is a great advantage; but this depends upon the nature of the material used: if weak and spongy, thickness is required; but in proportion to the greater strength and density of material, bands may be made thinner.

Preller's bands are in use all over the Kingdom, in different parts of Europe, India, Australia, South America, &c.

For hot climates the yellow leather is particularly suitable, and the grain never cracks in working even when the greatest power is applied.

All bands made of Preller's leather are warranted to be cut from the prime part of the hide (no shoulder being used), and are sewn with Preller's laces and twice stretched.

[1963]

RAWLINGS, JAMES, 10 *Carlton Hill East, N.W.*—Machine for cleaning boots with 2 brushes simultaneously, without inserting the hand.

[1964]

RENNIE, GEORGE, & SONS, 6 *Holland Street, Blackfriars, and Greenwich.*—Marine condensing engine for screw propellers, high and low pressure, with surface condensation. (*See page* 62.)

[1965]

RICHARDSON, THOMAS, & SONS, *Hartlepool.*—A pair of direct-acting inverted cylindrical marine condensing engines.

[1966]

RICHMOND, JOHN, *Hackney Wick Works, Victoria Park, N.E.*—Counting machines.

[1967]

RILEY, GEORGE, *South Lambeth.*—Patent helical refrigerator for brewers. Patent slotted false bottom for brewers.

[1968]

ROBERTS, RICHARD, & CO., 10 *Adam Street, Adelphi.*—Drawing and model of turbine.

[1969]

ROBERTS, WILLIAM, *Millwall, Poplar, E.*—Fire engines for house, factory, and general purposes. (*See pages* 60 *and* 61.)

[1970]

ROBINSON, WILLIAM, *Bridgwater.*—Machine for cleaning the inside of casks without un-heading, adapted for breweries, &c. (*See page* 61.)

RANSOMES & SIMS, *Ipswich.*—Portable double-cylinder steam engine, 20-horse power; portable steam crab, 5-horse power.

20-HORSE POWER DOUBLE-CYLINDER HIGH-PRESSURE STEAM ENGINE.

This engine is the largest of the exhibitors' standard series of portable steam engines, which are made from 3 to 20 horse power.

These portable steam engines are extremely simple, durable, and easy to manage; and are capable of application to almost all purposes where steam is required, such as working circular, horizontal, or vertical saws for cutting timber; for driving pumps for irrigation, mill-stones and mill gear, quartz-crushing machines, stampers, amalgamators, &c.; and are built for burning either wood or coal, a great desideratum in countries where coal is scarce.

The boiler, which is multitubular, is of the exhibitors' own make, and is constructed with especial reference to durability, on the same model as the most approved loco-motive boilers. The bulk of the plates are Low Moor, the others being best Staffordshire. Ample water-space is given round the fire box, and between the tubes, for the free circulation of the water, the escape of steam, and the settling of sediment. The boiler is tested by hydraulic pressure to 100 lbs. per square inch. It is fitted with a steam gauge, glass water-gauge, steam whistle, 2 gauge cocks, safety valve with spring balance, blow-off cock, &c. &c. and is lagged with wood, covered with sheet-iron. It is fitted with a lock-up safety valve when so ordered.

The chimney is furnished with a wire top, which extinguishes all sparks and prevents all danger of fire.

The crank shaft and connecting rods are of wrought-iron, and all small wearing parts are case-hardened.

The fly wheel is properly balanced, and can be hung on either end of the crank shaft.

The slide valve eccentric can readily be shifted to admit more or less steam, according to the amount of work to be performed, or to reverse the motion of the engine, if necessary.

The power is calculated at 45 lbs. pressure of steam in the boiler. Every engine is tested under steam before leaving the factory, and may be safely worked at 60 lbs. pressure, at which they give off double their nominal power, consuming, of course, fuel and water in the same increased proportion.

In estimating the power an engine will produce, the size of the cylinder is only one element, and by no means the most important, for it must be borne in mind that the power really depends upon the capability of the boiler to generate dry steam, as fast as the engine can utilise it. In a portable engine, the size of the boiler is limited by the condition that the engine must be easily portable; and Messrs. Ransome's engines are furnished with as large boilers as is compatible with that condition. The exhibitors have chosen a moderate sized cylinder, and a quick speed, in preference to a larger cylinder and a slow speed, as possessing, for this class of engine, very many substantial advantages; and it will be found in practice, that these engines will give off as much power, and cost as little to keep in repair, as any others of equal weight and portability, but furnished with larger cylinders.

These engines are all furnished with the following articles, viz. waterproof cover, tube brush, fire pricker, rake, shovel, screw spanners, oil can, large funnel, and spare gauge glass, which are included in the price quoted.

They are also sometimes fitted with a simple apparatus in the smoke box for heating the feed water. This economises the fuel considerably, and is not liable to get out of order.

RANSOMES & SIMS, *continued.*

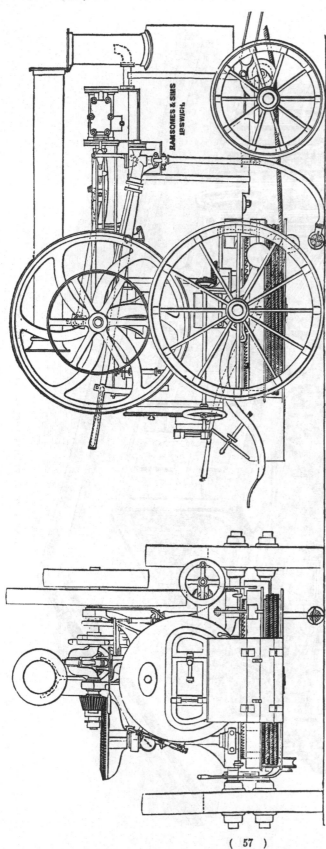

5-HORSE POWER PORTABLE STEAM CRAB,

A 5-HORSE POWER PORTABLE STEAM CRAB, with Biddell and Balk's patent boiler.

This crab is capable of raising about 25 cwt. at a rate of from 70 to 80 ft. per minute. It is especially designed for raising building materials; but if the winding gear is not required, it can be disconnected by shifting a clutch, and the engine can then, like any ordinary portable engine, be used for other purposes, such as driving a mortar mill, pumps, or circular saw, &c. The rope which generally passes through a snatch block, and over the swivel pulley on the fore carriage, is wound up on the winding drum, which is furnished with a ratchet wheel to retain it in its position, and also with a lever and rollers to enable the driver to cause the rope to coil properly. It is also furnished with a break, which is worked by the foot of the driver, and which serves for lowering and stopping suddenly. The release or this break is made self-acting by means of a counter-weight.

The engine is made to reverse to facilitate the starting, and for the purpose of unwinding the chain or rope on the drum, so as to facilitate the descent of the end of the chain or rope when empty.

Everything necessary for the working of this crab, can be done by the driver, without leaving his place.

RAVENHILL, SALKELD, & Co., *Glass House Fields, Ratcliff, and Orchard Wharf, Blackwall.*—Models of marine steam engines.

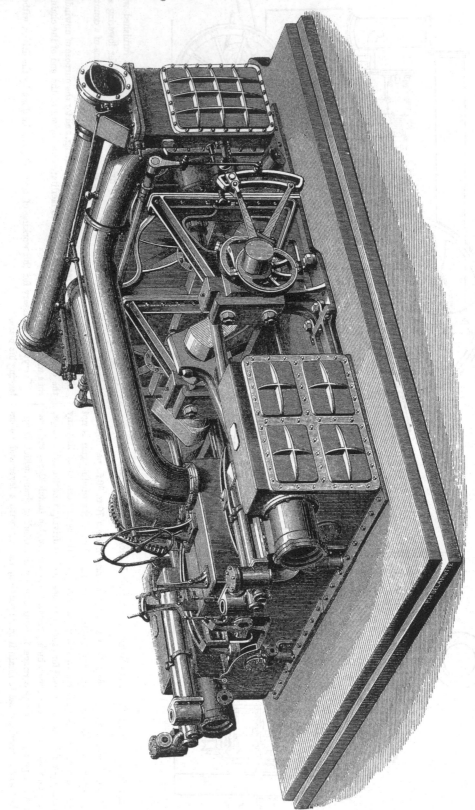

MARINE ENGINE, FROM A PHOTOGRAPH OF MODEL NO. 2.

RAVENHILL, SALKELD, & CO., *continued.*

1. MODEL OF ENGINES with feathering paddle wheels, of the Holyhead mail packets *Leinster* and *Connaught*, each of 720 horses nominal power.

This is an application of the oscillating cylinder to the largest class of marine steam engine, each cylinder 98 inches (eight feet two inches) internal diameter, weighed, when finished, upwards of twenty tons; the condenser weighed twenty-two tons. The engines were fitted with eight tubular boilers, having forty furnaces and 4176 tubes, giving a total length of four and three quarter miles of tubing, and the vessels attained an average speed at the official trial in Stokes Bay of eighteen knots or twenty-one miles an hour. The engines exerting an indicated power of 4,751 horses.

The first pair of engines with oscillating cylinders constructed by the exhibitors was fitted in the year 1838, and engines have since been manufactured by them upon this principle of the aggregate nominal power of 22,000 horses.

2. MODEL OF ENGINES with horizontal cylinders and double piston-rods of 500 horses nominal power for screw-propellers, such as are fitted by the exhibitors on board Her Majesty's 90-gun line-of-battle ships.

This model represents the plan of engines of the larger class made by the exhibitors for the British and foreign Governments, and is arranged so as to afford easy access to all the working parts.

The exhibitors were the first to introduce the double piston-rod engine into the British navy, engines of 300 horses nominal power so fitted having been made by them in the year 1845, since which time the following ships in her Majesty's service have been so fitted by them :—

Adventure.
Alacrity.
Alert.
Amphion.
Ariel.
Assurance.
Brunswick.
Centurion.
Charybdis.
Clio.
Coquette.
Dromedary.
Emerald.

Falcon.
Fawn.
Fox.
Gannet.
Glasgow.
Greyhound.
Jason.
Lapwing.
London.
Lyra.
*Narcissus.
Nelson.
Neptune.
Newcastle.
Pelican.
Pelorus.
Pioneer.
Racoon.
Rattlesnake.
Ringdove.
Roebuck.
**St. George.
Surprise.
Swallow.
Tamar.
Undaunted.
Victor.
Waterloo.
Wolverine.

This list refers only to engines made upon the double piston-rod plan, many other vessels in her Majesty's service having been fitted by the exhibitors upon various other systems, arranged according to the requirements of the service.

* Narcissus, 50-gun frigate, bearing the flag of Rear-Admiral Sir Baldwin W. Walker, Bart., K.C.B., late controller of the navy on the Cape of Good Hope station.

** St. George, 90-gun ship, on board which His Royal Highness Prince Alfred has been serving on the North American and West Indian station.

3. MODEL OF ENGINES of the same power as No. 2, arranged for surface condensation.

4. MODEL OF MARINE STEAM ENGINES, with inclined oscillating cylinders, designed for vessels having a small section with considerable rise of floor. Engines on this plan have been constructed by the exhibitors up to 240 horses nominal power. It is a light form of engine, and has given great satisfaction.

ROBERTS, WILLIAM, *Millwall*, *Poplar*, *E.*—Fire engines for house, factory, and general purposes.

ROBERTS'S HAND ENGINE IN RUNNING ORDER.

ROBERTS, WILLIAM,—*continued.*

DESCRIPTION OF W. ROBERTS'S PORTABLE FIRE-ENGINE.

A Pump having one suction at *a* through the side, and another in the bottom, which (the bottom) is screwed in the cylinder; also two deliveries, $a^1 a^1$, the air-chamber A^1 being screwed on the top, thus affording a ready means of access to the interior.

B B Levers with wood handles *b b*.

C Plank or bed, having an axle hooked at the ends *c c*, and at the end of the plank a plate C^1, with a latch C^2.

D D D Frames either of metal, or metal and wood combined.

E E Suspending rods working between the upright frames upon bolts at *e e*, the lower ends being hooked.

F Rod having a **T** foot, and connected with the frame at *f*.

G G Hose reel working upon the axle freely between the side frames *D D D D*, and upon which the hose is coiled.

H A box to carry the branch pipe, jets, spanners, and all the necessary tools. The suction hose can be carried upon the bed *C* at each side of the pump, or beside the box *H*, or both, if a long length is required.

I Canvas well, or cistern.

When used for horse traction, a seat is fitted to the frame to carry the driver and three firemen. The seat folds down upon the frame when not required.

Upon reaching a fire, the latch C^2 is thrown back, and the foot of the rod *F* withdrawn from the slot; this allows the handle end of the frame to rise; this action lowers the pump upon the ground, when the suspended rods *E E* being unhooked (the hose *g* being kept screwed to the pump), the reel is run towards the fire; when the necessary length being unwound, it is disconnected from the reel, the branch, spanners, &c., being in the box *H*, it can be got to work in a very few minutes. When done with, the whole can be quickly packed up and taken away.

THE FOLLOWING FIRE ENGINES, &C. ARE EXHIBITED:—

1. W. ROBERTS' PATENT FIRE ENGINE FOR 1 HORSE.

This engine will throw nearly as much water under pressure, as a brigade engine, with two-thirds the number of men to work it. It is about half the weight, and will pass through an opening one-third narrower.

Price £100 0

Hose and all gear extra.

2. W. ROBERTS' PATENT HAND FIRE ENGINE upon wheels, carrying its own hose and gear, can be run easily by 1 man. Price £50 0

3. W. ROBERTS' PATENT HAND FIRE ENGINE, will throw more than half the quantity of a brigade engine with one-third the number of men. Weight about 2 cwt.

Price £28 0

4. W. ROBERTS' IMPROVED HOSE REEL FOR 1 HORSE.

This reel will carry as much hose and gear as three brigade engines, and is specially adapted for use in cities or towns, having a constant supply of water at high-pressure.

5. W. ROBERTS' IMPROVED HOSE REEL FOR HAND WORK.

This reel will carry as much hose and gear as two brigade engines.

Price £15 0

Manufactured by Brown, Lenox, & Co., Millwall, Poplar, London.

ROBINSON, WILLIAM, *Bridgwater.*—Machine for cleansing the inside of casks without un-heading, adapted for breweries, &c.

PATENT CASK-CLEANSING MACHINES, for brewers, wine and spirit merchants and vinegar makers, &c.

These machines consist of two circular frames, one within the other. The outer one when set in motion revolves on its axis, and the inner one at the same time is moved in a circular direction, by lifts connected with each axle of the outer frame.

They can be worked by either hand or steam power; in shape and general construction they are exceedingly strong and durable; and being sent out in complete working order, no expense is incurred in fixing.

A cask, on being placed in the machine as shown in the diagram, speedily assumes a diagonal position, passing to the perpendicular, or head over head, and finally to the horizontal, thereby subjecting every part of the cask to the cleansing material, and rendering the labour, wear and tear of unheading the foulest cask unnecessary.

One machine worked by either hand or steam power will clean, of ordinary sweet casks 150 hogsheads or barrels, or 300 kilderkins, or firkins, per day, and one half that number of foul casks.

The superiority of these machines over any other yet introduced, consists in effectually cleansing a greater number of casks in the same time, and also in the capability of taking every size from a hogshead downwards.

These machines are therefore especially adapted for the use of brewers, wine and spirit merchants, vinegar makers, &c.

Price £30 complete for Steam Power.

References to a large number of firms now using them, together with testimonials of the highest character, may be had by applying to the patentee.

Prices from £22 to £35

RENNIE, GEORGE, & SONS, 6 *Holland Street, Blackfriars, and Greenwich.*—Marine condensing engine for screw propellers, high and low pressure, with surface condensation.

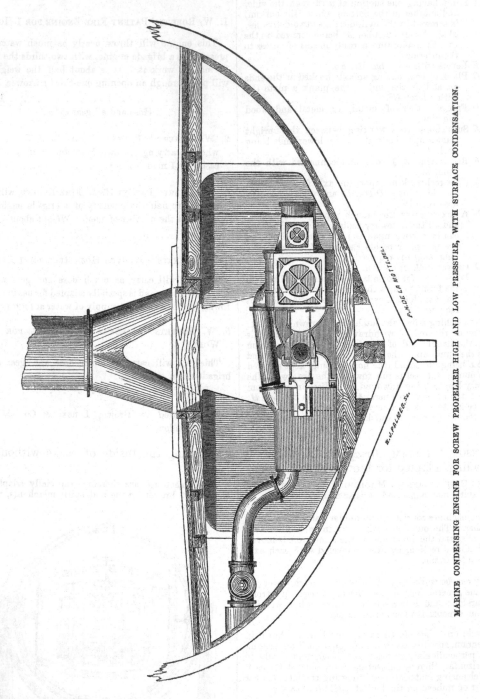

MARINE CONDENSING ENGINE FOR SCREW PROPELLER HIGH AND LOW PRESSURE, WITH SURFACE CONDENSATION.

P.M. DE LA MOTTE DEL.

W. J. PALMER Sc.

HIGH AND LOW PRESSURE MARINE CONDENSING ENGINE.

The above is an engraving of a high and low pressure, marine screw steam engine, placed in a vessel of war.

In class 8 is exhibited a working model of the above arrangement.

The advantage of 2 cylinders in direct-acting marine screw engines, is that of working steam expansively, whereby economy of steam and fuel is obtained, depending upon the pressure of steam, and the relative volumes of the high and low pressure cylinders. The engines are fitted with surface condensers, with copper tubes, and improved centrifugal pumps for circulating the water in condensers; these pumps being made on a double curvature principle; of least resistance to the flow of water occasioned by the centrifugal force generated by the angular velocity of the pump.

Engines on the above principle are fitted with boilers in proportion; apparatus for super-heating the steam, and feed-water heaters may be made to consume not more than 2 lbs. of coal per actual or indicated horse-power.

[1971]

ROSE, WILLIAM, 37 *Victoria Street, Manchester.*—Brigade fire engine and three patent portable fire engines, with fittings.

The exhibitor is a builder of fire engines, and manufacturer of fire-extinguishing apparatus. He is the sole builder of Hall's patent portable fire engine, also sole agent for Vaucher's patent woven hose, which will bear a pressure of 200 lbs. to the square inch. A large stock of brigade and portable engines with fittings on view at the depôt.

[1972]

ROUTLEDGE & OMMANNEY, *Salford, Manchester.*—Diagonal and double-acting engines, hydraulic-press pumps, self-acting boiler feeder; machine for cleaning brass turnings, &c.

[1973]

RUSE, CHARLES, 24 *Hereford Place, Commercial Road East, London, E.*—Two improved beer machines.

[1974]

RUSSELL, JOHN, & Co., *London, Wednesbury, Walsall, and Manchester.*—Wrought-iron tubes for boilers, gas, water, and steam.

The exhibitors are patent tube manufacturers, the original makers of wrought-iron gas tubes, and the inventors of the lap-welded tubes for locomotive and marine boilers. They also manufacture all kinds of tubes and fittings for gas, steam, or water; galvanised tubes and fittings; brass-work of all kinds for steam and gas; stocks, dies, and taps of all sizes.

The warehouses of John Russell & Co. are at 69, Upper Thames Street, and 5 Charles Street, Soho, London; and 35 Granby Row, Manchester; and the works, at Wednesbury, and at Walsall, Staffordshire.

All communications should be addressed, 69 Upper Thames Street, E.C. London.

[1975]

RUSSELL, JAMES, & SONS, *Wednesbury, and Upper Ground Street, Blackfriars, S.*—Iron tubes, iron and brass fittings.

Obtained honourable mention for lap-welded iron tubes at the American Exhibition; and a gold medal at the Paris Exhibition, 1855.

The exhibitors are the patentees and original makers of wrought-iron tubes.

[1976]

RUSSELL, J. SCOTT, *London.*—Three cylinder surface condensing marine steam engines. (*See page* 64.)

[1977]

RUSTON, PROCTOR, & Co., *Lincoln.*—Portable, fixed, and traction steam engines; flour and sawing mills.

[1978]

SALTER, GEORGE, & Co., *West Bromwich.*—Spring balances; dynamometers; spiral springs; pressure gauges; roasting jacks; bayonets; and swords. (*See page* 65.)

[1979]

SAMUELSON & Co., *Hull.*—Oil mill.

[1980]

SANDERS, FREDERIC, 473 *Oxford Street.*—Improvement in beer engine pumps, and spirit machines.

STOCKER'S IMPROVED PATENT BEER ENGINE, AND CRYSTAL SPIRIT FOUNTAINS.

These beer engines combine elegance, cheapness, and durability; and having no slot or sweep for the handle to work in, no dirt or grit can come in contact with the works. They are manufactured in mahogany, marble, and pewter. The improved crystal spirit fountain is of elegant design, and handsome ornament for licensed victuallers' counters.

RUSSELL, J. SCOTT, *London.*—Three-cylinder surface condensing marine steam engine.

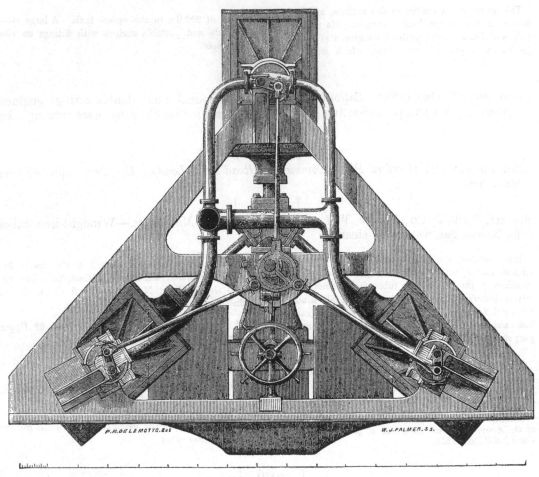

THREE-CYLINDER MARINE STEAM ENGINE.

THREE CYLINDER MARINE STEAM ENGINE AS ARRANGED FOR SCREW PROPELLER.

Collective power	100 horses.
Works expansively	with variable cut off.
Surface condensers	with India rubber packing.
Diameter of cylinder	30 inches.
Length of stroke	3 feet.

[1981]

SANDYS, VIVIAN, & CO., *Copper House Foundry, Hayle, Cornwall.*—18-horse power high-pressure horizontal steam engine.

[1982]

SCOTT, G., 35 *Page's Walk, Bermondsey.*—Portable engine, oscillating engine, surface condenser.

SALTER, GEORGE, & CO., *West Bromwich.*—Spring balances, dynamometers, spiral springs, pressure gauges, roasting jacks, bayonets, and swords.

The exhibitors are manufacturers of spring balances, vertical jacks, swords, bayonets, pressure gauges, pocket steelyards, steel springs, &c.

The following are exhibited, viz. :—

1. Small spring balances for locomotive and stationary engines.

2. Large spring balances for locomotive and stationary engines.

3. Patent pressure gauges in iron and brass cases, and gauges suitable for hydraulic presses.

4. Spring balances with straight movement for general weighing purposes.

5. Spring balances with circular movement for general weighing purposes.

6. Large dial weighing machines for railways, warehouses, &c.

7. Cheap dial weighing machines, registered pattern.

8. Patent quadrant pattern spring balances.

9. Patent counter spring weighing machine.

10. Spring letter balances.

11. Sportsman's pocket spring balances.

12. Spring balances for testing strength of cotton, yarn, gunlocks, &c.

13. Dynamometer for testing human strength.

14. Steel spiral springs.

15. Bayonets, sword bayonets, and cutlasses.

16. Pocket steelyards.

17. Extra strong vertical jacks.

[1983]

SEARBY, GEORGE, 2 *Crown Court, Threadneedle Street.*—Steam gauge.

SEARBY'S PATENT IMPROVED STEAM GAUGE (Mercurial). The advantages of this gauge are its cheapness, durability, and safety. It is manufactured on the only principle on which a steam gauge can act with certainty, and its internal arrangements, unlike all mechanical gauges, are in no danger of getting out of order. It also acts as a safety valve. It has been approved by many of the first engineers in the Kingdom.

[1984]

SHAND & MASON, *Blackfriars Road, London.*—Steam, brigade, military, and other fire engines, implements, &c. (*See pages* 68 *and* 69.)

[1985]

SHEPARD, EDWARD C., *Victoria Street, Westminster.*—Magneto-electric machine for electric light, and street lamp carburator. (*See page* 66.)

[1986]

SIEBE, DANIEL, 17 *Mason Street, Lambeth.*—Harrison's patent ice-making machine. (*See page* 67.)

SHEPARD, EDWARD C., *Victoria Street, Westminster.*—Magneto-electric machine for electric light, and street lamp carburator.

MAGNETO-ELECTRIC MACHINE, AND ELECTRIC LAMP, with an improved frotter, for producing a continuous electric light for lighthouses, steamers, signals, &c.

This machine possesses great advantages over all others in having a continuous frotter, which, with the improved electric lamp used with it, produces a continuous, steady, and uniform electric light, burning with unvarying intensity.

The beauty and brilliance of the electric light are undisputed. It shines through the midnight gloom, with a lustre, second only to that of the noonday sun ; and so pure and white is it, that all other flames assume a red tinge by contrast. It can be used for railway signals, for lighting mines, harbours, &c. and is of especial value for use on board steamers and sailing vessels ; materially reducing the risks of loss and damage from collision. It is invaluable for lighthouse use on dangerous coasts, where, for want of a light of sufficient power to reveal the hidden dangers, there has been such appalling loss of life and property.

STREET LAMP CARBURATOR. This apparatus effects a saving of one half the gas, and increases the brilliancy of the light. Over 2,000 carburators are already fitted to street lamps in London.

[1987]

SIEMENS, C. WILLIAM, 3 *Great George Street, Westminster.*—Regenerative gas engine and furnaces ; fluid meters.

[1988]

SIMPSON, G., *Glasgow.*—Pumps.

[1989]

SIMPSON, GEORGE, 315 *Oxford Street.*—Ash's piston freezing machine and wine cooler ; freezing vases ; refrigerators ; seltzogenes, &c.

 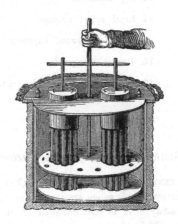

GEORGE SIMPSON is the sole manufacturer of ASH's PATENT PISTON FREEZING MACHINE and WINE COOLER, for producing, with or without ice, several kinds of dessert ices ready moulded for the table and blocks of pure ice, for icing wines, &c. The whole can be performed at one operation, or separately as desired. For hot climates this machine surpasses any other kind known.

Freezing vases, refrigerators or ice safes, butter coolers, and every article connected with the ice trade, seltzogenes for making soda water, &c. rotary knife cleaners, filters, and other patented inventions may be obtained from the exhibitor.

[1990]

SISSONS & WHITE, *Hull.*—Steam pile driver, simple, practical, economical, easily moved, and occupying small space. (*See page* 70.)

SIEBE, DANIEL, 17 *Mason Street, Lambeth.*—Harrison's patent ice-making machine.

PATENT IMPROVED ICE-MAKING MACHINE, capable of converting 24 cwt. (= 269 gallons) of spring or river water into blocks of solid ice without the use of chemicals, the ice being more or less transparent, in proportion to the relative quantity acted on at the same time.

The principle upon which this machine is constructed, is an application of the well known natural law, that, by evaporating fluids, the caloric contained therein passes off with the vapour, thereby reducing the temperature of the evaporating body. It will be seen on referring to the apparatus that science has been brought to the aid of nature, in the first place by the use of a volatile fluid as an evaporative agent; secondly, by a powerful pump, which, in its continued efforts to form a vacuum, assists the evaporation at a low temperature on the one hand, and by pressure with the assistance of water at an ordinary temperature, reduces the vapour again to a fluid on the other hand, thereby using and re-using the same volatile fluid without loss. In other words, the invention consists in the evaporation of volatile fluids *in vacuo,* at a low temperature, and condensing at a higher temperature, by pressure, and water at an ordinary temperature.

These machines can be made of any dimensions, the largest however at present in use produces 10 tons of ice daily. The manufacture of ice in tropical climates is the most important and successful operation, both in a sanitary and pecuniary point of view, to which refrigerating machinery on this principle has yet been applied, in many countries, where cooling drinks cease to be only a luxury, and become an actual necessity to Europeans; for it has been found impossible from the difficulties of transport, and loss in transit, to keep up the supply of natural ice, or to make it a remunerative speculation. These machines have been of the greatest service; for regardless of the high atmospheric temperature, the ice is formed daily on the spot where it is required for use, thereby avoiding all loss. It may not be out of place to remark here that one of the many machines now successfully at work is established nearly under the equator in Peru, supplying the neighbourhood with ice, an article rarely if ever seen in those regions before.

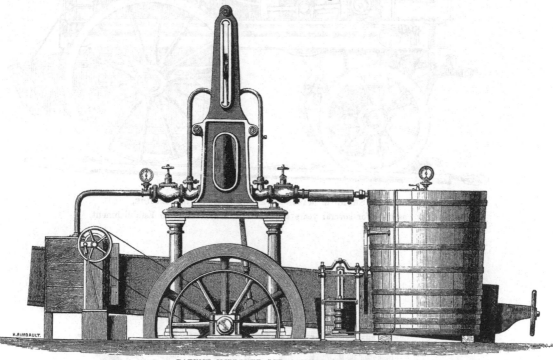

PATENT IMPROVED ICE-MAKING MACHINE.

The *cooling of hospitals,* and other buildings, is a subject which of late has attracted considerable attention, the formation of thermopathic sanitoriums in India has long been felt and acknowledged by eminent medical men, and has been discussed by the commission of inquiry into the sanitary condition of the army of India. The method proposed, is to reduce to, and retain the temperature at the required degree, by artificial means, on the converse of the principle by which buildings are warmed in this country. It has been proved by experiment that this is practicable, the inside temperature of a chamber having been reduced to within 6 degrees of freezing point, whilst the thermometer outside ranged at 90° Fahrenheit. How many valuable lives of our own public men who succumb to the climate of India might be saved, and their health as well as that of our army, secured at a comparatively small outlay when weighed against the benefit derived, it is impossible to form a just estimate.

The *cooling of wort* in breweries and distilleries is a process to which these machines, by reason of their immense refrigerative power and their capabilities to remove the caloric and lower the temperature of the wort to any desired degree are admirably adapted. It has been tested in Australia, both by direct action and by cooling a quantity of water to be used in the ordinary refrigerators and attemperators. It thus obviates the necessity of brewing only in winter in this country, and renders it possible to brew with success in any hot climate.

Salting and preserving meat, &c. is also materially assisted, as well as purity and wholesomeness secured, by the application of refrigerating machinery. The meat being removed, before congelation takes place, to a chamber kept at a low temperature, the salting trough being also kept at or near freezing point, the formation of animalculæ is entirely prevented, and salting may be carried on at any season of the year.

An admirable plan of preserving alimentary substances at a low temperature, has been proposed by Admiral Sir Charles Elliot, which, if carried out, would no doubt prove most valuable and beneficial.

It may be interesting to experimentalists to learn that 20° below zero (52° of cold), Fahrenheit, has been easily obtained and continued for some time.

SHAND & MASON, *Blackfriars Road, London.*—Steam, brigade, military, and other fire engines, implements, &c.

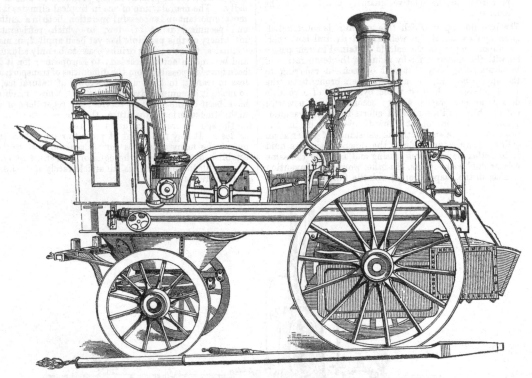

SHAND & MASON'S PATENT STEAM FIRE ENGINE,
as used for several years by the London Fire Engine Establishment.

BRIGADE FIRE ENGINE,
as made by Shand & Mason for the London Fire Engine Establishment.

SHAND & MASON, *continued.*

CAPTAIN FOWKES' PATENT MILITARY FIRE ENGINE,
as made by Shand & Mason for Her Majesty's War Department.

[1991]

SMITH, BROTHERS, & CO., *Hyson Green Works, Nottingham.*—Pressure and vacuum gauges patented by Sydney Smith, Nottingham.

[1992]

SOUL, M. A., 3 *Leadenhall Street.*—Salinometer for steam boilers using salt water (Long's patent).

[1993]

STEER, WILLIAM, *Crossland Street, Nottingham.*—Electro-magnetic engine.

[1994]

STEPHENSON TUBE COMPANY, THE, *Birmingham.*—Seamless locomotive, marine, steam, and other kinds of metal tubes ; calico printing rollers, &c.

[1995]

STONE, JOSIAH, *Deptford, Kent, S.E.*—3-throw ship's pump; double action ship's fire engines ; portable ship's fire engines.

[1996]

STONES, SETTLE, & WILKINSON, *King Street Brass Works, Hull.*—Brass works for engineers. (*See page* 71.)

[1997]

STRATFORD, WILLIAM, 6 *Edward Street, Mile End Road.*—Patent furnaces and bars. (*See page* 72.)

[1998]

STUBBS, W., 10 *Elliott Street, Liverpool.*—Registering machine for beer taps.

[1999]

STUBBS, WILLIAM, 1 *Union Street, Cleveland Street, Mile End.*—Specimens of coopering in wood, bone, and ivory.

SISSONS & WHITE, *Hull.*—Steam pile driver, simple, practical, economical, easily moved, occupies a small space.

SISSONS AND WHITE'S STEAM PILE DRIVER.

This machine supplies a deficiency which has long been felt, viz. :—something more expeditious and powerful than the common hand engine, and less ponderous and costly than those to which steam has hitherto been applied.

It is easily moved, and by a contrivance in the carriage part, can be transferred to other lines at any angle with great facility ; there is also an arrangement for readily altering the incline, to suit the various batters at which piles may have to be driven.

It requires 4 men to work it, and consumes 4 cwt. of coals in 10 hours.

Not least amongst its recommendations are its lightness and smallness of cost, as compared with the heavy and expensive steam drivers hitherto used ; and where staging is required the advantages are very great.

The total weight of the driver and boiler is 6 tons, including the ram and mountings, which are 22 cwt. ; it ordinarily falls 10 times in a minute, with a 5 ft. lift. The bottom framing of the driver is 7 ft. 3 in. square, and the boiler truck 5 ft. 6 in. square ; when in work the two are bolted together, and travel on the same tramway. Its comparative lightness, and the small space it occupies, make it capable of being worked in any position or circumstances in which a common hand machine can be *placed*, either on land or afloat.

By a different arrangement in the upright framing, piles can be driven in a tideway, down to a depth of 30 ft. below the stage on which the machinery stands, the ram driving quite down to the ground without using a "dolly," the dispensing with which is a great advantage.

It will be perceived from the annexed drawing, that the bottom framing is in two heights—the upper part revolving turntable fashion on the lower one. The machine can thus be faced round to any of the four sides.

The travelling wheels are castors, so that by lifting up each side with a lever the castors can be turned to run on a tramway at any angle.

It is moved by fastening the end of a rope ahead, passing it over a roller under the winch, and taking a turn round the barrel.

The pile is quickly pitched by attaching with a shackle a common chain to the pitched chain.

The height of the machine in the annexed drawing is 36 ft. and will pitch a pile 30 ft. long on ground the same level as that on which the machine stands ; this height is found to be sufficient for general use, but machines of greater height are made to order.

The ram is lifted by the horns passing through and projecting beyond the back of the upright guide pieces ; on this projection is fixed a frame and catches, which lay hold of the shoulder links of the pitched chain in its continuous revolution. The catches are closed by hand, and released, by striking against pins fixed in the back of the guide pieces.

The machine has been worked at the Vernatt's New Sluice, near Spalding ; the Harbour Works, Dublin ; the North Level Sluice, near Wisbeach ; at the Penarth Docks, Cardiff ; by Messrs. Smith & Knight, contractors, London ; at Messrs. Samuelsons' New Works, Hull ; the Jarrow Docks, Newcastle on Tyne, by Messrs. Jackson, Bean and Gow ; at the Surrey Canal Docks, Rotherhithe, London, by Messrs. Baker & Son ; at the Arsenal, Woolwich, by Mr. Lavers ; the new bridge at York ; the Bardney Bridge, Lincoln ; the Patent Slip, Genoa ; at Amsterdam, and several other places.

Extract from a report of a paper on pile driving, read before the Society of Civil Engineers, by Mr. W. F. Bryant, of the Westminster Bridge Works, Dec. 5, 1859 :—"Pile driving by steam power was next treated of, the author describing some of the principal machines which have been invented, preferring Sissons & White's, as being the most economical and practically useful."

STONES, SETTLE, & WILKINSON, *King Street Brass Works, Hull.*—Brass work for engineers.

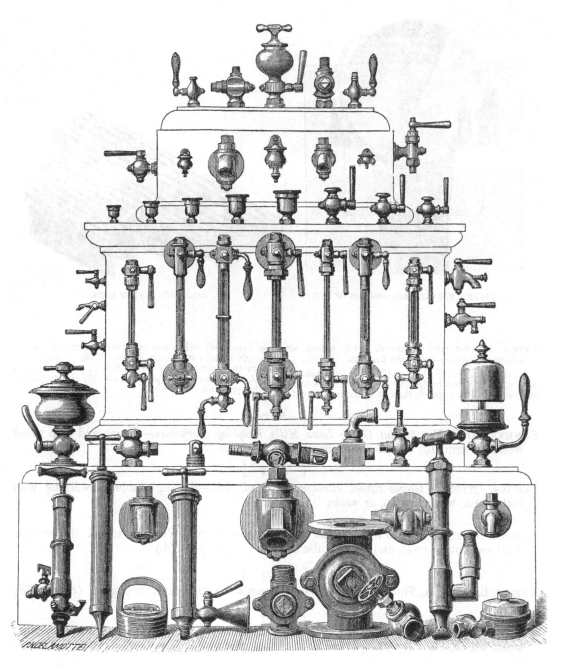

BRASS WORK FOR ENGINEERS.

STONES, SETTLE & WILKINSON exhibit a case of brass goods for steam engines and boilers, viz.: water gauges, gauge cocks, tallow cups and pumps, oil cups and syringes, steam whistles, steam taps and valves.

Price lists and drawings will be forwarded post-free on application.

STRATFORD, WILLIAM, 6 *Edward Street, Wentworth Road, Mile End Road.*—Patent furnaces and bars.

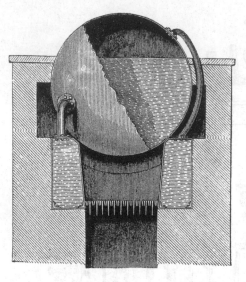

STRATFORD'S FURNACE AND SECTION OF BOILER.   STRATFORD'S FIRE BARS.

PATENT STEAM-BOILER FURNACE, with patent air-diffusing and smoke-consuming fire bars.

In this furnace, of which the figure is a transverse section, the walls and bridge, instead of being of fire-brick, are formed as water spaces, through which the feed water is introduced, and is compelled to traverse their entire length before entering the boiler. This, in conjunction with the patent bars shown, effects a great saving of fuel. The same principle is applicable also to various other purposes.

[2000]

SUMMERSCALES, W., & SON, *Coney Lane Mills, Keighley, Yorkshire.*—Brush and dash wheel washing, wringing, and mangling machine. (*See page* 73.)

[2001]

SYMONS, CYRUS, 2 *George Street, Blackfriars Road, S.E.*—Sewing machine, working with little noise, wear, trouble, or waste.

[2002]

TANGYE, BROTHERS, & PRICE, *Cornwall Works, Birmingham.*—Working model of hydraulic wool and cotton press, and a hydraulic ship jack. (*See page* 74.)

[2003]

TAPLIN, B. D., & Co., *Traction Engine Works, Lincoln.*—Patent traction engine. (*See page* 75.)

[2004]

TAYLOR, JAMES, & Co., *Britannia Engine Works and Foundry, Birkenhead.*—Traction engine; steam winch, with deck pumps; model of a steam crane, and of Stephenson's first locomotive.

[2005]

TENNANT, T. M., & Co., *Newington Works, Edinburgh, and Bowershall Iron Works, Leith.*—8-horse power upright portable steam engine; 6-horse power horizontal steam engine. (*See pages* 76 *and* 77.)

[2006]

THOMPSON & STATHER, *Green Lane, Sculcoates, Hull.*—Hydraulic press pumps.

[2007]

THORNEWILL & WARHAM, *Burton-upon-Trent.*—Pair of winding engines for colliery or other purposes.

SUMMERSCALES, W., & SON, *Coney Lane Mills, Keighley, Yorkshire.*—Brush and dash wheel washing, wringing, and mangling machine.

SUMMERSCALE'S WASHING, WRINGING, AND MANGLING MACHINE.

Obtained the silver medal at the Burnley Agricultural Show, August 30, 1860; also medals and prizes at the following agricultural meetings in 1861, viz.: Keighley, Newcastle-on-Tyne, Brigg, Darlington, Chester, Truro, Oxford, &c.; and the bronze medal at the Agricultural Meeting at Brussels in 1860.

A Drum inside of tub.

B Tub.

C Taps (two) to draw off water.

D Sycamore rollers, strongly hooped with iron, and capped with brass hoops.

E Drip board to bring the water back into the tub. This is water tight, and no slop whatever need occur.

F Mangling cloth, which travels the full length and is taken out when wringing.

G Wheels to remove the machine.

H Oscillating motion for dash wheel.

I Brushes for very dirty clothes (not used for blankets or flannels), rags, &c.; can be reversed at pleasure.

K Wheel thrown out of gear when mangling (in gear when washing).

PATENT COMBINED WASHING, WRINGING, AND MANGLING MACHINE, with a dash wheel or drum inside the tub, made to turn a circle with reversible action, by means of a tooth rack and pinion wheel, which are moved by the fly wheel being always turned in one direction.

The action for washing is thrown out of gear, by lifting a catch and moving the fly wheel shaft horizontally. This being done, the rollers are put into a working position for wringing and mangling. It will be seen on examining the drawing, that there is a mangling cloth attached, which can be put on in a few seconds for mangling. This machine will wash from 10 to 12 shirts at a time. The spring rests upon a patent bar, which increases its power considerably.

Full directions for use are supplied with each machine.

TANGYE, BROTHERS, & PRICE, *Cornwall Works, Birmingham.*—Working model of hydraulic wool and cotton press, and an hydraulic ship jack.

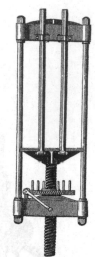

SCREW, COTTON, AND WOOL PRESS.

IMPROVED SCREW COTTON AND WOOL PRESS. Prices, from £45 to £150 0

WESTON'S PATENT PULLEY BLOCK. Sizes, to lift ½, 1, 1½, 2 and 3 tons. Prices, from £2 to . . . £5 10

ADVANTAGES.

1. One man can lift the weight specified.

2. The load cannot slip or run back, even if let go suddenly.

3. It requires no hoisting crab.

4. It is cheaper and safer than any other mode of doing the same work, and effects a very great saving in manual labour.

Upwards of 3,000 sets have been sold in about 9 months, and they are now in use in the works of all the leading engineers.

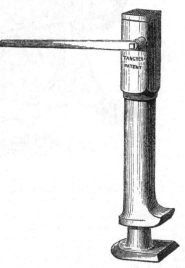

HYDRAULIC LIFTING JACK.

PATENT HYDRAULIC LIFTING JACK.

This jack is safer than any hitherto made, inasmuch as the lowering is under perfect control, being regulated by a screw; the foot and cylinder are also in one forging instead of being screwed together, or with the claw hung loosely over the head, as is the case with all other hydraulic jacks.

The cylinder and ram are made of the very best scrap iron. The prices, also, are as low as any in the trade.

Prices :—

To lift 4 tons	. £8 10	To lift 15 tons	. £18 10
ditto 6 do.	. 10 10	ditto 20 do.	. 22 10
ditto 8 do.	. 12 0	ditto 30 do.	. 25 0
ditto 10 do.	. 15 0	ditto 40 do.	. 27 10
ditto 12 do.	. 16 10	ditto 50 do.	. 32 10

Sole London agents for the sale of Weston's patent pulley blocks, S. & E. Ransome & Co. 31, Essex Street, Strand.

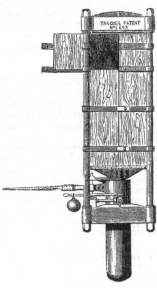

PATENT HYDRAULIC COTTON AND WOOL PRESS. Prices, from £55 to £200 0

Sole London agent for general machinery, Mr. S. Holman, 18, Cannon Street, E.C.

APLIN, B. D., & CO., *Traction Engine Works, Lincoln.*—Patent traction engine.

B. D. TAPLIN AND CO.'S NEW PATENT TRACTION ENGINE.

The above illustration represents one of B. D. TAPLIN & CO.'s TRACTION ENGINES of 16-horse power, with double cylinders and all the latest improvements : comprising their patent traction gear, also raising and lowering apparatus for regulating the height of water, when travelling up or down hill, or on irregular roads. The mode of steering is simple and effective.

The above engine is fitted with extra space for water and coals, sufficient for a 10 miles journey : and is built expressly for drawing heavy loads of 50 tons and upwards. It is suitable for Government works, contractors at home and abroad, mill, mine, and quarry owners, or for any other purpose requiring immense steam power.

Price £590 0

Further particulars sent post-free on application to B. D. Taplin & Co.'s traction engine works, Lincoln.

A 12-HORSE POWER ENGINE made on the same principle for farming purposes, such as steam ploughing, thrashing, grinding, sawing, &c. Price £425 0

Prices and particulars quoted for traction engines up to 50-horse power ; and also for waggons for contractors' purposes suitable for traction engines.

TENNANT, T. M., & Co., *Newington Works, Edinburgh, and Bowershall Iron Works, Leith.*—8-horse power upright portable steam engine; 6-horse power horizontal steam engine.

Elevation of 6-HORSE POWER HIGH-PRESSURE HORIZONTAL STEAM ENGINE, 8¼ in. cylinder, 20 in. stroke.

Price £60 0

Prices, with large tubular boiler and connexions, complete:—

3-horse power	£50
4 ditto	80
6 ditto	110
8 ditto	144
10 ditto	170
12 ditto	198
14 ditto	224
16 ditto	248

18-horse power	£275
20 ditto	300
25 ditto	375
30 ditto	450
35 ditto	490
40 ditto	520
45 ditto	585
50 ditto	650
60 ditto	800
70 ditto	950
80 ditto	1,150
90 ditto	1,300
100 ditto	1,500
120 ditto	1,800
140 ditto	2,200
150 ditto	2,400

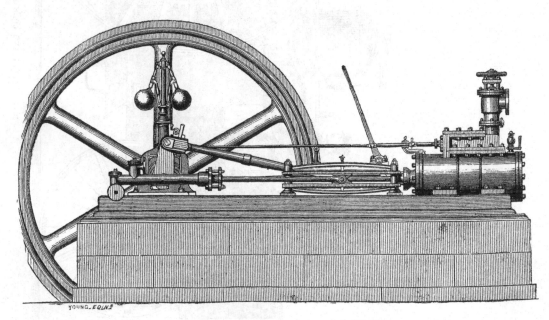

HIGH-PRESSURE HORIZONTAL STEAM ENGINE.

COMBINED HIGH AND LOW PRESSURE HORIZONTAL STEAM ENGINES, from 20-horse power upwards.

COMBINED HIGH AND LOW PRESSURE HORIZONTAL BEAM, up to 250-horse power.

LOCOMOTIVE ENGINES, from 9¼ in. cylinder upwards; water wheels and turbines; sugar, corn, saw, oil, and bone mills.

Designs and estimates prepared for machinery for home and foreign use.

TENNANT, T. M., & CO., *continued.*

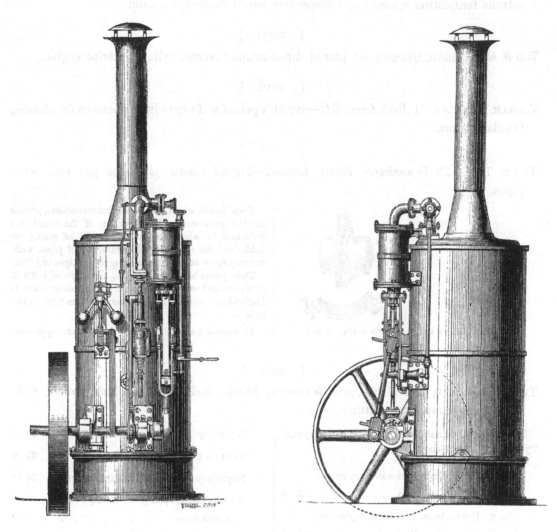

UPRIGHT STATIONARY STEAM ENGINE.

EIGHT-HORSE POWER UPRIGHT STATIONARY STEAM ENGINE, requiring no building-in, with large multi-tubular boiler, constructed to burn wood, or inferior coal, occupying a space 6 ft. by 5 ft. Price . . £160

4-horse power,	6½ in. cyl.	13 in. stroke	.	£105
6	ditto	8	ditto	14	ditto	. . 130
8	ditto	8¾	ditto	16	ditto	. . 160
10	ditto	9½	ditto	16	ditto	. . 190
12-horse power	2 ft. 8	ditto	14	ditto	. . 230
14	ditto	2 ft. 8¼	ditto	16	ditto	. . 250
16	ditto	2 ft. 8¾	ditto	16	ditto	. . 275
20	ditto	2 ft. 9½	ditto	16	ditto	. . 320

The above engines made portable on carriages and wheels, 10 per cent. extra.

PORTABLE STEAM CRANES, for wharf or railway, with above engines on wrought-iron carriage, pillar and jib, to hoist, lower, and turn round by steam.

To hoist 30 cwt.	£180
,, 40 cwt.	210
,, 60 cwt.	260
,, 80 cwt.	350

TRACTION OR ROADWAY ENGINES, with coal and water tanks, and steering apparatus complete :—

10-horse power.	£300
12 ditto	360
15 ditto	450
20 ditto	580

[2008]

TIZARD, WILLIAM LITTELL, 12 *Mark Lane, London.*—A surface refrigerator; an improved octuple fermenting apparatus; a suspension barrel washing machine.

[2009]

TOD & MC GREGOR, *Glasgow.*—A pair of direct acting inverted cylinder marine engines.

[2010]

TOPHAM, CHARLES, 31 *Bush Lane, E.C.*—Smith's patent self-expanding apparatus for cleaning tubular boilers.

[2011]

TRUSS, T. S., 53 *Gracechurch Street, London.*—Patent elastic joint for gas and water pipes.

ELASTIC JOINTS FOR GAS AND WATER PIPES.

These joints, which are simple and economical, provide for the expansion and contraction of the metal, the deflection to which continuous lengths of piping are liable, and the easy removal or insertion of pipes, without cutting or otherwise damaging either pipes or joints.

These joints have been tested to a pressure of 1,000 ft. of water; and are now applied to gas and water mains in England and on the Continent, varying from 2 in. to 4 ft. in bore.

They are also extensively in use for hot-water apparatus.

[2012]

TYLER, HAYWARD, & CO., 85 *Upper Whitecross Street.*—Soda-water machine, presses, well-engine, lift-pump, engine fittings.

The prices of manufactures exhibited by Hayward Tyler, & Co. are subjoined :—

No. 1. Patent beam soda-water engine, to make 200 doz. per diem £75 0

No. 2. Patent beam soda-water engine, to make 150 doz. per diem 70 0

No. 3. Patent beam soda-water engine, to make 100 doz. per diem 65 0

No. 1. Bramah's original soda-water machine to make 200 doz. per diem £65 0

No. 2. Bramah's original soda-water machine to make 150 doz. per diem 60 0

No. 3. Bramah's original soda-water machine to make 100 doz. per diem 55 0

A 5-in. hydraulic press for tinctures, &c. . 60 0

Press for proving girders 35 0

Naylor's pressure gauge in mahogany case . 16 10

A fire or manure pump, on wheels, with two 3½ inch barrels 7 0

Lift and force pumps on planks :—

	2 in.	2½ in.	3 in.	3½ in.	4 in.	4½ in.
Best . .	80/0	107/6	126/0	147/0	168/0	234/0
Good . .	74/0	85/0	105/0	127/6	147/6	
Common .	60/0	68/6	87/6	107/6	127/6	

For engine fittings, see price list issued by H. T. & Co. which, with price lists of all their manufactures, will be sent post-free on application.

[2013].

TYLOR, J. & SONS, *Warwick Lane, Newgate Street, London.*—Pumps, fire engines, steam fittings. (*See page* 80.)

[2014]

WALKER, THOMAS, & SON, 58 *Oxford Street, Birmingham.*—Steam boiler, alarm water gauges, and other machinery.

[2015]

WARD, F. O., *Hertford Street, Mayfair.*—Horizontal steam engine, combined with double-acting hydraulic power pumps, on cistern bed—new principle. (*See page* 81.)

[2016]

WARNER, JOHN, & SONS, *Crescent, Cripplegate, London.*—Water wheels, irrigators, ship manure pumps, fire engines. (*See page* 82.)

[2017]

WEBB & SON, *Comb's Tannery, Stowmarket, Suffolk; London office,* 11 *Leadenhall Street,* Mr. R. Pearce, Manager.—Leather; machine bands, buckets, and hose; glovers' leather, &c.

These machine bands are cut from level, and carefully selected oak-bark tanned English hides, and are manufactured throughout, to ensure a strength and durability which proves most satisfactory in all climates. The workmanship in single, double, and edged bands, is of the strongest description, and well adapted for heavy work. They are all thoroughly stretched by powerful machinery.

Considering the quality and price, these are the cheapest bands now manufactured.

The leather hose, buckets, rope, and thongs, are all of the best description.

The sole butts, glove and gaiter leather, calf skins and horse hides, both rough and dressed, with many other descriptions of leather, are tanned and manufactured upon the most improved principles by the exhibitors, who will forward prices and full particulars on application.

[2018]

WEIR, E., 142 *High Holborn.*—Washing, wringing, and mangling machines; cinder lifter, bread kneader.

[2019]

WENHAM, F. H., 1 *Union Road, Clapham Rise, S.*—A 10-horse power thermo-expansion steam engine, superheating the steam between the cylinders.

[2020]

WESTON, THOMAS ALDRIDGE, 31 *Essex Street, Strand, London.*—Improved pulley block and lifting apparatus.

[2021]

WHITE, JOSEPH, 7 *Trinity Street, Southwark, London, S.E.*—12 spring lever valve engine oil feeders, &c. (*See page* 83.)

[2022]

WHITMEE, JOHN, & CO., 70 *St. John Street, Clerkenwell, E.C.*—Crushers, cutters, mills and machines; Jolley's American provision safes, and refrigerators.

[2023]

WHITMORE & SONS, *Wickham Market, Suffolk.*—Improved steam engine; corn-mill machinery; engine details; framed drawings. (*See pages* 84 *and* 85.)

TYLOR, J., & SONS, *Warwick Lane, Newgate Street, London.*—Pumps, fire engines, steam fittings, soda-water machines, &c.

Obtained the Prize Medal, Great Exhibition, 1851; Dublin, 1853; and Paris, 1855.

J. TYLOR AND SON'S PATENT SODA-WATER MACHINES TO MAKE FROM 50 TO 360 DOZ. PER DIEM.

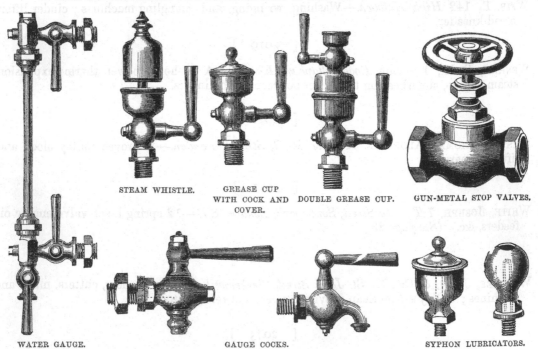

STEAM WHISTLE.   GREASE CUP WITH COCK AND COVER.   DOUBLE GREASE CUP.   GUN-METAL STOP VALVES.

WATER GAUGE.   GAUGE COCKS.   SYPHON LUBRICATORS.

Illustrated catalogues of steam engine and boiler fittings may be obtained post-free by application to J. Tylor & Sons, also catalogues of every description of brasswork for merchants, architects, builders, engineers, and plumbers.

WARD, F. O., *Hertford Street, Mayfair.*—Horizontal steam engine, combined with double acting hydraulic power pumps, on cistern bed—new system.

F. O. WARD'S PATENT IMPROVED HYDRAULIC PUMPING ENGINE.

This machine consists of a steam engine and four power pumps, horizontally disposed, the former on the top, the latter on the sides, of an elongated hollow bed, arranged to serve also as a cistern. The pumps are of a peculiar construction, which will be presently explained. Their disposition is such as to admit of a considerable increase of their length beyond that of ordinary hydraulic pumps, as will also be shown. They are worked on a plan believed to be novel and advantageous. The four pumps are coupled together in pairs, and the plungers of each pair are attached to the opposite ends of an intervening slide. This slide is moved to and fro, between horizontal guides, by one end of a connecting rod, whose opposite extremity is actuated by the crank of a driving shaft, transversely disposed across the oblique end of the cistern.

These arrangements are shown in the figure, which also exhibits the spur wheel and pinion, geared at 3 to 1, by which the pump-driving shaft just mentioned derives motion from the steam-engine shaft. The peculiar grouping of these parts upon and around the cistern is such, it will be observed, as virtually to compress two machines within the area which either would occupy alone. This valuable economy of floor space is effected by the super-position of the steam engine above the pumps ; yet it is not purchased (as might be expected) by any addition to the height of the machine, which could not conveniently be diminished, even were the pumps away.

To this advantage of compactness that of lightness is added, since the same casting which serves as bed-plate for the steam engine, affords also a stiff framing to the pumps, besides answering as water cistern for their supply. One casting thus replaces three ; and yet, compact as it is, its elongated form affords scope for a connecting rod of unusual length. This is a considerable advantage, enabling, as it does, pumps of increased length, to be worked with undiminished directness of thrust.

The economy hence resulting, in costly wrought-iron and brass work, is very great. For, on each side of the machine, one crank and one connecting rod are made to drive two long pumps, each equal in power to at least three ordinary short pumps, every one of which, on the old plan, would require a separate crank and connecting rod of its own.

We have therefore ten cranks and ten connecting rods saved out of twelve, and four sets of pump valves doing the work of twelve, with all the collateral advantages implied in these large economies. Thus, to take one example—whereas every set of six pumps on the old plan requires a costly 6-throw crank-shaft to work it, four long pumps, equal in power to two such sets, are driven on the new plan by one cheap, straight shaft, carrying one plain crank at each end.

Nor does this diminution in the number of cranks involve a less equable distribution of the resistance, seeing that the two cranks employed are set in such angular relation to each other, that when one pair of pumps is at dead-point, the other is at mid-stroke, and *vice versâ.*

With reference to power, indeed, it is beyond doubt that the gearing, which in this machine applies three rapid engine strokes to produce one slow pump stroke, brings power to bear against resistance far more advantageously, than when (as in certain direct-acting hydraulic pumps recently introduced) steam and water are made to travel at equal speed.

The remaining peculiarities relate chiefly to the internal fitting of the pumps. These have their inlet and outlet valves placed at one end, instead of, as usual, at opposite ends of the barrel. The water, therefore, enters and quits the pump at the same end, so that the annular water-way heretofore left between the plunger and barrel, ceases to be requisite. The plunger is accordingly turned to fit the barrel, so that no cavity remains to harbour air as usual, to the detriment of the vacuum or suction power of the pump, of which no less than one-third is sacrificed by the ordinary disposition of these parts. As to back-slip, or the reflux of water through the valves during their fall, this must needs be small in an engine having three times fewer than the ordinary number of valves, each making three times fewer than the usual number of lifts.

To sum up—this pumping engine, taken as a whole, is believed to present a series of advantages not heretofore combined in any one machine of its class. Its parts are few, simple, and light, easy and cheap to make and fit, in disposition singularly compact, in action very direct and efficient.

Though specially designed to work hydraulic presses, this engine is equally available, with slight modifications, for other kinds of pumping work. It may be obtained of Messrs. Wren & Hopkinson, machinists, Manchester ; who are appointed manufacturers under the patent.

WARNER, JOHN, & SONS, *Crescent, Cripplegate, London.*—Water wheels, irrigators, ship pumps, manure pumps, fire engines.

Obtained the Prize Medal in 1851.

The exhibitors are hydraulic engineers and manufacturers of fire engines, ship pumps, patent brass and iron pumps, garden engines, lamps, urns, braziery goods, plumbers' work, water-closets, steam and gas cocks, lead, tin, and copper pipe, imperial standard weights and measures.

Illustrated and priced catalogues can be had upon application.

No. 2 B. METHOD OF RAISING WATER from wells 50 to 200 ft. in depth by means of Warner's horse wheel frame and treble barrel pumps.

No. 13½. HYDRAULIC ENGINE for working pumps by steam power. Suitable for wells from 200 to 300 ft. in depth, or where a large quantity of water is required.

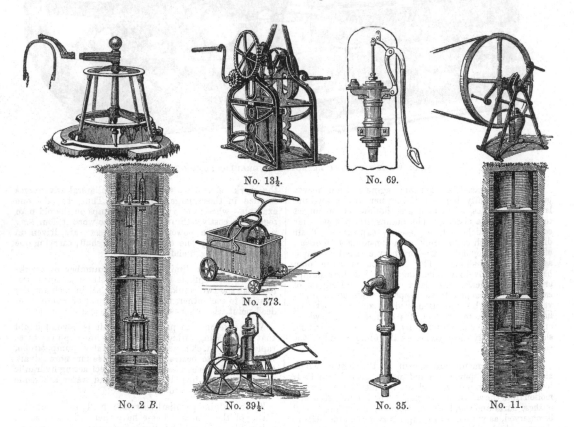

No. 13½.   No. 69.   No. 573.   No. 2 B.   No. 39½.   No. 35.   No. 11.

No. 11. FRAME AND PUMP for supplying the upper stories of houses from wells not exceeding 28 ft. in depth; to be worked by one man, or by steam power.

No. 5½.

No. 5½. WARNER'S FARM OR PLANTATION IRRIGATOR, mounted on a strong frame, will, with 1 horse, force upwards of 6,000 gallons per hour of water or liquid manure over the land, even at a high level. From its simple construction it is particularly adapted for

shipment abroad. This engine can also be used as an effective fire engine.

No. 573. WARNER'S PORTABLE FIRE ENGINE. Six men will throw 25 gallons of water per minute, to an altitude of 50 ft. It is particularly recommended for mansions, factories, and all large establishments.

No. 69. WARNER'S PATENT BRASS, VIBRATING STANDARD, LIFT AND FORCE PUMP, on wood plank. Recommended for its simplicity of construction and lowness of price.

No. 35. WARNER'S PATENT CAST-IRON LIFT PUMP for wells not exceeding 25 ft. in length.

No. 39½. WARNER'S DOUBLE-ACTION PUMP for raising and forcing water or liquid manure. This pump forms an effective fire engine for two men. The power required to work pumps on this principle, is only half that required by those of ordinary construction.

Brass and iron pumps of different sizes, and various patterns, are always kept in stock by John Warner & Sons.

WHITE, JOSEPH, 7 *Trinity Street, Southwark, London, S.E.*—12 spring lever valve engine oil feeders; 6 syphon machine ditto; 6 pyramid atmospheric ditto; 1 self-cleansing and filtering cistern; 3 improved filters; specimens of machine bands, laces, and spiral lathe bands.

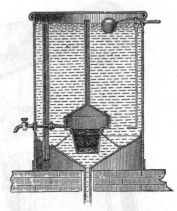

The IMPROVED SPRING LEVER VALVE OIL FEEDERS are extensively used by engineers and others for lubricating engines and machinery; and they are well adapted for filling oil lamps, both in private houses and on board ships, being unsurpassed for their lightness, strength, and durability. They are also very cleanly, and greatly economise time and oil, as not a single drop can escape except the air valve be pressed by the person using them, and they cost but a trifle more than the common oil can.

Price 16/6, 18/6, 20/6, 22/6, 25/0, 27/6, 31/6 per doz.

LAMP FEEDERS.

Pints, 2/6; 1½ pint, 3/0; quarts, 3/6 each.

WHITE'S IMPROVED SYPHON MACHINE OIL FEEDERS run more freely, and will not spill if upset; can be regulated from a large stream to the smallest drop, and the waste oil returns into the can.

Price 15/0, 18/0, 21/0, 22/6, 25/0, 30/0 per doz.

WHITE'S IMPROVED PYRAMID ATMOSPHERIC OIL FEEDERS.

Price 5/0, 6/6, 7/6, 9/0, 10/6, 12/6 per doz.

The B. H. CROWN LEATHER DRIVING BAND (Preller's Patent Leather Company, manufacturers,) are one-third thinner, stronger, and lighter than tanned leather, and will work double the time of any others before requiring repairs.

WHITE'S B. H. CROWN LEATHER SPIRAL LATHE BANDS are far superior to catgut or any other material; they will not slip, and can be lengthened or shortened at pleasure.

WHITE'S LEFT-HANDED COUPLING HOOKS AND EYES for 3-stranded gut, leather, or rope bands.

The SELF CLEANSING AND FILTERING CISTERN for household and manufacturing purposes (Rae's Patent). —T. W. Cowan, Kent Iron Works, Greenwich, manufacturer.

The cisterns complete in every respect from £2 upwards.

By the use of these cisterns the water drawn off for household purposes is always pure, the sediment being washed to the apex of the cone, whence it is drawn off. The water for culinary purposes passes upwards through the filter, and from thence to the kitchen. The cistern is washed by a suitable arrangement every time the water comes in from the main, effecting a saving of the pecuniary outlay attending the cleansing of the ordinary cisterns.

Filters from £1 upwards.

The cisterns and filters may be seen in operation at the offices of the agent, 7 Trinity Street, Trinity Square, Southwark, London, S.E.

[2024]

WILKINS, WILLIAM PICKFORD, *Ipswich.*—2-cylinder high-pressure and condensing steam-engine of 20-horse power; set of 4-inch 3-throw pumps; 4-inch and 2-inch improved stop valve.

A 2-CYLINDER HIGH-PRESSURE CONDENSING STEAM ENGINE of 20 horse-power, which will drive well and economically six pairs of 4 ft. millstones.

It has been successfully applied to pumping, and is well adapted for most other motive purposes, especially where economy of space and fuel is an object.

Price, with Cornish boiler of ample dimensions, and all fittings complete £550 0

A SET OF 3-THROW PUMPS for either hot or cold liquor, very compactly arranged, the valves of ready access, on a simple plan, and entirely of gun-metal. The whole mounted in strong cast-iron frame.

Price £65 0

TWO SPECIMENS OF WILKINS' STRAIGHTWAY STOP VALVE, with the screw, valve, and seating of gun-metal, well and strongly made for high pressures. Price 1*l.* per in.

[2025]

WILLIAMSON, W., 133 *High Holborn.*—Washing, wringing, and mangling machine.

MACHINE FOR WASHING CLOTHES AND FABRICS, with rollers attached for wringing and mangling.

These washing machines have been adopted by the Admiralty and War Department for naval and military hospitals. They are made of sizes from 1 ft. 6 in. long, 12 in. wide, up to 9 ft. long, 4 ft. 6 in. wide.

Nursery washing machines . . . £2 5 to £3 5
Ditto, and for wringing and
mangling . . . . 4 15 to 5 15

Domestic washing machines . .	£3 10 to	£8 10
Ditto, and for wringing and mangling	8 10 to	15 10
Machines for institutions and laundries	12 10 to	15 10
Ditto, and for wringing and mangling	22 10 to	26 10
Power washing machines . . .	30 0 to	100 0
Ditto, with steam engine attached	50 0 to	150 0

Whitmore & Sons, *Wickham Market, Suffolk.*—Improved steam engine; corn-mill machinery; engine details; framed drawings.

IMPROVED 35-HORSE POWER INDEPENDENT DOUBLE-CYLINDER CONDENSING ENGINE.

1. Improved Independent Iron Hurst and fittings for 2 pairs of stones.

(*See illustration on page 85.*)

The novelty and improvement, consist in the arrangement provided for driving the stones by belt in lieu of gearing, and also for the continuous discharge of the meal around the entire peripheries of the stones, into galvanized metal or wood reservoirs within iron cylinders, on which also the bed stones are supported and easily made to adjust. This is found to produce coolness of, and less injury to the meal; entirely prevents the unpleasantness and waste through stive, renders exhausting apparatus almost unnecessary, and admirably facilitates the process of good milling. The fittings for neck and step bearings of stone, spindles, centre irons, feeding, adjusting, and disengaging gear, are all of the most approved construction; the main features being simplicity and durability.

2. Improved Conveying and Elevating Machinery for meal and corn.

3. Improved and Newly-invented Direct-acting High-pressure Steam Engine of 10-horse power, with vertical crank shaft and horizontal fly wheel; the latter being situate within the bed plate of the engine.

(*See illustration on page 85.*)

This engine is particularly compact and simple, and well adapted for driving the foregoing machinery in mills on a small scale. The crank shaft is of solid forged iron with steel end revolving in steel step, and the governor, starting apparatus, &c. are all on an improved principle. The construction of the engine is such, that the usual friction is considerably reduced; it is easily accessible in all its parts; and requires little or no fixing.

WHITMORE & SONS, *continued.*

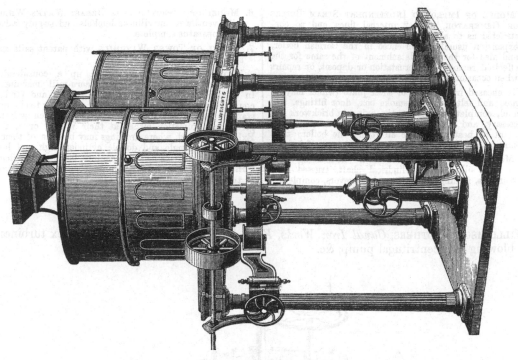

IMPROVED INDEPENDENT IRON HURST FOR TWO PAIR OF STONES.

DIRECT ACTING HIGH-PRESSURE 10-HORSE POWER STEAM ENGINE.

WHITMORE & SONS, *continued.*

4. **MODEL OF IMPROVED INDEPENDENT STEAM BOILER OR GENERATOR,** with 2 internal flues, and so constructed as to allow for the unequal contraction and expansion usually experienced in the Cornish boiler, and also for the easy detachment of the same for the effectual removal of incrustation or deposit, or repairs when occasion requires it.

It is encased in a jacket of sheet plate (which may be lagged and felted), with smoke box, door fittings, and funnel, complete; rendering the fixing in brickwork unnecessary, and providing at the same time an unusual amount of heating surface all round the boiler; thus ensuring safety, purity of steam, and economy of fuel.

5. **MODEL OF IMPROVED MACHINE** for dressing flour through silks, with cylindrical shaft, trussed reel, vibrators, feeding apparatus and conveyors, complete.

6. **MODEL OF IMPROVED IRON BREAST WATER WHEEL** with ventilated curvilinear buckets and supply valve, and apparatus complete.

7. **MODEL OF TOWER WINDMILL** with patent sails and self-winding tackle.

Whitmore & Sons have fitted up a considerable amount of the above improved corn-mill machinery in establishments on a large scale in this and foreign countries, the arrangements differing according to circumstances, particulars and testimonials of which will be given on application either at their Works, or at the Exhibition stand, where drawings may be seen of them; and of both single and double cylinder high and low pressure and condensing steam engines, boilers with patent furnaces, general mill, sawing, and agricultural machinery of their manufacture.

[2026]

WILLIAMSON BROTHERS, *Canal Iron Works, Kendal, Westmoreland.*—Patent vortex turbines, blowing fan, centrifugal pump, &c.

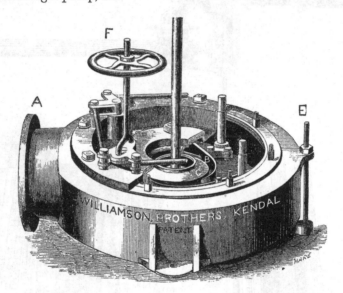

PATENT VORTEX TURBINE, an improved means of applying water-power.

The vortex turbine is equally adapted for high and low falls. Its peculiar advantages are :—

1. That the power is obtained with a slower velocity of the water, and consequently with less friction than in ordinary turbines.
2. Great steadiness and regularity of motion.
3. A thoroughly efficient means of adjustment to varying supplies of water. A large working model is exhibited in motion.

Williamson Brothers also exhibit in action—

A WHIRLPOOL BLOWING FAN,

A WHIRLPOOL CENTRIFUGAL PUMP; and

A PATENT VERTICAL-COLUMN STEAM ENGINE.

Descriptive circulars may be had on application to the makers.

[2027]

WILSON, JOHN C., & CO., 14A *Cannon Street, London, E.C.*—Portable steam sugar-cane mill. (*See page* 87.)

[2028]

WISE, FRANCIS, 22 *Buckingham Street, Adelphi, W.C.*—Feed water regulator, indicator, and alarm for steam boilers. (*See page* 88.)

[2029]

WOOD, ROBERT, & SONS, *Leeds.*—20-horse double-cylinder engine, steam pump, shafting, pulleys, and wheels.

WILSON, JOHN C., & Co., 14A *Cannon Street, London, E.C.*—Portable steam sugar-cane mill.

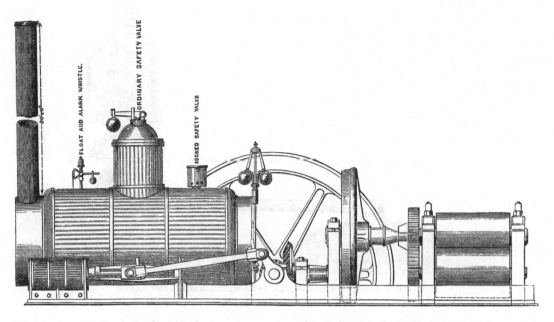

ELEVATION OF WILSON'S PORTABLE STEAM SUGAR-CANE MILL.

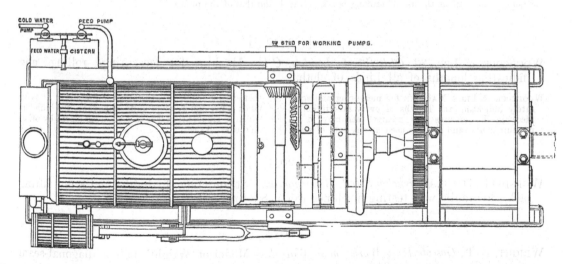

PLAN OF WILSON'S PORTABLE STEAM SUGAR-CANE MILL.

WILSON'S PATENT PORTABLE STEAM SUGAR-CANE MILL, with engine and portable boiler complete, on the same iron foundation-plate. Expensive brick foundations and setting of boiler dispensed with. No brick chimney required. An additional saving effected from the simplicity of erection and economy of fuel. Improved sugar machinery and apparatus of all kinds.

Cocoa nut and other oil machinery. Coffee and rice machinery. Flax steeping and scutching apparatus. Cotton cleaning and packing machinery, &c &c.

JOHN C. WILSON & Co., colonial engineers, 14A, Cannon Street, London.

International Exhibition. Machinery at work in Class VIII

WISE, FRANCIS, 22 *Buckingham Street, Adelphi, W.C.*—Feed water regulator, indicator, and alarm for steam boilers.

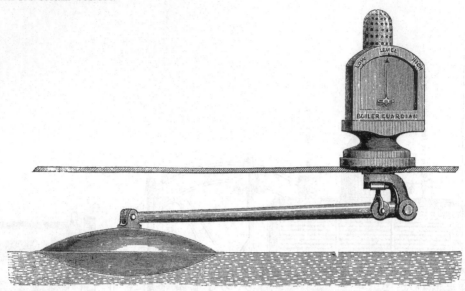

FEATHERSTONHAUGH AND WISE'S FEED-WATER REGULATOR GAUGE.

FEATHERSTONHAUGH & WISE'S PATENT BOILER GUARDIAN, OR SELF-ACTING FEED-WATER REGULATOR, GAUGE, AND ALARM.

By a simple arrangement connected with a float within the boiler, and without the use of stuffing boxes, cocks, or other complexities, this little apparatus regulates the action of the feed pump, indicates the water level, and should the latter from any cause fall below a certain point, it sounds an alarm, which calls timely attention to the fact. It is thus self-testing of its own efficiency and that of the pump.

[2030]

WOODCOCK & LEE, 33 *Old Street, London, C.E.*—Machine for measuring, rolling, and indicating lengths of all kinds of cloth.

WOODCOCK & LEE'S MACHINE for measuring and accurately indicating the lengths of every description of piece goods, ensures the greatest accuracy in all measurements, and may be used at any rate of speed that can be acquired with rollers and crank. It saves much time, and renders quite impossible the blunders which so frequently occur in measuring cloths by other methods. Price £10 0

[2031]

WORSDELL, THOMAS, *Berkeley Street, Birmingham.*—Steam crane, hydraulic and screw lifting jacks, hydraulic wire testing machines, &c. *(See page 89.)*

[2032]

WRIGHT, E. T., *Goscote Iron Works, near Walsall.*—Model of Wright's patent diagonal-seam steam boiler. *(See page 91.)*

[2033]

YARROW, A. F., *Barnsbury.*—Locomotive steam carriage.

[2034]

YORK & CO., 2 *Royal Exchange Buildings, London.*—High-pressure steam boiler.

[2035]

ZANNI, GEMMINIANO, 51 *Lamb's Conduit Street.*—Self-basting roasting apparatus.

WORSDELL, THOMAS, *Berkeley Street, Birmingham.*—Steam crane, hydraulic and screw-lifting jacks, hydraulic wire testing machines, &c.

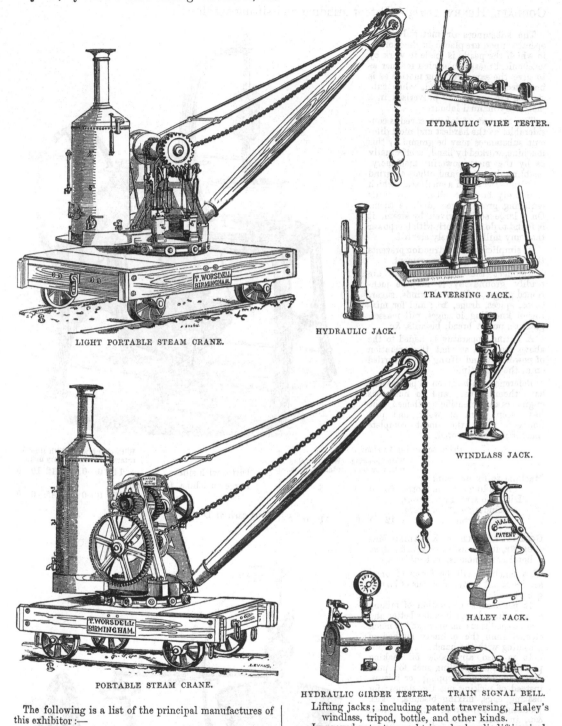

HYDRAULIC WIRE TESTER.

LIGHT PORTABLE STEAM CRANE.

TRAVERSING JACK.

HYDRAULIC JACK.

WINDLASS JACK.

HALEY JACK.

PORTABLE STEAM CRANE.

HYDRAULIC GIRDER TESTER.   TRAIN SIGNAL BELL.

The following is a list of the principal manufactures of this exhibitor :—

Contractors' locomotives, portable and other steam engines ; steam wharf, ship, and travelling cranes ; hand wharf, travelling, derrick, and other cranes ; steam and hand winches ; travelling crabs for gauntry frames ; pulley blocks, &c.

Engineers' tools, including self-acting screw-cutting and other lathes ; planing, slotting, drilling, screwing, punching and shearing machines ; ratchet braces, &c. ; nail-cutting, rivet, wood-screw, and general machinery.

Wrought-iron smiths' hearths, anvils, vices, &c.

Lifting jacks ; including patent traversing, Haley's windlass, tripod, bottle, and other kinds.

Improved patent wrought-iron hydraulic lifting jack, to raise from 4 to 200 tons.

Hydraulic machines to test bar iron, steel, chain cables, girders, and (small portable) for wire.

Railway bell signal apparatus, for communicating between guard and engine man, as supplied to various railway companies.

Rail-setting presses, jim-crows, and other contractors' tools.

Railway buffers, screw couplings and cramps, bolts and nuts, &c.

GOODALL, HENRY, *Derby.*—Patent grinding and sifting machine.

The substances or materials to be operated upon are placed in the mortar, in which the pestle is made to work by mechanical means in such a manner as to give the same rubbing motion as is imparted thereto by hand, when substances are ground or pulverised in a mortar by manual labour.

The amount of labour saved is considerable, as the hardest and most difficult substances may be ground by this machine, worked by hand, as effectually as by the most powerful machinery, enabling druggists and others to grind articles perfectly on a small scale which could only be done hitherto by mills requiring great power to drive them. On a large scale, driven by steam, it is found to do more work with less power than any mill previously erected.

Its simplicity of construction prevents the possibility of disarrangement.

Amongst the articles which are readily ground by it may be mentioned ginger, salts of all kinds, sugar, cocoa, spices, drugs, &c.; and for mixing or kneading lozenges, pill masses, glazier's putty, bread, biscuits, &c.

A sifting apparatus is added to the above machine, so that the operation of powdering and sifting may be carried on at the same time.

Reference can be given to parties who have them in use, and to numerous engineers and scientific gentlemen, who have seen them at work, and pronounced them the most complete machines ever invented.

Price according to size:

	With changing rotary motion.	With simple rotary motion.
Machine to fix on counter, including a 10-in. mortar	£6 0 0	£5 0 0
Ditto, on strong iron frame, including a 13-in. Wedgwood mortar	12 10 6	11 0 0

	With changing rotary motion.	With simple rotary motion.
Ditto, with sifter	14 0 0	12 10 0
Large machine for steam power	55 0 0	45 0 0

Nett cash at Derby.

GOODALL'S DOMESTIC KNEADING MACHINE, for the use of private families, hotels, club-houses, confectioners, &c.

This machine will be found of great service to all parties desirous of having good bread.

It performs the operation of mixing or kneading dough in a far better and more expeditious manner, and is much cleaner than the ordinary method of kneading with the hands.

It is also applicable for making potted meats, grinding suet for puddings, instead of chopping; or raisins, instead of picking out the stones; for beating eggs, mixing biscuits, cakes, puddings, powdering sugar, spices, &c., in short, for any purpose in the cooking department where kneading, grinding, or mixing is required.

They are made in sizes to mix from eight pounds of flour upwards.

Price according to size, a small machine, with tin bowl for making bread, £3 3s.; with stone mortar in addition, for making potted meats, &c. £4 4s.

May be procured from the inventor and manufacturer, H. Goodall, St. Peter's Street, Derby.

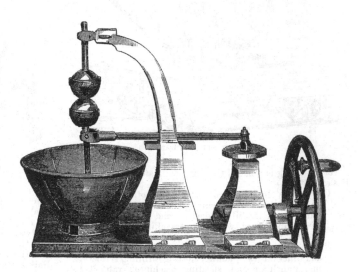

GOODALL'S DOMESTIC KNEADING MACHINE.

WRIGHT, E. T., *Goscote Iron Works, near Walsall.*—Model of Wright's patent diagonal-seam steam boiler.

DIAGONAL-SEAM STEAM BOILER.

MODEL OF A CYLINDRICAL STEAM BOILER, constructed with Wright's patent diagonal seams, by means of which, longitudinal joints are altogether avoided, and 40 per cent. additional strength is required.

It is but recently that attention has been directed to the inequality in the resisting forces in the transverse and longitudinal sections of cylindrical steam boilers. Mr. W. Fairbairn, in his book, "Useful Information for Engineers," says :—"If we refer to the comparative merits of the plates composing cylindrical vessels subjected to internal pressure, they will be found in this anomalous condition, that the strength in their longitudinal direction is twice that of the plates in the curvilinear direction. This appears by a comparison of the two orces, wherein we have shown that the ends of the 3 ft. boiler, at 40 lbs. internal pressure, sustain 360 lbs. of longitudinal strain upon each inch of a plate ¼ in. thick ; whereas plates of the same thickness have to bear, in the curvilinear direction, a strain of 720 lbs." And, it being a well ascertained fact, notwithstanding the vulgar notion to the contrary, that the ordinary or single riveted joint possesses but half the strength of the solid plate, it results that the longitudinal seams are the weakest parts of ordinary cylindrical boilers, however good the workmanship may be ; and that sound principles of construction demand the abandonment of the common method of making them.

CLASS IX.

AGRICULTURAL AND HORTICULTURAL MACHINES AND IMPLEMENTS.

[2071]

AMIES & BARFORD, *Peterborough.*—Portable steaming apparatus, registered sack elevator, water ballast land roller, clod crusher.

PORTABLE AND FIXED STEAMING APPARATUS.

STANLEY'S REGISTERED PORTABLE AND FIXED STEAMING APPARATUS.

Amies & Barford are the sole manufacturers of this celebrated and economical apparatus, of which 500 sets are now in successful use by eminent agriculturists in the United Kingdom for steaming food for cattle, and for domestic purposes. It is used in one hundred of the principal gaols, asylums, and other public buildings. It has obtained *every* first prize for which it has competed at the Royal and other agricultural shows during the last ten years, and is universally acknowledged to be the best, cheapest, and most economical apparatus extant.

The exhibitors are manufacturers also of prize clod crushers, land rollers, registered sack elevators, patent field stiles, Government outfall drainage pipes, &c. &c.

Descriptive illustrated catalogues will be forwarded on application.

[2072]

ASHBY, T. W., & Co., *Stamford*—Hay-maker ; rotating harrows ; meal grinding mill ; oil-cake breakers ; chaff and cane-top cutter for power; chaff-cutters ; 2½-horse power portable steam-engine ; horse rake ; wheel hand-rake ; horse power works ; saw-table, &c. (*See page* 2.)

ASHBY, T. W., & Co., *Stamford.*—Hay-maker; rotating harrows; meal grinding mill; oil-cake breakers; chaff and cane-top cutter for power; chaff-cutters; 2½-horse power portable steam-engine; horse rake; wheel hand-rake; horse power works; saw-table, &c

SMITH AND ASHBY'S PATENT HAYMAKING MACHINE.

The following implements are exhibited by T. W. Ashby & Co. :—

SMITH & ASHBY'S PATENT HAYMAKING MACHINE.

This haymaker has received 49 first-class prizes from the Royal and Provincial Agricultural Societies of England, Scotland, Ireland, France, Austria, and Holland. It is one of the strongest and most efficient haymakers in the world.

Price £15 15
Price, (T. W. A. & Co.'s new patent), simplified, and of extra power 16 16

SMITH & ASHBY'S PATENT STEEL-TOOTH HORSE RAKE, very greatly improved, has taken 38 prizes.
Price, with 26 teeth £8 5

SMITH & ASHBY'S PATENT WHEEL HAND-RAKE.
Price £2 0

IMPROVED 2½-HORSE POWER PORTABLE STEAM ENGINE, specially adapted for farm work and for the colonies, Price, with improvements £83 5
If with patent indicator, £3 extra.

1-HORSE GEAR WORKS. Price. £10 10

No. 3 PATENT SAFETY CHAFF CUTTER for steam-power.
Price £14 0

No. 4 PATENT SAFETY CHAFF CUTTER for horse-power.
Price £9 10 0

No. 6 CHAFF CUTTER for hand-power.   Price 6 0 0

No. 7a CHAFF CUTTER for small stables. Price 3 10 0

No. 8 CHAFF CUTTER   Do.   Price 2 12 6

IMPROVED OIL CAKE MILL No. 1.   Price 3 5 0
IMPROVED OIL CAKE MILL. No. 2.   Price 3 10 0

PORTABLE STONE GRINDING MILL, obtained first prize of the Royal Agricultural Society, 1860. Price £45 0

IMPROVED CIRCULAR SAW TABLE.   Price . £15 10

T. W. ASHBY & Co.'s PATENT ROTATING HARROW.
Single £3 0
Per pair, with draught bar complete. . . 6 6

PATENT STEEL SHIELD to prevent the wear of cranks; a patent staple beater drum and concave, and spring hanger for thrashing machines.

Descriptive catalogues will be sent free on application.

[2073]

AVELING, J. *Rochester, Kent.*—Agricultural locomotive engine (*See page* 3.)

[2074]

BAIN, McNICOL, & YOUNG, 29, *Cross Causeway, Edinburgh.*—Wire netting; fencing, and iron gates.

[2075]

BALL, WILLIAM, *Rothwell, Northampton.*—Agricultural cart, ploughs, &c.

[2076]

BAMLETT, ADAM C., *Middleton Tyas, Yorks.*—The Royal Agricultural Society's first prize manual delivery reaper, 1861.

[2077]

BARNARD, BISHOP, & BARNARDS, *Norwich.*—Patent root pulpers, Norfolk pig-troughs, patent mangles, iron garden and park chairs, galvanized iron wire netting.

PATENT ROOT PULPER.
PATENT ROOT PULPER for power.
Two varieties of IMPROVED NORFOLK PIG TROUGHS.
Universal SELF-ROLLING MANGLE.

Twelve varieties of Iron GARDEN AND PARK CHAIRS with spiral wire seats.
The Patent Gravitating British ROCKING LOUNGE.
Fifteen varieties of machine-made WIRE NETTING galvanized in the piece.

[2078]

BARRETT, EXALL, & ANDREWES, *Reading, Berkshire.*—Horse and steam thrashing machines engines, mills, and agricultural machinery. (*See pages* 4 *to* 6).

AVELING J., *Rochester, Kent.*—Agricultural locomotive engine.

This engine has an improved patent extra-large boiler, fitted with 37 2¾-in. tubes, external plates of the best Butterley iron, fire box and tube plates of Bowling iron, with extra stays for high-pressure; the fire grate measures 31 in. by 34 in. and is suitable for wood or coal fuel. The cylinder, 10 in. diameter, is surrounded by a jacket, and placed on the forward part of the boiler, by which arrangement priming in ascending steep inclines is prevented. The crank shaft is of Lowmoor iron. The engine is fitted with improved governor, reversing link

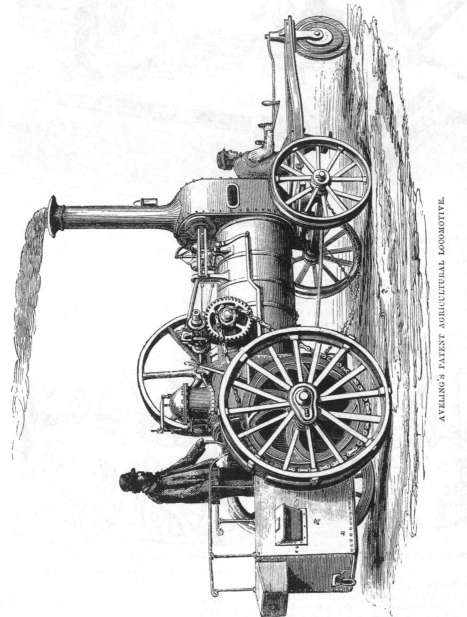

AVELING'S PATENT AGRICULTURAL LOCOMOTIVE.

AVELING'S PATENT AGRICULTURAL LOCOMOTIVE, to which the first prize gold medal was awarded at the late show in Mecklenburg-Schwerin.

Forty of these engines are now in constant use. Prices, particulars, and testimonials, and also estimates for the conversion of portable engines into locomotives, can be obtained on application to the manufacturers.

motion, patent tender and water tank, under foot-plate, driving chain and gear, steam-pressure gauge, extra lock-up safety valve, steam jet blower, firing tools and wrenches, driving wheels 5 ft. 6 in. diameter, 12 in. wide, patent steerage and screw break for descending inclines.

The engine is remarkable for its simplicity, and the great strength of its working parts. It is capable of drawing 10 tons up an incline of 1 in 6, and can be readily managed by any ordinary engine driver.

Price £420 0

BARRETT, EXALL, & ANDREWES, *Katesgrove Iron Works, Reading, Berkshire.*—Horse and steam thrashing machines, engines, mills, and agricultural machinery.

BARRETT, EXALL, AND ANDREWES' THIRTY-HORSE ENGINE.

A 30-HORSE POWER DOUBLE CYLINDER HIGH-PRESSURE EXPANSION ENGINE of the highest class, fitted with condensers (extra), giving full 30 per cent. of advantage in economy.

It is fitted with governor, pump, throttle and expansion valves, variable to any extent; a Lowmoor wrought crank shaft, mounted on a solid metal bed plate, planed and finished. It can be used without the condensers if there is not a good supply of water, but with that the consumption of fuel will be from 2 to 3 lbs. of coal per indicated horse power per hour.

It will be found at work in the Western Annex.

BARRETT, EXALL, AND ANDREWES' TEN-HORSE ENGINE.

A 10-HORSE POWER HORIZONTAL HIGH-PRESSURE FIXTURE ENGINE, fitted with governor, throttle valve, pump, and wrought-iron crank shaft; it is fixed upon a solid planed-metal bed-plate, which renders it self-contained.

The character of its construction enables it to be removed without disturbing the foundations, it works most economically, and has worked 3 hours with 140 lbs. of coal, which will be its average consumption. It will work expansively from one-third to full steam.

A 3-HORSE POWER HORIZONTAL HIGH-PRESSURE FIXTURE ENGINE, fitted same as the 10-horse power.

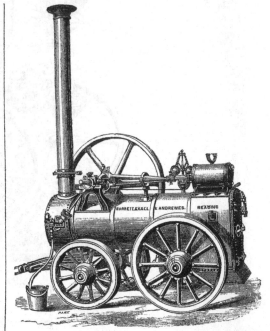

BARRETT, EXALL, AND ANDREWES' PORTABLE STEAM ENGINE.

AN 8-HORSE POWER HIGH-PRESSURE PORTABLE STEAM ENGINE.

The boiler has 20 square ft. of heating surface for each nominal horse-power. The fire box is of Lowmoor iron, and has an ample water space very accessible for cleansing. It is cased with wood and sheet-iron, and is supplied with governors, water gauge, steam-pressure gauge, steam whistle, gauge taps, pump, and partial water heater, the crank shaft is bent and of wrought-iron, and the engine has every modern improvement.

A 3-HORSE POWER HIGH-PRESSURE PORTABLE STEAM ENGINE.

BARRETT, EXALL, & ANDREWES, *continued.*

BARRETT, EXALL, AND ANDREWES' COMBINED THRASHING
MACHINE.

AN 8-HORSE POWER PORTABLE COMBINED THRASHING
AND FINISHING MACHINE.

It has the patent perforated beater drum 54 in. wide
which is made wholly of wrought-iron, as is also the
breasting. The straw shakers are of wood, and the screens
reciprocate with them, which prevents vibration. The
bearings are the patent spherical ones, which give absolute truth in setting the machine, never heat, and run
easy. It has the new patent corn elevator, which elevates
any description of grain without the ordinary cups, dresses
the corn whilst passing to the separating screen, by which
means a second blower and the usual barley aveller are
rendered unnecessary, and the machine is greatly simplified, and bearings, pulleys, and straps are dispensed
with, and the wear and tear greatly economised. It
finishes the corn for market.

AN 8-HORSE POWER PORTABLE COMBINED SINGLE-BLAST
THRASHING MACHINE, with patent perforated drum
48 in. wide. It is fitted with the patent spherical
bearings, and the new patent corn elevator.

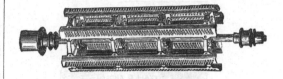

BARRETT, EXALL, AND ANDREWES' PATENT DRUM.

PATENT PERFORATED BEATER DRUM wholly of wrought-
iron, perforated at an angle of 45° to the square, and
bent in a mould to the requisite form for rubbing
out the grain. It will thrash all varieties of grain
without injury, and is adapted to any existing thrashing machine.

PATENT HORSE-POWER PORTABLE THRASHING MACHINE,
constructed for 3 light horses. The drum and breasting are entirely of wrought-iron; the latter formed of
separate bars with serrated faces, the ends of which
pass through slots in the sides of the machine and are
set nearer to, or carried further from, the drum, by
means of 2 revolving plates, having a continuous
grooved worm cut on their faces, which increase the
space between each as it is set to or from the drum,

BARRETT, EXALL, AND ANDREWES' PATENT THRASHING MACHINE.

and thus allow the larger corn room to escape. The
gear work is wholly inclosed in an iron cylinder, and
by the arrangement of the wheels the strain is equalized
and the friction reduced, whilst all accidents are effectually prevented.

The gear work gives a speed of about 100 revolutions
per minute, and the machine will thrash all kinds of
grain perfectly. It is most valuable for export, and
is in general use in all the corn-growing countries of
Europe.

BARRETT, EXALL, & ANDREWES, *continued.*

PATENT HAND THRASHING MACHINE, fitted with the patent wrought-iron drum and breasting, and contained in iron frame. It will thrash all kinds of grain and seed without injury.

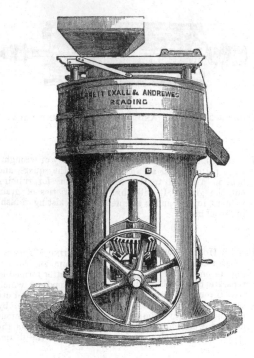

BARRETT, EXALL, AND ANDREWES' CORN MILL.

CORN-GRINDING MILL fitted with 48-in. French burr stones, the upper one revolving. An adjusting power is given so that it may be raised or lowered at pleasure. The whole mill is self-contained in a cylindrical iron frame, and is of the first class in all respects.

CIRCULAR SAW BENCH. The frame is wholly of iron, the bed plate is planed, fitted with metal fence, metal roller for the timber, boring apparatus, and fast and loose pulleys.

PATENT BAND SAW. The band or ribbon saw is passed over pulleys, and the table is of planed metal. It can be set for cutting wood at any angle or bevel.

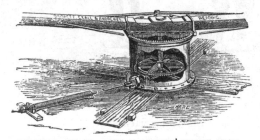

BARRETT, EXALL, AND ANDREWES' HORSE GEAR.

A 6-HORSE POWER PATENT GEAR. The whole of the machinery self-contained in an iron cylinder, and adapted for communicating power for thrashing, sawing, pumping, &c.

PATENT GEAR WORK, adapted for working a chaff cutter by 1-horse power with or without an intermediate motion; by using the latter the speed is doubled.

BARRETT, EXALL, AND ANDREWES' CHAFF CUTTERS.

CHAFF CUTTERS. The engraving represents a new and much improved series. The frames are of iron. The arrangement is simple, but very strong and effective, the working parts are protected; and they are adapted for manual, horse, and steam power.

GRAIN MILLS for crushing and splitting corn for horses and cattle. They will crush oats, barley, and linseed, and split beans and peas, with a minimum expenditure of power, and are adapted for hand, horse, or steam power.

SCREW LIFTING JACKS, or cotton screws. They are simple and strong; the screws are of the best forged iron, turned and chased in the lathe, and a corresponding nut is cut in the head of the frame which carries the weight.

BARLEY AVELLER, made entirely of iron. The barley is passed through a cylinder, in the centre of which is a spindle armed with knives, which remove the beard or awn from the grain. It is adapted for hand-power.

OIL-CAKE MILL. It has two sets of teeth revolving towards each other, so shaped as to take a firm grip of the cake, and break it small or large, as may be desired.

FLOUR-DRESSING MACHINE. The cylinder is fitted with wires of different meshes, with brushes revolving inside of it; thus four or more divisions of the meal may be produced at pleasure.

SLUICE VALVES, for either steam, water, or gas. The raising screws are square threaded and engine-turned, working in gun-metal nuts. The surfaces are brass faced, both in the sliding gates and seats.

HAY-MAKING MACHINE. It has two motions, a forward one for spreading the grass, and a backward one for lifting or turning it when nearly made. It has a simple arrangement with instantaneous action for giving it the desired motion, as also for elevating or depressing the rake barrels.

PATENT LEVER HORSE RAKE, very strong, simple, cheap, and novel. The outside frame used in all other rakes is dispensed with, the teeth are supported upon the axle, whilst a movable cleaner works inside the teeth and descends as they rise. It has a simple adjustment for altering the depth of the teeth while the rake is at work.

GORSE MACHINE, which first cuts up the gorse like fine chaff, and then crushes it between rollers running at different speeds. The prickles are completely destroyed, and the gorse reduced to a pulp fit for cattle food.

[2079]

BENTALL, EDWARD HAMMOND, *Heybridge, Maldon, Essex.*—Chaff cutters, corn and seed crushers, root pulpers, oil-cake mills. (*See pages* 8 *and* 9.)

[2080]

BEGBIE, JAMES, *Haddington, N.B.*—Adjustable sack holder and lifter; hand machine for sowing turnip seeds.

[2081]

BELL, GEORGE, *Inchmichael by Errol, Perthshire, N.B.*—Bell's reaping machine for two horses with patent sheaffer complete.

[2082]

BOBY, ROBERT, *Bury St. Edmunds, Suffolk.*—Machines for cleaning and separating grain, and improved wort pump for brewers. (*See page* 10.)

[2083]

BOOTHMAN, JAMES, *Gisburn Coates, near Skipton.*—Observatory bee-hive and feeding box, with ventilator for top.

[2084]

BOYD, JAMES, *Lewisham.*—Patent brush lawn-mower, self-cleaning, self-sharpening; shaft roller; tubular scythe handles.

[2085]

BROWN, WILLIAM, & CHARLES N. MAY, *North Wilts Foundry, Devizes.*—Portable steam engine, and patent sluice cock. (*See page* 11.)

[2086]

BURGESS & KEY, *London.*—Reaping, mowing, and thrashing machines, haymakers, horse rakes, carts, waggons, chaff-cutters, churns. (*See page* 12.)

[2087]

BURRELL, CHARLES, *St. Nicholas Works, Thetford, Norfolk, and* 69, *King William Street, City, London.*—Boydell's patent traction engine, &c. (*See pages* 13 *to* 16).

[2088]

BUSBY IMPLEMENT COMPANY, THE, *Bedale, Yorkshire.*—Ploughs, horse hoes, carts, and turnip tailers, &c.

[2089]

CAMBRIDGE, W. C., *Bristol.*—10-horse engine, steam cultivating tackle, clod-crushers, chain harrows, zinc riddles, washing boards.

[2090]

CARSON & TOONE, *Warminster.*—Prize chaff engines, Moody's turnip-cutters, horse gears, horse hoes, and cheese presses.

[2091]

CHANDLER, ROBERT, *Old Ford, Bow, Middlesex.*—Models of patent steam-cultivating apparatus for ploughing any shaped field.

DOUBLE-CYLINDERED ENGINE of 8-horse power, 3-tined cultivator, with all requisite adjuncts. Price, complete £500 0

DOUBLE-CYLINDER ENGINE, 10-horse power, 5-tined cultivator. Inclusive price £550 0

DOUBLE-CYLINDER ENGINE, 12-horse power, 5-tined cultivator. Inclusive price £600 0

Separate WINDLASS, with 5-tined cultivator, without engine, complete. Price £200 0

Further information may be obtained by applying to the exhibitor.

BENTALL, EDWARD HAMMOND, *Heybridge, Maldon, Essex.*—Chaff cutters, corn and seed crushers, root pulpers, oil-cake mills.

HORSE GEAR H. W. B.

INTERMEDIATE MOTION I.M.A.

BENTALL'S PATENT HORSE GEARS.

Descriptive Mark.

H. W. A. 1-horse	£7	7	0
H. W. B. 2-horse	8	8	0
H. W. C. 3-horse	11	11	0
H. W. D. 4-horse	12	12	0

BENTALL'S IMPROVED THRASHING MACHINES.

Bentall's patent 1-horse gear, intermediate motion, thrashing machine, pulleys, &c. £24 7 0

If fitted with travelling wheels and shafts . 29 12 0

Bentall's patent 2-horse gear, intermediate motion, thrashing machine, and pulleys 27 10 0

If fitted with travelling wheels and shafts . 32 15 0

Bentall's patent 3-horse gear, intermediate motion, thrashing machine, and pulleys 34 6 6

If fitted with travelling wheels and shafts . 41 13 6

Bentall's patent 4-horse gear, intermediate motion, thrashing machine, and pulleys 37 9 6

If fitted with travelling wheels and shafts . 44 16 6

CHAFF CUTTER C.D.D.

BENTALL'S PATENT PRIZE CHAFF CUTTERS.

Descriptive Mark.   Fitted with cast-iron legs.

C.C.X. for hand power,	7 in. mouth,	2 knives	2	5	0
C.D.A. ditto	7½ in. ditto	2 ditto	2	12	6
C.D.C. ditto	8 in. ditto	2 ditto	3	13	6
C.N.C. ditto	8 in. ditto	2 ditto	4	4	0

Fitted with wrought-iron legs.

C.D.D. for hand power, 9 in. mouth, 2 knives	5	5	0
C.D.E. { for hand or } 9 in. ditto 2 ditto	6	6	0
C.D.H. { horse power } 9 in. ditto 3 ditto	7	7	0
C.D.I. { for horse } 11 in. ditto 2 ditto	8	8	0
C.D.K. { or steam } 11 in. ditto 3 ditto	9	9	0
C.D.P. { power } 13 in. ditto 3 ditto	11	11	0

Fitted on a wood frame with wrought-iron legs.

C.W.D. for hand-power, 9 in. mouth, 2 knives	4	14	6
C.W.K. { for horse } 11 in. ditto 3 ditto	9	9	0
C.W.P. { or steam } 13 in. ditto 3 ditto	11	11	0

BENTALL'S INTERMEDIATE MOTION.

Descriptive Mark.

I.T.A. 1 and 2 horse, single gear £2 2 0

The following are fitted with brass bearings.

I.M.C. 1 and 2 horse, single gear	£3	3	0
I.M.D. 1 and 2 horse, double gear	4	4	0
I.M.A. 3 and 4 horse, single gear	3	13	6
I.M.B. 3 and 4 horse, double gear	4	14	6

ROOT PULPER R.P.A.

BENTALL'S PATENT PRIZE ROOT PULPERS.

Descriptive Mark.

R.P.D. barrel 9 in. diameter, 10 in. long .	£3	13	6
R.P.E. ditto 9 in. ditto 14 in. long .	4	14	6
R.P.C. ditto 12 in. ditto 10 in. long .	4	14	6
R.P.B. ditto 12 in. ditto 14 in. long .	5	15	6
R.P.A. ditto 12 in. ditto 20 in. long .	7	7	0

TURNIP CUTTER T.C.A.

BENTALL'S IMPROVED GARDNER'S TURNIP CUTTERS.

Descriptive Mark.

T.C.A. Gardner's turnip cutter as usually constructed. Price £4 10 0

T.C.A. fitted with Bentall's patent spout for separating the dirt &c. from the sliced turnips . . . 4 12 6

BENTALL, EDWARD HAMMOND, *continued.*

ROOT CUTTER R.C.A.

BENTALL'S PATENT ROOT CUTTERS.

Descriptive Mark.
R.C.A. to cut slices for beasts £4 4 0
R.C.A. to cut slices for beasts and finger-pieces for sheep. 4 14 6

OIL-CAKE MILL O.C.E.

BENTALL'S IMPROVED OIL-CAKE MILLS.

Descriptive Mark.
O.C.D. to break 1 size £2 2 0
O.C.A. to break various sizes 3 3 0
O.C.C. to break various sizes, or to dust . . 4 14 6
O.C.E. for horse or steam power, to break the hardest kinds of cake to various sizes, or to dust . 6 6 0

CORN AND SEED CHRUSHER R.S.A.

BENTALL'S PATENT PRIZE CORN AND SEED CRUSHERS.

Descriptive Mark.
R.S.A. for hand power. £5 5 0
  Ditto, with bean kibbler attached . . 6 16 6
R.S.B. for horse or steam power 8 8 0
  Ditto, with bean kibbler attached . . 10 0 0
R.S.C. for steam-power 12 12 0
  Ditto, with bean kibbler attached . . 14 0 0

BENTALL'S PATENT UNIVERSAL MILL.

Descriptive Mark.
U.M.A. adapted for kibbling oats, barley, beans, peas, Indian corn, &c. £5 5 0

BEAN KIBBLER B.K.C.

BENTALL'S PATENT BEAN KIBBLERS.

Descriptive Mark.
B.K.A. for hand-power. £2 12 6
B.K.C. ditto 3 3 0

BENTALL'S IMPROVED DRESSING MACHINES. Prices,
£6, £7, and £8 0 0

BENTALL'S BROADSHARE L.I.B.B.

BENTALL'S PATENT PRIZE BROADSHARE, CULTIVATOR, AND SUBSOIL PLOUGH.

Descriptive Mark.
B.I.B.F. heavy subsoil plough £4 4 0
B.I.B. 3-tined broadshare and subsoil . . 6 6 0
B.I.B.B. 5-tined ditto ditto . . 8 8 0
B.W.B. 3-tined ditto ditto wood beam 5 15 6
L.I.B.F. light subsoil plough 3 3 0
L.I.B.E. 3-tined broadshare and subsoil . . 5 15 6
L.I.B.B. 5-tined ditto ditto . . 7 7 0
L.I.B.C. 7-tined ditto ditto . . 8 8 0
L.I.B. 3-tined light broadshare 5 5 0
L.W.B. 3-tined ditto wood beam . 4 14 6

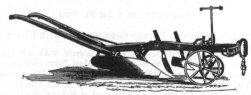

PATENT PLOUGH E.H.B.

BENTALL'S PATENT WOOD-BEAMED PLOUGHS.

Descriptive Mark.
E.H.A. fitted with curved handles, as a swing plough, including staff and coulter £2 13 6
E.H.B. fitted with straight handles, as a swing plough 2 13 6
  Ditto, fitted with 1 wheel 3 3 0
  Ditto, fitted with 2 wheels 3 10 0
Shares, per dozen 0 6 6

DYNAMOMETERS.

Bentall's traction dynamometer £35 0 0
Bentall's rotary dynamometer 50 0 0

BOBY, ROBERT, *Bury St. Edmunds, Suffolk.*—Machines for cleaning and separating grain, and improved wort pump for brewers.

Has obtained and still holds the First Prize and Silver Medal of the Royal Agricultural Society of England, and 30 other First-class Prizes.

4,000 of these machines have been sold in 5 years.

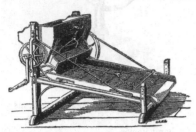

CORN SCREEN.

BOBY'S PATENT CORN SCREEN effectually separates all thin corns, stones, seeds, &c. from either barley, wheat, or sanfoin, and produces a sample that enables the merchant or farmer to obtain the highest market price for his corn.

To maltsters it is invaluable, as the duty on malt renders it necessary they should pay only on the best barley.

Screen No. 2, 50 bushels per hour . .	£7	0	0
Screen No. 1, 90 ditto . . .	9	0	0
Stone separator, extra	1	10	0
Screen No. 3, 150 bushels per hour . .	15	15	0

Boby's Patent Screen successfully competed at Norwich, in 1860, against a new patent self-cleaning and adjustable rotary screen, and obtained the silver medal.

CORN SCREEN WITH BLOWER.

BOBY'S IMPROVED PATENT CORN SCREEN, with blower.

This is the original well-known screen, with a simple blower in front of the hopper, the blast from which acts upon the grain as it falls from the hopper to the screen. This blower is driven, with a very small additional power, from the screen spindle, and is thrown out of use with the greatest facility by removing the strap, and the screen can then be worked alone.

This addition is very important, as it enables parties to separate most of the grown kernels, with the further great advantage obtained by the fact that the thin or tail corn is as clean and free from dust, &c. as the head sample. R. B. with the greatest confidence recommends this machine to all who are interested in obtaining a faultless sample of corn.

Screen No. 5, will screen 50 bush. per hour	£10	10	
Screen No. 6, ditto 90 ditto .	12	0	
Extra separators to the above	1	10	
Screen No. 7, will screen 150 bush. per hour	20	15	
Extra separator	2	5	
Extra pulley for power	1	0	

CORN-DRESSING MACHINE.

BOBY'S PATENT CORN-DRESSING MACHINE, with patent screen combined, enables any farmer or merchant to produce a sample of wheat or barley that will command the very highest price, as all the chaff, seeds, thin corns, &c. are effectually separated from the bulk.

Corn leaving a single-blast thrashing machine, has only to pass through Boby's dressing machine once, and a perfect separation will be found to have been made.

Price	£15	0	0
Dressing machine, with all the advantages of the above, but without patent screen	10	0	0

HAY-MAKING MACHINE.

BOBY'S NEW PATENT DOUBLE-ACTION HAY-MAKING MACHINE.

The inventor has effected everything in the above which is accomplished by any other implement of its kind, notwithstanding he has discarded more than 30 per cent. of the parts which usually compose them. In addition to this, the arrangement is such, that the revolving forks being placed a greater distance behind the carriage wheels, the weight of the machine has passed over the grass before the fork comes in contact with it, at the same time effecting a balance which leaves little or no weight on the back of the horse.

Amongst other advantages may be named its very ready and easy adjustment, the great facility with which its action can be reversed, coupled with the greatly diminished power required to draw it, and the simplicity with which every part of it can be removed or lucubrated and kept free from dirt, while it is impossible for it to choke.

The manufacturer offers to the public an implement which is at once lighter in weight, stronger in its parts, less likely to get out of order, and requiring less horse-power than any in the market.

Price £13 13

Catalogues will be sent post-free on application.

BROWN, WILLIAM, & CHARLES N. MAY, *North Wilts Foundry, Devizes.* London Agent: S. HOLMAN, 18, *Cannon Street.*—Portable steam engine, and patent sluice cock.

PORTABLE STEAM ENGINE.

EIGHT-HORSE POWER PRIZE PORTABLE STEAM ENGINE.

The accompanying woodcut represents an improved first-prize portable steam engine, which for simplicity, compactness, and economy in consumption of fuel, is not surpassed.

Among the many advantages this engine possesses over others, the following may be mentioned :—

The arrangement of the cylinder, being enclosed in a belt or jacket of steam, is kept, when working, at a high temperature, and thus the maximum advantage of using steam expansively is attained. All danger from fracture, through ice in the cylinder in frosty weather, is also obviated ; the ice melting as the steam gets up.

The lower part of the cylinder casting forms a steam chamber, from which the driest steam is taken off, directly into the valve case, without exposure to the cold air ; this is very important, as condensation and the liability to prime is thereby obviated.

The cylinder and all the working parts are fixed on the top of the boiler, as shown in engraving, so as to be well under the eye of the driver. The advantage of this arrangement must be obvious to all.

The simplicity of the arrangement is such, that any part is easy of access, in case of derangement, without interfering with or removing another.

The engine is furnished with an inside crank, which works between the bearings, so that the fly-wheel can be put on either side of the boiler, or a pulley of smaller dimensions fixed on its opposite side. One end of the shaft is also prolonged to fix a coupling on, for direct connexion with a windlass for steam cultivation or other purposes.

An improved patent steam-pressure gauge, without which no engine is complete, is furnished on every engine ; as well as a glass water-gauge, and gauge cocks of superior make, and a Salter's spring balance to safety valve.

The whole of the working bearings are of the best gun-metal, the guide bars are steel, and all working pins, as well as the nuts and screws, which are subject to much use, are case-hardened.

The boiler is made on the most improved plan to ensure durability and economy, together with as large an amount of heating surface as possible, upwards of 20 square ft. being allowed for each horse-power. The whole of the fire box is constructed of the best Lowmoor iron, or may be had of steel if preferred, at the same cost. The tubes are by the best makers, and are of such a length as to ensure no waste heat being emitted up the chimney. The barrel of the boiler is covered with a casing of hair felt and wood, and over all a covering of sheet iron, which gives a neat and finished appearance to the whole, and adds considerably to the facility of cleaning as well as economy in working.

The ash pan is fixed close round the fire box and fitted with a door which can be used as a damper, thus giving the driver full control over his fire. All live coals or cinders are also effectually prevented from falling out on the ground, so that the engine may be worked with the greatest safety even amongst straw or shavings.

BURGESS & KEY, *London.*—Reaping, mowing, and thrashing machines, haymakers, horse rakes, carts, waggons, chaff-cutters, churns.

PRIZE REAPING MACHINE.

Obtained the Council Medal at the Great Exhibition of 1851.

First prize consisting of 1,000 francs, a gold medal, and a great gold medal of honour, at Fouilleuse, near Paris, both in 1859 and 1860.

First prize at Goes, in Holland, in competition with Cuthbert's and Wood's.

First prize of the Highland Society, at Edinburgh.

First prize of the Yorkshire Society, at Hull.

First prize of the Lincolnshire Society, at Great Grimsby.

First prize of the North Northumberland Society.

First prize of the Hexham Society, at Haydon Bridge.

First prize of the Kent Society, at Ashford.

First prize of the Lancashire Society, at Southport.

Great gold medal of honour at Schwerin, and numerous other prizes and medals.

Royal Agricultural Society of England's first prize of £30, 1855. Ditto, £15 prize at Chelmsford, 1856.

First prize at Louth.

First prize at Hexham.

First prize of Highland Society's medal, at Lord Kinnaird's, Rossie Priory, North Britain.

Royal Agricultural Society's first prize, at Salisbury.

Australian medal of the Geelong Agricultural Society.

Austrian medal, at Vienna.

First-class diploma, at Pesth, Hungary.

First-class diploma, at Grossetto, Tuscany.

North Lancashire Agricultural Society, first prize and gold medal.

First-class diploma of the Central Society of Belgium, open to all nations.

BURGESS & KEY'S PRIZE REAPING MACHINE was introduced at the great Exhibition of 1851. Since that time it has been made entirely self-acting, and will now both cut and deliver the corn. In its improved form it has won the highest prizes and testimonials in all parts of the world.

The great advantage and superiority of Burgess & Key's screw-delivery reaper is fully attested by the enormous demand and use. They are in use on the royal farms in the United Kingdom, and on the Continent on the farms of the Emperor of the French, the Emperor of Russia, the Queen of Spain, the Grand Duchess Helena of Russia, the Grand Duke of Tuscany, Baron Ricasoli, Count Orloff Davidoff; and upwards of 3,000 are in use in the United Kingdom by the nobility and leading agriculturists.

Price £42 10

PRIZE MOWER AND COMBINED REAPING MACHINE.

BURGESS & KEY'S NEW PATENT MOWING MACHINE, adapted to cut all kinds of grasses, any required height; both wheels are geared and so placed that they do not travel on the cut grass; the driver can raise the knife so as to avoid obstruction, and the finger beam and knife are hinged so as to turn up when travelling along roads. It will cut about 1 acre of grass per hour.

This new machine is the result of 13 years' experience in the manufacture and use of reaping and mowing machines. During this time numerous prizes have been received by Burgess & Key for mowers, including the first prize of the Royal Agricultural Society of England at Canterbury and Warwick. They have also received the first prize

of £40 on 2 occasions in Holland, in competition with machines from all parts of the world. Numerous other prizes have also been received in various parts of the United Kingdom. It is a most simple and durable machine, and can be fully relied upon in the harvest field, as the greatest care is taken in its manufacture to ensure its working well. Price £25 as a mower.

The following are also exhibited :—

Horse rakes of superior construction.

Anthony's patent American churn, of which upwards of 10,000 have been sold.

Chaff-cutters of the very best description.

Turnip cutters; pumps and fire engines.

Lawn mowers; thrashing machines.

BURRELL, CHARLES, *St. Nicholas Works, Thetford, Norfolk,* and 69 *King William Street, City, London.*—Boydell's patent traction engine and endless railway, Fowler's patent steam ploughing apparatus, portable steam engines, combined and finishing thrashing machines, patent straw elevator, single and double corn mills, chaff machines, &c.

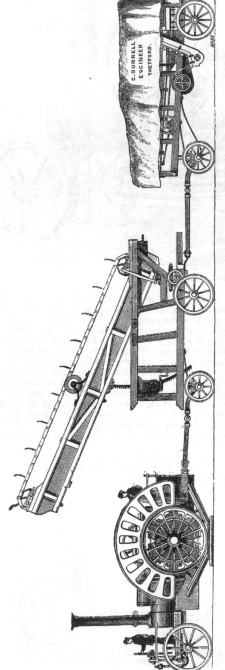

BOYDELL'S TRACTION ENGINE DRAWING A TRAIN OF THRASHING MACHINERY.

BOYDELL'S TRACTION ENGINE. This engine will draw heavy loads over soft sandy, rough, or hilly roads, or over country where no road exists. When not required for traction purposes, it is available for steam ploughing, as well as for every purpose to which an ordinary portable engine can be applied.

			£	s.	d.
Price of 10-horse power traction engine, with steering apparatus complete			£750	0	0
Price of thrashing machine with second dressing apparatus			115	0	0
Ditto ditto with patent reciprocating screen, and second dressing apparatus			120	0	0
Ditto patent straw elevator, to take the straw from the machine, and deliver it at any angle . . .			59	0	0

CHARLES BURRELL is the patentee and original manufacturer of the steam-power thrashing and dressing machines, and he exhibited the first of these machines at the Royal Agricultural Society's Show at York, in 1848; since which time he has directed his attention principally to the improvement of this class of machinery, and portable engines, and has so laid out his manufactory as to be enabled to produce work of the highest quality, and to execute orders with unusual promptness and accuracy.

Illustrated and descriptive catalogues, with numerous testimonials from agriculturists of high standing, and others who have used his engines and machinery, will be forwarded on application.

Orders or enquiries from English, foreign, or colonial houses addressed to the London offices or the Works will be promptly attended to, and estimates supplied with weights, measurements, and prices, inclusive or exclusive of packing and shipping charges, and delivered free to any dock in London, or any seaport in England or elsewhere.

BURRELL, CHARLES, *continued.*

ENGINE AND WINDLASS COMBINED.

This engine is so constructed that any parts requiring to be removed can be taken off when the steam is up, the fastenings being quite independent of the boiler. The windlass consists of a single sheave 5 ft. in diameter, round which the rope takes *half* a turn. The groove into which the rope passes is formed of a double series of small leaves, which on the least pressure clasp and hold the rope until it takes the straight line on the other side, when the clips freely open and liberate it. By this simple appliance all crushing and short bends, which are so detrimental to the profitable use of wire rope, are entirely avoided ; this, coupled with the fact that on each passage of the implement the rope is only bent twice, and then only round large diameters, will at once show this system of using wire rope to be most advantageous. The small leaves are made of chilled cast-iron, which is not liable to much wear, but when worn can be replaced at a trifling cost. The power is conveyed to the windlass by an upright shaft from the crank shaft.

PLAN OF WORKING.—On the headland is the engine and windlass, and directly opposite to them the anchor, which is self-moving, and between these the plough is pulled backwards and forwards, one end of the plough being alternately in the air and the other in its work, thus avoiding the necessity of turning at the headlands. The plough being constructed with patent slack gear, the rope is lengthened or shortened as the irregularity of the field requires, and at the same time both ropes are kept sufficiently tight to prevent them from trailing on the ground, by which means a great saving of draught is effected, the wear and tear (which must necessarily follow from the rope running on the ground) is entirely avoided without the least diminution of the power of the engine.

Any implement the farmer may deem it expedient to use may be substituted for the plough with a few modifications.

ROPE PORTERS.

These porters are placed along the fields at intervals of 40 yards, thereby keeping the rope entirely off the ground. | The outside ones are mounted on 3 wheels, so as to allow them to be moved by the rope.

BURRELL, CHARLES, *continued.*

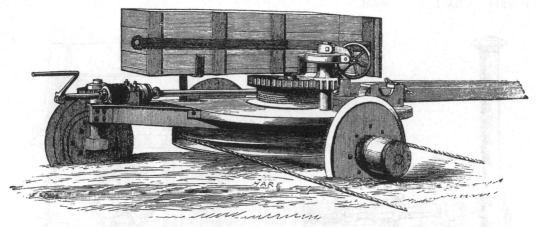

PATENT ANCHOR.

This anchor is made to resist the side strain of the implement worked, by the cutting into the ground of the disc wheels, and it is moved along the headland at pleasure, by the motion of the 5-ft. sheave, which is turned by the ploughing rope, and as the plough goes away from the anchor, the sheave winds up a rope stretched along the headland, and keeps the anchor opposite its work. The frame is made entirely of wrought iron. The steering of the disc enables it to be worked along a crooked headland. The box at the back is intended as a counterpoise to prevent the anchor being pulled over when doing very heavy work. This machine is managed by a boy, who also attends to shifting of rope porters.

PATENT BALANCE PLOUGH AND CULTIVATING MACHINE.

The above engraving represents the BALANCE PLOUGH AND CULTIVATING MACHINE, made of iron, and adjustable to different widths of furrow. The plough bodies and coulters are fixed on a bevel beam, and by altering their positions along the beam in either direction, a wider or narrower furrow is cut at pleasure, at the same time retaining the rigidity of a riveted frame which is so essential to the durability of a steam-going implement. A great many operations can be performed by this implement without much alteration being necessary. By removing the ordinary mould boards used for surface ploughing, and substituting short ones for scarifying, a tillage can be effected quite equal, if not superior, to digging, and leaving the land in a most desirable state for the action of the atmosphere. From the shares and mould boards being attached on the outside of the beam, all choking in very foul land is entirely obviated—a harrow can also be attached and drawn behind the plough if desired by the farmer. It can also be fitted with tines for cross cultivating.

10-horse set of ploughing and cultivating apparatus complete, consisting of engine with self-moving gear and windlass, a self-moving anchor and grappling anchor, 800 yards best quality steel rope, headland ropes, 18 rope porters, and snatch block, 4-furrow plough with scarifiers attached £780 0
12-horse set, ditto 825 0
14-horse set, ditto 875 0

PLOUGHING APPARATUS for attaching to ordinary portable engines of not less than 8 or 10 horse power.

8 or 10 horse self-moving windlass, a self-moving anchor and grappling anchor, 18 rope porters, 800 yards steel rope, headland ropes, and snatch block, 3 furrow plough with scarifiers attached . . . £365 0
8-horse power stationary windlass and ploughing apparatus, complete, consisting of driving shaft and windlass for engine, 2 self-moving anchors, 2 grappling anchors, 1,000 yards steel rope, headland ropes, and snatch block, 2 or 3 furrow plough, with scarifiers attached . . . 271 0

BURRELL, CHARLES, *continued.*

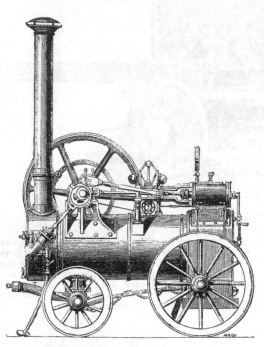

PORTABLE STEAM ENGINE.

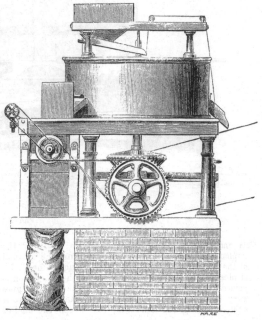

SINGLE FLOUR MILL, WITH DRESSING APPARATUS.

C. BURRELL'S PORTABLE STEAM ENGINES, of the most simple and approved construction, combining lightness and compactness with great strength and durability.

4-horse power, with cylinder 6½ in. diam.			.	£165	0
5	ditto	7	ditto	. 180	0
6	ditto	7¾	ditto	. 200	0
7	ditto	8½	ditto	. 215	0
8	ditto	9	ditto	. 230	0
10	ditto	10	ditto	. 270	0
10-horse power, with 2 cylinders		. . .	. 290		0
12	ditto	ditto	. . .	. 335	0
14	ditto	ditto	. . .	. 375	0

C. BURRELL'S IMPROVED SINGLE MILL, for grinding corn and dressing flour for household purposes at one operation.

With French burr stones, 36 in. diam.	.	.	£70	0
Ditto	ditto	42	ditto	. 85	0

If with Derbyshire stones, £10 less.

C. BURRELL'S IMPROVED PORTABLE DOUBLE CORN MILL, with dressing machine, mounted on a strong and suitable carriage, and fitted with wheels and shafts for moving from place to place. This mill is invaluable on large occupations, or in thinly populated districts.

Complete, as above, with 2 pairs of French burr millstones, 42 in. diameter . . . £165 0

Complete, as above, with 2 pairs of French burr millstones, 48 in. diameter . . . 180 0

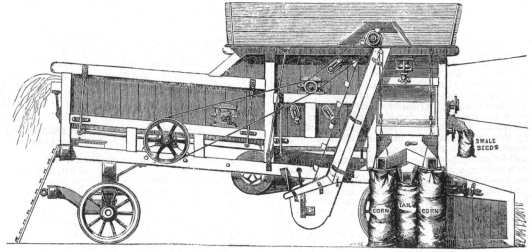

BURRELL'S IMPROVED THRASHING MACHINE.

With patent reciprocating screen, and second dressing apparatus . . . £120 0

[2092]

CHILDS, A. B., & OWEN, 481 *New Oxford Street.*—Grain separator, combining the action of the blast, riddles, and exhaust.

AMERICAN CORN OR FLOUR MILL SEPARATOR, combining the action of the blast, riddles, and exhaust. The most perfect machine extant.

Over 6,000 now in operation.

The flour miller of all others is most interested in this separator, owing to the large amount of grain he is constantly using. From vast competition, both at home and abroad, his profits are of necessity small, therefore his great aim should be to manufacture the sack of flour, from the least possible amount of wheat, and still preserve the quality. In thus giving the quantity with the quality,

the grain must be clean, and the cleaner the better, as it enables him to grind much more closely without specking the flour.

Corn factors, maltsters, brewers, farmers, and distillers, who have either water or steam as a motive power, will find its adoption of importance in proportion to their trade or traffic.

It first separates the corn according to bulk or size, by the means of the riddles and blast combined; and finely, by the exhaust, when every kernel of corn is weighed in air, and a division made agreeable to its weight or specific

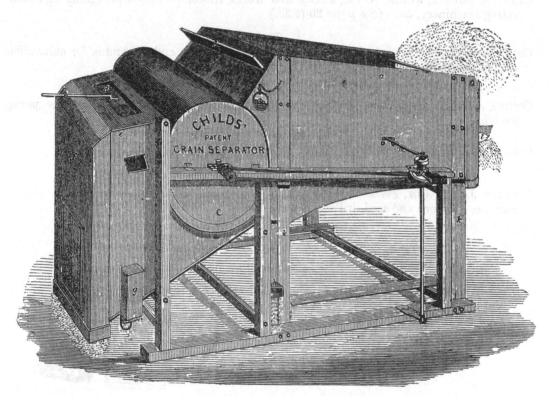

AMERICAN CORN OR FLOUR MILL SEPARATOR.

gravity. The operator is enabled to carry this division to an extreme nicety, from the simple though perfect means of regulating the machine, and thus remove all the imperfect, viz. sprouted, mouldy, shrunken, and weevil-eaten corn, the smut ball (without bursting), cliver, harriff seed, oats, garlic, cockle, &c.

The power required is merely nominal, being only sufficient to revolve the fan, and give the riddles a lateral motion.

Gentlemen calling at the office by appointment, can see the machine in operation.

LIST, &c. OF GRAIN SEPARATORS.

Size or No. of Machine.	Capacity or No. of bushels per hour.	Motion of fan per minute.	Diameter of driving pulley on machine.	Price of machine.
No. 1	200 to 400	525	9 in.	£70
No. 2	100 to 150	630	6 in.	50
No. 3	50 to 75	660	4½ in.	40
No. 4	30 to 50	700	4½ in.	30
No. 5	20 to 30	425	6 in.	20

The above machines delivered free upon the rail at Norwich.

[2093]

CLAY, CHARLES, *Stennard Iron Works, Wakefield, Yorkshire.*—Clay's patent cultivator and horse hoe.

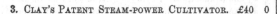

Obtained Prize at the Paris Exhibition, and also from the Royal Agricultural Society.

1. PATENT HORSE-POWER CULTIVATOR. The tines of this implement are raised backward, as in a horse-rake.

   Price, ranging according to the number of tines, from £6 5s. to £12 0

2. PATENT HORSE HOE. In this implement by a very simple arrangement, the width of cut can be instantly varied during the progress of the horses. Price £2 5
   Chain harrow, 15s. extra.

3. CLAY'S PATENT STEAM-POWER CULTIVATOR. £40 0

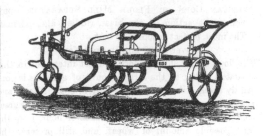

PATENT CULTIVATOR.

[2094]

CLAYTON, SHUTTLEWORTH, & Co., *Stamp End Works, Lincoln.*—Steam ploughing and cultivating machinery, &c. (*See pages* 20 *to* 23.)

[2095]

COLEMAN & SONS, *Chelmsford.*—8-horse power steam engine and apparatus for cultivation, cultivator, potato digger, and clod crusher. (*See page* 19.)

[2096]

COMINS, J., *South Molton, Devon.*—Self-cleaning clod crusher, set of drags, horse hoe, paring plough.

[2097]

CORNES, JAMES, *Barbridge Works, Nantwich.*—Prize chaff cutters.

[2098]

COULTAS, JAMES, JUN., *Perseverance Iron Works, Spittlegate, Grantham.*—Royal prize general purpose, and corn and seed drills, and horse hoe.

DRILL FOR GENERAL PURPOSES.

CORN AND SEED DRILL.

6-FT. 6-IN. 12-ROW GENERAL PURPOSE DRILL.
   Price £39 0
       With fore carriage steerage, £4 10s. extra.

The following highly valuable awards have been made for the above drills, showing an amount of success and increasing appreciation rarely obtained by any manufacturer. At the Royal Agricultural Society's Meeting, Leeds, 1861:—

   The first prize of £10 for the best corn and general purpose drill.
   The first prize of £10 for the best corn, seed, and root drill.
   The first prize of £5 for the best general purpose drill for small occupations.
   The first prize of £6 for the best drill for turnips and other roots.

6-FT. 6-IN. 12-ROW CORN AND SEED DRILL.
   Price £25 0

The first prize of £7 for the best drill for small seed and rye grass.
Small occupation corn and seed drill, highly commended.
Manure distributor, highly commended.

   In addition to which he has received a silver medal and 18 first prizes at other exhibitions in the short space of 5 years.

JAMES COULTAS JUN.'s IMPROVED HORSE HOE is the most efficient implement manufactured, being adapted for all kinds of corn and root crops, at any distance.
6-row general purpose £7 10

COLEMAN & SONS, *Chelmsford.*—8-horse power steam engine and apparatus for cultivation cultivator, potato digger, and clod crusher.

Coleman's Prize Cultivator has obtained upwards of 50 First Prizes, including the Prize Medal at the Great Exhibition, 1851.

PRIZE CULTIVATOR.

COLEMAN'S PATENT PRIZE CULTIVATOR combines in one implement the broadshare, grubber, and cultivator, and is the most efficient implement of its class for both spring and autumn work.

No. 5.   Price £7 0
No. 6.   Price 7 15
No. 9, with side levers.   Price 13 0

COLEMAN & SON'S IMPROVED HANSON'S PATENT POTATO DIGGER will raise potatoes cleaner, and with greater economy, than any other implement, and without injury to the crop.   Price £18 0

COLEMAN'S PATENT JOINTED CLOD CRUSHER accommodates itself to the undulations of the ground.   From the peculiar construction of its discs it is admirably adapted for abrading the surface of the soil, rolling young wheats, &c. and preventing the ravages of the wire worm.   Price £20 0

CLOD CRUSHER.

As Her Majesty's Commissioners were not able to allot to Messrs. Coleman sufficient space for a set of steam cultivating apparatus (Yarrow & Hilditch's patent), it is not exhibited.   It may be seen, however, at the Royal Agricultural Society's Show in Battersea Park.

Among the advantages of this system are the following:—

1. Direct action; no power being wasted by the use of pulleys between the implement and the engine.
2. Economy in first cost, and in expense of working.
3. The engine, being locomotive, can draw the entire set of tackle upon the roads, and is at the same time adapted for the general uses of a steam engine upon the farm.

Price lists will be forwarded post-free on application.

CLAYTON, SHUTTLEWORTH, & CO., *Stamp End Works, Lincoln; 78 Lombard Street, London; 125 Weiszgärber, Vienna; and Gegenüber dem Bahnhof, Pesth.*—Steam ploughing and cultivating machinery; portable steam engines for agricultural purposes; ditto for contractors, pumping and winding, &c.; ditto for sawing and general purposes; improved horizontal non-condensing fixed engines; improved combined thrashing and winnowing machinery; iron-framed mills for grinding all kinds of grain; flour-dressing machines; circular-saw tables; rack benches for large timber; improved pumping machinery, loam and mortar mills, &c. &c.

FIXED HORIZONTAL STEAM ENGINE.

Obtained a Prize Medal at the Great Exhibition, 1851; First Prize at the Royal Agricultural Society's Meeting, &c. &c.

FIXED, HORIZONTAL NON-CONDENSING HIGH-PRESSURE STEAM ENGINE, complete, with governors, starting valve, feed pump, fly wheel with turned-up rim, ample size, improved Cornish boiler with steam chest, furnace, safety valves, gauge cocks, blow-off cocks, connecting pipes, and every requisite to make same complete. Price, if in England, including man's time fixing:—

4-horse power complete £120 0

6-horse power complete	. . . . . .	£160	0
8 ditto	ditto	200	0
10 ditto	ditto	240	0
12 ditto	ditto	280	0
14 ditto	ditto	320	0
16 ditto	ditto	360	0
20 ditto	ditto	440	0

Link motion reversing gear if required.

SELF-ACTING SAW TABLE.

IMPROVED CIRCULAR-SAW TABLES.

Saw table, 24-in. saw	£15	0
Ditto 30-in. saw	20	0
Set of boring tools	1	10

Improved saw bench, 42-in. saw	£45	0
Self-acting ditto, complete	65	0
Set of trucks and railways	25	0
Improved rack bench	160	0

CLAYTON, SHUTTLEWORTH, & CO., *continued.*

IMPROVED OUTSIDE CYLINDER PORTABLE ENGINE.

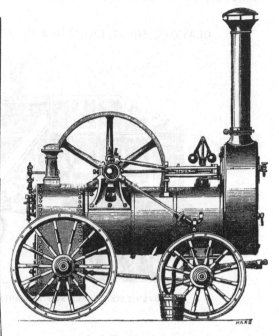

PATENT PORTABLE STEAM ENGINE.

CLAYTON, SHUTTLEWORTH, & CO.'S IMPROVED PORTABLE STEAM ENGINES FOR AGRICULTURAL PURPOSES, CONTRACTORS' USE, &c. &c.

Improved Portable Steam Engine.

4 horse power	. . . .	.	£165 0
5 ditto	. . . .		180 0
6 ditto	. . . . . .		200 0
7 ditto	. . . . .		215 0
8 ditto	. . . . . .		230 0
8 horse power, 2 cylinders	. . . .		250 0
10 ditto	1 cylinder	. . . .	270 0
10 ditto	2 cylinders	. . . .	290 0
12 ditto	2 cylinders	. . . .	335 0
14 ditto	ditto	. . . .	375 0
16 ditto	ditto	. . . .	415 0
18 ditto	ditto	. . . .	455 0
20 ditto	ditto	. . . .	495 0

4 horse power	. . . .	.	£170 0
5 ditto	. . . .	.	185 0
6 ditto	. . . . . .	.	205 0
7 ditto	. . . . .	.	220 0
8 ditto	. . . . .	.	235 0
8 horse power, 2 cylinders	. . . .		255 0
10 ditto	1 cylinder	. . . .	275 0
10 ditto	2 cylinders	. . . .	295 0
12 ditto	ditto	. . . .	340 0
14 ditto	ditto	. . . .	380 0
16 ditto	ditto	. . . .	420 0
18 ditto	ditto	. . . .	460 0
20 ditto	ditto	. . . .	500 0

Link motion reversing gear, £10 to £20 extra.

CLAYTON, SHUTTLEWORTH, & CO.'S PATENT STEAM-POWER THRASHING MACHINERY.

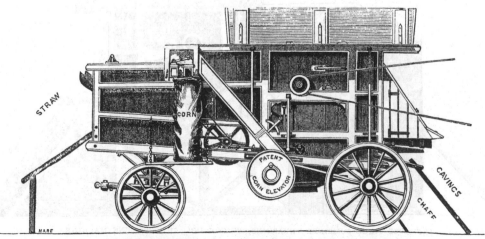

IMPROVED COMBINED THRASHING AND WINNOWING MACHINE.

CLAYTON, SHUTTLEWORTH, & CO., *continued.*

CLAYTON, SHUTTLEWORTH, & CO.'S PATENT STEAM THRASHING MACHINES.

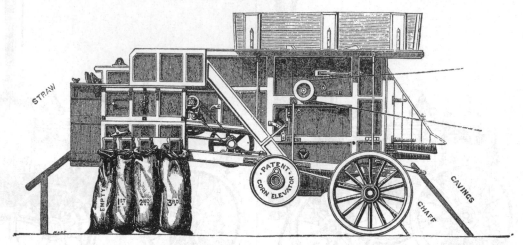

IMPROVED PATENT COMBINED THRASHING, WINNOWING, AND DRESSING MACHINE.

CLAYTON, SHUTTLEWORTH, & CO.'S IMPROVED COMBINED THRASHING AND WINNOWING MACHINE, with new patent elevator, single blast. Price £100 0

Clayton, Shuttleworth, & Co.'s improved patent combined thrashing, winnowing, and dressing machine, which thrashes the corn from the straw, dresses and cleans the same, and puts it in sacks ready for market.

Price, complete, with wood wheels and oil box, axles, drum 4 feet 6 in. wide, fitted with shafts or pole, and with a waterproof cloth cover £110 0

CLAYTON, SHUTTLEWORTH, & CO.'S IMPROVED FIXED COMBINED THRASHING, WINNOWING AND DRESSING MACHINE, with supporting frame, long spindle, and bearing, &c. Price, complete £118 10

Obtained first prizes at the Paris Exhibition of 1855; Royal Agricultural Shows at Lewes, Gloucester, Lincoln, Carlisle, and Chester; also the Yorkshire Shows, Bath and West of England, &c. &c.

The new patent corn elevator renders these machines perfect in their simplicity, thus dispensing with 6 driving pulleys, and 3 driving bands.

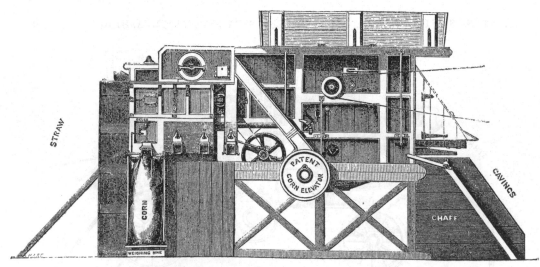

IMPROVED FIXED COMBINED THRASHING, WINNOWING, AND DRESSING MACHINE.

CLAYTON, SHUTTLEWORTH, & CO., *continued.*

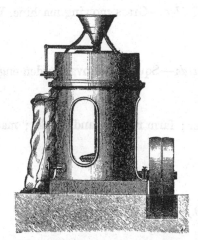

IMPROVED IRON-FRAMED CORN MILL.

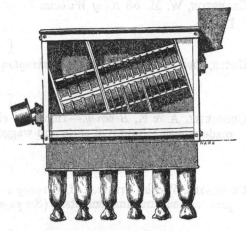

IMPROVED FLOUR-DRESSING MACHINE.

IMPROVED IRON-FRAMED CORN MILL, complete with driving pulley, and ready for immediate use.

		£	s.
Derby stones, 2 ft. 8 in.		£40	0
Ditto	3 ft.	55	0
Ditto	3 ft. 6 in.	65	0
Ditto	4 ft.	80	0
French bed, 2 ft. 8 in.		45	0
Ditto	3 ft.	60	0
Ditto	3 ft. 6 in.	70	0
Ditto	4 ft.	85	0
French bed and runner 2 ft. 8 in.		50	0
Ditto	ditto	3 ft.	65	0
Ditto	ditto	3 ft. 6 in.	75	0
Ditto	ditto	4 ft.	90	0

Fast and loose pulleys 50*s.* extra.

IMPROVED FLOUR-DRESSING MACHINE, with mahogany cylinder, including driving pulley.

	£	s.
Small, 12-in. cylinder	£25	0
Middle, 15-in. ditto	35	0
Large, 18-in. ditto	45	0
Crane for lifting runner stones	6	0
Complete set of tools for dressing millstones	4	10

Drawings and full particulars may be obtained on application.

CLAYTON, SHUTTLEWORTH, & CO.'s IMPROVED LOAM AND MORTAR MILL, which may be readily fixed and removed, and worked by an ordinary portable engine.

The pan is 5 ft. in diameter; and the two rollers are 3 ft. in diameter and 11 in. thick, giving an effective pressure of 50 cwt. on the material to be ground. The whole is firmly fitted and connected, and fixed on a timber frame. Scrapers are fitted to the vertical spindle, and every appliance is added to render this machine complete. The number already in use proves its efficiency.

	£	s.
Price	£85	0
For timber framing	5	0

CLAYTON, SHUTTLEWORTH, & CO.'s IMPROVED STRAW ELEVATOR (James Hayes, patentee).

Admirably adapted for taking the straw from the end of the shaker, when worked in conjunction with a portable thrashing machine. It absorbs very little power, and can be worked in any direction, varying from a straight line to right angles with the shaker. It is calculated to save the labour of 3 men, being capable of carrying the straw to a height of 20 ft. and upwards.

For straight delivery.

	£	s.
4 ft. 6 in. wide, 18 ft. long	£43	0
Ditto	20 ditto	48	0
Ditto	22 ditto	53	0

For delivery at any angle.

	£	s.
4 ft. 6 in. wide, 18 ft. long	£54	0
Ditto	20 ditto	59	0
Ditto	22 ditto	64	0

If fitted with wooden travelling wheels (which are strongly recommended), £5 extra.

CLAYTON, SHUTTLEWORTH, & CO.'s IMPROVED DOUBLE-BARREL PUMPS, for steam power.

These pumps are adapted for irrigation, or for pumping liquid manure in large quantities. They are intended to be worked by steam power (portable or fixed), and are capable of discharging 150 gallons per minute. They require no intermediate machinery, but may be worked direct, by a belt from a pulley fixed upon the engine fly-wheel shaft. The whole is fixed upon a strong iron frame, supported by two metal standards fitted with carriages and brasses. They have gutta-percha ball-valves, glands bushed with gun-metal, and an air vessel.

	£	s.
Price	£50	0

[2099]

CRANSTON, W. M., 58 *King William Street, London Bridge.*—Grass mowing machine, Wood's patent.

[2100]

CROSS, THOMAS WELLS, & CO., *Washington Works, Leeds.*—Square and oval garden engines.

[2101]

CROSSKILL, A. & E, *Beverley.*—Improved clod crusher; farm railway and trucks; machine-made wheels; improved carts and waggons.

[2102]

CROSSKILL, W., the Trustees of, *Beverley Iron Works, Beverley, Yorkshire.*—An assortment of prize agricultural implements. (*See pages 26 to 29.*)

[2103]

CROWLEY, MESSRS., & SONS, *Newport Pagnell, Bucks.*—General purpose cart; model cart; steam plough and apparatus.

[2104]

CUTHBERT, ROBERT, & CO., *Newton-le-Willows, Bedale.*—Patent reaping machines.

[2105]

DENNIS, T. H. P., *Chelmsford, Essex.*—Patent metallic horticultural building, or glazed structure. (*See page 25*).

[2106]

DORE, JOHN, 17 *Exmouth Street, Clerkenwell, London.*—Garden watering, rolling, and syringing machine, with registered spreader.

[2107]

DOWNIE, ROBERT, SEN., *Barnet, Hertfordshire.*—Improved open bee-hive and unicomb case, a substitute for bell glasses.

[2108]

DRAY, TAYLOR, & CO., 4 *Adelaide Place, London Bridge.*—Patent tubular iron gates.

[2109]

DRAY, W., & CO., *Farningham, Kent.*—Reaping machine, with drop platform.

[2110]

DRUMMOND, P. R., *Perth.*—Land-cleanser, which gathers, lifts, and carts stones, felt, corn, &c. without hand.

[2111]

EATON, JOHN, *Thrapstone.*—Patent turnip thinner and horse hoe combined; circular sheep crib, lifting jacks, &c.

[2112]

FENN, ROBERT, *Rectory, Woodstock.*—Bee-hive, adapted for cottagers, on the depriving system, without destroying the bees.

It was said by Dr. Johnson, that "the next best to knowing a thing was to know where to find it." Those who are desirous of becoming acquainted with an economical system of bee-keeping, will find one set forth in Nos. 639, 652 *(old series)*, and 4, 10, 21, 22, 29, 39, 40, 42, 43, 46 *(new series)*, of the *Journal of Horticulture, Cottage Gardener, and Country Gentleman*, published at 162 Fleet Street, E.C.

DENNIS, T. H. P., *Chelmsford, Essex.*—Patent metallic horticultural building, or glazed structure.

PH. DELAMOTTE del.   W.J.PALMER Sc.

METALLIC HORTICULTURAL BUILDING.

PATENT METALLIC HORTICULTURAL BUILDINGS, manufactured by T. H. P. Dennis, horticultural builder and hot-water engineer, High Street, Chelmsford.

These buildings are constructed of iron, and by the introduction of malleable fittings, the several parts are brought together with such facility as to overcome the only obstacle hitherto existing to their universal adoption.

The cost of these structures will defy competition even by the perishable wooden houses, whilst in increased strength and durability, shadowless frames, and illimitable forms, their advantages are so obvious, that they cannot fail to secure the patronage of those who require the highest order of conservatory, or the useful and profitable forcing house. They are all correctly fitted previous to leaving the Works, and can be erected by an ordinary mechanic in a few hours (screws and bolts being entirely dispensed with), by which the undesirable lengthened presence of workmen is obviated. Their extreme portability is of no small advantage, and however long they may have been fixed, their removal and re-erection can be accomplished without injury to any of the framing. They can be transmitted as low-rated freight. Provision for their extension has been carefully studied, and can be accomplished without alteration to any existing structure.

Every front light can be made to open and swing upon the mullion of the house, and the roof ventilation has no limits.

The condensed water from the roof is carried outside the building, thereby preventing the decay which follows when it is allowed to accumulate upon the eaves plate.

Every one is now supposed to be aware that glazed iron roofs, judiciously arranged, have not the least tendency to break the glass, and for those who still have this erroneous impression, the patentee wishes to explain that the most evident causes of fracture arise either from the glazier extending the laps of the glass so far upon the preceding square, that in cold weather ice is formed to such an extent, from the quantity of moisture necessarily retained, as to break the glass by its expansion, or from the bars of which the roof is composed being so irregularly spaced that the glazier is often compelled to introduce the glass with the bars on either side pressing tight against it, thus causing fracture, whilst other panes necessarily fall short of the width, and certain leakage is the result. These well known evils have been successfully overcome by the application of distance pieces between the bars, by which each glass-space is rendered equidistant from top to bottom and throughout any extent of surface, ensuring a water-tight and perfect roof.

Several of the parts are arranged so that iron roofs can be applied to wooden structures when preferred.

CROSSKILL, W. THE TRUSTEES OF, *Beverley Iron Works Beverley, Yorkshire.*—An assortment of prize agricultural implements.

THREE-HORSE REAPING MACHINE.

The Trustees of W. CROSSKILL's NEW 3-HORSE REAPING MACHINE, with self-delivery. Awarded the first prize of £14 by the Royal Agricultural Society of England at Leeds, 1861. Price £37 0

The Trustees of W. CROSSKILL's IMPROVED NORWEGIAN HARROW. Awarded the silver medal by the Royal Agricultural Society of England. Price . . £15 15

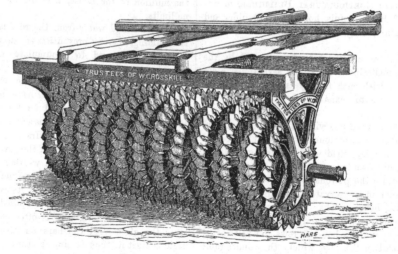

SELF-CLEANING CLOD CRUSHER AND ROLLER.

The Trustees of W. CROSSKILL's IMPROVED PATENT SELF-CLEANING CLOD-CRUSHER AND ROLLER. Awarded the special gold medal, 37 sovereigns, and 2 silver medals by the Royal Agricultural Society of England. Price, £16 10s. or with travelling wheels £18 10

CROSSKILL, W. THE TRUSTEES OF, *continued.*

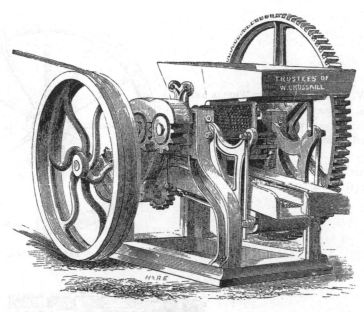

BONE MILL.

The Trustees of W. CROSSKILL'S IMPROVED BONE MILL. Awarded the first prize of £5 by the Royal Agricultural Society of England at Canterbury, 1860. Price £85 0

The Trustees of W. CROSSKILL'S IMPROVED BONE-DUST MILL. Awarded the first prize of £5 at Chester, 1858, and £10 at Canterbury, 1860, by the Royal Agricultural Society of England. Price £80 0

LIQUID MANURE DISTRIBUTOR.

The Trustees of W. CROSSKILL'S IMPROVED LIQUID-MANURE DISTRIBUTOR OR WATER CART. Awarded the first prize of £6 by the Royal Agricultural Society of England at Leeds, 1861. Price £17, or with portable pump and pipe . . . . £22 15 Apparatus for watering 4 rows of turnips, 15s. extra; and portable tripod stand for pump, 15s. extra.

CROSSKILL, W. THE TRUSTEES OF, *continued.*

PATENT WHEEL.

The Trustees of W. CROSSKILL's PATENT WHEELS AND AXLES for carts, waggons, &c. Awarded 2 silver medals, for combining good workmanship with cheapness, by the Royal Agricultural Society of England.

These articles are manufactured by machinery, and celebrated for strength, durability, and easy running.

LIGHT SPRING CART.

The Trustees of W. CROSSKILL's LIGHT SPRING CART. Awarded the prize of £2 for the best cheap market cart | by the Royal Agricultural Society of England, at Leeds 1861. Price £12 15

ONE-HORSE CART.

The Trustees of W. CROSSKILL's IMPROVED ONE-HORSE CART. Awarded the first prize of £6 by the Royal | Agricultural Society of England, at Leeds, 1861. Price £13, or with harvest ladders £14 1

CROSSKILL, W. THE TRUSTEES OF, *continued.*

The Trustees of W. CROSSKILL's MODEL 1-HORSE CART. Awarded the first prize of £5 by the Royal Agricultural Society of England, at Newcastle-on-Tyne. Price £14 5s. or with harvest shelvings. £15 15

The Trustees of W. CROSSKILL's YORK PRIZE 1-HORSE Cart. Price £11 10s. or with harvest raves . £13 10

PAIR-HORSE WAGGON.

The Trustees of W. CROSSKILL's PAIR-HORSE WAGGON. Awarded 4 first prizes by the Royal Agricultural Society of England. Price £29 10

The Trustees of W. CROSSKILL's ARCHIMEDEAN ROOT WASHER. Awarded the silver medal by the Royal Agricultural Society of England. Price . . £5 10

The Trustees of W. CROSSKILL's IMPROVED PATENT PIG TROUGH. Price £3 5

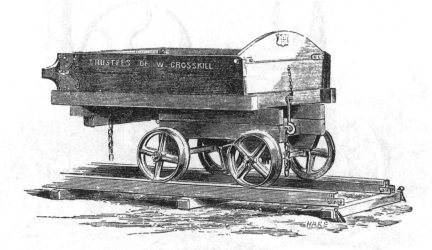

PORTABLE FARM RAILWAY.

The Trustees of W. CROSSKILL's PATENT PORTABLE FARM RAILWAY. Awarded 2 silver medals by the Royal Agricultural Society of England. Price 4s. per running yard. Trucks to tip sideway or endway, £5 10s. each.

Fowler, John, Jun., 28 *Cornhill, London, E.C.; and Steam Plough Works, Leeds, Yorkshire.*— Steam ploughs.

PLAN OF WORKING.

On the left headland is the engine and windlass, and directly opposite to them the anchor, which is self-moving, and between these the plough is pulled backwards and forwards, one end of the plough being alternately in the air and the other in its work, thus avoiding the necessity of turning at the headlands.

LOCOMOTIVE ENGINE.

The above is an engraving of a Locomotive Engine, adapted for steam ploughing. The clip drum for hauling the plough is placed under the boiler.

FOWLER, JOHN, JUN., *continued.*

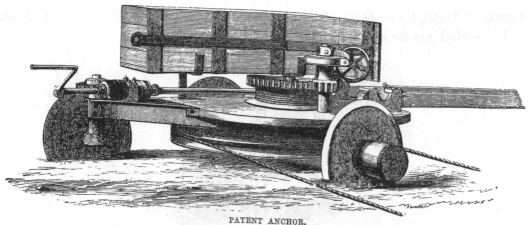

PATENT ANCHOR.

This anchor is made to resist the side strain of the implement worked, by the cutting into the ground of the disc wheels, and it is moved along, the headland at pleasure by the motion of the 5-ft. sheave, which is turned by the ploughing rope.

PATENT BALANCE PLOUGH.

The above engraving represents the PATENT BALANCE PLOUGH, made of iron, and adjustable to different widths of furrow.

PATENT BALANCE CULTIVATOR.

This is an engraving of the PATENT BALANCE CULTIVATOR. It will take a breadth of 6 ft. at each bout.

This Apparatus has gained every Prize for which it has competed.

[2113]

FERRABEE, JAMES, & CO., *Stroud, Gloucestershire; and* 75 *and* 76A *High Holborn, London W.C.*—Machines for mowing lawns.

LAWN MOWER FOR TWO MEN.

FERRABEE'S PATENT LAWN MOWERS.

M 1. The "Handy Lawn Mower," which a lady may use with ease £4 10	M 4. 19-in. for 2 men 7 0	
M 2. 16-in. for 1 man 5 5	M 5. 22-in. ditto 7 10	
M 3. 16-in. for man and boy. 6 10	M 6. 26-in. pony machine 10 10	
	M 7. 28-in. horse machine 16 0	
	M 8. 36-in. ditto 20 0	

[2114]

FERRYMAN, E. *Mendep Place, Oundle, Northamptonshire*—Patent self-kneading lever churn.

[2115]

FOWLER, JOHN, JUN., 28 *Cornhill, London, E.C.*—Steam ploughs.   (*See pages* 30, 31.)

[2116]

FRY, A. & T., *Temple Gate, Bristol.*—Cart, and American horse rake.

ONE-HORSE CART.

IMPROVED 1-HORSE CART for agricultural purposes, fitted with hay frame, tyres 3¼ in. wide.

Price, delivered in London £15 15

PATENT TUBULAR IRON AMERICAN HORSE RAKE, capable of putting 3 acres of hay into winrows per hour. These rakes are only a few pounds heavier than the wooden ones, which are already so well known. Price according to size.

[2117]

GARRETT, RICHARD, & SON, *Leiston Works, Suffolk, England.*—A selection of the most approved agricultural machinery.

GARRETT, RICHARD, & SON, *continued.*

Obtained the Council Medal in 1851 ; Gold Medal of Honor, Paris, 1855 ; and First-Class Gold Medal, Vienna, 1856 ; also 50 gold and silver medals from the different agricultural societies of Europe. R. G. & Son have in addition to these received an unprecedented number of money prizes, amounting to £1,200, and commendations almost without limit.

ESTABLISHED A.D. 1778.

GARRETT AND SON'S IMPROVED PORTABLE STEAM ENGINE.

The firm of RICHARD GARRETT & SON solicit the attention of noblemen, land owners, and farmers of all nations (who are desirous to improve agriculture), to their engines, machines, and implements, which are constructed upon the most scientific principles, of first-class workmanship, aided by the most modern mechanical appliances to facilitate the manufacture in both wood and metals, all which materials are selected with a view to the utmost durability, which can best be appreciated by those who have recently inspected their works.

R. G. & Son respectfully invite all who may desire to form their judgment upon a sound basis, to avail themselves of the convenience which the Eastern Counties Railway now affords, for an inspection, which cannot

GARRETT, RICHARD, & SON, *continued.*

fail to induce them to patronise implements and machines of such superior manufacture and perfect finish.

All machinery and implements of R. G. & Son's manufacture may be seen in practical use on the farm annexed to the works, and adjoining the Leiston Railway Station.

The widely spread, and rapidly increasing demand throughout Europe for steam thrashing machines (in which this firm has retained the precedence for improvements during the last half century) has urged upon them the necessity for producing a machine, more simple and effective in performing all the operations necessary to separate the corn from the straw, and make it a clean and perfect sample for sale. This is now done in one process, without any waste, and with very little manual labour, by the combined thrashing and dressing machine, described in the following pages.

Richard Garrett & Son also exhibit their well known standard implements and machines, viz. portable and fixed steam engines, horse-power thrashing machines, dressing machines, grinding mills, drills, and horse-hoes; adapted for all methods of cultivation, which will be briefly described in the following pages. Detailed catalogues and particulars of shipment, with estimates of cost of delivery to any part of the world, may be obtained on application to Leiston Works, or at their stand, in Class 9, at the International Exhibition.

In consequence of the extensive connexions of this firm, shipments are made in full cargoes, by vessels freighted direct from the works to many of the principal European ports, and this arrangement saves their customers the heavy charges usually incurred for packing and incidental shipping expenses, and insures the machinery being delivered in a sound and perfect condition. When required a competent man can be sent at a moderate expense, to instruct others in the use and management of the machines in work.

GARRETT & SON'S IMPROVED PORTABLE STEAM ENGINE. Price, with travelling wheels, complete, varying according to power, from £170 to £420 0

A marked improvement will be found not only in the appearance, but also in the practical working qualities of R. G. & Son's improved portable steam engines since the exhibition of 1851, at which, in common with one other exhibitor only, they received the council medal for the portable steam engine then exhibited. So great has been the demand for their improved portable steam engines in connexion with the patent thrashing machines, and also for contractors' purposes, that an entire new set of workshops, covering about two acres of ground, have been added to the Leiston Works since the year 1851. These shops are fitted with the latest and most approved mechanical tools and appliances for producing this class of machinery of the very best description, mechanically constructed and arranged, with every part perfectly true, and thoroughly well finished, and also at a first cost so moderate as to enable R. G. & Son to defy competition. With these facilities at their command, Garrett & Son are in a position to execute orders to any extent promptly.

These engines are adapted for working a thrashing and dressing machine of large dimensions, a circular or vertical sawing machine, a stone grinding mill, a set of steam cultivating apparatus, &c. &c. The two main points R. G. & Son have taken especial pains to carry out in the improvements introduced into their portable engines, are—

Economy in the consumption of fuel; and the arrangement of the working parts with a view to durability and facility for renewing the wearing parts when necessary. And in these very important points, R. G. & Son have been eminently successful, and can with confidence refer to any of their portable engines now in use in almost every county of England, and in many parts of Ireland, Scotland, Europe, and the Colonies.

GARRETT AND SON'S IMPROVED FIXED STEAM ENGINE, WITH HORIZONTAL CYLINDER.

GARRETT & SON'S IMPROVED FIXED STEAM ENGINE, with horizontal cylinder. Price, complete, 4 to 20 horse power, £120 to . . . £440 0

This form of engine is now generally preferred to those with vertical cylinders, being more compact in form, occupying less space, is fixed with greater facility and at less cost, and more easy of removal.

It is fixed on a metal foundation plate, and the various working parts being easily accessible, the adjustment and repairs are done with facility, the boiler is on the Cornish principle of an improved construction, and will be found very economical in generating steam, exceedingly strong and durable. Every part required for working these engines, including all the requisite fittings, to the end of the fly-wheel shaft, are sent with each engine.

D 2

GARRETT, RICHARD, & SON. *continued.*

GARRETT AND SON'S DOUBLE-CYLINDER STEAM PLOUGHING ENGINE AND TACKLE.

RICHARD GARRETT & SON have recently arranged for the manufacture of steam cultivating apparatus, with the latest improvements, under Messrs. Howard's various patents. Full detailed particulars will be given on application.

GARRETT, RICHARD, & SON, *continued.*

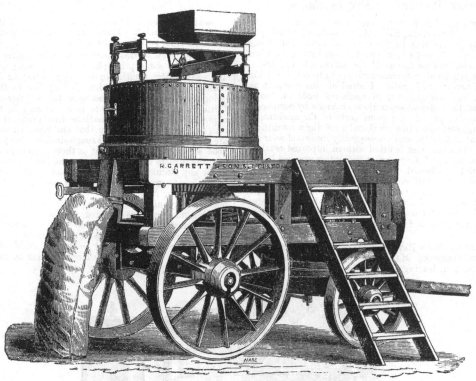

GARRETT AND SON'S IMPROVED STONE MILL FOR GRINDING WHEAT.

GARRETT & SON'S IMPROVED STONE MILL for grinding wheat for flour, and other corn for feeding purposes. Price, with pair of French burr or peak stones, £44 to £90 10

This mill is adapted for grinding every description of farm produce expeditiously and economically, and is fitted with an improved apparatus for adjusting the stones so as to grind to any degree of fineness required An improved wrought-iron crane is sent with it which is used for the purpose of raising the upper stone when required to do so for dressing.

GARRETT & SON'S PATENT COMBINED THRASHING AND DRESSING MACHINE, for steam or water power, on the new principle introduced by R. G. & Son, in 1859, and secured by letters patent. Price, according to power, complete for travelling, from £85 to £130 0

The cut given below shows one of R. G. & Son's new patent machines described in their Illustrated Catalogue as "A2, 'and explains the new and improved construction introduced into this class of machines, as compared with those on the old principle. The main difference between

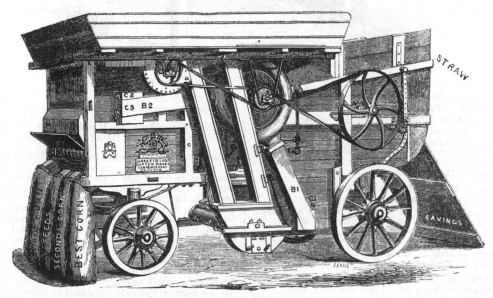

GARRETT AND SON'S PATENT COMBINED THRASHING AND DRESSING MACHINE.

GARRETT, RICHARD, & SON, *continued.*

the two is, that in the patent machines, all the blast necessary for dressing the grain is produced at one place by a fan (marked A in the cut), and conducted through the trunks B1 and B2, where it comes into contact with the grain, and by means of the valves C1, C2, and C3 can be regulated instantaneously, so as to suit every kind and description of grain. Instead of this compact and convenient arrangement, in machines made on the old system, the different separations are made by complicated machinery, placed in various parts of the machine, requiring great power to work it, and involving a considerable amount of friction, and consequently increased wear and tear. This machine is fitted with an improved revolving screen, by which four perfect separations are made, and the grain delivered into four sacks, viz. best corn, seconds ditto, tail corn, and seeds; the chaff (a very great improvement) being delivered quite free from seeds. This machine will deliver the corn rough-dressed instead of finishing it for market when required, and is adapted to the power of an 8-horse engine.

GARRETT & SON's PATENT COMBINED THRASHING AND DRESSING MACHINE, for rough-dressing grain, *i. e.*, leaving the sample so that by once passing through a finishing-dressing machine, it is fit for market. Price according to power, complete for travelling, from £80 to . . . £120 0

No cut is given of this machine, as it is constructed on precisely the same principle as the foregoing machine, and only differs from it in having two instead of three blasts. The same description that applies to the foregoing machine (excepting merely so far as it refers to the third blast) applies equally to this. It may be thought superfluous to bring out a machine that professes only to rough-dress the corn, seeing that the finishing machines are adapted for being used as rough-dressers, doing equally good work in either case; but as there are a large number of agriculturists who will not have the finishing machines under any circumstances, R. G. & Son found it desirable to bring out a machine on their new patent principle, for rough-dressing only, and without the necessary additional machinery required for finishing the sample.

These machines are adapted for the power of a 5 to 8-horse engine, and are described in R. G. & Son's Illustrated Catalogue (in English) as the B machine.

It is essentially important to bear in mind, that this machine will under ordinary circumstances be found to finish the sample for market.

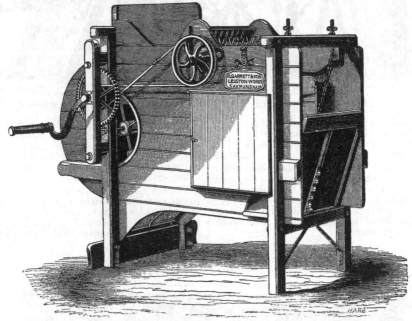

GARRETT AND SON'S IMPROVED CORN-DRESSING MACHINE.

GARRETT & SON's IMPROVED CORN-DRESSING MACHINE. Price, complete £9 0

This machine is more particularly adapted for the purpose of dressing corn when the chaff, broken straw, ears, leaf, and rubbish are all mixed, and for separating the inferior corn from the best. It will dress all kinds of grain or small seeds, and is fitted with a spiked roller for chaffing the corn when in a very rough state.

GARRETT & SON's IMPROVED CORN-DRESSING MACHINE, OF SMALL SIZE. Price £6 10

This machine is for the purpose of dressing all kinds of corn or small seeds in a perfect manner after being roughly sifted. It requires but a small amount of power to work it, say one man and a boy, and it will perform a large quantity of work in proportion to the power, cleaning all kinds of corn and small seeds perfectly.

GARRETT & SON's HORSE-POWER BOLTING THRASHING MACHINE. Price, complete, with travelling wheels,

3-horse power	£50 0
4-horse power	57 10

R. G. & Son's Thrashing Machine was the only one included in the award of a Council Medal, in 1851.

These machines were introduced by R. G. & Son about 25 years ago, in order to supply a demand then and still considered to be of the greatest importance, viz. the delivery of the straw quite uninjured, and fit either for use on the farm, for thatching purposes, or for sale; and the peculiar form of drum fitted to these machines answers for this purpose most satisfactorily, as the straw is delivered by it quite straight and unbroken. For thrashing barley by horse-power, this form of machine stands unrivalled, as it in no way injures the germ of the seed. Agriculturists residing in the vicinity of large towns where the straw can be profitably sold, have found this machine a valuable acquisition. The "bolting" drum has also been applied to the combined machines for steam-power, and is used extensively in every part of England.

GARRETT, RICHARD, & SON, *continued.*

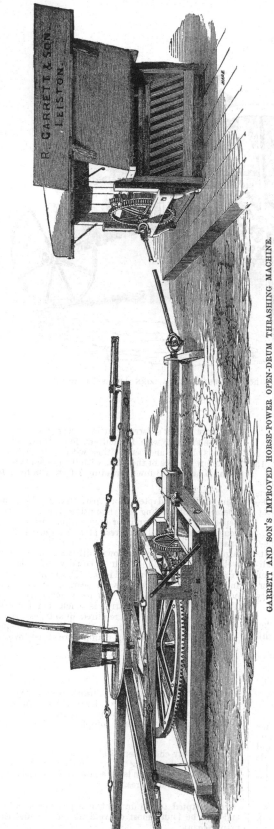

GARRETT AND SON'S IMPROVED HORSE-POWER OPEN-DRUM THRASHING MACHINE.

The above engraving represents the machine set down for work, with the jointed spindle and bridge to connect them.

GARRETT & SON'S IMPROVED HORSE-POWER OPEN-DRUM THRASHING MACHINE, adapted expressly for being worked by small colonial or foreign horses. Price, complete, with travelling wheels,

2-horse power	£36 0 0
4-horse power	48 0 0

R. G. & Son were the only exhibitors who received the Council Medal for thrashing machines in 1851.

They have also received for their horse-power thrashing machines, awards at the Great International Exhibitions of England, France, and Germany, and numerous other prizes (in all some 50 medals and £1,200 in specie).

This machine has been brought out expressly for the purpose of suiting the special requirements of the colonial and foreign farmers; the travelling wheels are made extra broad and strong, fitted with iron axles, with hollowed boxes to carry the grease, especially adapted for travelling over rough roads or uncultivated ground, and properly proportioned to the power of the horses for effectually working the same. The woodwork is extra seasoned by being specially dried, and will not be the least injured in the hottest climates, and the machine is constructed throughout with the view of economising cost of freight by packing in the smallest possible compass.

The working parts are precisely similar in construction to the ordinary open-drum thrashing machines of R. G. & Son's manufacture, of which 3,500 have been sold during the last 30 years.

GARRETT, RICHARD, & SON, *continued.*

GARRETT AND SON'S IMPROVED ELEVEN-ROW SUFFOLK LEVER CORN AND SEED DRILL.

GARRETT & SON'S IMPROVED ELEVEN-ROW SUF-
FOLK LEVER CORN AND SEED DRILL. Price of
the drill for 9 rows spreading 5 feet, to 13 rows
spreading 7 feet, £21 10s. to £26 15

This drill is extensively used at home and abroad for
the purpose of drilling in rows at any distance apart,
wheat, barley, beans, peas, and other grain, and by
changing the delivery barrel, turnips, mangold wurtzel,
and other seeds.

An improvement has been made in the fore-steerage,
rendering it easier of management, and preventing its
proper working being affected by clods and other inequalities of the surface. It is adapted for every description
of soil, for flat or hilly lands, and will be found to perform in the most efficient and economical manner every
operation for which a drill can be employed.

** For price of the different wearing parts, extras, &c.
see R. G. & Son's Illustrated Catalogue (in English),
page 5.

GARRETT & SON'S IMPROVED SMALL OCCUPATION
LEVER CORN AND SEED DRILL. Price of the
drill for 7 rows spreading 4 feet, to 10 rows
spreading 5½ feet, £16 to £20 0

This drill is similar to the preceding one so far
as it is adapted for drilling all kinds of corn and
seed, but as it is constructed on a smaller scale, and
the frame, and also the various wearing parts, are made
lighter, it is not adapted for drilling such large quantities
as the full size Suffolk corn drill, being more suitable for
small light land farms.

** For price of the different wearing parts, extras, &c.
see R. G. & Son's Illustrated Catalogue (in English),
page 6.

GARRETT & SON'S IMPROVED GENERAL PURPOSE
LEVER DRILL, with 11 coulters, for drilling all
kinds of corn and seed, either with or without
manure. Price of the drill, with 9 rows spreading 5 feet, to 13 rows spreading 7 feet, £38 to £46 0

This drill is adapted for performing the various
operations of seeding and manuring the soil. It will
deposit all kinds of grain or seed, either with or without
compost or artificial manures, at any required distances
apart, and at any depths. It will drill with perfect
regularity in going up or down hill, also on side hills as
well as on flat land, and it is equally well adapted for
lands ploughed flat or in ridges. The jointed iron lever
introduced by R. G. & Son some years since, is a valuable
adjunct to the drill, and has been generally adopted.
The Chambers' patent barrel is calculated for drilling
artificial manures in any required quantities, say from
2 to 60 bushels per acre ; and the drill, when required,
can be used for distributing manure broadcast, over grass
or corn lands.

GARRETT & SON'S IMPROVED THREE-ROW ECONO-
MICAL SEED AND MANURE DRILL, for turnips
and other seeds with manures on flat or ridge
lands. Price of the drill complete, for 3 rows £16 10

This is a very cheap, serviceable, and efficient drill, for
the purpose of drilling in rows on either flat or ridge
ploughed lands, all kinds of seeds with artificial or any
light pulverized manures.

It is adapted to the draught of a pony or small horse,
and will be found most convenient for use and easy of
management.

GARRETT, RICHARD, & SON, *continued.*

CHAMBERS' PATENT BROADCAST MANURE DISTRIBUTOR.

CHAMBERS' PATENT BROADCAST MANURE DISTRI-
BUTOR. Price complete, spreading 7½ feet,
£16 10s. to £19 0
*Included in the award of the gold médaille d'honneur,
Paris, 1855, and the first gold medal, at Vienna, 1857.*
This machine is in extensive use both at home and
abroad, for the purpose of distributing either broadcast
or in rows, all kinds of artificial manures, such as guano,
blood manures, salt, nitrate of soda, &c. which it delivers
in the most even manner, in quantities varying as required
from 2 to 100 bushels per English statute acre.
GARRETT'S PATENT HORSE-HOE. Price of the
horse-hoe, complete, £16 to £22 0
This well-known implement has met with an unparalleled
success, both in competing for prizes, and in being
brought into practical use, as it has won every prize for
which it has contended, and after twenty years' practical
test in all parts of the world, is universally admitted to
be as thoroughly efficient and useful an implement as
there is in use.

This implement will hoe in an effectual manner every
variety of drilled root or grain crop, at the rate of 10 to
15 English statute acres per day, and at a cost of not
exceeding 6d. per acre. It will work effectually on
uneven ground; the hoes are kept a uniform depth in the
ground, and the weeds are effectually destroyed, however
uneven the surface of the ground may be. The steerage is
a valuable addition to the hoe, as it enables the attendant
to steer the hoes to the greatest nicety, and does away
with any risk of cutting up the plants. The new
patented arrangement for regulating the position of the
hoe blade will be found superior to any other, and very
effectual. A grass seed engine is attached to this hoe for
the purpose of sowing grass seeds broadcast, while hoeing
spring corn, delivering the same in any required quantities
by means of revolving brushes.

GARRETT'S PATENT HORSE-HOE.

*** *For Description in* FRENCH, GERMAN, DUTCH, ITALIAN, *and* SPANISH, *see Appendix, pages* 113 *to* 116.

[2118]

GIBBONS, PHILIP & HENRY PHILIP, *Wantage, Berkshire.*—A portable combined double-blower thrashing machine.

[2119]

GRAY, JAMES, *Danvers Street, Chelsea.*—Elegant span-roof conservatory; tubular boiler, patent valve.

JAMES GRAY carries on the business of a horticultural builder in all its branches. The building department is managed by first-rate practical men, and the heating is under his own special care. He has been honoured during the past year with the erection and heating of the immense ranges of glass structures in the gardens of His Grace the Duke of Hamilton, at Hamilton Palace, Scotland; and also those in the gardens of the Right Hon. the Earl of Craven, at Combe Abbey, Warwickshire, where his works may be seen, as also in many others of the principal gardens in the country.

[2120]

GRAY & SONS, *Belfast.*—Agricultural machinery.

[2121]

GRAY, JOHN & CO. *Uddingston, near Glasgow.*—Agricultural implements, machines, and engine. (*See page* 43.)

[2122]

GREEN, THOMAS, *Smithfield Iron Works, Leeds, and Victoria Street, Holborn, London.*—Green's patent lawn-mowing machines. (*See page* 44.)

[2123]

HALKETT, PETER, 142 *High Holborn.*—Guideways; entire steam agriculture; connexion shown between fields and homestead.

THE GUIDEWAY SYSTEM OF STEAM CULTIVATION has been described in detail in a paper read by the exhibitor before the Society of Arts, in December, 1858, and published in its Journal. Persons who are desirous of knowing more of the system than can be shown by the model, or explained in the catalogue, are referred to that paper. It is sufficient here to state that the system consists in laying down at intervals of fifty feet or more, permanent guideways or rails, upon which a locomotive cultivator is supported and guided. This cultivator carries the motive power, and has attached beneath it the various implements that are required for the agricultural operations.

By this system every kind of field work can be most efficiently performed, a large concentration of power is obtained, and with very few hands a great amount of acreage is completed in a day, the soil is deeply and thoroughly worked and comminuted, and a fine tilth is the result, which is never trodden on by the foot of men or horses, or compressed by the wheels of carts and implements.

The cost of laying down the permanent way is £20 per acre, but this outlay is much more than compensated by the great economy of the operations, while large profits will be realized by the much increased produce raised by the great superiority of the cultivation.

The following is a list of the operations which the exhibitor has performed by steam:—ploughing, subsoiling, grubbing, rolling, clodcrushing, harrowing, finely comminuting the soil, drilling seed, hoeing crops, reaping corn, carting, watering.

No. 1. The model shows how the fields of a farm so laid down are connected with the homestead, and how the engine-power is brought to the barn for thrashing, &c., and how the trucks carry the produce and manure to and from the homestead.

No. 2. This drawing shows a modification of the guideway system in which the travelling cultivator is drawn by a rope from a distance, the engine power being stationary on the ground.

No. 3. This drawing shows a modification of the system especially suitable for our colonies and for countries where land can be cheaply obtained in large tracts, but where labour is scarce. The motive power used is a traction engine, travelling on a narrow strip of grass land. Much wider cultivators are used, much lighter rails, and the cost of the system is reduced to £2 10s. per acre. From a series of experiments, it appears that with the present traction engines one apparatus could plough 60 acres per day, and harvest (on the Australian system), the ears of 400 acres of wheat, performing other operations with equal celerity. One man only for the annual cultivation would be required for every ten now employed. The facility of superintending a few well-paid men with machinery, would enable large capitalists to embark in the business of agriculture; and one most important point for such countries, is that crops, as well as being raised, would be harvested by these few men. In Spain, Hungary, Poland, and Russia, the system would be a very valuable one; and in Australia and in America, where land rises in value with great rapidity when brought into cultivation, the ratio at which a certain amount of capital would increase annually if embarked in this patent, would be without doubt a very great one. The whole outlay would be often far more than paid in the first year's crop, and the field for the enterprise would be practically unlimited.

Persons desirous of further information are referred to the exhibitor's agent, Mr. Edward Weir, 142 High Holborn.

[2124]

HANCOCK, J. & F. & CO., *Tipton Green, Furnaces, Staffordshire.*—Pulverising plough; butter machines; steam plough; windlass for hauling implements. (*See page* 45.)

[2125]

HAYWOOD, JAMES, JUN., *Phœnix Foundry, Derby.*—Cast-iron ornamental vases and chairs.

The exhibitor manufactures portable and fixed steam engines of all sizes, combined portable thrashing machines, grinding mills, chaff cutters, &c.

GRAY, JOHN, & Co., *Uddingston, near Glasgow.*—Agricultural implements, machines, and engine. *Obtained a Prize Medal at the Great Exhibition, 1851.*

No 1. 8-HORSE POWER PORTABLE STEAM ENGINE for agricultural and other purposes, £225.

No. 2. PORTABLE THRASHING AND FINISHING MACHINE, delivering corn in bags ready for market, £110*

No. 3. 2-HORSE PLOUGH, for general purposes, strongly framed of wrought iron, can be used as a swing or wheel plough,—as swing plough, 95/; as wheel plough, 102/6; with improved steel mouldboard, 7/6 extra.

1 2-HORSE SWING PLOUGH, 85/: with wheel, 92/.
5. TURNWREST PLOUGH, 220/.
6. DEEP-SOIL PLOUGH, with steel mould, 150/.
7. GRAY'S IMPROVED ANGLED IRON HARROWS, 84/.
8. IMPROVED ZIG-ZAG IRON HARROWS, 3 in set, 85/.
9. LEVER SUBSOIL PULVERISER, 2 or 3 horses, 140/.

10. 3-HORSE FIELD GRUBBER, improved leverage. 210/.
11. DRILL GRUBBER, for pulverising between drills of root crops, with improved bridle, 85/.
12. IMPROVED CHAFF CUTTER, for power, 200/.
13. IMPROVED OAT BRUISER, for power, 220/.
14. IMPROVED 2 AND 3 HORSE YOKES, &c.

GREEN, THOMAS, *Smithfield Iron Works, Leeds, and Victoria Street, Holborn, London.* — Green's patent lawn-mowing machines.

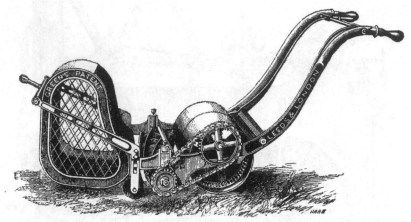

LAWN-MOWING MACHINE.

These machines now stand unrivalled, having improvements of the most important character (which have been secured by Her Majesty's royal letters patent for 1862).

By the use of these machines lawns can be brought to a state of perfection, unequalled by any other means; they are simple, durable, and effective, and are made in sizes suitable for the smallest plots, or lawns of the greatest extent.

Manufacturer of portable steam engines, sawing and wood-working machinery, iron roofing, fire-proof, and horticultural buildings.

Heating and ventilating apparatus on the most approved principles.

Entrance gates, field gates, iron and wire fencing, palisading, plain and ornamental iron and wire work of every description.

Hare, rabbit, poultry, and game netting.

Thomas Green keeps agricultural and horticultural implements of every description in stock.

Illustrated price lists will be sent free on application.

GREEN'S NEW PATENT GARDEN ROLLER.

GREEN'S NEW PATENT GARDEN ROLLER.

These garden rollers are made in two parts, and are free to revolve on the axis, the outer edges are rounded off or turned inwards, thus avoiding the unsightly marks always left by the use of the old form of roller; they can be used by the most unskilful workman, with the greatest certainty of producing a beautifully even surface, whether used upon lawns or on gravel paths, and for the bowling green and cricket field, are really indispensable. They are manufactured of the best materials, and finished in a superior manner.

Illustrated price lists will be sent free, on application to Thomas Green, Smithfield Iron Works, Leeds; and No. 2 Victoria Street, Holborn Hill, London, E.C.

HANCOCK, J. & F. & CO., *Tipton Green Furnaces, Staffordshire.*—Pulverising plough; butter machines; steam plough; windlass for hauling implements.

THE FOLLOWING ARTICLES ARE EXHIBITED :—

1. HANCOCK & CO.'s PATENT PRIZE PULVERISER PLOUGH.

This implement is used for making a seed-bed at one operation, on any kind of soil. It is a triple trenching plough, one share working below the other, and instead of ploughing land into clods, ploughs it into a good tilth, and makes a perfect seed-bed, every yard it works.

The turn-furrows are removable, by which any depth of soil may be turned, or none at all. It is worked from 3 to 12 in. deep, with 2, 3, or 4 horses, just in the same way as the common plough.

Price £6 10

2. Is a modification of the above for steam-power.

It is arranged to work without turning at the headlands, and takes a double furrow and makes a seed-bed at one operation.

Price £25 0

3. HANCOCK'S PATENT REGULATING WINDLASS.

This arrangement will double the hauling power of the engine without stopping up hill, and through irregular, heavy, or wet patches of ground ; the hauling drums are mounted on, and keyed to independent shafts, and each drum is acted on by two speed "clutches." By this contrivance, a common 8-horse power engine will be enabled to master all the difficulties of steam tillage.

Price £100 0

4. HANCOCK'S PATENT PRIZE BUTTER MACHINE separates all traces of acid and milk from butter without touching it with the hand. It also cools it, and will make it crimp in the hottest weather. Price . £2 12

Electro-plated £5 0

No. 2 ditto 2 2

Electro-plated 4 0

HAYWARD TYLER & CO., 85 *Upper Whitecross Street, London.*—Garden engines, conservatory pump, syringes, fountain jets.

The prices of the manufactures exhibited are subjoined:—

Garden engine with 28-gallon oak tub, and registered spreader £5 15 0

Small garden engine, with galvanized iron tub to hold 12 gallons 2 10 0

Oval galvanized iron tub engine, to hold 16 gallons 4 4 0

Small ditto ditto . . 12 gallons 2 10 0

Ditto ditto . . 10 gallons 1 15 0

Conservatory engine 3 3 0

Conservatory pump 2 10 0

3½-inch double fire or manure pump, on wheels 7 0 0

Lift pump, on oak plank, 2½ inch . . 5 12 6

Read's patent garden syringe 0 15 0

Common garden syringe, with rose and jet £0 7

Strong garden syringe, with rose and jet . 0 9

Strong garden syringe, with improved rose and jet 0 12

Small or ladies' syringe 0 6

FOUNTAIN JETS.

Prince of Wales feathers, large, 7/6, small, 5/0.

Crown . . . . ,, 7/6 ,, 5/0.

Barker's mill . . . . . ,, 17/6 ,, 6/6.

Convolvulus . . . . ,, 8/6 ,, 5/0.

Sheet . . . . ,, 8/6 ,, 5/0.

Reverse sheet . . . . . ,, 8/6 ,, 5/0.

Fountain basket with jet ,, 15/0 ,, 10/0.

[2126]

HENSMAN, WILLIAM, & SON, *Linslade Works, Leighton Buzzard, Beds.*—Patent ploughs, prize corn and seed drills, &c.

Obtained Prize Medal at the Exhibition of 1851.

THE "WOBURN DRILL."

WILLIAM HENSMAN & SON recommend to the attention of agriculturists their pair-horse steerage, corn, seed, and manure drill, known as the WOBURN DRILL. In addition to the medal of the Great Exhibition, and prizes at the meetings of various societies in intermediate years, it obtained in 1861 the prizes of the Royal Agricultural Society of England, of the Bath and West of England, and of the West Middlesex Societies.

This implement is adapted to all kinds of grain and seeds, and may be used on any land, possessing this great advantage—that the corn-hopper is self-acting, and delivers the seed with as great regularity when traversing a hilly district, as when employed on a level plain. The delivery of seed is most accurate, and exceedingly easy of management; the coulters can be set to any distance apart; and the steerage is the most complete yet introduced.

8-coulter cup drill as above, complete for corn
    and seeds. Price £20   0
6-coulter cup drill as above, complete for corn
    and seeds. Price 18   0

W. Hensman & Son also request attention to their improved land presser, which obtained the first prize of the Royal Agricultural Society at the Leeds Meeting, 1861. It is fitted with drill and hoes, so as to press the land, drill the corn, and cover it in at one operation. It is a very efficient implement on light lands.

    Price, with drill and hoes £13 10

An illustrated catalogue may be obtained post-free by application.

[2127]

HEREMAN, SAMUEL, 7 *Pall Mall East, London.*—Sir Joseph Paxton's patent hothouses for the million.

These HOTHOUSES are made of the very best seasoned red deal, and as the sashes are much stronger than those generally used in ordinary old-fashioned structures, they will, if properly erected, stand as permanent buildings proportionably longer. Whilst thus suited for a permanency, they are also particularly adapted for persons having temporary tenures, as they can be so fixed that at the expiration of a tenancy they may with ease be packed up and removed like any other furniture. Their moderate cost places them within the reach of all. As span-roofs, they are adapted for orchard houses, vineries, pineries, and indeed every horticultural purpose; and to all who have walls already standing they offer immense advantages, as they can be formed into ranges of lean-to houses, with facility and at an extremely low cost.

[2128]

HILL & SMITH, *Brierley Hill, Staffordshire.*—Patent continuous iron fencing and hurdles, prize wrought-iron land rollers.

[2129]

HOLMES & SONS, *Norwich, Norfolk.*—Prize corn, seed, and manure drills; portable engine; thrashing machine; seed sheller. (*See page* 48.)

[2130]

HORNSBY, R. & SONS, *Spittlegate Iron Works, Grantham.*—Improved patent prize portable and fixed steam engines, agricultural implements and machinery. (*See pages* 49 *to* 60.)

[2131]

HOWARD, JAMES & FREDERICK, *Britannia Iron Works, Bedford.*—Steam ploughs, steam cultivators, ploughs, harrows, horse rakes, hay-making machines. (*See pages* 62 *and* 63.)

[2132]

HUGHES & SONS, *Dover Road, and* 29 *Mark Lane, London.*—French burr millstones; flour-dressing and grain-cleaning machines.

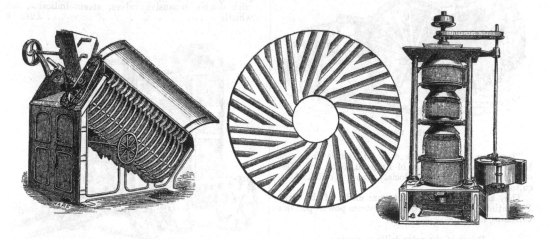

J. HUGHES & SONS have an extensive stock of French millstones, made from a very superior quality of burr, obtained from quarries recently discovered in France, and which, for workmanship and finish, cannot be excelled.

J. H. & Sons are also importers of Cologne stones, Granite peak, and grindstones direct from the quarries. Have always on hand at Mark Lane an assortment of flour-dressing machines, patent grain cleaners and bolting machines, best machine wire for cylinders, screen and kiln wire, square and round smut wire, wove by machinery; bolting cloths; best leather millbands; iron and brass sheave blocks; metal provers and staffs; corn and flour sacks; measures; machine brushes, all bristle; sack and heaving barrows; mill bills; and all other articles used in mills.

Hughes & Sons employ none but the best workmen; and, employing no travellers, avoid the necessarily high expenses connected therewith; and are thus in a position to give their customers all the advantages from this circumstance, as well as those arising from adequate capital, combined with practical judgment.

All goods are delivered free to wharves or railways in London.

[2133]

HUGHES, HENRY, *Regent Street, Loughborough.*—Improved bee-hive.

1. The hive is completely impervious to the weather.
2. The interior of the hive is less acted upon by changes of temperature, hair felt being inserted between the inner and outer casings.
3. The depriving hive is fitted with bars, by which each comb can be examined or removed at pleasure.

4. An improved feeding trough, by which the bees can be fed at the top of the hive at any time and under any circumstances.
5. The hive is mounted on a stand of iron, which will keep it free from the attacks of vermin.

[2134]

HUMPHRIES, EDWARD, *Pershore, Worcestershire.*—6-horse portable steam engine, and two combined finishing thrashing machines.

Manufacturer of the celebrated COMBINED THRASHING MACHINES, which have obtained the first prize at the Bath and West of England's Society's Meetings for six years in succession. Also the £20 prize at the Royal Agricultural Show at Canterbury. Price . . £93 0
Illustrated priced catalogues on application.

HOLMES & SONS, *Norwich, Norfolk.*—Prize corn, seed, and manure drills; portable engine; thrashing machine; seed sheller.

Obtained Prize Medals at the Exhibition of 1851.

These exhibitors maintain their high position in the manufacture of corn and seed, seed and manure drills, and manure distributors. At the Royal Agricultural Society's last meeting at Leeds they had awarded to them the Society's Four Prizes.

For the best seed and manure drill, ridge or flat—
The highest prize of £10.
For their best corn and seed drill—The prize of £5.
For their manure distributor—The prize of £3.
For their small seed drill—The prize of £3.

No drill trials having taken place since the year in which H. & Sons had awarded them the Three First prizes for the best corn drill; 3 prizes for manure distributor; and prizes for seed and manure drill.

CORN DRILL.

They now have received over 60 awards from the Royal Agricultural Society of England, Bath and West of England Agricultural Society, and Norfolk Agricultural Society.
They also obtained the first-prize medal of the Great Exhibition of 1851 for the best steam-power portable thrashing machine.

Prices of the prize drills as above :—

No. 1. The Leeds and Salisbury prize
manure distributor £14 14

No. 2. The Leeds and Salisbury prize corn
drill, 11 rows at 6 in. wrought iron levers 23 12
Fore carriage steerage to fit drill for broad
work 4 0

No. 3. The Newton Abbots prize small
occupation corn drill 18 0
Fore steerage to ditto, extra 4 0

No. 4. Improved general purpose drill for ⎫
8 rows of corn without manure, 6 in. apart ⎪
6 ditto corn with manure, 8 in. ditto ⎪
4 ditto turnip or rape seed, 13½ in. ditto ⎬ 32 10
3 ditto ditto 18 in. ditto ⎪
2 ditto swedes or mangold on the ridge, ⎪
27 in. apart ⎪
Having extra sets of levers and coulters for ⎪
each purpose. ⎭

No. 5. The Leeds prize seed and manure
drill, ridge or flat, with ridging rollers
and double-action wrought-iron levers . 25 0

No. 6. A small hand drill for seed without
manure 2 0

Upwards of 4,000 drills have now been manufactured by Holmes & Sons, a fact, which is an additional guarantee of their being approved.

H. & Sons had the honour of receiving at the Great Exhibition of 1851, the first-prize medal for the BEST PORTABLE STEAM-POWER THRASHING MACHINE, and they would now call especial attention to their

No. 7. NEW COMBINED PORTABLE MACHINE, which separates the chaff from the corn, delivers the chaff into large bags, cleansed from dust and seeds, as well as the corn into the sacks. This arrangement effects a considerable saving in labour; the chaff is more easy of removal, and there is much less waste of corn than by any other arrangement. Fitted with patent beaters, and on wood travelling wheels. Price . . £100 0

No. 8. IMPROVED 8-HORSE POWER PORTABLE STEAM ENGINE. For durability, efficiency, first-class workmanship, and small consumption of fuel, these engines are gaining a very high reputation. They are fitted with double expansive valves, steam indicator, and whistle £245 0

PORTABLE SEED SHELLER.

No. 9. PRIZE PORTABLE SEED SHELLER (New machine, received special award at Cardiff).

From the very many testimonials H. & Sons have received from gentlemen who have been supplied with these machines, there is no question about their being very far in advance of any other machine for the purpose now before the public.
On wood wheels £55 0

No. 10. IMPROVED CIRCULAR SAW TABLES, fitted up with every regard to wear and steadiness of the saw when in work. The large benches, with self-acting feed motion to the saw, are the most complete and portable that can possibly be. (See H. & Sons' Catalogue, pages 29—31.)

Strong table, with 36-in. saw and rollers at
end £20 0
Extra for boring apparatus 3 10
Extra for loose pulley and throwing-out lever 1 10
If fitted with improved fence to cut at any
angle, extra 1 10

No. 11. HOLMES & SONS' CORN DRESSING AND WINNOWING MACHINE, of which 3,000 have been manufactured, still hold their position. They are simple, turn easy, and make a good sample. Price . . £9 9

Hornsby, Richard, & Sons, *Spittlegate Iron Works, Grantham.*—Improved patent prize portable and fixed steam engines, agricultural implements and machinery.

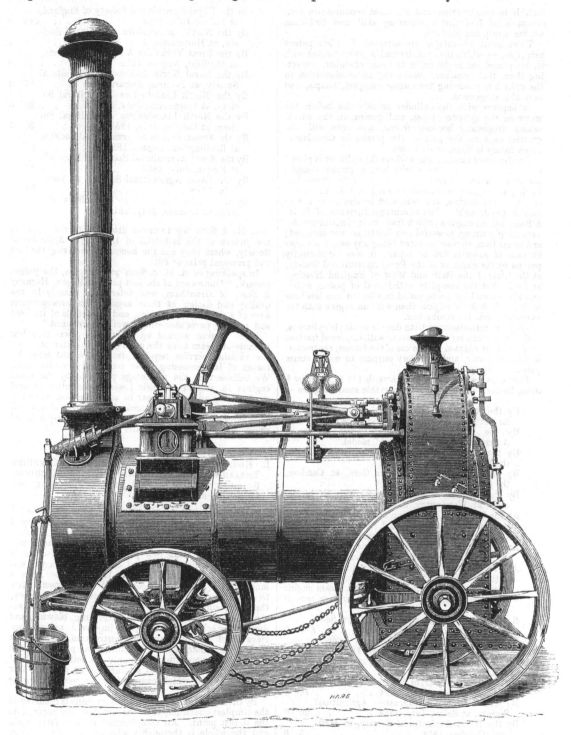

R. HORNSBY AND SONS' PATENT PORTABLE DOUBLE-CYLINDER STEAM ENGINE.

Portable Steam Engines have been a leading manufacture with R. Hornsby & Sons since their introduction, and they can refer with confidence to the proud position taken by them at every competition, in proof of their unapproachable excellence.

Their superiority has now been tested, not only by public trials at the principal meetings throughout the world, but by successful increasing use in the farm-yards of every country, where for years past they have proved themselves the best in principle, the simplest and most

HORNSBY, RICHARD, & SONS, *continued.*

durable in construction, and the most economic in consumption of fuel that engineering skill and first-class workmanship can produce.

These great advantages are attained by their patent principle, on which the cylinder and pipes connected with it, are placed inside the boiler or steam chamber, protecting them from weather, preventing all condensation in the cylinder, rendering the engine compact, simple, and easy of management.

In engines with the cylinder outside the boiler, the water in the cylinder, pipes, and pumps, in the winter season, frequently becomes frozen, and even with the greatest care on the part of the person in attendance, much injury is done, or time lost.

On the other hand, in cases where the cylinder is placed for protection inside the smoke box, a greater complication and weight of parts is necessary, by the smoke box having to be made sufficiently strong to take the whole strain of the engine, and increased in size so as not to impede the draught. The advantages therefore of R. H. & Sons' patent engines (which from their construction are exceedingly strong, powerful and light), are not obtained; and so far from engines so fitted being any easier of access in case of examination or repair, it was satisfactorily proved by the engineer of the Royal Agricultural Society, at the trials of the Bath and West of England Meeting, at Bath, that the complete withdrawal of piston, slides, and all the working parts, could be effected in a less time with R. H. & Sons' engine, than with an engine with the cylinder inside the smoke box.

They are manufactured with double or single cylinders, with link motion reversing gear, or with improved fraction gear; and are adapted for the use of the farmer, contractor, builder, exporter, and for every purpose to which steam power can be applied.

The following prizes have been awarded to R. Hornsby & Sons, for their improved patent portable steam engines:—

By the Royal Agricultural Society of England, at its last trial at Chester, the first prize of £25.
By the Imperial Royal Agricultural Society of Austria, at Vienna, the gold medal.
By the Hungarian Agricultural Society, at Pesth, the highest diploma of merit.
By the Agricultural Society of Gers, at Condom, the gold medal.
By the Manchester and Liverpool Agricultural Society, at Warrington, the first prize.
At the Universal Exposition at Paris, 1856, the first prize of £24 and gold medal, for the best portable steam engine for agricultural purposes.
At the Universal Exposition at Paris, 1855, the medal of honour for the best portable steam engine.
At the Great Exhibition of the industry of all nations, held at the Crystal Palace, Hyde Park, London, 1851, for the best portable steam engine for agricultural purposes, the first prize or council medal.

By the North Lincolnshire Agricultural Society, Boston, August, 1855 . . . . . £20 0
By the Bath and West of England Agricultural Society, at Tiverton, June, 1855 . . 10 0
By the Royal Agricultural Society of England, at Lincoln, July, 1854 . . . . 20 0
By the Bath and West of England Agricultural Society, at Bath, June, 1854 . . . . 10 0
By the Selby and Tadcaster Agricultural Society, at Selby, July, 1854 . . . . . 20 0
By the Herts Agricultural Society, at Hertford, October, 1854 . . . . 5 0
By the Great Yorkshire Agricultural Society, at York, August, 1853 . . . . . . 12 10
By the North Lincolnshire Agricultural Society, at Gainsboro', July, 1853 . . . . 20 0
By the Royal Agricultural Society of England at Gloucester, July, 1853 . . . . . . 10 0
By the Bath and West of England Agricultural Society, Plymouth, June, 1853 . . . . 15 0

By the Royal Agricultural Society of England, at Lewes, July, 1852 . . . . £40 0
By the North Lincolnshire Agricultural Society, at Horncastle, July, 1852 . . . . 7 0
By the Great Yorkshire Agricultural Society, at Sheffield, August, 1852 . . . . . 15 0
By the Royal North Lancashire Agricultural Society, at Preston, August, 1852 . . . 5 0
By the North Lincolnshire Agricultural Society, at Horncastle, July, 1852 . . . . 20 0
By the North Lincolnshire Agricultural Society, at Caistor, July, 1851 . . . . . 20 0
By the Great Yorkshire Agricultural Society, at Bridlington, August, 1851 . . . . 15 0
By the Royal Agricultural Society of England, at Exeter, July, 1850 . . . . 50 0
By the Royal Agricultural Society, at York, July, 1848 . . . . 50 0
By the North Lincolnshire Agricultural Society, at Lincoln, July, 1848 . . . . . 20 0

R. H. & Sons beg to direct attention to the report of the judges at the last trial of the Royal Agricultural Society, where they had the honour of receiving the first and principal prize of £25.

In speaking of R. H. & Sons' prize engine, the judges remark, "Our award of the first prize to Messrs. Hornsby & Sons, of Grantham, was determined mainly by the quality and design of their engine. Its arrangements were of a superior description, and the details of its fixed and working parts exceedingly well proportioned."

"The engine worked up to its full power at a less pressure of steam than the others, and is better fitted for the variable service required from it on the farm, by reason of its possessing fuller command over its work. We believe that the advantage possessed by the other engines in respect to their lower consumption of fuel would be found to disappear in actual service, when the appliances for reducing the area of their fire-grates would be removed. The fire-grate of Messrs. Hornsby & Sons' engine was in its ordinary state."

R. HORNSBY & SONS' IMPROVED PATENT COMBINED THRASHING, SHAKING, AND DRESSING MACHINES, for either portable or fixed purposes—preparing corn for market at one operation.

R. Hornsby & Sons' thrashing machine was awarded the first prize of £20 by the Royal Agricultural Society of England, at its last trial of thrashing machines at Chester, being the highest prize awarded for thrashing machines.

R. Hornsby & Sons' new patent combined machine is introduced by them with great confidence from the improvements it offers over any yet brought out. After a long series of experiments conducted with a special view to simplifying the machine and increasing its efficiency, they have succeeded in completing an entirely new arrangement of thrashing, shaking, elevating, blowing, and dressing machinery, having an improved patent corn elevator of the most perfect construction, new patent shaker, and other important features, and performing the whole of these operations with only one belt, and without gearing; a result hitherto attained by no machine ever introduced, and which cannot from its very nature be surpassed.

By these great improvements the machine is rendered of the simplest character, the wear and tear are reduced to the lowest point. All complication is done away with, and the whole is thoroughly adapted for doing a large amount of work with little power, and at the least cost to the farmer.

The following advantages amongst others may be particularly referred to:—

Extreme lightness.—Each machine weighing nearly one ton less than hitherto, and being therefore much easier of removal.

Great simplicity.—The screws are entirely dispensed

HORNSBY, RICHARD, & SONS, *continued.*

with, and replaced by vibrating boards ; and, from the compactness and convenience of the internal arrangement, these boards are rendered much shorter than ordinarily, and therefore involve less wear and tear.

The utmost durability of parts.—This is obtained by several important features, to which R. H. & Sons invite special notice. It is well known to all concerned in such machinery how troublesome and expensive a part the shaker crank is, in every machine yet introduced, and how much time is lost, and cost frequently incurred, by the necessity for its renewal. In *these* machines, by patented improvements, the cranks are case-hardened, and consequently will give fully double the wear of any others, and a proportionate advantage be realized by their use. The newly-invented and patented cranks, by which the vibrating boards are worked, are also case-hardened, which is impossible in the ordinary manufacture. The boards are hung on patent globular links, which work easily, and retain the oil for a much longer period than in any others, and are driven by R. H. & Sons' patent rods, constructed of flexible wood, which need no attention, require no oil-

ing, and springing with their own movement, cannot possibly wear out or break.

Renewal of the cranks of the shaker and vibrating boards.—This is an important feature, when the cost of replacing either of the above-mentioned cranks is considered. In these machines the patented cranks are constructed in parts (case-hardened), so that when one bearing is worn, it can be removed and replaced without necessitating a new crank.

New improved shaker.—The machines are fitted with their improved patent "differential shaker," the action of which is entirely new and of the most effective character—shaking the straw with the least possible power, and in the most perfect manner.

New improved patent corn elevator and cleanser.—This novel corn elevator is exceedingly simple, requires the least amount of power, does away with a number of belts and wearing parts, and will be found the most perfect in operation. In fact, having in a finishing machine only one belt, which is less than in any machine yet brought before the public.

R. HORNSBY AND SONS' IMPROVED PATENT COMBINED THRASHING, SHAKING, AND DRESSING MACHINE.

With the numerous improvements in detail which have for some time occupied the attention of R. H. & Sons, and the extraordinary lightness, simplicity, durability, and efficiency of the machine as described, they can confidently bring it before the notice of the public as the most perfect in principle, and the most effective in operation ever brought out.

The following first prizes have been awarded to R. H. & Sons for their thrashing machines, at the meetings of the Royal and other Agricultural Societies :—

By the Royal Agricultural Society of England, at its last trial of thrashing machines, at Chester, the first prize of £20, being the highest prize awarded for thrashing machines.

The first prize of £10 by the Royal Agricultural Society of England, at Lincoln.

The first prize of £10 by the Highland and Agricultural Society of Scotland, at Glasgow.

The highest diploma of merit of the Hungarian Agricultural Society, at Pesth.

The gold medal of the Imperial Royal Agricultural Society of Austria, at Vienna.

The first prize of £20 at the Royal Agricultural Meeting, at Carlisle.

The medal of honour at the Universal Exposition, at Paris.

By the Manchester and Liverpool Agricultural Society, at Warrington, the first prize.

And numerous other first prizes from the Great Yorkshire, North Lincolnshire, and other Agricultural Societies.

The performance of R. H. & Sons' prize machine at the Royal Society's trials was noticed as follows in the report in *The Times* of July 26 :—" Messrs. Hornsby and Sons have succeeded in turning out a machine which on trial was found to 'thrash perfectly,' leaving no grain in the ear, and none being wasted, as is too often the case, in either the chaff, which was well separated, the straw, which sustained little or no damage in the operation, or the pulse. The machine seemed to do its work 'in all respects thoroughly.' The straw underwent a strict examination at many hands, but no waste was discovered ; and several practical farmers expressed themselves in terms of unqualified satisfaction at the result, having narrowly watched the process throughout."

HORNSBY, RICHARD, & SONS, *continued.*

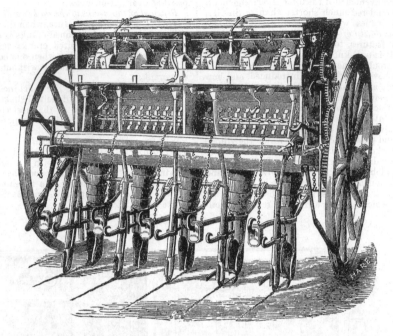

HORNSBY AND SONS' PATENT PRIZE DRILL.

R. HORNSBY & SONS' PATENT PRIZE DRILLS for corn and seeds of all descriptions, with or without manure.

Since the important improvements introduced under recent patents, these drills are undoubtedly by far the best ever brought before the public. In other drills there is great imperfection in the most essential point, namely, the conveyance of the corn or seed to the ground. R. Hornsby & Sons' patent flexible india-rubber tubes remedy this defect, and supply a perfect conductor in the place of the clumsy and inefficient tin cups in ordinary use, the delivery through which must necessarily be irregular, especially in windy weather. The patent tubes constitute the greatest improvement ever introduced in drills, simplifying and rendering them more efficient. The seed being delivered down one elastic tube, neither wind nor rain has the least effect on it as it passes through the continuous tube with the greatest precision into the channel made by the coulter : all bouncing of the seed from one cup to another, which must be the case in drills where tins are used, is entirely done away with.

The general purpose drill is capable of drilling all descriptions of corn and seeds, with or without manure, in any required quantities, and at any distance apart. It is alike suitable to the various requirements of all farms.

The patent corn and seed drills are suitable to all methods of cultivation, will work upon any soil, and deposit corn and seeds at any distance apart. They are constructed with improved slides, for regulating the quantity of seed at the pigeon holes. The feed of every coulter can, by these means, be increased or diminished without stopping the drill ; also, with two coulter bars, to equalize the pressure upon each coulter.

The patent prize drills for general purposes have received the following prizes :—

By the Royal Agricultural Society of England, R. H. & Sons' last trial at Salisbury, the first prize.
By the Universal Agricultural Exposition, at Paris, June, 1856, the first prize of £10 and the gold medal.
By the Universal Agricultural Exposition, at Paris, 1855, the medal of honour.

In addition to 9 other first prizes from the Royal Agricultural Society of England at its meetings at Lincoln, Lewes, Norwich, York, Newcastle-upon-Tyne, Shrewsbury, Derby, Bristol, and Liverpool, and upwards of 100 from the Great Yorkshire, North Lincolnshire, and othe Agricultural Societies.

The patent corn and seed drills have received the following prizes:—

The council medal at the Great Exhibition of the industry of all nations, held at the Crystal Palace, Hyde Park, London, 1851.
The first prize of £10, and the gold medal, at the Universal Agricultural Exposition, at Paris.
The medal of honour at the Universal Agricultural Exposition, at Paris.

In addition to 5 other first prizes from the Royal Agricultural Society of England at its meetings at Carlisle, Lincoln, Gloucester, Lewes, and Exeter ; and upwards of 100 first prizes from the Royal North Lancashire, Bath and West of England, Great Yorkshire, and other Agricultural Societies.

R. Hornsby & Sons also manufacture patent drills for small occupations, patent drills for turnip or mangold wurtzel with manure, patent ridge drills of various kinds, and every description of drill for depositing corn seeds and manure on ridge or flat ground ; to all of which an immense number of first prizes have been awarded by the various agricultural societies.

The Great Exhibition PATENT PRIZE CORN-DRESSING MACHINE.

R. Hornsby & Sons have received 9 first prizes from the Royal Agricultural Society of England, and upwards of 150 from other agricultural societies, for their patent corn-dressing machines, being a far greater number than has been awarded to any other machine. The first prize of £5 was also awarded to R. H. & Sons by the Royal Agricultural Society of England, at its last trial at Chester ; the medal of honour at the Universal Exposition, at Paris, in addition to 8 other first prizes from the Royal Agricultural Society of England at its meetings at

HORNSBY, RICHARD, & SONS, *continued.*

SIDE-VIEW OF THE GREAT EXHIBITION PRIZE CORN DRILL, WITH STEERAGE.

Carlisle, Lincoln, Gloucester, Lewes, Exeter, Norwich, York, and Newcastle-upon-Tyne, and upwards of 150 from the Great Yorkshire, Bath and West of England, North Lincolnshire, and other societies.

ADELAIDE AGRICULTURAL SOCIETY.—From the *South Australian Advertiser*, Wednesday, December 26, 1860.— "The judges awarded the prize to a Hornsby's machine exhibited by the Messrs. Tuxford, and remarked that it is capable of being worked with a saving of labour in consequence of having a spiked roller on the top to feed the machine. It also divides the wheat better than any other machine, thus occasioning some saving in time."

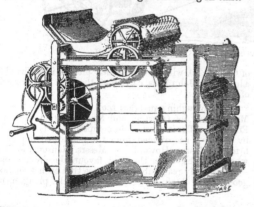

THE GREAT EXHIBITION PRIZE CORN-DRESSING MACHINE.

This improved and powerful dressing machine may be fitted with, or without a spiked roller working through a grating, so arranged as to form a hopper; and is easily adjusted to suit corn either in rough chaff or any other state. It is also fitted with a double shaking screen at bottom, which more effectually cleans the corn from all kinds of small seeds than a fixed one. For the second time over the strap is taken off, which puts the roller out of action; and a board placed in front of the grating makes it an excellent machine for finishing the corn for market. About 3,000 of these machines have been sent out, including a very large number to the colonies.

R. HORNSBY & SONS' IMPROVED ROYAL WARWICK CHAMPION PATENT PRIZE PLOUGHS, which last year,

considerably exceeded their past successes, and gained an immense number of champion and first prizes, including 6 in competition with Messrs. Howard, of Bedford, and 6 with Messrs. Ransomes & Sims, of Ipswich.

Since their first introduction, they have surpassed any known implement in their success at every competition, and have rapidly taken their position as the best, simplest, and most efficient implements for the farmer's use. The principal advantages of construction and arrangement for which these ploughs are remarkable, and which have gained them their high reputation, are—

1st. The *beam handles and frame* are one *solid continuous* piece of wrought-iron work, by which the usual cumbrous cast-iron body is dispensed with, and the utmost lightness, strength, and durability secured.

2d. The *lever neck* is of wrought-iron, for giving the share more or less "pitch," and more or less "land" as may be desired. The joint is globular or spherical, and is therefore of immense strength, and does not allow of the least dirt working in.

3d. The *slipe or slade* is of patent arrangement, and serves for the frame or body, to which the lever neck is attached. It is of great durability, and will wear considerably longer than those in ordinary use.

4th. The *simple* and *perfect* mode by which the breast is fastened, and can be expanded or contracted at pleasure.

5th. The *shares*, which are made in different forms, suit any and every description of soil. They are constructed to work accurately in conjunction with the breast, and by simply changing them, the plough can be made to give any class of ploughing to suit all localities, and each variety of land.

6th. The *coulter* and coulter fastenings, which are simple and effective, will cut in any position, as may be necessary to suit the variation in the share.

7th. The *wheels* are of novel construction, and have been recently patented. The old-fashioned long axle is done away with, and for it are substituted *discs*, working in suitable recesses, which prevent any oscillation of the wheel. Great steadiness is thus attained; the discs may be easily and cheaply replaced if necessary: they retain the oil and make the wheels self-lubricating, and are of great durability, as breakage of the wheel is rendered almost impossible.

R. Hornsby & Sons invite special attention to the following abstract of the successes of their Warwick royal

HORNSBY, RICHARD, & SONS, *continued.*

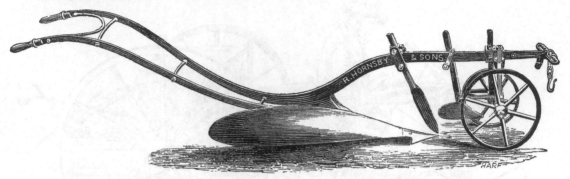

R. HORNSBY AND SONS' CHAMPION PATENT PRIZE GENERAL PURPOSE PLOUGH.

champion ploughs with their ploughman last season, in competition with those of Messrs. Howard, Ransomes, Page, Ball, and others.

R. Hornsby & Sons' ploughman, George Brown, has during 1861 been judged at 13 competitions, and taken 12 champion prizes, of which 6 have been gained, in competition with Messrs. Howard's ploughman Purser, and 6 in competition with Messrs. Ransomes and Sims' man Powell, viz. :—

At Berkeley and Thornbury, Gloucester, August 28, beating Messrs. Howard's man, with their latest improved plough.

At Sparkenoe, Leicestershire, September 11, champion class, beating Messrs. Howard's and Ransomes' men, with their latest improved ploughs.

At Sparkenhoe, separate trial for straight furrows, beating Messrs. Howard's and Ransomes' men, with their latest improved ploughs.

At Huntingdon, October 15, beating Messrs. Howard's and Ransomes' men, with their latest improved ploughs.

At Highnam, Gloucester, October 17, special silver medal, beating Messrs. Howard's and Ransomes' men, with their latest improved ploughs.

At Weobley, Hereford, November 14, sweepstakes of

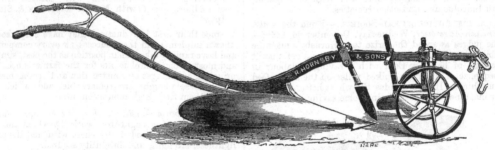

R. HORNSBY AND SONS' CHAMPION PATENT PRIZE LIGHT LAND PLOUGH.

£22, beating Messrs. Howard's and Ransomes' men, with their latest improved ploughs.

At Bennington, September 27, beating Messrs. Ransomes' man, with their latest improved plough.

R. H. & Sons' ploughman and their Warwick royal champion plough have never this season been beaten by local men ; and having carried off more prizes, in competition with others, than any in England, are alone entitled to the name of champion.

Ross MEETING.—R. H. & Sons' ploughman could not be present ; but the following, from the *Hereford Times*

of October 30, will show that their plough, in the hands of a local man, John Rees, carried off the palm for superiority of work :—

"The ploughing matches took place on the farms of Mrs. Barrett, of Overton, and Mr. John Cadle of Over-Ross farm. The interest of the competition was heightened by the presence of men from the celebrated implement firms of Messrs. Howard and Messrs. Ransome & Sims, whose ploughs, under the most favourable circumstances —as it is only reasonable to suppose that each firm takes good care to obtain the services of ' crack ' ploughmen—

R. HORNSBY AND SONS' CHAMPION PATENT PRIZE PONY PLOUGH.

were thus brought into competition with those of our well-known local makers, the Messrs. Kell, which are very extensively used in the district. The result of the contest was the awarding of the prize to Messrs. Howard's champion ploughman, Frederick Purser ; but, when we state upon the authority of very competent judges, that the work of Purser and the majority of the ploughmen was not *very* first-rate, the contest will hardly be looked upon as a test of the comparative merits of the ploughs. If

such is to be the result, the palm of victory must unquestionably be awarded to Messrs. Hornsby, for the best ploughing of the day was that of John Rees, in the service of Mr. Dowle, near Chepstow, the champion ploughman of that district. He was, however, not entitled to take the prize, for the simple fact that he ploughed only half the allotted quantity of land, none having been apportioned to him on the ground, in consequence, we believe, of some informality or error in

HORNSBY, RICHARD, & SONS, *continued.*

the entry. As affording a test, however, of the plough-man's skill, his work was universally pronounced to be immeasurably superior to that of any other competitor, and but for the 'chapter of accidents' disqualifying him, he would have taken the champion's prize."

WILLITON AND DUNSTER PLOUGHING MATCH, October 80, 1861.—Special prize of £5 for champion ploughmen.— John James Warren, of Milverton, with one of R. Hornsby & Sons' ploughs, beat Messrs. Howard's ploughman, Purser.

The following are a few of the matches at which R. Hornsby & Sons' ploughs have last year been successful. A large number might be added from all parts of the Kingdom, for which they cannot find space.

Berkley and Thornbury, champion prize, against Messrs. Howard's man.
Sparkenhoe, champion, farmers' sons' cup, and other prizes, against Messrs. Howard's & Ransomes' men.
Huntingdon, champion prize, against Messrs. Howard's & Ransomes' men.
Highnam, Gloucester, champion prize, against Messrs. Howard's & Ransomes' men.
Weobley, Hereford, champion prize, against Messrs. Howard's & Ransomes' men.
Hitchin, champion prize, against Messrs. Howard's & Ransomes' men.
Bennington, champion prize, against Messrs. Ransomes' man.
Evercreech, champion, first, second, and third prizes.
St. Neots, champion prize, against Messrs. Howard's man.
Derby, first prize.
Grantham, champion and other prizes.
Sleaford, champion and other prizes.
Waltham, champion and other prizes.
Wellingboro', champion and other prizes.
Metz, France, first prize.
West Riding of Yorkshire, first prize and extra ditto.
Bingham, Notts, champion prize, farmers' sons' cup, and other prizes.
Worksop, Notts, first prize.
Duloe, Cornwall, first prize.
Witheridge and Thelbridge, first prize.
Halberton, first and second prizes.
Upton, farmers' sons' cup.
Wellington, Wivelscombe, and Milverton, silver cup.

Bideford, champion prize and prize for the plough.
Chepstow, champion prize, by a local man, against the man who ploughed for Messrs. Howard at Hitchen and Luton.
Yarnscombe, two first prizes.
Huntsham, champion prize.
West Buckland and Bradford, silver cup, gained by Stephen Morgan.
Ropsley, first prize and silver cup.
Horncastle, champion, farmers' sons' cup, and the first prizes in every class.
Caistor, champion prize.
Alford, champion prize.
Wenlock, all the prizes.
Greasley and Silston, Notts, farmers' sons' cup.

ROYAL AGRICULTURAL SOCIETY OF ENGLAND, Meeting at Warwick, July, 1859.— R. Hornsby & Sons have pleasure in referring to the unequalled success which has attended their competition, in the trials for ploughs at the above meeting, and beg to submit the award of the judges, viz. :—

For the best plough for light land:—First prize awarded to R. Hornsby & Sons ; second ditto to Ransomes & Sims ; third ditto to J. & F. Howard.
For the best plough for heavy land:—First prize awarded to R. Hornsby & Sons ; second and third equally divided between Ransomes & Sims and J. & F. Howard.

GREAT PLOUGHING CONTEST AT STANLEY, PERTHSHIRE, confined exclusively to makers.—At the great competition of ploughs at Stanley on the 7th and 8th March, 1860, under the auspices of His Grace the Duke of Athole, K.T. President of the Highland Society of Scotland, R. H. & Sons had the honour to receive two prizes and one commendation, for their improved patent general purpose and light land ploughs, which were so successful at the Royal Agricultural Society's meeting at Warwick in July last.

The Warwick general purpose prize ploughs of Messrs. Howard, of Bedford, and Ransomes & Sims, of Ipswich, were amongst the competitors, and were worked by their champion ploughmen ; but the superior position taken by R. H. & Sons will be seen from the fact that they were the ONLY English makers, *the whole of whose ploughs were selected for the second day's trial*, and each of which received an award.

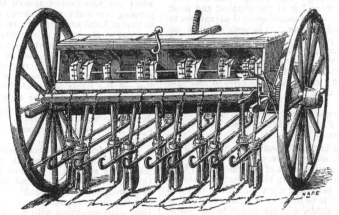

R. HORNSBY AND SONS' PATENT DRILL FOR SMALL OCCUPATIONS.

PATENT DRILLS FOR SMALL OCCUPATIONS, to which the first prize of £5 was awarded by the Royal Agricultural Society of England, at its last trial at Salisbury ; the first prize of £5 by the Bath and West of England Agricultural Society, at Plymouth ; the first prize of £5 by the Royal Agricultural Society, at Carlisle ; the

first prize of £5 by the Bath and West of England Society, at Tiverton.

These drills combine most of the advantages of R. Hornsby & Sons' patent corn drills, for which so large a number of prizes have been awarded.

The patent india-rubber seed conductors are used instead

HORNSBY, RICHARD, & SONS, *continued.*

of tins, the advantages of which are now too well known to require further comment. They are fitted with iron levers, each acting independently: they will deposit any quantity or description of corn, and by simply changing the cup barrel, are equally well adapted for drilling turnip, mangold wurtzel, or cole seed, at any required depth in the soil, or at any distance apart in the rows, from 6 in. and upwards, for the various crops of corn and seed.

R. HORNSBY AND SONS' IMPROVED FIXED STEAM ENGINE.

IMPROVED FIXED STEAM ENGINES, to which the prize of £10 was awarded at the last trial of the Royal Agricultural Society, at Chester.

This engine, the design and workmanship of which are mentioned in the report of the judges of the Royal Agricultural Society as being "very good," and the consumption of fuel as "low," is of the simplest and most serviceable character, and presents especial advantages to persons requiring such a power. It is made complete on a metal foundation plate, easy of removal, and supplied with cylindrical Cornish boiler, of suitable size and strength. The whole is made of the best materials, and executed in a superior style of workmanship; all complete to the end of fly wheel shaft, exclusive of carriage, masonry, and brick-work.

By Her Majesty's royal letters patent—R. HORNSBY & SONS' IMPROVED PATENT WASHING, WRINGING, AND MANGLING MACHINES.

R. Hornsby & Sons having for some time been satisfied that a really simple and effective washing machine was amongst the first domestic wants of the community, have devoted considerable attention to the subject, and have at length succeeded in perfecting a machine which they can unhesitatingly assert to be the best and most efficient, to wash thoroughly, quickly, without injury to the linen, with the least possible quantity of water, and at the least cost for fuel.

The following is a brief description of the principle and action of the machine:—

This washing machine, which may be fitted either with or without the wringing and mangling apparatus, consists of a tub or vessel, of well-seasoned wood, and first-class workmanship, the inside of which is covered with ribs, and at the bottom of which is a patented hollow bridge.

The clothes, soap, and water are prepared in the usual way, and when the vessel is charged it is swung backwards and forwards in such a manner as to bring the top quite perpendicular at every movement. By the action, some of the air and water rush to and fro between the spaces of the bridge, and inflate or spread the clothes, which are also rubbed against the bridge by the water pressing over and through them; the process being similar to that of hand-washing, viz. rubbing the clothes both in and out of the water, submerging them, and by a sort of syringeing action, removing every particle of dirt. On opening the vessel, the clothes are never found either rolled into a mass or swimming on the surface, but always thoroughly opened to the action of the water, which passes through the fabric of the clothes. The water is used as hot as possible, and as the tight lid confines the steam, clothes are subject to a thorough boiling and steaming during the washing operation. The machine is well got up, is very compact and portable, as well as convenient to use, and easy to work.

These machines have been severely tested for a length of time, both at hotels and in private families, and the results are such as to warrant R. H. & Sons in saying that nothing of more general utility has for a long time been brought before the public. The beautiful cleanliness and improved appearance of the linen when washed by them—the quickness with which a large and heavy wash can be got through without discomfort or annoyance—the great saving in the fire required—and the fact that articles which the most careful hand-washing would injure, are washed with perfect safety, and that clothes of every kind receive no damage in the process—are sufficient evidence that their machine supplies the long-wanted desideratum, and must be adopted by almost every class.

The principal advantages of R. H. & Sons' washing machine may be briefly summed up as follows:—

HORNSBY, RICHARD, & SONS, *continued.*

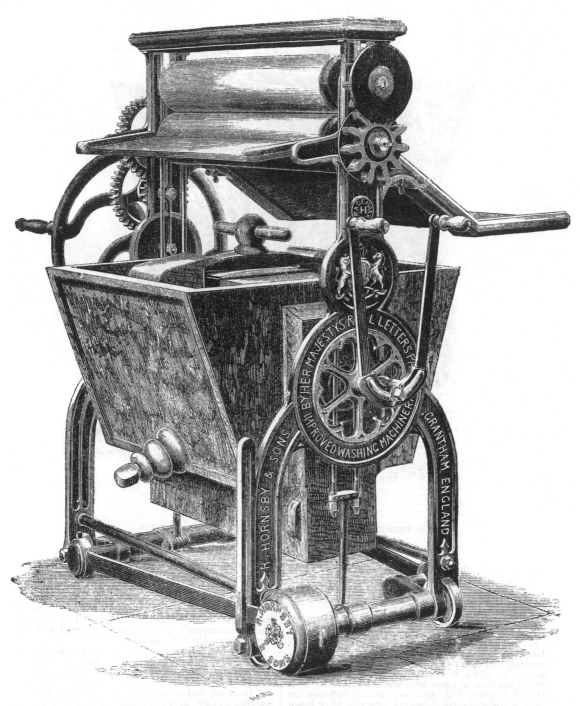

R. HORNSBY AND SONS' IMPROVED PATENT WASHING, WRINGING, AND MANGLING MACHINE.

1st. *It is thoroughly effective and perfect in operation.* The machine is really what it professes to be, and not a mere vessel in which the linen is swum, and therefore scarcely washed at all. This efficiency is attained by the improved form of the tub or vessel—the construction of their patent hollow bridge, as before described—and by their PATENT DOUBLE HANDLE, by which the Machine is worked. This handle—which will be seen in the en-

graving—gives TWICE THE MOTION to the Machine that is obtained in any other, and produces so violent a cleansing action, that the linen is thoroughly washed in an incredibly short time, without hand-rubbing.

2d. *It is simple,* containing nothing that can by any possibility injure the most delicate article—requires no management, and is easily worked.

3d. *It is compact in arrangement,* occupying less

HORNSBY, RICHARD, & SONS, *continued.*

R. HORNSBY AND SONS' IMPROVED PATENT WASHING MACHINE.

room than any other—having the washing vessel placed immediately under the rollers, so that the water, in wringing, falls directly into it, and large articles are drawn out and wrung without labour in lifting from the vessel to the rollers.

4th. *It is fitted with an improved tub*, to be used in blueing or rinsing, which offers many advantages over similar appliances offered to the public. When performing either of the above operations, the lid of the washing vessel is removed, and this tub fitted in its place under the rollers, to receive the blue or rinsing water, which can then be taken away without interfering with the other portion of the machine. It is of sufficient size to answer every purpose; does not form part of or lessen the size of the washing vessel; will be found useful for a variety of domestic purposes; and is included in the price of the machine.

5th. *It is fitted with an improved patent lid*, which prevents leakage when in motion, and confining the air, steam, and water, gives them a full and very important action on the clothes.

6th. *It is remarkably easy to work*, having their improved patent pendulous motion, by which the tub is swung from side to side in working with the least amount of power. The patent balance box underneath the tub is partly filled with sand, and assists the motion, causing the machine to be readily worked by any ordinary domestic servant.

7th. *It is of first-class material and workmanship;* every portion of both wood and iron work is as good as can be made; the rollers are well-seasoned sycamore or beech, brass-capped; and every part is finished with due regard to strength and durability.

BY ITS USE the whole of the washing of any family may be done *without inconvenience*, in an exceedingly *short space of time, without extra assistance.* A GREAT SAVING in fuel and in soap will be effected. The linen will be greatly improved by the bleaching action of the machine, and will wear much longer than when subjected to the destructive friction of hand-rubbing. When the combined machine is used, the wear of the clothes will also be enhanced by the wringing being effected without the usual injury to the fabric; and the entire performance will be such as to render it *an essential* in every well-regulated and economically managed family.

R. Hornsby and Sons have a large number of testimonials as to the efficiency of their washing, wringing, and mangling machines, but space will only allow of the publication of the following. They will be happy to furnish any number on application.

HORNSBY, RICHARD, & SONS, *continued.*

"*Haggerstone, Beal, Northumberland,*
Nov. 23, 1861.
"GENTLEMEN,
"I HAVE very great pleasure in bearing testimony to the efficient manner in which the washing, wringing, and mangling machine does its work, which I ordered from you when at the Royal Agricultural Society's Show, held at Leeds in July last. I consider it a very great saving of time, labour, fuel, &c. when compared with hand-washing, as we generally wash once in four weeks, which took two women two days ; on each time now, since using the machine, the whole of the operation is completed in about eight hours ; the clothes are equally well washed ; the wringing is much better done, and not so much injured as in twisting by the hand; the mangling is beautifully done ; indeed, the linen altogether has a beautiful clean appearance when finished by the machine.
"The machine itself is a compact and convenient piece of workmanship, and occupies very little room in the wash-house.
"The short experience that I have had of the machine, I would not now be without one—as you know I did not receive it until October last.
"I am, sirs, yours truly,
"J. MAIN."

"MRS. HOPPER has great pleasure in testifying to the simplicity and efficiency of the washing and wringing machine manufactured by Messrs. Hornsby & Sons. The linen washed by means of it, is done in much less time, with a great saving of labour and fuel, and with comparatively little, if any, injury arising from the process. The appearance of the linen, too, is very much improved, as compared with what can be attained when the washing is done by hand, or by the numerous imperfect washing, wringing, and mangling machines hitherto in use. For heavy articles, such as blankets, counterpanes, &c. this machine is indispensable. From her own experience, no less than from that of her friends, Mrs. H. can testify that no previous machine, whether in respect to compactness, convenience, or general efficiency, has given such unqualified satisfaction as Messrs. Hornsby's.
"*Houghton-le-Spring, Durham,*
"*Nov. 25, 1861.*"

"*Barnsley, Nov. 25, 1861.*
"GENTLEMEN,—Having received your machine a short time ago at my residence, I can testify to its worth. As a washing, wringing, and mangling machine, it is superior to anything I ever saw or used ; and the saving of time, labour, and expense is truly wonderful.
"I am, gentlemen, yours, &c.
"JOHN GREEN, Agent.
"Be kind enough to send a few prospectuses as soon as convenient, and I can then obtain an order for a larger machine—one to wash, wring, and mangle."

"*Terrington, Nov. 25, 1861.*
"DEAR SIRS,—I have great pleasure in recommending the machine supplied by R. Hornsby & Sons, as an efficient washer, wringer, and mangler ; saving a great deal of time and labour, when compared with the original method of washing by the hand.
"Yours truly,
"MATTHEW JOHNSON."

"*Agricultural Implement Works,*
Gloucester, Nov. 26, 1861.
"GENTLEMEN,—I am glad to acquaint you that your new patent washing machine I purchased of you at Leeds Show, works to my entire satisfaction, as I find that one woman can thoroughly wash as many clothes in four hours, with your patent machine, that used to take two women nearly two days to do, and with less than half the soap, and without the least damage to the finest material ;

and I am sure they only require to be known, and you will have a great demand for them. I would not be without one if they were double the price.
"I am, gentlemen, your obedient servant,
"WILLIAM SNOWDEN."

"*Blows Hall, Ripon, Nov. 25, 1861.*
"GENTLEMEN,—The washing, &c. machine, which you supplied, fully answers our expectations. The saving of time, fuel, and labour, is great ; and owing to its simplicity, it may be put into the hands of any one using reasonable care.
"I am, gentlemen, yours truly,
"WILLIAM HARLAND, Jun."

"*To MR. E. HEADLY,*
Corn Exchange Street, Cambridge.
"SIR, As you wish to know how I like the Hornsbys' patent washing machine, which I had of you, I must say I very much approve of it. It makes the linen a beautiful colour, and, by what I have seen, I believe it will not injure the linen so much as washing by hand. I can get it done in half the time, and would not be without it for double its value.
"Yours respectfully,
"E. WINTERS.
"*Chesterton, Cambs. Nov. 2, '61.*"

"*Melton Mowbray, Nov. 23, 1861.*
"GENTLEMEN,—Each and every washing machine I have sold, I am pleased to say, fully answers your representations. I am informed by one of the purchasers he had a good deal of prejudice to overcome with his servants, that, after a trial of some months, experience has proved it to possess great advantages in saving of time and labour. The wringing and mangling having a compound leverage, any amount of pressure can be attained without risk of breakage ; the wood-work is carefully put together, and well seasoned—a very important advantage in such articles. As a whole, it is a valuable article, and well worthy the attention of parties who have any amount of washing to get through.
"I am, gentlemen, yours respectfully,
"WARREN SHARMAN."

"*Albion Foundry, Pitts' Lane, Mill Street,*
"*Kidderminster, Nov. 19, 1861.*
"GENTLEMEN,—The washing machine I had from you is the best I have seen for economy and simplicity ; its saving in labour is quite 50 per cent.; also the same amount in fuel and soap—three important items. Since we have had it, we have done without the assistance of a washerwoman—another important saving, as all our washing for 9 individuals is done by our ordinary servant girl. I am sure all persons not using them sacrifice a great amount every year.
"I am, gentlemen, yours respectfully,
"JOHN TURTON."

"*King's Lynn, Nov. 20, 1861.*
"GENTLEMEN,—I beg to inform you that my family have used the washing machine I had of you these last three months, and like it remarkably well ; I cannot speak too highly of it. I lent it to a friend of mine ; they also will not wash any more without one. Please to send me one, price £8 8s. as soon as possible.
"I think no family would be without one if they did only know the use of such valuable machines—both saving time, fuel, and a great deal less injury to the clothes than hand-washing. In fact, our washing was done at less than half the usual expense, and in one-quarter the time.
"Believe me, yours most respectfully,
"FREDERICK SAVAGE."

HORNSBY, RICHARD, & SONS, *continued.*

"*Swan Lane, Upper Thames Street,*
"*London, Nov.* 20, 1861.
"DEAR SIRS,—Respecting the efficacy of your washing machines, I would state the one we (Carter & Co.) ordered at the Leeds Show, I sent to my own residence, and having had the opportunity of trying a great number of them before the public, I can with confidence state it is decidedly the best, as it both does a larger amount of work with less labour, and most effectually cleans the linen without in any way injuring it.
"I am, dear sirs, yours truly,
"C. CARTER."

"*Brent House, near South Shields,*
"*Nov.* 26, 1861.
"GENTLEMEN,—I beg to say that the washing machine (size R.) which you have sent me has given great satisfaction; it is an exceedingly well constructed and powerful implement, and is found to facilitate the labour of washing very much; the quantity of clothes which previously required a long day to wash, is now done in a few hours by your machine. The mangling apparatus is also very effective.
"I am, gentlemen, yours obediently,
"WM. ANDERSON."

"*Polbathic, Cornwall,*
"*Nov.* 28, 1861.
"GENTLEMEN,—The two washing machines I had from you answer remarkably well; one I have sold, the other (seeing its value) we have kept for ourselves. The machine possesses advantages over several others I have seen. 1st.—It is so remarkably simple, that any person, in two or three minutes, can understand its working. 2d.—It is very easy to work. A boy or girl, of 12 or 15 years of age, could work it all day. 3d.—It washes very clean, without the least injury to the clothes, however fine or delicate. 4th.—The wringing and mangling process is certainly complete. The peculiar, though simple, way in which the levers are brought to bear on the rollers, I should think, will never be surpassed.
"I herewith forward you two testimonials which I have received, and expect to give you an order for two more machines very shortly.
"I am, yours respectfully,
"WM. BRENTON."

"*November* 21st, 1861.
"SIR,—The machine I had from you far exceeds anything I expected; I have given it a fair trial, and am convinced that a woman and a girl will perform as much work as six good washerwomen. I can now do my washing with one-third of soap and fuel I formerly used. I think the machine complete.
"MR. W. BRENTON." "Yours respectfully,
"J. PARKER."

"*Trebole Farm, St. Germans.*
"THE washing machine I had from you on trial last washed remarkably well. It certainly effected a considerable saving of soap, fuel, and labour. The washing was well done, and the wringing and mangling process is excellent. "Yours, &c.
"MR. BRENTON." "S. P. PAIGE."

"*Hanley, Staffordshire, Dec.* 12, 1861.
"DEAR MADAM,—The washing, wringing, and mangling machine (R.) with which you supplied me, gives every satisfaction.
"Its simplicity of construction, superior workmanship, immense saving of time and labour, the ease with which it is worked, and the small space occupied, make it the most convenient and efficient machine I have ever seen, and I can strongly recommend it to those who have not

already availed themselves of the assistance of so valuable a machine.
"Believe me, madam, yours truly,
"To MRS. T. MELLARD, "BOYCE ADAMS.
Agricultural Implement Depôt,
Uttoxeter."

"*Mail Hotel, Grantham, Dec.* 3, 1861.
"GENTLEMEN,—The washing machine I had from you for use at this hotel has now been in operation for some time, and I have great pleasure in bearing my gratified testimony to its thorough efficiency and value.
"Our large wash, which has hitherto had to be got through with great inconvenience by the ordinary mode of hand-rubbing, is now accomplished by the machine, with an immense saving of time, labour, fuel, &c.; and so perfectly satisfied am I of the advantage I derive from it, that I cannot speak too highly of it, and would not be without it on any consideration.
"No hand-rubbing can produce such clean and beautifully-bleached linen as is obtained by the simple action of the machine. It will wash clothes of any description without the slightest injury—but with a great saving to the fabric; it is of first-rate workmanship, very compact and convenient, easy to work, and can be managed by any one, and the wringing and mangling rollers with which it is fitted, are the very perfection of their kind. The tub for blueing or rinsing which you send with it is also very valuable, and in fact my only wonder is that any one can continue to do their washing without the use of a similar machine.
"I have tried many machines, but have found none to equal it, and it will afford me great pleasure to answer any inquiries, or to bear any testimony to its value.
"I am, gentlemen, yours truly,
"MARY TURGOOSE."

"*Croft House, Marsh, near Huddersfield,*
"*Nov.* 30, 1861.
"GENTLEMEN,—I have much pleasure in bearing my testimony to the general usefulness of your washing, wringing, and mangling machine. My servants inform me that by using it the saving of time and labour is *very considerable*, and that it effectually washes the linen without the slightest injury to the lightest fabric. It is the most compact and convenient machine of the kind I know of, and the workmanship is thoroughly good. I can recommend it with every confidence.
"Yours very truly,
"DAVID SYKES."

In concluding this notice of the principal manufactures exhibited by them at the International Exhibition, R. Hornsby & Sons desire to draw the attention of the many home, foreign, and colonial agriculturists and merchants who may inspect their machinery and implements, to the high position taken by them during many years, and to the extraordinary facilities they possess for the supply of every description of first-rate goods, that may be required for any part of the world. They refer with confidence to the successful introduction of their manufactures into France, the Austrian dominions, Hungary, Spain, Russia, Sweden, Australia, New Zealand, South America, Canada, and almost every quarter of the globe, and beg to assure those who may favor them with inquiries or orders they are prepared and determined to spare no efforts to keep their implements in advance of all competition.
Illustrated catalogues, containing prices, drawings, and full particulars, may be had at their stand in the Exhibition, or by application to
RICHARD HORNSBY & SONS,
Spittlegate Iron Works,
Grantham, Lincolnshire,
England.

[2135]

HUNT & PICKERING, *Goulding Works, Leicester.*—Corn crusher mills, root pulpers, oil cake breakers, ploughs, rakes, whippletrees, &c. (*See pages 64 and 65.*)

[2136]

HUNT, T. R. & R., *Earls Colne, Essex.*—Steam-power machine for hulling clover and trefoil seed.

[2137]

HUNTER, PHILIP, 64 *Nicolson Street, Edinburgh.*—Latest improved churns, dairy utensils, ornamented Scotch cooper work, &c.

CHURNS AND DAIRY UTENSILS.

PHILIP HUNTER's registered PRIZE CHURN is unrivalled in simplicity, durability, and cheapness. Having bestowed the greatest portion of his time during the last 25 years to the consideration of the construction of churns, the exhibitor can state with confidence that his churn possesses advantages over any other at present in use. He exhibits the following articles of Scotch carved wood and cooper's work:—

Luggies, quaighs, table whiskey casks, punch-bowls, toddy ladles, trenchers, butter coolers, &c.

[2138]

JAMES, ISAAC, *Tivoli, Cheltenham.*—Liquid manure distributor and pump.

LIQUID MANURE DISTRIBUTOR.

This apparatus has obtained no fewer than 17 prizes at the shows of various agricultural societies. It possesses the very great advantage of incapacity of derangement. It can be applied with equal facility to the distribution of liquid manure, to irrigation, and all similar purposes.

[2139]

KAY, THOMAS, *Holbeck Moor Pottery, near Leeds.*—Horticultural pots, garden pots, fern cases, bordering for garden walks.

[2140]

KEMP, MURRAY, & NICHOLSON, *Stirling, N.B.*—Combined reaper and mower; 2-horse self-cleaning grubber.

KEMP, MURRAY, & NICHOLSON'S COMBINED REAPING AND MOWING MACHINE is unsurpassed for simplicity, efficiency, and durability, and has been awarded a prize at all the places where it has been in competition, including the highest award of the Highland and Agricultural Society of Scotland, for 1861. Catalogues containing full description and numerous testimonials to be had free on application.

Combined reaping and mowing machine	£24	0
Machine for reaping only	23	0
Either of the above with shafts extra	1	0

KEMP, MURRAY, & NICHOLSON'S TENNANT'S SELF-CLEANING 2-HORSE GRUBBER (or cultivator), is the best implement now in use for cleaning and pulverising the soil, and the highest commendations have been awarded it by numerous extensive agriculturists. It can be readily taken to pieces and packed into small compass, thus rendering it peculiarly adapted for export. Price,

With 2 wheels	£5	5
With 1 wheel	5	0

[2141]

KENNAN & SONS, *Dublin.*—Improved iron fences and erecting tools, log saws, lawn mowers, and root blasters. (*See page 66.*)

[2142]

KINGSTON, SAMUEL, *Spalding.*—Rotary cupola beehive on Nutt's principle.

[2143]

LEACH, GEORGE, *Leeds.*—Models of patent steam mole for cultivating land, or pulverising the soil.

HOWARD, JAMES & FREDERICK, *Britannia Iron Works, Bedford.*—Steam ploughs, steam cultivators, ploughs, harrows, horse rakes, haymaking machines.

Thirty-five First Prizes have been awarded to James and Frederick Howard by the Royal Agricultural Society of England; also the Prize Medal at the Great Exhibition of all Nations in 1851, the gold Medal of Honour at the Paris Exhibition in 1855, and the gold Medal of Honour at the Vienna Exhibition in 1857.

HOWARD'S PATENT STEAM PLOUGHING AND CULTIVATING APPARATUS.

1. HOWARD'S PATENT STEAM CULTIVATING APPARATUS, consisting of engine, windlass, wire-rope, cultivator, anchors, pulleys, &c. complete.

2. HOWARD'S PATENT PLOUGH for steam-power.
3. HOWARD'S PATENT CULTIVATOR for steam-power.
4. HOWARD'S PATENT HARROW for steam-power.

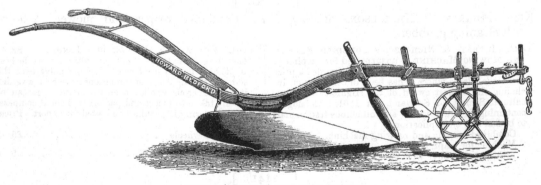

HOWARD'S PATENT CHAMPION PLOUGH.

5. HOWARD'S PATENT 4-HORSE WHEEL PLOUGH.
6. HOWARD'S PATENT 3-HORSE WHEEL PLOUGH.
7. HOWARD'S PATENT 2-HORSE WHEEL PLOUGH.
8. HOWARD'S PATENT 2-HORSE SWING PLOUGH.
9. HOWARD'S PATENT 1-HORSE WHEEL PLOUGH.

10. HOWARD'S PATENT DWARF OR PONY PLOUGH.
11. HOWARD'S PATENT POTATO-DIGGING PLOUGH.
12. HOWARD'S PATENT SUBSOIL PLOUGH.
13. HOWARD'S PATENT RIDGING PLOUGH.
14. HOWARD'S PATENT 4-HORSE HARROWS.

HOWARD, JAMES & FREDERICK, *continued.*

15. HOWARD'S PATENT 3-HORSE HARROWS.
16. HOWARD'S PATENT 3-HORSE JOINTED HARROWS.
17. HOWARD'S PATENT 2-HORSE HARROWS.
18. HOWARD'S PATENT 2-HORSE JOINTED HARROWS.
19. HOWARD'S PATENT 1-HORSE HARROWS.

20. HOWARD'S IMPROVED 4-HORSE TUBULAR IRON WHIPPLETREES
21. HOWARD'S IMPROVED 3-HORSE TUBULAR IRON WHIPPLETREES.
22. HOWARD'S IMPROVED 2-HORSE TUBULAR IRON WHIPPLETREES.

HOWARD'S PATENT HAYMAKING MACHINE.

23. HOWARD'S PATENT 2-HORSE HAYMAKING MACHINE. | 24. HOWARD'S PATENT 1-HORSE HAYMAKING MACHINE.

HOWARD'S PATENT HORSE RAKE.

25. HOWARD'S PATENT HORSE RAKE, large size.
26. HOWARD'S PATENT HORSE RAKE, middle size.
27. HOWARD'S PATENT HORSE RAKE, small size.
28. HOWARD'S DYNAMOMETER, for testing the draught of ploughs, &c.

Full particulars of the above, and other implements manufactured by JAMES & FREDERICK HOWARD, Britannia Iron Works, Bedford, may be had free on application.

HUNT & PICKERING, *Goulding Works, Leicester.*—Corn crusher mills, root pulpers, oil cake breakers, ploughs, rakes, whippletrees, &c.

CORN CRUSHER.

DISC ROOT PULPER.

1. An improved corn crusher or kibbling mill, for crushing beans, peas, oats, barley, Indian corn, wheat, &c. Two solid steel rollers with fluted surfaces are made to pass each other at different velocities, by which the corn is crushed with little power, and to any required size.

They are made in the following order :—
No. 3, for one man, will crush per hour (making 50 revolutions per minute), 2 bushels of beans, 7 pecks of oats. Price £4 7 6
No. 4, for two men, will crush per hour (making 50 revolutions per minute), 5 bushels of beans, 3 bushels of oats. Price £5 5 0
No. 5, for power only, will crush per hour (100 revolutions per minute), 22 bushels of beans, 12 bushels of oats. Price £7 0 0
No. 5 B, as No. 5, but mounted in brass and highly finished. Price . £8 8 0
Size exhibited No. 4.

2. An improved disc root pulper, for reducing roots to a pulp for feeding cattle, possesses the following points—

1st. Strength, with efficiency and simplicity, no gearing or wheels required.
2d. Can be set to produce any size pulp.
3d. The cast-steel knives are removable, to allow of sharpening, &c.
4th. Requires less power than any pulper publicly tested.
5th. Prices moderate.
No. 6 KS, for one man, will pulp 8 cwt. per hour. Price £4 10 0
No. 8 KS, for two men, or power, 12 cwt. per hour. Price £5 5 0
No. 12 KS, for two men, or power, 1½ to 4 tons per hour. Price £6 0 0
Pulleys extra according to size.
Size exhibited No. 8 KS.

3. An improved oil-cake breaker, for breaking oil cake in pieces for feeding beasts, sheep, lambs, calves, &c. Is so arranged that it can be instantly set to any of 9 different gauges, without altering the depth of gearing ; has movable hopper for pieces, and screens the dust in falling.

No. 3, for one man, with wood hopper . £3 5 0
No. 4, ,, ,, iron ,, . . 3 10 0
No. 5, ,, ,, ,, ,, for cotton cake 3 12 6
Size exhibited No. 4.

OIL-CAKE BREAKER.

HUNT & PICKERING, *continued.*

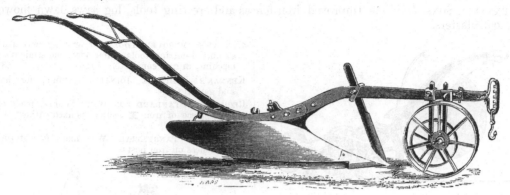

PATENT IRON PLOUGH.

4. A patent iron plough, possessing many important improvements, being the result of a quarter of a century's practical experience in plough manufacture.

1st. Every part is readily accessible and simple.

2d. Proportionate strength is obtained throughout, without any obstructions upon the beam.

3d. A new and novel mode of pitching the share.

4th. Every movement is given to the wheels, by two screws and clips.

5th. The patent oil box wheels exclude grit and retain the oil.

6th. The coulter is *straight* and readily sharpened, every required movement being given by a novel adaptation of the circular wedge.

W. H. A. Light land plough with 2 wheels.

Price	. . . .	£4	5	0
Extra for steel furrow turner		0	6	0
Ditto patent wheels		0	5	0
W. H. B. General purpose plough . .		4	10	0
Extra for steel furrow turner.		0	7	6
Ditto patent wheels		0	5	0

Plough exhibited is W. H. B.

5. A pair of Russell's patent oil box plough wheels, by which the grit is entirely excluded from the axle, and the oil retained; a valuable addition to the plough.

Price, land wheel, 4s. furrow wheel, 6s. 6d. complete with axles.

A section is also exhibited showing the construction.

IRON SACK BARROW.

6. A set of improved link whippletrees, so arranged that the strain is equally divided throughout, enabling them to bear double the resistance of the ordinary whippletree. Price, per set of 3 . . 12s. 6d.

7. An improved couch-grass or twitch rake. A simple arrangement by which great strength is obtained, without much weight; is readily repaired with new teeth, and may be said to be everlasting. The teeth are of solid steel. Price,

Without handles 3s. 0d.
Handled . . 3 6

8. An improved iron sack barrow, in which both axle and wheels revolve independent of each other, enabling them to be turned upon a barn floor without injury.

No. 3, general size,
12s. 6d.
No. 4, large size 13s. 6d.

9. A Leicester garden seat; composed entirely of wrought iron and wood. The ends or supports are made from one piece of iron curved to the required shape, on which are bolted the back and seat which are of wood, thus forming one of the most simple seats ever produced. Each seat is finished equal to coach painting.

Prices from 18s. upwards.

LEICESTER GARDEN SEAT.

KENNAN & SONS, *Dublin.*—Improved iron fences and erecting tools, log saws, lawn mowers, root blasters.

KENNAN'S IMPROVED TOOLS for erecting wire fences, strainer, knotting tools, collar vice, and straightening machine, in case complete. Price £3 10

KENNAN'S PORTABLE JOINTED LADDER, for house and garden use.

REGISTERED STANDARD FOR WIRE FENCES, made of a single piece of iron **I** section, is fixed without wood or stone blocks.

REGISTERED TANGENTIAL WINDERS, for straining wire fences on iron posts.

KENNAN'S LAWN MOWERS, with registered tilt gear.

When the box *E* has been filled with cut grass, it is raised and emptied into the large box at the back of the machine by simply depressing the handle *A*, which moves the segment *B*, gearing into the pinion *C*, and so raises the arm *D*, upon which the front box *E* is hung. The large grass carrier can be quickly over-set or lifted off the machine.

KENNAN'S APPARATUS FOR BLASTING ROOTS AND STUMPS OF TREES. The set includes perforated screw plug bent lever handle suitable screw auger, and a mould for cartridges.

This presents the simplest, most economical and efficient mode of breaking up roots or stumps of trees, so as to facilitate their removal or conversion into fuel. With 5 ounces of blasting powder, a large root may be split in about 8 minutes.

The apparatus complete in case with fuze for 50 roots, price £2 10

LOG-SAWING MACHINE.

KENNAN'S LOG-SAWING MACHINE, worked by one man.

The saw frame is hung on jointed bars, fitted with box springs at *A* and *B*, which materially assist the back stroke, and take all strain off the wrist. The frame is strongly made of seasoned timber, and no part is liable to get out of order. The blocks to be cut are secured by an iron cramp, and short pieces are supported by a movable rest.

Price, complete as above £2 15

[2144]

LEE, CHARLES, 12 *Warwick Crescent, Kensington.*—Water barrow of light draught ; runner for box barrows ; greenhouse ventilator.

[2145]

LESLIE, BRADFORD, 2 *Abercorn Place, St. John's Wood.*—Model of a pump for irrigation in India, worked by wind.

[2146]

LIPSCOMBE, FREDERICK, & Co., 233 *Strand, near Temple Bar.*—Improved fountain jets for aquariums, conservatories, gardens.

[2147]

LOVEY, EDWARD, *Ponsnooth, Perran Wharf, Cornwall.*—Beehives.

[2148]

MAGGS & HINDLEY, *Bourton, Dorset.*—Agricultural machinery.

[2149]

MAPPLEBACK & LOWE, *Birmingham.*—Draining tools, and agricultural implements.

[2150]

MARRIOTT, JOSEPH, *Gracechurch Street.*—Apiary, working bees, unicomb observatory pivot hives, humane and glass beehives.

[2151]

MESSENGER, THOMAS GOODE, *Loughborough.*—Patent triangular tubular boiler, hinged valve and indicator.

The advantage of this patented boiler is an immense surface exposed to the direct action of the fire by entirely surrounding it, the greater part being placed *over* the fire; it is exceedingly economical in fuel, and very quick in action, and the whole surface can be cleaned at any time.

The superiority of this patented hinged valve for steam, water, or gas, consists in its great simplicity, impossibility of becoming fast, its thorough efficiency, great durability, and unparalleled cheapness.

References and prices for these and general horticultural building will be supplied upon application.

[2152]

MILFORD T., & SON, *Wheel Works, Thorverton, Cullompton, Devon.*—Carts and waggons for agricultural purposes.

The IMPROVED 1-HORSE CART has obtained 11 first prizes and 2 silver medals from the Royal and the Bath and West of England Societies in 1853, '4, '5, '6, '7, '8, and 1861. Price £14 0
Harvest shelvings £1 extra.

The IMPROVED 2-HORSE WAGGON has obtained 9 first prizes and a medal from the Royal and the Bath and West of England Agricultural Societies in 1852, '3, '4, '5, '7, '8, and 1861. Price, complete . . . £28 0

[2153]

MOODY, CHARLES PETERS, *Holway, Sherborne, Dorset.*—Patent field gate formed of machine-made duplicate parts.

[2154]

MORTON, H. J., & CO., *Basinghall Buildings, Leeds.*—Manufacturers of corrugated iron roofs and buildings, and patent cable-strained fences.

[2155]

MUNN, MAJOR, *Throwley, Kent.*—Model of a beehive.

[2156]

MUSGRAVE, BROTHERS, *Ann Street, Belfast.*—Iron stalls for cattle.

[2157]

NEIGHBOUR, GEORGE, & SONS, 127 *Holborn, W.C. and* 149 *Regent Street, W.*—Beehives, bees at work.

LIGURIAN BEES AT WORK IN GLASS HIVES.

NEIGHBOUR'S IMPROVED BEE HIVES for taking honey without the destruction of the bees.

These bee hives may be viewed in full operation at the International Exhibition of 1862, Class 9.

Drawings and detailed lists will be forwarded on receipt of 2 postage stamps.

Applications for stocks of Ligurian bees, &c. may be made to Geo. Neighbour & Sons, 127 High Holborn or 149 Regent Street, London.

[2158]

NICHOLSON, WILLIAM NEWZAM, *Trent Iron Works, Newark.*—Hay machines, horse rakes, oil-cake crushers, sack lifters, garden rollers, &c.

Obtained two Prize Medals at the Exhibition of 1851.

W. N. Nicholson's factory is situated on the river Trent, by which there is communication to the iron and coal districts of Staffordshire, Derbyshire, and Yorkshire, and with the great shipping ports of Hull, Gainsborough, &c. The Great Northern and Midland Railways intersect at Newark, the latter line adjoining his Works; and by both these important lines there is direct access to the iron and coal districts, as well as to the ports of London, Liverpool, Hull, Bristol, &c. so that goods are forwarded to every part of England and Scotland on all narrow gauge lines without unloading.

AGRICULTURAL MACHINES EXHIBITED :—

1. NICHOLSON'S PATENT HAY-MAKING MACHINE, with patent tubular iron shafts, and wire screen for protecting the horse from the hay in windy weather, and other recent improvements.

To this machine has been awarded the first prize of £4 by the Royal Agricultural Society at the great quadrennial trial of this class of machines at Salisbury, in 1857, and another prize of £4 at the Leeds trial in 1861, besides numerous other first prizes in England, France, Ireland, &c.

Several varieties are made, having the double action and reverse motion, as the one exhibited. The ordinary 1-horse machine, price £15 ; a stronger machine, suitable for water-meadows and uneven lands, at £16. A 2-horse machine, easy to work in the heaviest crops with a pair of light horses, having the forks divided into 4 sets; and a similar 1-horse machine.

W. N. Nicholson has also just introduced a new cheap 1-horse hay machine without the reverse motion, price only £10.

All these are usually made with the patent tubular iron horse-shafts, but can be had with wood shafts, or for the Continent or colonies, with pole for pair of horses or oxen.

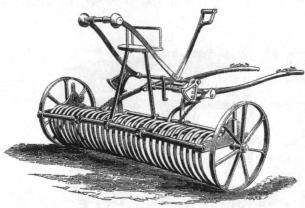

a most simple and effective manner, and it has met with the approval of the first mechanicians. The frames are made of iron in one piece, ensuring remarkable strength and firmness. They are made either with one or two pairs of crushing rollers, according to the purpose for which they are required. Prices from £3 10*s.* to £10 10*s.*

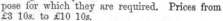

2. NICHOLSON'S PATENT HORSE RAKE, with seat, having patent tubular iron shafts and frame, and wrought-iron wheels. Several varieties of this labour-saving implement are made both with and without the seat, at prices varying from £7 10*s.* to £12 0

The absence of wood in their construction is of great importance, as it ensures strength and durability with lightness, as the tubular iron shafts are not liable to rot, decay, or breakage as those made of wood. The carrying wheels being of wrought-iron, cannot be broken as the ordinary cast-iron wheels. The addition of the seat is an important improvement, saving nearly one-half the labour in using, whilst the general construction and performance of the rake is not excelled by any.

3. NICHOLSON'S PATENT OIL-CAKE BREAKER. Six first prizes have been received from the Royal Agricultural Society by W. N. Nicholson for this class of machine.

The method of driving the breaking rollers so as to obtain great variation in the sizes broken, is contrived in

4. NICHOLSON'S REGISTERED SACK-RAISING BARROW by a simple arrangement allows a man to raise a filled sack the proper height for taking on the back, thereby enabling 1 man to do the work for which 3 are usually required. Two sizes are made at £2 10*s.* and £3 10*s.*, the larger one being also available for merchandise in bales, or other packages.

5. NICHOLSON'S IMPROVED WINNOWING MACHINES are made in 4 varieties, for large and small occupations. They are unequalled in simplicity and efficiency, and for the excellent sample they are capable of producing. A corn-elevating apparatus can be had with them, by which the clean corn is lifted into the sack without extra labour.

6. NICHOLSON'S PATENT DOUBLE-CYLINDER GARDEN ROLLER is the most perfect and effective instrument of its class, and being finished by machinery, is less liable to disarrangement than any. It is used with less labour, and can be turned round without injuring either grass or gravel.

Full particulars and catalogues of W. N. Nicholson's various inventions and manufactures, including stoves, &c. shown in Class 31, may be had free on application.

[2159]

NIXEY, W. G., 12 *Soho Square.*—Patent garden labels, patent money tills, specimens of refined black-lead.

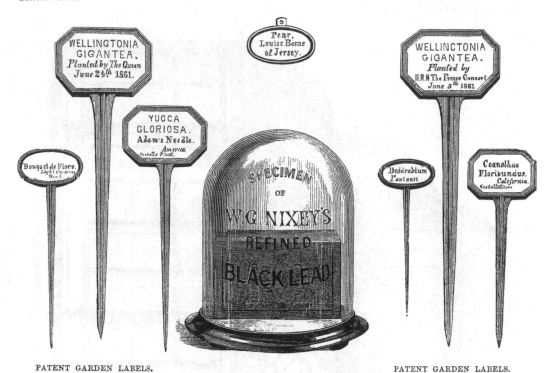

PATENT GARDEN LABELS. PATENT GARDEN LABELS.

W. G. NIXEY'S PATENT GARDEN LABELS, composed of iron and glass hermetically sealed, are imperishable and indestructible by time or weather. Patronised by Her Majesty the Queen.

PATENT MONEY TILL.

W. G. NIXEY'S PATENT MONEY TILLS for the prevention of fraud and error, causing mutual satisfaction between employer and employed.

W. G. NIXEY'S CHEMICAL PREPARATION OF BLACK LEAD for polishing stoves and ornamental iron-work without waste or dust.

[2160]

ORMSON, HENRY, *Stanley Bridge, King's Road, Chelsea, London, S.W.*—Conservatory, hot-water tubular boilers, &c.

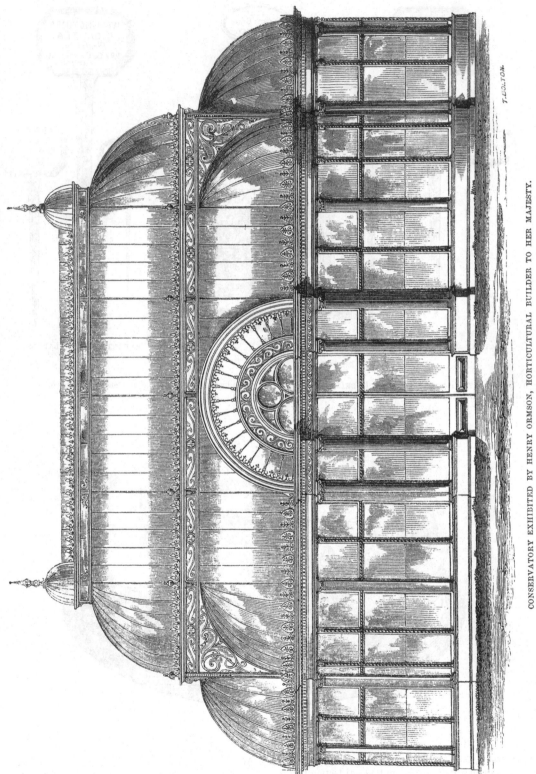

CONSERVATORY EXHIBITED BY HENRY ORMSON, HORTICULTURAL BUILDER TO HER MAJESTY.

ORMSON & SON, *continued.*

ORMSON'S PATENT JOINTLESS TUBULAR BOILER, an original invention, is now fully acknowledged to surpass the old-fashioned jointed tubular boilers, which have been made for the last 20 or 30 years by all other manufacturers. The advantages of this patent will be manifest to every person's understanding from the following facts. For instance, one of the old-fashioned jointed boilers with 50 tubes would have 100 joints made of rope-yarn and cements exposed to the direct action of the fire; whereas in this patent there is not one joint so exposed. It should be fully understood that as the old-fashioned jointed tubular boiler

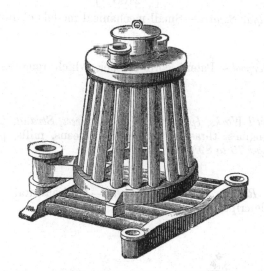

PATENT JOINTLESS TUBULAR BOILER.

increases in size, and in number of tubes and joints, its liability to leakage also increases, and hence the reason why Ormson's one-boiler system and patent jointless tubular boilers have become so universally adopted on account of their superior power, great safety, and economy:—

Ormson's No. 1 boilers are heating upwards of
250 ft. of pipe.

Ormson's No. 2 boilers   ,,   ,,   600   ,

Ormson's No. 3 boilers   ,,   ,,   1300   ,,

Ormson's No. 4 boilers are heating upwards of
3000 ft. of pipe.

Ormson's No. 5 boilers   ,,   ,,   5000   ,,

These boilers can be made in larger sizes if required, to heat 12,000 or 15,000 ft. of pipe.

Henry Ormson, horticultural builder to Her Majesty, and hot-water apparatus manufacturer to the Commissioners of Her Majesty's royal palaces and public buildings, and to the Royal Horticultural Society, Stanley Bridge, King's Road, Chelsea, London, S.W.

[2161]

PAGE, E., & Co., *Victoria Iron Works, Bedford.*—Ploughs, horse hoes, horse rakes, chaff cutters, harrows, &c.

2162]

PETTITT, WYATT JOHN, *The Apiary, Dover.*—Bee-hives; Major Munn's bar-frame hive.

[2163]

PHILLIPS, GEORGE, *Harrow-on-the-Hill, Middlesex.*—Improved collateral beehives, composed of wood, glass, and zinc.

[2164]

PICKSLEY, SIMS, & CO., *Leigh, near Manchester.*—Agricultural implements. (*See page* 73.)

[2165]

PRIEST & WOOLNOUGH, *Kingston-on-Thames.*—Horse hoes, turnip, manure, and corn drills. (*See page* 74.)

[2166]

PRINCE & CO., 4 *Trafalgar Square.*— Small mechanical models of inventions.

[2167]

RANKIN, R. & J., *Liverpool.*—Patent corn cleaner, which removes smut and all impurities from the grain.

[2168]

RANSOMES & SIMS, *Orwell Works, Ipswich ;* 31 *Essex Street, Strand, London;* 23 *Water Street, Liverpool.*—Steam engines, thrashing machines, screens, mills, ploughs, and agricultural machinery. (*See pages* 75 *to* 89.)

[2169]

READ, RICHARD, 35 *Regent Circus, Piccadilly.*—Horticultural engines, machines, and syringes, of every description.

HORTICULTURAL ENGINES, MACHINES, AND SYRINGES, manufactured only by Richard Read, instrument maker (by appointment) to Her Majesty.

[2170]

REEVES, ROBERT & JOHN, *Bratton, Westbury, Wilts.*—Liquid manure drills, manure distributors, patent corn manure, and turnip drills, &c. (*See page* 90.)

PICKSLEY, SIMS, & CO., *Leigh, near Manchester.*—Agricultural implements.

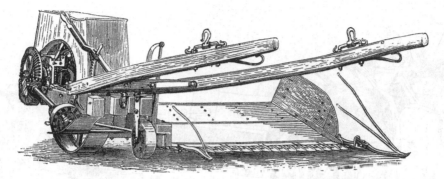

REAPING MACHINE FOR CUTTING GRAIN AND ARTIFICIAL GRASSES.   (Price £25.)

This machine obtained the

First prize at the Royal Agricultural Society's Show at Leeds, 1861.

First prize at the East Lothian Agricultural Society's Show at Haddington, 1861.

First prize at the North Lonsdale Agricultural Society's show at Ulverstone, 1861.

First prize at the Leyland Agricultural Society's Show at Leyland, 1861.

The first prize was also awarded to P. S. & Co. at Leyland, for their combined reaping and mowing machine (Bamlett's patent), price £35, in competition with Woods'.

Since taking the above prizes, however, P. S. & Co. have incorporated several important improvements suggested by the experience of last season, and they can now warrant the machine as the champion reaping machine for manual delivery.

Priced catalogues may be obtained post-free on application at the Works, as above.

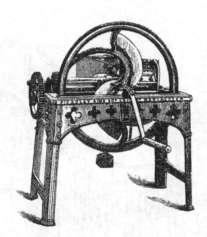

CHAFF CUTTER.   (No. 1A, price £4.)

P. S. & Co. have in 1861 with their celebrated machines taken upwards of 50 first-class prizes at the Royal and other principal agricultural shows in England, France, and Australia.

Chaff cutters . . . . . from £2 5 to £25 0

Oat and bean mills . . . from 3 10 to 15 0

Grinding mills . . . . from £8 10 to £15 0

Turnip cutters . . . . from 3 0 to   6 10

Turnip pulpers . . . . from 3 10 to   8 10

Lawn-mowing machines . from 5 0 to   7 10

P. S. & Co. are the sole and exclusive makers of Bamlett's patent reaping machine.

PRIEST & WOOLNOUGH, *Kingston-on-Thames.*—Horse hoes, turnip, manure, and corn drills.

PRIEST & WOOLNOUGH'S IMPROVED CORN DRILL, with ore carriage steerage attached, is adapted to all the requirements of a farm for depositing wheat, barley, beans, peas, turnips, mangold, Indian corn, or maize, clover, and any other grain or seed.

PRIEST & WOOLNOUGH'S IMPROVED DRILL for turnips and manure.—*Obtained the Royal Agricultural Society of England's prize at Leeds,* 1861.

This drill is for the purpose of depositing turnips, or mangold wurtzel with guano, superphosphate, or other highly concentrated manures.

PRIEST AND WOOLNOUGH'S PATENT FIRST PRIZE HORSE-HOE.

Obtained the Royal Agricultural Society's prizes at Salisbury, 1857, *and at Leeds,* 1861; *a special medal at Vienna,* 1857, *and gold medal at Paris,* 1860.

PRIEST & WOOLNOUGH'S PATENT LEVER HORSE-HOE for light lands and small occupations.

This implement is adapted for hoeing between the rows of drilled crops of every description, either on the level surface or on ridges. It is made a corresponding width

with the drill it is to follow, and will hoe at once as many rows as were drilled.

PRIEST & WOOLNOUGH'S DRILL for light land or small occupations.

This drill is the same in principle as the corn drill before described, but made altogether lighter.

RANSOMES & SIMS, *Orwell Works, Ipswich ; 31 Essex Street, Strand, London ; 23 Water Street, Liverpool.*—Steam engines, thrashing machines, screens, mills, ploughs and agricultural machinery.

SECTION I.—PLOUGHS.

PATENT TRUSSED - BEAM IRON PLOUGH, marked Y W B, made principally of wrought-iron, and intended for ordinary ploughing with two or more horses. The annexed cut represents this plough in the form in which it is ordinarily used for prize ploughing.

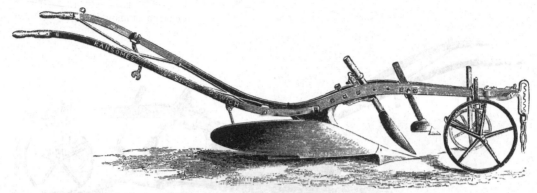

RANSOMES AND SIMS' PATENT TRUSSED-BEAM IRON PLOUGH Y W B.

This plough forms one of a series of four ploughs, three of which obtained prizes at the last ploughing match of the Royal Agricultural Society of England, at Warwick, in 1859. This series of ploughs is modelled after our well-known ploughs, the Y L and Y R C, which have received the following prizes from the Royal English Agricultural Society, &c. &c. :—

The prize of £10 and silver medal, as the best heavy land plough ; also to the same plough, a prize of £10 and silver medal as the best light land plough, at the Royal Agricultural Society's meeting at Southampton.

A prize of £10 at the Royal Agricultural Society's meeting at Northampton.

The council medal of the Great Exhibition with this plough as made by Busby.

The first prize at the meetings of the Royal Agricultural Society at Lewes, 1852, at Lincoln, 1854; and again at the Carlisle meeting, 1855; as the best plough for general purposes.

The divisional prize at the Bath and West of England meeting at Tiverton, 1855.

The prize for deep ploughing at the Royal Agricultural Improvement Society of Ireland, Carlow meeting, 1855.

This is the medium-sized plough of that series, two smaller sizes, and one larger being made.

The *handles* are of sufficient length to give perfect command over the ploughs.

The *beam* is on their patent trussed principle, by which greater rigidity and strength are secured, with the same weight of metal, than can be obtained by an ordinary solid beam. This construction of the beam also permits the coulter to be placed quite centrally, so that it does not require to be necked, and is therefore more easily kept in its proper position than when it is necked, which it must always be in solid-beam ploughs.

The *wheels* are carried on one cross bar, the advantage of which is, that they can be more firmly fixed in any desired position, and can be more quickly shifted, than when they are carried on two separate bars, also that the whole of the wheel fastenings are rendered extremely simple, without omitting any adjustment, that can possibly be required for either the land or furrow wheels.

The *draught* is taken directly from the head, for the result of very careful experiment has convinced R. and S. that in a properly constructed plough, the draught bar is quite needless and often very injurious.

The *share* is fixed to a wrought-iron movable lever neck, which allows it to be set with more or less pitch, as may be required, and the arrangements for fixing the neck in the desired position are very simple and effective. The form of the share and of the mould board is the result of a series of most careful experiments on a variety of soils. They will be found to leave the furrow slice neatly turned over at an angle of 45°, leaving the upper edge full and sharp. The ploughs are very steady in work, and leave the furrow bottom clean and square.

The *skim coulter.*—This is a very useful addition to a plough, particularly when ploughing clover, ley, or stubble. It just pares the surface, and turns into the bottom of the furrow all long grass, weeds, stubble, &c. which are then completely buried and enrich the soil by their decomposition. A weight attached by a chain to the coulter, is often used when the grass is long to assist in drawing everything cut by the skim coulter into the furrow, so that it may be thoroughly covered. This chain is also useful when ploughing in manure.

The plough as shown in the above cut is fitted to cut a rectangular furrow slice, and to deposit it so that the angle A B C is a right angle, and the side A B equal to the side B C, as shown in the accompanying diagram. This is the ordinary form of ploughing, which is considered in England most advantageous for producing a crop ;

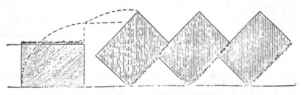

RECTANGULAR FURROW SLICE.

RANSOMES & SIMS, *continued.*

but if desired the same plough may be fitted with a mould board and share suitable to cut a furrow of a trapezoidal section, and deposit it so that the side A B is equal to B C, and the angle A B C is somewhat less than a right angle, as represented in the accompanying diagram.

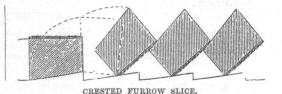

CRESTED FURROW SLICE.

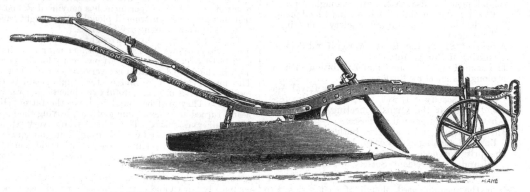

RANSOMES AND SIMS' PLOUGH Y W B FOR PRODUCING CRESTED FURROWS.

This form of furrow is usually termed a "crested" furrow. It possesses the advantage of exposing rather more surface to the atmosphere than the rectangular furrow, but to set against this, there is one-ninth less soil moved when ploughing with the same depth, than on the previous plan, in consequence of the furrow bottom being inclined to the land side, instead of at right angles to it, and the horses must travel two miles further per acre to plough the same depth of furrow than is necessary on the rectangular system.

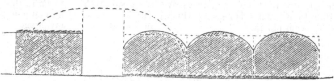

RANSOMES AND SIMS' PLOUGH Y W B FITTED AS A KENT PLOUGH.

This plough may also be fitted with a mould board and share, as shown above, which cuts a furrow slice of a rectangular section, and turns it completely bottom upwards, as shown in the annexed diagram, thus exposing the lower soil to the fertilising action of the atmosphere, and burying all the surface vegetation so that it decomposes and enriches the soil.

KENTISH FURROW SLICE.

Ransomes & Sims have mould boards suitable for producing each of the above-described forms of furrow, on either heavy or light land, and being interchangeable one with another at pleasure this plough becomes a very complete implement.

By removing the ordinary body from the plough and

RANSOMES & SIMS, *continued.*

also the furrow wheel, and substituting a ridging body (which may be done in a few minutes), this plough becomes a convenient moulding or ridging plough, as shown below, which is useful for setting out land, for ordinary ploughing, for ridging or moulding up beet-root, potatoes, or other plants sown on the ridge, and for opening water furrows.

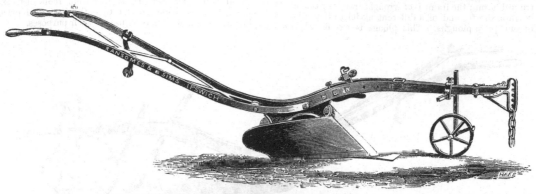

RANSOMES AND SIMS' PLOUGH Y W B, WITH RIDGING BODY.

By removing the mould boards from this body the plough is adapted (as shown in the engraving), for breaking up the subsoil after the plough, the land being thus stirred from 12 to 14 in. deep.

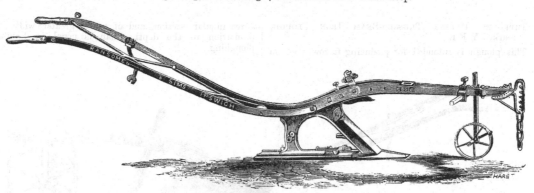

RANSOMES AND SIMS' PLOUGH Y W B, WITH A SUBSOIL BODY.

By attaching to the frame a pair of open-ribbed mould boards (as shown below), the plough is adapted for raising potatoes, which operation it performs in a superior manner, leaving fewer in the ground than when raised with a fork, and not damaging the potatoes.

Although the practice of using the same implement for

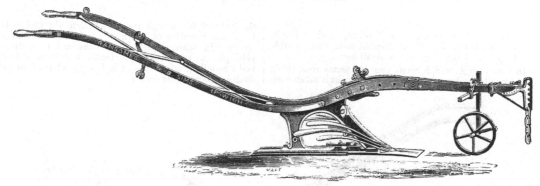

RANSOMES AND SIMS' PLOUGH Y W B, WITH A POTATO BODY.

various dissimilar purposes, where the land under cultivation is of sufficiently large extent to justify the use of a special implement for each purpose, cannot be recommended, yet inasmuch as this plough in each of its forms is perfectly complete, and will perform each operation thoroughly, there are occasions when it will be convenient and desirable to employ it as an interchangeable implement in the manner described above.

PATENT TRUSSED-BEAM IRON PLOUGH, marked Y X.

This is the smallest of the Warwick series of prize ploughs (described above), and is suitable for two horses on light and mixed soils.

RANSOMES & SIMS, *continued.*

PATENT TRUSSED-BEAM IRON PLOUGH, marked Y F L.

This plough is fitted with Ransomes & Sims patent trussed beam; the frame is of wrought-iron, and the mould board is shorter and of a different model to that adopted in our prize ploughs. This plough is extensively used abroad in cases where the land is newly cleared, as in consequence of the great strength of the beam and frame, it is well suited for resisting strains from roots of trees, stones, &c., and being shorter than the improved prize ploughs, it is more handy in use for such purposes.

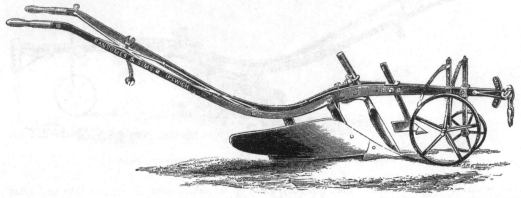

RANSOMES AND SIMS' PLOUGH Y F L.

IMPROVED PATENT TRUSSED-BEAM IRON PLOUGH, marked Y F R.

This plough is intended for producing furrow s ces of a rectangular section, and of much greater width in proportion to the depth, than is usual in English ploughing.

RANSOMES AND SIMS' PLOUGH Y F R.

IMPROVED SOLID-BEAM IRON PLOUGH, marked B F S, constructed principally of wrought-iron, and suitable for use with one large or two small horses.

In this plough, which forms one of a series comprising one smaller and two larger sizes, the draught is taken from the body of the plough instead of from the end of the beam, the intention of which is to relieve the beam from strain. In other respects, it is constructed on similar principles to the ploughs previously described, and will make equally good work, but it is only adapted for producing a rectangular furrow.

RANSOMES AND SIMS' PLOUGH B F S.

RANSOMES & SIMS, *continued.*

This series of ploughs is also provided with ridging and subsoil bodies as previously described with the Y W B, and two sizes of it are also made as double furrow ploughs for turning two furrows at the same time, and thus saving one man, as shown in the subjoined diagram.

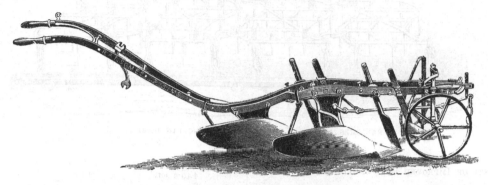

RANSOMES AND SIMS' DOUBLE-FURROW PLOUGH.

IMPROVED SOLID-BEAM IRON PLOUGH, marked T C.

This plough is constructed entirely of wrought-iron with the exception of the mould board, and it will turn a furrow from 6 to 10 inches deep, and from 9 to 15 inches wide. It is strong enough to resist the strain of 10 horses or 12 or 14 bullocks; but it may be easily worked by 2 or 3 horses. The short beam and great length of handle give the holder a great command over it, and it is extremely suitable for breaking up new and rough land.

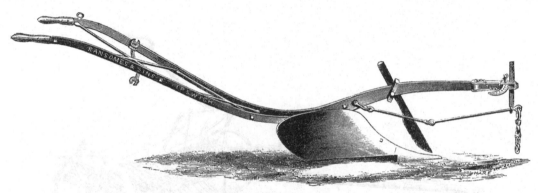

RANSOMES AND SIMS' PLOUGH T C.

IMPROVED PLOUGH, with wood beam and handles, marked W V R L.

In this plough, the beam and handles are of well-seasoned timber. The draft is taken from the body of the plough, which is designed on the same model as those of the exhibitors' best iron ploughs, and will produce equally good work; but being partially of wood, it is not so expensive in the first cost.

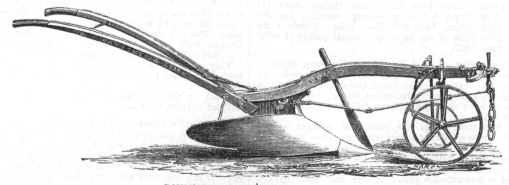

RANSOMES AND SIM'S PLOUGH W V R L.

RANSOMES & SIMS, *continued.*

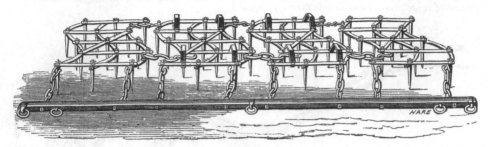

RANSOMES AND SIMS' IMPROVED EAST ANGLIAN HARROW.

SET OF IMPROVED EAST ANGLIAN HARROWS.

The beams are diagonally braced; the teeth tracks equally distant. The teeth will not shake loose in work; the harrows will not run over each other in rough work, and are fitted with hind hooks to draw the contrary way, so as to give a lighter finish in seed harrowing. They are made in the following sizes, and all with 5 rows of teeth :

	4-Beam Harrows.
No. 1. Light harrows.	Cover 9 feet 3 inches.
4 to a set.	
3 ,,	7 ,, 0 ,,
2 ,,	4 ,, 8 ,,
No. 2. Medium Harrows.	3-Beam Harrows.
4 to a set.	10 feet 0 inches.
3 ,,	7 ,, 6 ,,
2 ,,	5 ,, 0 ,,
No. 3. Heavy Harrows.	3-Beam Harrows.
4 to a set.	10 feet 0 inches.
3 ,,	7 ,, 6 ,,
2 ,,	5 ,, 0 ,,

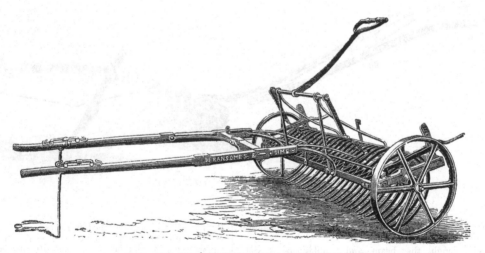

RANSOMES AND SIMS' IMPROVED IRON HORSE DRAG-RAKE.

IMPROVED IRON HORSE DRAG-RAKE. Highly commended at the Royal Agricultural Society's meeting, Lincoln, 1854. Prize, Paris Exhibition, 1856. Obtained the First Prize of the Royal Agricultural Society at Salisbury, 1857.

Horse rakes are used for collecting hay, corn, stubbles, twitchgrass, &c. for raking in clover and grass seeds, and as weed extirpators on young cereal crops, for which purposes they are of the highest utility, performing the work more cheaply and thoroughly than can be done in any other way.

Each tooth swings independently of the other, so that the whole set readily adjusts itself to uneven land. The frame is also furnished with side levers, so that the rake can be used with one wheel in the furrow if necessary. There is an arc on the shafts, so that the teeth may be set to penetrate the ground, or skim the surface lightly, by which they are prevented from collecting the soil or rubbish with the corn, or from pulling up the young clover when raking barley. Each alternate tooth can be raised out of work when desired, so as to form a coarse rake. This is useful when raking twitch, or other weeds, brought to the surface by harrowing. The wheels are of iron, and capped to prevent the admission of dirt.

They are made in three sizes, and with steel or iron teeth. The steel teeth are much the lightest and most durable.

A lad can clear the rake of its load instantaneously, without stopping the horse.

PATENT SELF-BALANCING HORSE DRAG-RAKE.

In this rake the driver rides on a seat so arranged that the weight of his body partly counterbalances the teeth. The rake is cleared by the driver depressing the footboard, and as the driver rides, much more land can be got over in a day, than with those rakes in which he has to walk behind.

RANSOMES & SIMS, *continued.*

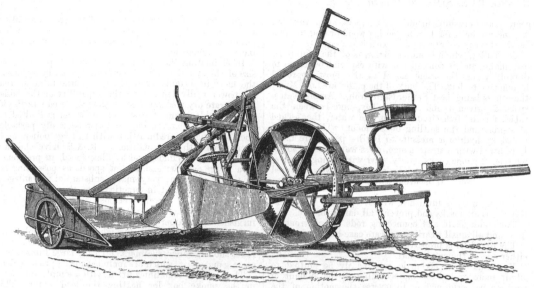

RANSOMES AND SIMS' PATENT SELF-RAKING VICTORIAN REAPER.

MODEL OF RANSOMES' PATENT SELF-RAKING VICTORIAN REAPER.

This machine is suitable for cutting any description of cereal crop, and delivering it at the side of the machine in neatly-formed sheaves. The automatic delivery is extremely simple, and capable of variation according to the weight of the crop.

It consists of a series of rakes and arms which revolve round a vertical shaft under the guidance of an irregular waved ring or cam, in such a manner that they bring the grain forward to the knives, and when cut deliver it at the side of the machine. This machine has been very severely and successfully tested by practical farmers under very varying conditions.

RANSOMES AND SIMS' EIGHT-HORSE POWER PORTABLE HIGH-PRESSURE STEAM ENGINE.

8-HORSE POWER PORTABLE HIGH-PRESSURE STEAM ENGINE.

This engine is one of Ransomes & Sims' standard series of portable steam engines which are made from 3 to 20 horse power, and with single or double cylinders.

Portable steam engines are extremely simple, durable, and easy to manage ; and are capable of application to all the requirements of a farm ; such as driving thrashing machines and all machinery for the preparation of food for stock, steam ploughing and tilling ; and also for working circular, horizontal, or vertical saws for cutting timber— for driving pumps for irrigation—millstones and mill

RANSOMES & SIMS, *continued.*

gear, quartz-crushing machines, stampers, amalgamators, &c. and are built for burning either wood or coal, a great desideratum in countries where coal is scarce.

The boiler, which is multitubular, is of the exhibitors' own make, and is constructed with especial reference to durability, on the same model as the most approved locomotive boilers. The bulk of the plates are Low Moor, the others being best Staffordshire iron. Ample water-space is given round the fire box, and between the tubes, for the free circulation of the water, the escape of steam, and the settling of sediment. The boiler is tested by hydraulic pressure to 100lbs. per square inch. It is fitted with a steam gauge, glass water-gauge, steam whistle, 2 gauge cocks, safety valve with spring balance, blow-off cock, &c. &c. and is lagged with wood covered with sheet-iron. It is fitted with a lock-up safety valve when so ordered.

The chimney is furnished with a wire top which extinguishes all sparks and prevents all danger of fire.

The crank shaft and connecting rods are of wrought-iron, and all small wearing parts are case-hardened.

The fly wheel is properly balanced, and can be hung on either end of the crank shaft.

The slide valve eccentric can readily be shifted to admit more or less steam, according to the amount of work to be performed, or to reverse the motion of the engine, if necessary.

The power is calculated at 45lbs. pressure of steam in the boiler. Every engine is tested under steam before leaving the factory, and may be safely worked at 60lbs. pressure, at which they give off double their nominal power, consuming, of course, fuel and water in the same increased proportion.

In estimating the power an engine will produce, the size of the cylinder is only one element, and by no means the most important, for it must be borne in mind, that the power really depends on the capability of the boiler to generate dry steam as fast as the engine can utilise it. In a portable engine the size of the boiler is limited by the condition that the engine must be easily portable, and these engines are furnished with as large boilers as is compatible with that condition. R. & S. have chosen a moderate sized cylinder and a quick speed, in preference to a larger cylinder and a slow speed, as possessing for this class of engine very many substantial advantages, and it will be found in practice that these engines will give off as much power, and cost as little to keep in repair, as any others of equal weight and portability, but furnished with larger cylinders.

These engines are all furnished with the following articles, viz. waterproof cover, tube brush, fire pricker, rake, shovel, screw spanners, oil can, large funnel, and spare gauge glass, which are included in the price quoted.

They are also sometimes fitted with a simple apparatus in the smoke box for heating the feed water. This economises the fuel considerably, and is not liable to get out of order.

RANSOMES AND SIMS' TEN-HORSE PORTABLE ENGINE WITH BIDDELL AND BALK'S PATENT BOILER.

A 10-HORSE POWER PORTABLE STEAM ENGINE, with Biddell & Balk's patent boiler. Biddell & Balk's patent boiler obtained the only prize which the Royal Agricultural Society offered for the best steam boiler in the year 1858.

These patent engines are made in various sizes from 5 to 14 horse power. They are suitable for every purpose to which a portable engine is usually applied, especially for steam cultivation and use in foreign countries where repairs are difficult. The patent boiler offers the greatest facilities for keeping the inside perfectly clean and free from mud, and thereby avoiding waste of fuel and risk of burning the plates. The boiler, as shown in the subjoined woodcut, is so constructed that the fire box, tubes, and tube plates, can be taken out all in one piece and put in again with facility. This is effected by using screws and bolts instead of rivets for the connexion of the above-mentioned parts with the shell of the boiler, the surfaces making the steam-tight joints being faced.

The great advantages of this boiler are, facility of cleaning, inspection, and repairs, so that when circumstances compel the use of bad water, the evil consequences of the same may be avoided by a frequent cleaning.

The exhibitors remark upon these boilers in comparison with the usual form of locomotive boilers,

"That the principal deposit takes place in parts of the boiler where it does the least amount of injury, and not in the parts immediately exposed to the action of the fire.

"That accurate experiments have shown a favourable result in point of economy of fuel, as compared with the ordinary boiler.

"That they are better adapted for burning wood, and

"That being higher above the ground they will travel better over rough roads."

RANSOMES & SIMS, *continued.*

RANSOMES AND SIMS' FIFTEEN-HORSE POWER HORIZONTAL STATIONARY HIGH-PRESSURE STEAM ENGINE.

A 15-HORSE POWER HORIZONTAL STATIONARY HIGH-PRESSURE STEAM ENGINE.

The above engraving represents one of Ransomes and Sims' standard series of high-pressure stationary engines, which are made in various sizes from 4 to 20 horse power, and which have been awarded the following important prizes, viz. :—

A prize of £10 awarded by the Royal Agricultural Society of England at the Lewes Meeting, 1852 ; the first prize of £20, by the same Society, at the Lincoln Meeting of 1854 ; and was again awarded the first prize of £20 at the Carlisle Meeting, 1855 (for the 8-horse power engine).
Prize, Paris Exhibition, 1856.

These engines are made of the best materials, and first-class workmanship. They are exceedingly simple in construction and compact in form. All the parts are easy of access, and afford every facility for adjustment or repairs. They are principally supported on a very strong cast-iron frame, and may be either erected on a stone or brick foundation, or be carried on two wood sills.

The crank shaft and connecting rod are of the best wrought-iron ; the slide valve is on the best principle ; the feed pump is very simple, and not liable to be put out of order ; there is a governor of the best construction for controlling the speed of the engine, and an improved regulating valve. The boiler is on the Cornish principle, and is perfectly safe and easy to manage ; the fire is placed in an internal circular flue, and the flame passes through and along each side of the boiler to the chimney. By this arrangement, all sediment contained in the water can collect underneath the fire-flue. The boiler is fitted with a good safety valve, glass water-gauge, and everything necessary for its safe and efficient working, and is very economical in the consumption of fuel. They may be fitted with lock-up safety valve if desired.

These engines are peculiarly adapted for driving fixed thrashing machines and barn machinery, or for sawing, pumping, driving corn mills, or any other purpose for which steam power is required.

Engines of this construction are well adapted for grinding corn ; they can readily be attached to ordinary millstones in wind or water mills, and are, therefore, well worthy the attention of millers who may wish to ensure the means of grinding, at all times, with economy and regularity.

When economy of fuel is important these engines are furnished with an apparatus in the foundation plate for heating the feed-water, by which one-seventh of the fuel and one-tenth of the water that would otherwise be used are saved.

PORTABLE CORN MILL on iron frame, fitted with a pair of 24-in. French burr stones, the lower one of which runs.

This mill is suitable for grinding any substance to which millstones are usually applicable, and will perform its work more rapidly than the generality of such small mills.

PORTABLE CORN MILL on wood frame, fitted with a pair of 36-in. French burr stones and dressing apparatus.

This forms one of a series of portable corn mills comprising the following sizes:—30-in. 36-in. 42-in. and 48-in.

These mills are suitable for any purposes to which millstones are usually applied.

They consist of a pair of French or English millstones, mounted on a strong timber frame, as shown in the woodcut.

When fitted with French burr stones and a dressing apparatus, which can be neatly attached to the framework without adding to the bulk of the mill, they will produce the finest flour for household purposes.

For grinding barley-meal, bruising oats, or splitting beans, the English stones are quite sufficient, but French stones are preferable for producing fine flour, and much more durable.

RANSOMES & SIMS, *continued.*

When required for grinding Indian corn, it is recommended that the corn should be first split in a Biddell's patent bean cutter, which can be readily fixed on the top of the stone case. By this plan the stones wear much longer without dressing, and will grind faster.

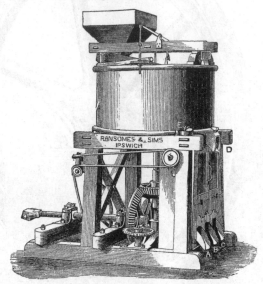

RANSOMES AND SIMS' PORTABLE CORN MILL.

If so ordered, they can be fitted with a small crane for easily turning over the top stone when it requires dressing. They may be driven by means of a horse-gear, by a portable or fixed steam engine, or by water power.

PATENT COMBINED DOUBLE-BLAST STEAM THRASHING, RIDDLING, STRAW-SHAKING, WINNOWING, BARLEY AWNING, AND FINAL DRESSING MACHINE, marked A 1.

This machine is intended for thrashing wheat, rye, barley, oats, and other grain and dressing them ready for market, which operations it performs in the best manner. The drum is 54 in. clear width, and is fitted with patent reversible wrought-iron beaters, which do not break or injure the grain, and which will thrash barley so that it will malt perfectly. The grain is fed into the machine longitudinally, so that the straw is not bent in thrashing, and consequently it leaves the machine without being injured. After the corn has passed through the drum, the straw is carried into the patent shaker, in which part this machine radically differs from any other yet brought out. The objects to be obtained by a good shaker are threefold :—

1st. To separate the straw so that any grain remaining in it may be retained in the machine.

2d. To carry back the grain and short straws thus separated to the dressing part of the machine.

3d. To carry away the straw so that it may be easily removed from the tail of the machine.

The old reciprocating or crank shakers only accomplish the first and third of these requirements, and that at a great expense of power and wear and tear. To accomplish the second desideratum, additional apparatus is necessary, which generally consists of heavy reciprocating riddles set at an angle under the shaker, or some other equally cumbrous contrivance.

The patent rotary shaker with which these machines are fitted, accomplishes the above three requirements in a perfectly satisfactory manner.

The rollers are so placed that the spikes of one roller nearly touch the circumference of the adjacent rollers, and also the board which forms the bottom of the shaker. The rollers revolve at equal speed, so that as the straw leaves the drum it is shaken continually in a jerking

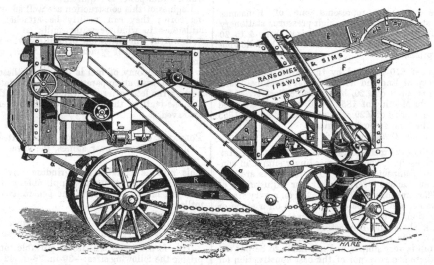

RANSOMES AND SIMS' PATENT COMBINED DOUBLE-BLAST STEAM THRASHING, RIDDLING, STRAW-SHAKING, WINNOWING, BARLEY-AWNING, AND FINAL DRESSING MACHINE.

manner, which exactly resembles the action of hand-shaking by means of a fork. The spikes on the rollers nearly touching the bottom board of the shaker, all the short straws and grain are helped back at each revolution of the shaker towards the riddling apparatus. As there is no reciprocating movement about the shaker, the power required to work it is very small, and as the riddling apparatus is, by the shaker, so much relieved from the work which it usually has to do, it also is very light,

and requires but small power to drive it, whilst the wear and tear are consequently reduced to the lowest point. A series of riddles is arranged in the machine; their use is to sift the straws and ears from the grain and chaff; and they differ in size of mesh to be used according to the grain to be thrashed. Whilst the corn is passing through the riddling apparatus, it is subjected to a blast produced by the fan, and this blast is made stronger or weaker by opening or closing the doors at the ends of

RANSOMES & SIMS, *continued.*

the fan box. The chaff is blown towards the back of the machine. After the thrashed grain has passed through the riddles, the clean corn is carried down to the elevator bottom, whence it is carried up by the elevators, dropped into the barley awner through which it passes into a chob-cleaner or white-coater, which effectually strips the husk from the kernels to which it may still be adhering. If the drum and concave are properly set, this machine leaves very few chobs or unthrashed ears.

The grain then passes over the sieves, which are arranged as in a common dressing-machine, and is simultaneously operated upon by a blast which removes all the dust, dirt, and seeds, and leaves the grain perfectly bright and clean.

It next passes into the patent adjustable rotary screen, which separates the thin kernels from the best corn, and leaves the sample ready for market, dressed in a better manner than it could have been done by hand.

It will be evident that a machine which has to finish wheats of different kinds and sizes of kernels, also oats, barley, rye, &c. and take out from the bulk the light corn, must either be furnished with several screens, or its one screen must be capable of considerable and ready adjustment. After much consideration, R. & S. have succeeded in producing and patenting a screen which may be readily and quickly adjusted to suit different kinds of grain, and which, by a simple arrangement is self-cleaning, so that it never can become blocked up by the grain lodging between the wires.

By means of this screen two separations are made, viz. good corn and light or chicken corn. The amount of distance between the wires is regulated at pleasure, so that the owner of the machine, or the person who has hired it, can set it to take out as much or as little tail corn as he pleases.

The A 1 machines are all fitted with this screen, and the advantage which it gives them over the old plan, of having a separate screen for each kind of grain, is very great.

This machine may also be fitted with a patent apparatus (carried on the same framework), by which the straw, as it leaves the machine, may without any manual labour at the machine be carried to any convenient distance, within a limit of 50 feet, and formed into a stack not exceeding 27 feet in height.

These machines are made in several sizes, all of which stand in the first rank for simplicity, durability, economy of power in proportion to the work done, and excellence in all the operations which they profess to perform. In order to meet the rapidly increasing demand for these articles, Ransomes and Sims have recently erected a costly series of tools for preparing all the woodwork for them, so that the whole machine is constructed with an accuracy and solidity entirely unattainable by hand-work, and as only thoroughly seasoned wood and the best materials are used, the utmost durability is thus ensured.

These machines are sold at the lowest prices which are compatible with the above qualifications, and Ransomes & Sims desire to draw attention to the fact that the customer has their reputation to guarantee the excellence of their machines, for they cannot afford to send out any which are not in every respect of the first class; also, that so-called cheap machinery can in general only be produced by the use of inferior material and unsound workmanship, and speedily shows itself to be the most expensive, often entailing on the unfortunate purchaser a permanent outlay and annoyance.

These machines have been awarded prizes and medals as follow:—

> At the great trial of thrashing machines by the Agricultural Society of Belgium, during the second week of April, 1858, at Brussels, to the No. 1 and also to the A 1 machine, a special prize, grand gold medal of honour.
> At Vienna, 1857, a gold medal.
> At Pesth, 1857, the highest diploma of merit awarded.
> At Amsterdam, 1857, a gold medal.
> At Paris, 1856, a first-class gold medal and 300 francs.
> At Rotterdam, 1858, first-class gold medal.
> At St. Petersburg, 1860, a gold medal.
> At Schwerin, 1862, from the German Landowners' Union, the gold medal of honour.

PATENT COMBINED DOUBLE-BLAST STEAM THRASHING, RIDDLING, STRAW SHAKING, WINNOWING, AND BARLEY AWNING MACHINE, marked B 1.

This machine is similar in its general construction to the A 1 previously described, but it has no screen. Except the final screening, it performs the same operations as the A 1, and is strongly recommended in all cases where screening is not absolutely necessary.

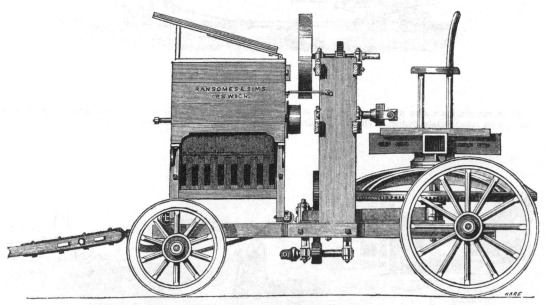

RANSOMES AND SIMS' IMPROVED PORTABLE THRASHING MACHINE.

Ransomes & Sims, *continued.*

Improved Portable Thrashing Machine, suitable to be worked by horses or oxen without unloading the horse-works, as shown on page 85.

This machine thrashes various kinds of grain perfectly and without injury, leaving the grain and chaff together. The barn-work or thrashing part only requires to be unloaded; the driving gear remaining upon the 4-wheel carriage around which the horses walk. The grain when thrashed is dressed by hand at any convenient period. From its simplicity and the facility with which it can be transported and used, this machine is well adapted for mountainous and other countries where repairs are difficult and the roads bad.

The exhibitors also manufacture portable horse-power

PORTABLE HORSE-POWER THRASHING MACHINE.

thrashing machines mounted on 2-wheel carriages, as shown in woodcut.

Before using these machines the horse gear and thrashing drum must be both unloaded and fixed as below.

These machines are also made to drive with a strap through an intermediate motion. The annexed drawing represents one on this construction as fixed for work, and with a winnowing machine attached.

RANSOMES & SIMS, *continued.*

NEW PATTERN DRESSING MACHINE, No. 3.

This machine is fitted with a spiked roller. It will dress rough grain just as it comes from thrashing machines which have no riddle or blower. By throwing off the strap, lifting out the spiked roller, closing the toothed plate, through which it works, with an iron cover, and changing the sieves, the machine is prepared for dressing grain as usually delivered from the single-blast steam thrashing machine. By setting the machine as above, but taking out all the sieves, it may be used as a blower or as a malt screen. Extra sieves for seed dressing are sent when ordered, and when they are used the screen must be closed.

PATENT SELF-CLEANING AND ADJUSTABLE ROTARY SCREEN, with stone separator.

This machine will separate thin and light grain from a sample of barley, wheat, &c. making a perfect sample without leaving good grain with the tail.

The distances between the wires can be altered, so that more or less light grain may be removed as desired, and the screen is therefore equally applicable to grain grown on different soils, or in different climates or seasons.

This screen is perfectly self-cleaning, so that it is always equally effective. It has no brushes either inside or outside, nor any washers or cleaners passing between the wires, and is therefore free from the objections to which screens so constructed are liable in wear.

The action of this screen is continuous. It is therefore subjected to less strain in working, and requires less power, than those in which the action is backwards and forwards.

To merchants and maltsters this screen is invaluable, being from its adjustability applicable to foreign as well as home-grown barley, and for one season's growth as well as another. To farmers it is also invaluable, enabling them so to dress their barley that it shall command the highest price, and to use for feeding purposes the light corn, which, if not separated, would lower the value of the whole sample.

BIDDELL'S PATENT BEAN CUTTER, for splitting hard or soft beans, peas, and Indian corn.

> Obtained the silver medal of the Royal Agricultural Society at Gloucester; the silver medal of the Yorkshire Agricultural Society at York, 1853; and a second-class medal at a meeting of the Royal Agricultural Improvement Society of Ireland, at Killarney, 1853.

It is well known that neither solid-roller mills nor mill-stones will split beans unless they are in good condition, on account of their sticky nature when damp. A stone or other foreign substance passing into a solid-roller mill generally damages the rollers, which are costly and difficult to repair.

In Biddell's patent bean cutter these defects are entirely remedied. The barrel or cutting roller is hollow, and is formed by a number of separate triangular steel cutters, arranged around the circumference of two end rings, and so set that there is more clearance at their back than at the cutting edge, therefore the mill can never choke, no matter what may be the size or condition of the beans. Each tooth has three separate cutting edges, which can be successively used, and when all are worn out they may be easily replaced with new teeth by an ordinary labourer at a very small cost, viz. 7s. 6d. for a complete set. The amount to which the beans are crushed is governed by a screw, and care must be taken not to set the cutting plate so close that it touches the barrel.

They are also well adapted for cracking peas or Indian corn, and have been successfully applied over a pair of stones for cracking the Indian corn before grinding, which enables the stones to do their work more quickly, and with less power. For this use they are mounted over the stone box, and the split corn passes from the bean cutter spout into the hopper of the corn mill.

They are made in two sizes, No. 1 and No. 2, and are mounted on colun n, as shown in woodcut, or on a bracket.

With the No. 1 mill one man can cut 3 bushels of beans per hour; two men, 5 bushels per hour. If driven by horse or steam-power, at a speed of 150 revolutions per minute, it will cut 24 bushels per hour.

The No. 2 mill, which is extra strong, and with brass bearings for power only, when driven at 150 revolutions per minute, will cut from 25 to 30 bushels of beans per hour.

RANSOMES & SIMS, *continued.*

BIDDELL'S NEW PATENT STEEL OAT MILL.

PATENT COMBINED BEAN AND OAT MILL, AND OIL-CAKE BREAKER, No. 17.

In Biddell's patent oat mill, the roller has the cutting edge formed of pure steel, supported at the back by cast-iron. This enables us to harden the steel as much as can be done by fire and water, for the cast-iron not being susceptible of hardening by the same process, we get the toughness of the soft material supporting the keen cutting edge of the harder metal. Thus a very durable and excellent article is produced, and at a cheaper rate than could be done by the old process of making the cutting barrels of wrought-iron, and then case-hardening them, an operation which was attended with much risk and expense. The other process, of making them of cast-iron and case-hardening them, produced an apparently good article, but a very worthless one really, as the hardening was only skin-deep, and soon wore away.

This mill consists of three distinct mills on one frame, viz. a Biddell's patent bean mill which will cut and crush about 3 bushels of beans per hour; a Biddell's patent oat mill, adapted for cutting and crushing from 3 to 5 bushels of oats per hour; and a No. 4 oil cake breaker, for breaking and screening linseed or rape cake.

PATENT COMBINED STEEL MILL for beans and oats, on iron stand, No. 10.

The combined mills consist of the working parts of the above mills mounted on the same spindle and frame, by which the efficiency of each mill is maintained, but the cost of two frames is saved, and the space occupied is also less.

Some thousands of these patent mills are in use, and giving the greatest satisfaction.

BIDDELL'S PATENT UNIVERSAL MILL, No 18.

This mill forms one of a series, and consists of a smooth-roller bruising mill and a Biddell's patent bean cutter mounted upon the same frame, and which may be used simultaneously or separately at pleasure.

The smooth-roller mill is intended for bruising oats, linseed, malt, or barley. It consists of two cast-iron rollers of equal diameters and widths, mounted on a strong frame, and to do the same work requires *less power* than those which are constructed with one *large* and one *small* wheel, which was proved by the trial before the Royal Agricultural Society at Chester, in 1858, where one of these mills, in competition with all the best mills on

RANSOMES & SIMS, *continued.*

that construction, was awarded the first prize as the best oat and linseed crusher.

The patent bean cutter is intended for breaking beans, peas, Indian corn, &c. They are made of various sizes, and suitable for hand, horse, steam, or water power.

BIDDELL'S PATENT CAM CHAFF CUTTER, No. 3.

This chaff cutter obtained the first prize of the Royal Agricultural Society of England, at Chester, in 1858, as the best chaff cutter for hand power.

This machine is fitted on iron frame with wrought legs, and cuts two lengths of chaff, $\frac{5}{16}$ in. and $\frac{3}{8}$ in. It is adapted to cut a large amount of chaff with but little consumption of power.

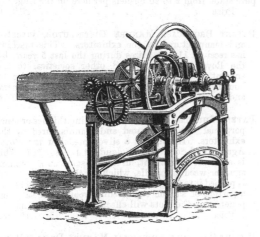

UNIVERSAL CHAFF CUTTER, No. 7, with rising and falling rollers, for hand, horse, steam, or water power.

This machine may be worked by one man, with handle at A; by two men, with handles at A and B; by horse or steam power, through a crotch or pulley on spindle A.

It is simple in construction, and not liable to get out of order. Except the wood feeding trough, it consists entirely of metal.

The knives are as easily sharpened and set as in an ordinary chaff cutter.

This machine is furnished with a patent lever for instantly stopping the rollers, in case the hand of the feeder should be drawn in. The few cog wheels in this machine are cased over, so that no danger can arise from them.

It cuts two different lengths, viz. $\frac{1}{8}$ in. and $\frac{1}{4}$ in. or a greater variety if so ordered.

Of $\frac{1}{8}$-in. chaff one man will cut about 3 cwt. of hay, two men 5 cwt. one horse 10 cwt. per hour.

This machine will cut any substance to which machines of this class are usually applied, such as hay, straw, clover, hop-bines, sorghum, cane trash, &c.

BIDDELL'S NEW PATENT ROOT CUTTERS.

One of these machines obtained the first prize of the Royal Agricultural Society of England at Chester, in 1858.

They are made in different sizes, and to cut roots either into slices for beasts, into finger-pieces for sheep, or into thin shreds for fermenting.

This machine is fitted with knives for cutting slices $\frac{1}{8}$ in. thick, the entire width of the root. It is also fitted with cross-knives, which are easily thrown in and out of work, and which cut finger-pieces $\frac{3}{4}$ in. wide and $\frac{5}{8}$ in. thick.

They effectually cut the last piece. The roots do not hang up as in other machines. The hopper is divided into three parts, so that by filling one or all, the machines may be worked either by a boy or a man as is most convenient.

REEVES, ROBERT & JOHN, *Bratton, Westbury, Wilts.*—Liquid manure drills, manure distributors, patent corn manure, and turnip drills, &c.

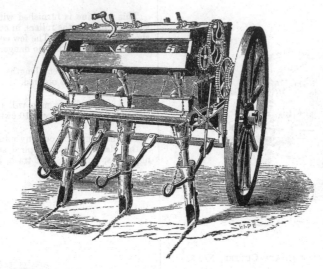

PATENT LIQUID-MANURE AND SEED DRILL.

PATENT LIQUID-MANURE AND SEED DRILL. Invented by Thomas Chandler of Aldbourne, and manufactured by the exhibitors. This drill has received 42 prizes from the Royal Agricultural and other societies, including the first prize at the Royal Agricultural Show at Leeds, 1861; first prize and gold medal at the Paris Universal Exhibition, 1856; and prize medal at the great International Exhibition of England, 1851. Price £25 0

CHAMBER'S PATENT DROP LEVER, manufactured by the exhibitors, for dropping liquid manure and seed in bunches. Price £2 10

A 7-ROW SMALL OCCUPATION CORN DRILL. This drill is suitable for small light land farms. It will sow all kinds of grain and seed. Price £14 10

PATENT MANURE AND SEED DRILL for sowing manure with mangolds and turnips, invented and manufactured by the exhibitors. This drill received the prize at the Royal Agricultural Show at Leeds, 1861. It is suitable for drilling general compost manure, or for artificial manure in its pure state, from 3 to 60 bushels per acre. Price £16 0

PATENT ECONOMICAL MANURE AND SEED DRILL, invented and manufactured by the exhibitors.

This drill has received the following prizes during the last 5 years:—

First prize at the Bath and West of England Show at Newton, 1857.
Silver medal at the Royal Agricultural Society of Engand's Show at Salisbury, 1857.
The prize at the North Lincolnshire Show at Louth, 1857.
First prize at the Bath and West of England Show at Cardiff, 1858.

First prize at the Yorkshire Show at Hull, 1859.
First prize at the Highland Agricultural Show at Edinburgh, 1859.

First prize at the Highland Agricultural Show at Dumfries, 1860.
A prize at the Royal Agricultural Show at Leeds, 1861; and the first prize at the Highland Agricultural Show at Perth, 1861.

It is adapted for sowing artificial manures in their pure state, from 2 to 20 bushels per acre on the ridge. Price £12 0

PATENT BROADCAST MANURE DISTRIBUTOR, invented and manufactured by the exhibitors. This machine has received 11 first prizes during the last 5 years, by the Royal Agricultural and other societies, to the amount of £54; it also received the first prize or honourable acknowledgment at the German Farmers' and Foresters' Show at Schwerin, 1861. Price £10 0

PATENT THISTLE DESTROYER, for killing thistles or other perpetual weeds, invented and manufactured by the exhibitors. It obtained a silver medal at the Royal Agricultural Society of England, at Leeds, 1861. This is a simple implement, used the same as a common spud or weeding paddle, which, at the same time as it is pushed into the ground to cut off the weed, discharges a portion of salt on the bleeding root. The salt thus penetrating the roots will effectually destroy them. Price . . . . 10s. 9d.

PATENT COMBINED ARTIFICIAL MANURE DRILL & HORSE HOE, invented by H. & T. Proctor of Bristol, and manufactured by the exhibitors. This implement is adapted for hoeing the plants and, at the same time, to deposit a small portion of suitable manure to carry out the growth of the plant in its last stages.

R. & J. R. can with confidence recommend the above class of implements, having received during the last month a large number of testimonials as to the efficiency of their patent manure drills and manure distributors, which they will be happy to forward post-free on application.

[2171]

RICHMOND & CHANDLER, *Salford, Manchester.*—Chaff cutters and machinery for the preparation of food for cattle, &c.

The long experience of Richmond & Chandler in the manufacture of chaff cutters, corn crushers, &c. enables them to produce machines of a superior construction and style of workmanship at as low prices as any house in the trade. The chaff cutters have for years taken the prize of the Royal Society of England, also the Societies of Ireland, Scotland, France, Holland, Russia, and elsewhere ; they are fitted with toothed rollers, rising mouths, stop motion, and the patent steel mouthpiece, which being harder than the knives, prevents their constant friction wearing it hollow or uneven.

These machines are thus described in Stephen's Book of Farm Implements and Machines, edited by R. S. Burn:—"The admirable workmanship which characterises these unrivalled machines may be cited as a good example of what agricultural mechanism ought in all cases to be."

CHAFF CUTTER.

CHAFF CUTTERS.		
No. 58	£2	10
No. 57	3	15
No. 1A	4	10
No. 59	5	10
No. 3C	7	0
No. 4	10	0
No. 5	15	0
No. 6	16	0
No. 7	20	0

CORN CRUSHERS.		
No. 1	5	5
No. 2	6	10

No. 3	10	0
No. 4	14	0
No. 5	24	0

TURNIP CUTTERS.		
No. 1	3	0
No. 2	4	10

STEAMING APPARATUS.

Price £6 9s. to	16	4

BREAD KNEADING MACHINES.

Price £5 to	50	0

Catalogues may be obtained free on application.

[2172]

ROBEY & CO. *Lincoln.*—Ten-horse double-cylindered traction engine, and double-blast thrashing machine.

TRACTION ENGINE.

The above wood-cut represents one of Robey & Co.'s double-cylinder patent traction engines, and a No. 1 finishing machine, as they appear when travelling from one place to another.

Manufacturers of portable and fixed steam engines, patent steam ploughing tackle, traction engines, thrashing machines, corn mills, saw benches, &c. &c. Descriptive, illustrated, and priced catalogues free by post.

MESSRS. R. & CO.'S 10-HORSE DOUBLE-CYLINDERED TRACTION ENGINE can be used for all agricultural purposes, and when required to be moved from farm to farm, will take a thrashing machine or any other agricultural implement, without horses.

The silver medal was awarded to R. & Co. for their ploughing and general purpose engine at the Royal Show held at Leeds, 1861.

PATENT HIGHWAY LOCOMOTIVE.

[2173]

ROWLEY, J. JEPHSON, *Rowthorne, Chesterfield.*—Hedge clipping machine, combining a grass mower and manual delivery reaper.

[2174]

ROWSELL, SAMUEL, *Buckland St. Mary, near Chard, Somerset.*—Patent tubular-iron horse rake (American) ; field and entrance gates.

[2175]

RUSTON, PROCTER & CO., *Lincoln.*—Eight-horse power portable engine, and combined finishing and thrashing machine (*See page* 94).

[2176]

ST. PANCRAS IRON WORK COMPANY, THE, *Old St. Pancras Road, London, N.W.*—A conservatory, and glass walls.

[2177]

SAMSON & JEWELL, MESSRS., *St. Heliers, Jersey.*—Combined paring and breaking cultivator in lieu of skim plough.

Obtained the Prize Medal of the Royal Agricultural Society at Leeds, July, 1861 ; and the First Prize of the Jersey Royal Agricultural Society, May, 1861.

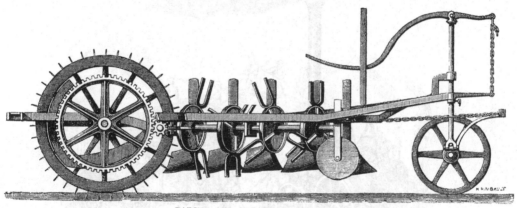

PARING AND BREAKING CULTIVATOR.

This implement is guaranteed to work in all kinds of soil, and will do from 3 to 4 acres per day, performing the work of the skim plough, harrow, and scarifier, at one operation. It also acts as a turn-furrow plough to deposit all kinds of seed. It may be used as a subsoil plough, as all the working parts are disconnected. The makers can guarantee a saving of 50 per cent. by its use.

Application can be made to Messrs. Samson & Jewell, Jersey ; W. H. Samson, Underhill, Iden, Sussex ; and Mr. S. Jewell, Nursling, Hants ; A. Lewis De Jongh, Bishopsgate Street Within, London ; Messrs. Tasker & Sons, Andover, Hants ; or to Edward Parsons Fowler, travelling agent.

Prices :—
Double-action steam plough . . . . . 55 gs.
Single-action plough, drawn by 4 horses . 25 gs.

[2178]

SAMUELSON, BERNHARD, *Banbury, and* 76 *Cannon Street West, London.*—Harvesting and food-preparing machinery. (*See page* 95.)

[2179]

SCOTT, THOMAS, 18 *Parliament Street.*—Self-regulating drinking trough for cattle and sheep.

[2180]

SCOTT, THOMAS, *Newcastle, county Down, Ireland.*—Carrot-seed bearding and dressing machine ; grass-seed separating apparatus.

RUSTON, PROCTOR, & CO., *Lincoln, and Kennet Wharf, 67 Upper Thames Street.*—Eight-horse power portable engine, and combined finishing and thrashing machine.

R. P. & Co. will exhibit during the season in Classes VIII. and IX. their celebrated prize portable steam engines and thrashing machines, which have in public competition won the highest honours, as will be seen from the following list of prizes which have recently been awarded them for their tested excellence :—

St. Petersburgh, 1860. *The Gold Medal and Diploma o Merit.*

Burnley, August, 1860. *Five Pounds and Silver Medal.*

Gothenborg, 1860. *Two Prize Medals.*

Bolton, September, 1860. *The First Prize and Silver Medal.*

Whitchurch, 1861. *The First Prize.*

Mecklenburg-Schwerin, 1861. *The Two First Silver Medals.*

Chorley, 1861. *Ten Pounds and Silver Medal.*

Ashton-under-Lyne, 1861. *The Two Silver Medals.*

Belfast, 1861. *The Silver Medal.*

And numerous other money awards and high commendations.

These engines and thrashing machines are now in extensive use throughout Europe. They are especially remarkable for their extreme simplicity of arrangement, strength of construction, high finish, economy in working, and general efficiency for all the purposes of their construction.

PORTABLE ENGINE AND FINISHING AND THRASHING MACHINE AT WORK.

Ruston, Proctor, & Co. are prepared to execute orders without the least delay for their improved portable and fixed engines, from 2 to 50 horse-power ; thrashing machines, flour mills, portable or fixed ; sawing benches, timber mills, bone mills, steam pumps, mortar mills, winding gear, &c.

CIRCULAR SAWING BENCH.

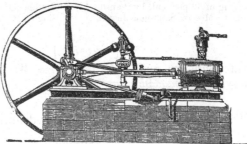

FIXED STEAM ENGINE.

PORTABLE FLOUR MILL.

Illustrated catalogues, with prices and descriptions, may be had on application at the stand, or at the Sheaf Iron Works, Lincoln, or Kennet Wharf, 67 Upper Thames Street, London.

SAMUELSON, BERNHARD, *Banbury, and* 76 *Cannon Street West, London.*—Harvesting and food-preparing machinery.

HARVESTING AND FOOD-PREPARING MACHINERY.

SAMUELSON'S SELF-RAKING REAPING MACHINE. (R. C. Ransome & Samuelson's combined patents.)

PATENT SELF-RAKING REAPING MACHINE, which deposits the grain in sheaves, clear of the track of the horses, by means of revolving rakes. Power required, 2 horses walking at the ordinary farm pace. Width of cut, 5 ft.
Price for full crops £38 0
If with 4 arms, for light and continental crops 36 0

SAMUELSON'S PATENT MEADOW-MOWING MACHINE is distinguished for the flexibility of the cutting apparatus, which enables it to follow the undulations of uneven ground, and to avoid contact with any obstacles. Made of various widths for the draught of 1 or 2 horses.
Price, according to width of cut, £20 to . . £23 0
The same with reaping attachment for cutting grain as well as grass, £26 to £29 0

SAMUELSON'S HAND-RAKING REAPING MACHINE, with side and back delivery, for 1 and 2 horses.
Price, according to width of cut, £16 to . . £17 17

SAMUELSON'S (MAINWARING'S & BOYD'S PATENTS), LAWN-MOWING MACHINES, with Mainwaring's silent wheels, and Boyd's self-cleaning apparatus, for rolling, cutting, and collecting grass on lawns at one operation. Power, varying according to width, from that of a boy, to a light horse. Price, from £5 to £15 15

SAMUELSON'S IMPROVED HORSE RAKE, with steel teeth, width 7½ ft. and 8½ ft. Prices, £8 and . . . £8 10

SAMUELSON'S PATENT GARDNER'S TURNIP CUTTERS, for cattle, calves, sheep, and lambs, on iron and wood frames, cutting the last flat piece of each root to the proper size. By reversing the motion, they are made to cut for cattle or sheep, and by adding the so-called lamb-plates, the cattle knives are made to cut for calves, and the sheep knives for lambs.
Prices, £4 to £6 18 6

SAMUELSON'S & CORBETT'S PATENT ROOT-PULPING MACHINES, for reducing roots so as to be fit for mixing with chaff, cake, or corn. Prices, £4 10 to . . £8 8

SAMUELSON'S IMPROVED IRON-FRAME CHAFF CUTTERS. The two sizes exhibited are adapted specially for export owing to the small compass into which they can be packed.
Price £3 and £5 0
Other sizes from £2 15s. to 13 0

SAMUELSON'S PATENT OIL-CAKE BREAKERS for feeding only, and for feeding and manure, £2 to . . . £8 10

SAMUELSON'S NEW PATTERN GARDEN ROLLERS, £1 12s. 6d. to £4 0
SAMUELSON'S IMPROVED GARDEN ENGINES, £4 to 5 0

[2181]

SELLAR, GEORGE, & SON, *Huntley, Aberdeenshire.*—General plough for home and colonial use; large plough, ridging plough.

GENERAL PURPOSE PLOUGH, for home and colonial use; with improved long steel mould board and wrought-iron frame and share. Constructed not only for producing the finest style of ploughing on cultivated farms, but for standing the rougher work of reclaiming land in the colonies.

These ploughs have carried an immense number of prizes at shows and ploughing matches at home; while in Australia they have been more successful than those of any other maker.

Price £4 15

With land wheel, as exhibited, 7s. 6d. extra.

LARGE PLOUGH for deep ploughing, with wrought-iron frame and share. This plough has carried the first prize at every competition where it has been put forward.

Price £6 0

With steel, instead of cast-iron, mould board, 18s. extra.

RIDGING OR DRILL PLOUGH, with drill gauge,—specially suited for turnip culture, being adapted for effectually covering in the manure, and topping the ridge so as to leave the finest of the soil where the seed is deposited.

Price £4 5

With steel, instead of cast-iron, mould boards, 20s. extra.

[2182]

SHANKS, A., & SON, *Arbroath;* Sole Agents for London, J. B. BROWN & Co., 18 *Cannon Street, City.*—Improved lawn mowers.

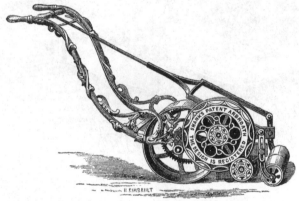

NEW HAND MACHINE.

NEW SMALL HAND MACHINE.

SHANK'S NEW PATENT LAWN MOWING, ROLLING, COLLECTING, AND DELIVERING MACHINE for 1862. (With silent motion if specially desired.)

SHANK'S NEW HAND MACHINE mows the grass wet or dry, on lawns uneven or otherwise, in a much neater manner than the scythe, and at half the expense.

The exhibitors, in introducing their patent lawn mowers to the public for this season, do so with that confidence in their merits and superiority which the eminently successful result of their long and continued efforts to improve fully entitle them to have. The effect of the improvements of previous years has been so much appreciated by the practical gardener, that a large and steady increase in the number of machines sold has every year taken place, every one of which, so far as known, has given the greatest satisfaction. Further improvements have been introduced into the machines for this season, to which reference is respectfully made.

The patentees have brought out an entirely new hand machine, which combines, in addition to the improvements of last year's machine, other improvements of importance, with a new and tasteful design, which has been duly protected by registration. The new models have been produced without regard to expense, and the

patentees have been particularly careful in observing that all the parts possess sufficient strength and firmness to enable them to stand satisfactorily the tear and wear of out-door work, and the rather rough handling these machines are sometimes subject to from the labourer unskilled in machinery. A. S. & Son are gratified in being able to offer to the public a machine, which is not only a graceful ornament to the flower garden, but the most perfect and the most easily worked lawn mower that has ever been in use. The cutter (the most important part of the machine) is this season still further strengthened. The bearings, and everything tending to increase friction, and consequently the draught, are carefully constructed to reduce the friction to the smallest possible amount. The whole of the machines (horse, pony, and hand sizes) are fitted with care and precision, and possess the great advantages of ability to mow on uneven lawns without injury to the turf, of having loose rollers for ease in turning, of having wheels properly guarded, and so placed as to give an equal balance to either side of the machine. No change of rollers is necessary in mowing a verge or close to a flower bed.

The patentees feel convinced that the advantages possessed by their new machine over all others of its class will ultimately lead to its being the

SHANKS, A., & SONS, *continued.*

only, or almost the only, lawn mower in practical use.

Prices :—

SHANKS' PATENT HORSE MACHINE.

No. 1. Width of cutter, 48 in.	. . . .	£28	0
No. 2. ditto 42 in.	. . . .	26	0
No. 3. ditto 36 in.	. . . .	22	0

Drawn by a horse.

No. 4. Width of cutter, 30 in.	. . . .	£19	0

Drawn by a horse or strong pony.

Patent delivering apparatus for Nos. 1 and 2, extra £2; for Nos. 3 and 4, extra £1 10s.; silent movement, extra £1; boots for horses' feet, per set, £1 4s.

SHANKS' PATENT PONY AND DONKEY MACHINE.

No. 5. Width of cutter, 30 in.	. . . .	£15	15
No. 6. ditto 28 in.	. . . .	14	10

Drawn by a pony.

No. 7. Width of cutter, 25 in.	. . . .	£12	10

Drawn by a donkey.

Patent delivering apparatus for Nos. 5 and 6, extra £1 10s.; for No. 7, extra £1 5s.; Silent movement, extra 12/6; boots for pony, per set £1 1s.; boots for donkey, per set 16/0.

SHANKS' NEW PATENT HAND MACHINE, for pushing or drawing, separately or together.

No. 8. Width of cutter, 24 in.	. . .	£9 0 0
No. 9. ditto 22 in.	. .	8 7 6

Easily worked by 2 men.

No. 10. Width of cutter, 19 in.	. . .	7 12 6

Easily worked by a man and a boy.

No. 11. Width of cutter, 16 in.	. . .	6 17 6

Easily worked by a man.

No. 12. Width of cutter, 13 in.	. . .	6 2 6

Easily worked by a boy.

Patent delivering apparatus, if attached to the hand machine, extra, £1 5s. 6d.; Silent movement, extra, 7/0.

SHANKS' NEW PATENT SMALL HAND MACHINE for 1862 is made on the same elaborate and graceful model as their other new hand machine, and is specially intended to be used by ladies and gentlemen for recreation or amusement in the flower garden, and for small gardens where no regular gardener is kept.

The machine is fitted with one roller which is fixed to the shaft, and is so light and easily worked that no draw-rod is necessary. No person having a lawn, however small, should be without one of these useful machines. Nothing looks better in a garden than well-kept grass, and it is not possible to keep the grass well without a machine. The patentees particularly wish it to be borne in mind that these machines are not like toys, more for ornament than use. They are constructed to stand the tear and wear of many years; so that in point of economy, as well as in beauty of work, the mowing machine is unquestionably a most useful horticultural invention.

SHANKS' NEW PATENT SMALL HAND MACHINE for pushing only.

Prices :—

No. 16. Width of cutter, 16 in.	. . . .	£6	5
No. 17. ditto 14 in.	. . . .	5	15
No. 18. ditto 12 in.	. . . .	5	5

Easily worked by a boy.

Silent movement, extra 4/0.

The above are net cash prices, and include carriage to most of the principal railway stations and shipping ports in the Kingdom.

The first practical gardeners of the day who have devoted their attention to examining all the different lawn mowers, do not hesitate in recommending Shanks' machine as the best mower for general use.

The patentees have had the honour of supplying their patent mowing and rolling machine to Her Majesty, for the Royal gardens at Kew, Windsor, Buckingham Palace, Hampton Court, Osborne, and Balmoral; to His Majesty the Emperor of the French; His Royal Highness the Prince of Prussia; His Excellency the Belgian Minister; the Right Hon. Lord Palmerston; His Grace the Duke of Bedford; His Grace the Duke of Sutherland; His Grace the Duke of Buccleuch; and to most of the principal nobility and gentry in the Kingdom. Their machines are also in operation in many of the botanic, and in many hundreds of other gardens in the Kingdom, as well as in almost every country throughout the world, where their merits have been fully proved, and their success established.

The machines are warranted to give ample satisfaction, and if not approved of they may be at once returned.

Sole agents for London, J. B. Brown & Co. 18 Cannon Street, City.

[2183]

SHARPE, BENJAMIN, *Hanwell Park, Middlesex.*—Grass harrows, by which grass and other crops are greatly increased.

[2184]

SMITH, GEORGE, 31 *St. John's Square, Clerkenwell.*—Enamelled garden labels.

[2185]

SMITH, WILLIAM, *Kettering, Northamptonshire.*—Patent horse hoe; winnowing machine; patent sugar machine; patent currant machine.

[2186]

SMYTH, JAMES, & SON, *Peasenhall, Suffolk; Witham, Essex; and Dieppe, France.*—Patent drilling and sowing machines. (*See page* 98.)

[2187]

SNOWDEN, WILLIAM, *King's Cross, London, and Gloucester.*—Paring plough. (*See page* 98.)

[2188]

STANLEY, JOHN, M., & Co., *Midland Works, Sheffield.*—Ornamental octagon conservatory.

[2189]

STEEVENS, W. 6, *Godolphin Road, Hammersmith.*—Steam plough for ploughing, cultivating, and tilling.

SMYTH, JAMES, & SONS, *Peasenhall, Suffolk; Witham, Essex; and Dieppe, France.*—Patent drilling and sowing machines.

PATENT LEVER CORN DRILL, WITH FORE CARRIAGE STEERAGE.

SNOWDEN, WILLIAM F. *King's Cross, London, and Gloucester.*—Paring plough; pares ¾ to 4 inches deep; turns the turf over and cuts it into lengths.

PARING PLOUGH.

SNOWDEN'S PATENT CHAMPION PARING PLOUGH, which has gained the Royal Agricultural Society's of England prizes at Chester and Warwick; the Royal Agricultural Society's of Ireland prizes at Londonderry, and 13 other prizes. It is the only implement that will pare any land from ¾ in. to 4 in. thick, turn the turf over, cut it into 1 or 2 ft. lengths as may be required.

Price complete £6 16 6
Ditto, for stubble 5 5 0

Testimonial from the Right Hon. the Earl of Essex.

"*Cassiobury, Watford,*
"*Nov. 23, 1861.*

"SIR,—I have delayed giving you a report of your paring plough until I had had a fair trial of it, which I have now had, and it has been most satisfactory. I have used it for paring old spongy turf in nearly 5 miles of woodland drives, averaging 12 ft. wide. It performed its work admirably in 12 days at a cost of 9*s.* per day, £5 8*s.* I have no hesitation in saying that had I done it by hand labour it would have taken 10 men at least

5 or 6 weeks at a cost of £30 or £40. I have also used it for paring good sound turf for removal, and it did the work admirably. Next autumn I shall chiefly use it for paring and cleaning stubbles, and I feel sure it will be most efficient.

"You are welcome to make any use you please of this report.
"I am, yours faithfully,
"ESSEX."

CHAFF CUTTER.

SNOWDEN'S PATENT CHAFF CUTTER, HOP-BINE AND CANE-TOP CUTTER, will cut any length from ¼ in. to 3 in. long by only shifting a pin, and is more simple, more durable, and does more work with less labour than any other. Price,

For steam or horse-power £9 19 6
No. 2, for hand-power 8 8 0
No. 3, small for ditto 5 5 0

[2190]

TASKER & SONS, *Waterloo Iron Works, Andover, Hants.*—Combined thrashing and winnowing machine.

PATENT COMBINED THRASHING MACHINE for preparing all kinds of grain for market in one operation.

This machine will thrash every description of grain completely, separating the corn, straw, caving, and chaff from one another, delivering them in the places assigned to each. The drum beaters are less liable to split the corn than any other kind, perfectly cleaning the straw no matter what its condition may be.

The straw shakers possess all the advantages resulting from the use of 2 cranks; one only being used, the amount of tossing of the straw can be regulated as circumstances dictate. The short straws (or caving) are separated from the corn and chaff by the vibrating riddles. The winnowing apparatus is fitted with one fan and suitable screens for dividing the corn from the chaff, and conducting the former to the patent corn elevator to be elevated into the separating screen.

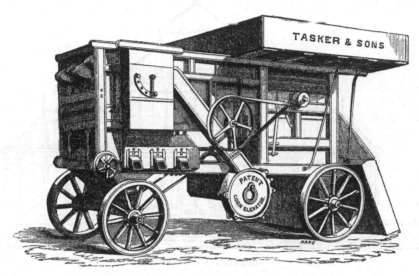

THRASHING MACHINE.

This machine is 3 ft. 4 in. wide, and can be driven with either a 3 or 4-horse power engine. It does its work equally well with the larger machines (requiring 8-horse power engines to work them), preparing the corn in one operation fit for market. In the construction of it none but the very best materials and workmanship have been employed, complication of working parts have been reduced to the smallest minimum, economising the expense in repairs, reducing friction, and consequently the power required to work it. The revolving separating screen is placed immediately across the machine, behind the drum, the several samples of corn on leaving it are conducted to either side of the machine, by suitable spouts, into the sacks placed to receive them. This arrangement removes the large projecting box in which the separating screen is usually placed from the side of the machine, not only giving it a neater appearance, but preventing all accidents to the same when being removed from place to place. The separating screen is novel in construction : its form is conical, the object being that as the corn passes down it, its diameter increasing, the amount of screening surface the corn has to pass over becomes greater. In itself it is of the strongest form, the longitudinal bars are drilled, giving the required distances of the wires encircling it, and which wires are passed through these holes instead of being bound on the outside, as usually the case ; from this it is obvious the mesh of the screen cannot be altered by the encircling wires shifting on the longitudinal bars. The bearings (which are supplied with lubricators) are almost all external, and so open the internal parts of the machine to view.

These small machines will be found of great advantage to persons whose extent of occupation precludes their employing engines of greater power than 4 horse, yet the same results are obtained, and the engine can be used for driving mill-stones, bruising mills, chaff cutters, and other barn machinery.

Price of engine for above machine, 4-horse power £165 0

The advantages of the corn elevator in this machine, are :—

1. It elevates any description of grain in any quantity, without the use of the ordinary tins.

2. It dispenses with the second blower, as the corn is dressed in its passage to the separating screen, from which it is delivered in different samples fit for the consumer.

3. It greatly simplifies the machine, inasmuch as the barley, horner, 2 fans, 2 sets of elevators, tins, 6 straps, and 17 pulleys are dispensed with, thereby economising wear and tear to a considerable amount.

Prices :—

3 ft. 4 in. wide in drum, suitable to be driven by a 3-horse engine £84 0

3 ft. 9 in. or 4 ft. wide in drum, suitable to be driven by a 4 or 5-horse engine 93 0

4 ft. or 4 ft. 6 in. wide in drum, suitable to be driven by a 6 or 7-horse engine . . . 110 0

4 ft. 6 in. wide in drum, suitable to be driven by a 7 or 8-horse engine 120 0

[2191]

TAYLOR, J., & SONS, *Kensall Green, London, W.*—Conservatory, double chambers and improved horizontal tubular boiler, furnace doors, &c.

J. TAYLOR & SONS call the attention of the nobility and gentry to the very superior manner in which they erect conservatories, vineries, forcing, fruit and plant houses of every description, combining the most modern improvements with elegance of design and durability of material and workmanship. They undertake the arrangement for heating with hot water, on the most improved and economical principles, churches, man-

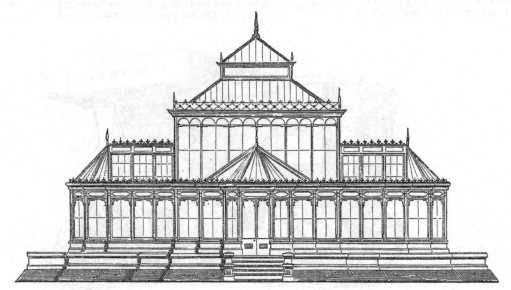

ORNAMENTAL CONSERVATORY.

sions, public buildings, baths, horticultural buildings, &c.

J. T. & Sons' ventilating apparatus supersedes any now in use as being the most simple and effective.

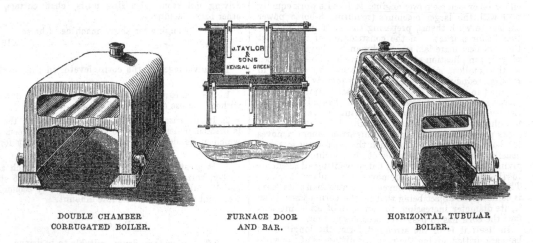

DOUBLE CHAMBER CORRUGATED BOILER.

FURNACE DOOR AND BAR.

HORIZONTAL TUBULAR BOILER.

[2192]

TEGETMEIER, W. B., *Muswell Hill, N.*—Movable frame and permanent observatory beehives.

In these hives each comb is in a separate frame, which may be removed alone, thus every comb is under control, and full honey-combs, or brood-combs for rearing queens, may be withdrawn. The virgin honey is removed in top boxes.

The observatory hive has each side formed of 4 plates of glass, and preserves its temperature through the winter.

Price £1 1

[2193]

THOMPSON, HENRY ATWOOD, *Lewes, Sussex.*—Entrance and other gates, &c. (*See page* 102.)

[2194]

TURNER, E. R. & F., *Ipswich.*—Portable steam engine, &c. (*See pages* 103 *to* 105.)

[2195]

TUXFORD & SONS, *Boston, Lincolnshire.*—Portable steam engines, road locomotives, thrashing, stacking, grinding, and sawing machinery.

[2196]

TYE, JOHN, *Lincoln.*—Double mill, French stones, and governors. (*See page* 106.)

[2197]

TYLER, HAYWARD, & CO., 85 *Upper Whitecross Street, E.C.*—Garden engines, conservatory pump, syringes, fountain jets.—(*See page* 45.)

[2198]

UNDERHILL, W. S., *Newport, Salop.*—Corn elevator, thrashing machine, &c. (*See pages* 107, 108.)

[2199]

WALLIS & HASLAM, *Basingstoke.*—2-horse and 3-horse portable thrashing machines, flour mill, ploughs, harrows, patent spherical bearings. (*See page* 109.)

[2200]

WARNER, JOHN, & SONS, *Crescent, Cripplegate, London.*—Garden engines, pumps, syringes, fountains, fumigators for graperies. (*See page* 110.)

[2201]

WEEKS, JOHN, & COMPANY, *King's Road, Chelsea.*—Improved boiler, ornamental heating stacks, models of conservatories, &c.

[2202]

WEIR, EDWARD, 142, *High Holborn.*—Spirit draining levels with French and English scales, churns, and irrigating pumps.

[2203]

WILKINSON, WRIGHT, & CO., *Boston, Lincolnshire.*—Steam engines, &c. (*See page* 111.)

[2204]

WILLISON, ROBERT, *Alloa, N.B.*—Ventilator for vineries, lift and force pump.

[2205]

WOODBOURNE, JAMES, *Park Iron Works, Kingsley, near Alton, Hampshire.*—Improved machine for packing hops.

[2206]

WOODS & COCKSEDGE, *Suffolk Iron Works, Newmarket.*—New iron horse-gear, &c. (*See page* 112.)

[2207]

YOUNG, J. & T., *Vulcan Foundry, Ayr.*—Drill for mangold wurzel and turnip seed.

YOUNG'S REGISTERED DOUBLE-DRILL TURNIP AND MANGOLD WURTZEL DROP-SOWING MACHINE, drops the seed continuously or at almost any required distances apart; and is so constructed as to work effectually, however unskilful the person may be who is attending it. Being made wholly of cast and malleable iron, it is not liable to be damaged either by the weather or rough usage, and can sow in damp weather when most other machines must stop. By using this machine, a saving of one-half the seed is effected; and the plants can be thinned for from 1/6 to 2/0 less per acre than if the common machine were used. Price, at the Works . . . £6 0

This machine has gained 6 first prizes and several silver medals at some of the leading agricultural exhibitions.

THOMPSON, H. ATWOOD, *Lewes, Sussex.*—Entrance and other gates, drainage instruments, &c.

ENTRANCE GATES.

SET OF ENTRANCE GATES, WITH CAST-IRON PIERS.

These gates are constructed on a principle which gives great strength with an ornamental appearance. They are made of iron and wood, the latter forming a trussed framing which is tied by the former, so that it is impossible for them to drop from wear or ill-usage. The mountings are a peculiar combination of levers, whereby the gates will close themselves without jar; and the latches have locking fastenings. They may be had of any size, and finished in any style.

10 ft. centre gate, grained oak	£6 10
2 side gates, £4 10s. each	9 0
Eccentric lever mountings	4 10

CAST-IRON PIERS FOR ENTRANCE GATES, cheaper and more durable than stone, and quite equal to it in appearance, weighing nearly ½ ton each. Price, each . . £7 7

9-FEET ENTRANCE GATE AND CAST-IRON PIERS, on the same construction as preceding, but with smaller piers.

Price of gate	£4 10
Eccentric lever mountings	1 10
Iron piers, weighing about 5½ cwt. each . .	4 15

9-FT. FIELD GATE, on the same principle, with hangings complete.

Price	£2 10
Wood posts, 7 ft. square, to match, each . .	0 8

CAST-IRON CAPS AND MOULDINGS of various sizes for wood posts. Per pair,

7 in.	£1 0
8 in.	1 5
9 in.	1 10
10 in.	1 15
12 in.	2 0

Gold medal and 100 francs awarded to H. A. Thompson, for drainage levels, at the Exposition Universelle de Paris, 1856; and 2 silver medals by the Royal Agricultural Society of England.

ECONOMIC DRAINAGE LEVEL. A cheap and accurate instrument, well adapted for all ordinary purposes in draining, and may be used by a labourer of ordinary intelligence. Price £1 18

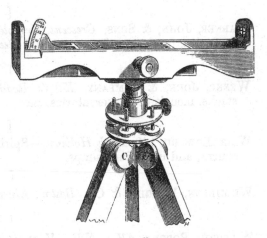

"IMPROVED ECONOMIC" DRAINAGE LEVEL.

IMPROVED ECONOMIC DRAINAGE LEVEL, similar in principle to Economic, but entirely of metal, with a brass mounted spirit tube of the best quality. Price £2 15 0

TELESCOPE DRAINAGE LEVEL. This instrument is constructed entirely of brass, with a telescope attached; it is simple and accurate. Price, including polished case and levelling staff £5 10

WORKMAN'S BEVEL, a modification of the plumb level, to enable an inexperienced workman to work accurately to any gradient. Price £0 18 6

GRADUATED LEVELLING STAFF, with sliding vane. Price £0 12 6

Mahogany SLIDING STAFF, 9½ ft. 2 2 0

Cross STAFF, with brass head 0 15 0

Ditto, with compass 1 0 0

LAND CHAINS, OFF-SET STAFFS, &c.

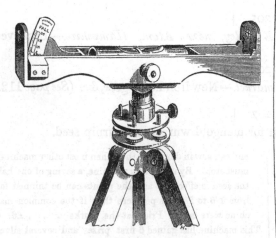

"ECONOMIC" DRAINAGE LEVEL.

TURNER, E. R. & F., *Ipswich.*—Portable steam engine, thrashing and dressing machine, corn crushing mills, &c.

Obtained the First Prizes of the Royal Agricultural Society of England for their Corn Crushing Mills in 1849, 1853, 1854, 1855, and 1860; the Society's Silver Medal for Thrashing Machine in 1860; Prize Gold Medal and 150 Francs at the French Universal Exhibition of Agriculture, 1856; Large Silver Medal at the Imperial Exhibition of Agriculture, at Vienna, 1857; 3 Medals in Silver-gilt and Silver, and 750 Francs, for Steam Engine, Thrashing Machine, and Corn Crusher from the Royal Agricultural Society of East Flanders, 1861.

E. R. AND F. TURNER'S PORTABLE STEAM ENGINE.

1. A PORTABLE STEAM ENGINE OF 4-HORSE POWER.

The cylinder is 6¼ in. diameter; length of stroke, 10¼ in. The fly wheel, which serves also as a driving pulley, is 4 ft. 4 in. diameter, and makes 140 revolutions per minute. The crank shaft is of wrought-iron, it admits of the fly wheel hanging on either end, and of an additional driving pulley. The feed pump has two delivery valves and a tap to regulate the quantity of water. The pump valves, slide valve, cylinder, piston, and every part of this engine are so constructed, as to afford the greatest facility for repairs or adjustment. The boiler is of the ordinary locomotive form; it is strongly stayed and proved both with steam and water at a high pressure; it has ample heating and evaporating surface, as well as water space. The fire-box is of Low Moor iron, the tubes are 20 in number and 2¼ in. diameter. There is a plug at the top of the fire box, which fuses at a low temperature, thus preventing any serious accident arising from the water getting too low. The exterior of the boiler and cylinder are clothed with wood, and neatly cased over with sheet iron. The engine is mounted on a run of 4 wood wheels with iron axles, and improved locking gear, and is furnished with a pair of horse shafts, wheel skid, tube cleaner, firing tools, and waterproof cover.

Price £155 0
Steam Gauge £3 extra.

2. A COMBINED PORTABLE THRASHING AND DRESSING MACHINE.

The "drum" or thrashing part is 4 ft. wide, and is with the breastwork so constructed as to effectually thrash out the grain without injuring it. The straw passes to the end of the machine over a set of frames worked by a crank motion, forming an effective straw shaker, through which the corn, which would otherwise be carried along by the straw, falls on to the riddle board beneath. The bulk of the corn descends from the thrashing part on to the riddle board, and passes on to a wooden riddle perforated and channelled, through which the corn and chaff fall, whilst the short straws are conveyed to the ground. After passing through the riddle, the chaff is separated by the blast of a fan, and the chobs removed by a wire screen, the corn descending into a box whence it is elevated into the barley horning barrel, which carries it to the other side of the machine and delivers it on to a sieve box to be again sifted and winnowed, all foreign substances being thus removed. It then falls on to a patent screen, which separates the small corn from the large, and delivers the bulk a perfect sample into the sack ready dressed for market.

This machine is worked with ease by the 4-horse power steam engine previously described, thrashing and dressing

TURNER, E. R. & F., *continued.*

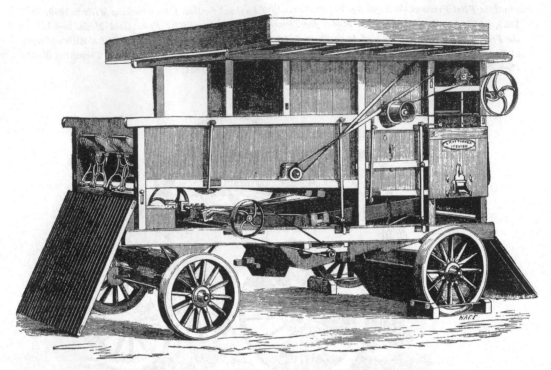

E. R. AND F. TURNER'S COMBINED THRASHING AND DRESSING MACHINE.

from 30 to 40 quarters of corn per day, doing equally well wheat, barley, rye, and other grain.

Price £80 0

The manufacturers make machines similar to the above, but without the finishing dressing apparatus, at a cost of £65. These machines may be worked by their 3-horse power engines.

3. A CORN, SEED, AND MALT CRUSHING MILL (No. 8).

It consists of a large wheel or roller 4 ft. diameter by 6 in. wide, working in contact with a smaller roller of equal width. Between the surfaces of these rollers the corn, &c. is crushed, pressure being applied to the small roll by means of a screw and spring. By this arrangement several advantages are gained, the large roller acting also as a fly-wheel insures regularity of motion; in conjunction with the small roll its capacity for crushing is fully as great as if in conjunction with a roller of its own size, thus economizing both cost and space. The smaller roller

is also more readily influenced by the springs, the elastic pressure thus gained tending to promote the efficiency of the work done, as well as increasing the durability of the surfaces.

Price £18 18
Pulley for power, £1 2s. extra.

4. A CORN, SEED, AND MALT CRUSHING MILL, WITH BEAN MILL COMBINED (No. 8 B).

The crushing part of this mill is precisely like that of the preceding one. The bean mill consists of a pair of metallic plates, the one fixed to the mill frame, the other revolving with the spindle; they are readily adjusted, are very durable, and easily renewed. This mill admits of being used simultaneously for crushing corn and grinding beans.

Price £24 0
Pulley for power, £1 2s. extra.

TURNER, E. R. & F., *continued.*

. A CORN, SEED, AND MALT CRUSHING MILL (No. 1).

The large roll is 3 ft. 10½ in. diameter by 4 in. wide, in other respects the description of article 3 applies to this mill.

Price £12 0
Pulley, 17s. 6d. extra.

6. A CORN, SEED, AND MALT CRUSHING MILL, WITH BEAN MILL COMBINED (No. 1 B).

The crushing part of this mill is like article 5; the description given of bean mill with article 4 applies to this.

Price £15 0
Pulley, 17s. 6d. extra.

7. A CORN, SEED, AND MALT CRUSHING MILL (No. 2),

On the same principle as the preceding, the large roll being 3 ft. 2 in. diameter by 3½ in. wide. This mill may be used by hand as well as by horse or steam power.

Price £8 0
Pulley, 15s. extra.

8. A CORN, SEED, AND MALT CRUSHING MILL, WITH BEAN MILL COMBINED (No. 2 B),

On the same principle as those already described, the crushing part corresponding in size with article 7.

Price £10 10
Pully, 15s. extra.

9. A CORN, SEED, AND MALT CRUSHING MILL (No. 7),

For hand power, the large roller 2 ft. diameter by 4 in. wide; it is furnished with a fly wheel, in other respects the description of the preceding articles apply to it.

Price £6 10

10. A CORN, SEED, AND MALT CRUSHING MILL, WITH BEAN MILL COMBINED (No. 7 B).

Price £8 8

11. A CORN, SEED, AND MALT CRUSHING MILL (No. 6),

For hand power, the large crushing roller is 1 ft. 6 in. diameter by 2½ in. wide.

Price £4 15

12. A CORN, SEED, AND MALT CRUSHING MILL, WITH BEAN MILL, COMBINED (No. 6 B),

In other respects like article 11.

Price £5 15 6

TYE, JOHN, *Lincoln.*—Double mill fitted with two pairs, French stones, governors.

The exhibitor manufactures portable and fixed corn-grinding mills, flour-dressing machines, with wood and iron cylinders; silk machines, for dressing flour; improved barley mills, for making pearl barley for agricultural purposes; smut machines, French burr stones, &c.; and deals in Derbyshire greystones, mill chisels, &c.

He also constructs and erects waterwheels in the most approved manner.

The following engraving represents a fixed corn-grinding mill on a metal frame, with 2 pairs of French stones, 4 ft. diameter, and governors attached for regulating the stones when at work, which makes it easy to manage.

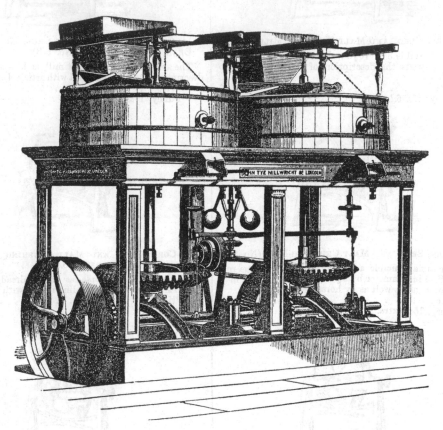

FIXED CORN-GRINDING MILL.

It offers an advantage over any other, as it is portable, and can be set to work without being fixed to the walls, or fastened to the floors of a building. It is so constructed that it can be driven either by steam, wind, or water power.

J. Tye's portable and fixed corn-grinding mills are offered to the public, as unrivalled in the combination of advantages which they possess. For quality of material, strength of construction, high finish, and economy of working, they successfully maintain the first rank. They are admirably adapted for the foreign trade.

In addition to several prizes and medals, J. Tye has received a great quantity of flattering testimonials, both from home and abroad, testifying to the superiority of his mills.

Illustrated and priced catalogues can be had on application at the Works.

John Tye's improved corn-grinding mills received the prize at the North Lincolnshire Agricultural Show, held at Louth, 1857.

Highly commended by the judges, at the Meeting of the Royal Agricultural Society of England, held at Chester, July, 1858.

The prize at the North Lincolnshire Agricultural Show, held at Grimsby, 1859.

The silver medal at the Agricultural Show, held at Melbourne, Australia, 1859.

The silver medal at the Manchester and Liverpool Agricultural Society's Meeting, held at Liverpool, September, 1859.

The prize at the Meeting of the Royal Agricultural Society of England, held at Canterbury, 1860.

Also the prize at the North Lincolnshire Agricultural Show, held at Horncastle, July, 1860.

J. T. intends exhibiting at the Royal Agricultural Society's Meeting at Battersea Park, where mills may be seen at work in the trial yard.

UNDERHILL, W. S. *Newport, Salop.*—Corn elevator, thrashing machine, field and barn implements, patent game and poultry fences.

WROUGHT-IRON CULTIVATOR.

3-HORSE OR 7-TINED WROUGHT-IRON CULTIVATOR, invented, improved, and manufactured by the exhibitor, is now universally used, and wherever introduced during the last 17 years, it has superseded all others, including Ducie's, Coleman's, Howard's, &c. It is made entirely of wrought-iron, mounted on high wheels; the draft is light; it turns easily on the headlands, and is an implement well suited for exportation. Price £6 0 0

LIGHT IRON GENERAL PURPOSE PLOUGH, marked *A* 3, improved and manufactured by the exhibitor. It is a good serviceable plough, easy of draught, and does its work neatly and well. Price £3 5 0

RIDGING PLOUGH, improved and manufactured by the exhibitor. The manner of adjustment is very simple and effective, and is fitted with the manufacturer's new pattern fore-end mould boards and cast shares. Price £3 0 0

SET OF HARROWS, improved and manufactured by the exhibitor; suitable for general purposes. Price £3 10 0

SET OF CHAIN HARROWS of a medium size, improved and manufactured by the exhibitor, for collecting twitch and dressing turf land. They are self-relieving. Price £3 0 0

HORSE HOE, improved and manufactured by the exhibitor; of very simple construction, and great strength; made entirely of wrought-iron. Price £1 5 0

GRUBBER, improved and manufactured by the exhibitor. Chiefly used for working between the rows of potatoes, turnips, &c.; it is very light and strong, and fitted with an extra set of tines, to work as a horse hoe. Price £2 5 0

RYE-GRASS DRILL, improved and manufactured by the exhibitor, and fitted with a new slide invented by him, and is warranted to sow every kind of seed with the greatest regularity, and without clogging. It is the only implement that can be depended on for sowing with certainty and regularity. Price . . £3 10 0

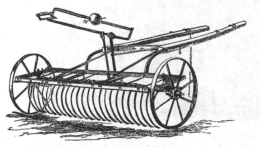

LEVER HORSE RAKE.

LEVER HORSE RAKE, improved and manufactured by the exhibitor, fitted with the oval tooth and the sliding balance ball, both of which are the invention of the exhibitor, and much approved. It is constructed wholly of wrought-iron, firm and strongly made, and is well suited for exportation. Price . . £6 0 0

GAPPING DRILL, invented by Mr. John Phillips, of Brockton, Newport, and manufactured by the exhibitor. Most useful for sowing seeds in rows, when liable to be injured by the fly, &c. Price . £0 8 6

CHEESE PRESS, invented by Henry Bruckshaw and manufactured by the exhibitor; it is simple, strong, and durable, occupies small space, has no levers, weights, or wheels, as those in ordinary use; it is easy to manage, not liable to get out of order, and well adapted for exportation. Price £2 0 0

WROUGHT-IRON COW CRIB, improved and manufactured by the exhibitor, 4 ft. square; it is well made, strong, and handsome. Price £1 10 0

IRON CATTLE TROUGH, manufactured by the exhibitor, 3 ft. square. Price £1 0 0

CAKE BREAKER, invented and manufactured by the exhibitor; a most compact and effective implement, occupying small space, and the only one of its kind where both sets of rollers are adjustable. Price £4 10 0

ROLL OF FENCE.

SPECIMENS OF IRON FENCES—game, poultry, sheep, park, tree, &c.—invented, patented, and manufactured by the exhibitor. This is a most compact and light style of fencing, and the mode of connecting the horizontal bars with the vertical rods is original, (*see illustration*); the horizontal bars are notched by machinery, and the whole forms the strongest, most compact, and durable fence ever offered to the public; it is easily fixed, and if broken by accident, simple to repair; it packs closely, and is neat when fixed. It is strongly recommended for exportation. Price, from 6*d.* to 5*s.* per yard.

UNDERHILL, W. S., *continued.*

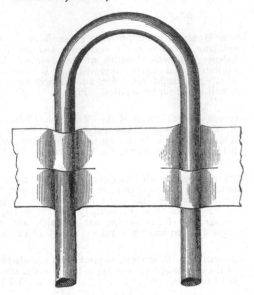

CONNEXION OF FENCING BARS AND RODS.

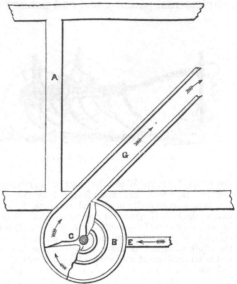

GRAIN ELEVATOR.

SHEEP RACK.

PATENT SHEEP RACK, invented and manufactured by the exhibitor, made wholly of wrought-iron, and mounted on wheels as shown in illustration. This is the cheapest article of the kind ever offered, and has had a silver medal awarded at the Leeds Show of the Royal Agricultural Society. Price £1 10 0

PATENT GRAIN ELEVATOR. This is a new implement, invented and patented by John and Henry Bruckshaw, and W. S. Underhill, of Newport, and manufactured by the exhibitor. It is applicable for raising grain in mills, corn stores, or unloading grain vessels, and will lift any quantity of grain in any required direction, and when applied to a thrashing machine acts as elevator, barley peeler, and smutter. It requires to be seen at work to be fully appreciated. Amongst its advantages over the ordinary tins are—that it dresses the corn in its passage to the separating screen, and when applied to thrashing machines, dispenses with half the ordinary number of straps and pulleys, and can be fixed to any machine of any maker.

A is the main frame of the machine; *B* the case of the elevator; *C* the flyers or blades fixed on the fan shaft *E* of the first winnowing machine, or in any other more convenient position.

THRASHING MACHINE.

THRASHING MACHINE, improved and manufactured by the exhibitor, fitted with every known improvement, including the patent grain elevator. It finishes the grain ready for market. Price, complete . £100 0 0

WALLIS & HASLAM, *Basingstoke.*—2-horse and 3-horse portable thrashing machines flour mill, ploughs, harrows, patent spherical bearings.

TWO-HORSE POWER PORTABLE THRASHING MACHINE.

1. 2-HORSE POWER PORTABLE THRASHING MACHINE, loaded for travelling with patent spherical bearings and wrought-iron drums, and breasting and patent beaters.

The 4-horse power machine, similar to this, but stronger, obtained the Royal Agricultural Society's first prize of £20 at the Canterbury meeting, in 1860. These machines are made very strong, and are well adapted for exportation. The prices and sizes are as follows :—

			Fixture.	Portable.
			£ s.	£ s.
1 h. p. with 18-in. barn-work	26 0	30 0
2 ditto 24-in. ditto	35 0	40 0
3 ditto 30-in. ditto	41 10	48 10
4 ditto 42-in. ditto	51 0	58 10

	Fixture.	Portable.
	£ s.	£ s.
4-horse power, if with a separate 2-wheel carriage for the conveyance of the barn-work or thrashing part, extra		4 10
4-horse power, with 54-in. barn-work, and a separate 2-wheel carriage for the barn-work when made portable £55 0		67 10
An additional draft pole to any of the above machines, complete with fittings to adapt them for countries where the horses are small, extra 1 10		1 10

2. THREE-HORSE POWER PORTABLE THRASHING MACHINE, on 4 wheels.

This machine has been designed to meet the requirements of those purchasers who prefer a machine which can be set down for work without being taken off the travelling wheels. It requires no fixing, the wheels being merely let into the ground 3 or 4 inches. It is shown fitted with a pole, as used in the south of Russia and some other parts, but may be had fitted with shafts at the same price. Two sizes of this class of machine are made, viz. :—

				£ s.
3 horse power, with 30-in. barn-work	.	.	£50 0
4 ditto with 42-in. ditto	.	.	60 0
4 ditto with 48-in. ditto	.	.	62 0
4 ditto with 54-in. ditto	.	.	64 0

An additional draft pole to either of the above machines, complete with fittings, to adapt them for countries where the horses are small, extra 1 10

3. PORTABLE FLOUR MILL, with a pair of 3-ft. stones, fitted complete with fast and loose driving pulleys. These mills have a strong cast-iron frame, which carries the stones above it, and contains the necessary gearing inside. They are made in parts, which are carefully fitted together so as to be easily taken to pieces for packing and transport. Those fitted with 2 ft. 6 in. stones, and smaller, may be had either with pulleys or with a universal joint to adapt them to be driven by horse-power gear-works.

4. IRON PLOUGH adapted for general purposes on light land, fitted with 1 wheel and steel breast, marked W.H.B.
   Price, with cast-iron breast £3 7 6

5. IRON PLOUGH, adapted for general purposes on both light and heavy land, fitted with 2 wheels and patent screw stumps, to allow the man to alter the depth without stopping, and with steel breast, marked W.H.F.
   Price, with cast-iron breast £4 17 6
   Extra to either if with steel breast . . . 0 7 6

6. PATENT EXCELSIOR HARROWS, marked X 10. Price, per set of 3, with whippletree £4 4

7. SET OF SELF-RELIEVING CHAIN HARROWS, 5 ft. wide, and 7½ ft. long. Price complete £2 5

8. PAIR OF PATENT SELF-ADJUSTING SPHERICAL BEARINGS.

WARNER, JOHN, & SONS, *Crescent, Cripplegate, London.*—Garden engines, pumps, syringes, fountains, fumigators for graperies, &c.

Obtained Prize Medal at the Great Exhibition, 1851.

No. 547. WARNER'S OAK OVAL TUB GARDEN ENGINE, with registered spreader.
These engines, in oak tubs, are made to hold either 14 or 24 gallons; in galvanized iron tubs, 10, 16, or 24 gallons.

No. 579½. WARNER'S IMPROVED SWING WATER BARROW to hold 35 gallons.
By its use the gardener will save much time and labour where much watering is done with the water-pot.

No. 585. FOUNTAIN DESIGN—Prince of Wales' feathers.

No. 587. FOUNTAIN DESIGN—mushroom pattern.

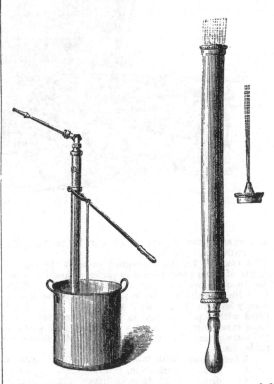

No. 546. WARNER'S GARDEN ENGINE, recommended for orchard houses and conservatories.
A large variety of fountain designs for lawns or conservatories can be seen at the manufactory.

BRASS SYRINGES of various sizes for greenhouses.
Illustrated and priced catalogues may be obtained by application.

WILKINSON, WRIGHT, & CO., *Boston, Lincolnshire.*—Steam engines, thrashing machines, stacking machines or straw carriers, saw tables.

PORTABLE STEAM ENGINE.

The above illustration represents a PORTABLE STEAM ENGINE, with horizontal cylinder and ordinary tubular boiler, as manufactured by MESSRS. WILKINSON, WRIGHT, & CO.; the workmanship is of first-rate quality, and in the engine will be found to contain all the latest improvements.

The above engine is of 7 horse-power. Price . . £200

PATENT PRIZE STACKING MACHINE OR STRAW CARRIER.

The silver medal was awarded to this implement at the Royal Agricultural Show at Leeds, 1861; also the silver medal at Mecklenburgh Schwerin, 1861.

This implement is applicable to stacking hay, straw, or other similar produce. May be worked by hand, horse, or steam power, will carry the material to be stacked to any reasonable distance or height, and in any direction. Price, with a range of 15 yards and 30 ft. high £31 10

WOODS & COCKSEDGE, *Suffolk Iron Works, Stowmarket.* New iron horse-gear, grinding mills pulper, &c.

IRON PRIZE HORSE-GEAR.

NEW IRON PRIZE HORSE-GEAR, fixed for driving agricultural and other machinery. Price . . . £13 13
The above sketch represents W. & C.'s new iron horse-gear, fixed for driving a chaff engine, prize root pulper, and crushing and grinding mill, &c.

CORN-GRINDING MILL.

PORTABLE CORN-GRINDING MILL, fitted with French burr stones for grinding agricultural produce, 20 in.
Price, £21 to £75 0
ROLLER CRUSHING AND GRINDING MILL, for crushing oats, barley, linseed, malt, and also for grinding or splitting beans, peas, Indian corn, &c.
Price, £5 15s. to £13 13

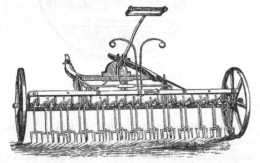

PHILLIPS' PATENT POPPY AND WEED EXTIRPATOR.

PHILLIPS' PATENT POPPY AND WEED EXTIRPATOR AND LEVER HARROW, for exterminating poppies and other

weeds, for passing over root crops, and harrowing in small seeds. Obtained 2 silver medals of the Royal Agricultural Society of England, and several special prizes.
Price £8 15 0

IMPROVED OIL-CAKE BREAKERS, breaks 2 sizes.
Price £3 7 6

PATENT ROOT PULPER.

IMPROVED PATENT ROOT PULPER. Awarded several first prizes of the Royal Agricultural Society. Price for hand-power £4 15 0

This implement when fixed to horse or steam-power will pulp into a fine regular mince 5 to 7 tons per hour.

MODELS.

SALISBURY FIRST-PRIZE AGRICULTURAL CART for general purposes. Price £15 10 0

MITCHELL'S NEW PATENT COMBINED HARROW, SEED DRILL, AND HORSE HOE. Price, £7 to . . £15 0

NEW STEEL'D TEETH HORSE RAKE, &c.
Price £6 15 0

The celebrated SALISBURY FIRST-PRIZE GENERAL PURPOSE CART; the body made of the best seasoned English timber, oak bottom, and thick plank sides, strongly bolted together. The harvest frames are made with strong wrought-iron joints, to take to pieces. The wheels are made with every recent improvement, of the best dry well-seasoned timber. The arms and axles are of the best scrap iron, fitted with caps, and are warranted. Price £15 10 0

HAYWOOD, JAMES, JUN., *Phœnix Foundry, Derby.*—Cast-iron ornamental vases and chairs.

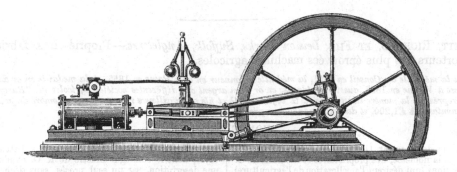

STATIONARY STEAM ENGINE

Manufacturer of stationary steam engines of all sizes, from two-horse power upwards, with or without expansive gear ; steam boilers and boiler fittings ; portable steam engines and combined thrashing machines, of two, three, and four-horse power, the first ever constructed of these sizes to thrash, winnow, and bag the corn in one operation ; patent combined thrashing and finishing machines, to prepare the corn ready for market ; improved portable grinding mills, with either French or Grey stones, from 18 in. to 48 in. diameter ; flour-dressing machines ; improved chaff cutters, with wood or iron frames, constructed to pack in small space for exportation ; improved horse gear, made entirely of iron ; every description of saw machinery ; liquid manure and other pumps.

FOUNDRY DEPARTMENT.

Light and heavy castings of every description ; wrought and cast iron roofs, bridges, girders, and tanks, and every kind of smith's and founder's work in general ; railway chairs, switches, and crossings ; machinery and sugar mill castings ; windows, stoves, cooking ranges ; cast-iron ornamental vases and chairs, and other ornamental castings. The contracts executed at these works comprise some of the largest railway bridges in the kingdom, including the one over the Westminster Road, London, on the South Western Railway, which is 90 feet span ; a great number of iron roofs for railway stations ; many covered markets, including the one at Manchester, which is the largest in England ; the whole of the iron roofing and castings required in the erection of the Enfield Small Arms Factory ; and other important works.

James Haywood, Jun. much regrets that Her Majesty's Commissioners were unable to afford space for the exhibition of his machinery in the International Exhibition ; specimens will be found in the Agricultural Department at the Crystal Palace, Sydenham ; and at the Royal Agricultural Society's Show, to be held in London in June.

APPENDIX TO CLASS IX.

GARRETT, RICHARD, ET FILS, *Leiston Works, Suffolk, Angleterre.*—Propriétaires, fabricants et exporteurs des plus éprouvées machines agricoles.

Ont reçu la médaille du Conseil en 1851, la médaille d'honneur en or à Paris en 1855, et la medaille en or de première classe à Vienne en 1857, aussi 50 médailles en or et en argent des différentes sociétés agricoles de l'Europe, comme les représente la montre illustrée de R. G. et fils. R. G. et fils ont d'ailleurs reçu un unique nombre de prix d'argent se montants à £1,200, et des recommandations presque sans limites.

FONDÉ EN 1778.

La raison de RICHARD GARRETT et FILS sollicite l'attention de la noblesse, des propriétaires et des fermiers de toutes nations (qui désirent l'amélioration de l'agriculture) pour leur machines et instruments aratoires comme exposés dans Classe No. 9, et qui l'on trouvera construits d'après les maximes les plus scientifiques et d'une fabrication du premier rang, assisté par les applications mécaniques les plus nouvelles pour faciliter le travail en bois et métaux, quels matériaux sont tous choisis avec égard à la plus grande solidité, ce que peuvent apprécier le mieux ceux, qui ont dernièrement visité leur fabrique.

R. G. et fils invitent respectueusement tous ceux, qui aient le désir de faire leur jugement sur une base saine, de profiter de la commodité, qui présente à présent le chemin de fer par les comtés de l'est, pour une telle visite, qui ne manquera pas d'occasionner leur patronage d'instruments et de machines d'une fabrication si supérieure et si parfaitement finis.

Les machines et les instruments fabriqués par R. G. et fils peuvent être vus en fonction pratique sur le ferme, qui est annexée à la fabrique et avoisine la station du chemin de fer de Leiston.

La demande, qui c'est répandue et s'augmente rapidement par toute l'Europe pour les machines à battre à vapeur (pour les perfectionnements desquelles cette raison à Leiston Works a pendant le dernier demi-siècle continuellement tenu le premier rang) leur fit voir la nécessité de la production d'une machine, qui serait capable d'achever, d'une manière plus simple et efficace, les opérations nécessaires pour préparer le blé et rendre l'échantillon propre et parfait pour vente, tout ce que fait à présent leur machine combinée à battre et nettoyer, dont le catalogue illustré de R. G. et fils contient une description, par un seul procès, sans déchet et avec très peu d'ouvrage manuel.

Richard Garrett et fils exposent aussi leur instruments et machines types bien connus, savoir :—Machines à vapeur portatives et fixes, Machines à battre à manège, Machines à nettoyer le grain, Moulins à moudre, Semoirs et Houes à cheval, qui sont adaptés pour toutes les méthodes de culture, et dont ce catalogue donne une brève description. On peut obtenir des catalogues détaillés avec des reseignements complets sur le transport par eau, et des devis du prix de livraison à toutes parts du monde, en s'adressant à Leiston Works ou à leur place dans Classe No. 9 de l'exposition internationale.

En conséquence des connexions étendues de cette fabrique l'on envoie des cargaisons entières par bâtiments affrétés directement de la fabrique à beaucoup des ports principaux de l'Europe, et par cet arrangement les acheteurs épargnent les grandes dépenses, qui résultent ordinairement de l'emballage et des frais casuels d'embarquement, et les machines sont delivrées en bon état. L'on envoie aussi sur demande, à une dépense modique, un homme capable d'instruire dans l'emploi et le maniement des machines.

On peut obtenir des Catalogues en Anglais et différentes langues étrangères franco, aussi des plans, des dessins et des calculs des machines fabriquées par R. G. et fils en s'adressant à leur place dans Classe No. 9, ou directement à Leiston Works, Suffolk.

Garrett Richard und Sohn, Leiston Works, Suffolk, England. — Patent Inhaber, Fabrikanten und Ausführer der erprobtesten Landwirthschaftlichen Maschinen.

Erhielten die Council-Medaille in 1851, die goldene Ehren-Medaille in Paris, 1855, und die goldene Medaille erster Classe in Wien, 1857, nebst 50 goldenen und silbernen Medaillen der verschiedenen landwirthschaftlichen Gesellschaften Europas, wie sie R. G. und Sohns illustrirter Aushängebogen darstellt. R. G. und Sohn haben ausserdem noch eine beispiellose Anzahl Geldpreise, £1,200 betragend, nebst zahllosen Empfehlungen erhalten.

Gegründet im Jahre 1778.

Die Firma Richard Garrett und Sohn erlaubt sich die Aufmerksamkeit der Edelleute, Gutsbesitzer und Landwirthe aller Nationen (die sich für die Verbesserung des Ackerbaues interessiren) auf ihre in Classe No. 9 ausgestellten Maschinen und Geräthe zu leiten, welche man nach den wissenschaftlichsten Prinzipien gebaut und aufs Vollkommenste gearbeitet finden wird, unterstützt durch die neuesten mechanischen Anwendungen zum Erleichtern der Arbeit in Holz und Metallen, welche Materialien alle mit Rücksicht auf größt mögliche Dauerhaftigkeit auserlesen sind; dies werden aber Diejenigen am besten schätzen können, welche vor Kurzem ihre Werkstätten besucht haben.

R. G. und Sohn laden aufs Höflichste alle Diejenigen ein, welche ihr Urtheil auf einer gesunden Basis zu bilden wünschen, sich der Bequemlichkeit zu bedienen, welche die Eisenbahn der Ost-Provinzen jetzt für einen solchen Besuch gewähren, der nicht fehlen kann zur Begünstigung von Geräthen und Maschinen solcher vorzüglichen Fabrikation und vollkommenen Arbeit Veranlassung zu geben.

Alle von R. G. und Sohn gefertigte Maschinen und Geräthe können auf dem an die Fabrik und Leiston Station grenzenden Pachthofe in praktischem Betriebe gesehen werden.

Das sich über ganz Europa weit verbreitete und schnell zunehmende Begehr für Dampfdreschmaschinen (in welcher Branche diese Firma zu Leiston Works für Verbesserungen während des letzten halben Jahrhunderts fortwährend den Vorrang gehalten hat), hat sie von der Nothwendigkeit überzeugt eine Maschine zu erzeugen, welche auf einfachere und wirksamere Weise die zur gehörigen Zubereitung des Korns und zum Hervorbringen eines reinen und vollkommenen

Musters, zum Verkauf, nöthigen Operationen ausführen könne, und dies geschieht jetzt durch einen einzelnen Prozeß, ohne Verwüstung und mit sehr wenig Handarbeit, mittelst ihrer combinirten Dresch- und Reinigungs-Maschine, welche in R. G. und Sohns illustrirtem Catalog beschrieben steht.

Richard Garrett und Sohn stellen ebenfalls ihre wohlbekannten Normal-Geräthe und Maschinen aus, nämlich:— Transportirbare und feststehende Dampfmaschinen, Göpel-Dreschmaschinen, Kornreinigungsmaschinen, Mahlmühlen, Säemaschinen und Pferdehacken, welche allen Arten von Anbau angemessen sind und in diesem Catalog beschrieben stehen. Umständliche Cataloge mit vollständiger Auskunft, betreffend Verschiffung und Anschläge der Lieferungskosten nach irgend einem Theile der Welt, sind zu haben, wenn man sich an ihre Fabrik in Leiston, oder ihren Stand, in Classe No. 9 der Ausstellung aller Nationen wendet.

In Folge der ausgebreiteten Verbindungen dieser Fabrik finden Verschiffungen ganzer Ladungen Statt, mittelst Schiffe, welche direkt von der Fabrik aus nach vielen der Haupthäfen Europas befrachtet werden, und diese Einrichtung erspart ihren Kunden die schweren Kosten, welche gewöhnlich aus Verpackung und zufälligen Verladungsspesen entstehen, und sichert dabei die Lieferung der Maschinen in einem unverdorbenen und vollkommenen Zustande; wenn man es wünsche, wird auch gegen billige Berechnung ein sachkündiger Mann gesandt, um in dem Gebrauche und der Handhabung der Maschinen zu unterrichten.

Illustrirte Cataloge, auf Englisch und in verschiedenen fremden Sprachen, sind frei zu haben, auch Pläne, Zeichnungen und Kostenanschläge der von R. G. und Sohn gefertigten Maschinen, wenn man sich an ihren Stand in Classe No. 9, oder direkt an Leiston Fabrik, Suffolk, wende.

GARRETT, RICHARD, OG SÖN, *Leiston Works, Suffolk, England,*—Patenthavere, Fabrikanter og Exporteurer af de meest sögte Agerdyrkningsmaskiner.

Erholdt Council-Medaillen i 1851, Guld-Æresmedaillen i Paris 1855 og Guldmedaillen af förste Classe i Vien 1857, samt 50 andre Guld- og Sölv-Medailler fra de forskjellige Agerdyrknings-Selskaber i Europa (see R. G. og Söns illustrerte Ark).—R. G. og Sön have desforuden erholdt et uhört Antal af Pengepræmier, der belöbe sig til £1,200, og tallöse Anbefalinger.

ANLAGT I AARET 1778.

Firmaet RICHARD GARRETT og SÖN anmoder Adelsmænd, Landeiere og Landmænd af alle Nationer (der maatte interessere sig for Agerdyrkningens Fremgang) om at skjenke deres i Classe No. 9 udstilte Maskiner og Redskaber, deres Opmærksomhed, hvilke man vil finde construerte efter de videnskabeligste Principer og af fortrinligt Arbeide, understöttet af de nyeste mekaniske Anvendelser for at lette Forarbeidelsen af baade Træ og Metaller, alle hvilke Materialier blive udsögte med Hensyn til yderst mulig Varighed, hvilket de ere bedst istand til at vurdere, som have for kort Tid siden besögt deres Fabrik.

R. G. og Sön indbyde ærbödigst alle de, som maatte önske at bygge deres Omdömme paa en sund Basis, til at benytte sig af Leiligheden, som "Eastern Counties" Iernbanen nu yder for et saadant Besög, idet ikke kan andet end formaa dem til at begunstige Redskaber og Maskiner af saa udmærket Fabrikation og fuldkomment Arbeide.

Alle de Maskiner og Redskaber, som R. G. og Sön forfærdige, kunne sees i praktisk Brug paa Avlsgaarden, der staaer i Forbindelse med Fabbrikken og stöder op til Leiston Iernbane-Stationen.

Den over hele Europa vidt udspredte og hurtigt tiltagende Efterspörgsel om Damp-Tærskemaskiner (for hvis Forbedring dette Firma i Leiston bestandig har holdt Forrangen gjennem det sidste halve Aarhundrede) drev dem til at indsee Nödvendigheden af at frembringe en Maskine, der var simplere og virksommere i Udövelsen af de nödvendige Operationer for at tilberede Kornet og frembringe en reen og fuldkommen Pröve for Salg,

og dette fuldbringes nu in in en eneste Proces og med meget lidet Haandarbeide ved Hjælp af deres kombinerede Tærske- og Rensemaskine, der staaer beskreven i R. G. og Söns illustrerte Catalog.

Richard Garrett og Sön udstille ogsaa deres velbekjendte Mönster-Redskaber og Maskiner, nemlig bevægelige of faste Dampmaskiner, Tærskemaskiner med Hesteværk, Rensemaskiner, Malemöller, Saaemaskiner og Hestehypper, der passe for al Slags Dyrkning og ere beskrevne i denne Catalog. Udförlige Cataloger med omstændelig Underretning om Udskibning, samt Overslag over Omkostninger for Aflevering overalt i Verden, kunne erholdes, naar man henvender sig til "Leiston Works" eller til deres Stade i Classe No. 9 paa International-Udstillingen.

Som Fölge af denne Fabriks udbredte Forbindelser, finde Udskibninger Sted i hele Ladninger med Skibe, der befragtes umiddelbart fra Fabrikken til mange af Europas fornemste Havne; dette sparer dens Kunder de svære Udgifter, de ellers i Almindelighed paadrage sig for Pakning og tilfældige Indskibningsomkostninger, og sikkrer desuden Maskinernes Aflevering i sund og fuldkommen Tilstand, og ifald det skulde være nödvendigt, kan en kompetent Mand blive sendt, imod moderat Betaling, for at give Underviisning i Brugen og Behandlingen af de respektive Maskiner.

Illustrerte Cataloger paa Engelsk og i forskjellige fremmede Sprog kunne erholdes frit, ogsaa Planer, Tegninger og Beregninger, R. G. og Söns Maskiner vedkommende, naar man henvender sig til deres Stade i Classe No. 9 eller umiddelbart til "Leiston Works" Suffolk.

GARRETT, RICHARD, É HIJO, *Leiston Works, Suffolk, Inglaterra.*—Autorizados con patente, fabricantes, y exportadores de la maquinaria de agricultura, con todas las ultimas y mas recientes mejoras.

Obtuvieron la Medalla del Consejo en 1851; la Medalla de Oro de Honor, en Paris 1855; y la Medalla de Oro de Primera Clase, en Viena 1857; como tambien 50 Medallas de oro y de plata de las diversas Sociedades de Agricultura de Europa, como consta de su Carton Ilustrado. Ademas las dichas medallas, R. G. é Hijo han recibido un numero sin ejemplo de premios en dinero, del importe en junto de £1,200, y encomios casi ilimitados.

ESTABLECIDOS A.D. 1778.

La casa de Richard Garrett é Hijo, llaman la atencion de los Senores Proprietarios, y Agricultores de todas las naciones (que quieran mejorar la agricultura), a sus maquinas é utensilios que se exponen en la Clase No. 9, los cuales se hallarán construidos segun los principios mas cientificos, de trabajo de primera calidad, con el auxilio de los medios mecanicos mas modernos para facilitar la fabrica en madera y metales. Sus materiales son escogidos para asegurar la mayor durabilidad, cosa que podran confirmar aquellos que han inspeccionado recientemente sus Fabricas.

R. G. é Hijo invitan á los Senores que quieran satisfacerse por medio de una inspeccion personal, á que se valgan de las facilidades que les ofrece el Ferro Carril Eastern Counties, cuya Estacion de Leiston está muy proxima de sus Fabricas. Los Senores Visitadores hallarán en las mismas, utensilios y maquinas todo de primera clase y perfeccion ; y paraque su inspeccion sea mas satisfactoria, se podrá ver en operacion y actividad de servicio, en la Tierra que esta junta á los Talleres y unida á la Estacion de Leiston todas las maquinas é utensilios de la fabrica de R. G. é Hijo.

Como la demanda en toda la Europa para Maquinas de Vapor para trillar está ahora tan extendida y siempre en aumento, y como la casa de R. G. é Hijo á Leiston Works ha tenido constantemente la precedencia durante medio siglo, tocaron la necesidad de producir una maquina mas simple y efectiva en hacer todas las operaciones precisas para separar el grano de la paja, y convertirlo en una muestra limpia y perfecta para el mercado. Esto se hace ahora por un proceso, sin desperdicio ninguno, y con poquisimo trabajo manual, por medio de su Maquina combinada para Trillar y Preparar como se describe en el Catalogo Ilustrado.

R. G. é Hijo exponen tambien los utensilios y maquinas reconocidas que siguen, á saber : Maquinas de Vapor locomobiles y fijas, Maquinas para Trillar con fuerza de caballos, Maquinas para preparar Grano, Molinos para Muler, Taladros, Azadas trabajados por caballos, ajustadas para todos los metodos de cultivacion y que se describen en las paginas del Catalogo. El mismo Catalogo, con todos los detalles y particularidades con respecto al embarque, y los presupuestos de gastos de entrega en todas las partes del mundo, podran obtenerse en Leiston Works ó al mostrador de los Fabricantes, Clase 9, en la Exposicion Internacional.

Por motivo de las muchas relaciones de los Fabricantes, se hacen consignaciones de sus articulos, en cargos redondos y completos por buques que se fletan directamente de los talleres pora muchos de los Puertos principales de Europa. Por este medio, sus clientes no tienen que hacer los fuertes gastos que hay que hacer ordinariamente para gastos de embalage y otros incidentes á la navigacion, y la maquinaria se entrega, con mas seguridad, en una condicion sana y perfecta. Ademas, si es menester, se puede mandar, con pocos gastos, un artesano competente para enseñar á otros el uso y el modo de hacer trabajar las maquinas.

Catalogos Ilustrados, en lengua Inglesa y otras estrangeras, se obtienen gratis ; como tambien, los Planos, Diseños y Aprecios de la Maquinaria de Richard Garrett é Hijo en su Mostrador en la Exposicion Internacional, Clase 9, ó directamente de su Establecimiento, Leiston Works, Suffolk.

GARRETT, RICHARD, E FIGLIO, *Leiston Works, Suffolk, Inghilterra.*—Patentati, fabbricanti, ed esportatori della più perfezzionate macchine d' agricultura.

Ottennero la Medaglia del Consiglio in 1851 ; la Medaglia d'Oro d'Onore di Parigi in 1855 ; e la Medaglia d'Oro di Prima Classe, di Vienna in 1857 ; come pure 50 medaglie d'oro e d'argento dalle diverse Società Europee d'Agricultura, indicate nella Carta Illustrata dei medesimi. Oltre queste, R. G. e Figlio hanno ricevuto Premj in denaro che ammontano alla somma complessiva di £1,200, e commedazioni quasi senza limite.

STABILITI IN 1778.

La casa di Richard Garrett e Figlio ha l' onore di richiamare l' attenzione dei Sigri. Proprietarj di tutte le Nazioni (i quali s' interessano nell' agricultura) alle sue macchine ed utensili che si espongano nella Classe No. 9. Si troveranno i medesimi fabbricati secondo i principi più scientifici, di lavoro di prim' ordine, dietro i mezzi meccanici più moderni onde facilitarne la manifattura tanto del legno che dei metalli ; e coloro che hanno esaminato recentemente le sue Fabbriche sapranno dire che i materiali di cui si serve sono scelti per le loro durevoli qualità.

R. G. e Figlio pregano i Sigri. amatori i quali desiderano formarne un giudizio, di valersi della commodità che offre loro la Ferrovia Eastern Counties ; ed i Fabbricanti rimangono persuasi della preferenza che sarà accordata alle lor macchine, dopo tale ispezione in consequenza della superiorità della manifattura e della perfezione delle medesime.

Tutte le macchine ed utensili della Fabbrica de' Suddetti potranno vedersi in operazione sui Terreni attigui alle Fabbriche, e contigui pure alla Stazione della Ferrovia di Leiston.

Il favore che sempre più godono in Europa i Trebbiatoj a Vapore (pei quali, durante mezzo Secolo, la Casa di R. G. e Figlio ha avuto la più grande riputazione) ha imposto loro la necessità di produrre una Macchina, più semplice ed efficace ad eseguire tutte le operazioni necessarie onde estrarre il Grano dalla Paglia, e renderlo pulito ed atto alla vendita. Questo si effettua adesso per via de

un solo processo, senza guasto, e con pochissimo lavoro manuale, dalla sua macchina complessiva da trebbiare e preparare, come si troverà descritta nel suo Catalogo.

I medesimi Fabbricanti espongono pure le loro Macchine a Vapore locomobili e fissi, Trebbiatoj a forza di Cavalli, Macchine da Preparare il Grano, Molini da Maccinare, Succhielli, e Zappi da Cavallo ; che si adattano a tutti i metodi di coltivazione e che si descrivono con tutti i dettagli nei Cataloghi. Questi ultimi, come pure le notizie necessarie quanto alla Spedizione, e Stima delle spese di consegna in tutti i paesi, potranno ottenersi a Leiston Works, oppure alla loro mostra all' Esposizione Internazionale, Classe 9.

Per ragione delle relazioni estese dei Fabbricanti, le consegne si fanno in esclusive cariche, da vascelli noleggiati direttamente dalle Fabbriche a diversi dei Porti principali di Europa, e in tal modo risparmiano ai loro clienti le forti spese in che s' incorrono generalmente per l' imballaggio e l'espedizione delle medesime ; garantisce poi la consegna delle macchine in uno stato sano e convenevole ; e al bisogno viene mandato un uomo capace con poca spesa onde insegnare ad altri l' uso e modo di usare le macchine.

I Cataloghi Illustrati, in Inglese ed in Lingue Straniere, si danno gratis ; e si possono ottenere alla mostra dei Sigri. R. G. e Figlio, Classe 9, oppure alle lor Fabbriche, Leiston Works, Suffolk, Piani, Disegni, e Valutazioni delle loro Macchine.

LONDON :
R. CLAY, SON, AND TAYLOR, PRINTERS,
BREAD STREET HILL.

Printed in the United States
By Bookmasters

Printed in the United States
By Bookmasters